Elements of Cartography

Elements of Cartography

Fifth Edition

Arthur H. Robinson
University of Wisconsin–Madison
Lawrence Martin Professor Emeritus of Cartography

Randall D. Sale
University of Wisconsin–Madison
Professor Emeritus of Geography

Joel L. Morrison
*U.S. Geological Survey**
(formerly, Professor, University of Wisconsin–Madison)

Phillip C. Muehrcke
University of Wisconsin–Madison
Professor of Geography

*this book was prepared by the author as a private individual without any contribution of government facilities, funds, materials or information which is not already available or that may be made available to the general public upon request to the U.S. government or U.S.G.S.

John Wiley & Sons

New York • Chichester • Brisbane • Toronto • Singapore

Text Design: Loretta Saracino-Marotto
Cover Design: Karin Gerdes Kincheloe
Cover Illustrator: Intergraph Corporation
Production Supervisor: Miriam Navarro
Copy Editor: Elizabeth Hovinen
Illustrations Designer: Onno Brouwer
Map Coordinator: Lisa Heard

Library of Congress Cataloging in Publication Data

Main entry under title:

Elements of cartography.

Revision of: Robinson, A.H. Elements of cartography. 4th ed. ©1978.
Includes index.
1. Cartography. I. Robinson, Arthur Howard, 1915–
GA105.3.E43 1985 526 84-11860
ISBN 0-471-09877-9

Printed in the United States of America

10 9 8 7 6 5 4 3 2 1

Preface

Cartography is in transition. Where the changes will lead is uncertain, but change in the discipline is pervasive, and the rate of change seems to be accelerating. Many of the changes are the result of very rapid and substantial development in the technology available to cartography. But, equally important, a conceptual maturation of the discipline itself has evolved. Technological changes alone would be relatively easy to deal with, but the maturation of the discipline coupled with the rapid changes in technology demand that practitioners learn not only ways of doing new things, but additional ways of doing old things with new equipment. Many practitioners must also adopt a new awareness of why cartography exists and develop an appreciation for its growing usefulness.

The revision of a textbook during such a time of transition is a challenge. Until recently, many people outside the field have viewed cartography as essentially a technology, and any book purporting to serve satisfactorily as a text for a technology was supposed to be simply a recipe book. But as cartography has matured into an independent field, its basic principles have received increasing attention with the result that the field of cartography has developed to the stage where it is possible to talk with some confidence about basic theoretical principles that guide the mapping process. In fact, today the choice between the mapping alternatives that technological advances have made possible depends more on an understanding of mapping requirements than on technical training itself. Thus, more than a recipe book is needed because the discipline is at the point where a balanced treatment must be given both cartographic theory and cartographic practice. This fifth edition of *Elements of Cartography*, then, is an attempt to provide that balance. Not only has a new, fourth author joined the team, but many other changes have been made as well.

Because so many conceptual and technological changes have taken place since the publication of the fourth edition six years ago, much new material has been added. Many of the appendices of earlier editions have been dropped as being unnecessary in a world of computers and calculators. In response to many requests, a chapter on the representation of the land surface, missing from the fourth edition, has been added. A glossary has been compiled, consisting of terms from related sciences that a cartographer should know. Many of these terms are used in the text of the fifth edition. Cartographic terminology itself is well defined in the text and can be found by referring to the index. The manipulation of cartographic data that appears in image form has received greater attention in the fifth edition as has the systemization of the graphic elements used in creating a map image.

While the field of cartography has been experiencing change, cartographic curricula have remained rather conservative and have lagged somewhat behind the research frontiers. Many institutions still offer only one or two courses, and many students take such courses primarily in the hope of gaining a marketable skill. To accommodate the wide variety of situations in which *Elements* is used, the authors have divided the text into four parts. Part One is general and introductory, consisting of three chapters introducing the student to the nature, the history, and the technology of cartography. These three chapters may serve as an introduction to the field for any student. The remaining three parts may be used together for a one-year course or individually, each as the basis for a separate semester or quarter course. Part Two deals with fundamental operations and principles of cartography, such as map scale, coordinate systems and projections, data manipulation and generalization, and the basic aspects of map design from a more theoretical point of view. Part Three treats data manipulation, generalization, and map symbolization in a more practical manner. Part Four details map production and reproduction procedures as practiced today. The continuing move to the use of computer technology is incorporated throughout the text wherever appropriate.

The authors are deeply indebted to many people without whom the production of the fifth edition would have been made more difficult, if not impossible. Research findings published by many scholars and the expertise of colleagues were immensely useful as were the many comments and suggestions made directly to the authors. Onno Brouwer prepared many and coordinated all of the hundreds of illustrations that appear in the fifth edition. June Bennett helped immensely with typing and managing distribution of copies. The many students at the University of Wisconsin and elsewhere who were subjected to the organization and reorganization of the material in this book over the years deserve much credit for their feedback to its authors. Finally, as always in the production of such works, the families of the authors must sacrifice. Collectively to our wives and, for some of us, our children, the authors express their appreciation and thanks for their support and help.

J.L.M.
A.H.R.
R.D.S.
P.C.M.

Contents

Elements of Cartography

PART ONE

Introduction to Cartography

1

The Nature of Cartography

An appropriate question at the beginning of this book is: What is cartography? To answer we will first look at where cartography fits in our endeavor to know and communicate. The beginnings of that struggle are lost in prehistory, but we can confidently assume that the earliest ways used utterances and drawings to create the mental images involved in understanding objects and their relationships. From these sounds developed the spoken and written natural languages of today and the sketches evolved into the variety of present-day graphics.

A significant part of peoples' interest focuses on their surroundings, and the desire for imagery of the spatial organization of things, particularly things in the environment, seems to be as normal as breathing. It can be simple and elementary, as when one is concerned with basic relations, such as inside or outside, near or far, in front or behind, or it can be quite sophisticated as when it involves abstract concepts like the distribution of air pollution. Other animals, primitive people, and young children probably construct unique and situational kinds of spatial images, while experienced adults obviously are capable of highly rational spatial constructions. These images formed in our brains are concerned with the spatial relationships among things or ideas and with the spatial forms of entire distributions. If one person were to attempt to communicate with someone else by describing verbally such relationships or forms, we can only hope that the description would evoke a more or less similar image if all conditions were favorable. That would, however, be much more likely to happen were we to provide a visual representation of the image. This graphic representation of spatial relationships and spatial forms is what we call a *map,* and, very simply, cartography is the making and study of maps in all their aspects.

Written and verbal languages allow us to develop ideas and concepts and express them in a variety of ways, ranging from tightly structured scholarly dissertations to literary creations and dramatics which evoke emotional responses. The use of the written and spoken language, sometimes called "literacy" and "articulacy," is a way of developing, manipulating, analyzing, expressing, communicating various sorts of ideas and beliefs. Mathematics, which has been referred to as "numeracy," is a way of symbolizing and dealing with the relationships among abstractions, sets, numbers, and magnitudes. Just as literacy can range from dealing with highly emotive novels to staid scientific matters, and mathematics from abstract relationships to very precise and accurate calculations, in similar fashion graphics, a fourth way of communicating concepts and relationships, deals with a variety of image presentational methods. These methods range from drawing and painting to the construction of plans and diagrams. The term "graphicacy" denotes this form of communication. Cartography is an important branch of graphics, since it is an extremely efficient way of manipulating, analyzing, and displaying, and thus expressing, ideas, forms, and relationships that occur in two- and three-dimensional space.

In the broad sense, cartography includes any activity in which the presentation and use of maps is a matter of basic concern. This includes teaching the skills of map use; studying the history of cartography; maintaining map collections with the associated cataloging and bibliographic activities; and the collection, collation, and manipulation of data and the design and preparation of maps, charts, plans, and atlases. Although each activity may involve highly specialized procedures and require particular training, they all deal with maps, and it is the unique character of the map as the central intellectual object that unites those cartographers who work with them.

Maps Are a Necessity

People must have assistance in observing and studying the great variety of phenomena that concern them. Some things are very tiny, and we must use complex electronic and optical means (such as the microscope) to enlarge them in order to understand their configuration and structural relationships. In contrast, some geographical phenomena are so extensive that we must somehow reduce them to bring them into view. Cartography consists of a group of techniques fundamentally concerned with reducing the spatial characteristics of a large area—a portion or all of the earth, or another celestial body—and putting it in map form to make it observable. The same techniques can be used to enlarge microscopic things to make them easily visualized. Although it is uncommon to refer to these activities as cartography, the resulting images are sometimes called maps. Just as spoken and written language allows people to express themselves beyond the restriction of having to point to everything, a map allows us to extend the normal range of vision, so to speak, and makes it possible for us to see the broader spatial relations that exist over large areas or the details of microscopic particles.

Even an ordinary map is much more than a mere reduction. It is a carefully designed instrument for recording, calculating, displaying, analyzing, and in general understanding the interrelation of things in their spatial relationship. Nevertheless, its most fundamental function is to bring things into view.

Maps range in size from the tiny portrayals that appear on some postage stamps to the enormous mural-like wall maps used by civilian and military security groups to keep track of events and forces. They all have one thing in common: to add to the geographical understanding of the viewer. All beings live in a temporal and spatial environment in which everything is related to everything else in one way or another. Since classical Greek times, curiosity about the geographical environment or milieu has steadily grown in one civilization or another, and ways to represent it in a meaningful way have become more and more specialized. Today there are many different kinds of mapmaking, and the objectives and methods involved seem very different. It is important to realize, however, that all maps have the same basic objective of serving as a means of communicating spatial relationships and forms; therefore, however dissimilar the maps may seem, the cartographic methods involved are fundamentally alike.

The rapidly growing population of the earth and the increasing complexity of modern life, with its attendant pressures and contentions for available resources, has made necessary detailed studies of the physical and social environment, ranging from population to pollution, from food production to energy resources. The geographer, the planner, historian, economist, agriculturalist, geologist, and others working in the basic sciences and engineering long ago found the map to be an indispensable aid.

A large map of a small region, depicting its land forms, drainage, vegetation, settlement patterns, roads, geology, or a host of other detailed distributions, makes available the knowledge of the relationships necessary to plan and carry on many works intelligently. The ecological complexities of the environment require maps for their study. The building of a road, a house, a flood-control system, or almost any other constructive endeavor requires prior mapping. Smaller maps of larger areas showing things such as flood plain hazards, soil erosion, land use, population character, climates, income, and so on, are indispensable to understanding the problems and potentialities of an area. Maps of the whole earth indicate generalizations and relationships of broad earth patterns with which we may intelligently consider the course of past, present, and future events.

Basic Characteristics of Maps

The radar map on the television weather report showing precipitation and storms seems very unlike the map in the travel brochure proclaiming the "glories of ancient Greece"; yet, they have much in common.

All maps are concerned with two fundamental elements of reality: locations and attributes at locations. Locations (L) are simply positions in a two-dimensional space, such as places with the coordinates *x,y*. Attributes (A) at locations are some qualities or magnitudes, such as languages or temperatures. From these two basic elements many relationships can be formed. Some examples are:

$L_1 - L_2$ relationships among locations when no. attributes are involved, such as the distances or bearings between origins and destinations needed for navigation;

$L_1(A_1,A_2,A_3)$ relationships among various attributes at one location, such as temperature, precipitation, and soil type;

$(L_1)A_1 - (L_2)A_1$ relationships among the locations of the attributes of a given distribution, such as the variation of precipitation amounts from place to place;

$(L_1)A_1A_2 - (L_2)A_1A_2$ relationships among the locations of derived or combined attributes of given distributions, such as the relation of per capita income to educational attainment, as they vary from place to place.

All sorts of topological and metrical properties of relationships can be identified and derived, such as distances, directions, adjacency, insidedness, patterns, networks, interactions. A map is therefore a very powerful tool.

Most maps are reductions, and thus the map is usually smaller than the region it portrays. Each map has a defined dimensional relationship between reality and the map; this relationship is called *scale* and is of primary importance. Because of the relative "poverty" of map space the scale sets a limit on the information that can be included and on the manner in which it can be delineated.

All maps involve transformations of various kinds. A common geometric one is to transform a spherical surface (essentially the shape of the earth) to an easier surface to work with, such as the screen of a monitor or a flat map sheet, by a systematic transformation called a *map projection*. The choice of a map projection affects how a map should be used. It is often convenient to employ on maps referencing systems called *plane coordinate grids*. These coordinate systems assist the map user in calculating distances and directions from the map, two metric properties of common interest. Coordinate systems depend on map projections for their accuracy.

All maps are abstractions of reality. The real world is so intricate and wonderfully complex that merely reducing it or putting a small part of it in image form would make it even more confusing. Consequently, maps ordinarily portray data that have been chosen to fit the use of the map, and these data are subjected to a variety of operations, such as classification and simplification, to enhance their comprehension.

All maps employ signs to designate the elements of reality. Even the responses of the various systems of remote sensing, which produce or yield kinds of maps, are signs in that the display of their spectral sensitivities is different from that which would result from our direct observation if that were possible. The designated meanings of the signs constitute the symbolism of cartography. Relatively few of the symbols used on maps have universal meanings, just as relatively few words have universal meanings in all written and spoken languages. Some maps use unique symbolization schemes, while some map series use many conventional signs, which make the map readable by persons who do not speak the same natural language and thus find it difficult to communicate verbally.

All maps portray data by using various kinds of marks, such as lines, dots, colors, tones, patterns, and so on, which require the user constantly to compare the symbols with those in a legend. Unlike the words we read in the natural language or languages we know, which usually convey information to us without our paying much attention to their appearance, the marks of a graphic display are often very noticeable. Whether the marks are on a luminous cathode ray tube or on a piece of paper, their selection and the way they are assembled (how the map is designed graphically) greatly affects the communicability of the map, that is, the way a viewer will organize the data it presents.

CLASSES OF MAPS

The number of possible combinations of scales, subject matter, and objectives is astronomical; consequently, there is an almost unlimited variety of maps. Nevertheless, there are recognizable groupings of objectives and uses for maps, which permit us to catalog them to some degree. One of the problems in doing so is distinguishing between the objectives of the mapmaker and the responses of the viewer. As observed by R. A. Skelton, "Maps have many functions and many faces, and each of us sees them with different eyes."* The chart that serves the utilitarian purpose of providing bearings, depths, and coastal positions for a ship's navigator will, for another, conjure up visions of coconut palms and an idyllic beach life. We cannot, therefore, reasonably predict responses; we can be quite sure only of the cartographer's objectives. In order to provide a basis for the appreciation of the similarities and differences among maps and cartographers, we will look at maps from three points of view: (1) their scale, (2) their function, and (3) their subject matter.

Scale

The dimensions of reality must necessarily be changed to the proportion that will accomplish the objectives and serve the function of the map. The proportion or ratio between the map dimensions and those of reality is called the *map scale;* the various ways map scale can be stated and portrayed are treated in Chapter 4. Here it is necessary only to point out that the ratio between the size of a map and the size of the area it represents can range from very small to very large. When a small sheet is used to show a large area (such as a map of the United States, or even the world on a sheet the size of this page), that map is described as being a *small-scale* map. If a map the size of this page showed only a small part of reality (for example, less than 1 km^2), it would be described as a *large-scale* map.

*R. A. Skelton, *Maps: A Historical Survey of Their Study and Collecting*. Chicago: University of Chicago Press, 1972, p. 3.

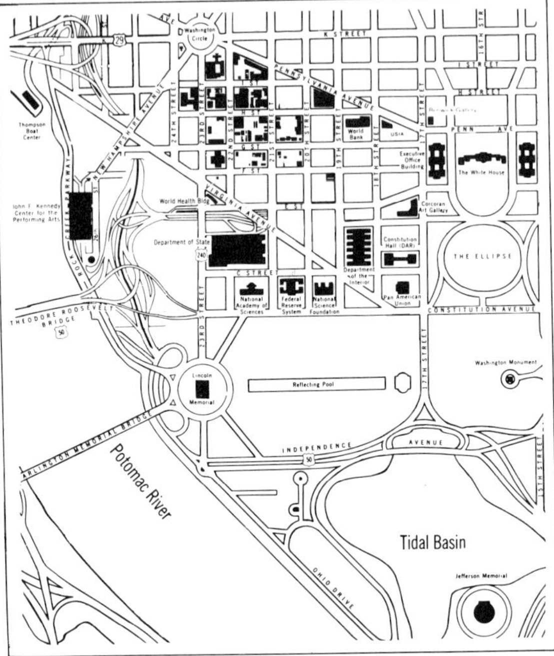

FIGURE 1.1 An example of a large-scale map. (From *George Washington University Bulletin*, courtesy, George Washington University.)

The terms *large* and *small* when combined with scale refer to the relative sizes at which objects are represented, not to the amount of reduction involved. Accordingly, when comparatively little reduction is involved and things such as roads and other features are shown with considerable magnitude, the map is termed a large-scale map (Fig. 1.1). When great reduction has been employed, as for a small-scale map, most of the smaller features on the earth cannot be shown at a size proportional to the amount of reduction, but must be greatly magnified and symbolized to be seen at all. Consequently, reality must be portrayed selectively and with considerable simplification on small-scale maps (Fig. 1.2). On the other hand, although selection is also characteristic of large-scale maps, such maps can portray many aspects of reality in the actual proportion of the amount of reduction employed.

There is no general consensus on the quanti-

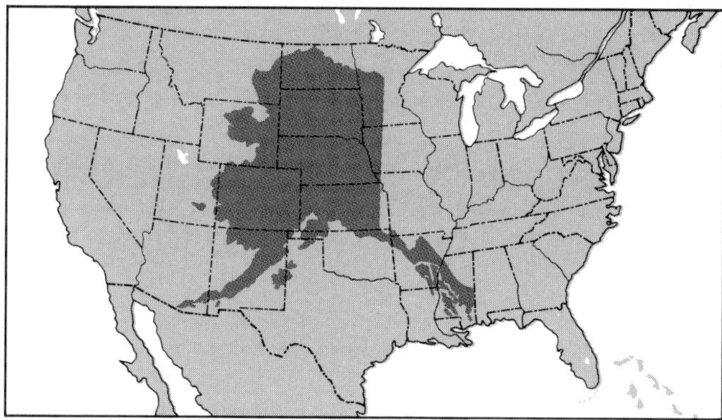

FIGURE 1.2 An example of a small-scale map. (After F. Van Zandt, "Boundaries of the United States and the Several States," U.S. Geological Survey Professional Paper 909, Washington, D.C.: U.S. Government Printing Office, 1976.)

tative limits of the terms small, medium, and large scale, and there is no reason why there should be, since the terms are hardly more than relative. Most cartographers would agree, however, that a map with a reduction ratio of 1 to 50,000 or less (for example, 1 to 25,000) would be a large-scale map, and maps involving ratios of reductions of 1 to 500,000 or more (for example, 1 to 1,000,000) would probably be considered small-scale maps.

Function

As we have just seen, the range from large scale to small scale is a continuum with no clear divisions separating the classes of map scale. Similarly, if we try to divide maps into classes based on their function, we find a great difference between extremes, but the transition along the range from one class to another is a gradual instead of an abrupt change. We can recognize three main classes of maps: *general* maps, *thematic* maps, and *charts*.

General Maps. General, or reference, maps are those in which the objective is to portray the *spatial association* of a selection of diverse geographical phenomena. Things such as roads, settlements, boundaries, water courses, elevations, coastlines, and bodies of water are typically chosen for portrayal on general maps.

Large-scale general maps of land areas are usually called *topographic* maps (Fig. 1.3). They are issued in series of individual sheets and are very carefully made, usually by photogrammetric methods, by national or other public agencies. Maps of much larger scale are required for site location and other engineering purposes, and they employ only ground survey methods. Great attention is paid to their accuracy in terms of positional relationships among the items mapped. In many cases they have the validity of legal documents and are the basis for boundary deter-

FIGURE 1.3 A section of a modern topographic map. (From USGS 1:24,000 Farmersville, Ohio, quadrangle, 1974.)

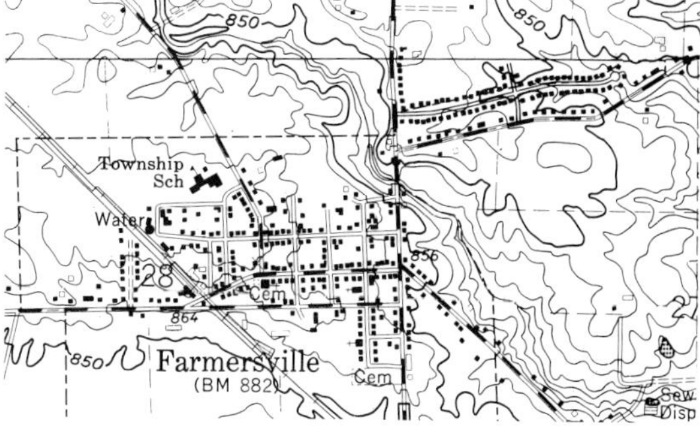

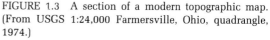

mination, tax assessments, transfers of ownership, and other such functions that require great precision. In the United States and other countries, official map accuracy standards have been established for such general, large-scale maps. The map accuracy standards apply only to the metrical horizontal and vertical qualities of the maps, not to nonmetrical aspects such as labeling blunders, incompleteness, or being out of date.

Small-scale general maps are typified by the maps of states, countries, and continents in atlases. These maps portray the array of phenomena similar to those shown on the large-scale general maps but, because they are small scale and the symbolization and representation must be greatly generalized, they cannot attain the standard of positional precision striven for on the large-scale maps.

Thematic Maps. Thematic maps are quite different from general maps. Whereas general maps attempt to portray the positional relationships of a variety of different attributes on one map, thematic maps concentrate on the spatial variations of the form of a single attribute or the relationship among several. In thematic maps the objective is to portray the form or structure of a distribution, that is, the character of the whole as consisting of the interrelation of the parts. There is no limit to the subject matter of thematic maps, and they

FIGURE 1.4 Distribution of course offerings in cartography by state. Data from the Mapping Sciences Education Data Base. AK = Alaska, DC = District of Columbia, HI = Hawaii, PR = Puerto Rico. "Relative importance" is the number of cartography courses as a percentage of all mapping sciences courses in that state. (From Dahlberg, 1981, courtesy Richard E. Dahlberg.)

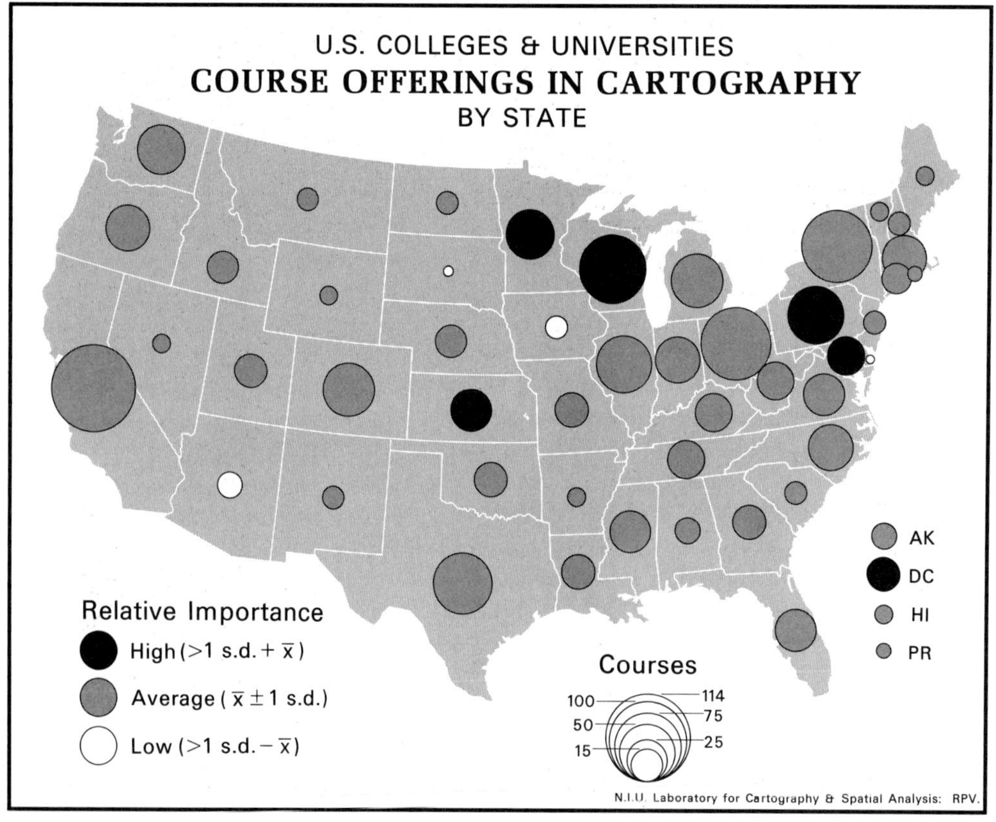

range in appearance from the satellite cloud cover image to the shaded map of election results. They are typified by maps of average annual precipitation or temperatures, populations, atmospheric pressure, and average annual income (Fig. 1.4).

Just because a map deals largely with a single class of phenomenon does not necessarily mean that it is a thematic map. Maps showing the diversity of soils, bedrock geology, or population density can be properly classified as general maps if the primary objectives are simply to show the locations of types of soils, rock, or population density at particular places. On the other hand, maps made from the same data may employ methods of symbolization that focus attention on the structure of the distribution, and they would then be properly called thematic maps.

Thematic maps are commonly small-scale maps largely because many geographical distributions occur over considerable areas, and to portray their essential structure requires great reduction. Nevertheless, this tends to be relative; when the area of interest is a city, for example, maps intended to show the structure of individual phenomena may be of relatively large scale (Fig. 1.5). At small scales accuracy in thematic mapping is less a matter of concern for the precision of individual map positions than it is for the truthfulness of the portrayal of the basic structural character of the distribution.

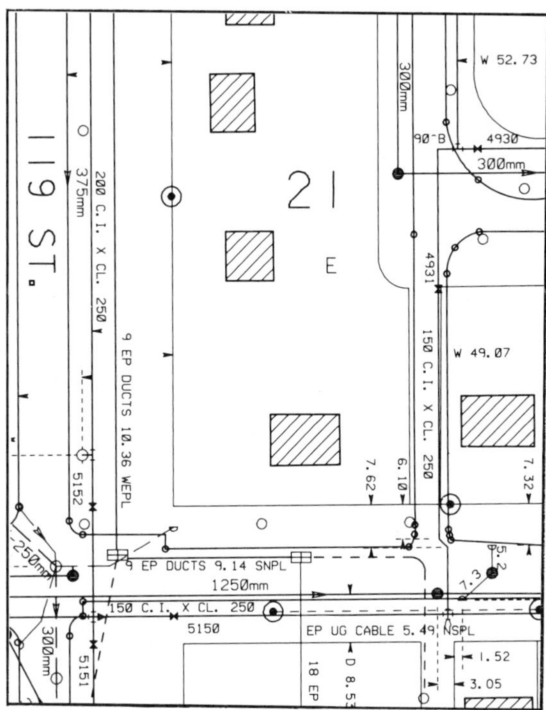

FIGURE 1.5 A portion of a 1:1,000 (large-scale) utility cadastre. (Reproduced by permission from E. A. Kennedy and R. G. Ritchie, "Mapping the Urban Infrastructure," *Proceedings of Auto-Carto V*, copyright 1983 by the American Congress on Surveying and Mapping and the American Society of Photogrammetry.)

Charts. Maps especially designed to serve the needs of navigators, nautical and aeronautical, are called charts. Although it is an oversimplification, one distinction is that maps are to be looked at while charts are to be worked on; courses are plotted, positions determined, bearings are marked, and so on, on the chart. It should be noted that navigators also use general maps. The marine equivalent of the topographic map is the bathymetric map.

There are many varieties of charts. *Nautical* charts include sailing charts for navigation in open waters, general charts for visual and radar navigation offshore using landmarks, coastal charts for near-shore navigation, harbor charts for use in harbors and for anchorage, as well as small-craft charts. All show, precisely located, such things as soundings, coasts, shoal waters, lights, buoys, and radio aids (Fig. 1.6). Their scales vary depending upon the detail necessary; unlike topographic maps, chart series are not made at a uniform scale. Chart design is focused on producing something accurate and easy to read and to mark on.

There are two types of aeronautical charts, those for visual flying and those for instrument navigation. *Aeronautical* charts for visual flying are similar to general maps that show a selection of recognizable features, such as cities, roads, railroads, and so on, as well as other significant elements, such as airports and beacons (Fig. 1.7). Charts for instrument navigation include radio fa-

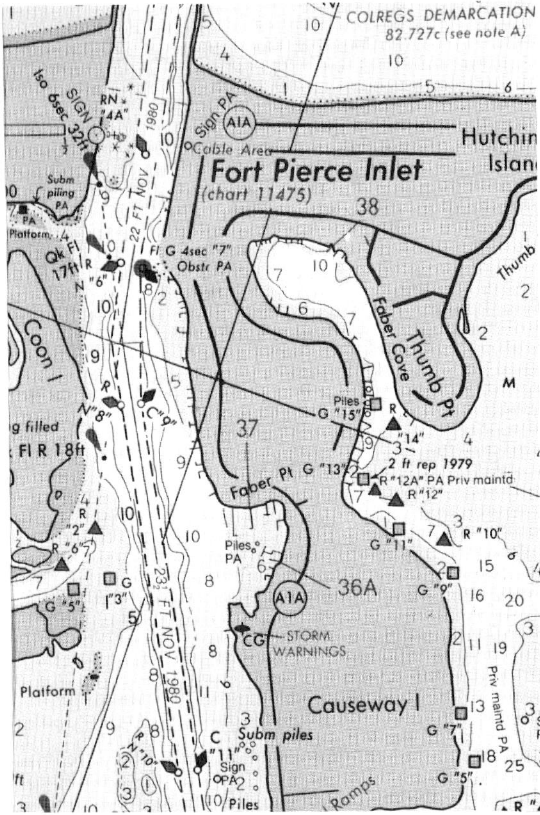

FIGURE 1.6 A section of the Fort Pierce–Fort Pierce Inlet areas on Nautical Chart 11472, Inset 1, Side A, by the National Ocean Survey. Soundings in feet at mean low water.

cility en route charts, high altitude en route charts, terminal arrival charts, taxi charts, and others.

Although it is not called a chart, the familiar road map is really a chart for land navigation. It supplies information about such things as routes, distances, road qualities, stopping places, and hazards, as well as incidental information such as regional names and places of interest.

It should be strongly emphasized that, while there are "pure" general maps, thematic maps, and charts, a majority tend to combine functions to some extent. For example, the green printing often seen on topographic maps shows the distribution of the forested areas, and the representation of terrain shows the structure of the landsurface form. Thus topographic maps, basically general maps, may have thematic components. Similarly, most thematic maps include a selection of boundaries, cities, rivers, and other features, so that the user can more easily fix the location of the subject distribution. Many charts are more "functionally specific," so that they tend to be less duplicative of functions.

Subject Matter

The variety of geographical phenomena and the myriad uses to which maps may be put combine to cause an enormous variety. Although they may all be classed as large-scale or small-scale and

FIGURE 1.7 A portion of a VFR (Visual Flying Rules) Terminal Area Chart—Kansas City, by the National Ocean Survey.

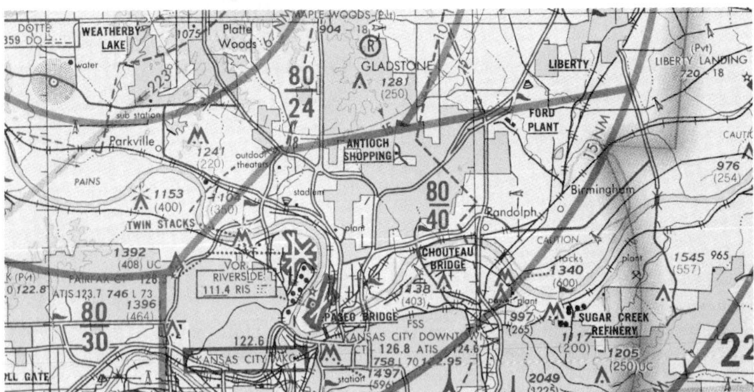

can usually be placed somewhere in the continuum from general to thematic, it is useful also to group maps on the basis of their subject matter. Several important categories may be recognized.

Probably among the earliest "permanent" maps were drawings to accompany the cadastre, the official register or list of property owners and their landholdings; these drawings were called *cadastral maps,* and they showed the geographic relationships among the various parcels. They are common today, and they record property boundaries in much the way they did several thousand years ago (Fig. 1.8). A principal use of cadastres is to provide a basis on which to assess taxes—which may account in part for the fact that cadastral maps have always been with us.

FIGURE 1.8 A portion of a typical 1:1,000 (large-scale) Cadastral map. (Reproduced by permission from E. A. Kennedy and R. G. Ritchie, "Mapping the Urban Infrastructure," *Proceedings of Auto-Carto V*, copyright 1983 by the American Congress on Surveying and Mapping and the American Society of Photogrammetry.)

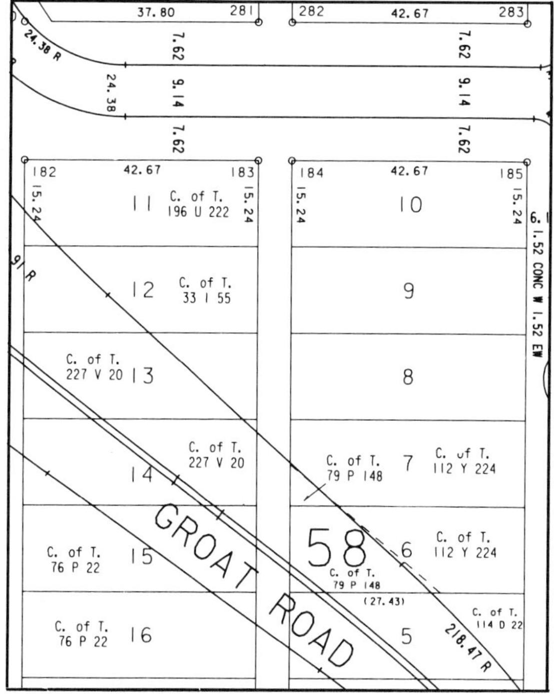

Closely allied to cadastral maps, but more general in nature, is a category of large-scale general maps called *plans*. These are detailed maps showing buildings, roadways, boundary lines visible on the ground, and administrative boundaries. Plans of urban areas are likely to be very large scale. In countries that have a well-integrated mapping program, such as that of the Ordnance Survey of the United Kingdom, the very large-scale plans form the basis for the topographic map series.

There is no limit to the number of classes of maps that can be created by grouping them according to their dominant subject matter. Thus, there are soil maps, geological maps, climatic maps, population maps, transportation maps, economic maps, statistical maps, and so on without end. Such categories are useful only in that any one such class will have numerous similarities in the cartographic treatment of the substantive material and the associated problems. It is a mistake, however, to think that all such maps are alike. There is likely to be more difference between a large-scale map of surface bedrock in an area and a small-scale map of plate tectonics in the Atlantic basin—both geological maps—than there would be between a soil map and a vegetation map of a given area. Cartography is independent of subject matter.

CONCEPTIONS OF CARTOGRAPHY

A subject as complex and important as cartography is bound to have many interesting dimensions worthy of special attention. Five such focuses of attention within the field are singled out for further consideration here. This selection is made on the basis of the emphasis placed on mapping costs, map accuracy, essential mapping activities, communication effectiveness, and the aesthetic aspects of maps. These concerns should not be viewed as contradictory, but rather should be seen as the focal points for the ancillary activities and the subjects of the more active discussions that take place in the field. All of the material presented in this book is relevant in some degree to one or more of these dimensions of the field. Although it is somewhat misleading to assign simple, one-word descriptive labels to these various facets, we do so anyway primarily for convenience of discussion.

Geometric Focus

One widely held conception of cartography we will term the geometric view of the discipline. The aim in this case is to create a cartographic model of reality that is primarily for metrical (measurement) use and analysis. Counts and measurements taken from the map, we hope, will approximate closely those that would be attained were the same analysis carried out directly in the mapped environment. Applications may involve tasks with rather simple aims, such as measurements of position, direction, distance, area or volume, or counts of features. On the other hand, applications may become very involved in support of navigation or complex engineering procedures, such as the siting of dams, airports, toxic waste dumps, or land communications/transportation corridors.

Emphasis in this conception is placed on metrical accuracy (Fig. 1.9). There is a concern for uniform and high quality in data collection, manipulation, and portrayal. Indeed, attempts are often made to adhere to rigid map accuracy standards. A broad user group is assumed, and it is also assumed that these map users will be able to read and analyze accurately whatever the cartographer produces.

The land maps that result from the geometric viewpoint tend to be large scale and to portray distributions of the physical features of the earth or other celestial bodies. Topics mapped commonly include the landform surface, hydrography, political boundaries, transportation/communication routes, and so forth. Nautical and aeronautical charts employ a wide variety of scales to fit the situations in which they are used and the precision required in the data to be obtained from them.

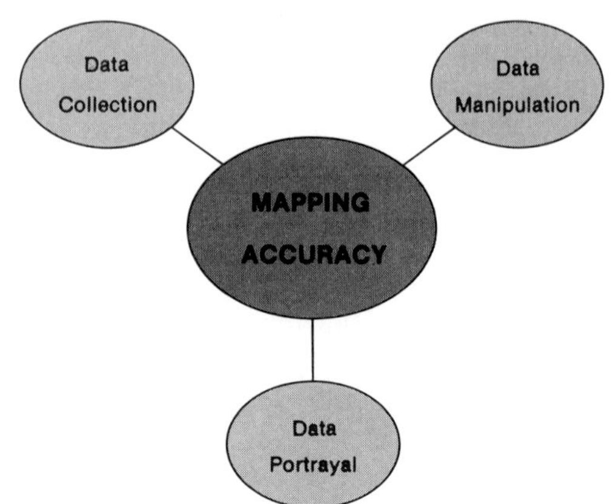

FIGURE 1.9 Basic characteristics of the geometric focus.

Many topographic map and navigational chart series are produced by government agencies under this conception of the field. The design of these maps tends to be traditional, neat, and clean in appearance. Rarely is the design experimented with or changed.

Within the geometric conception of cartography social or cultural themes, such as population, per capita income, health of inhabitants, and so on are de-emphasized. Quite possibly this de-emphasis is simply the result of difficulty in accurately collecting and portraying information about these kinds of distributions. With the emphasis placed so strongly on the truthfulness of the map product in analytical terms, the mapping of sociocultural phenomena may merely be thought best avoided.

Technologic Focus

In this view cartography is considered as a technology for producing maps, and maps are considered to be a storage medium for spatially ordered information. From the perspective of this conception, cartography is viewed as a series of processes concerned with data collection, map design, production, and reproduction. Research is directed at improving mapping efficiency (Fig. 1.10). Thus, attention is focused mainly on technical innovation and the streamlining of the stages

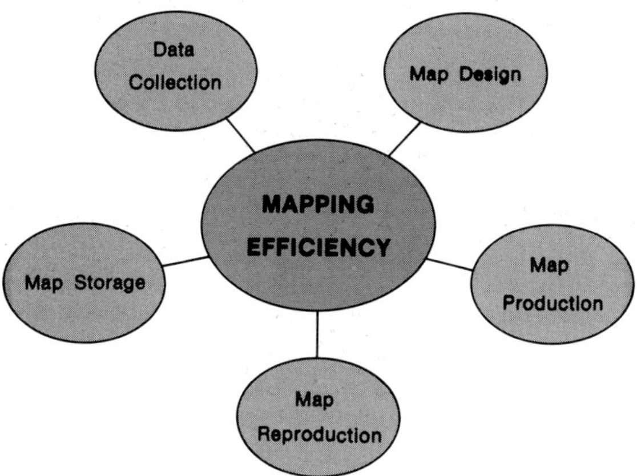

FIGURE 1.10 Basic characteristics of the technological focus.

in map preparation, a primary aim being to increase production speed and volume while lowering per unit costs.

Those who take a technological view of cartography can point to many concrete advances in the field. Manual skills were largely replaced by optical-mechanical technologies, and these in turn largely gave way to photo-chemical technologies. Currently electronic methods are revolutionizing the field. Thus the modern cartographer is advised to pick and choose among a wide variety of methods, making decisions on the basis of available time, labor, and capital resources.

The technological conception is evident in the planning and execution of many map series. Some initial emphasis is given to the effects of data manipulations, map design aesthetics, map accuracy, and map effectiveness, but once production has begun, practically no further attention is given to these topics. One result is that the symbolic systems used on the maps tend to be complex and standardized.

The use for which the product is intended also receives little direct attention. Similarly, the readability of the map is not given serious consideration; it is assumed that some-day someone will have cause for using any map that is produced, and at that time the mapped information will be decipherable and sufficient.

Presentation Focus

The presentation focus stems from concerns about what cartographers do and the relationship between cartography and each of the other mapping sciences (geodesy, surveying, remote sensing, photogrammetry) and associated disciplines. This model emphasizes map design as the central focus or core activity of the field (Fig. 1.11).* The cartographer's role is first to determine the map content and then generalize and represent this information with symbols in a well-balanced layout. The entire process of the production of the map is carefully planned.

The presentation concept is exclusive in the sense that it establishes limits to the discipline. The characteristics of geographical data, technology, map conditions, graphic arts, and visual perception receive attention only as they may be needed to fit the requirements of the map design. These topics and the information and processes that they encompass are not in a direct way part of cartography but instead provide linkages and channels of information flow between cartography and a wide array of important support disciplines.

Indeed, this recognition of the extensive ties between cartography and other disciplines is probably the most valuable characteristic of the core activity model. It makes clear the fact that cartography draws its strength from a diversity of sources. The physical, natural, and social sciences provide the necessary raw data for mapping. The cognitive sciences create an awareness of the needs and limits of human visualization, which can be addressed through map design ideas. The engineering and technological professions and trades provide the means for executing the designed map in an efficient manner.

Artistic Focus

The three conceptions of cartography discussed so far are all rather clinical in nature. They sug-

*E. S. Bos, "Another Approach to the Identity of Cartography," *ITC Journal*, 1982–2, pp. 104–8.

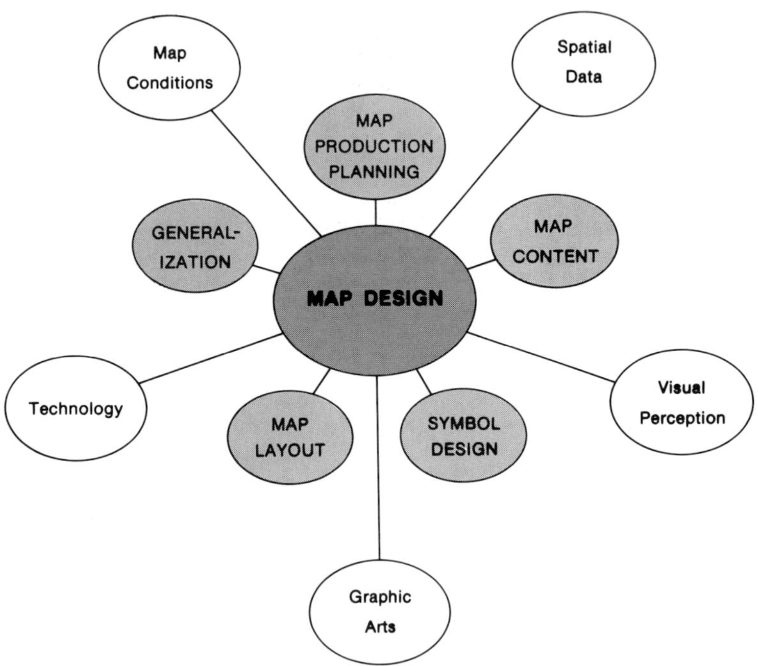

FIGURE 1.11 Basic characteristics of the presentation focus.

gest that the mapping process and the effects of using maps can be rigorously defined and manipulated. Although rational analysis and logical step-by-step procedures are important, by themselves they can lead to rather sterile portrayals, which may fail to convey a realistic impression of the mapped environment in spite of being technically correct. What may be missing is an artistic dimension (Fig. 1.12).

A primary aim in the artistic focus is to employ an understanding of such visual qualities as color, balance, contrast, pattern, line and shape character, selection, exaggeration, and other graphic characteristics to create forms and associations that evoke appropriate impressions and sensations. Perspective views, enhancement of sphericity, variations of scale, unusual orientations, and nontraditional treatments are techniques that can capture and portray significant mental images (Fig. 1.13). Occasionally a unique perceptual space can be employed as a framework in place of the usual rigid Euclidean space. Sometimes the aim is simply to bring a semblance of vitality to an otherwise lifeless map. This artistic focus seems to involve holistic, intuitive processes, which fall largely beyond the realm of verbal or numerical reason. However, practice in learning to see effectively and a great deal of experience in artistic expression are important.

The artistic conception of cartography is purposefully vague with respect to mapping rules or

FIGURE 1.12 Basic characteristics of the artistic focus.

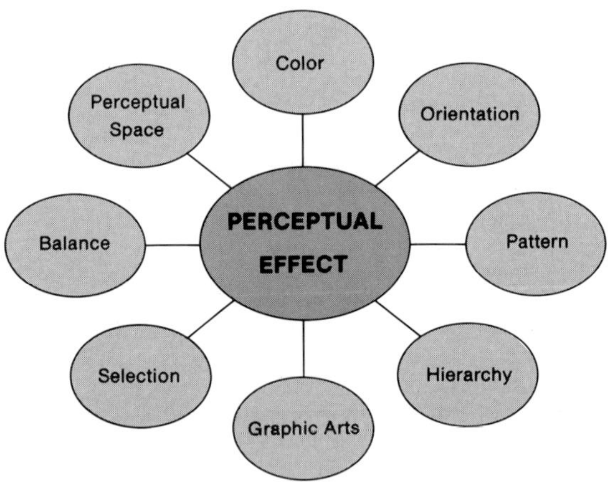

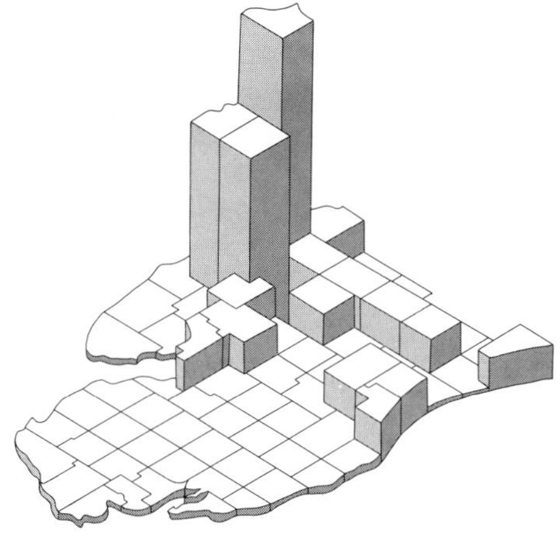

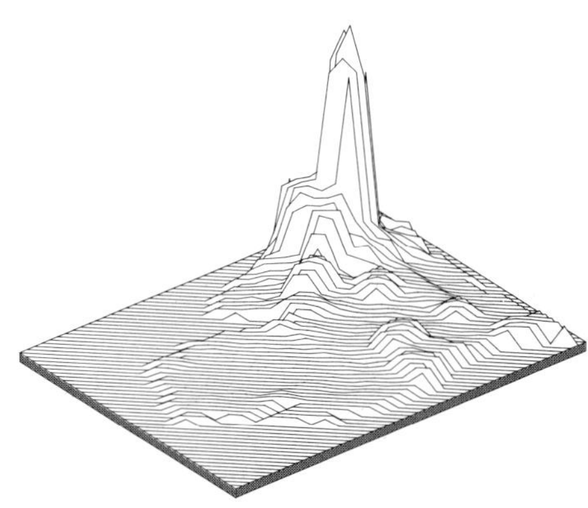

FIGURE 1.13 Nontraditional ways of portraying population density in the lower peninsula of Michigan. Perspective views of three-dimensional models as seen from the northwest. (From R. E. Groop and P. Smith, "A Dot Matrix Method of Portraying Continuous Statistical Surfaces," *The American Cartographer* 9, (1982):123–30. Courtesy R. E. Groop).

guidelines. The emphasis is on creative expression and whatever seems to work with the situation at hand, rather than on following established conventions. Standards of accuracy in this kind of cartography are not metrical. Rather, their quality is to be judged by the subjective responses of the viewer. Innovation and variety are characteristic.

Communication Focus

The communication focus identifies the principal task of cartography to be the effective communication of information via the use of a map. It is based on the belief that graphics (including maps) play an important role in human thought and communication, which in many ways is comparable to that served by natural languages and mathematics/statistics.

From a communication perspective, making and using a map are treated equally, and an attempt is made to design maps in such a way that the user's ability to retrieve information from them will be enhanced (Fig. 1.14). This aim is particularly appropriate in the case of thematic mapping, where the goal is to create a general impression of a phenomenon's spatial distribution or form rather than to provide information about individual places. In thematic mapping there is considerable latitude in the application of the principles of cartography, and the need for accuracy or precision is limited by the user's ability to obtain information or create impressions from the map. As a consequence, thematic portrayals tend to be relatively forceful, exhibiting the clean, crisp graphic characteristics that simplicity makes possible.

With the communication focus the mapping process is seen as a series of information transformations, each of which has the power to alter the appearance of the final product (Fig. 1.15).* In data collection the environmental information is distorted through the filters of ground survey, census, remote sensing, or compilation procedures. Through generalization, mapping further modifies these data by the abstraction processes of selection, classification, simplification, and symbolization. Finally, the use of the map leads to the distorting effects of map reading, analysis, and interpretation. The point here is that there are many possible maps of the same geographical information, each of which will possess certain communication advantages and limitations. The

*W. R. Tobler, "A Transformational View of Cartography," *The American Cartographer* 6 (1979): 101–6.

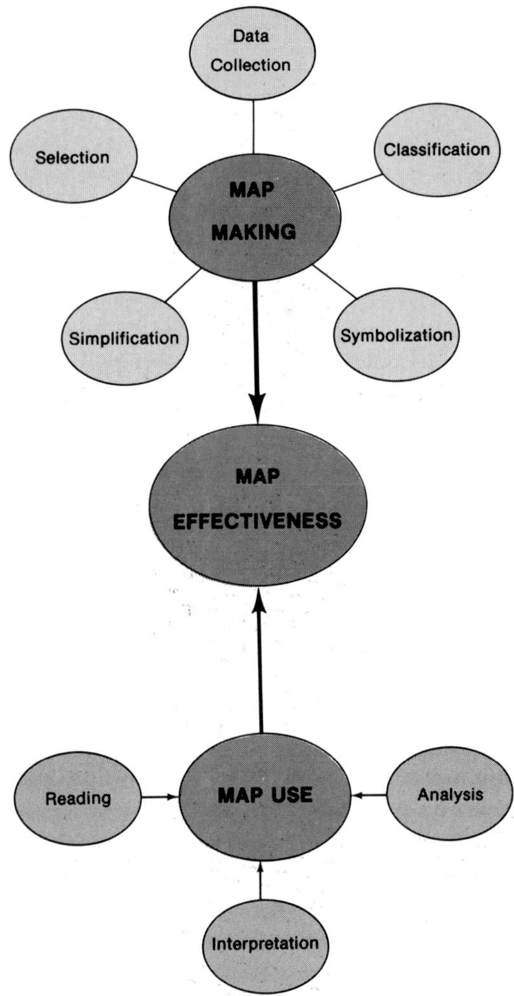

FIGURE 1.14 Basic characteristics of the communications focus.

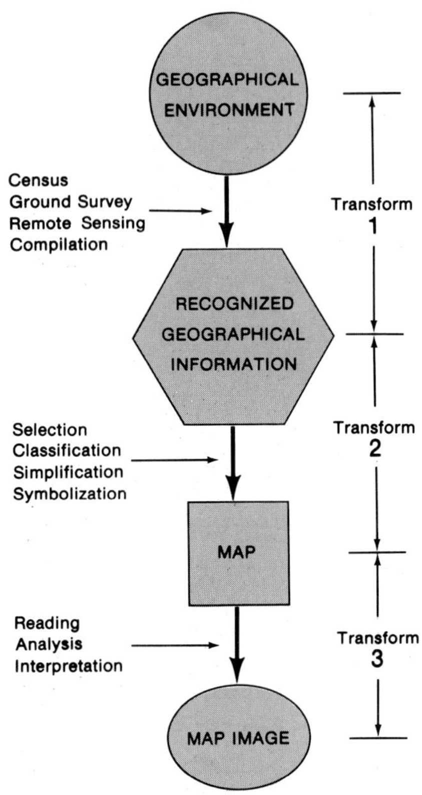

FIGURE 1.15 Fundamental information transformations in cartography.

cartographer's task is to explore the ramifications of each mapping possibility and to select the most appropriate for the intended communication.

What the communication conception emphasizes, then, is that the mapping effect may in some contexts be as valuable as the metrical accuracy of mapping in geometric terms might be in other contexts. In its view, the great power of the mapping process lies in its ability to provide fresh, insightful perspectives, sometimes even distorted ones, on our environment. Furthermore, the effect of a map is in large part a function of the user's skill, experience, and perceived needs.

THE SCOPE OF CARTOGRAPHY

A professional normally works primarily within one of the main areas of emphasis of the field, but regardless of which one that may be, the trained cartographer must know about a broad scope of information and processes. We can liken cartography to a kind of drama played, at a minimum, by two performers with two stage properties. The actors are (1) the cartographer, and (2) the map user. In some instances the same person can take both parts, but the parts are played sequentially. The properties are (1) the map, and

(2) what we may call the "data domain," which is simply a term to encompass all potential information that may be put on a map, read from a map, or be ancillary to either process.

The normal sequence is for the cartographer to select data from the data domain, to process those data into a map format, then for the user to observe and respond to or use the data. This identifies four categories of processes in cartography:

1. Collecting and selecting the data for mapping
2. Manipulating and generalizing the data, designing and constructing the map
3. Reading or viewing the map
4. Responding to or interpreting the data

It follows that the five focuses or conceptions of cartography differ in terms of which of the processes each stresses. The geometric conception emphasizes both the data collection and the map construction processes; the technological view stresses the construction processes; the presentational focus concentrates on the design aspects; the artistic emphasis is primarily concerned with the responses of the viewers; and the communication viewpoint focuses on the design and reading processes. Each conception contains elements of some or all of the other processes, as well as the processes on which it concentrates. Thus, even though some models explicitly stress some of these activities and de-emphasize others, none denies the fact that a cartographer should know about all of them.

The communication model is by far the least restrictive of these five conceptions of cartography. It not only includes all mapping activities, but also embraces the full range of the mapping sciences and relies heavily on input from many associated disciplines. At the same time, the cartographic method is understood to be perfectly general. This means that in addition to geographers, who have long been primary users of cartographic products, the map is seen throughout the sciences, engineering professions, and humanities as being a valuable means for organizing, analyzing, and expressing data and concepts. For these reasons, the communication conception of cartography is used whenever appropriate to organize and present material in this book.

Specifically, this book is organized into four parts. This first part examines the nature of cartography, its history (Chapter 2), and its technologies (Chapter 3). The second part concentrates on some basic aspects of cartography, including scale (Chapter 4), map projections (Chapter 5), manipulating and generalizing data (Chapter 6), and the graphic elements of map design (Chapters 7 to 9). Part III emphasizes data collection (Chapter 10) and generalization (Chapters 11 to 15). Part IV includes map construction and printing (Chapters 16 to 18).

SELECTED REFERENCES

Balchin, W. G. V., "Graphicacy," *The American Cartographer* 3 (1976): 33–8.

Board, C., "Cartographic Communication," *Cartographica* 18 (1981): 42–78.

Bos, E. S., "Another Approach to the Identity of Cartography," *ITC Journal,* 1982-2, pp. 104–8.

Dahlberg, R. E., "Educational Needs and Problems Within the National Cartographic System," *The American Cartographer* 8 (1981): 105–13.

Guelke, L. (ed.), "The Nature of Cartographic Communication," *Cartographica,* Monograph No. 19. Toronto: University of Toronto Press, 1977.

Kadmon, N., "Cartograms and Topology," *Cartographica* 19 (1982): 1–17.

Keates, J. S., *Understanding Maps.* London: Longman Group Limited, and New York: Halstead Press, John Wiley & Sons, 1982.

Kember, I. D., "Some Distinctive Features of Marine Cartography," *The Cartographic Journal* 8 (1971): 13–20.

Morrison, J. L., "Towards a Functional Definition of the Science of Cartography with Emphasis on Map Reading," *The American Cartographer* 5 (1978): 97–110.

Muehrcke, P., "Maps in Geography," *Cartographica* 18 (1981): 1–41.

Petchenik, B. B., "From Place to Space: The Psychological Achievements of Thematic Mapping," *The American Cartographer* 6 (1979): 5–12.

Robinson, A. H., "The Uniqueness of the Map," *The American Cartographer* 5 (1978): 5–7.

Robinson, A. H., and B. B. Petchenik, *The Nature of Maps: Essays Toward Understanding Maps and Mapping*. Chicago: University of Chicago Press, 1976.

Salichtchev, K. A., "Cartographic Communication: Its Place in the Theory of Science," *The Canadian Cartographer* 15 (1978): 93–9.

Thompson, M. M., *Maps for America,* 2d ed. Washington, D.C.; U.S. Geological Survey, 1982.

Tobler, W. R., "A Transformational View of Cartography," *The American Cartographer* 6 (1979): 101–6.

2

The History and the Profession of Cartography

When someone first arranged symbols to represent geographical facts in a reduced, two-dimensional space, it was an achievement in abstract thinking of a very high order, since it enabled people to discern directional relationships, observe associations, and discover geographical structures that could only be known when mapped. No one knows when the first map was made, but mapmaking is probably at least as old as communication by written language.

The story of mapmaking is not only an account of the endeavor to understand our environment, but also a reflection of people's attitudes, beliefs, and priorities at various times. A general awareness of the changes and developments during the long history of mapmaking provides the cartographer with an understanding of the roots of the profession, but perhaps even more important, it provides a background that serves to emphasize how profound are the developments of the present day.

The history of cartography is a field of study to which a great many scholars have devoted their lives. It has an extensive literature, ranging from the very scholarly to the popular. In our brief review we can only touch the high points, and the interested reader will find a number of basic sources listed in the Selected References at the end of the chapter.

THE HISTORY OF MAPMAKING

No one knows when the first cartographer prepared the first map, which was probably a crude representation of locations drawn in soft earth or sand or scratched on a rock. One of the older authentic maps to survive is a clay tablet nearly 5000 years old showing mountains, water bodies, and other geographic features in Mesopotamia. It is thought that portions of the valley of the Nile were carefully mapped in ancient times in order to recover property lines after the annual floods. Figure 2.1 is an early, large-scale Mesopotamian city plan.

FIGURE 2.1 An ancient plan (map) of a city in Mesopotamia on a clay tablet. (Photo courtesy of Norman J.W. Thrower.)

The maps made by technologically primitive peoples of recent times have been studied, and the evidence suggests that mapmaking is a common and well-developed skill, albeit with major differences from the maps made by technologically more advanced groups. All peoples seem to develop an appreciation of the spatial distribution of topographic phenomena and employ whatever materials are at hand to represent vital matters such as routes and hunting grounds. Eskimos use driftwood, pebbles, and bone; South Sea islanders use reeds, shells, and leaves; and Indians use birch bark and sand (Fig. 2.2). Although it is difficult to generalize from the meager data, it appears that such peoples are especially conscious of topological relationships (relative positions and so on) and employ a different system of distance scaling from that normally used in most modern mapping. Distances

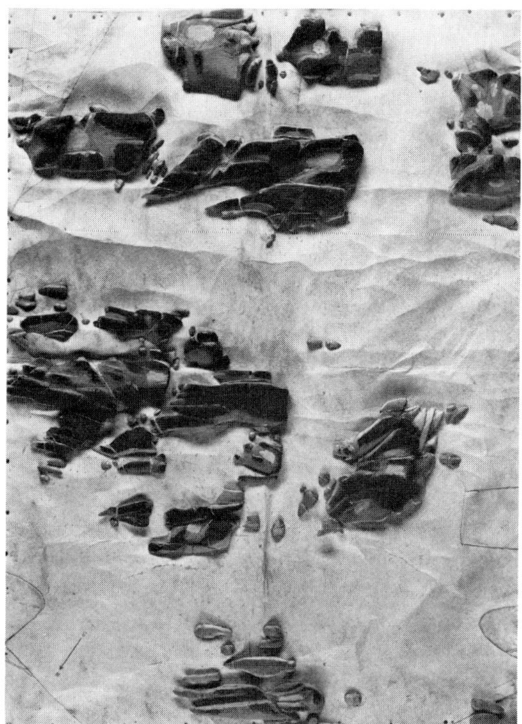

FIGURE 2.2 A map, carved from driftwood, colored and tied with gut to sealskin, showing eighty-three islands and reefs, lakes, swampy ground, tidal areas, and such in an area of about 70 square miles in Disco Bay, Greenland. Made by an Eskimo hunter who had never seen any other map. (From the Library of Congress collection.)

are often thought of in terms of relative travel time, regardless of the actual earth space involved. Both this method of scaling and their topological concerns are eminently practical.

Topographical maps, which portray detailed features of a limited area, seem to have been the earliest maps to have been made in all societies. They certainly would have been the most useful and practical in the era before there was widespread travel and before people began to be curious about the earth as a whole. Only a few fragments of such maps have survived from the ancient periods in Eastern and Western cultures.

Cartography in the Eastern World

From the evidence at hand, it appears that topographical maps matured earlier and attained a higher level of appreciation in the Eastern world than in the Western. The oldest civilization in Asia developed in China and spread from there to other parts of the continent. There is evidence of astronomical knowledge as early as the first historic dynasty, the Shang dynasty, which ended in the eleventh century B.C. There are literary references to geography and maps at an early date, such as the following excerpt from the chapter called "Fragment on Maps" written sometime between about 480 and 100 B.C.

> *All military commanders must first examine and come to know maps. They must know thoroughly the location of . . . defiles, streams . . . , famous mountains, passable valleys, arterial rivers, highlands, . . . the distances of roads, the size of the city . . . , famous and deserted towns, and barren and fertile land. They should thoroughly store up [in their minds] the relative location of the configurations of the terrain. Then afterwards they can march their armies and raid towns. In the disposition [troops] they will know [what lies] ahead and behind, and will not lose the advantages of the terrain. This is the constant value of maps.**

In 1973 three maps of the second century B.C. were discovered in a tomb from the Han dynasty at Ma-wang-tui in the city of Chang-sha in Hunan Province. The maps are manuscript on silk and have greatly deteriorated, but two of the maps, one topographic and one focused on military matters, have been reconstructed. The topographic map shows some thirty streams, twenty roads, many mountain ranges, and identifies over one hundred items by name (Fig. 2.3). The maps are quite sophisticated in that the concept of scale

*From W. Allyn Rickett, *Kuan-Tzu: A Repository of Early Chinese Thought*, Vol 1, (Hong Kong: Hong Kong University Press, 1965), p. 234. Quotation supplied by Professor Mei-Ling Hsu.

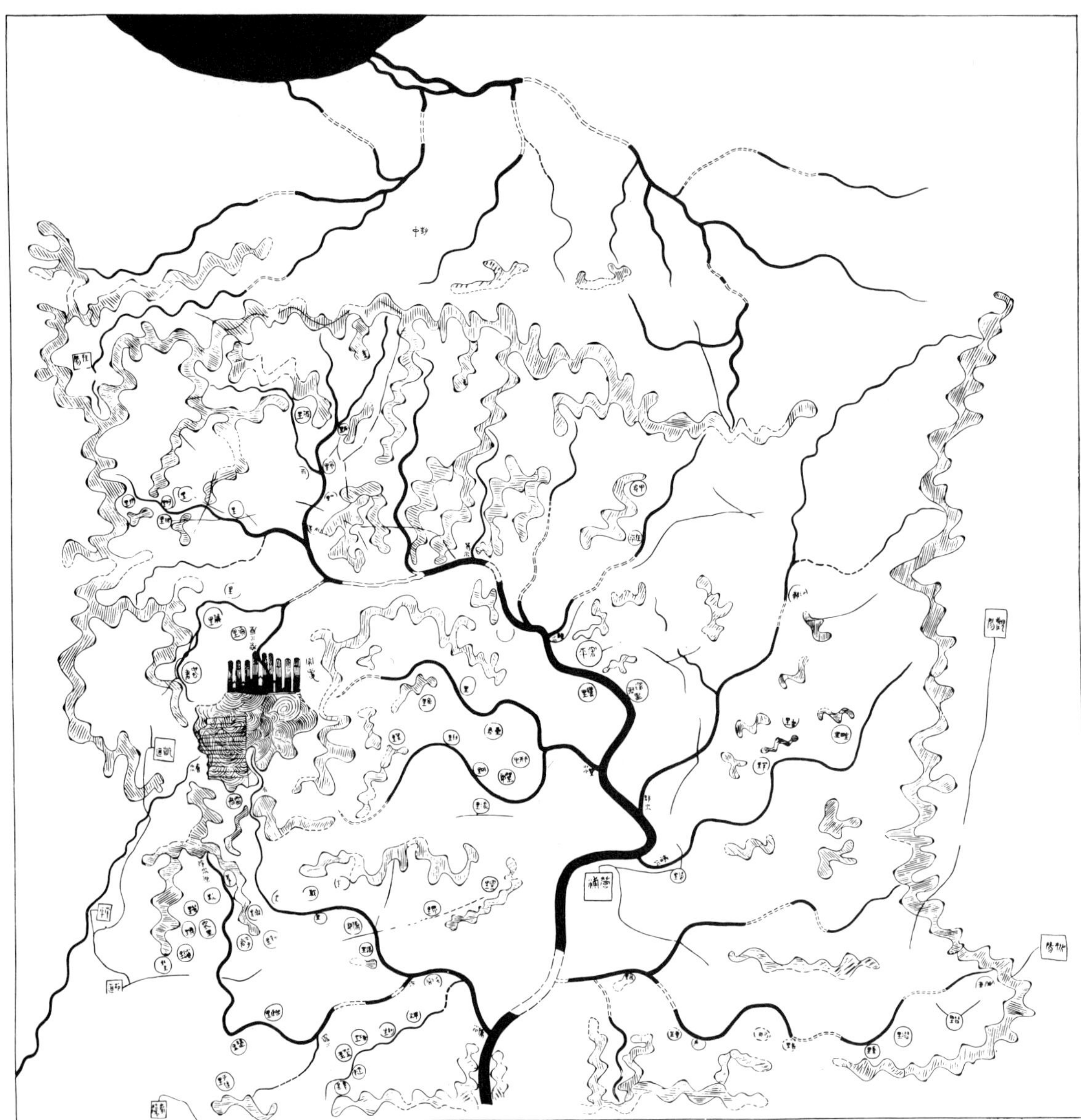

FIGURE 2.3 The reconstructed Han topographic map which covers the south central part of present Hu-nan Province and adjacent areas. The original map is 96 by 96 cm. The wavy, shaded areas are symbols for mountain ranges, and the black lines are rivers. South is at the top. (From *Wen Wu* [1975], No. 2. Reproduced with the permission of the Cultural Relics Publishing House, Beijing, Peoples Republic of China. Courtesy of Mei-Ling Hsu.)

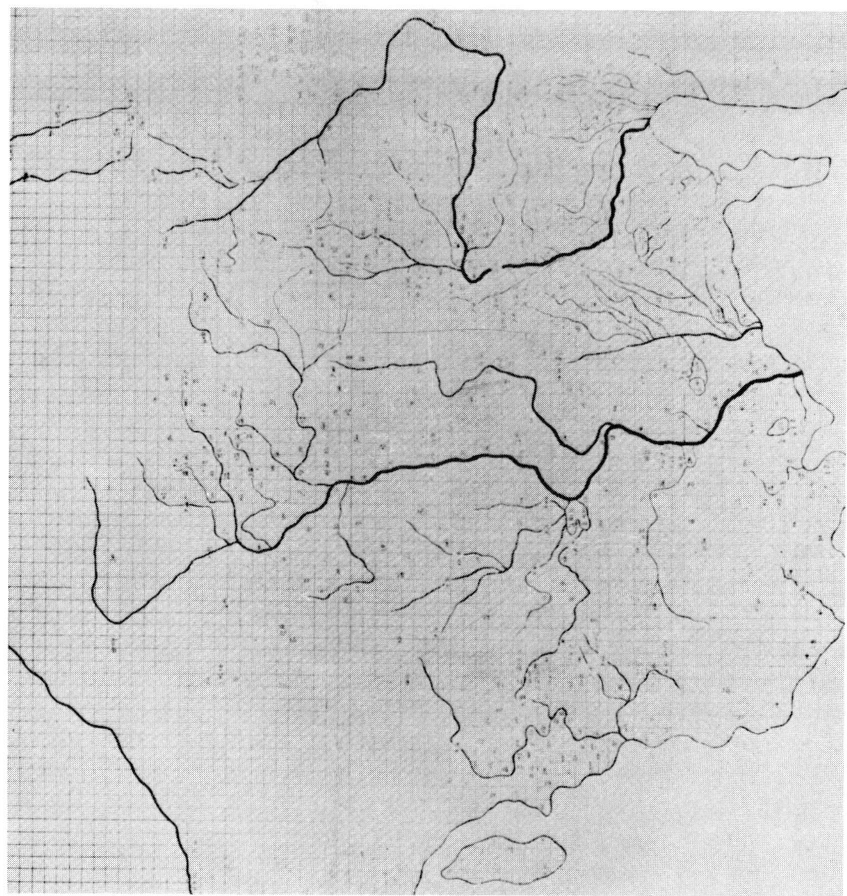

FIGURE 2.4 The earliest extant, small-scale map of China, the "Yü chi tu," carved in stone about 1137 by an unknown cartographer of the Sung dynasty showing the characteristic square grid. The legend states that the side of each square represents 100 *li*. (The *li* is the Chinese mile; one *li* = ca. 0.5 km.) The map is remarkably accurate, as comparison with a modern map will show. (Photo courtesy of Norman J.W. Thrower.)

is clearly understood, the symbolism is well developed, and an orthogonal viewpoint (perpendicular to the map) is assumed. The scientific, technological, and artistic qualities of the maps suggest that they came from a long tradition.*

One of the more distinctive features of early Chinese cartography is the use of a square grid over the map area (Fig. 2.4). The lines are not latitude and longitude lines, and there is no real evidence of the use of projections by the early Chinese mapmakers. The square grid is just that, a grid made up of equally spaced, orthogonal straight lines. Each map specified the length of the sides of the squares in earth distance, and in that sense it was a statement of scale. With it, it was easy to work out straight-line distances by

* Mei-Ling Hsu, "The Han Maps and Early Chinese Cartography," *Annals of the Association of American Geographers*, 68, no. 1 (1978): 45–60.

the use of the Pythagorean theorem, a relationship long understood by Chinese mathematicians. In compiling from one map to another on a different scale, the square grid would be a useful guide. Although the square grid is similar to the rectangular reference plans used after the sixteenth century to specify the locations of places on maps and to the modern plane coordinate systems, there is no evidence that it was employed for those purposes in China.

A number of early Chinese scholars and cartographers should be noted. The square grid may have been introduced by Chang Heng, an astronomer who flourished in the second century A.D. The so-called father of Chinese cartography is Pei Hsiu (A.D. 224–271) who was minister of public works during the Chin dynasty. In a preface to his now-lost map "Yü kung ti yü tu" (Map of the Tribute of Yu), he enunciated six principles to be followed in mapmaking: The first three are that scale should be defined by using graduated divisions, that a square grid should be used to depict correct locations, and that right-angled triangles should be used to derive distances. The last three deal with the problems of determining positions in uneven terrain and locating them properly on the plane surface of the map. Pei Hsiu's principles were followed by other analytically minded cartographers in later periods, but as has often been the case, their maps have disappeared even though descriptions of them exist. Notable among these is the large "Hai nei hua yi tu" (Map of both Chinese and Barbarian Peoples within the Four Seas) by Chia Tan (730–805) and the "Yü tu" (Terrestrial Map) by Chu Ssu-pen (1273–1337). Although Chu Ssu-pen's map was later lost, it had been expanded and put in the form of a forty-five-map atlas by Lo Hung-hsien (1504–1564) entitled *Kuang yü tu* (Enlarged Terrestrial Map), which went through seven editions.

In some ways significant to cartography, China was considerably more advanced technologically than the Western world. The compass was used much earlier there, paper was invented in China in the second century A.D. (it was not really used in the Western world until about 1000 years later), and the first printing of a map in China occurred about A.D. 1155, 300 years earlier than in Europe.

By the seventeenth century, European influences began to make themselves felt in Chinese cartography, particularly through the work of Jesuit missionaries. Thereafter the mapmaking of the Eastern and Western worlds ceased to be very distinctive.

Early Cartography in the Western World

The extension from a concern with the local area to the contemplation of the character of other lands or even the nature of the whole earth is very large. No one knows when this took place in the Western world, but we do know that by the time of Aristotle (384–322 B.C.), the earth was recognized as being spherical from evidence such as differences in the altitudes of stars at different places, from the fact that shorelines and ships seemed to "come over" the horizon as one moved across the sea, and even from the assumption that the sphere was the most perfect form.

By the second century B.C. the system of describing positions on the earth by latitude and longitude and the division of the circle into 360° had become well established. The development of civilization allowed more frequent travel, and a greater interest in faraway places was accompanied by increased thought about ways of presenting the relationships of areas on maps. Estimates of the size of the earth were made by the ancient scholars Eratosthenes (ca. 276–195 B.C.) and Posidonius (ca. 130–50 B.C.) from angular observations of the sun and stars in the eastern Mediterranean area (see Fig. 4.1). The methods used were entirely correct, but the required assumptions and the precision of their observations were not. Nevertheless, although we cannot be absolutely sure of their results in terms of modern units of measure, the estimates seem not to have been very far off. Both apparently overestimated the size by only 12 to 15 percent.

From the occasional references to maps in the classical Greek literature we may infer that mapping was not considered to be anything unusual. Consequently, although we may assume that many maps were made during the time of the classical Greeks, none of the actual maps appears to have survived. Fortunately for our understanding of the cartography of the ancient period, there is, however, a record that apparently clearly reflects the advanced stage to which cartography had developed by the end of the Greek period; these are the writings attributed to Claudius Ptolemy (ca. A.D. 90–160).

In the ancient Carthaginian trading city of Tyre on the southern part of the coast of present-day Lebanon, a cartographer-geographer, Marinus, a contemporary of Ptolemy, wrote and made maps. Unfortunately, Marinus's materials are lost, but Ptolemy set out to correct Marinus and, by good fortune, what are thought to be Ptolemy's writings have survived. Ptolemy lived and worked in Alexandria, Egypt, which had become the intellectual center of the Western world. There was a great library at Alexandria as well as a community of scholars. Ptolemy brought together into a book, called simply *Geography,* all that was apparently then generally known concerning the earth. The *Geography* included, among other things, a treatise on cartography.

He described how maps should be made; he gave directions for dealing with the problem of presenting the spherical surface of the earth on a flat sheet; he recognized the inevitability of distortion in the process; and in many other ways he described what was known about cartography at that time. Ptolemy also made a list of numerous places in the world that he had learned about from the writings of others and from travelers, and he gave his best estimate of their positions in latitude and longitude terms. To illustrate this he is thought to have provided a series of maps of local and larger areas (Fig. 2.5); this is not certain, however, because if he did, they have since been lost. He did give a detailed account of how they were made. Ptolemy's writings were lost to the European world for more than 1000 years, but fortunately they were preserved by the Arabs and later came to light again in Europe. From the descriptions the Ptolemaic maps were reconstructed, and they had a profound influence on European geographical and cartographical thinking during the Renaissance.

Medieval Cartography

The Greeks and Romans apparently were technologically rather primitive, and the "classical" period is classical only in philosophical and intellectual matters. So far as we can tell, the writings of Marinus and Ptolemy's *Geography* were unique, and there is no reference to any other such activity for many centuries. The Romans did considerable surveying of landholdings, and there are references in Roman literature to administrative and engineering types of maps but, as the Roman Empire declined, concern with the general theory and practice of cartography all but died out.

Ptolemy's concern in the second century A.D. with projections and systematic ways of making maps to show geographical facts was succeeded in following centuries by a quite different attitude. The map became an artistic and didactic device to illustrate scriptural theory about the nature of the earth, and objective geographical thinking about faraway places was replaced by fancy and whimsy. Maps of the "known world" (known as *mappae mundi* in Latin) were produced but, whereas during the earlier Greek period these had been based on observation and reason, they now came to be but media for preserving the results of fanciful speculation and literal interpretations of biblical passages. Quite a variety were prepared containing various symbolic representations of the earth, such as rectangular maps probably based on the scriptural reference to the "four corners of the earth," and

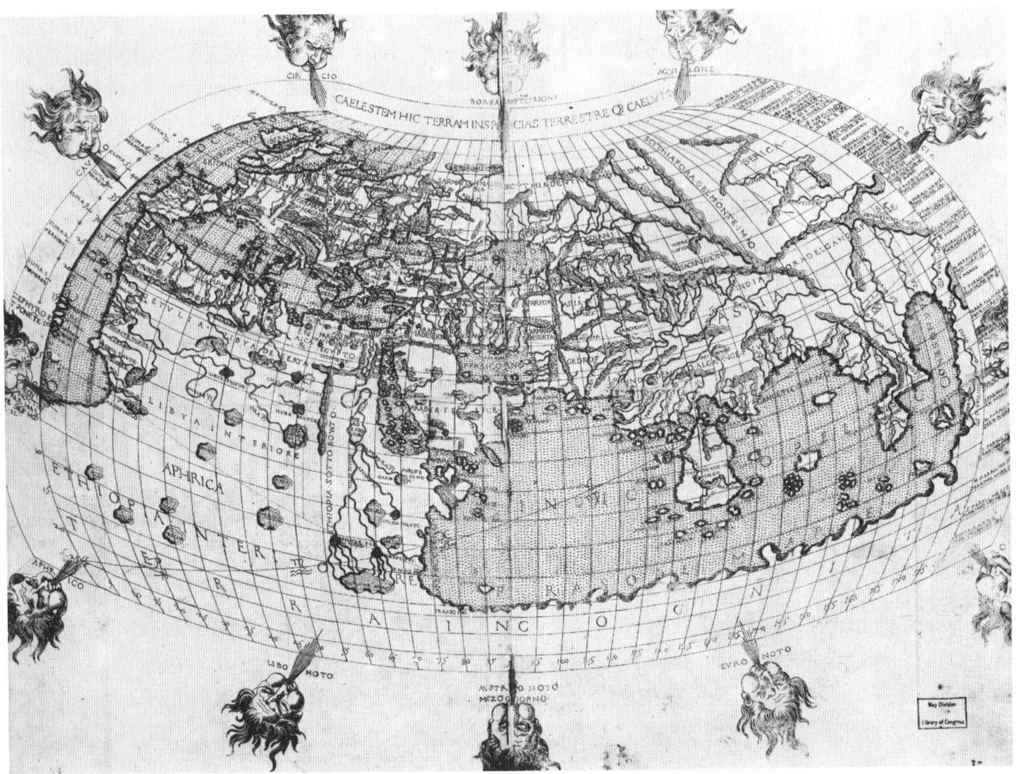

FIGURE 2.5 One of the maps constructed from Ptolemy's written directions and descriptions. The area shown extends from the Atlantic on the west to beyond the closed Indian Ocean. This "world map," reflecting the knowledge of the second century A.D., provided what was thought to be up-to-date information about the known world when the written works attributed to Ptolemy were reintroduced to Europe in the fifteenth century. (From Berlinghieri's edition of Ptolemy's *Geography*, Florence, 1482. From the Library of Congress collection.)

circular maps with the holy city of Jerusalem at the center. These maps reflected the threefold division of the earth among Shem, Ham, and Japheth, the sons of Noah. They are called T in O maps, because they were designed with the Mediterranean as the upright part of the T, the Don and Nile rivers as the crosspiece, and the whole inside a circular ocean (Fig. 2.6).

The farthest area that was known at all was, of course, the Orient. It became traditional to locate Paradise in the difficult-to-reach, far eastern area and to put it at the top of the map. From this practice we have derived the term *to orient* a map—that is, to turn it so that the directions indicated are understood by the reader—but today to orient a map means to arrange it so that map and earth directions correspond or, alternatively, so that north is at the top.

On these maps were often placed mythical places, beasts, and dangers such as the kingdom of the legendary Gog and Magog, who were nonbelieving menaces to the Christian world. Most of these maps were very small and served as illustrations in manuscript books. This kind of cartography hung on for a long time, and some of the later maps that follow this general pattern are very detailed and ornate. One of the larger is the map from the Hereford Cathedral in England (Fig. 2.7).

The earlier part of the medieval period in the

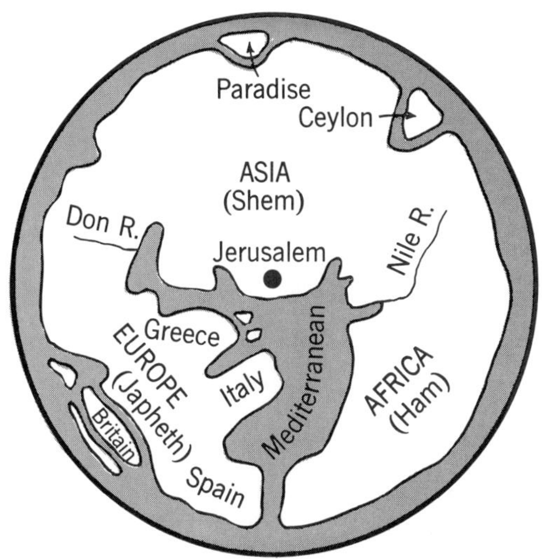

FIGURE 2.6 The common layout of a T in 0 map. On such maps Jerusalem was usually located at the center and the farthest east, where Paradise was thought to be, was placed at the top. The known world was surrounded by the river Oceanus. Compare with Figure 2.7.

Western world is called the Dark Ages, and it was not until after the first millennium of the Christian era that enlightenment began to show in the field of cartography and geography. Three advances can be noted: (1) the contributions of the Arabs, (2) the growth of interest in faraway places, and (3) the remarkable development of portolan charts.

The Arabs had preserved the writings of Ptolemy, had translated them, and had made several attempts to measure the size of the earth in the ninth century A.D. The culmination of Arab cartography was reached in the work of Idrisi (A.D. 1099–1164), an Arab scholar at the court of King Roger of Sicily. His world maps, made in A.D. 1154 and 1161, were a great advance over contemporary non-Arab works (Fig. 2.8).

The first few centuries of the second millennium A.D. saw a change taking place in intellectual standards that was reflected in the cartography. The transformation began slowly, but the several Crusades, the travels of missionaries and merchants, such as the Polos, and the generally increased movement of peoples and goods began to awaken an interest in the outside world; the fancy and superstition characteristic of the Dark Ages began to recede. With the translation into Latin, and then the printing of the writings and maps attributed to Ptolemy (items that had lain dormant for 1000 years) in the fifteenth century, a new interest in geography and cartography developed in Europe.

One of the more remarkable occurrences in the history of mapmaking is the sudden appearance in the thirteenth century of quite accurate sailing charts called portolan charts. They were the products of the experience of a large amount of open-water and coastwise sailing in the Mediterranean and adjacent areas (Fig. 2.9). Characteristically, they were covered by a systematic series of unlabeled, intersecting, straight, directional lines radiating from wind, and later, compass roses. The lines and shapes of the bounding coasts of the seas on these charts are surprisingly accurate but, unfortunately, this accuracy did not seem to extend to maps of the land behind the coasts until much later. Although many portolan charts must have been made, none earlier than the late thirteenth century has survived.

The medieval period in cartography may be said to have come to an end during the fifteenth century.

The Renaissance in Western Cartography

The Age of Discovery, centering on the monumental achievements of Columbus, da Gama, Cabot, Magellan, Elcano, and others at the end of the fifteenth and the beginning of the sixteenth centuries, kindled such an interest in the rapidly expanding world that mapmaking and publishing soon became a lucrative calling. By the latter half of the sixteenth century, the profession was generally in good standing and was well supported, even though its products were still far from being first-class examples of objective scientific thought.

One of the circumstances that contributed greatly to the rapid advance of cartography was

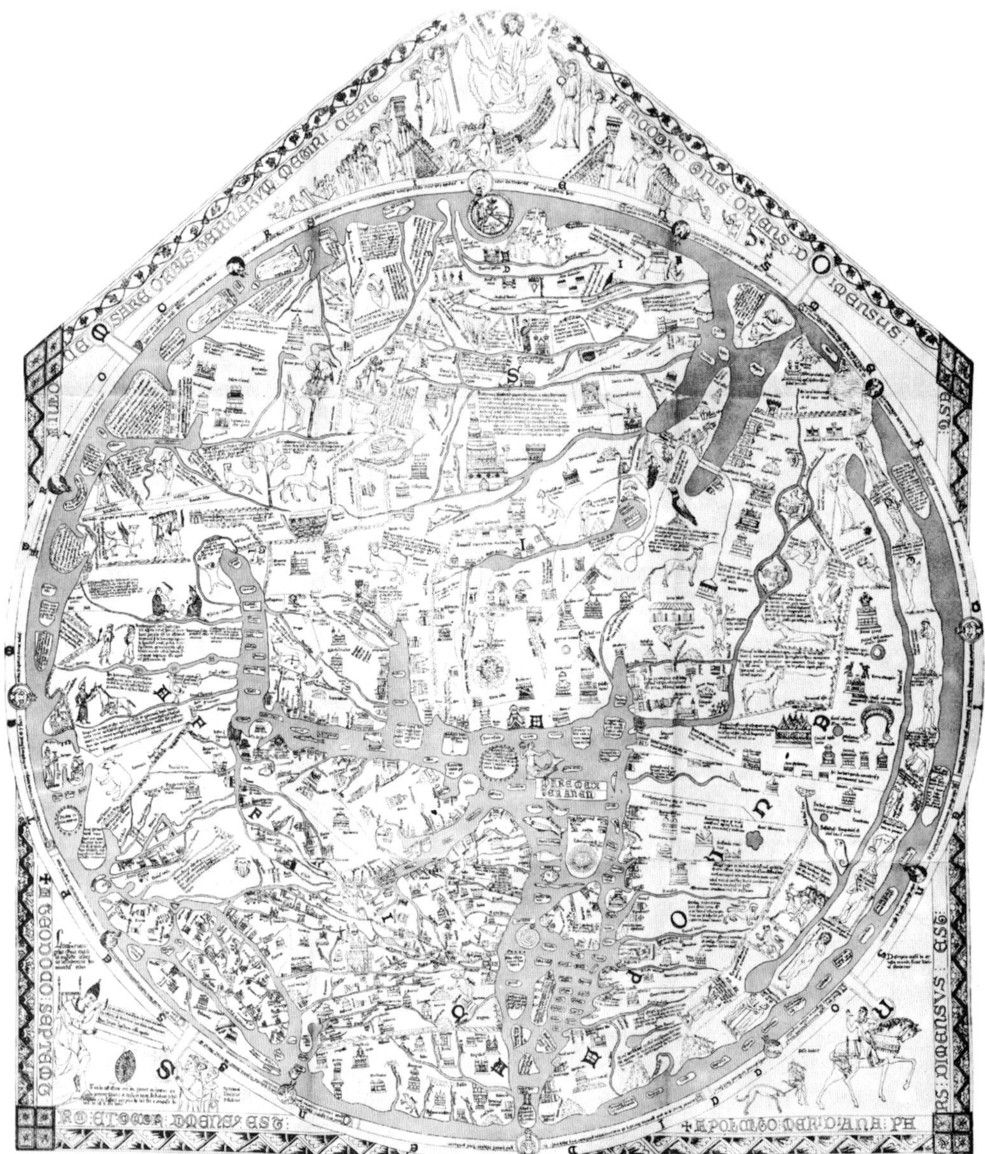

FIGURE 2.7 The Hereford map. This map, 1.34 by 1.65 m, was made in the thirteenth century to serve as an altarpiece in the cathedral at Hereford, England, where it has been preserved. It illustrates the degree to which much cartography had become figurative and allegorical compared to that in the time of Ptolemy, 1000 years earlier. The map is oriented with east at the top and Jerusalem at the center. (Photo from the Library of Congress collection.)

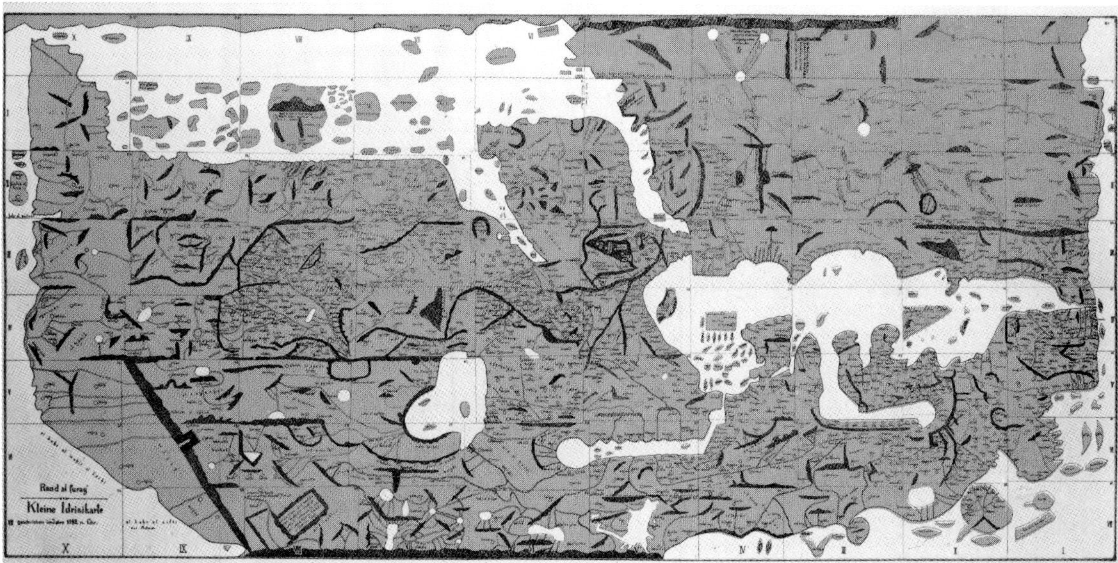

FIGURE 2.8 A redrawing, by Konrad Miller, of the smaller of two world maps by Idrisi made in A.D. 1161 as reduced from the original seventy-three sheets. The original is oriented with south at the top, a common practice at that time. Here it is displayed with north at the top so that the geographical relationships can be more easily recognized, since we are now used to seeing most maps oriented that way. (From the Library of Congress collection.)

the invention in Europe shortly after 1450 of printing and engraving, which soon made possible the reproduction of maps in numerous copies. Previously, each copy of a map had to be laboriously hand drawn. Great map-publishing houses such as Mercator, Blaeu, Hondius, and others in Holland and France rose and flourished. Their maps were primarily general maps containing not much more than coastlines, rivers, cities, and occasional crude indications of mountains. Fancy and intricate craftsmanship was popular, and the maps were richly embellished with ornate scrolls, compass roses, and drawings of people, animals, and ships (Fig. 2.10). Except for religious and some navigational data, the mapping of any geographic information beyond what we today call base data was unknown. The mapping of most other kinds of geographical distributions had to wait for the inquiring minds who would want such maps.

Between 1600 and 1650 a new and fresh attitude toward cartography arose. For the first time since the end of the classical era 1500 years earlier, accuracy and the scientific method became fashionable again. This attitude, replacing the dogmatic and unscientific outlook that was more or less dominant during the long Dark Ages, made itself evident in a number of ways in cartography.

In the second half of the seventeenth century the French Academy of Sciences was founded with one of its avowed purposes being the improvement of charts and navigation. This, of course, included cartography among its important concerns. Precise navigation in the open ocean was becoming a serious problem, and its solution depended on an accurate determination of the size and shape of the earth and on the development of a practical method for determining the longitude. The necessity for increased mobility in military actions also made desirable the development of land survey methods.

The French Academy began by measuring carefully the arc along a meridian and, with the new technique of triangulation, it began to po-

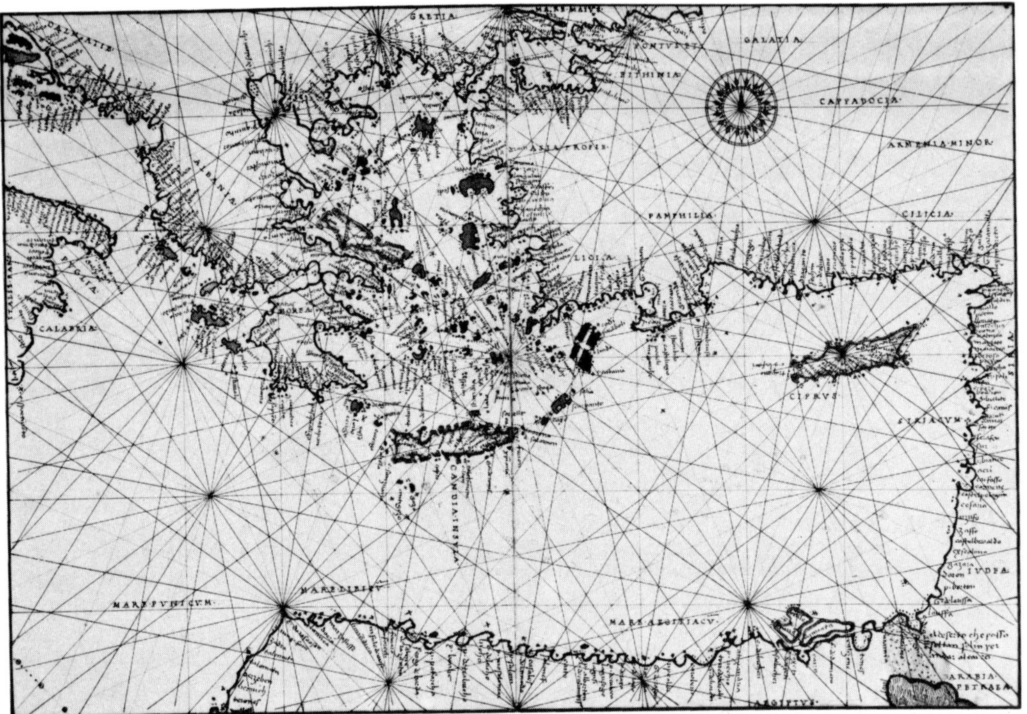

FIGURE 2.9 A section of one of the later portolan charts. Note the detail and relative accuracy of the delineation of the coasts of the eastern Mediterranean. The maps are drawn on vellum and decorated in various colors including gold. (From Plate 12 of the manuscript atlas prepared by Battista Agnese of Venice about 1543, now in the Library of Congress.)

sition more accurately the outlines of France (Fig. 2.11). Because of differences noted in the lengths of degrees along the meridian and the behavior of the pendulum at various latitudes, the question arose as to the precise shape of the earth. During the first half of the eighteenth century, French expeditions were sent to Peru and Lapland to measure other arcs along meridians. Their determinations settled once and for all that the polar radius was shorter than an equatorial radius.

In the middle of the eighteenth century the French initiated a detailed topographic survey of their country at a scale of 1:86,400 (about $1\frac{1}{4}$ mi. to the inch) and almost completed it prior to the end of the century (Fig. 2.12). Harrison's chronometer for longitude determination was perfected in England in 1765, and there were many evidences of curiosity about the earth. But perhaps the most notable and significant cartographic trend was the realization by many that their fund of knowledge about the land behind the coastlines was quite erroneous. Even the administrators and rulers of countries, particularly in Europe, became aware that it was impossible to govern (or fight wars) without adequate maps of the land. Soon other great national topographic surveys of Europe were established, such as that of England in 1791, and the relatively rapid production of topographic maps followed.

The problem of representing the land-surface form more precisely arose and, almost as quickly, devices such as the hachure and contour were developed. By the last half of the nineteenth century, a large portion of Europe had been covered by topographic maps. These maps were expensive to make, however, and did not have a wide

FIGURE 2.10 The Hondius map of America. This map was included in the Hondius-Mercator Atlas of 1606. The original is in the Newberry Library, Chicago. (Courtesy Rand McNally and Company.)

distribution. But they were the foundation on which all future cartography of the land was to be based.

Another development was the official production of charts. As noted above, as early as the thirteenth century nautical charts were in use to assist the mariner, but their production was largely in private hands. By the early sixteenth century in both Spain and Portugal, official organizations were established to supervise chart making, but it was not until the nineteenth century that the great modern governmental hydrographic offices were established to produce the detailed charts needed for safe use of the oceans. These counterparts to the great land surveys produced navigational charts, bathymetric maps ("topographic" maps of the sea bottom), and a variety of special purpose oceanic maps.

The Introduction of the Metric System. Before the beginning of the nineteenth century, every country had its unique system of weights and measures. This caused enormous difficulty, because no one knew precisely the relationship of one national unit to another (for example, the French *toise* to the English yard). Consequently, it was difficult to compile maps and to convert the scales of maps.

Earlier attempts by scholars to institute a new unit of length were given considerable impetus by the proposal in 1791 of the French Academy of Sciences and by the National Assembly's approval of its plan to employ the length of a meridian quadrant in fixing a universal unit.* The

*Georg Strasser, "The Toise, the Yard and the Metre—The Struggle for a Universal Unit of Length," *Surveying and Mapping* 35 (1975): 25–46.

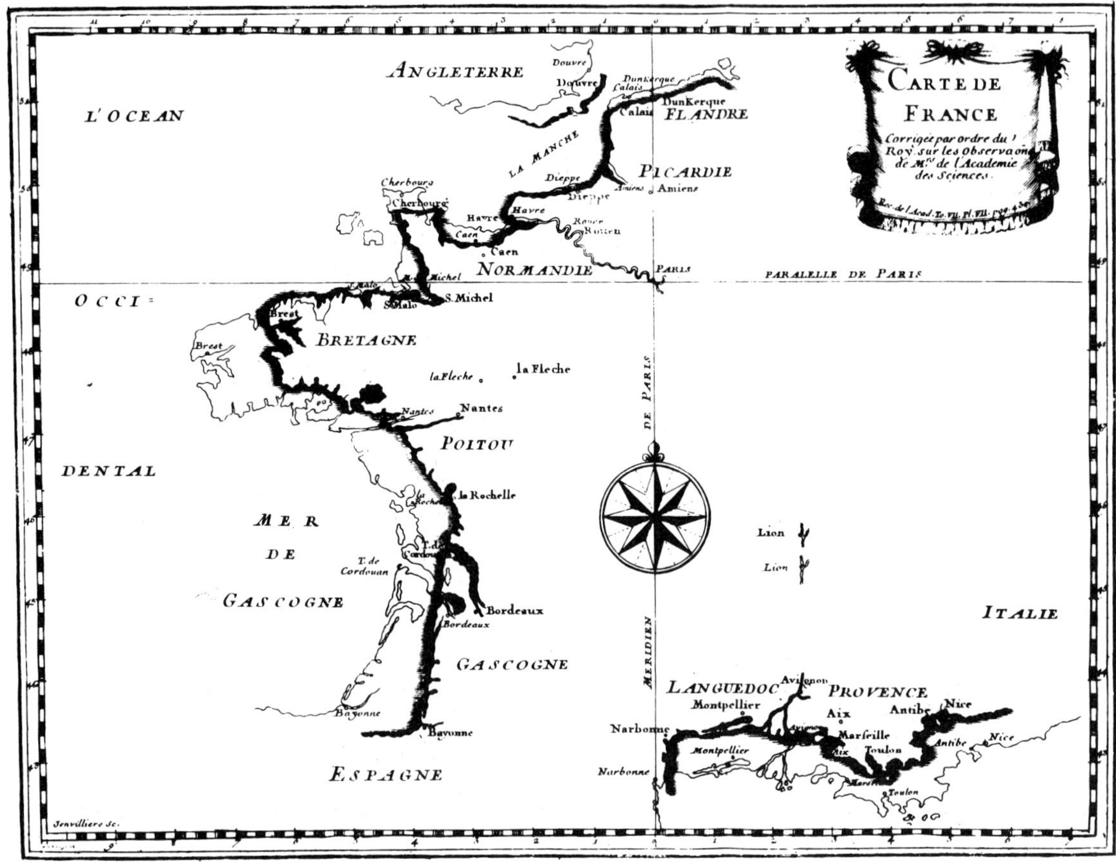

FIGURE 2.11 The map published by the French Academy of Sciences showing the more accurate 1693 coastlines of France (shaded) compared with their delineation on a 1679 map by the highly respected cartogrpaher Sanson.

Revolution in France first delayed and then promoted what was to become the International System of Units, in which the basic unit of distance is the meter, originally defined as 1/10,000,000 part of the arc distance from the equator to the pole. The metric system provided a kind of natural universal unit to which every other unit could be compared. Soon thereafter, in the early nineteenth century, map scales began to be stated as fractions or proportions in which one unit on the map represented so many of the same units on the earth, for example, 1:100,000. When map scales are stated in this way, conversions are easy, because such a proportion is independent of any one kind of unit of measurement. The use of the metric system led to "round number" scales such as 1:25,000 instead of odd scales such as 1:63,360 (1 in. on the map represents 1 mi. on the earth).

All the world, except for a few areas including the United States, have officially adopted the metric system with its easy decimal relationships. It is used in all scientific inquiry and, even though not mandatory in civilian use, it has been legal in the United States for many years. The military agencies adopted it years ago, and most maps made in the United States now are either "metric" or include graphic scales that show metric units. Ultimately the United States will officially "go metric."

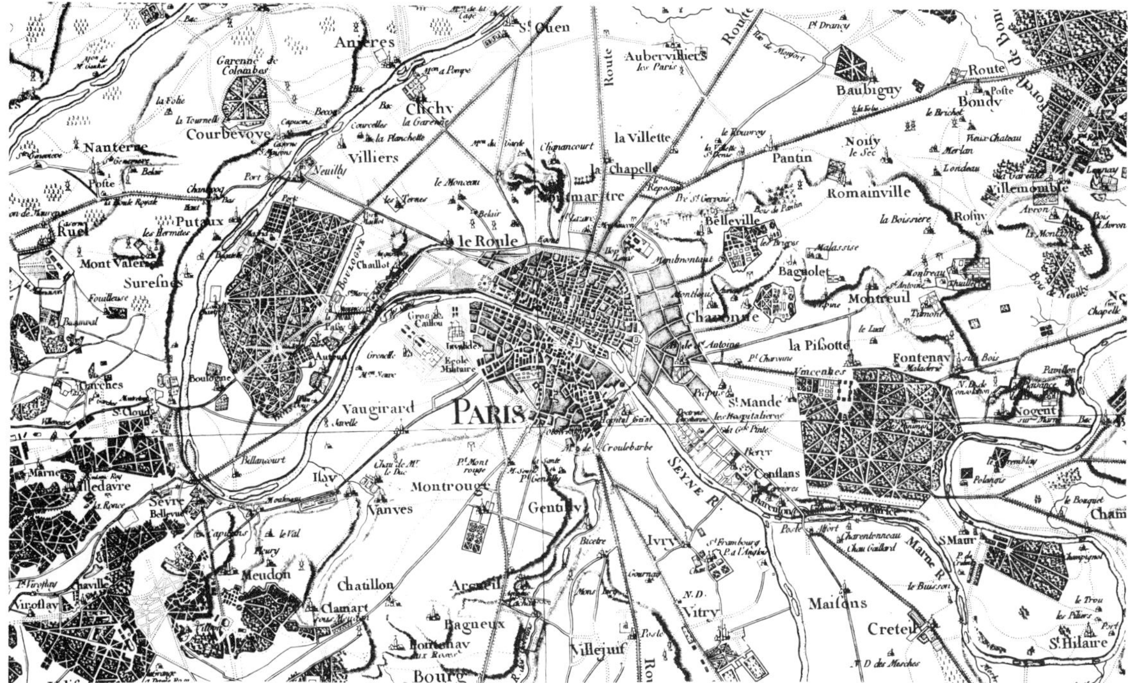

FIGURE 2.12 A section (reduced 50 percent) of the Paris sheet of the first French topographic survey map, called the "Carte de Cassini." The map was concerned with horizontal positions (planimetry), not with elevations.

The Rise of Thematic Cartography

Until the eighteenth century most maps simply showed where places, rivers, coasts, and boundaries were located. Beginning slowly in the latter part of the seventeenth century, in addition to the general small-scale and large-scale topographic maps and navigational charts, a second great class, thematic maps, was added to the repertoire of cartography. Thematic cartography had sketchy beginnings and had to wait for the broadening of scientific inquiry and the study of people and their institutions, which grew rapidly, especially in the nineteenth century. This required much mapping. A purely thematic map is quite different from a general atlas map, a navigational chart, and a topographical map. It is a kind of graphic "geographical essay," since its main objective is to portray geographical relationships regarding particular distributions; it is concerned with such facts as densities, relative magnitudes, gradients, movements, and various other environmental and geographical aspects of earthly phenomena. Often general maps have some "thematic" content, and the data of a thematic map are usually displayed on a background base of general locational data.

The variety of subject matter mapped after thematic cartography developed was very great. Of particular interest in the history of thematic mapping is Edmund Halley's publication in 1701 of a map showing the distribution, as then known, of the variation of the compass, that is, the angular difference between true north-south and the alignment of the compass needle (Fig. 2.13). The famous astronomer-cartographer portrayed this by connecting points of equal variation (also called declination) with lines. Later named *isogonic lines,* these Halleyan lines or "curve lines," as he called them, led to the widespread use for many purposes of this kind of line symbol (isarithm or isoline) in later times.

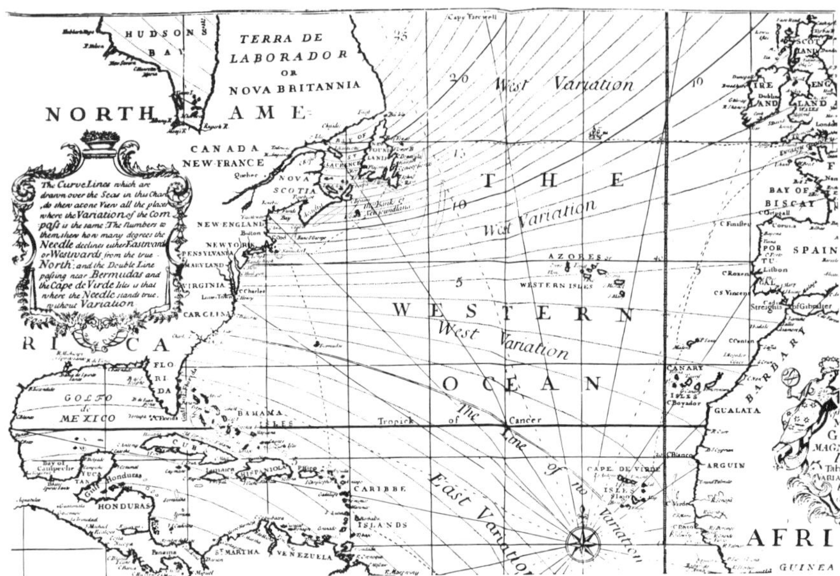

FIGURE 2.13 A portion of Halley's isogonic chart of 1701 showing "curve lines" (isarithms or isolines), which are intended to connect points with the same variation of the compass. Note the compass rose in the lower right with its radiating rhumb lines. (From the Library of Congress collection.)

FIGURE 2.14 Part of a map of Ireland, made in 1837 by H. D. Harness, showing passenger movement by regular public conveyance. The map was made to help plan where to locate railway lines.

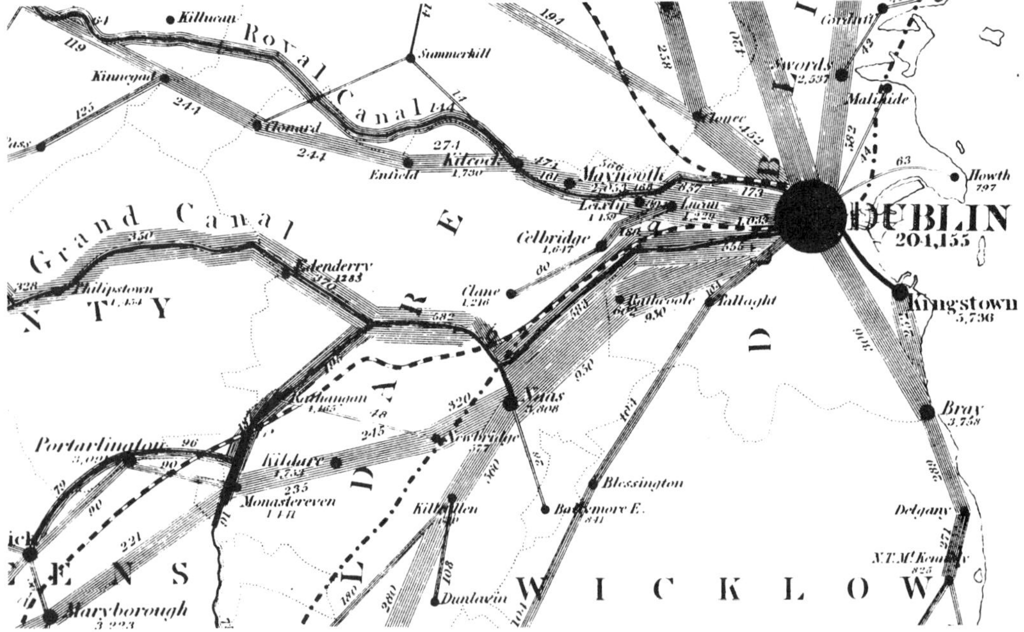

Many other new kinds of ways of showing thematic data were devised, especially in the first half of the nineteenth century. They range from variable shading to show the incidence of cholera or criminal activity to proportional flow lines to portray the transportation of goods and people by the newly developed road and railway systems (Fig. 2.14).

Most important to the development of thematic cartography was the branching out of science into a number of separate fields; this was in contrast to its previous state, which was a kind of all-inclusive complex of physical science, philosophy, and general geography. The more exact studies such as physics, chemistry, mathematics, and astronomy had progressed far; but earth and life scientists who were concerned with certain classes of earth distributions, such as the physical geographer, geologist, meteorologist, and biologist, and the investigators whom we now call "social scientists," were just beginning their studies. The taking of censuses and the recording of all sorts of statistics from weather observations to crime rates was generally initiated during the early part of the nineteenth century, and this also had a significant effect on the development of thematic cartography. Most of the rapidly growing earth, life, and behavioral sciences needed maps and, in general, they required the smaller-scale, compiled, thematic maps.

The Growth of Modern Cartography

Many factors helped to promote an acceleration in cartography during the nineteenth century. They include the development of lithography, which made possible the easy and inexpensive duplication of drawings; the invention of photography; the development of color printing; the rise of the technique of statistics; the growth of mass transport, such as the railroad; and the rise of professional scholarly societies.

By the beginning of the twentieth century, the thirst for knowledge about the earth had led to remarkable strides in all aspects of mapmaking. The colored lithographic map was fairly common; a serious proposal to map the earth at the comparatively large scale of 1:1,000,000 had been made; and some great map-publishing houses such as Bartholomew in Great Britain, Justus Perthes in Germany, and Rand McNally in the United States had come into being.

Cartography has probably advanced more, technically, since the beginning of the twentieth century than during any other period of comparable length. It is probably correct that the number of maps made since 1900 is greater than the production during all previous times, even if we do not count the many millions made for military purposes. Many factors have combined to promote this phenomenal growth; we will mention only a few.

The development of the airplane was very significant to modern cartography. It operated as a catalyst in bringing about the demand for more mapping; at the same time, the airplane made possible photogrammetry (mapping from photographs). The need for smaller-scale coverage of larger areas, such as the aeronautical chart, promoted larger-scale mapping of the unknown areas. Furthermore, the earth seen from the vantage point of an airplane in flight appears somewhat like a map, and the air traveler frequently develops an interest in maps. The rapid developments connected with space exploration and satellite technology are having a similar effect, but at a different level. Remote sensing, mapping of the moon and planets, and precise measurements of the earth and of relative positions on it from orbital observation are just a few of the developments.

Photography, in which rapid advancements were made during the latter half of the nineteenth century, has also had enormous significance for mapping. The potential of photographs for photogrammetric uses was evident almost immediately after the invention of the camera, and the advantages of aerial over terrestrial photographs were obvious. Today, photogrammetric methods are extensively used in the preparation of nearly all published large-scale topographic maps. Furthermore, techniques for removing the perspective displacement in an air photograph were developed, and a different kind of map with a photographic base, called an orthophotomap, is becoming increasingly available. In an orthophotomap the portrayal is the "natural" photographic image of the land supplemented by various symbols.

The union of photography with the printing processes in the latter part of the nineteenth century provided a relatively inexpensive means of reproducing an image on almost any material. From then on technical developments in the photolithographic and photoengraving fields were rapid and continuous, and they became inextricably woven into the methods of cartography. Today, high-speed, multicolor lithographic offset presses are capable of handling any kind of cartographic problem. Modern manual map construction with its scribing, open window negatives, photo lettering, and so on, is as different from the mapmaking of 50 years ago as is the automobile from the horse and buggy.

The rapidly increasing population of the earth, the phenomenal growth of urban centers, and the associated technological and industrial development have also had a powerful impact on cartography. The need to protect the environment and the demand for environmental information by planners and administrators have greatly promoted all classes of mapping. The urban, rural, resource, industrial, or other kinds of planner-engineers are usually concerned with relations in the spatial dimension; for this they must have maps.

Probably the most profound changes in the broad field of cartography occurring at present are a result of the continuing development of electronic techniques associated with telecommunications, automation, and computer applications. The hardware, consisting of satellites, computers, digitizers, plotters, printers, lasers, and the cathode ray tube together with the associated software, the programs for accomplishing particular tasks, are literally revolutionizing mapmaking. In many applications these remarkable mechanical and technical advances simply make the solutions to old problems easier. Cartography is a body of theory and method for dealing with problems of recording, analyzing, and communicating geographical information; the computer, remote sensing, electronic typography, and the other innovations are clearly aids to better, more efficient mapmaking.

Electronics and the computer are more than merely technical developments that make traditional mapmaking easier and faster. They are having a major effect on the field itself in several ways. One such consequence is a change in our concept of the map; instead of conceiving of it as a sheet with markings on it, we now include video displays to which we can add or subtract data as desired. Of tremendous importance is the development of data bases from which map displays may be constructed. How these are structured to make their cartographic use most efficient and how they can be kept current with changes in the data are primary tasks in cartography. These kinds of problems are particularly important in data bases involving land records and associated data, such as the myriad of elements integrated in an urban complex.

In order to promote the development of what will be an extremely useful National Digital Cartographic Data Base for the United States, it is necessary to modernize the Public Land Survey System on which much locational data are based, such as landownership and natural-resource inventory data. The problems and tasks involved in developing such a multipurpose cadastre are presently being carefully considered.*

THE PROFESSION OF CARTOGRAPHY

Mapmaking is a very old activity, but in the Western world, it did not attain the status of a profession until the fifteenth century. At that time,

*National Research Council, Committee on Integrated Land Data Mapping, *Modernization of the Public Land Survey System* (Washington, D.C.: National Academy Press, 1982).

with the burgeoning of exploration, the translation of Ptolemy's *Geography* into Latin, and the application to maps of the newly developed printing processes, a class of scholar-technician mapmakers came into being. As we noted earlier, in Spain and Portugal, countries deeply committed to overseas colonial development and trade, official bureaus were established to promote chart making and navigation methods. Later, large commercial trading institutions in several countries, notably Holland and England, maintained similar, but private, mapping offices. After the introduction of topographic mapping in the eighteenth century, government agencies concerned with official large-scale mapping of the land were established. Today, most countries support large government agencies with primary responsibilities for official mapping and charting. Many other agencies, organizations, and individuals are active in cartography. The great technical advances in the processes of mapmaking, the growing need for education and research in cartography, and the importance of maps as documents, analytical tools, and communicative devices has caused cartography to become a broad, diverse field.

The Major Divisions of Cartography

Cartography is usually thought to consist of two classes of operations. One is concerned with assembling and making available geographical data and the preparation of a variety of general maps used for basic reference and operational purposes. This category includes, for example, large-scale topographic maps of the land, hydrographic charts, and aeronautical charts. The other division has to do with the preparation of an even larger variety of maps. This category includes the usually small-scale thematic maps of all kinds, as well as atlas maps, road maps, maps to accompany the written text in books, and the maps in planning and other governmental offices. Within each category there is also considerable specialization such as may occur among the survey, design, production, and reproduction phases of a topographic map. All such divisions and activities blend one into another, however, and sharp compartmentalization rarely occurs.

The first-mentioned category of cartography works primarily from data obtained by field or hydrographic survey or by satellite and photogrammetric methods. Things such as the shape of the earth, height of sea level, land elevations, precise distances, and detailed locational information are of fundamental concern. Electronic and photogrammetric instruments and remote sensing are an integral part of this sort of cartography. Generally, this group includes the great national survey organizations, the oceanographic and aeronautical charting agencies, and most military mapping organizations, which altogether employ thousands of cartographers. They produce an enormous number of maps. For example, the National Ocean Survey in the United States, only one of the several federal mapmaking agencies, issues more than 40 million copies of charts each year.

The other category, which includes thematic cartography, draws on the basic work of the first group but is mostly concerned with the communication of general information and with the effective portrayal of relationships, generalizations, and geographical concepts. The specific subject matter may be aimed at history, economics, urban planning, rural sociology, engineering, and many other areas of the physical and social sciences—there is no limit. The data for such maps are usually compiled from diverse sources, ranging from institutional reports to information obtained by remote sensors in a satellite, such as LANDSAT.

It should not be inferred that the practical considerations that tend to separate the cartographers basically concerned with large-scale maps, survey, and photogrammetric methods from those concerned with small-scale maps and compilation necessarily create a well-defined void between the two groups. The contrary is true. The fundamental cartographic problems of each group are conceptually similar. For example, the concepts on which the delineation of landforms by contours depends are the same as those on which the delineation of an abstract statistical surface by isopleths depends. Most of the principles on which cartographic techniques are based are equally applicable in either division of the field. Instead, the major distinction is in the methods of acquiring the data to be mapped; these are commonly different. On the other hand, the cartography, that is, the conception, the designing, and the production of the map, as distinct from the gathering of the data, is fundamentally the same in both divisions.

Other Cartographic Operations

The major conceptual-operational divisions of the field outlined above do not take into account some other fundamental areas of cartographic activity, which, in the United States, tend to cut across the large-scale, survey-oriented and the small-scale, compilation-oriented groups. There are several such areas: private mapmaking companies, commercial survey and mapping companies, planning agencies, and cartographic information organizations.

Commercial cartography is made up of a large number of companies, from small, local institutions to large, complex organizations with national and international markets. In the United States there are more than 100 such firms, and they employ over 1000 cartographers; the majority of them are technicians, and the rest are compilers, editors, and researchers. The products of such companies are diverse, ranging from street and plat maps to specialized atlases and road maps. The maps are either prepared from material supplied by the consumer, such as a street plan for a town, or they are compiled by research staffs in major mapmaking companies or in the cartographic departments of publishing companies.

Many commercial survey and mapmaking firms do contract mapping for industries, planning departments, and federal, state, and municipal agencies. This often involves air survey. The work ranges from cadastral-type mapping for private and public appraisal or tax assessment needs to site analyses for urban and industrial development. Some of these companies are quite sophisticated in their application of advanced electronic and computer aids to surveying and cartography.

Most cities, many states, and even some regional groupings of areas have planning departments that make quantities of large-scale and small-scale maps. An example of a large-scale map would be the detailed mapping of an area to be redeveloped; population and land use maps for zoning and other aspects of general urban planning are representative of smaller-scale maps. Cartographers employed by planning agencies often work with remote sensing imagery and computers.

The enormous number and diversity of maps that have been made in recent years, the great variety of activities that require maps, the need to provide reference assistance to those who need to know what is available, and the necessity for maintaining collections of maps has led to the rapid growth of cartographic information services. Many cartographers are involved in this aspect of cartography, which ranges from the map divisions of large libraries and the numerous smaller ones in federal mapping agencies, universities, and colleges to the map information bureaus maintained by the federal and many state governments. The professional activities in this area of cartography are less concerned with mapmaking than with the communication of information about maps, their availability, reliability, cost, and so on, although such organizations often publish index maps showing the variety of coverages, scales, and types of maps. A great deal of cataloging and bibliographic work is involved. The two main federal units in this area are the Geography and Map Division of the Library of Congress* and the National Cartographic Information Center (NCIC), a cooperative activity of

*Library of Congress, Geography and Map Division, Washington, DC 20540

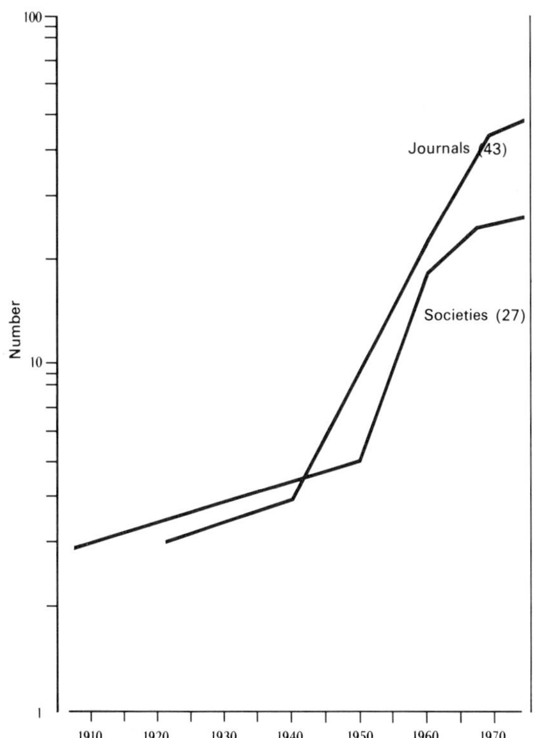

FIGURE 2.15 Cumulative numbers of cartographic societies and journals, to 1975. Note that the period of fastest growth has been since 1940. The y axis is a logarithmic scale so that the slope of the curve shows the relative rate of growth. (From data compiled by John A. Wolter.)

federal mapping agencies maintained by the U.S. Geological Survey.*

Societies and Journals

Since the mid-1940s, the field of cartography has grown rapidly in all aspects. The establishment of professional societies has been particularly notable. A main function of professional societies is to disseminate information; one way to accomplish this is by the publication of a journal. Consequently, the number of journals has grown along with the societies. There are nearly thirty societies around the world and more than sixty journals, newsletters, and other serials are now being published (Fig. 2.15). The major journals in English are listed in the Selected References.

Because cartographers have much in common with geographers, surveyors, and photogrammetrists, some societies include these interests. For example, in the United States the American Cartographic Association is a member organization of the American Congress on Surveying and Mapping (ACSM), which encompasses surveyors as well as cartographers. There are also cartographic associations in English-speaking Australia, Britain, Canada, and New Zealand and in most of the other leading countries of the world.

Cartography is represented at the international level by the International Cartographic Association (ICA), an organization in which countries, not individuals, hold membership. Some sixty countries now are members. The ICA is engaged in cooperative activities with the Federation Internationale des Geomètres (FIG), the International Association of Geodesy (IAG), the International Society for Photogrammetry and Remote Sensing (ISPRS), the International Federation of Library Associations (IFLA), and the International Geographical Union (IGU).

* U.S. Geological Survey, NCIC, 507 National Center, Reston, VA 22092. Tel: (703) 860-6039.

SELECTED REFERENCES

Bagrow, L., *History of Cartography,* revised and enlarged by R. A. Skelton. London: C. A. Watts and Co., Ltd., 1964.

Bickmore, D. P. (ed.), "Perspectives in the Alternative Cartography, Cartographic Computing Technology and its Applications," *Cartographica* 19, no. 2, Monograph 28 (1982).

Brown, L. A., *The Story of Maps.* Boston: Little, Brown and Company, 1949.

Chu, G. H-Y., "The Rectangular Grid in Chinese Cartography," unpublished master's thesis, University of Wisconsin–Madison (1974).

Crone, G. R., *Maps and Their Makers: An Introduction to the History of Cartography,* 5th ed. Folkestone, Kent, England: Dawson, Archon Books, 1978.

Guelke, L. (ed.), "Maps in Modern Geography, Geographical Persperctives on the New Cartography," *Cartographica* 18, no. 2, Monograph 27 (1981).

Harvey, P.D.A., *The History of Topographical Maps.* London: Thames and Hudson, 1980.

Hsu, M-L., "The Han Maps and Early Chinese Cartography," *Annals of the Association of American Geographers,* 68, 1 (1978): 45–60.

Robinson, A. H., *Early Thematic Mapping in the History of Cartography.* Chicago: University of Chicago Press, 1982.

Robinson, A. H., and B. B. Petchenik, *The Nature of Maps: Essays Toward Understanding Maps and Mapping.* Chicago: University of Chicago Press, 1976, pp. 108–23.

Stephens, J. D., "Current Cartographic Serials: An Annotated International List," *The American Cartographer* 7, no. 2 (1980): 123–38.

Strasser, G., "The Toise, the Yard and the Metre—The Struggle for a Universal Unit of Length," *Surveying and Mapping* 35 (1975): 25–46.

Taylor, D.R.F. (ed.), *The Computer in Contemporary Cartography,* Vol. 1 of *Progress in Contemporary Cartography.* New York: John Wiley & Sons, 1980.

Thrower, N.J.W., *Maps and Man.* Englewood Cliffs, N.J.: Prentice-Hall, Inc., 1972.

A select list of cartographic serials that publish articles in English and the country of their present publication:

The American Cartographer (USA)

The Cartographic Journal (United Kingdom)

Cartographica (Canada)

Cartography (Australia)

International Yearbook of Cartography (Federal Republic of Germany)

New Zealand Cartographic Journal (New Zealand)

Imago Mundi: A Review of Early Cartography (United Kingdom)

The Map Collector (United Kingdom)

Bulletin, Society of University Cartographers (United Kingdom)

Bulletin, Special Libraries Association, Geography and Map Division (USA)

World Cartography (United Nations, New York)

3

Technology of Cartography

TECHNICAL ADVANCES IN CARTOGRAPHY

MANUAL TECHNOLOGY
OPTICAL-MECHANICAL TECHNOLOGY
PHOTO-CHEMICAL TECHNOLOGY
ELECTRONIC TECHNOLOGY
- Hardware
- Software
- Data

IMPACT OF CHANGING TECHNOLOGY

TECHNOLOGICAL INTEGRATION
COST OF CHANGE
ENORMOUS AMOUNTS OF DATA
SHIFTING STANDARDS
IMPROPER APPLICATIONS BY NONCARTOGRAPHERS

This chapter considers the impact of technological advances on the evolution of the modern map. The previous discussion has dealt with the close association between conceptual developments in society and the history of cartography and mapping. Traditionally there has been a strong relationship between mapping and the prevailing state of technological development, and throughout the centuries cartographers have been quick to borrow and adopt technological innovations. How data for mapping are selected, how map design is conceived, and how maps are produced and reproduced have always strongly reflected the technological achievements of the period.

One reason for this close correlation between technological advances and cartographic achievement is that each succeeding generation of cartographers has had to face the same two goals. Society has been unrelenting in its demand for maps that are more timely, accurate, and complete, and at the same time there has been a continual demand for greater accessibility to lower-cost maps. These dual forces have been at play regardless of cartographic developments in the previous generation. Indeed, to a large extent it has been the constant struggle to meet these goals that has led to the kinds of maps we know today.

TECHNICAL ADVANCES IN CARTOGRAPHY

A wide array of technical developments has been important in cartography. Advanced knowledge of the principles of mechanics, optics, chemistry, metallurgy, electromagnetism, and electronics has found application in the mapping process. To a large extent this knowledge has been made possible by continual refinement in engineering and manufacturing processes. Indeed the two, understanding and engineering, have gone hand in hand. An advance in one usually soon leads to an advance in the other. This see-saw growth in knowledge and technology has meant that cartographers in each new generation have had the benefit of better tools, machines, and materials than the last. Today the products of technology that find cartographic application are of greater sensitivity, accuracy, speed, and durability; are easier to handle and use; and are generally of much higher quality than ever before. As a result contemporary cartographers enjoy an unprecedented flexibility in their mapping endeavors.

Although constant adjustment to technological change has long been a fact of life for the cartographer, two fundamental trends are evident. First, the magnitude of change brought about by each new technology seems to be greater than the impact of the one that came before. Secondly, the rate of change seems to be accelerating (Fig. 3.1). *Manual techniques* based on simple hand-held tools dominated mapmaking for thousands of years. Then, beginning with the development of the printing press and movable type, in the twelfth century in China and the fifteenth century in the Western world, *optical-mechanical technology* came to dominate mapping activities. For the past 175 years cartographers have been integrating *photo-chemical technology* into their work. Currently the revolution in *electronic technology* is rapidly affecting the way cartographers think and work. The profession is fast becoming a high-technology field, and there seems to be no end in sight to possible cartographic applications of this latest technology.

In the eyes of those who view the field from a short-term perspective, the magnitude of the changes due to electronic technology has been sufficient to give rise to a false dichotomy. Thus, distinction is sometimes made between manual cartography (apparently including optical-mechanical and photo-chemical technologies) and automated or computer-assisted cartography. This division implies that the two are quite different. On the contrary, what has occurred recently has happened several times before. That is, machines (in this latest case, electronic) have been introduced to perform some of the necessary time-consuming, exacting, and repetitive operations. It is worth emphasizing that in this case, as before when new technologies were introduced, basic

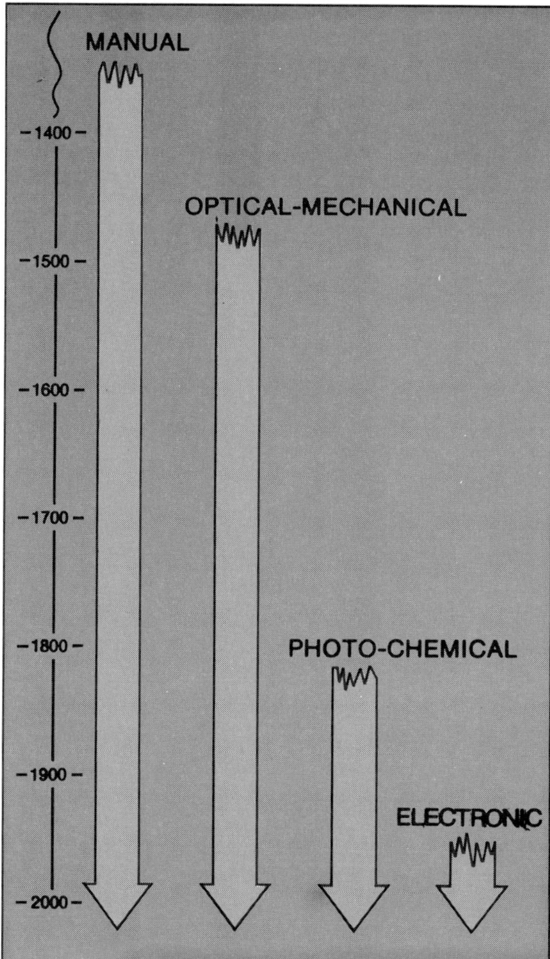

FIGURE 3.1 Mapping in the Western world has undergone four major technological revolutions since the time when cognitive images were first transcribed into tangible cartographic products. Each technology has had a dramatic impact, both in terms of the scope and depth of the discipline, and with respect to the ease, speed, and cost of mapping. The timing is somewhat different in the Eastern world.

cartographic processes have not been substantially changed. What has happened is that the way these processes are carried out has in many respects been modified to take advantage of computer technology. The apparent, but not real, dichotomy is enhanced because the knowledge of electronic technology is, for the most part, in the hands of a different set of specialists from the group that knows the other technologies. Thus, what has seemed to some to be a division is nothing more than procedural developments resulting from the introduction of the latest new technology into an existing field.

It is also important to point out that only rarely has one technology completely replaced another. Over the years the cartographer has simply been given more options from which to select. Thus, the modern cartographer is privileged to pick and choose procedures from the full array of the technologies so far devised. Cartographers also can take advantage of nearly every conceivable combination or hybrid technology. Since this means that there are many alternative ways to make a map, the cartographer is increasingly gaining the opportunity to tailor the mapping process to the situation at hand. The particular approach adopted for a given mapping project should represent a balance between available time, acceptable costs (labor, equipment, materials), and type and quality of the product (map) desired. For this reason great differences in mapping practices currently exist among various mapping establishments.

Clearly, by introducing more things to consider, technological innovations serve to make the cartographer's decision-making responsibilities more rather than less difficult. The real value of technological advance is that it opens to the cartographer options or alternatives that can lead to more effective and efficient mapping. But to take advantage of technology's potential benefits requires a thorough understanding of the central role that technology plays in the cartographic process.

The chapters that follow treat this topic in great detail. The aim at this point is merely to provide a general overview of the subject to help to put the discussion in subsequent chapters into useful perspective. To this end the next sections of this chapter are organized around the broad headings taken from Figure 3.1. In each section the impact of technology is assessed with respect to all phases of the mapping process. These include the capture of raw data for mapping, the arrangement or compilation of these data into a draft map format, the design or specification of graphic factors that will give the map its visual character, the physical production or execution of the map design in an appropriate medium, and the duplication or reproduction of the original map.

Manual Technology

Manual mapping procedures were dominant during the longest period in the recorded history of cartography. Craftsmen became incredibly skillful in the era when only hand-held tools, such as brushes, quills, and the stylus, were available to work on papyrus, silk, parchment, and even in clay and metal. The miniature *mappae mundi* in medieval illuminated manuscripts and the later portolan charts were intricate and colorful. As a matter of fact, it is suggested that mapmaking became a recognized profession when the demand for skillfully made charts and other navigational instruments burgeoned in the fourteenth and fifteenth centuries.

Manual techniques did not go out of fashion with the development of printing. Persons skilled in making relief woodcuts (nonprinting areas carved away) and intaglio copper plates (printing areas incised) carried on the manual tradition (Fig. 3.2). To these groups were added the lithographic and wax engraver in the nineteenth century. Some of the great names in the history of cartography are associated with the "handmade map." Mercator was a skilled engraver; Abraham Ortelius, who produced the first modern atlas, started out as a map colorist; and August Petermann, who became a leading German cartographer-geographer and who founded a leading periodical *Petermanns Geographische Mitteilungen,* began his career as a skilled engraver-lithographer.

FIGURE 3.2 A *Formschneider* (one who carves woodcuts) at work in front of a window. In those days there was no satisfactory substitute for daylight. From Amman and Hans Sachs, *Eygentliche Beschreibung aller Stände auf Erden*, Frankfurt, 1568. (Courtesy of the John M. Wing Foundation on the History of Printing of the Newberry Library, Chicago.)

Today, thousands of years after the first extant cartographic map, manual procedures are still responsible for a sizable portion of mapping activity. Continual improvements in materials, tools, and techniques have helped to ensure a place for cartographers who are skilled in manual procedures, even in modern mapping activities. Indeed, for many purposes, it is difficult to match the speed, flexibility, and skill of the human craftsperson and decision maker in the mapping process. Thus, rather than disappearing with the introduction of new technologies, manual methods have been incorporated into the new ways.

Optical-Mechanical Technology

The second generation of major technological innovation in cartography involved applications of the principles of optics and mechanics. Lenses greatly enhanced human perception, and light projection devices substantially reduced the labor and improved the accuracy of image transfer

FIGURE 3.3 Printing from a copperplate engraving with the rolling press was hard work. From Abraham Bosse, *De la manière de graver . . .* , Paris, 1745. (Courtesy of The John M. Wing Foundation on the History of Printing of The Newberry Library, Chicago.)

operations. Machine power augmented and magnified human muscle power (Fig. 3.3). The result was a major increase in the speed and efficiency of the mapping process, with a commensurate reduction in mapping cost. Since this was just what cartographers wanted, mechanical technology was widely adopted, ranging from engraving machines for producing closely spaced, parallel lines to engine-driven presses. Maps became far more accessible to a much wider audience of potential users than was possible when production was limited to manual technology in which every map, original and copy, had to be drawn by hand. The adoption of mechanical technology also started a trend toward greater need for more expensive capital equipment compared to what was required for manual methods.

The impact of mechanical technology on the different phases of the mapping process has varied considerably, however. The reproduction phase has been by far the most drastically affected. Map compilation and production procedures in large part have become slave to the choice of subsequent map reproduction method. In fact, the introduction of mechanical technology has changed the way the cartographer must think about the mapping process.

Photo-Chemical Technology

The development of lithography and photography and the application of etching techniques to mapmaking in the early nineteenth century stimulated the third major technological revolution in cartography. Lithography, which was first called chemical printing, produced duplicates from a flat surface by employing the mutual repulsion of oil and water to form a printing image (see Chapter 17). It was much cheaper and more manageable than copperplate engraving and led to color printing by mid-century. Today lithography is the most widely used method for printing maps.

Photography, first applied to mapmaking in the mid-nineteenth century, rapidly had a major impact on the field. Field photography, begun in earnest in the late 1800s and currently referred to as *environmental remote sensing,* gave cartographers a dramatic new map form, the *photomap.* The variety and uses of this image-based map form now seem unlimited. The photographic process also provided cartographers with a powerful new technique for carrying out map compilation, production, and reproduction tasks in the laboratory. Some chores characteristic of manual and mechanical mapping practice, such as image enlargement or reduction and the transfer of a guide image from the surface of one material to that of another, have been greatly modified. Other tasks, such as hand engraving of printing plates, have been altogether eliminated. Still other chores, such as those associated with the processing and preparation of photographic images (for example, aerial photos) for reproduction, are new additions to the cartographer's list of responsibilities.

The basic components of the photographic process are similar in both remote sensing (field) and laboratory applications. There must be an object to be photographed, a light source, a light-sensitive medium, a device for controlling exposure, and a means for chemically developing the photographic image. Actual equipment and procedures used in the laboratory differ in significant ways from those used in the field, however. In laboratory work, called *graphic arts photography,* artwork replaces the ground scene as the object to be copied. Illumination is achieved with a light source that can be precisely regulated rather than by the sun. And devices for making controlled exposures can use light-reflection as well as light-transmission principles. These factors combine to provide the cartographer working with graphic arts photography with a wide array of materials and procedures.

The fact that photo-chemical technology could be so effectively applied in the field as well as in the laboratory has had diverse consequences in the mapping process. Unlike mechanical technology, which has a strongly unbalanced influence on the different phases of mapping, the impact of photo-chemical technology is quite evenly distributed. Indeed, it has become an integral part of all phases of mapping. Even today new applications are constantly being discovered.

Even more than mechanical technology, photo-chemical technology has forced the cartographer to think backward through the stages of the mapping process. To a large degree reproduction method alters production procedures. For example, map artwork would be correctly scribed right-reading for letterpress printing, whereas it should be scribed in reverse (wrong-reading) for lithographic printing. Similarly, production method alters compilation procedure. Thus, on a compilation worksheet designed to be traced in subsequent ink drafting, the use of colored pencils on paper might be the best means of distinguishing one category of feature from another. In contrast, a worksheet to be used as a guide image in scribing might better be constructed in black ink on drafting film in order to facilitate the photographic transfer of the image to the scribecoat surface. What these examples suggest is that in order to take full advantage of the alternative mapping strategies made possible by photographic technology, the cartographer must plan ahead carefully. Failure to do so generally leads to inefficiency and added costs in time, labor, and materials.

Electronic Technology

Cartographers began to explore the power of electronics in the early 1950s and now find themselves in the midst of a high-technology revolution (Fig. 3.4). When one considers that computer-assisted cartography is less than 30 years old, its effects on the field are revolutionary. An entirely new technology has been invented and implemented. It has concentrated on doing by machine many operations that were formerly painstakingly done by hand. The increased flexibility and capabilities of the new technology will enable even further extensions of the field which, at present, are only partially conceived. The jargon of the electronic age has become part of the cartographer's language. This fourth major technological revolution to affect cartography promises to alter the mapping process as nothing before. The analog image, that is, the visible two-dimensional map, that characterized previous technologies is being replaced by a digital record

FIGURE 3.4 Multipurpose electronic work stations, such as this multiple-screen television weather station system, make it possible to map the same region in a variety of ways simultaneously. The electronic gadgetry permits dramatic special effects, such as animation, zooms, fades, and a host of other image enhancements and manipulative possibilities. (Courtesy Color Graphics Weather Systems, Inc.)

or file in which all locations and characteristics are coded in a binary system. In the process, the way that cartographers think about maps and go about data collection, compilation, production, and reproduction is being changed dramatically. The conventional graphic map is no longer the only, or ultimate, product to be generated in the cartographic process.

There are two important effects of these changes. One is simply that the use of computer-assisted methods requires that cartographers become familiar with the new techniques, since it is unlikely that a cartographer who is expert in another aspect of cartography can immediately interact with computer technology. In some operations the transition is easier than in others. For example, the process of line digitizing (recording the *x*-*y* positions along a line) is similar to scribing. In some instances they are almost identical, since a visible line is created to show the cartographer what part of a given line has been digitized. However, the concomitant entry of necessary auxiliary information into the digital record has no counterpart in the scribing operation, and in fact, the process of digitizing demands that the cartographer learn something about the structures of digital data files.

A second effect has to do with the time lag between technological development and its full-scale implementation. Technology has advanced relatively fast compared to its incorporation in efficient cartographic systems. Although the directors of cartographic units may have long known that many operations can be carried out with the assistance of computer technology, the lack of capital and/or the exact hardware configurations to make an efficient system for a specific cartographic unit may not have been available.

Complete automation of the conventional mapping processes is far from being attained. The changes brought about by electronic technology are to date quite uneven. In some aspects the developments have been profound, but in others relatively little has occurred. Initial attention was directed at integrating electronics with older technologies. This was highly successful and resulted in a period of imitation and replication rather than true innovation. More recently significant steps have been taken toward fully electronic mapping. Already this has led to the creation of photo image maps based on digital data sensed from nonvisible parts of the electromagnetic spectrum and displayed as dynamic/interactive video images. Although other innovations will undoubtedly soon follow, *computer-assisted cartography* is still a better term than *automated cartography* to describe the present state of affairs. But clearly the strongest force being felt in the field is pushing toward further automation, so this may soon change.

Computer-assisted mapping involves the development and integration of the three components of any computer system (Fig. 3.5). There are the machines that perform the operations, or *hardware*; there are the instructions that tell the machine what to do (what, when, where, how), or *software;* and there are the *data* to be manipulated by the machines under software control. The success of computer-assisted mapping rests on the skill of the cartographer and the development and application of each of these com-

FIGURE 3.5 Progress in computer-assisted mapping depends upon the development and integration of the hardware, software, and data components of the system. Of course, the central character is the cartographer.

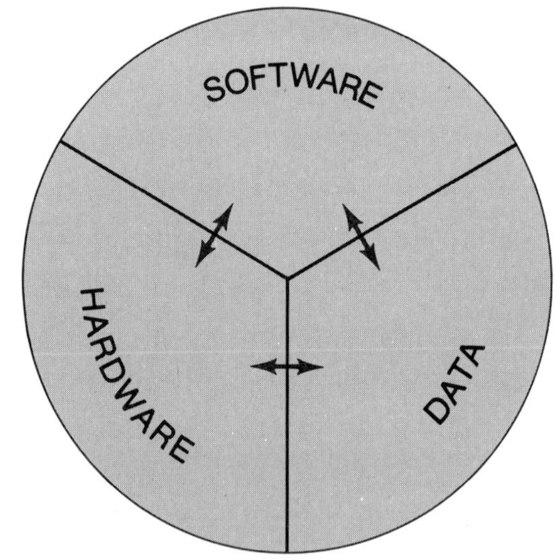

puter system components within a cartographic environment.

Hardware. The hardware aspects of a computer-assisted mapping system are outlined in Figure 3.6. At the heart of the system is a high-speed digital computer or *central processing unit* (CPU). Its function is to receive data in digital form and process them into a form that the cartographer needs as output. Control units range widely in speed, power, and function. Relatively small capacity *microcomputers* (or microprocessors) are fast bringing the potential for computer-assisted mapping into the office and home, at least in limited scope. Medium capacity *minicomputers* are generally dedicated to a few special functions and have become standard equipment in even relatively small mapping establishments (Fig. 3.7). Very large *mainframe* computers are designed to handle many varied tasks from a variety of data entry devices in seemingly simultaneous fashion and are now routinely used to support daily operations in large mapping agencies and in cartographic research establishments.

In support of the CPU are a host of peripheral devices for data entry, storage, and output (see Fig. 3.6). Most of these devices serve a wide range of users, but some are of special interest in computer-assisted cartography. These include data entry devices, called digitizers, and special graphic output devices such as printers, plotters, and display screens suitable for map presentation. Since these digitizers and graphic output devices have become important mapping tools, they will be discussed in detail in subsequent chapters.

Software. The software component of a computer mapping system is somewhat less tangible but no less important than the hardware component. Computers require explicit instructions. Thus, sets of machine instructions, called *programs,* are needed to carry out the processing and reformatting (restructuring) of data that cartographers heretofore have had to do by nonelectronic means. Before such programs can be written, each cartographic task must be analyzed and then

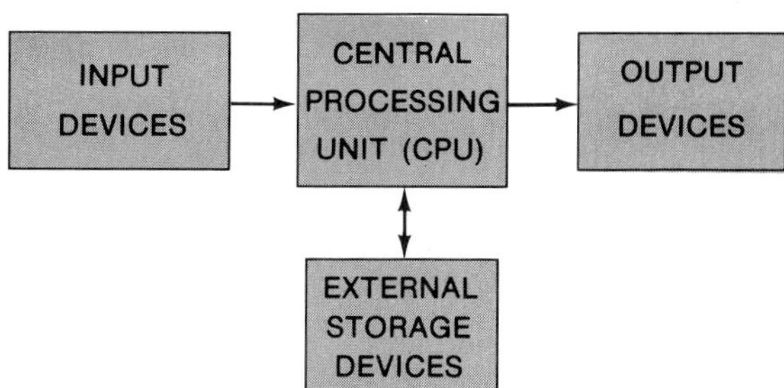

FIGURE 3.6 The hardware configuration of a standard computer system consists of data input devices, a central processing unit (CPU), external storage devices, data output devices, and appropriate linkages for communicating data from one unit to another.

structured as a logical step-by-step procedure, known as an *algorithm.* Most common procedures used by cartographers have been translated into software programs written in special *computer languages,* such as FORTRAN or Pascal. Indeed, many mapping procedures have been programmed more than once. Since this usually has been done by individuals working independently at different locations, the programs are commonly written using slightly different algorithms, computer languages, and machine specifications.*

Although individual mapping programs have been marketed commercially for some years, general purpose mapping packages, which can handle a variety of chores, are just now becoming commercially available.**

*One of the more frustrating aspects of computer-assisted mapping is that programs written for use on one manufacturer's equipment may have to be extensively rewritten before they will function properly on another company's computer.

**Recently marketed mapping software packages that fall into the general purpose category include: ODYSSEY, an extremely sophisticated set of integrated mapping programs designed for use on large-capacity computers (Laboratory for Computer Graphics and Spatial Analysis, Harvard Graduate School of Design, 520 Gund Hall, Cambridge, MA 02138); SAS/GRAPH, a package of general purpose graphics programs designed for use on medium capacity computers, including some mapping options (SAS Institute Inc., SAS Circle, Box 8000, Cary, NC 27511); and MICROMAP II, a multipurpose mapping package designed for use on microcomputers (Morgan-Fairfield Graphics, 1923 First Avenue, Suite 300, Seattle, WA 98101).

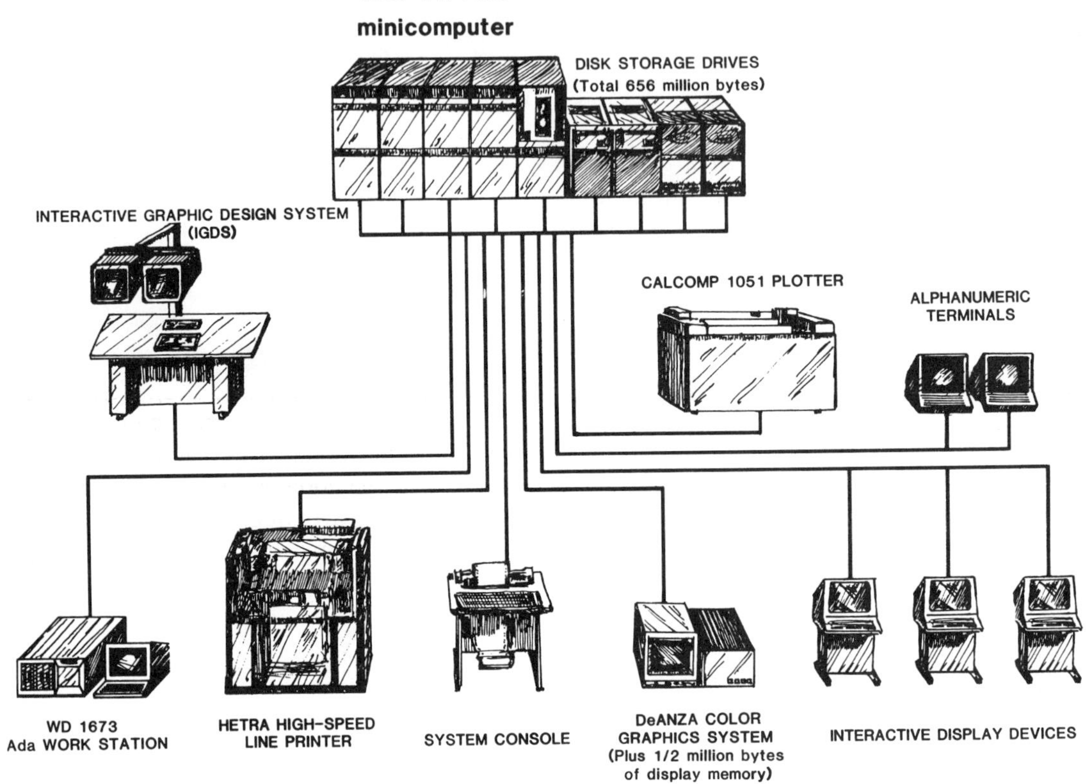

FIGURE 3.7 The sophisticated computer graphics laboratory in the Department of Geography and Computer Sciences at the U.S. Military Academy is formulated around a minicomputer that is lavishly supported with peripheral hardware of special mapping interest. (Courtesy U.S. Military Academy.)

Today's well-rounded cartographer is routinely involved with these "canned" (prewritten) mapping programs and should possess sufficient computer literacy to be able to understand the cartographic implications of each algorithm being used. Not all possible or useful programs have been created, of course, which means that modern-day cartographers will also be directly involved in computer programming tasks in the course of performing day-to-day duties. Thus, the professional cartographer should have a working knowledge of at least one computer language.

Data. Computer hardware and software are of little value to cartographers unless the data to be mapped are available in computer-compatible form. This necessity points out the importance of the data component of a computer mapping system and helps explain why data entry devices such as digitizers have assumed such a prominent role in modern mapping. But there is more to making data computer-compatible than merely transforming them into digital records. The raw numbers must be edited, labeled, tagged with the necessary reference information, and formatted and structured in a fashion that preserves their spatial character and facilitates ease of future use. Once created, the resulting *files* or *data bases* must be kept up to date and otherwise maintained to serve the needs of cartographers both now and in the future.

To date, the data component of computer mapping systems has been the weakest of the three major parts. Data base construction has lagged far behind hardware advances and software development. Progress is being made at a more rapid rate than in the past, however, and a number of government agencies, including the U.S. Geological Survey, have recently committed themselves to building and maintaining data bases that will be of great value to cartographers for some years to come.

IMPACT OF CHANGING TECHNOLOGY

Technology historically has had a changing and far-ranging impact on cartography, and as already pointed out, the pace and magnitude of change seems to be accelerating. It is neither easy nor very meaningful to categorize the changes that are taking place as being good or bad. To try to do so would be to miss the point that change reflects evolutionary adjustments to the needs of a dynamic society of map users.

What *is* useful is to consider how the cartographer's life is being influenced by the changes that are taking place. Several topics are of special interest to someone about to enter the field of mapping. When considering these topics, it is important to keep in mind that in the upheaval of the new replacing or adding to the old the primary, overall goal is not simply to make maps faster and more inexpensively with less human labor, but also to make maps that are more effective triggers of human thought and better communicators of environmental concepts and information.

Technological Integration

The power, operating speed, and functional capabilities of modern computers can handle a vast array of cartographic chores, but at this time several things are still lacking. One is the availability of adequate files of computer-compatible data to suit the broad range of mapping scales and topics. The other is sufficiently versatile software to enable the cartographer to perform the host of data and graphic manipulations involved in the mapping process. Although the supply of digital data and cartographic software is improving steadily, it will take some time before data and programs are available to cover the full range of mapping tasks and be workable on each of the computer systems currently in use.

Already, however, electronic technology is having a melding influence on cartography. For one thing, applications of electronics are blurring past distinctions among various mapping technologies. Cartographers have worked with three broad generations of electronic devices during the last quarter century (Fig. 3.8). Beginning in the mid-1950s there was a mating of digital and mechanical technologies. A mating of digital and photographic technologies followed in the early 1960s. And, finally, fully digital mapping had its origins in the mid-1960s. Currently machines representing every possible hybrid of the various technologies are being used. The integration with the most sophisticated modern devices is so complete that it has become meaningless to try to identify or separate the contribution of the individual technologies.

Mapping procedures have become just as difficult to categorize by technology as the machines that are used in mapping. Old boundaries are fast breaking down. Since certain strengths

FIGURE 3.8 Electronic map production devices have gone through three broad developmental stages during the past quarter century. Although all three generations of equipment are currently in use, the trend is toward more fully digital units.

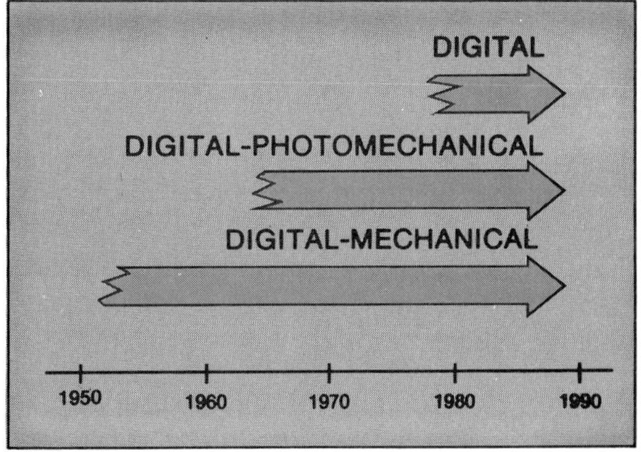

and weaknesses are inherent to each technology, modern cartography commonly involves procedures based on several approaches. Manual, optical-mechanical, photo-chemical, and electronic activities may all be integrated in an attempt to attain a mapping goal. The idea is to blend procedures based on the different technologies in such a way that each will contribute its own special advantages with respect to speed, cost, accuracy, and other desirable qualities.

In similar fashion, electronic technology is creating fuzzy boundaries between the stages in the mapping process. Previously it made sense to think about mapping in terms of the distinct phases of data gathering, compilation, production, and reproduction. But if data are collected directly in digital form, the stages of compilation, production, and reproduction in some cases tend to merge into one operation. A mere touch of a button (switch) moves the cartographer from one phase to the next. This breakdown of old boundaries is particularly true when only a single copy of a map is made for the benefit of a single user, or when a map image is telecommunicated to several display terminals for direct viewing and is then erased.

This blurring of mapping technologies, procedures, and stages is changing the way maps are made and in the process is altering the relationship between mapmaker and map user. When maps were manually created and only one or a few copies were ever made, the map user was normally in some way directly involved with their construction. But during the period when sophisticated mechanical and photo-chemical technologies dominated mapping activities, the map user tended to be left out. Professional cartographers were responsible for designing, constructing, and reproducing maps and map series, which were then stored in depositories and libraries to be made available later to potential users. Until recently, for the most part, the map user had to make do with whatever printed maps were available.

Today, electronic technology is fast making it possible for the map user to be directly involved, especially if mapping is being done on a display screen or with microcomputer-driven plotting devices. The result, of course, is that the knowledgeable user can better tailor mapping decisions to requirements at hand and need not rely on the judgment of a professional cartographer, who may know neither what a map is to be used for nor how sophisticated the map user might be. Thus electronic technology is also changing the role of the professional cartographer. Rather than pursuing the goal of building up the stock of printed maps for potential use, cartographers will increasingly need to concern themselves with making environmental data and mapping programs more accessible to potential mapmakers. There will also need to be a major new emphasis on raising the level of cartographic sophistication among these professional map users, as well as the populace at large, if the promise of modern mapping is to be realized.

Cost of Change

Changing anything involves a cost. Time, labor, materials, and equipment all have a price. Thus, in order to save mapping time and reduce the dependence on skilled labor, cartographers have had to increase equipment and materials expenditures. When the substitution is carried to the extreme, capital outlays can be enormous. Pieces of photo processing equipment commonly cost from $5,000 to $20,000, computer peripheral devices from $5,000 to $50,000, and integrated computer systems from $250,000 to $1,500,000. The largest mapping agencies may have several duplicates of each machine.

Technological change has led to an increasing difference between small and large mapping establishments. In the past the same chores were carried out regardless of the size of the shop. Workers could move freely between large and small establishments and generally carry on what they had been doing. The need for retraining was minimal. But this is no longer the case. Small mapping establishments today are more likely to rely on the older technologies, while large establishments tend to base their operations on the latest technologies. Although individuals may be able to move relatively freely between establishments of about the same size because the establishments use similar technologies, they cannot move between those of vastly different sizes and technologies without undertaking a major retraining effort.

An important consequence is that the price of changing technology may include enormous education and retraining expenditures. Indeed, continuing education is fast becoming a fact of life for the cartographer. This may be done in-house, through workshops associated with professional societies, or by returning to institutions of higher education.

Enormous Amounts of Data

Changing technology has led to the capture, processing, storage, and retrieval of ever larger volumes of data suitable for mapping. Effective and efficient management of these data has become a major concern to cartographers. Questions involving security, updating frequency, access, charges to users, and archival procedures all raise difficult problems. Standards need to be established, and someone needs to assume responsibility for seeing to it that these standards are met. Until these things are accomplished, the future character of mapping is uncertain.

The concerns associated with the handling of immense amounts of data seem only to be accentuated with modern digital methods. Vast amounts of data have already been lost due either to a lack of interest or to improper archiving that has permitted deterioration of the electronic records. It will take someone with broad authority and deep commitment to resolve the issue. Fortunately, there are signs that some of the large government mapping agencies, such as the U.S. Geological Survey, are assuming greater responsibility for the creation, preservation, and distribution of digital data files. But much remains to be accomplished before cartographers can feel confident that data for potential future maps are being well managed for the general benefit of the mapping community.

Shifting Standards

Change of any sort is likely to be accompanied by a shift in standards. There has been a continual push for improved map accuracy standards, for example, as technology has progressed. As the mapping process has been speeded up by new technologies, there has been increasing demand for the mapping of shorter-lived phenomena. There is even evidence that the static map presentations that have been with us for centuries are giving way to animated maps, such as those made possible by repetitive weather satellite coverage. Many other examples could be cited.

There is always the danger, of course, that a new procedure or technique will be thought to be good and acceptable merely because it is quick and easy. Mapping convenience and mapping effectiveness are quite different, however. This point has been dramatized recently by the numerous crude maps that have been produced with computer assistance by individuals lacking even the most rudimentary knowledge of sound mapping principles. Although knowledgeable scientists and professionals who also happen to be skilled map users may have been able to obtain something from these maplike presentations, for many others it is more likely that the cartographic message was misinterpreted or lost. The mystique of the latest mapping technology is only worth so much. Beyond that point the usefulness of the map rests on how well good mapping technique has been incorporated into its design. Convenience is no substitute for sound cartographic judgment.

Improper Application by Noncartographers

There is far more to mapping than a mere mastery of technologies. A skilled cartographer also knows a great deal about human thought and communication, the other mapping sciences, and the

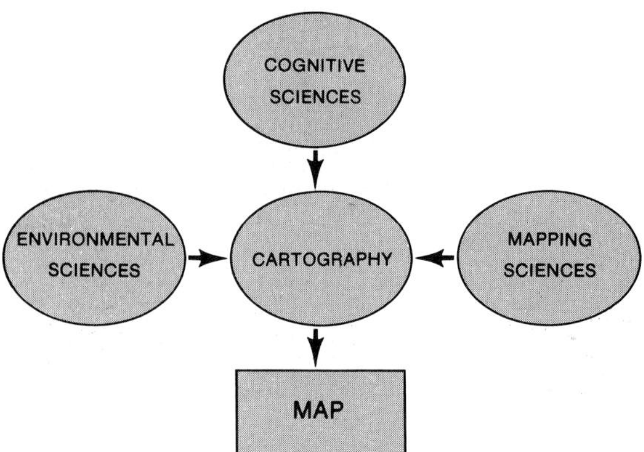

FIGURE 3.9 Skilled cartographers not only will have mastered the principles of mapmaking but also will have a firm grasp of the various environmental sciences, understand the cognitive processes of thought and communication, and be familiar with the other mapping sciences, such as geodesy, photogrammetry, remote sensing, and surveying.

disciplines involved with the environmental issues being mapped (Fig. 3.9). Being a computer graphics specialist does not make one a cartographer because a map is far from being a neutral document. Although it is not obvious in the data, to a rather large degree a map reflects the human conception of the mapped environment. It takes skill and experience to express cartographically the essential characteristics of environmental data in an effective manner.

As mapping becomes easier and more accessible to the untutored, the need for basic cartographic literacy will increase. Without such education we risk ending up with more and more maps that say less and less about our environment. We begin to address this need for developing cartographic awareness and responsibility in the following chapters.

SELECTED REFERENCES

Bickmore, D. P. (ed.), "Perspectives in the Alternative Cartography: Cartographic Computing Technology and its Applications," *Cartographica,* Monograph No. 28. Toronto: University of Toronto Press, 1982.

Boyle, A. R., "Cartography in 1990," *Proceedings of Auto-Carto IV,* Reston, Va.: American Congress on Surveying and Mapping, 1 (1979): 40–7.

Dudycha, D. J., "The Impact of Computer Cartography," *Cartographica* 18 (1981): 116–50.

Dutton, G. H., and Nisen, W. G., "The Expanding Realm of Computer Cartography," *Datamation* 24 (1978): 134–42.

Guelke, L., and Lai, Poh Chin, "Computer Cartography in Historical Geographical Research," *The Canadian Cartographer* 27 (1983): 207–22.

Lange, J. C., and Shanahan, D. P., *Interactive Computer Graphics Applied to Mechanical Drafting and Design.* New York: John Wiley & Sons, 1984.

Moellering, H., "Strategies of Real-time Cartography," *The Cartographic Journal* 17 (1980): 12–5.

Monmonier, M. S., *Computer-Assisted Cartography: Principles and Prospects.* Englewood Cliffs, N.J.: Prentice-Hall, Inc., 1982.

Morrison, J. L., "Changing Philosophical-Technical Aspects of Thematic Cartography," *The American Cartographer* 1 (1974): 5–14.

Robinson, A. H., Morrison, J. L., and Muehrcke, P.C., "Cartography 1950–2000," *Transactions of the Institute of British Geographers,* New Series 2 (1977): 3–18.

Taylor, D.R.F. (ed.), *The Computer in Contemporary Cartography.* New York: John Wiley & Sons, 1980.

Thompson, M. M., *Maps for America,* 2d ed. Washington, D.C.: U.S. Government Printing Office, 1982.

PART TWO

Theoretical Principles of Cartography

4

The Spheroid, Map Scale, Coordinate Systems, and Reckoning

THE EARTH

Even two thousand years ago most people knew that, if we disregard things like hills and valleys, the earth is spherical. Of course, some people thought it was flat, and some still do, but such beliefs become more difficult with satellite images and orbits. To assume the earth to be a true sphere was and is good enough for many kinds of mapmaking, but modern accurate topographic maps and charts require a more careful consideration of the shape of the earth.

The Shape of the Earth

The continents and mountains that rise above sea level are minor irregularities of the spheroidal earth and are quite insignificant relative to the size of the earth, although not, of course, to human beings. Even the flattening, that is, the difference between the polar and equatorial radii, which results from the rotation of the earth, is comparatively small. If the earth were reduced to the size of a large ball 1 m in diameter (a little more than 1 yd.), the polar flattening would total less than 3.5 mm (about $\frac{1}{8}$ in.), while the highest mountain would scarcely be measurable without special equipment and would certainly not be noticeable to most of us.

Even though the departures of the shape of the earth from a perfect sphere are relatively very small, they are important in the mapping process, since they affect the observations of surveyors and the accuracy with which their data can be transferred to maps.

The geographical framework on which all maps are based requires the systematic transfer of the geometric relationships observed on the spheroidal surface of the earth to the plane surface of the map.* Several factors are involved. First, the uneven distribution of terrestrial mass affects the direction of gravity, which determines the horizontal and vertical at places, upon which depend many locational observations. The irregular spheroidal shape which takes into account the variations in gravity is called the *geoid.* Second, for mapping purposes the observations made on the geoid must be transferred to a regular geometric reference surface, called an *ellipsoid,* which incorporates the flattening and closely approximates the geoid. Third, the three-dimensional geographical relationships of the ellipsoid must be transformed to the two-dimensional plane of the map by any of several procedures called *map projections*. The determination of geoidal characteristics and appropriate ellipsoids are concerns of the field of applied mathematics called *geodesy*. Brief descriptions of the geoid and ellipsoid appear below. The subject of map projections is part of cartography and will be introduced in the next chapter.

*Strictly speaking, we may make maps on surfaces other than a plane. The representation on a reduced globe and that of a three-dimensional terrain model are both maps. But 99 percent of all maps are on a plane (or the near-plane of a monitor screen), because for most purposes, it is the most convenient.

Geoid

The figure of the earth is unique, and therefore it is called the *geoid,* meaning earthlike. It is the shape that would be approximated by an undisturbed mean sea level in the oceans and the level of the water in a series of sea-level canals crisscrossing the land. It is technically defined as an equipotential surface, that is to which the direction of gravity everywhere is perpendicular. Because of variations in the distribution of the mass and density of the earth's constituents, the geoid generally rises over the continents and is depressed in oceanic areas. It also has various other bumps and hollows that depart from an "average smoothness" by as much as 60 m. The geoid is a very significant factor in mapping because all observations on the earth are, of course, made on the geoid. Because the geoid is irregular and therefore the direction of gravity is not everywhere toward the center of the earth, great efforts are required to correct for the deflections of the vertical so that measurements of distances on the surface will be consistent with those determined by astronomic observations.

Ellipsoid

The geoid is also deformed by the rotation of the earth. Because it spins on an axis, the geoid is bulged somewhat in the equatorial area and flattened a bit in the polar regions. A circumference through the poles is elliptical. The actual amount of the flattening is about 21.5 km, the difference between the polar and equatorial radii; the equatorial radius is, of course, the larger.

For precise mapping a regular geometric reference surface must be used. Observations on the geoid are transferred to the regular form that most closely approximates the geoid. This is an *ellipsoid of revolution,* that is, a figure produced by an ellipse rotated around its minor (smaller) axis.

The amount of the polar flattening (ellipticity or oblateness) is given by the ratio $f = (a - b)/a$ where a is the equatorial semiaxis and b is the polar semiaxis. It is usually expressed as $1/f$ and for most useful mathematical ellipsoids the value of $1/f$ is close to 297 or 298.

Because of the earth's oblateness, the flatter portion of the geoid is in the polar sector and the more rapid curvature is in the equatorial. Since a considerable amount of navigation is based on the angle between some celestial body and the horizon plane (or a perpendicular to it—the vertical), it is apparent that complications result from this departure from a true sphere. Consequently, whenever charts are being prepared for navigation or for determining or plotting exact courses and distances from one place to another, it is necessary to take the oblateness into account. In most cases of small-scale mapping it may safely be ignored.

Size of the Earth

About 250 B.C. Eratosthenes, a Greek scholar living in Egypt, calculated the size of the earth and apparently came close to the figures we now accept. Eratosthenes estimated that the distance between Alexandria and Syene (modern Aswan) in Egypt was 5000 stadia, about 925 km (575 mi.), since that Greek unit of measure is thought to equal about 185 m. He determined the arc distance to be 7° 12′ by observing the angle above the southern horizon of the noon sun at Alexandria to be 82° 48′ when the noon sun was supposed to be vertical at Syene (Fig. 4.1). Arc distance 7° 12′ is $\frac{1}{50}$ of a full circle, so 50 times the estimated distance between Alexandria and Syene equaled 250,000 stadia. Thus, he found the earth's circumference to be 46,250 km (approximately 28,750 mi.), which is only about 15 percent too large. Eratosthenes' method was correct, but his measurements and assumptions were somewhat erroneous; fortunately, the errors tended to compensate one another.*

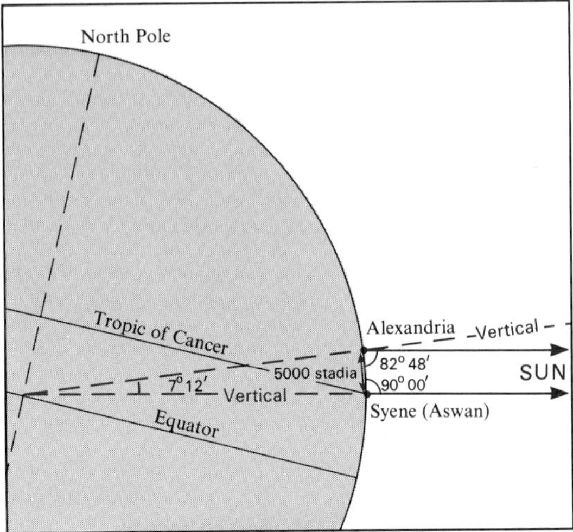

FIGURE 4.1 The relationships that Eratosthenes used for calculating the circumference of the earth. See text for explanation.

In recent times the dimensions of various ellipsoids have been calculated with great care in the same way from various astro-geodetic arcs

*The early estimates by Eratosthenes and one other were recorded by Ptolemy in his book *Geography*. Ptolemy, recognizing the errors but not their compensation, reported "corrected" values that reduced the earth's circumference by nearly a fourth. Unfortunately, or fortunately depending on one's viewpoint, the estimates in *Geography* were generally accepted. If Columbus had known the true size of the earth, he probably would have not dared set sail to find the Indies by going west; in that case, perhaps the latest colonization of the Americas might have been from Asia instead of Europe.

TABLE 4.1 Geodetic Reference System 1980 (GRS 80)

Equatorial Semi-axis (a)	Polar Semi-axis (b)	$1/f$
6,378,137 m	6,356,752 m	298.257

TABLE 4.2 Dimensions of the Earth (Based upon GRS 80)

	Kilometers	Statute Miles (U.S.)
Equatorial diameter	12,756.3	7,926.4
Polar diameter	12,713.5	7,899.8
Equatorial circumference	40,075.1	24,901.5
Radius of the sphere of equal area* (approximate)	6,371	3,959
Area of the earth (approximate)	510,064,500 km^2	196,937,000 mi^2

*The sphere of equal area is a true sphere which has the same surface area as the ellipsoid. All radii of a sphere are identical.

measured with precision. These have differed because the parameters used were derived for limited areas. Recently the International Association of Geodesy has adopted new dimensions for the reference ellipsoid, called the Geodetic Reference System 1980 (GRS 80) given in Table 4.1. The GRS 80 ellipsoid will be used as the basis for a new reference mapping surface in the United States called North American Datum 1983 (NAD 83).

Given an ellipsoid, various other dimensions of the earth that are useful in cartography may be calculated. Table 4.2 gives dimensions of the earth based upon GRS 80.

The Great Circle

The shortest distance between two points is a straight line; however, on the earth it is obviously impossible to follow such a straight line. The shortest "straight line" course over the surface between any two points on a sphere is the arc on the surface directly above the true straight line. This arc is formed by the intersection of the spherical surface with the plane passing through the two points and the center of the earth (Fig. 4.2). The circle established by the intersection of such a plane with the surface divides the earth equally into hemispheres and is called a *great circle*.

FIGURE 4.2 The trace of the intersection of a plane and a sphere is a great circle whenever the plane includes the center of the sphere. Any two points on a sphere, such as A and B, can be connected by a great circle arc, and that arc is the shortest route over the surface between the two points.

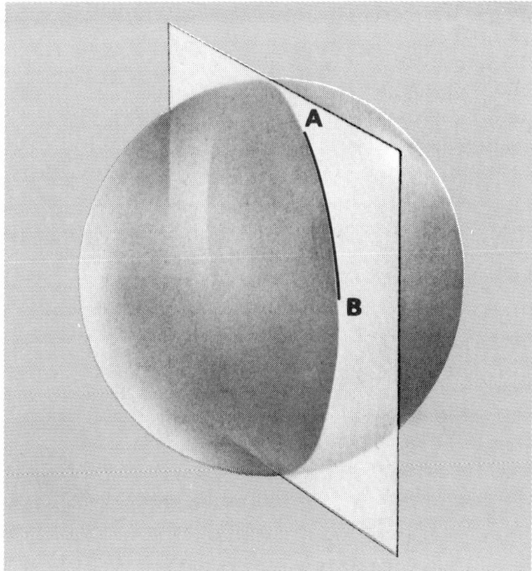

Great circles bear a number of geometrical relationships with the spherical earth that are of considerable significance in cartography and map use.

1. A great circle always bisects another great circle.
2. An arc of the great circle joining them is the shortest course between two points on the spherical earth.
3. The plane in which any great circle lies always bisects the earth and hence always includes the center of the earth.

Because a great circle is the shortest distance between two points on a spherical surface, air and sea travel, insofar as is possible or desirable, move along such routes as do radio signals and certain other electronic impulses. Consequently, many maps must be made on which great circles are shown by straight lines or simple curves.

MAP SCALE

Maps are necessarily smaller than the areas mapped, and consequently, in order to be usable, they must state the ratio or proportion between comparable measurements. This ratio is called the *map scale* and should be the first thing of which the map user becomes aware. The scale of a map may be shown in many ways: It can be specifically indicated by some statement or graphic device, and it may be shown indirectly by the spacing of the parallels and meridians and even subtly by the size and character of the marks on the map.

Map scale is an elusive thing because, by the very nature of the necessary transformation from the sphere to the plane, the scale of a map must vary from place to place and will even vary in different directions at a point.

Statements of Scale

The map scale is commonly thought of as being the ratio between a distance on the map and the corresponding distance on the earth, with the distance on the map always expressed as one. The map scale may be expressed in the following ways.

Representative Fraction (RF). This is a simple fraction or ratio. It may be shown either as 1:1,000,000 or 1/1,000,000; the former is preferred. This means that (along particular lines) 1 mm or 1 cm or 1 in. on the map (the numerator) represents 1,000,000 mm, cm, or in. on the earth's surface (the denominator). It is usually referred to as the "RF" for short. The unit of distance on both sides of the ratio must be the same.

Verbal Statement. This is a statement of map distance in relation to earth distance. For example, the RF 1:1,000,000 denotes a map on which 1 mm represents 1 km or about 1 in. to 16 mi. Many older map series were commonly referred to by this type of scale, for example, the "one-inch" or "six-inch" maps of the British Ordnance Survey (1 in. to 1 mi., 6 in. to 1 mi.).

Graphic or Bar Scale. This is a line placed on the map, often in the legend box or margin of the sheet, that has been subdivided to show the map lengths of units of earth distance. One end of the bar scale is usually subdivided further, so that the user may measure distances more precisely (Fig. 4.3).

Area Scale. This refers to the ratio of areas on the map to those on the earth. When the transformation from the sphere to the plane has been

FIGURE 4.3 Examples of graphic or bar scales. Often, the left end of the bar is subdivided into smaller units to provide more precise estimation of distances.

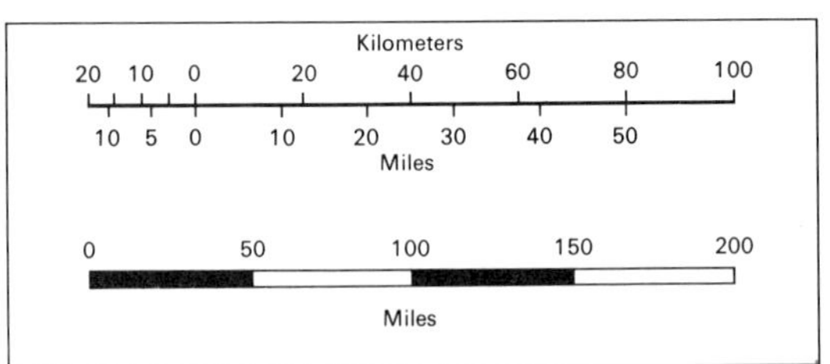

made so that all area proportions on the earth are correctly represented, the stated scale is one in which 1 unit of area (square centimeters, square inches) is proportional to a particular number of the same square units on the earth. If the area scale were expressed, it would be shown, for example, either as $1{:}1{,}000{,}000^2$ or as 1 to the square of 1,000,000. Usually, however, the fact that the denominator of the map scale is squared is assumed and not shown. The area scale can also be shown graphically by a square representing a stated number of square km or square mi.

Scale Factor

It is not possible to transform the spherical surface to a plane without differentially "stretching" or "shrinking" the spherical surface in the process. This means that the stated scale, the RF, will fit only at selected points or along particular lines; elsewhere the actual map scale will be either larger or smaller than the given RF. This is true to some degree in *all* flat maps. The statement of the relation between the given RF and the actual scale value is called the *scale factor* (SF).

Perhaps the simplest way to appreciate the concept of the SF is to imagine the necessary reduction and transformation of the spherical surface as being accomplished in two stages: (1) the reduction of the earth sphere to a globe map of a selected scale, and (2) the transformation of that spherical map to a plane map. The stated RF of the plane map will then be the RF of the globe and is called the *principal* (or nominal) RF. The real RF will be the *actual* scale on the map and will, of course, vary from place to place.

The SF may be computed by the following formula.

$$\text{SF} = \frac{\text{actual scale}}{\text{principal scale}}$$

This expresses the SF as a ratio related to the principal scale as unity (one). A SF of 2.0 would mean that the actual scale was twice the principal scale, which would be the case if, for example, the actual scale were 1:15,000,000 and the principal scale were 1:30,000,000. (Remember that the RF is a fraction and the larger the denominator, the smaller the scale.) Similarly, a SF of 0.50 would show that the actual scale was half that of the principal scale, as would be the case if, for example, the actual scale were 1:60,000,000 and the principal scale were 1:30,000,000.

Scale factors of the magnitudes used for these illustrations occur only on some small-scale maps. On large-scale maps the SF at various places will vary only slightly from one (unity). For example, on large-scale maps employing the transverse Mercator projection, the SF magnitudes within a 6° longitude zone may vary only from 0.99960 to 1.00158.

Determining the Scale of a Map

Sometimes it is necessary to determine the scale of a map or the scale of a particular part of the map, since, as observed previously, the scale can never be the same all over a flat map.

The map scale along a particular line may be approximated by measuring the map distance between two points that are a known earth distance apart and then computing the scale. Certain known distances are easy to use, such as the lengths of degrees of latitude or longitude (see Appendix B). Be careful that the measurement is taken in the direction the scale is to be used; frequently the distance scale of the map will not be the same in all directions from a point.

If the map scale referring to areas is desired, a known area on the earth (see Appendix B) may be calculated or measured on the map and the proportion determined. Remember that scales based on the relation between areas are conventionally expressed as the square root of the number of units on the right of the ratio. Thus, if the measurement shows that 1 square unit on the map represents 25,000,000,000,000 of the same units of the earth, it would be recorded as the square root, 1:5,000,000, which normally would approximate the linear scale.

Transforming the Map Scale

Frequently the cartographer is called on to change the size of a map, that is, to reduce or enlarge it. The mechanical means of accomplishing this are dealt with in a later chapter, but the problem of determining how to change it in terms of scale is similar to the problem of transforming one type of scale to another. A cartographer who can develop a facility with this sort of scale transformation will have no difficulty enlarging or reducing maps.

There is, of course, no problem in transforming decimal metric scales. U.S. Customary units are more bothersome; the essential information is that there are 63,360 in. in 1 mi. (statute). With this information we can change each of the linear scales (RF, graphic, stated) previously described to the others. Examples follow.

If the RF of the map is shown as 1:75,000:

EXAMPLE 1.

To determine the stated scale:

Metric.

The cm/km scale will be:

1. 1 cm (map) represents 75,000 cm (earth), and
2. 75,000 cm = 0.75 km
3. 1 km/0.75 km = 1.333
4. Therefore 1.333 cm represents 1.0 km

U.S. Customary.

The in./mi. scale will be:

1. 1 in. (map) represents 75,000 in. (earth), and
2. 1/75,000 = x/63,360, and
3. x = 0.845
4. Therefore, 0.845 in. represents 1.0 mi.

EXAMPLE 2.

To construct a graphic scale, a proportion is established as:

Metric.

1. 1.333 cm:1.0 km::x cm:10 km
2. x = 13.33
3. 13.33 cm represents 10 km, which may be easily plotted and subdivided

U.S. Customary.

1. 0.845 in.:1.0 mi.::x in.:10 mi.
2. x = 8.45 in.
3. 8.45 in. represents 10 mi., which may be easily plotted and subdivided

EXAMPLE 3.

To determine the RF:

Metric.

If the graphic scale shows by measurement that 1 cm represents 50 km, then:

1. 1 cm represents 50 × 100,000 cm, or
2. 1 cm to 5,000,000 cm and, therefore,
3. The RF is 1:5,000,000

U.S. Customary.

If the graphic scale shows by measurement that 1 in. represents 75 mi.:

1. 1 in. represents 75 × 63,360, or
2. 1 in. to 4,752,000 in. and, therefore,
3. The RF is 1:4,752,000

Changing the scale of a map that has an area scale is accomplished by converting the known area scale and the desired area scale to a linear proportion.

Common Map Scales

Maps are made with an infinite variety of scales. An experienced map user learns to associate an approximate level of generalization and accuracy with particular scales, and the RF then becomes a kind of index of precision and content. It is helpful to translate the RF mentally into common units of measure. For example, on a map at a scale of 1:1,000,000, 1 mm represents 1 km and $\frac{1}{16}$ in. represents about 1 mi. Table 4.3 is a list of some of the more common map scales.

COORDINATE SYSTEMS

Locating points relative to one another requires use of the concepts of direction and distance. These can only be specified in terms of some system; primitive peoples probably did so in relative terms, using aids such as the directions of the rising and setting sun, forward and backward, left and right, and so on, and they probably expressed distance in terms of travel time—all of these being reckoned with respect to one's location.

Two types of systems are now in general use. The older, the geographical coordinate system employing latitude and longitude, was first put into practice by Greek philosopher-geographers before the beginning of the Christian era. It is the primary system, since it is used for all basic locational reckoning, such as navigation and fundamental surveying.

The second system, called *plane rectangular coordinates* or just *plane coordinates,* is also old, at least in its basic form. It was a standard feature of Chinese cartography after being included in the six principles of mapmaking by Pei Hsiu in the third century A.D. In modern form the plane coordinate system evolved from Cartesian coordinates applied to military needs, but it has since become extremely useful for a variety of other uses.

Any system must be based upon some reference point. From such a starting point, the locations of all other points can be stated in terms of a defined direction and distance from it.

Geographical Coordinates

On a motionless spherical surface there would be no natural starting point, but the earth both rotates on an axis and revolves around the sun in a regular way. Because the positions of the

TABLE 4.3 Common Map Scales and Their Equivalents

Map Scale	One Centimeter Represents	One Kilometer Is Represented by	One Inch Represents	One Mile is Represented by
1:2,000	20 m	50 cm	56 yd.	31.68 in.
1:5,000	50 m	20 cm	139 yd.	12.67 in.
1:10,000	0.1 km	10 cm	0.158 mi.	6.34 in.
1:20,000	0.2 km	5 cm	0.316 mi.	3.17 in.
1:24,000	0.24 km	4.17 cm	0.379 mi.	2.64 in.
1:25,000	0.25 km	4.0 cm	0.395 mi.	2.53 in.
1:31,680	0.317 km	3.16 cm	0.500 mi.	2.00 in.
1:50,000	0.5 km	2.0 cm	0.789 mi.	1.27 in.
1:62,500	0.625 km	1.6 cm	0.986 mi.	1.014 in.
1:63,360	0.634 km	1.58 cm	1.00 mi.	1.00 in.
1:75,000	0.75 km	1.33 cm	1.18 mi.	0.845 in.
1:80,000	0.80 km	1.25 cm	1.26 mi.	0.792 in.
1:100,000	1.0 km	1.0 cm	1.58 mi.	0.634 in.
1:125,000	1.25 km	8.0 mm	1.97 mi.	0.507 in.
1:250,000	2.5 km	4.0 mm	3.95 mi.	0.253 in.
1:500,000	5.0 km	2.0 mm	7.89 mi.	0.127 in.
1:1,000,000	10.0 km	1.0 mm	15.78 mi.	0.063 in.

other celestial bodies are thus predictable, one's location can be calculated if one has some way of telling time and an *ephemeris,* an astronomical almanac containing tables that list daily apparent positions of celestial bodies. The geographical coordinate system was devised to make possible a statement of location, and the two poles, where the axis of rotation intersects the earth's surface, provide places on which to base the system. Specifying a location on the earth requires determining a north-south distance, called latitude, and an east-west distance, called longitude.

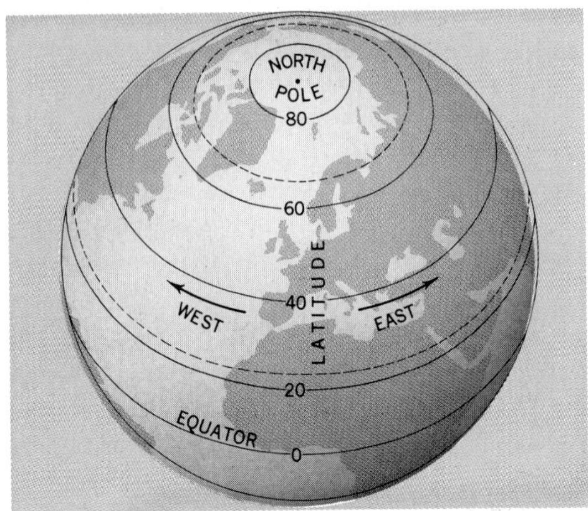

FIGURE 4.4 The parallels of latitude (showing distance north-south) specify the directions east-west. (From Trewartha, Robinson, Hammond, and Horn, *Fundamentals of Physical Geography*, 3d ed., New York: McGraw-Hill Book Company, 1977.)

Latitude

The system of locating oneself in a north-south position depends on the regular curvature of the earth's surface. One's latitude may be defined as the angle between a normal (perpendicular) to the surface and the plane of the equator at that place. This is accomplished by observing the altitude (angle above the horizon) of some celestial body and then, with the aid of the data in an ephemeris, calculating the desired angle in the north-south direction. Because the earth is spherical, change of position along a north-south line is accompanied by a change in the angular elevation of celestial bodies in relation to the horizon plane on the earth. When the celestial bodies lie in the plane of the arc, such as Polaris or the noon sun, there is a one-to-one relationship; that is, for each degree of distance along a north-south arc, the elevation above the horizon of the body will change by 1°. Any star can be observed, and appropriate corrections can be made to obtain a corresponding zenith angle.

This simplifies the problem somewhat, because the earth rotates on its axis, and most of the celestial bodies therefore also seem to move while the observer is moving from one place to another. The fundamental fact, however, is that north-south position can be determined by measuring the vertical angle between horizontal and a celestial body.

To use this relationship for a spherical coordinate system was natural. The ancients imagined an infinite number of circles around the earth parallel to one another (Fig. 4.4). The one dividing the earth in half, equidistant between the poles, was named, as might be expected, the equator. The series north of the equator is called north latitude, and the series south of the equator is called south latitude.

Length of a Degree of Latitude. In the usual system of angular measurement, a circle contains 360°; and a half-circle contains 180°. Consequently, there are 180° of latitude from pole to pole. The quadrant from the equator to each pole is divided into 90°, and the numbering starts from 0° at the equator and goes by degrees, minutes, and seconds to 90° at each pole. Latitude is always designated as north or south.

Degrees of latitude are very nearly the same length, but not quite. Because the earth is an oblate spheroid (flattened in the polar areas), a north-south line (a meridian) has less curvature near the poles and more near the equator. Therefore, to observe a 1° difference in the altitude of a celestial body requires a longer traverse along the meridian in the polar regions. Consequently, degrees of north-south arc on the earth are not quite the same lengths in units of uniform surface distance, but vary from about 110.6 km (68.7 mi.) near the equator to about 111.7 km (69.4 mi.) near the poles. This difference of about 1 km in 110 is of little significance in small-scale maps, but is important on large-scale maps and charts. A complete table of lengths is included in Appendix B.

Longitude

Only position north-south on the earth is established by latitude. The transverse component, longitude, or distance east-west is provided by an infinite set of great circles, called *meridians,* arranged perpendicularly to the parallels. Unlike the equator in the latitude system, no meridian has a natural basis for being the starting line from which to reckon distance east-west in degrees, minutes, and seconds of longitude. From a given meridian, selected as a starting line, east-west position is designated by the angular distance along the parallel circle in the latitude system (Fig. 4.5).

Prior to the middle of the eighteenth century, only latitude could be easily reckoned with any precision. Distance east and west depends on time differences and, for easy reckoning, requires that one know the times of day at the two places at the same instant. Without accurate timepieces that can be carried or instantaneous communication, this can only be done by very elaborate astronomical observation and calculation. Through the years the result was considerable error in east-west location, which was one of the contributing factors to the glorious error of the fifteenth century, the idea that the distance from Europe westward to Asia was less than half what it actually is.

When the precise determination of longitude became critical for navigation, generous prizes for its solution were offered. A variety of suggestions were put forward, ranging from the observations of a celestial timepiece, such as the behavior of the satellites of Jupiter, to the employment of the variations (declination) of the magnetic compass. When the chronometer (a very accu-

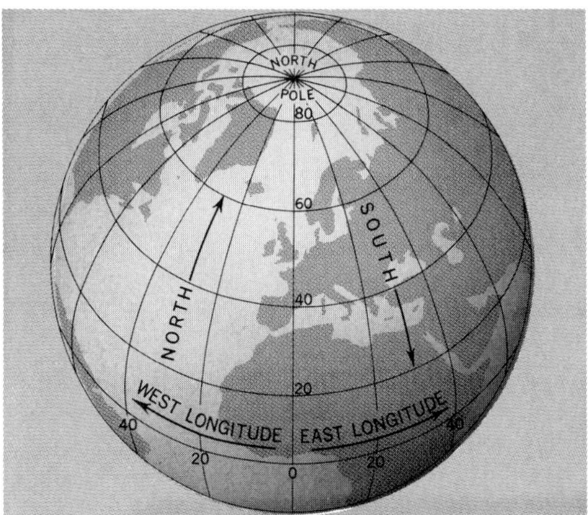

FIGURE 4.5 The meridians of longitude (showing distance east-west) specify the directions north-south. (From Trewartha, Robinson, Hammond, and Horn, *Fundamentals of Physical Geography*, 3d ed., New York: McGraw-Hill Book Company, 1977.)

rate clock) was developed by Harrison and others in the middle of the eighteenth century, the problem was solved.

Because all parallels are concentric circles, they all rotate at the same angular speed—360° per day or 15° per hour. By carrying a clock showing the accurate time somewhere else, the difference between that time and local sun time in hours, minutes, and seconds can be converted to the difference in longitude between the two places by arithmetic. Today this is accomplished by time signals broadcast at regular intervals, as well as with the aid of chronometers.

Length of a Degree of Longitude. The equator is a great circle but, as we go toward the poles, all other parallels become smaller and smaller circles; yet each is divided into 360°. Therefore, each east-west degree of longitude becomes shorter with increasing latitude and is finally reduced to nil at the poles. The relationship between the length of a parallel (the circumference of a small circle) and the circumference of a great circle (such as the equator or a meridian circle) is the circumference of the great circle multiplied by the cosine of the latitude of the parallel; stated another way, the length of a degree of longitude = cosine of the latitude × length of a degree of latitude.

A table of cosines will show that

$$\cos 0° = 1.00$$
$$\cos 30° = 0.87$$
$$\cos 60° = 0.50$$
$$\cos 90° = 0.00$$

Thus, at 60° north and south latitude, a degree of longitude is half as long as a degree of latitude. Table 4.4 illustrates the decreasing lengths of the degrees of longitude from the equator toward the pole. A more complete table is included in Appendix B.

The Prime Meridian. The meridians are all alike, and any one can be chosen as the meridian of origin from which to start the numbering for longitude. The choice became, as might be expected, a problem of international consequence. Numerous countries, each with patriotic ambition, wished to have 0° longitude within its borders or as the meridian of its capital. For many years each nation published its own maps and charts with longitude reckoned from its own meridian of origin. The result was much confusion.

During the last century, many nations began to accept the meridian of the observatory at Greenwich near London, England, as 0° and, in 1884, it was agreed on at an international conference. Today this is universally accepted as the prime meridian, but some maps still show two sets of meridians, one numbering system based on a local meridian and the other on the Greenwich system.

Since longitude is reckoned as either east or west from Greenwich (to 180°), the prime meridian is somewhat troublesome because it divides both Europe and Africa into east and west longitude. The choice of the meridian of Greenwich as the prime meridian establishes the "point of origin" of the geographical coordinate system in the Gulf of Guinea. The opposite of the prime meridian, the 180° meridian, is more fortunately located; its position in the Pacific provides a convenient place for the international date line. Days on earth must begin and end somewhere and only a few deviations from 180° are needed in that sparsely populated region to keep from separating inhabited areas into time zones with different days.

TABLE 4.4 Lengths of One Degree of Parallels

Latitude	Kilometers	Statute Miles
0°	111.321	69.172
10°	109.641	68.129
20°	104.649	65.026
30°	96.448	59.956
40°	85.396	53.063
50°	71.698	44.552
60°	55.802	34.674
70°	38.188	23.729
80°	19.394	12.051
90°	0	0

Rectangular Coordinates

On a limitless plane surface there is no natural reference point; that is, every point is like every other point. An arbitrary system of location on a plane surface has long been used by establishing a "point of origin" at the intersection of two conveniently located, perpendicular "axes." The plane is then divided into a grid by an infinite number of equally spaced lines parallel to each axis. The position of any point on the plane with reference to the point of origin may then be stated by indicating the distance from each axis to the point, measured in each case parallel to the other axis, and expressed to any desired precision. In the familiar rectangular coordinate system (for example, cross-section paper) the "horizontal" distance is called the X value or *abscissa,* and the distance perpendicular to it is called the Y value or the *ordinate* (Fig. 4.6).

The geographical coordinate system is useful for large areas, and the measurement of distances and directions in angular measure in degrees, minutes, and seconds can hardly be improved on. But it is often cumbersome. With the increas-

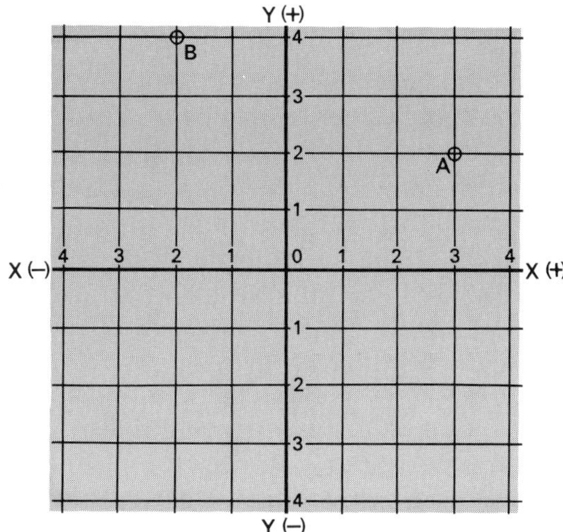

FIGURE 4.6 A rectangular coordinate system. The origin is 0. The abscissa values are −X, 0, +X and the ordinate values are −Y, 0, +Y. The position of point A is 3,2; the position of point B is −2,4. In the geographical coordinate system, Y values correspond to latitude north and south and X values to longitude east and west of the origin (0° latitude and 0° longitude). In designating position in a plane rectangular coordinate system, the X value is always given first; in the geographical coordinate system, latitude is usually given first.

ing range of artillery in World War I it became more and more difficult to arrive at accurate azimuth (bearing or direction) and range (distance). To simplify the problem, the French constructed a series of local plane, rectangular coordinate grids on maps. Since the formulas of plane geometry are far simpler than those of spherical geometry, other nations quickly followed suit and, between World Wars I and II, a great many systems of plane rectangular coordinates were devised and put into use. By now the use of rectangular grid systems is almost universal.

The basic procedure is as follows: first, a map is made by transforming the spherical surface to a plane (by a system of map projection), preparing the map on the plane, and placing a rectangular plane coordinate grid over the map. The two sets of straight, parallel grid lines are equally spaced, perpendicular to one another. To locate a position we need only specify the *X* and *Y* coordinates to whatever degree of precision we desire in decimal divisions of whatever earth distance units are used. This is much simpler than degrees, minutes, and seconds of latitude and longitude.

To simplify the reckoning of position, only the upper right-hand part of a plane coordinate system is used (Fig. 4.6), so that both sets of coordinates are positive, and therefore there will be no repetition of numbers east and west or north and south of the axes. Normally, the origin of the numbering is assumed to be outside the map area to the lower left.

One reads a grid reference in the same way a point is located on cross-section paper. In rectangular map coordinates, the *X* value is always given first and is called an *easting;* the *Y* value is called a *northing*. A rule of thumb is that when using grid references, we must always "read right, up." Looking at Figure 4.7 will show that point P can be given an easting value of 14.5 and a northing value of 20.1 by decimal subdivision of the squares. A grid reference is given as an even set of numbers run together, so to speak, with

FIGURE 4.7 A portion of a rectangular grid. If the squares are 1 km on a side, then point P may be located to within a 10 m² with the grid reference 14562011.

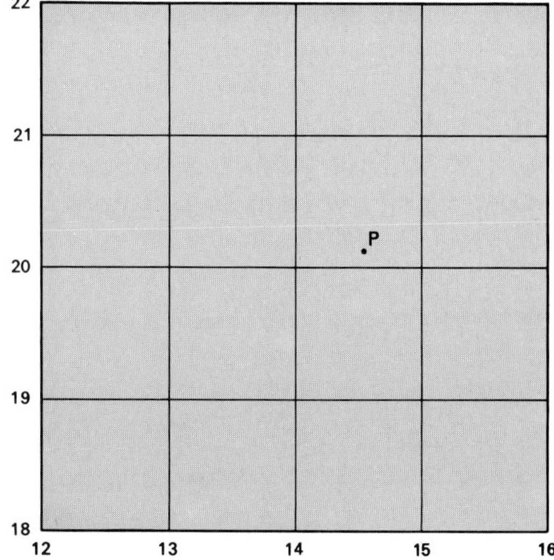

the first half of the group being the easting and the second half being the northing. Decimals are not shown, but are deduced from the numbering of the lines; thus, the grid reference for point P would be 145201. With lesser precision it would be 1420, and with greater precision it would be 14562011. If each square in Figure 4.7 represented 1 km^2 (1000 m on a side), the reference 145201 would be a statement that point P lies within a 100-m square, the point read being the location of the southwest corner of the 100-m square. Adding an additional digit would narrow the position to within a 10-m square.

Normally, plane coordinates are used only on large-scale maps, since the distortions that result from the transformation of the spherical surface to the plane make small-scale maps undesirable for detailed reference and calculation. Most large-scale topographic maps show one or more systems of plane coordinates, and some areas (for example, Great Britain) have adopted a national system of plane coordinates. It should be noted that small-scale reference maps sometimes employ a letter-number indexing system to help locate map data, but that is not a plane coordinate system.

One important characteristic of a plane coordinate system is the way the scale factors (SF) are arranged. This depends upon the system of map projection. In the early part of this century, a variety of projections were used for plane coordinate systems, but today most plane coordinate systems are based on only three map projections: the transverse Mercator, the polar stereographic, and Lambert's conformal conic. These will be described in the next chapter, but the distribution of the SF is quite simple in all three.

The UTM System

Although individual countries may develop particular systems suitable to their needs, one system that is commonly used is the Universal Transverse Mercator (UTM) grid system. The grid system and the projection on which it is based have been widely adopted for topographic maps, referencing of satellite imagery, natural resources data bases, and other applications that require precise positioning.

In the UTM grid system the area of the earth between 84°N and 80°S latitude is divided into north-south columns 6° of longitude wide called *zones*. These are numbered from 1 to 60 eastward, beginning at the 180th meridian. Each column is divided into quadrilaterals 8° of latitude high. The rows of quadrilaterals are assigned letters C to X consecutively (with I and O omitted) beginning at 80°S latitude (Fig. 4.8). (Row X is 12° latitude extending from 72°N to 84°N to cover all land areas in the northern hemisphere.) Each quadrilateral is assigned a number-letter combination. As always, in giving a grid reference, one reads right, up. Each quadrilateral is divided into 100,000-m zones designated by a system of letter combinations.

Within each zone the meridian in the center of the zone is given an easting value of 500,000 m. The equator is designated as having a northing value of 0 for the northern hemisphere coordinates and an arbitrary northing value of 10 million m for the southern hemisphere.

For the UTM the transverse Mercator projection is employed, so that along each north-south grid line (only the grid line in the center of a zone is a meridian) the SF is constant, but it varies in the east-west direction. Along the center grid line of each UTM grid zone the SF is 0.99960 (smaller scale), and at the margins of the widest part of a north-south column (equator) including overlap, about 363 km distant from the center, the SF is 1.00158.

The UPS System

In the UPS grid system, used instead of the UTM system in the polar areas, each circular polar zone is divided in half by the 0°–180° meridian. In the north polar zone the west half (west longitude) is designated grid zone Y, the east half as Z. In the south polar zone the west longitude half is designated A, the east half B (Fig. 4.8).

In the polar areas the northings and eastings of

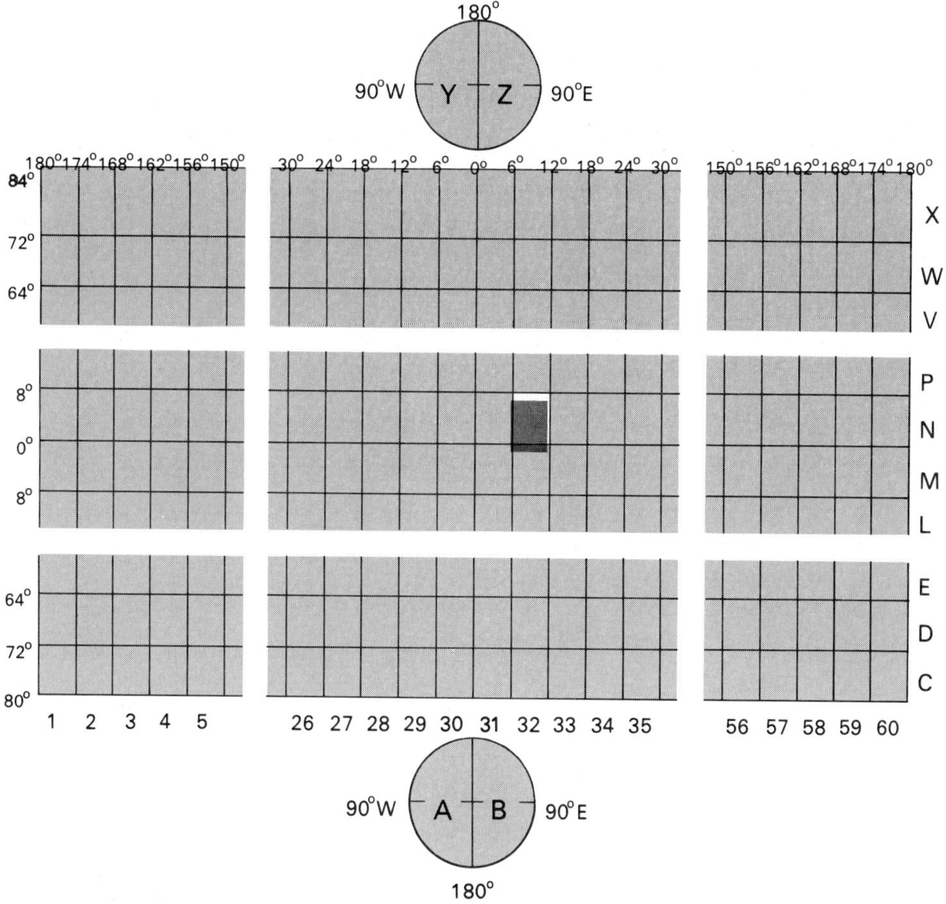

FIGURE 4.8 The system of UTM and UPS grid zone designations. Each quadrilateral is identified by its column number and row letter (I and O are omitted). The tinted zone is 32N. The UPS grid zones in the polar areas are sectors on each side of the 0°–180° meridians.

a grid system must be arbitrarily assigned. In both zones the 2 million–m easting coincides with the 0°–180° meridian line. The 2 million–m northing coincides with the 90°E–90°W meridian line. Grid north is parallel to true north along the 0° meridian and, therefore, also to true south along the 180° meridian. The UPS zones are divided into 100,000-m squares like the UTM.*

For the UPS the stereographic projection centered on the pole is employed. This application of the projection arranges the scale so that the SF's are constant along the parallels (circles in this case), but they vary from parallel to parallel. At the pole the SF is set at 0.994, at approximately 81° latitude it is 1.0 and increases to 1.0016 in the vicinity of 80° latitude.

State Plane Coordinates

In order to provide the convenience of plane coordinates and a way to ensure the permanent recording of the location of original land survey

*A complete description of the UTM and UPS grid systems is given in Department of the Army, *Grids and Grid References*, TM 5-241-1, Headquarters, Department of the Army, Washington, D.C., 1967.

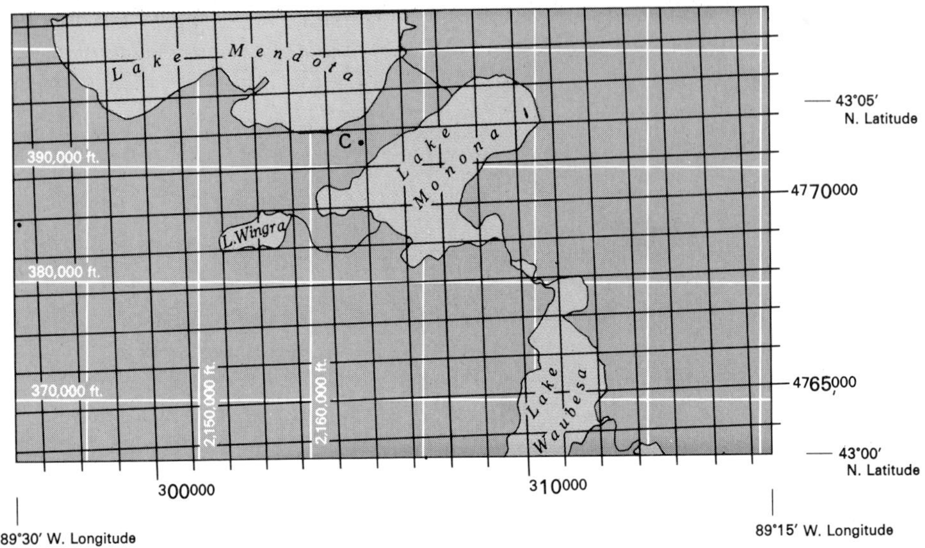

FIGURE 4.9 The three systems of coordinates as they appear on part of the USGS Madison, Wisconsin, quadrangle, 1:62,500. The ticks in the bottom and right margins show the graticule, that is, the spherical coordinates, which conform to the edges of the figure; the white lines show the 10,000-ft grid lines of the Wisconsin Coordinate System, South Zone; and the straight black lines inside the figure show the 1000-m UTM grid lines.

monuments, the U.S. Coast and Geodetic Survey (now the National Ocean Survey) worked out a system of plane rectangular coordinates for each of the states on either the transverse Mercator or Lambert's conic projection that are specifically tied to locations in the national geodetic survey system. To keep the inevitable scale variation to a reasonable minimum, a state may have two or more overlapping zones, each of which has its own projection system and grid. The units used are feet. The large-scale topographic maps published by the U.S. Geological Survey carry tick marks showing the locations of the 10,000-ft. grid.*

Figure 4.9 shows both the 1000-m UTM grid lines of Zone 16 and the 10,000-ft. grid lines of the Wisconsin Coordinate System South Zone, extended over the southern third of the area included on the Madison, Wisconsin, Quadrangle, 1:62,500, U.S. Geological Survey topographic map. The three different orientations (the graticule, that is, the net of parallels and meridians, the UTM grid, and the grid of the Wisconsin Coordinate System South Zone) result from each set being differently arranged. In the graticule all lines are true north-south or east-west. In the UTM Zone 16 the meridian of 87°W longitude is central and is the north-south axis about which the rectangular grid is arranged.* In the Wisconsin Coordinate System the north-south axis of the rectangular grid is 90°W longitude. Accordingly, in the area shown on Fig. 4.9 it will be seen that

*A complete description of the state plane coordinate systems is given in Hugh C. Mitchell and Lansing G. Simmons, *The State Coordinate Systems*, U.S. Coast and Geodetic Survey, Special Publication No. 235. (Washington, D.C.: U.S. Government Printing Office, 1945.)

*The 100,000-m squares of the military UTM grid system are not used in the application of the UTM system on U.S. Geological Survey topographic maps. The numbering is based on the equator and the central meridian of each zone, as described earlier.

the UTM grid deviates slightly from a true north-south orientation as does the Wisconsin Coordinate System in an opposite direction.

On published topographic maps the positions of the grid lines on the two sets are shown only by hard-to-distinguish ticks around the margins of the map. Point C in Figure 4.9 is the location of the state capitol building in Madison, Wisconsin. Grid references of C would be read as follows.

1. UTM SYSTEM. The full UTM coordinates to locate Point C (in a 100-m square) would be 3058 for the easting and 47716 for the northing, but the initial small digits (3 in the eastings and 47 in the northings in Fig. 4.9) are not used in giving a grid reference. Therefore, the grid reference of 058716 would locate Point C within a 100-m square; this reference is the location of the southwest corner of the square.
2. WISCONSIN COORDINATE SYSTEM. On a $7\frac{1}{2}$-minute, 1:24,000 scale topographic map, 1/100 in. represents 20 ft., so that the reading of a coordinate position to that precise value would be quite useless, since the paper the map is printed on would be subject to greater distortion with changes in humidity. State plane coordinates are surveyed to the fraction of a foot in the field, and then are fully given. Point C in Figure 4.9 is located at approximately 2,164,600 ft. east and 392,300 ft. north in the Wisconsin Coordinate System, South Zone.

RECKONING

The determination of directions, distances, and areas on the earth is done at various levels of precision. On the one hand they may be determined with the aid of maps and charts using the information provided by coordinate systems. Except for directions and distances between places relatively close together on large-scale maps, the derivation of such data by direct measurement on maps is subject to considerable error because of the inevitable variation in scale factors. On the other hand, the detailed, precise measurement of such things as property boundaries and railways is done by surveyors, while the establishment of accurate geographical positions and long distances is accomplished by geodetic engineers using sophisticated instruments and complex techniques. Navigators of aircraft and oceangoing vessels are similarly concerned to establish location, course, and speed. All such activities require an understanding of the concepts of direction and distance, as well as their derivative, area, as they apply on the spherical earth.

Direction

Directions on the earth are entirely arbitrary, since a spherical surface has no edges, beginning, or end. By definition, then, north-south is along any meridian and east-west is along any parallel; because of the arrangement of the graticule, these two directions are everywhere perpendicular except, of course, at the poles. The directions determined by the orientation of the graticule are called geographic or *true* directions as distinguished from two other kinds of direction, *grid* direction and *magnetic* direction.

It is apparent that when a rectangular grid is placed over the graticule of a map, in most places the "north" direction of the grid will not coincide with true north as specified by the graticule (see Fig. 4.9). Therefore, most detailed topographic maps specify the discrepancy in degrees and minutes between grid north and true north at the center of the sheet.

The needle of the magnetic compass aligns itself with the total field of magnetic force; in most parts of the earth this is not parallel with the meridian.* Consequently, there is usually a difference between true north and magnetic north that is called *compass variation* or *magnetic decli-*

---

*Contrary to the belief of the uninformed, the compass needle does not point directly toward the magnetic pole(s), except in the sense that if one were to "follow the compass" he would ultimately arrive at the magnetic pole, but it would be by a devious route.

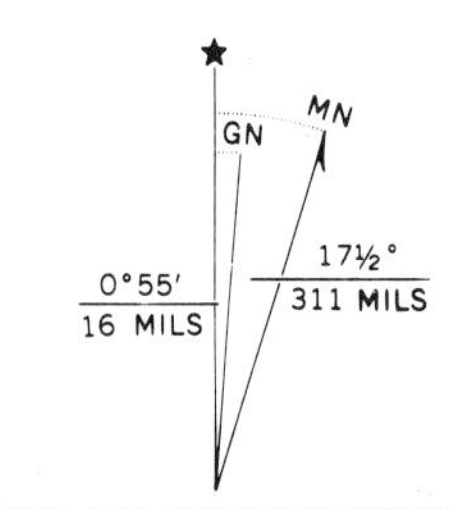

FIGURE 4.10 Arrows showing the differences among true north, grid north, and magnetic north at the center of USGS Paradise East, California, quadrangle, 1:24,000 (1980). A *mil* is an angular division equal to 1/6,400 part of the circumference of a circle.

nation. The difference between true north, grid north, and magnetic north is usually shown with a diagram on detailed maps (Fig. 4.10). Furthermore, the magnetic field changes slowly so that the variation value is likely to be correct only for the date the map was issued. Often a statement of the amount of annual change in variation is included.

FIGURE 4.11 A typical compass rose from a modern nautical chart published by the U.S. National Ocean Survey. The rose is positioned on the chart so that 0°–360° shows true north. Note how magnetic north and the annual variation are indicated.

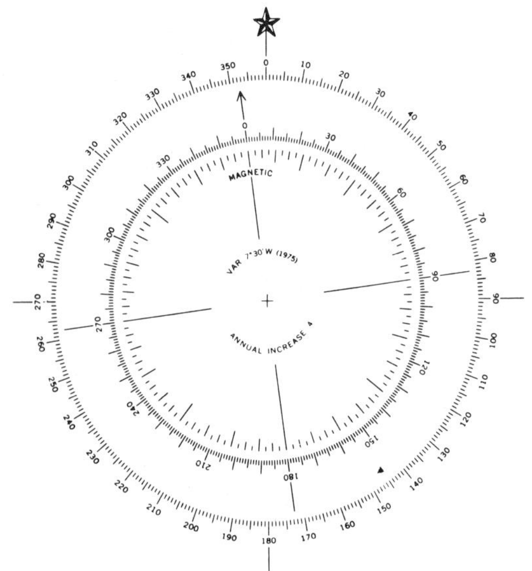

Prior to the regular use of the magnetic compass, mariners in the Mediterranean area often specified directions by typical winds; these were shown diagramatically on a wind rose, a star-shaped device with eight or sixteen labeled points. From this developed a compass card in which the 360° circle was divided into thirty-two points (1 point = $11\frac{1}{4}°$). It became common practice on charts to include a similar representation, called a compass rose, to help the navigator reckon position and plot the course. Divisions finer than a point later became necessary, and today the compass rose on charts usually consists of a circle divided into degrees (Fig. 4.11).

The direction of a line on the earth is called many things: bearing, course, heading, or azimuth. Their meanings are essentially the same, differing largely in the context in which they are used. The two of importance in cartography are azimuth and bearing.

Azimuth. As is apparent from observing a globe, the directions on the earth, established by the graticule, are likely to be constantly different if we move along the arc of a great circle. Only on a meridian or on the equator does direction remain constant along a great circle. It is convenient to be able to designate the "direction" the great circle has at any starting point toward a destination. This direction is reckoned by observing the angle the arc of the great circle makes with the meridian of the starting point. The angle is described by the number of degrees (0 to 360°), reading clockwise, usually from north.

Since arcs of great circles are the shortest courses between points, movement along them is of major importance. Hence many maps are constructed so that directional relations are maintained as far as possible.

The computation of azimuths in the geographical coordinate system is quite involved and is ordinarily not needed except in geodetic work. In plane rectangular coordinates the grid azimuth (Az_G) from A to B is

$$\tan Az_G = \frac{E_2 - E_1}{N_2 - N_1}$$

in which E_1 is the easting value for A and E_2 is the easting value for B; N_1 is the northing value for A and N_2 the northing value for B. When the tangent angle equivalent to Az_G is found, it must then be employed as an angular measure clockwise from grid north. If B lies in the northeast quadrant from A, then the tangent value $= Az_G$ $=$ the grid azimuth; if B is in the southeast quadrant, then $Az_G = 180° - \tan$; if B is in the southwest quadrant, then $Az_G = 180° + \tan$; and, if B lies in the northwest quadrant, then $Az_G = 360° - \tan$.

Rhumb Lines. A bearing is the direction from one point to another, usually expressed in relation to the compass rose either in a fashion such as northeast or as north 45° east. A great circle is the most economical route to follow when traveling on the earth. A pilot of an aircraft may do this by following a radio beam, but it is practically impossible otherwise, except when travel is along a meridian or the equator, because directional relations constantly change along all other great-circle routes. This is illustrated in Fig. 4.12. Because a course of travel must be directed in some manner, such as by the compass, it is not only inconvenient but impracticable to try to change course at, so to speak, each step.

A bearing which intersects meridians at a constant oblique angle is called a *rhumb line* or *loxodrome*. As a matter of fact, rhumb lines are complicated curves, and if one were to continue along a rhumb line, one would spiral toward the pole along a spherical helix with the pole the limit.

In order for ships and aircraft to approximate as closely as possible the great circle route between two points, movement is directed along rhumb lines that approximate it. Such a course is planned to begin on the great circle and shortly return to it, then depart and return again, as shown

FIGURE 4.12 How azimuth (direction) is given. The drawings show a great circle arc on the earth's graticule. The drawing on the right is an enlarged view of the center section of the drawing on the left. The azimuth, from the starting point, of any place along the great circle arc to the northeast is stated as the angle between the meridian and the great circle arc, reckoned clockwise from the meridian. Notice that the great circle arc intersects each meridian at a different angle.

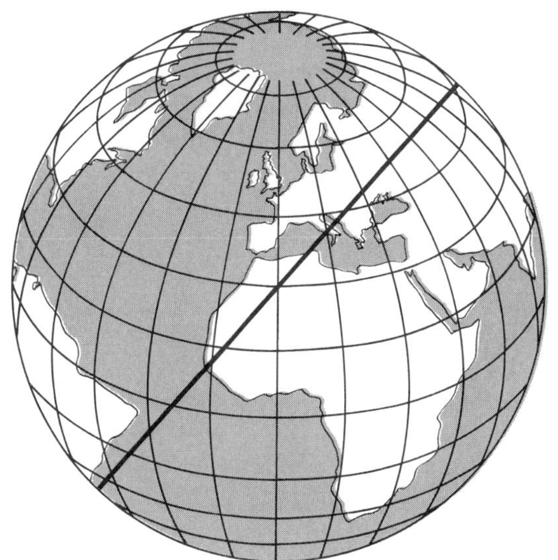

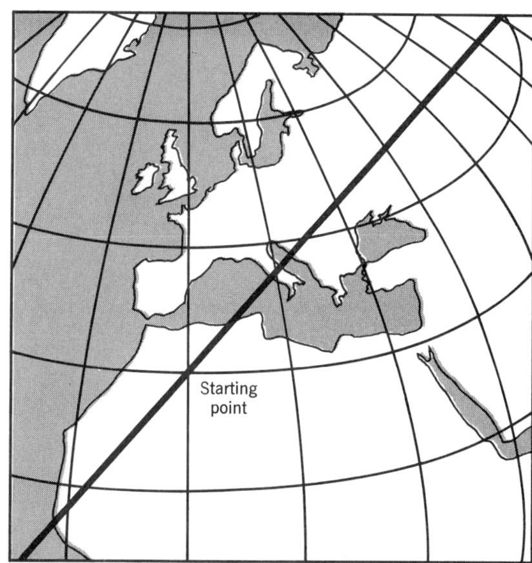

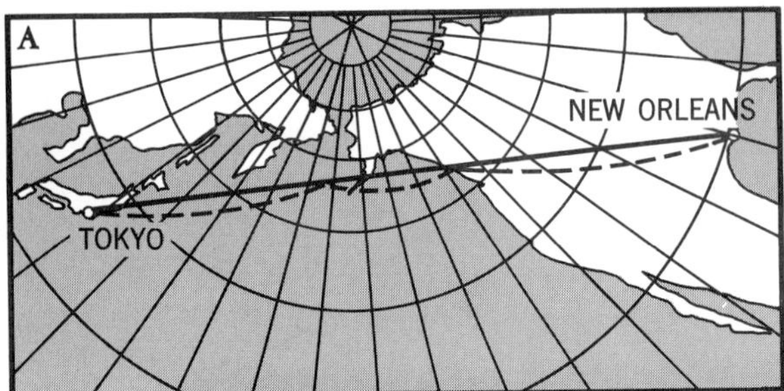

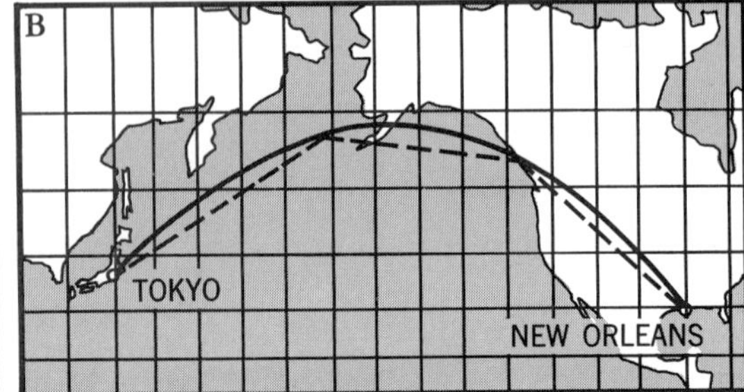

FIGURE 4.13 Two maps showing the same great circle arcs, and rhumb lines. Map (A) is constructed so that the great circle arc appears as the shortest distance between Tokyo and New Orleans, that is, as a straight line, while the rhumbs appear as somewhat longer "loops." This is their correct relationship. In map (B) the representation has been reversed by constructing the map in such a way as to make the rhumbs appear as straight lines, which "deforms" the great circle arc into a longer curve on the map.

in Fig. 4.13. This procedure is similar to following the inside of the circumference of a circle by a series of short straight-line chords. It is not the same, of course, since the great circle is the "straight line" and the rhumb lines are curved lines and are actually longer routes.

Orientation. A map sheet when looked at or held in the hand has an upper and lower edge; if the sheet represents part of the earth in some way (picture or map) it is natural to think of what is shown in the upper part of the representation as being farther away. In the allegorical medieval *mappae mundi* ("world maps") Paradise was logically located at the top and placed in the most remote part of the known earth—the Orient, or the Far East. From this and the practice of arranging churches with the chancel and main alter to the east came the term "orientation," which among other things has come to mean determining one's bearings and arranging things with the points of the compass.

Several centuries ago it became the practice to arrange maps with north instead of east at the top, and it has now become so strongly established that we think of "up north" and "down south." Australia and New Zealand being "down under," and "upper" Michigan and "lower" California are examples of the unconscious adjustment to this convention. Needless to say, since there is no up or down along a spherical surface, there is no reason why a map cannot be oriented in some other way. Since we think of the top of a sheet as "away" from us, it is apparent that in some instances orienting a map in the direction of interest or movement, if any, may well be more useful. On the other hand, having most maps oriented to north probably helps most users because the orientation is familiar.

Distance

Distances on the earth's surface are always reckoned along straight lines ("as the crow flies"), which means along arcs of great circles unless otherwise indicated. Because no map, except one on a globe, can represent the distances between all points correctly, for longer distances it is necessary to refer to a globe or to a table of distances, or to calculate the length of the great-circle arc between two places. A piece of string or the edge of a piece of paper can be used to establish the great circle on a globe. If the scale of the globe is not readily available, the string or paper may be transferred to a meridian and its length in degrees of latitude ascertained. Since all degrees of latitude are nearly equal and are approximately 111 km (69 mi.), the length of the arc in conventional units can be determined.

The arc distance D on the sphere between two points A and B, the positions of which are known, can be calculated using ordinary trigonometric functions by the formula

$\cos D = (\sin a \sin b) + (\cos a \cos b \cos P)$ in which

D = arc distance between A and B
a = latitude of A
b = latitude of B
P = degrees of longitude between A and B

When D is determined in arc distance, it may be converted to any other convenient unit of measure. Note that if A and B are on opposite sides of the equator, the product of the sines will be negative. If P is greater than 90°, the product of the cosines will be negative. Solve algebraically.*

The plane grid distance D_G, between two points A and B on a rectangular coordinate system can be calculated by the formula

$$D_G = \sqrt{(E_A - E_B)^2 + (N_A - N_B)^2}$$

in which E_A and E_B are the eastings of points A and B, respectively, and N_A and N_B are the respective northings.

There are, of course, many different units of distance measurement that have been used on maps in the past, and it is sometimes necessary to refer to encyclopedias or other sources to obtain data for conversion. Most maps today use metric units, and nautical charts are converting depth measures to meters from the traditional 6-ft. fathom. Civilian mapping agencies in the United States are now changing to the metric system, but the conversion is costly and will take time. The relations between measures of the International (metric) System and the U.S. Customary System are given in the conversion table in Appendix A.

Area

The curved surface of the spherical earth makes difficult the reckoning and representing of areas. To arrive at the area of a polygonal segment of the surface of a sphere (a closed figure bounded by arcs of great circles) is relatively easy, but the earth is not a perfect sphere, and furthermore, most of the areas in which we would be interested, aside from small landholdings, are extremely irregular (for example, continents, countries, states, counties, and the like), with complex boundaries or coastlines. Consequently, the easiest way to determine the sizes of such areas is to map them first and then measure the area enclosed. This can be done in various ways, such as counting squares on a graph-paper overlay, by using a device called a *planimeter,* which measures an area by tracing its boundary, or by computer from the digitized boundary. Of course, the map must be one in which the transformation from the spherical surface to the plane surface was done in such a way that the sizes of areas are correctly represented.

*Calculation using haversines $\left(\frac{1 - \cos}{2}\right)$ avoids the difficulty caused by negative values.

SELECTED REFERENCES

Bossler, J. D., "The New Adjustment of the North American Datum," Article 17, "Datum Parameters," *ACSM Bulletin,* November 1979.

Bowditch, N., *American Practical Navigator,* H.O. Pub. No. 9, 1966 Corrected Print, U.S. Navy Hydrographic Office (now Defense Mapping Agency, Topographic/Hydrographic Center). Washington, D.C.: U.S. Government Printing Office, 1966.

Burkhard, R. K., *Geodesy for the Layman,* Aeronautical Chart and Information Center (now Defense Mapping Agency, Aerospace Center), St. Louis, Mo., 1968.

Department of the Army, *Grids and Grid References,* TM 5-241-1, Headquarters, Department of the Army, Washington, D.C., 1967.

Maling, D. H., *Coordinate Systems and Map Projections.* London: George Philip and Son Ltd., 1973.

Mitchell, H. C., and L. G. Simmons, *The State Coordinate Systems,* U.S. Coast and Geodetic Survey (now National Ocean Survey), Special Publication No. 235. Washington, D.C.: U.S. Government Printing Office, 1945, revised 1974.

U.S. Naval Observatory, *The American Ephemeris and Nautical Almanac,* Washington, D.C.: U.S. Government Printing Office, issued annually.

5

Map Projections

The earth is spherical, and a simple way of mapping it without distortion is to map it on a globe, or if a much larger scale is desired, on a spherical segment of a globe. All that has been changed is the size (scale), since relative distances, angles, and areas, as well as such things as azimuths, rhumbs, and great circles, are all retained without any additional change. Maps on globes are almost indispensable for appreciating global strategic and geopolitical relationships. On the other hand, globes have some practical disadvantages: They are expensive to make, difficult to reproduce, cumbersome to handle, awkward to store, and difficult to measure and draw on. Less than half is visible at any one time. All of these drawbacks are eliminated when a map is prepared on a flat surface, which is the way most maps are made.* Nevertheless, constructing a map on a flat surface does require an important operation in addition to altering scale: The spherical surface must be transformed to a plane (flat) surface. The system of transformation is called a *map projection.***

In the broadest sense any portrayal of anything on any surface other than that of the thing itself is a projection. If one assumes that the term "map" encompasses all sorts of photographs, perspective representations, satellite views, side-looking radar images, and so on, then the subject of map projections can involve some extremely complex geometry. In this chapter our concern with projections will be limited to the types of transformation suitable as a framework for what we can call, for the lack of a more precise term, the standard or common kind of map. Such a map, whether it is drawn by hand or by computer, or is created by transforming the perspective geometry of an air photograph, involves a basic assumption, namely, that the viewer has an orthogonal relationship (looking straight down) with all parts of the earth's surface and to the map portraying it.

The geometry of the earth's surface is important in many ways. We are concerned with distances between places; accurate areas of cities, counties, states, and properties are needed for many purposes; and directions of electronic signals, winds, and headings for navigation are often required. For these and many other kinds of data, we use flat maps instead of globes, and the necessity of map projections adds a complication. It is a fact that a spherical surface cannot be transformed to a plane without in some way modifying the surface geometry. But happily, it is also a fact that there are a great many transformations that retain one, or even several, of the geometric qualities of the globe. The significance of the various geometric qualities that can be retained depends, to a considerable degree, on the extent of the region being mapped. Some transformation systems are suitable for representations of the whole earth, when our interest is likely to be in global topological relationships rather than azimuths or precise distances. At the other end of the scale range, with maps of countries, states, or small areas, we may be very much concerned with geometric qualities, such as the maintenance of angular consistency or the most efficient fit of a rectangular coordinate system so as to minimize the scale factor variation.

Some transformations can make the geometry of the map more useful than the globe for certain purposes. A good example is provided by the projection introduced in 1569 by Gerardus Mercator, a Flemish cartographer and mathematician. Mercator's projection was devised to provide a solution to a serious problem facing navigators at that time.

The sixteenth century was an exciting time of exploration and greatly expanded sea travel. Two new continents had been found; one of Magellan's ships had succeeded in circling the globe; and ships were setting out "to all points of the compass." One of the major trials of early na-

*The image on a monitor screen is assumed to be flat.

**The term "map projection" stems from the fact that the earliest transformations were constructed by employing systems of parallel, converging, or diverging lines to depict the one figure on the other. In geometry this is called projection. Most map projections are actually mathematical transformations.

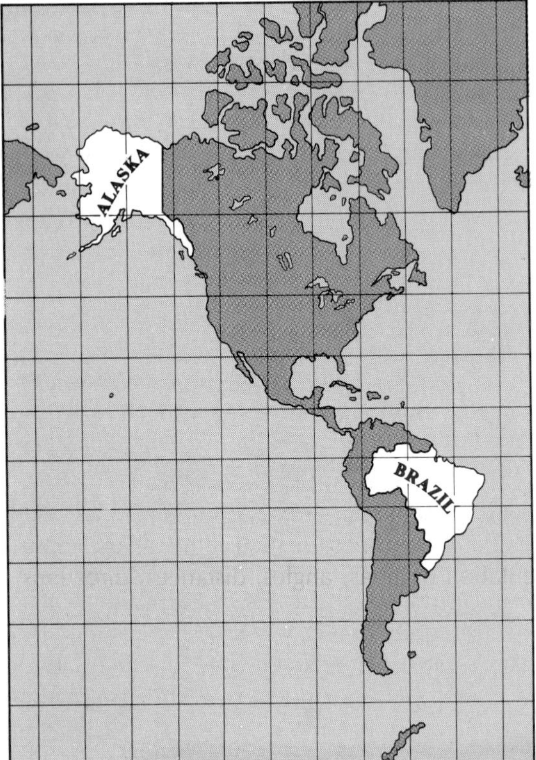

FIGURE 5.1 How Mercator's projection enlarges areas in higher latitudes. Note that Alaska in North America appears about the same size as Brazil in South America, whereas in fact Brazil is more than *five* times larger.

vigators was that, although they had a rough idea of where lands were and had the compass to help them, they had no way of determining the bearing of courses that, with any degree of certainty, would take them to their destinations.

It was pointed out earlier that a mariner must sail along a line of constant bearing because there is no other course that can be readily maintained. The solution was to transform the spherical surface to a plane in such a way that a straight line on the resulting chart, *anywhere in any direction,* was a line of constant bearing. Thus the mariner need only draw a straight line (or a series of straight lines) from the starting point to the destination and, if appropriate allowances were made for currents, winds, and compass declinations, the mariner had a reasonably good chance of arriving somewhere near the destination. Due north and due south are lines of constant bearing, but all rhumbs are parts of a complex curve on the globe, a spherical helix; by making them all appear as straight lines, for purposes of *navigation,* Mercator's projection has extremely desirable geometric characteristics.

Mercator's projection also provides an example of how a transformation can be useful for one purpose but unfit for another. A navigator is not much concerned with the shapes and relative sizes of land areas, but people concerned with geography and world affairs are. In order to show all rhumbs as straight lines, Mercator's projection severely strains the spherical surface, so that regions in middle and higher latitudes appear both misshapen and grossly enlarged. For example, on a Mercator chart Alaska and Brazil appear to be about the same size, but in fact, Brazil is five times larger (Fig. 5.1). For non-navigational purposes, then, Mercator's projection has quite unsuitable geometric characteristics.

Cartography and Map Projections

As one who is supposed to "know all about maps" a cartographer ought to be thoroughly familiar with the topic of map projections, but the subject covers a wide range, and few cartographers can be versatile in all its aspects. For example, the formulation of new map projections for specific purposes can be very complex, usually requiring considerable mathematical background, and such specialization is uncommon. On the other hand, a cartographer must understand the effects various transformations have on the representation of angles, areas, distances, and directions, so that proper allowances can be made when doing all sorts of measurements on maps, or to put it another way, to keep from making mistakes. (For example, one should not measure areas on Mercator's projection!) Furthermore, cartographers frequently transfer data from one map to another on a different projection, and knowing the distortion characteristics of each is an aid to accuracy. Cartographers occasionally must also choose a projection for a particular map or map series, and this requires arriving at a best fit between the characteristics of a projection and the objectives of the proposed map or maps. There is an unlimited variety of such problems, ranging from what kind of projection is most suitable for portraying some statistical data to which projections will allow easy subdivision for a series of sectional maps.

Until recently a cartographer also had the responsibility of calculating and drawing the selected projection, both tedious and painstaking tasks. Happily, with the advent of computers and plotters, this arduous operation, including the plotting of essential geographical data, can now be done mostly by machine at relatively low cost. Most mapmaking organizations have the necessary equipment and programs, for example the CAM software package; for those that do not, drawn projections and plotted base data are obtainable at reasonable cost.* The ease with which such fundamental cartographic operations can be performed enhances the cartographer's primary task, the *selection* of the proper projection. In a very real sense, the greater the number of options, the more important is the choice.

INTRODUCTION TO MAP PROJECTIONS

An infinite number of map projections is possible, and one might get the feeling that the subject is overwhelming. Nevertheless, although the mathematics of actually devising a system of transformation with certain characteristics may be formidable, the characteristics themselves seldom are. It is the characteristics that concern most cartographers.

In all its ramifications, the subject of transformation from the sphere or spheroid to a plane is much more complex than can be treated in a book of this sort. There is a large literature on the topic; a few selections are listed at the end of the chapter, and the interested reader will find in each of them references for further exploration. In the following sections we will introduce the subject of projections primarily from the point of view of those attributes that affect their usefulness. The actual methods of construction will be left to the specialist and the computer.

Although the characteristics of projections are generally described in such terms as the representation of areas, angles, distances, directions, whether parallels and meridians are straight or curved, and so on, these are all functions of scale relationships. Thus, we can approach the subject through an analysis of what happens to the scale at various points when the spherical surface is transformed to a plane surface.

Transformation and Scale Factor

The best way to understand the transformation process and the role of scale is to proceed as suggested in Chapter 4, namely, to assume that the earth has been mapped on a globe of the size chosen for the ultimate flat map, no matter how large or small. As explained there, this reference globe will have a given representation fraction (RF), called the principal scale, derived by dividing the radius of the earth by the radius of the globe. On the reference globe the actual scale anywhere will be the same as the principal scale, and by definition the scale factor (SF = actual scale ÷ principal scale) will be 1.0 everywhere. When all or part of the globe map is transformed to a flat map by some system of projection, the actual scale at various places on the map will be larger or smaller than the principal scale because the sphere and the plane are *nonapplicable,* that

*As of the date of publication, plotted projections, with or without geographical data, are available from the University of Wisconsin Cartographic Laboratory, 383 Science Hall, University of Wisconsin, Madison, WI 53706. Its *Projection Handbook* ($7.50) lists twenty-two projections, most of which are available in equatorial, polar, and oblique cases, on computer paper or drafting Mylar up to about 84 cm. World Data Base II (6,000,000 points of coastlines and political boundaries) suitable for scales of 1:1,000,000 or smaller can also be plotted on the projections.

is, one cannot be transformed to the other without stretching, shrinking, or tearing. Consequently, in some way the SF will always vary from place to place on a flat map.

To visualize what happens, imagine a pattern of equally distant points on the globe map, and the establishment of corresponding, or homologous, points on the flat map. The system employed to specify the positions of the points on the flat map constitutes the method of projection. Since the two surfaces are not applicable, distance relationships among the points on the flat map must be modified. Since angular and areal relationships are functions of relative distances, alterations of these relationships are bound to occur. Consequently, it is impossible to devise a system of projection for the globe to the plane such that any figure drawn on the one will appear exactly the same on the other. Nevertheless, by suitable arrangement of the SF variations, it is possible (1) to retain some angular relationships, *or* (2) to retain relative sizes of like figures. If these particular qualities are not wanted in a projection, but some other geometric attribute of the spherical surface is desired, such as straight line azimuths from one point to all others, then most angular relationships will usually be changed, and areas of like regions on the two surfaces will not have a constant ratio to each other.

In order to understand these stated facts as they apply to map projections, it is necessary to realize that SF values may occur at a point and may be different in different directions at a point.

To demonstrate graphically the first of these propositions, imagine an arc of 90°, as in Figure 5.2, projected orthographically to a straight line tangent at a. If *a, b, c, . . . j* are the positions of 10° divisions of the arc, their respective positions after projection to the line tangent at *a* are indicated by points *a, b′, c′, . . . j′*. Line *aj′* therefore represents line *aj*. The drawing clearly shows that the intervals on the straight line, starting at the point of tangency (a), become progressively smaller as *j′* is approached. If SF = 1.0 along the arc then, on the projection of the arc, the SF is gradually reduced from 1.0 at *a* to 0.0 at *j′*. The rate of change is graphically indicated by the diminution of the spaces between the points. Since it is a continuous change, it is evident that every point on *aj′* must have a different SF.

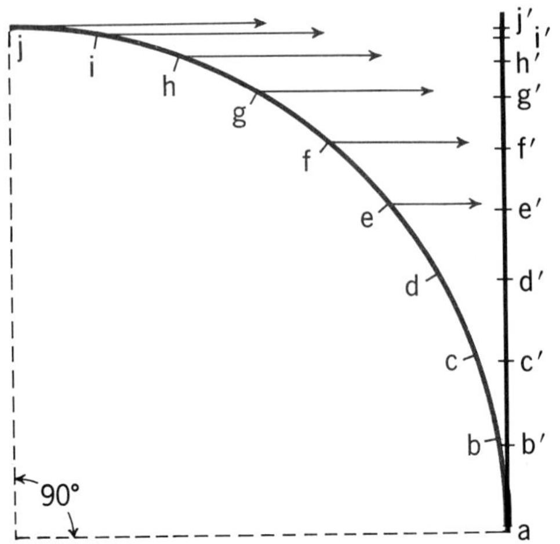

FIGURE 5.2 Orthographic projection of an arc to a tangent straight line. See text.

In order to visualize that scale at a point may be different in different directions, imagine a rectangle *abcd* and its projection to *ab′c′d,* as in Figure 5.3A, so that side *ad* coincides in *abcd* and its projection *ab′c′d*.

Figure 5.3B is an orthogonal view of rectangle *abcd* and its projection superimposed with *ad* of each coincident. If the SF of *abcd* is assumed to be 1.0 and the length *ad* is the same in each rectangle, there has been no change in scale in that direction. However, since the length *ab′* is half the length *ab* and since it is evident from the method of projection that the change has been made in a uniform fashion, then the SF along *ab′* must be 0.5. Furthermore, by projection, line *ac* has become line *ac′*. The ratio of lengths *ac′* to *ac* constitutes the SF along *ac′*, and it is evident that it is neither the 1.0 ratio along *ad* or the 0.5 ratio along *ab′*; it is somewhere in between. Any other diagonal from point *a* to a position on the side *bc* would have its corresponding place of intersection on *b′c′*. The ratio of lengths of sim-

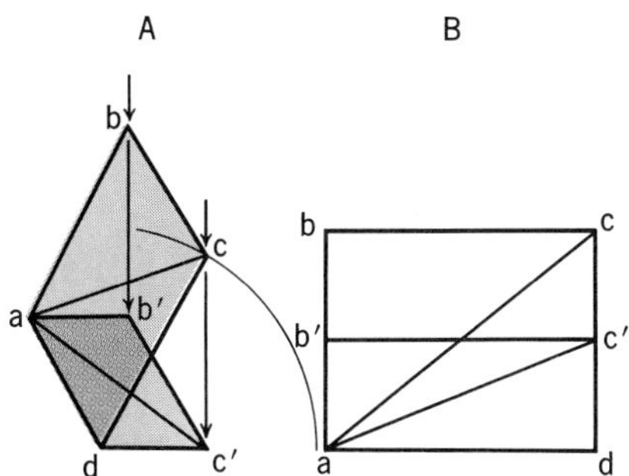

FIGURE 5.3 Projection of rectangle *abcd* to rectangle *ab'c'd* with side *ad* held constant. In drawing (A) the perspective view shows the geometric relation of the two rectangles. Drawing (B) shows the relation of the two rectangles when they are each viewed orthogonally (that is, perpendicularly to their surfaces).

ilar diagonals on the two rectangles would be different for each such line. Hence, the scale at point *a* in rectangle *ab'c'd* is different in every direction.

The values of the SF in different directions at each point on a map projection and the changes in SF from point to point provide the bases for analyzing, to a considerable degree, what the system of map projection has accomplished for, or done to, the geometric realities of distances, directions, angles, and areas on the sphere.

Tissot's Theorem

On the globe, as on any spherical surface, there is at each point an infinite number of paired orthogonal (intersecting at right angles) directions, such as N-S with E-W, NE-SW with NW-SE, and so on. When a transformation to the plane is made, the paired orthogonal directions on the sphere will, of course, be represented by paired directions on the map projection, but the pairs will not necessarily remain orthogonal. The theorem formulated by the French mathematician, A. Tissot, states that *whatever the system of transformation, there is at each point on the spherical surface at least one pair of orthogonal directions which will also be orthogonal on the projection.* The paired directions retained as orthogonal on the projection may be called the *principal directions.* Although what those directions may happen to be is generally of no particular interest, the SF in the principal directions at the point is important. As explained earlier, the SF will usually be more or less than 1.0 in different directions at each point on a projection, and the maximum deviations of the SF will be in the two principal directions. The greater value of the SF deviation is called *a* and the lesser, *b*. With those values one can calculate the angular and area distortions caused by the system of transformation for any point.*

The Indicatrix

Tissot employed a graphic device, which he called the *indicatrix,* to illustrate the concepts of angular and areal distortions that occur at points as a consequence of transformation. The indicatrix is based upon the scale factors *a* and *b*.

At any point on the globe map the SF is the same in every direction; therefore, the SF $= a = b = 1.0$ everywhere. The value 1.0 is a magnitude rather than a dimension, since a point is infinitely small. However, in order to display the changes that take place, Tissot represented the point by a finite circle with a radius of 1.0. In any system of transformation, the magnitudes *a* and *b* will ordinarily be larger or smaller than 1.0. In some projections (called *conformal*) $a = b \neq 1.0$ at most places, but in all other projections at the majority of points $a \neq b$. In those cases the indicatrix circle is transformed to an ellipse which is defined by *a* as the semimajor axis and *b* as the semiminor. By an analysis of the geometric changes in the indicatrix that occur when the circle is transformed to an ellipse, we can determine how much angular distortion and

*The mathematics of the derivation of *a* and *b* is fully explained in Derek H. Maling, *Coordinate Systems and Map Projections* (London: George Philip and Son, Ltd., 1973), pp. 62–67.

increase or decrease in area representation has occurred at any point on the projection.

As an example of this method of evaluating the changes produced by a transformation, an indicatrix is shown in Figure 5.4. The gray circle represents a point on the globe at O. The SF at O in every direction is 1.0, so the circle is constructed with $OA = 1 =$ the radius r. In the system of projection the directions OA and OB are the principal directions on the projection, that is, the pair of orthogonal directions on the globe that are retained as orthogonal in the projection.

The system of projection transforms the circle of the indicatrix to the tinted ellipse with a semimajor axis of OA' and a semiminor axis of OB'. The value of $OA' = a = 1.25$, so the SF has been increased in that direction, while the value of $OB' = b = 0.80$, showing a reduction of scale in that direction.

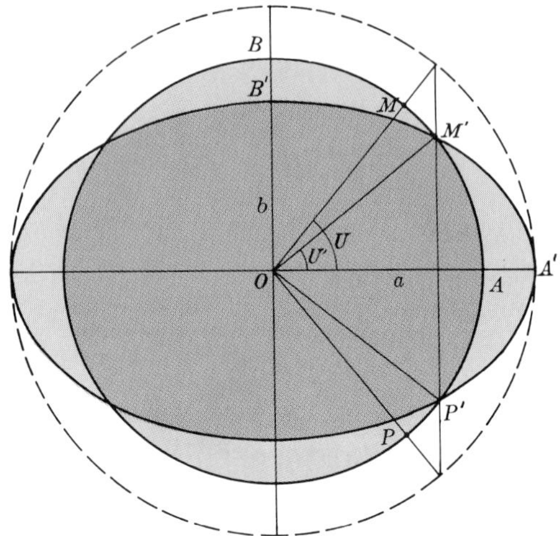

FIGURE 5.4 The indicatrix shows an equal-area transformation of the circle to which the following indices apply: $OA = OB = 1.0$. The corresponding indices of the ellipse are: $OA' = a = 1.25$, and $OB' = b = 0.80$; therefore, $ab = S = 1.0$, which shows that there has been no change in area representation. Angle $MOA = U = 51°21'40''$ has been transformed to $M'OA' = U'$ $38°39'35''$; $U\text{-}U' = 12°42'05''$ and $2\omega = 25°24'10''$. (After Marschner.)

Angular Deformation. To analyze first the angular distortion that has taken place at O, it is necessary to understand that all points on the circumference of the circle shown in Figure 5.4 will have their counterparts on the ellipse. Point B has been shifted to point B', and A to A', and it is evident that no angular change in these directions has taken place, since angle $BOA =$ angle $B'OA'$. All other points on the arc between B and A will, when projected to the ellipse, be shifted a greater or lesser amount in their direction from O. The point subjected to the *greatest* deflection is identified in the circumference of the circle with M, and it has its counterpart on the ellipse in point M'. The angle MOA on the globe thus becomes $M'OA'$ in the projection. If angle $MOA = U$ and angle $M'OA' = U'$, then $U - U'$ denotes the *maximum* angular distortion within one quadrant. The value of $U - U'$ is designated as *omega* (ω). If an angle such as MOP were to have its sides located in two quadrants and if they were to occupy the position of maximum change in both directions, then the angle in question would be changed to $M'OP'$ and would thus incur the maximum deflection for each quadrant. Consequently, the value of 2ω denotes the *maximum angular distortion* that may occur at a point. All other angular distortions at point O would be less than 2ω. Since the values of ω will range from 0° in the principal directions to a maximum somewhere between them, it is not possible to state an average angular distortion.

Area Change. Changes in the representation of areas may or may not occur as a consequence of the transformation of the circle into an ellipse by the projection system. If there has been a change, its magnitude can be readily established by comparing the areal "contents" of the circle with that of the ellipse. The area of a circle is $r^2\pi$, whereas the area of an ellipse is $ab\pi$, where a and b represent the semimajor and semiminor axes. Therefore, since the axes of the ellipse are based on the original circle whose radius was unity, and since π is constant and can be ignored, the product of ab compared to unity (one) expresses how much the representation of the area has been changed. The product of ab is designated as S.

Distortion in Map Projections

Regardless of the system employed to transform the spherical surface to a plane, the geometrical relationships on the sphere cannot be entirely duplicated. Angles, areas, distances, and directions are subject to a variety of changes, and there are many other characteristics that may or may not be duplicated in map projections, such as parallel parallels, converging meridians, perpendicular intersection of parallels and meridians, poles being represented as points, and so on. The major alterations, however, are those having to do with angles, areas, distances, and directions. In order to provide an understanding of the basic consequences of these alterations, a brief resume of their characteristics follows.

Representation of Angles. The compass rose appears the same everywhere on the globe surface (except at the poles); that is, at each point, the cardinal directions are always 90° apart, and each of the intervening directions is everywhere at the same angle with the cardinal directions.

It is possible to retain this property of angular relations to some extent in a map projection. When it is retained, the projection is termed *conformal* or *orthomorphic,* and both words imply "correct form or shape." It is important to understand that these terms apply to the directions or angles that obtain *at points.* The attribute of correct shape does not apply to regions of any significant dimension.

On the sphere or a globe map, by definition the SF is 1.0 in every direction at every point. In any transformation distortion of some kind must occur, and the SF must vary from point to point. It is possible, however, to arrange the stretching and compression so that at each point on a conformal projection $a = b$, but ab will not necessarily equal 1.0. When this condition obtains, *all* directions around each point will be represented correctly, and the parallels and meridians will intersect at 90°. It must be emphasized that this desirable quality is limited to directions at points and does not necessarily apply to directions between distant places. It is also important to realize that just because a projection shows perpendicular parallels and meridians, it does not necessarily have the property of conformality.

Representation of Areas. It is possible to retain in a map projection the representation of areas so that all regions on the projection will be represented in correct relative size. When this characteristic is retained, the projection is said to be *equal-area* or *equivalent.* This property is obtained by arranging the SF in the principal directions so that the product of $ab = S = 1.0$ everywhere.

On such a projection $a = b$ can occur at only one or (at the most) two points or along one or two lines. At all other places $a \neq b$. Hence, angles around all such points will be altered.

It is evident that the scale requirements for conformality and for equivalence in a map projection are contradictory, and consequently, no projection can be both conformal and equivalent. Thus all conformal projections will present similar earth regions with unequal sizes and all equal-area projections will deform most earth angles.

The representation of areas and angles are the most important for a great many cartographic representations. Two other aspects of representation need to be considered, however, for a sound understanding of what must happen when the one surface is transformed to another. One of these concerns distances.

Representation of Distances. Needless to say, any map projection represents all distances "correctly," provided we know the scale variations involved. As generally understood, however, distance representation is a matter of maintaining consistency of scale; that is, for finite distances to be represented "correctly," the scale must be *uniform* along the extent of the appropriate line joining the points being scaled and must be the same as the principal scale on the reference globe from which the projection was made, that is, have an SF of 1.0. The following are possible.

1. Scale may be maintained along one or more parallel lines, but only along the lines. When this is done, the lines are called *standard lines.*
2. Scale may be maintained in all directions from *one* or *two* points, but only from those points. Such projections are called *equidistant.*

Representation of Directions. Just as it is impossible to represent all earth distances with a consistent scale on a projection, so also is it impossible to represent all earth directions correctly with straight lines on a map. It is true that we can arrange the SF distribution so as to show rhumbs or arcs of great circles as straight lines. But no projection can show *true direction* in the proper sense that all great circles are shown as straight lines which have the same angular relations to the graticule of the map that they have with the graticule of the globe. For example, the oft-stated assertion that Mercator's projection "shows true direction" applies only to the fact that constant bearings are shown as straight lines. Such an assertion is erroneous, since true direction on a sphere is along a great circle, not along a rhumb (Fig. 5.5).

If we think of a correct direction as being a great circle shown as a straight line on the map which has the proper azimuth reading with the meridian of the starting point, certain representations are possible.

1. The great circle arcs between all points may be shown by straight lines for a limited area, although the angular intersections of the great circles with the meridians (azimuths) will not be shown correctly. To do this causes such a strain, so to speak, on the transformation process that it is not possible to extend it to even a hemisphere.
2. Great circle arcs with correct azimuths may be shown as straight lines for all directions from *one* or, at the most, *two* points. Such projections are called *azimuthal.*

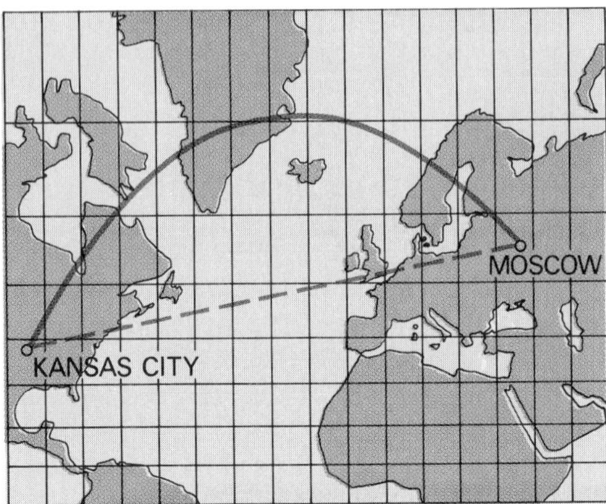

FIGURE 5.5 The great circle (solid line) and the rhumb (dashed line) between Kansas City in the United States and Moscow in the U.S.S.R. as they appear on Mercator's projection. The great circle arc shown is the "true direction," since it is the shortest route from the one point to the other.

The Analysis and Depiction of Distortion

There are several approaches to comparing one projection with another in terms of the amount and distribution of the distortion. Some are entirely graphic and simply provide a visual display of the amount and location of the deformation. Generally more useful are the quantitative measures 2ω and S.

On all conformal projections, $a = b$ everywhere on the projection and, when $a = b$, the value of 2ω is 0°. Hence there is no angular deformation at points on conformal projections but, because the values of a and b vary from place to place, the product of ab, that is, S, will also vary from place to place. Consequently, all conformal projections enlarge or reduce areas relative to one another, and S at various points provides an index of the degree of areal change.

On equal-area projections the scale relationships at each point are such that the product of

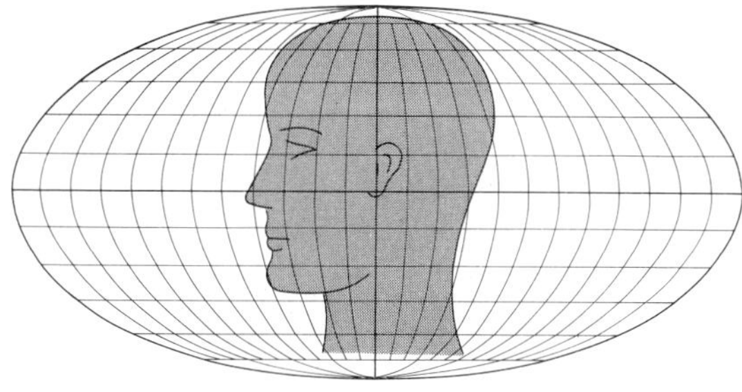

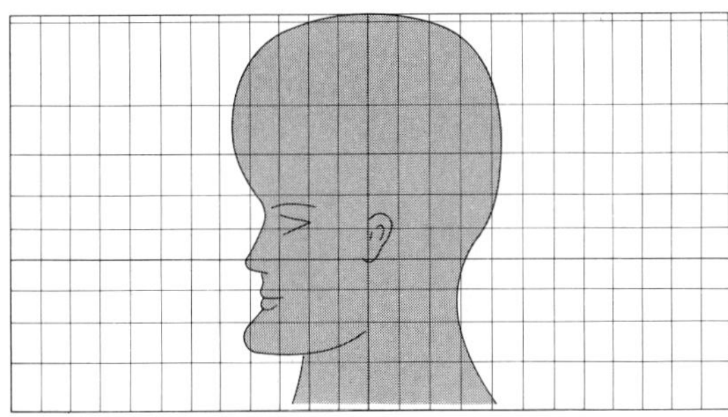

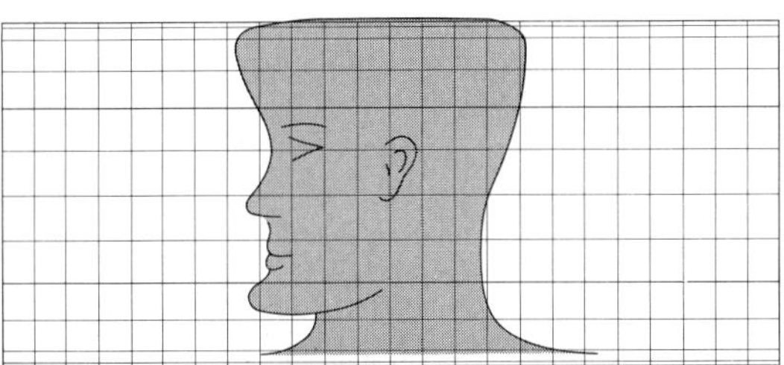

FIGURE 5.6 A head drawn on one projection (Mollweide's) has been transferred to Mercator's projection (center), and to the cylindrical equal-area projection with standard parallels at 30° (bottom). Just because the profile looks most natural on Mollweide's projection, that projection is not necessarily "better." The natural profile could have been drawn on any one and then plotted on the others.

ab always equals 1.0. Any difference between the values of *a* and *b* produces a value of 2ω greater than 0°. Consequently, all equal-area projections deform angles, and the value of 2ω at various points provides an index of the degree of angular distortion.

On all projections that are neither conformal nor equal-area, *a* will not equal *b* nor will the product of *ab* = 1.0. Therefore, on such projections, both the values of *S* and 2ω will vary from place to place.

The arrangement or pattern on a projection of the values of *S* and 2ω may be shown by plotting lines of equal value as illustrated later in this chapter. Many projections may differ in detail but be derived from the same basic geometric model, such as a cone or a cylinder. Commonly, all the projections in a class will have similar patterns of distortion.

It is also possible to derive the mean value of the distortion for either the entire projection or for only a portion of the projection, such as the land areas. A comparison of mean values for several similar projections is helpful in evaluating their relative merit.

Tissot's method of analysis is limited to values at a point. It does not provide much help in evaluating another kind of distortion, which exists in all projections, namely, changes in the distance and angular relations among widely spaced points or areas such as continents. To date, this kind of distortion has not been found to be commensurable; consequently, graphic means have been employed to help show the alteration that takes place with respect to larger spatial relations.

Various devices have been employed to this end, such as a head plotted on different projections to illustrate the elongation, compression, and shearing of larger areas (Fig. 5.6). The deformation of directions can be effectively demonstrated by plotting various great circles in different parts of a projection (Fig. 5.7). Another useful procedure has been to cover the globe with equilateral triangles or a pattern of equally spaced dots and then reproduce the same triangles or dots on the different projections.

The Arrangement of the Graticule

The surface of a globe is uniform insofar as its geometry is concerned. At any point, the surface curves away in all directions at the same rate. In

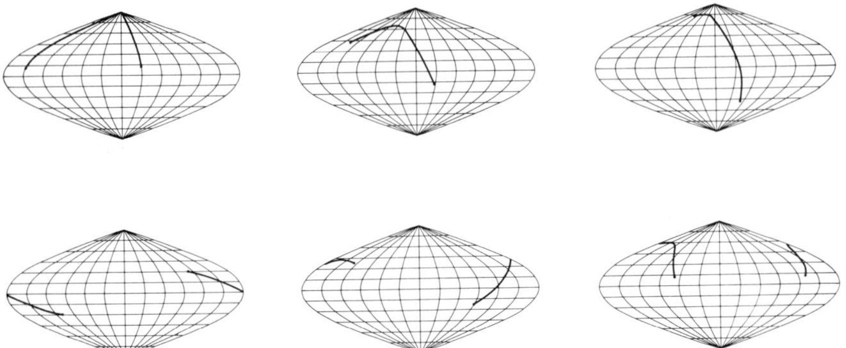

FIGURE 5.7 Selected great circle arcs on an equatorial case of the sinusoidal projection showing their departures from straight lines. Each uninterrupted and interrupted arc is 150° long. (Courtesy W. R. Tobler).

order to have something to provide positive identification of location on such a limitless, uniform space, we use the earth's coordinate system. It is so useful that we find it difficult to think of the earth's surface without automatically including the graticule. Consequently, we tend to conceive of a map projection as a representation of the graticule. Furthermore, because the concept of cardinal directions is so important to people in their thinking about the earth, it is ordinarily sensible to arrange a projection so as to present these directions to good advantage. This is done by making significant directions, such as east-west parallels or north-south meridians, appear as straight lines or arcs.

It is important, however, to realize that whatever way the system of projection may be arranged or oriented with respect to the globe, any system of projection is the transformation of a spherical *surface* to a plane *surface*. The graticule merely provides a handy series of reference points. In whatever way the system of projection may be positioned with respect to the globe, the characteristics of the projection are the same; the pattern and amounts of distortion do not change. Figure 5.8 shows how the graticule appears when the sinusoidal projection is centered at various places. They are all the same system of projection.

On the other hand, it is conventional for the transformation system to be applied to the globe in such a way that the graticule is displayed in a simple fashion. This is accomplished by arranging the system so that the distortion is symmetrical around particular points or lines, such as a pole, a meridian, or one or two parallels. When a SF of 1.0 is held constant along such lines, they are called *standard lines*. They constitute reference lines that define the principal scale and from which the SF departs in other parts of the projection.

The concept of standard parallels and standard meridians will be used in the next section to help describe the conventional forms of projections. Remember, however, that the arrangement of the transformation system so that standard great circles coincide with the equator or meridians, or so that standard small circles coincide with parallels, is merely a convenience; it is not a necessity.

MAP PROJECTIONS

There are an infinite number of ways to transform a spherical surface to a plane and, of course, the variety of cartographic objectives is unlimited. Fortunately, a number of systems of map projection combine several useful characteristics, with

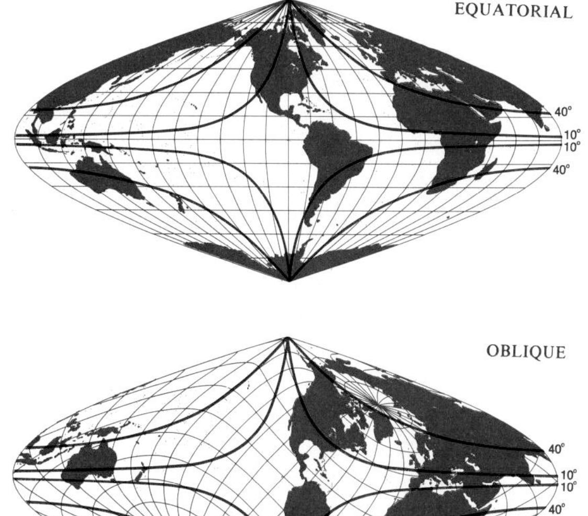

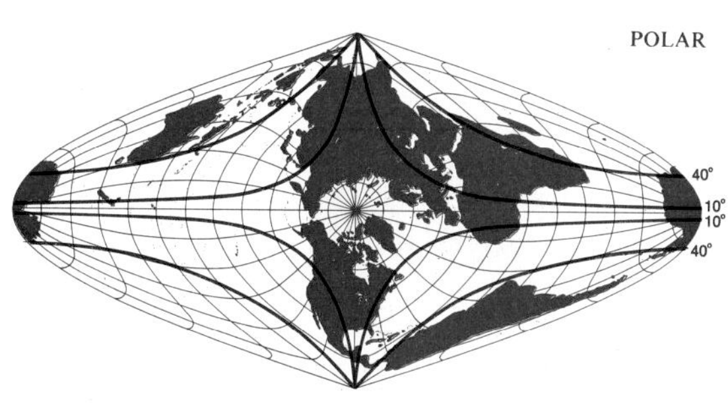

FIGURE 5.8 Different centerings of the sinusoidal projection produce different appearing graticules. Nevertheless, the arrangement or pattern of the deformation is the same on all, since the same system of transformation is employed. Isarithms of distortion are 2ω.

the result that relatively few map projections are in common use. The versatility of the computer and plotter have made it much easier to obtain a good match between map objective and projection, but the advantage of familiarity has tended to restrict the use of transformation systems to those that are better known.

The number of diverse factors that may influence the choice of a map projection is surprising. The geographer, historian, and ecologist are likely to be concerned with relative sizes of regions. The navigator, meteorologist, astronaut, and engineer are generally concerned with angles and distances. The atlas mapmaker often wants a compromise, the illustrator is usually limited by a prescribed format, and the maker of a series of maps is interested in how the individual maps may be made to fit together well.

Because the differences among map projections necessarily involve matters of distortion, there is a tendency to think of map projections as necessary but poor substitutes for the globe surface. On the contrary, in most instances projections are advantageous for reasons in addition to the face that it is cheaper to make a flat map than a globe. Map projections enable us to map distributions and derive and convey concepts that would be either impossible or at least undesirable on a globe. Furthermore, the notion that one projection is by nature better than another is unwarranted. There are systems of projection for which no useful purpose is now known, but there is no such thing as a bad projection—there are only good and poor choices.

The Employment of Map Projections

All flat maps are made on map projections. In many cases the projection in use has become standard because of a particular functional requirement, or by international agreement, or by reason of convention. For example, Mercator's projection is the obvious choice for nautical charts, and Lambert's conformal conic projection was selected as the framework for international aeronautical charts (in the middle latitudes), since it combines conformality with relative ease in scaling distances and plotting courses. On the other hand, there are no such given prescriptions or conventions for the innumerable maps that are made for atlases and portrayal of geographical aspects of all kinds of data on wall maps, separate maps, and the maps in texts, books, encyclopedias, reports, advertising, and the like.

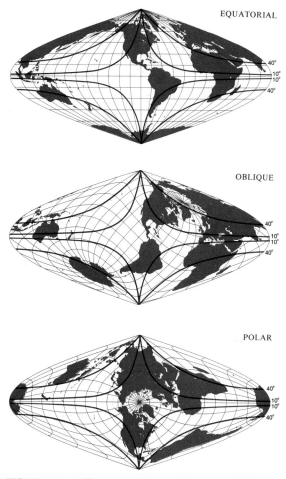

There are many factors to keep in mind when choosing a system of projection. No specific formulas can be given that will lead to the right selection because each map is a complex mixture of objectives and constraints, but a few generalizations can be drawn to exemplify the complex nature of this fundamental choice that the cartographer must make. There are several general rules or guidelines that should be kept in mind when selecting a projection. The first is that the inherent distortion increases away from those parts of the projection where the SF = 1.0, sometimes called the origin of the projection. Therefore, a good match between the shape of the region being mapped and the shape of the area of low distortion on the projection is desirable. Certain general classes of projections have specific arrangements of the distortion, and the knowledge of these patterns helps considerably in choosing and using a particular system.

Maps to be made in series, such as sets for atlases or even topographic series, have different requirements from those made as individual maps. For example, a most useful attribute of some aspects of some projections is the fact that any portion can be cut from the whole and provide a segment that is in itself a relatively good selection for a smaller area, with good symmetry and deformation characteristics. Most projections in which the meridians are straight lines that meet the parallels at right angles satisfy this requirement.

A great many maps demand more from the map projection than merely one or a combination of the special properties of projections (equivalence, conformality, and azimuthality). Projection attributes such as parallel parallels, localized area distortion, and rectangular coordinates frequently are significant to the success of a map. For example, a map of a distribution that does not require equivalence may have a concentration of the information in the middle latitudes. In that case, a projection that to some extent expanded the areas of the middle latitudes would be a great help by allowing relative detail in the significant areas. Any small-scale map of temperature distributions over large areas is made more effective if the parallels are parallel; it is even more expressive if they are straight lines that allow for easy north-south comparisons. The overall shape of an area on a projection may be of great importance. Many times the format (shape and size) of the page or sheet on which the map is to be made is prescribed. By using a projection that fits a format most efficiently, a considerable increase in scale can often be obtained, which may be a real asset to a crowded map.

The Classification of Projections

The usual categorization of projections is based on general geometric characteristics. Conceptually, the spherical surface is transformed to a "developable surface," which is a geometric form capable of being flattened, such as a cone or a cylinder (both of which may be cut and laid out flat) or a plane (which is already flat). Conventionally, the axis of the globe is aligned with the axes of the cylinder and cone (see Fig. 5.9) so that the graticule lines will be simplified. In a projection based on a cone, meridians converge in one direction and diverge in the other; on the opened-up cylinder, meridians and parallels are straight, perpendicular lines. Projections on a plane are not so conventionally aligned, and no generalizations can be made about their appearance. Such a constructional grouping of projections results in categories called cylindrical, conic, azimuthal (plane), pseudocylindrical, and miscellaneous (those based on no geometric form). Whatever the terminology employed, the grouping is not strictly a classification but a listing. Other classifications or listings of projections include the parametric, the appearance of the graticule, and the relation of the spherical surface to the plane (secant, tangent, transverse, or oblique).

The subject of the classification of projections has been much investigated and several proposals for its accomplishment have been made, none of which seem to have gained general accept-

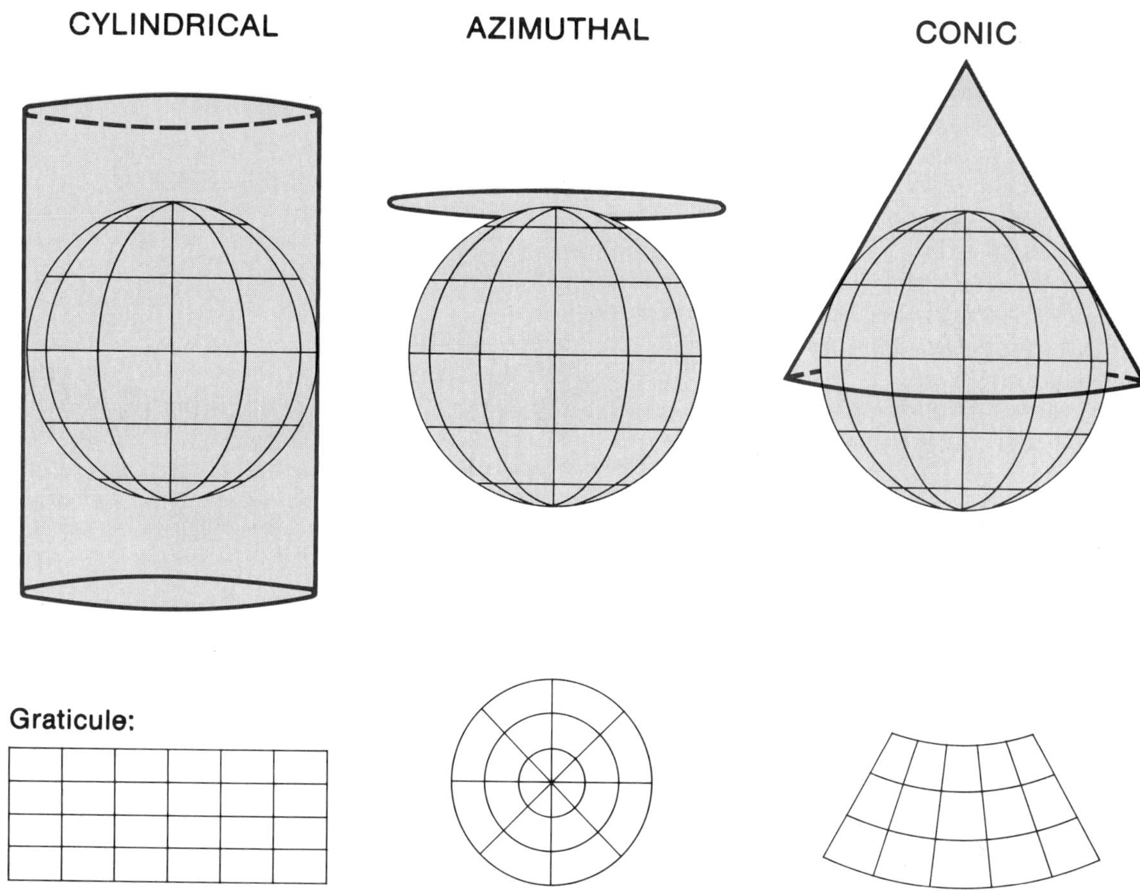

FIGURE 5.9 The developable surfaces to which the earth's surface may be "projected" and the appearances of the graticules when the transformations are arranged conventionally.

ance. The topic is of considerable interest from a theoretical point of view, but it is too complex to be dealt with here.*

Some Attributes of Projections

The primary elements in the choice of projections are generally their major property or attribute, such as conformality, equivalence, azimuthality, reasonable appearance, and so on. In most instances one property or attribute is generally of sufficient significance that it is usually the first distinguishing characteristic with which the cartographer begins to make a selection. The major properties have been defined earlier.

Secondary elements of considerable significance are the amounts and arrangements of the distortion. *Mean distortion,* either maximum angular or area (2ω or S), is the weighted arithmetic mean of the values that occur at points over the projection. When derived for similar areas on different projections, a comparison of the mean distortion values provides one index of the relative efficiency of the forms of projection.

The pattern of distortion is the arrangement of

*Refer to D. H. Maling, *Coordinate Systems and Map Projections* (London: George Philip and Son Ltd., 1973), pp. 82–108.

either S or 2ω values on a projection. It is most easily symbolized and visualized by thinking of the quantities as representing the z values of an S or 2ω third dimension above the surface. Then isarithms (contours) of this surface will show the arrangement of the relative values and, by their spacing, the gradients or rates of change. Certain classes of projections have similar patterns of distortion. These are diagrammed in Figures 5.10 to 5.12. Within the shaded areas the heavy lines are the standard lines; the darker the shading, the greater the deformation.

An azimuthal pattern occurs if the transformation takes place from the globe to a tangent or an intersecting (secant) plane. The trace of the intersecting plane and sphere will, of course, be circular. Lines of equal distortion are concentric around the point of tangency or the center of the circle of intersection (see Fig. 5.10).

A conical pattern results if the initial transformation is made to the surface of a true cone tangent at a small circle or intersecting at two small circles on the globe. Lines of equal distortion parallel the standard small circles (Fig. 5.11).

FIGURE 5.10 Azimuthal patterns of deformation. (A) The pattern when the plane is tangent to the sphere at a point, and (B) the pattern when the plane intersects the sphere. In B the trace of the intersection will, of course, be a small circle on the globe, not necessarily a parallel of latitude. The standard point in A and the standard line in B, where the SF = 1.0, are shown in red.

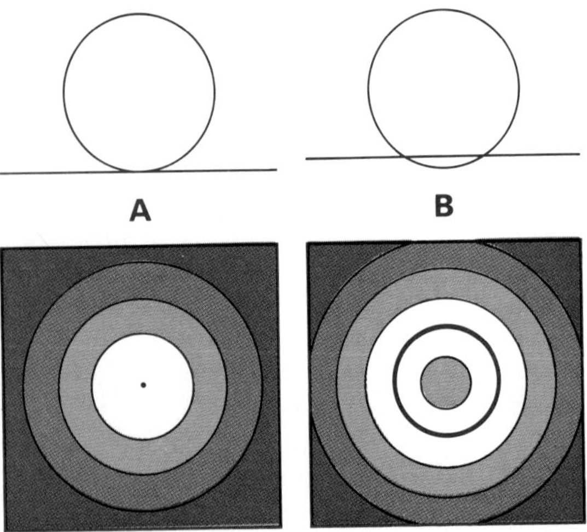

A cylindrical pattern occurs on all map projections that, in principle, are developed by first transforming the spherical surface to a tangent or secant cylinder. In all cases the lines of equal distortion are straight lines parallel to the standard lines (see Fig. 5.12). Distortion will increase away from the standard lines, and the greatest gradient will be in the directions normal (at right angles) to the standard line.

A fourth group of projections, called *pseudocylindrical,* a term meaning similar to or abnormal cylindrical, has patterns of distortion less systematic than azimuthal, conical, or cylindrical. In the conventional or equatorial case, these projections usually have straight-line parallels that vary in length, with curved meridians equally spaced along them. There have been dozens of such projections proposed; some are widely known and often used. None are conformal or azimuthal, but many are equal-area.*

The pattern of angular distortion (2ω) varies among the pseudocylindrical projections, but as can be expected, it increases away from the central areas with (in the equatorial case) maximum distortion in the polar regions.** As with all projections, the graticule may be arranged with respect to the pattern of distortion in any desired way.

There are a great many projections that can only be classed as "miscellaneous," since each one tends to be unique. Many have been devised for particular purposes for which they were admirably suited, but because of familiarity, have been put to other uses. Some projections with no great merit, such as Gall's cylindrical or Van der Grinton's, for some reason seem to have caught the fancy of mapmakers for a while and have been used when some other projection may have been a better choice.

*John P. Snyder, "A Comparison of Pseudocylindrical Map Projections," *The American Cartographer* 4 (1977): 59–81.

**The patterns on the equivalent pseudocylindrical projections shown later in the chapter in Figure 5.19 are typical of this class.

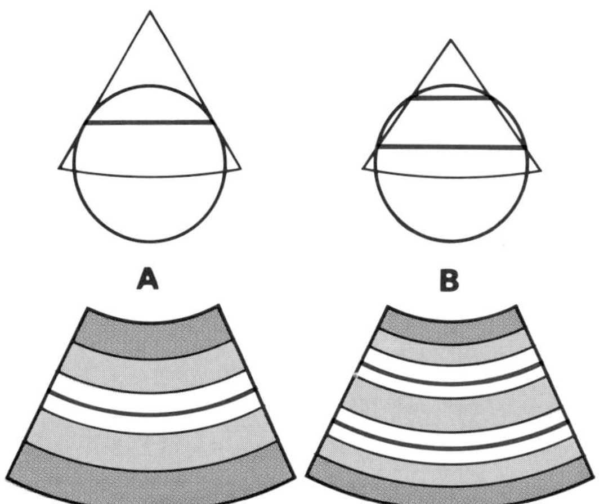

FIGURE 5.11 Conical patterns of deformation. (A) The pattern when the cone is tangent to one small circle, and (B) The pattern when the cone intersects the sphere along two small circles. The small circles may or may not be parallels of latitudes. Standard lines, where the SF = 1.0, are shown in red.

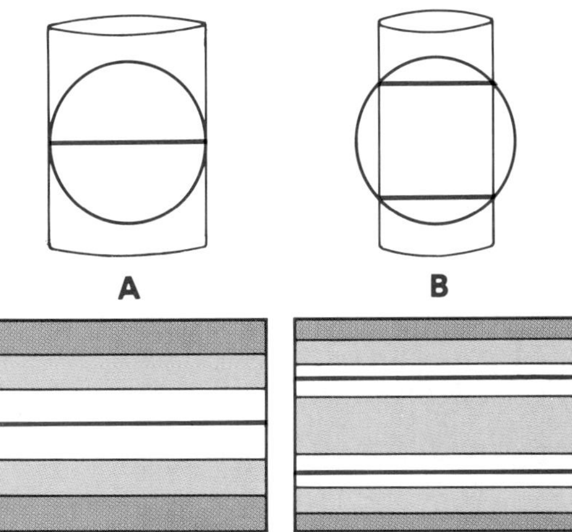

FIGURE 5.12 Cylindrical patterns of deformation. (A) The pattern when the cylinder is tangent to a great circle with one standard line where the SF = 1.0 (red). (B) The pattern when the cylinder is secant and the standard lines are two parallel small circles. The tangent great circle or the parallel small circles need not be a meridian or parallels of latitude.

Many projections used for medium and smaller scale maps, and especially those that fall in the miscellaneous class, can be evaluated quite effectively just by "looking"—by comparing the inherent attributes of the graticule on the globe with the way they appear on the projection. The following lists the more important visual characteristics of the earth's coordinate system as portrayed on a globe.*

1. Parallels are parallel.
2. Parallels when shown at a constant interval are spaced equally on meridians.
3. Meridians and great circles on a globe appear as straight lines when viewed orthogonally, which is the way we look at a flat map.
4. Meridians converge toward the poles and diverge toward the equator.
5. Meridians when shown at a constant interval are equally spaced on the parallels, but their spacing decreases from the equator to the pole.
6. When both are shown with the same intervals, meridians and parallels are equally spaced near the equator.
7. When both are shown with the same intervals, meridians at 60° latitude are half as far apart as parallels.
8. Parallels and meridians always intersect at right angles.
9. The surface area bounded by any two parallels and two meridians (a given distance apart) is the same anywhere between the same parallels.

*In the following list several slight variations (due to the earth being a spheroid rather than a sphere) have been approximated in order that the important relationships may be easily appreciated: in No. 2 there is a variation of about 0.12 percent; in No. 6 there is a difference of about 0.7 percent; and in No. 7 there is a difference of about 0.17 percent.

Conformal Projections

Maps that are to be used for analyzing, guiding, or recording motion and angular relationships require the employment of conformal projections. In these categories fall the navigational charts of the mariner or aviator, the plotting and analysis charts of the meteorologist, and the general class of topographic maps. Many of the uses to which topographic maps are put do not require conformality but, since topographic maps serve a wide variety of engineering purposes, they are often made on a conformal framework.

The concept of distortion on a conformal projection is a matter of definition because, in one sense, nothing is distorted, since all angular relationships at each point are retained (that is, $a = b$ everywhere). All that changes is the SF, and one point is as "accurate" as another; only the scales are different. Thus we can refer to the principal scale as "correct," and the other parts of the projection will be merely relatively enlarged or reduced. On the other hand, variation in S is obviously distortion, and the systematic change of S from place to place can result in noticeable distortion of shapes if the map covers an area of considerable extent.

There are four conformal projections in common use: Mercator's, the transverse Mercator, Lambert's conformal conic with two standard parallels, and the stereographic. Much the best known is Mercator's projection because of its long history of usefulness to the mariner. The stereographic projection, familiar to the classical Greeks, is also widely used. In recent times Lambert's conformal conic and the transverse form of Mercator's have become more popular among cartographers.

Mercator's projection (Fig. 5.13) is probably the most famous map projection ever devised. It was introduced in 1569 by the famous Flemish cartographer specifically as a device for nautical navigation, and it has served this purpose well. Conceptually, it is a cylindrical projection, but it must be mathematically derived. It has the property that all rhumbs appear as straight lines, an obvious advantage to someone trying to proceed along a compass bearing. Except for the meridians and the equator, great circle arcs do not appear as straight lines, so Mercator's projection does not show "true direction"; but such courses can be easily transferred from a projection (gno-

FIGURE 5.13 Mercator's projection. Values of lines of equal scale exaggeration compared to the value $S = 1.0$ at the equator are 25 and 250 percent.

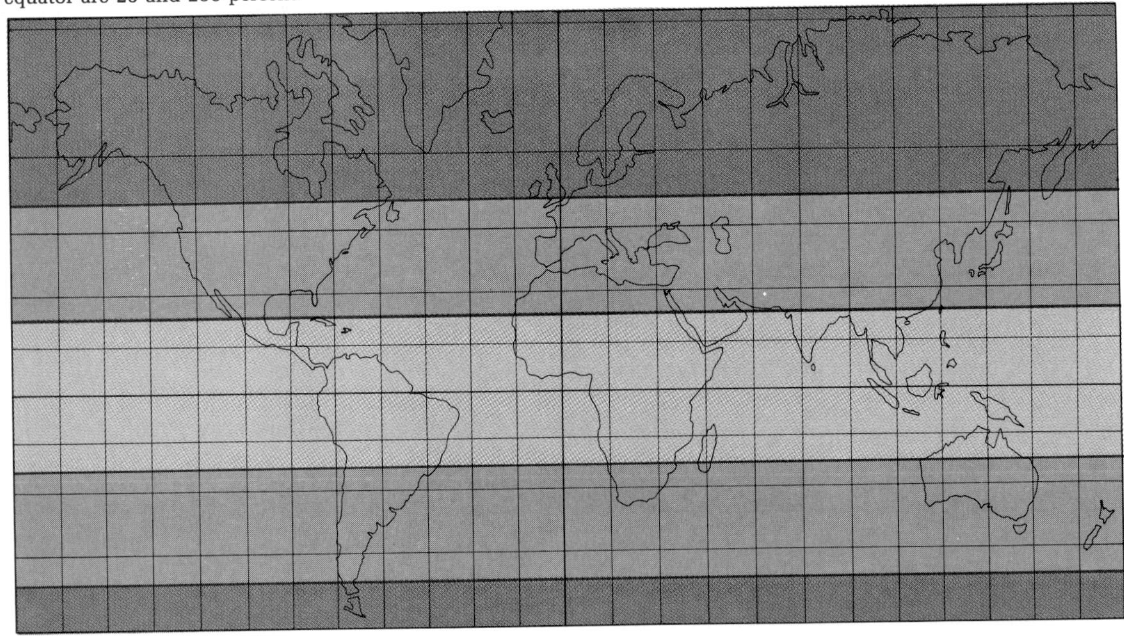

monic) that does so. The great-circle course can then be approximated by a series of straight rhumbs. It is apparent that Mercator's projection enlarges areas at a rapidly increasing rate toward the higher latitudes, so it is of little use for purposes other than navigation.

In the normal form of Mercator's projection the standard line is the equator along which the SF is 1.0. The rate of change of the SF values is relatively small for the lower latitudes, meaning that a zone along the equator is well represented. For example, even at 10° latitude the SF would only be about 1.016. Since the earth is spherical and the equator is like any other great circle, it is apparent that we can twist Mercator's projection (or the sphere) 90° so that the standard line becomes a meridian (a great circle) that takes the place of the equator (Fig. 5.14). When this is done, the projection is called the transverse Mercator.

FIGURE 5.14 (A) The conceptual cylinder for the normal form of Mercator's projection is arranged parallel to the axis of the sphere, resulting in the equator (0°) being tangent and thus the standard line. (B) To develop the transverse Mercator projection, the cylinder is turned so that a meridional "great circle" is tangent and a meridian, in this case 80°W, becomes the standard line.

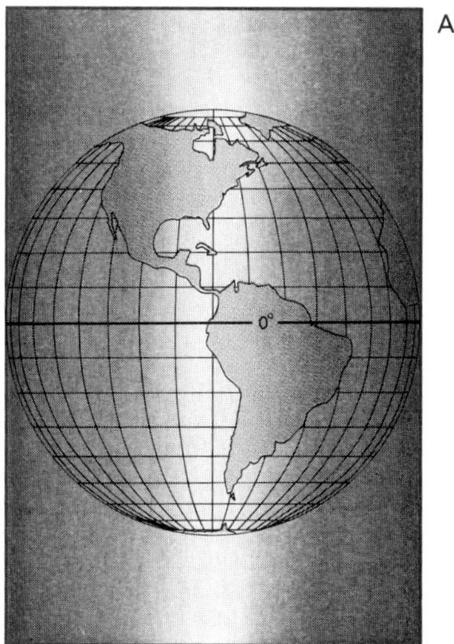

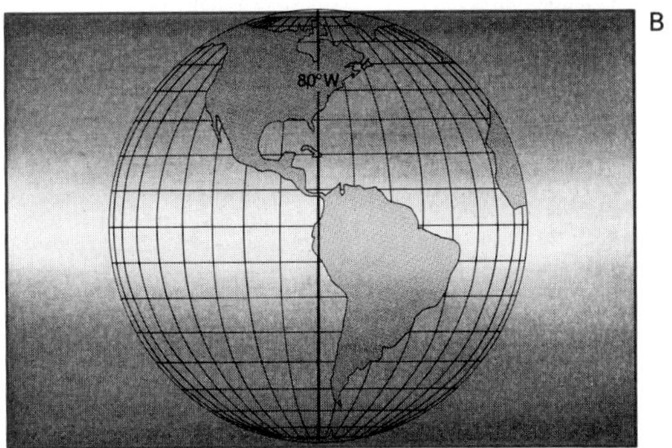

The transverse Mercator projection is conformal, but because most of the parallels and meridians are curves, it does not have the attribute that all rhumbs are straight lines. Consequently, since scale exaggeration increases away from the standard meridian, it is useful for only a small zone along that central line (Fig. 5.14). Lines of equal-scale difference (parallels of the graticule in the normal aspect of Mercator's projection) are small circles parallel to the central meridian in the transverse Mercator projection. In recent years the projection has become widely used for topographic maps, for example in the United Kingdom and the United States, and as a base for the Universal Transverse Mercator (UTM) plane coordinate system.

In order to reduce the variation in SF in the zone along the central meridian in the transverse Mercator projection used for the UTM grid reference system, the projection is made secant. By so doing the principal scale is defined as being 1.0 along two small circles paralleling the central meridian instead of making the central meridian a single standard line. Figure 5.15 shows that this procedure makes the SF <1.0 between the two standard lines and >1.0 outside them, thus reducing the overall deviation of the scale factors from 1.0 on the projection.

Lambert's conformal conic projection with two standard parallels in its normal form has concentric parallels and equally spaced, straight meridians that meet the parallels at right angles (Fig. 5.16). The SF is <1.0 between the standard parallels and >1.0 outside them. Area distortion be-

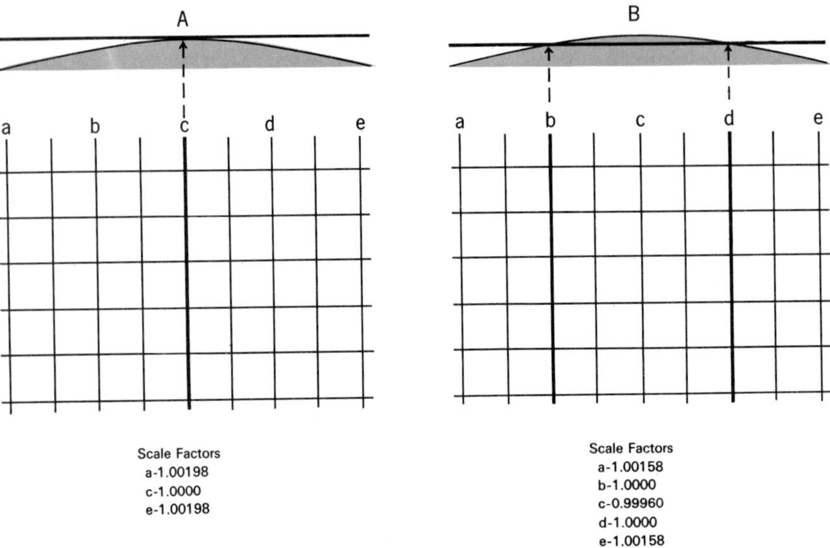

FIGURE 5.15 On the transverse Mercator projection system used for UTM rectangular grid reference purposes, instead of a single line being standard, as in (A), two are made standard, as in (B). In (B) the standard lines, b and d (small circles parallel to the central meridian) are placed 180 km east and west of the central meridian. The east-west zone shown here, that includes an overlap area, would be approximately 725 km wide or about 6.5° of longitude near the equator.

tween and near the standard parallels is relatively small; thus the projection provides exceptionally good directional and shape relationships for an east-west latitudinal zone. Consequently, it is used for air navigation in intermediate latitudes, for topographic maps, and for meteorological charts.

The conformal stereographic projection (Fig. 5.23A) also belongs in the azimuthal group. The distortion, that is, variation in *S*, is arranged symmetrically around the center point. This is an advantage when the shape of the area to be represented is more or less compact. When the projection is made secant, as for the polar areas in the UPS grid system, the area within the standard line has an SF <1.0; outside, the SF is >1.0.

Equal-Area Projections

For the majority of maps used for instruction and for small-scale general maps, the property of being equal-area commands high priority. The comparative extent of geographical areas, such as nations, water bodies, and regions of geographical similarity (vegetation, population, climate, and so on) are matters of obvious significance.

Many of our general impressions of the sizes of regions are acquired subconsciously. Because non-equal-area projections have been so frequently used for instructional and small-scale general maps in the past, most people have erroneous conceptions of the comparative sizes of regions, such as that Greenland is considerably larger than Mexico (nearly the same size) or that Africa is smaller than North America (Africa is more than 2 million sq. mi. larger). For some maps the property of equivalence is more than a passive factor. Some kinds of symbolization require near equivalence in order to portray relative densities properly. An example is illustrated in Figure 5.17.

The choice among equal-area map projections depends on two important considerations.

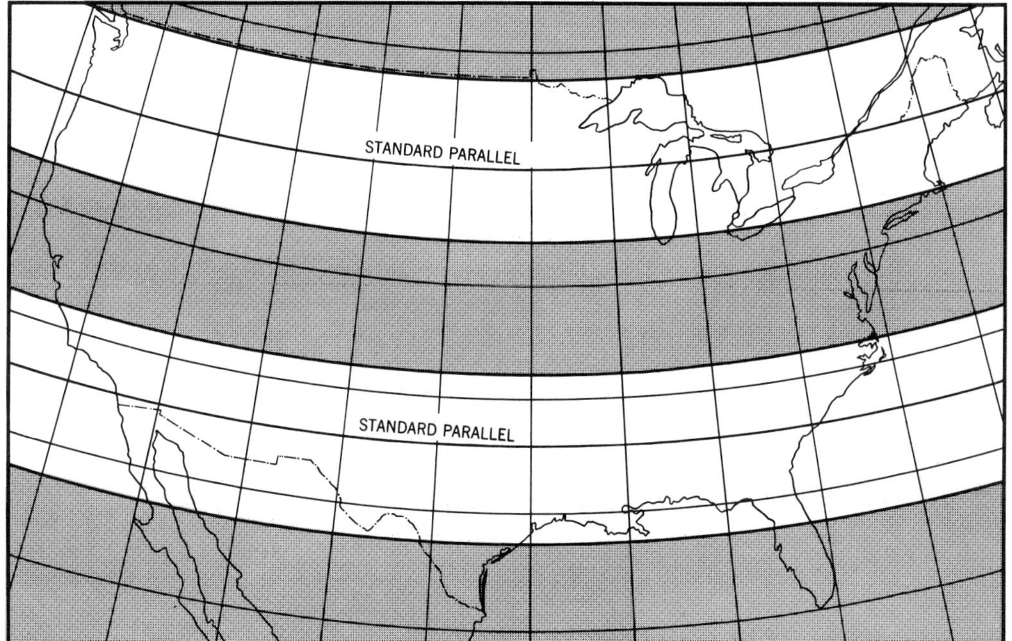

FIGURE 5.16 Lambert's conformal conic projection with two standard parallels. Values of lines of equal scale exaggeration (S) are 2 percent.

1. The size of the area involved.
2. The distribution of the angular deformation.

FIGURE 5.17 These two areas, France, left, and the Malagasy Republic (the island of Madagascar), right, are nearly the same size. On the non-equal-area projection from which these two outlines were traced, France is shown larger than it actually is in comparison to the Malagasy Republic. The same number of dots has been placed within each outline, but the apparent densities are clearly not the same.

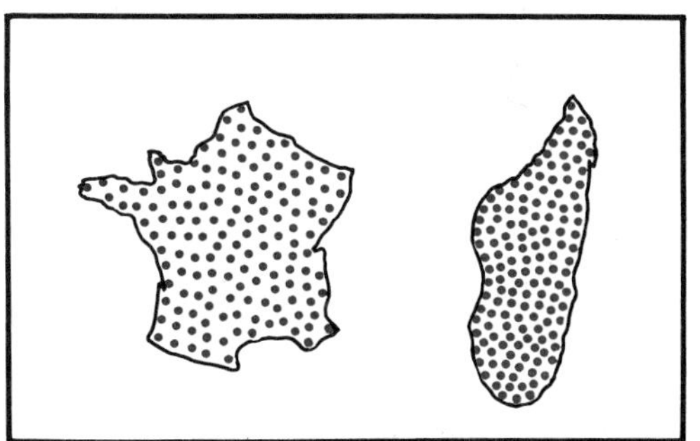

There are a great many possibilities from which to choose; a cartographer who keeps these two elements in mind will rarely make a bad choice.

As a general rule, the smaller the section to be represented, the less significant is the choice of an equal-area projection. Because there are a great many ways to transform the globe surface to a plane while maintaining the requirement for equivalence ($ab = 1.0$ everywhere), there are a relatively larger number of well-known equal-area projections than azimuthal and conformal projections. Because the entire surface area of the earth can be plotted within the bounding lines of a plane figure of almost any shape, there is also a large number of pseudocylindrical "world projections," many of which are equal-area.

It is apparent that for a world map the pattern of deformation is a matter of paramount concern. The "better" portions of such projections can also be used for larger-scale maps of continents or even for areas of smaller extent.

A few representative types of common equal-area projections are shown, along with brief notes on their employment. Most of the illustrations include isarithms of 2ω to show the overall pattern of deformation.

Albers' conic projection (Fig. 5.18) conventionally has two standard parallels along which there is no angular distortion. Because it is conically derived, distortion zones are arranged parallel to the standard lines. Any two small circles reasonably close together may be chosen as standard, but the closer they are, of course, the better will be the representation in their immediate vicinity. Because of the low distortion value and, when conventionally arranged, its neat appearance with straight meridians and concentric arc parallels that meet the meridians at right angles, this is a good choice for a middle-latitude area of greater east-west extent and a lesser north-south extent. Outside the standard parallels, the SF along the meridians is progressively reduced. Parallel curvature ordinarily becomes undesirable if the projection is extended for much over 100° longitude. Its obvious superiority for maps of the conterminous United States on which to plot and study geographical distributions has led to its selection as the standard base map by many government agencies such as the Bureau of the Census.

Lambert's equal-area projection (Fig. 5.23B) is azimuthal as well as equivalent. Since distortion is symmetrical around the central point, which can be located anywhere, the projection is useful for areas that have nearly equal east-west and north-south dimensions. Consequently, areas of continental proportions are well represented on this projection.

The cylindrical equal-area projection (Fig. 5.19A), when conventionally arranged, is capable of variation like Albers' conic, that is, the projection employs two standard parallels. The two parallels may "coincide," so to speak, and

FIGURE 5.18 Albers' equal-area conic projection of the United States. Values of lines of equal maximum angular deformation (2ω) are 1°.

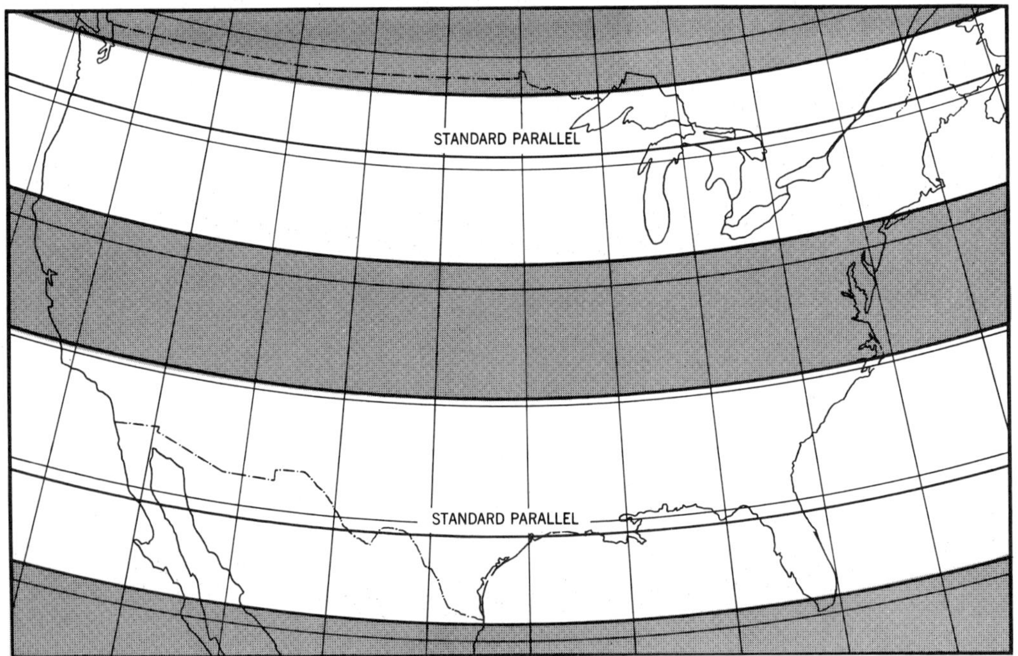

be a great circle (the equator), or they may be any two others so long as they are homolatitudes (the same parallels in opposite hemispheres). Distortion is arranged, of course, parallel to the standard small circles. Although for a variety of reasons this projection looks "peculiar" to many people, it does in fact provide, when standard parallels just under 30° are chosen, the least *overall* mean angular distortion of any equal-area world projection.

The sinusoidal projection (Fig. 5.19B), when conventionally oriented, has a straight central meridian and equator, along both of which there is no angular distortion. All parallels are standard, and a merit of this projection is that when a constant interval is used, they appear to be equally spaced on the projection, giving the *illusion* of proper spacing, so that it is useful for representations where latitudinal relations are significant.* The sinusoidal is particularly suitable when properly centered for maps of less-than-world areas, such as South America, for which the distribution of distortion is especially fortuitous.

Mollweide's projection (Fig. 5.19C) does not have the excessively pointed polar areas of the sinusoidal, and thus it appears a bit more realistic. In order to attain equivalence within its oval shape, it is necessary to decrease the north-south scale in the high latitudes and increase it in the low latitudes. The opposite is true in the east-west direction. Shapes are modified accordingly. The two areas of least distortion in the middle latitudes make the projection useful for world distributions when interest is concentrated in those areas.

Eckert's IV (Fig. 5.19D) projection, when conventionally arranged, represents the poles by lines half the length of the equator instead of by points. Accordingly, the polar areas are not so compressed in the east-west direction as on the preceding two projections. This takes place,

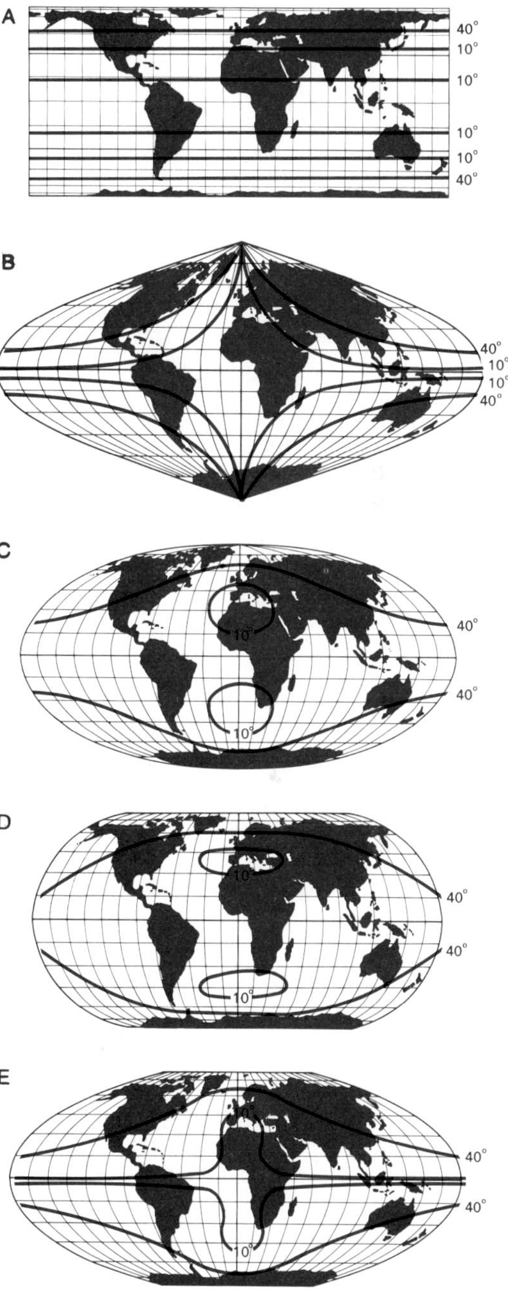

FIGURE 5.19 A few of the many equivalent world map projections: (A) cylindrical equal-area with standard parallels at 30°N and S latitude; (B) sinusoidal projection; (C) Mollweide's projection; (D) Eckert's No. IV projection; and (E) flat polar quartic projection. The black lines are values of 2ω.

*It is an illusion because the distances between parallels on the projection are properly measured along the meridian, not simply perpendicular to the parallels on the map.

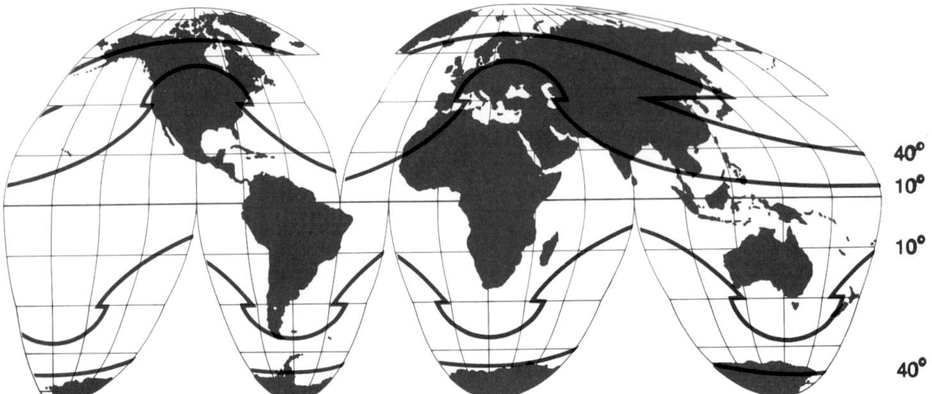

FIGURE 5.20 Goode's Homolosine projection is an interrupted union of the sinusoidal projection equatorward of approximately 40° latitude and the poleward zones of Mollweide's projection. The black lines are values of 2ω.

however, at the expense of their north-south representation. As in Mollweide's, the equatorial areas are stretched in the north-south direction. The distribution of distortion is similar to that of Mollweide's.

The flat polar quartic projection by McBryde and Thomas (Fig. 5.19E) employs a line one-third the length of the equator for the pole (when conventionally arranged). This provides for a more reasonable portrayal of the polar areas than either the sinusoidal or the Mollweide.

Several projections for world maps have been prepared by combining the better parts of two. The best known of these is Goode's Homolosine (Fig. 5.20), which is a combination of the equatorial section of the sinusoidal and the poleward sections of Mollweide's; thus, it is equal-area.* The two projections, when constructed to the same area scale (as from the same reference globe), have one parallel of identical length (approximately 40°) along which they may be joined. The projection is usually used in interrupted form (see below) and has been widely employed in the United States. Its overall quality, as shown by a comparison of mean values of deformation, is not appreciably better than Mollweide's alone.

Interruption and Condensing. When either the continents or the seas are the areas of primary interest, it is possible to display one or the other (lands or seas) to better advantage on a projection by (1) interrupting the projection, and (2) condensing the map. Interruption allows one to repeat the better parts of the projection, as in Goode's Homolosine projection (Fig. 5.20). Condensing is accomplished by deleting the unwanted sections of the projection in order to attain greater scale for the areas of interest within a fixed page format (Fig. 5.21).

Azimuthal Projections

As a group, azimuthal projections (sometimes called zenithal) have steadily increased in prominence. The advent of the airplane, the development of radio electronics and satellites, the mapping of other celestial bodies, and the general increase in scientific activity all have contributed to this development. It has occurred be-

*Mollweide's projection is sometimes called the *homolographic,* hence the combined form *homolo + sine.*

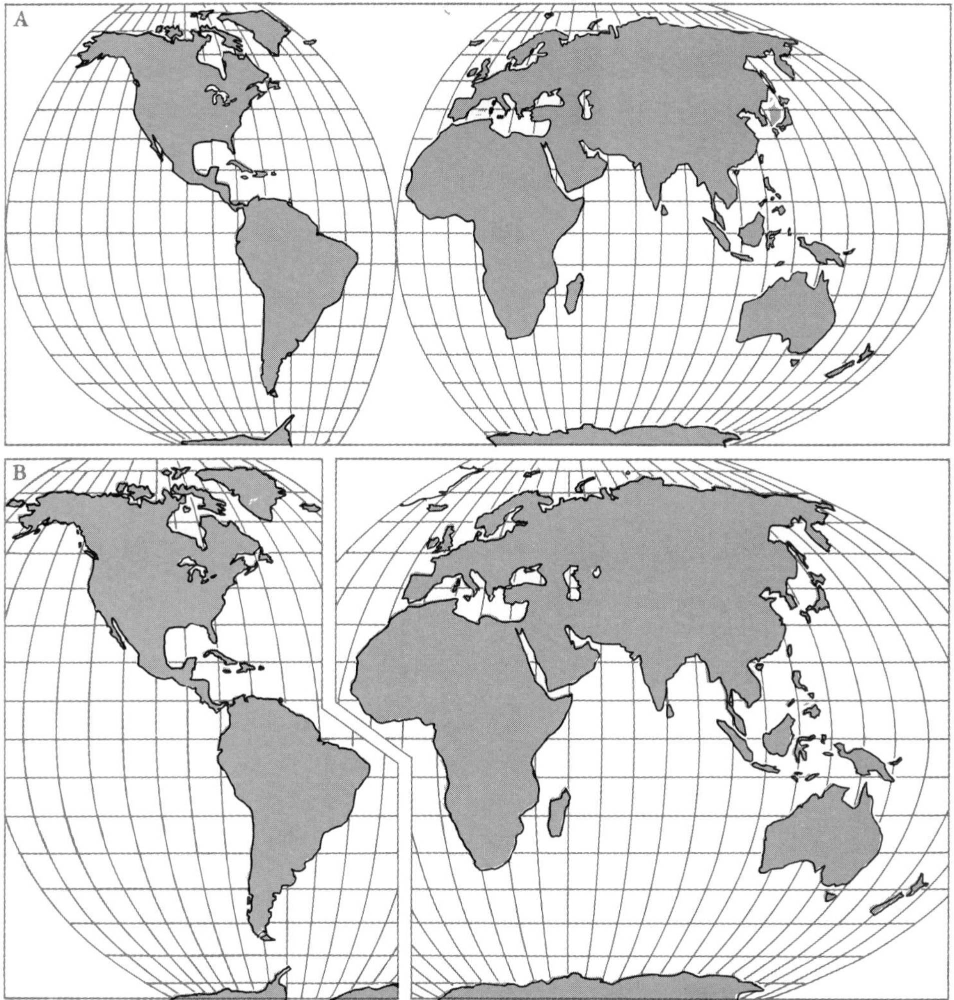

FIGURE 5.21 (A) an interrupted flat polar quartic equal-area projection of the entire earth. By deleting unwanted areas, (B), additional scale is obtained within a limiting width. The areas of concern could just as well be the ocean areas.

cause azimuthal projections have a number of useful qualities not shared by other classes of projections.

All azimuthal projections are "projected" on a plane that may be centered (made tangent or secant) to the globe anywhere. A line perpendicular to the plane at the center point of the projection will necessarily pass through the center of the globe. Consequently, the distortions characteristic of all azimuthal projections are symmetrical around the chosen center. The variation of the SF in all cases changes from the center at the same rate in every direction. If the plane is made tangent to the sphere, there is no distortion of any kind at the center; if it is made secant, the distortion will be least along a small circle. Furthermore, all great circles passing through the center of the projection will be straight lines and will show the correct azimuths from and to the center in relation to any point. It should be emphasized that only azimuths (directions) from and to the *center* are correct on an azimuthal projection.

At the center point all azimuthal projections with the same principal scale are identical, and the variation among them is merely a matter of the way the scale changes along the straight great circles that radiate from the center. Figure 5.22 illustrates this relationship. The fact that the distortion is arranged symmetrically around a center point makes this class of projections useful for areas having more or less equal dimensions in each direction, or for maps in which interest is not localized in one dimension. Because any azimuthal projection can be centered anywhere and still present a reasonable-appearing graticule, the class is more versatile than others.

FIGURE 5.22 Comparison of segments of five azimuthal projections, in this case all centered at the pole. Meridians (great circles), when shown at a constant interval, in all cases appear as equally spaced straight lines radiating from the center. The important thing to note is that the only variation among the projections is in the spacing of the parallels; in other words, the only difference among them is the radial scale from the center.

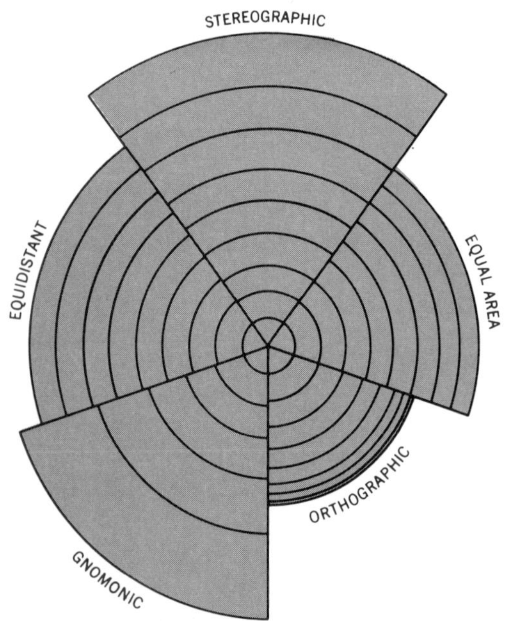

An infinite number of azimuthal projections is possible,* but only five are well known: the stereographic, Lambert's equal-area, the azimuthal equidistant, the orthographic, and the gnomonic (Fig. 5.23).

The conformal stereographic (Fig. 5.23A) and Lambert's equal-area (Fig. 5.23B) projections were considered in the sections that treated those two major properties. The stereographic, however, is unique among projections in that it has the attribute that any circle or circular arc on the globe will plot as a circle or circular arc on the projection.** It is possible, therefore, to plot the ranges of such things as radio waves or aircraft merely with a compass, as directed in the footnote.

The azimuthal equidistant projection (Fig. 5.23C) has the unique quality that the linear scale is uniform along the radiating straight lines through the center. Therefore, the position of every place is shown in simple relative position and distance from that *center*. Directions and distances between points whose great circle connection does not pass through the center are not shown correctly.

Any kind of movement that is directed toward or away from a center, such as radio impulses and seismic waves, is well shown on this projection. The projection has an advantage over many of the other azimuthal projections in that it is possible to show the entire sphere (Fig. 5.24).

*Any vertical air photograph is a perspective azimuthal projection.

**Since all great circles through the center of the projection are, of course, straight lines, we must define these as circles with a radius of infinity to make this statement not open to argument! In any case, any small circle on the earth within the limits of the projection can be drawn with a compass on the projection. This is done by first locating the ends of a diameter of the wanted circle along a straight line radial (great circle) through the center of the projection. The *construction* center of the circle is found by halving the diameter. The actual center of the circle on the earth and the construction center for that circle on the projection do not coincide except at the center of the projection.

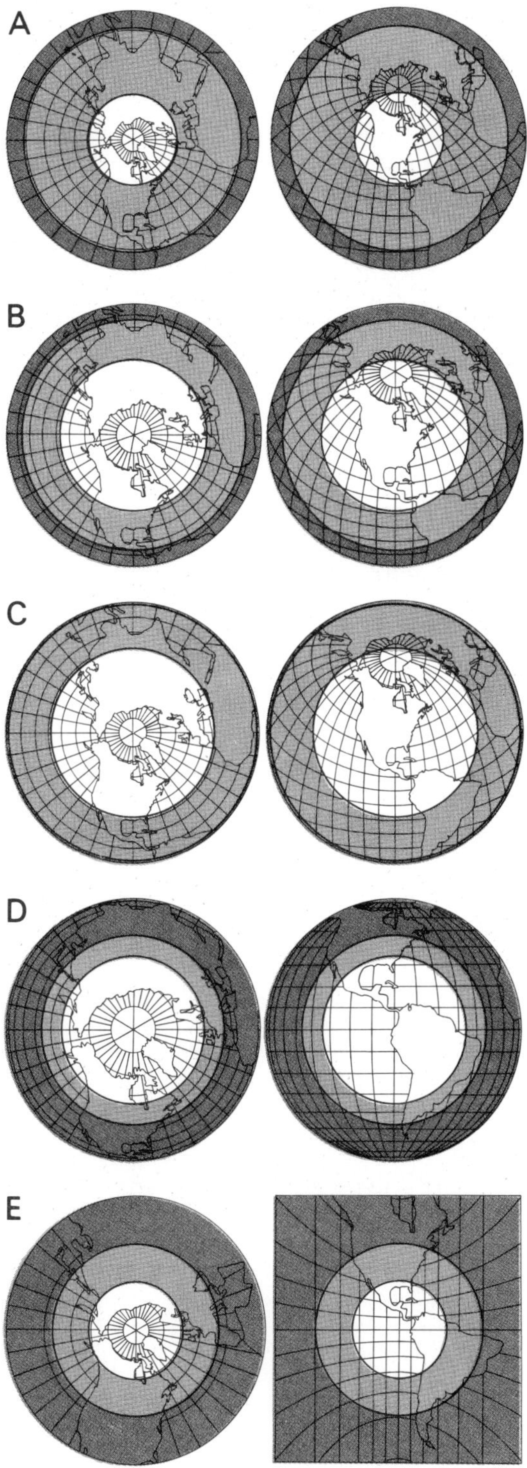

FIGURE 5.23 The five well-known azimuthal projections: (A) stereographic; (B) Lambert's equal-area; (C) azimuthal equidistant; (D) orthographic; and (E) gnomonic. In (A) the zones of areal exaggeration (S) are less than 30 percent, 30 to 200 percent, over 200 percent; in (B), (C), (D), and (E), zones of 2ω are 0° to 10°, 10° to 25°, over 25°.

Most azimuthals are limited to presenting a hemisphere or less. Since the bounding circle on this world projection is the *point* (antipode) on the sphere opposite to the center point, shape and area distortion in the peripheral areas are extreme.

The orthographic projection (Fig. 5.23D) looks like a perspective view of a globe from a considerable distance, although it is not quite the same. For this reason it might almost be called a visual projection in that the distortion of areas and angles, although great around the edges, is not particularly apparent to the viewer. On this account it is useful for preparing illustrative maps wherein the sphericity of the globe is of major significance.

The gnomonic projection (Fig. 5.23E) has the unique property that all great-circle arcs are represented as straight lines anywhere on the projection. The projection is useful in marine navigation, since the navigator need only join the points of departure and destination with a straight line on a gnomonic chart, and the location of the great-circle course is displayed. Because compass directions constantly change along most great circles, the navigator transfers the course from the gnomonic graticule to one on a conformal projection and then approximates it with a series of rhumbs.

Other Systems of Map Projection

More than 250 different systems of map projection have been devised and described in the literature of mathematical cartography. New ones are proposed with regularity, either to fit a particular need or merely because the transformation of a spherical surface to a plane with given constraints is an intriguing mathematical prob-

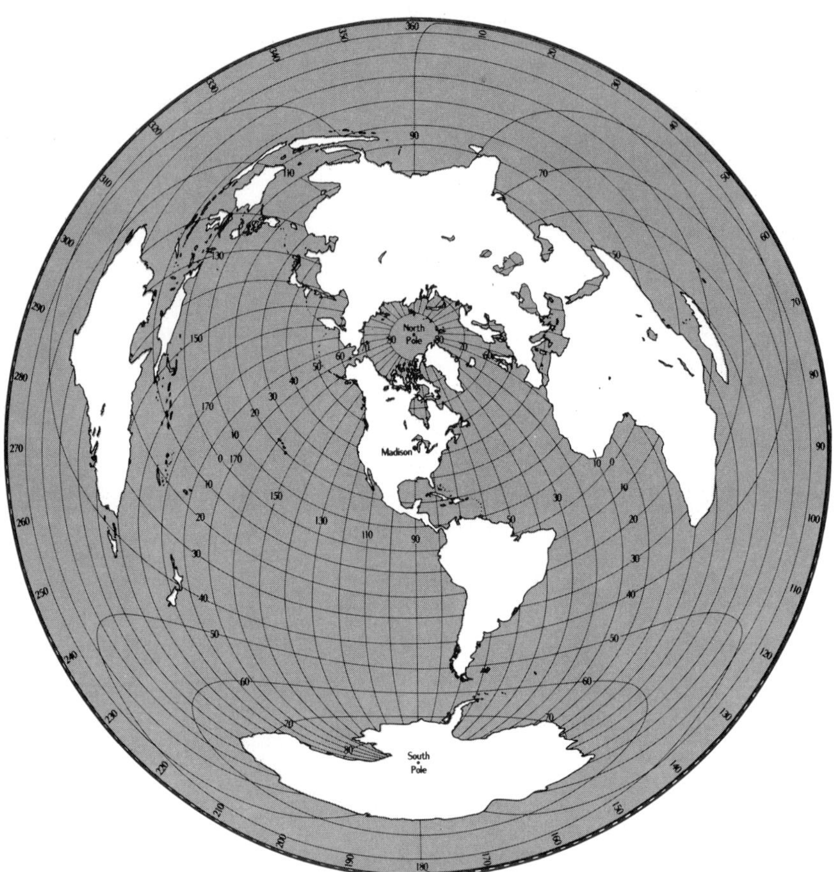

FIGURE 5.24 An azimuthal equidistant projection of the world centered on Madison, Wisconsin, prepared for geophysical uses. Note the enormous deformation (S and 2ω) at the outer edge. The bounding circle is the *point* (antipode) on the globe opposite the center point of the projection. (University of Wisconsin Cartographic Laboratory).

lem. It is impossible in this textbook to do more than describe briefly the widely used projections. To range more widely, it is necessary to refer to technical treatises, some of which are listed in the selected references at the end of this chapter.*

Most of the well-known and widely used projections are those we have already described, which have specific properties, but there are a few that do not and yet have some attributes that make them useful for particular purposes. Those include the plane chart and the simple conic with two standard parallels.

The plane chart, sometimes called the *equirectangular projection,* is one of the oldest and simplest of cylindrical map projections. It was widely used for navigational charts until superseded by Mercator's projection. It is useful for city plans and for base maps of small areas. It is easily constructed and, for a limited area, has small distortion. All meridians and any chosen central parallel are standard. The projection may be centered anywhere.

*A most useful reference and introduction to the complexity of the great variety of projections is D. H. Maling, "The Terminology of Map Projections," *International Yearbook of Cartography* 8 (1968): 11–64.

There are several possible arrangements of scale which may be used to produce conic projections with two standard parallels. These projections are similar in appearance to Albers' and Lambert's conic projections, but they do not have their special properties. They do not distort either areas or angles to a very great degree, if the standard parallels are placed close together and provided the projections are not extended far north and south of the standard parallels. They have been widely used in atlases for areas in middle latitudes.

To provide examples of the kinds of problems that have led to the need for new projections and the kinds of solutions that have been proposed, brief descriptions of three are presented here. The problems and their solutions are purposely chosen to be as different as possible in order to illustrate the wide range of cartographic requirements. The three are the polyconic projection, the pseudocylindrical Robinson's projection, and the space oblique Mercator projection.

The Polyconic Projection. The factors to be considered in choosing a projection for a set of large-scale topographic maps for a country appear simple: conformality is desirable, distortion should be minimal, and adjacent maps should fit together. A relatively small region, such as Great Britain, can be mapped in its entirety on one projection (transverse Mercator), and then the individual map sheets are sections of the whole. In such an arrangement the sheets fit together perfectly, and because the entire area is not large, the variation in the SF is well within reasonable limits, even at the margins. That kind of cartographic luxury is limited to small countries. A larger area can employ one projection, such as Lambert's conformal conic with two standard parallels or the transverse Mercator, and then repeat it, so to speak, with different standard parallels or standard meridians for successive north-south or east-west zones. Each of the zones would in effect be a single map from which separate sheets could be sectioned, but there would be bothersome junction-overlap regions where the zones meet. Only the sheets within one zone would fit together. Most of the modern topographic map series are based on these kinds of plans.

Some survey organizations, which were initiated at a time when conformality was judged less important than it is now, employed a system in which each sheet was based on its own projection in such a way as to allow easy fit of sheets in one direction or another. Such is the polyconic projection used until recently for the standard topographic maps of the United States.

The polyconic projection, devised by the first superintendent of the U.S. Coast and Geodetic Survey, Ferdinand Hassler, and later adopted by the U.S. Geological Survey for the topographic map series, is best explained by describing its construction in the conventional case. A straight-line central meridian is standard, that is, the parallel intersections on it are correctly spaced. Every parallel is standard, since each one is constructed on the tangent line of its own cone, hence the name *polyconic*. The arcs for the parallels do not have a common center, and therefore they are not concentric, but diverge from one another away from the central meridian. All the other meridians are curves correctly spaced on the parallels. The distortion characteristics are easily summarized (Fig. 5.25): SF = 1.0 east-west everywhere; SF = 1.0 north-south on the central meridian; SF = slightly >1.0 north-south away from the central meridian, but on a 7.5-minute topographic quadrangle, the deformation at the east-west margins is inconsequential. Topographic quadrangles in a north-south column will fit together exactly. Although in theory, those in an east-west row would not fit, in fact they do. Within each sheet the scale departures are so small that the outer meridians hardly depart from being straight lines. However, adjacent east-west rows of quadrangles would slowly diverge.

Robinson's Projection. In 1961 Rand McNally and Company commissioned a new pseudocylindrical projection for world maps at all scales. In their view all the then-available projections for uninterrupted world maps had serious shortcomings, especially in the general appearance of the

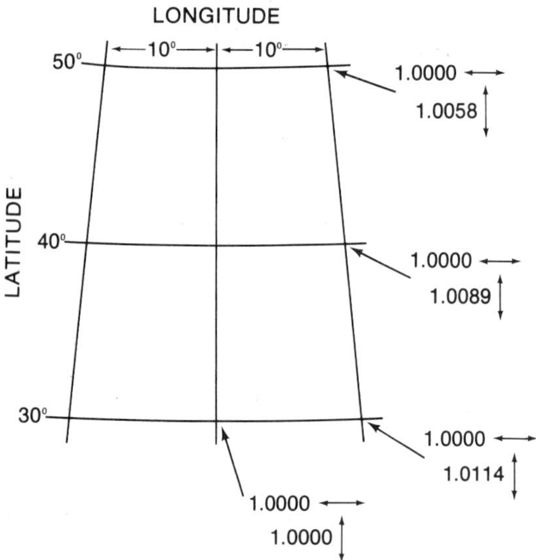

FIGURE 5.25 The distribution of scale factors on a polyconic projection in the vicinity of 40° latitude. North-south SF values away from the central meridian are approximate. Note that the section of the projection which was used for a standard 7.5-minute quadrangle topographic map would be $\frac{1}{8}$ degree east-west and north-south along the central meridian.

land masses. The constraint of equivalence puts a great strain on the representation of shapes on world maps, and cylindrical world maps tend to appear overly diagrammatic. The ultimate objective was a projection for world maps that minimized the appearance of angular and area distortion. The result is the projection shown in Figure 5.26.

The pattern of angular distortion (2ω) is similar to those on other pseudocylindrical projections (Fig. 5.19), but includes a significantly larger area of minimum values in the central regions than do the equivalent projections. Values of S vary according to latitude and increase away from the parallels of 38°N and S (in the conventional case), which may be considered the standard parallels when the projection is constructed for a globe of the same surface area. As so defined, more than three quarters of the earth's surface is shown on the projection with less than a 20 percent departure from its true size at scale.

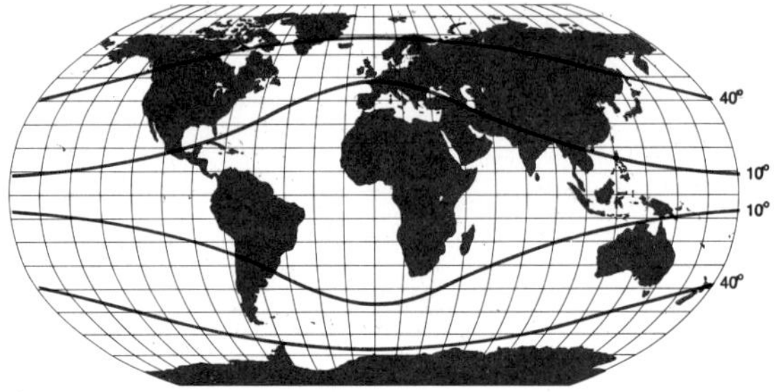

FIGURE 5.26 Robinson's projection. The black lines are values of 2ω. The value of S would be 1.0 at approximately 38°N and S latitude. Isolines of values of S would parallel the parallels.

Space Oblique Mercator Projection. Heretofore we have considered a map projection to be a framework to which geographical data are subsequently fitted. The development of continuously recording scanning devices in orbiting satellites, in a sense, turns this concept around. A scanning device, such as LANDSAT, continuously records data in digital mode, which is subject to more than a dozen geometric corrections before printout as imagery. The geometrically corrected imagery is on a cylindrical class map projection with unique characteristics resulting from the relative motions of the scanning device, the orbiting spacecraft, and the earth. Figure 5.27 illustrates the general relationships involved.

The space oblique Mercator projection is essentially conformal, but it is significantly different from the transverse Mercator projection. The groundtrack of the satellite becomes the central line of the projection along which the SF is nearly 1.0. Instead of being a simple great circle on the conceptual sphere, the ground track is slightly curved, and it crosses the equator at an angle (Fig. 5.27). The curved groundtrack results from the fact that as the satellite moves along its orbit, the earth is also rotating. The inclination of the orbit to the plane of the equator makes the limit of the groundtrack in the polar areas about 81°N and S. The space oblique Mercator projection is not perfectly conformal, but within the scanning swath the departure is negligible. The graticule and the UTM plane coordinate grid can be fitted

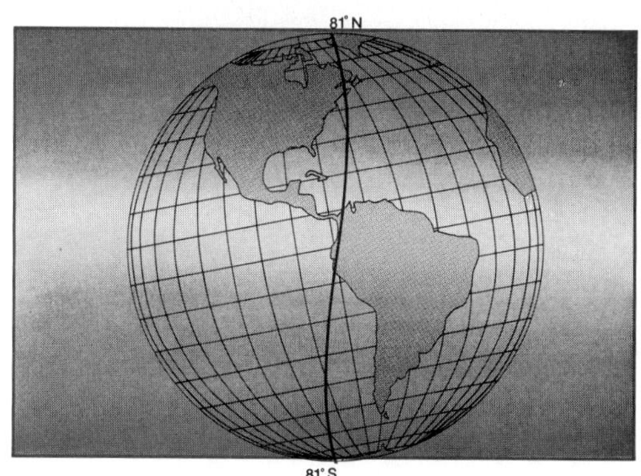

The central line (LANDSAT groundtrack), along which the SF is essentially 1.0, of the space oblique Mercator projection is slightly curved and oblique to the equator. It crosses the polar areas at about 81°N and S latitude. The conceptual basis for the projection is similar to that for the transverse Mercator projection, but the central line is not a great circle (compare Fig. 5.14).

to LANDSAT imagery with no detectable anomalies.

Map projections have always served as a base on which to record geographical information, allowing subsequent measurement and calculation. With the introduction of first aerial and then satellite photography, data were recorded for the first time in a map projection mode capable of being transformed by photogrammetric methods into more desirable projectional forms. Photography is static, however, and the transformational problems did not include time. The development of continuous orbital scanners with sufficient geometric fidelity for mapping introduces dynamics into the transformation process. This holds the prospect of continuous real-time mapping, which will, of course, still be on map projections.

SELECTED REFERENCES

Colvocoresses, A. P., "Space Oblique Mercator," *Photogrammetric Engineering* 40 (1974): 921–26.

Dahlberg, R. E., "Evolution of Interrupted Map Projections," *International Yearbook of Cartography* 2 (1962): 36–53.

Hsu, Mei-Ling, "The Role of Projections in Modern Map Design," *Cartographica,* Monograph 27. vol. 18, no. 2 (1981): 151–86.

Keuning, J., "The History of Geographical Map Projections Until 1600," *Imago Mundi* 12 (1955): 1–24.

Maling, D. H., *Coordinate Systems and Map Projections* London: George Philip and Son Ltd., 1973.

Marschner, F. J., "Structural Properties of Medium and Small Scale Maps," *Annals of the Association of American Geographers* 34 (1944): 1–46.

McBryde, F. W., and P. D. Thomas, "Equal-area Projections for World Statistical Maps," Special Publ. No. 245. Washington, D.C.: U.S. Coast and Geodetic Survey, 1949.

Robinson, A. H., "The Use of Deformational Data in Evaluating Map Projections," *Annals of the Association of American Geographers* 41 (1951): 58–74.

———, "Interrupting a Map Projection: A Partial Analysis of Its Value," *Annals of the Association of American Geographers* 43 (1953): 216–26.

———, "A New Map Projection: Its Development and Characteristics," *International Yearbook of Cartography* 14 (1974): 145–55.

Snyder, J. P., "A Comparison of Pseudocylindrical Map Projections," *The American Cartographer* 4 (1977): 59–81.

———, "Map Projections Used by the U.S. Geological Survey," U.S. Geological Survey Bulletin 1532. Washington, D.C.: U.S. Government Printing Office, 1982.

Tobler, W. R., "Geographic Area and Map Projections," *The Geographical Review* 53 (1963): 59–78.

———, "A Classification of Map Projections," *Annals of the Association of American Geographers* 52 (1962): 167–75.

6

Processing and Generalizing Geographical Data

GEOGRAPHICAL VARIABLES AND THEIR SPECIFICATION

CLASSES OF GEOGRAPHICAL VARIABLES

Place or Positional Data
Linear Data
Areal Data
Volumetric Data

CONTINUITY AND SMOOTHNESS
SCALING GEOGRAPHICAL VARIABLES

Nominal Scales
Ordinal Scales
Interval Scales
Ratio Scales

BASIC STATISTICAL CONCEPTS AND PROCESSES

ABSOLUTE AND DERIVED DATA

Averages
Ratios
Densities
Potentials

APPROPRIATE AVERAGES AND INDEXES OF VARIATION

Mode—Variation Ratio
Median—Decile Range
Arithmetic Mean—Standard Deviation
Standard Error of the Mean

GENERALIZATION

The Elements and Controls of Generalization

THE ELEMENTS OF GENERALIZATION

Simplification
Classification
Symbolization
Induction

THE CONTROLS OF GENERALIZATION

Objective
Scale
Graphic Limits
Quality of Data

Earthly phenomena occur in bewildering numbers and intricate arrays. One obvious way to make sense out of this confusing complexity is to make maps. The cartographic process requires that after the variables to be mapped have been selected, they must be processed in order to make an effective portrayal. These range from the calculation of ordinary statistical measures, such as averages and ratios, to the execution of several somewhat more complex operations, such as simplification and classification, collectively called *cartographic generalization*.

As we have seen in the preceding two chapters, even the rather straightforward procedures of cartographic reduction and transformation involve a variety of possibilities. In order to deal logically with the options open to the cartographer in processing and generalizing data, we shall begin by considering the essential nature of geographical variables and then survey the statistical procedures appropriate to dealing with them. Finally, we will examine how the intellectual process of generalization functions in cartography.

GEOGRAPHICAL VARIABLES AND THEIR SPECIFICATION

Anything that is anywhere, whether material such as a road, or immaterial such as a religious adherence, is a geographical or spatial phenomenon that has location and therefore can be mapped. Because geographical variables are so diverse and often quite complex, in order to prepare meaningful maps, one must understand clearly the essential nature of the geographical data.* This involves several aspects.

*The study and mapping of other celestial bodies, such as the moon and Mars, has caused a minor problem in terminology. Since ancient times the systematic study of earthly spatial relationships has been known as *geography* (Greek *geo*, earth, + *graphein*, to draw, write). Now that we are similarly concerned with the surface characteristics of other celestial bodies, it is a bit awkward to use earth-based terms, such as "geographical" or "geological," to refer to their study. Nevertheless, we shall continue to do so in the belief that such terms are now essentially generic rather than earth-specific.

Foremost is the locational relationships among the data, or what is often called their *geographical ordering;* the display of this is what cartography is all about. Geographical ordering is an inherent quality of all spatial data, and we can take it for granted, even though establishing position is not always easy. Equally important is a systematic approach to the description of the geographical variables and the specification of categories within classes. Without a logical way of dealing with their qualitative and quantitative characteristics, there would be nothing but confusion.

Classes of Geographical Variables

We can recognize four basic categories of geographical phenomena: place or positional, linear, areal, and volumetric.

Place or Positional Data. A point is a nondimensional location. Conceptually, place or positional data are things that exist at individual locations. There is a great variety of such kinds of data, ranging from a depth sounding to a road intersection. At a higher level of abstraction, we may conceive of a city at a location. Even though the city may cover a considerable area, it can be endowed with the attribute of being at a different place from another city. Even some summary characteristic of a considerable region, such as the average annual production of a state or country, can be thought of as being represented by some central place within the boundaries. The essential characteristic of place data is the conception of its existence at a single location, however abstract that conception may be.

Linear Data. Some kinds of data are linear, the distinguishing characteristic being that they are one-dimensional. Even though a phenomenon may have significant width, such as a road or a river, its course and relative length are the dominant attributes that allow us to think of it as a line. There are many kinds of linear data, ranging from the invisible boundary between two areas of different administration or the coastline separating land from water, to the routes followed by transported materials or spreading ideas.

Areal Data. Conceptually areal data are two-dimensional, and a primary concern is the areal extent of the phenomenon. Even though a region may be long and narrow, as an area, its linearity is not its prime attribute. Like the other classes of geographical variables, the conceptual range of areal data is wide. Attributes such as national sovereignty, dominant language or religion, climatic type, and vegetation character all fall in the class of areal data.

Volumetric Data. Geographical volumes are three-dimensional in concept. Such data may range from a mental construct (for example, the population of a city as a "quantity" of people), or they can be tangible (like the volume of precipitation that falls on an area or the tonnage of freight moved by rail). We conceive of geographical volumes in different ways. The population of a city or the gross national product of a country are simply sums, whereas many geographical volumes are thought of as quantities spread out over some base level, or *datum,* and extending above or below it, such as land or water in relation to sea level. Conceptually, a volume can be quite abstract, such as geographical density, that is, so many units of some phenomenon per unit area.

We tend not to be very systematic in the way we deal with geographical variables because we often put the same item in different categories, depending on how we may be thinking about it. For example, we may conceive of New York City as a place (in contrast to Philadelphia), or as an area (a particular administrative region, as contrasted with an adjacent region), or as a "volume" of humanity. Nevertheless, any geographical phenomenon can be put in one of these four categories, and some phenomena may be transformed conceptually from one to another.

Continuity and Smoothness

Some geographical arrays are discrete or discontinuous in that a distribution may be composed of individual items at particular locations, and the intervening areas are empty of the item. Such would be the case, for example, of individual houses, industrial plants, cities, or routes of movement. In contrast, other distributions are continuous in that no area is empty. It is inconceivable, for example, that temperature or the surface of the solid earth (above and below sea level) can *not* exist anywhere. Spatial data that intrinsically are discrete can be transformed conceptually into continuous data. For example, people are discrete, so that population would have to be classed as discontinuous, but if numbers of people are related to the areas that they occupy by applying the density concept (number of persons per square kilometer), the ratio becomes continuous, since all areas then must have some density value, even those where the value is zero.

Geographical distributions are also either smooth or nonsmooth. *Smooth phenomena* are those in which the differences from place to place are transitional rather than abrupt. For example, atmospheric pressure is found to vary more or less gradually from place to place and the pressure between two places close together will be intermediate. On the other hand, some things, such as land use categories and statistics of gross national productions, change abruptly at the boundaries between classes.

Commonly, areal data tend to be nonsmooth and volumetric data tend to be smooth; however, just as in the case with continuity, some kinds of distributions can conceptually be either. For example, density can be thought of as volumetric in which the value changes are abrupt at the boundaries of the areal units used in its calculation. But, more abstractly, it can also be conceived as forming a smooth, uneven, statistical surface with sloping ups and downs from place to place.

Scaling Geographical Variables

When dealing cartographically with place, linear, areal, and volumetric data, it is necessary, of course, to determine the locations of the variables. That provides the spatial attribute or geographical ordering that is the fundamental function of the map, but it is not enough. It is also necessary to differentiate within the classes of data. A map that shows all locations of rivers, roads, boundaries, and railways as lines but fails to show which is which would not be very useful. For cartography, the most efficient method of describing the observed characteristics and the specification of the categories within a set of variables is one that involves four levels of precision. Such a method is called a *scaling system,* and the four scales, in increasing order of descriptive efficiency, are *nominal, ordinal, interval,* and *ratio.*

Nominal Scales. Nominal scales are employed when we distinguish among a set of things only on the basis of their intrinsic character, that is, the distinctions are based only on qualitative considerations without any implication of a quantitative relationship.

Examples of nominal differentiation of place data are Chicago, a gravel pit, and the north magnetic pole; an example of areal data is a land use class; an example of linear data is a river or road; and an example of volumetric data is a maritime air mass or a foreign-born population.

Although we can conceive of particular geographical volumes on a nominal scale, they cannot be mapped as volumes without employing a higher order scale (ordinal, interval, or ratio). Because the ordinary map is only two-dimensional, a volume without any quantitative attribute can only be treated as place, linear, or areal data. For example, a volume of foreign-born population might be mapped as existing in a city (place data) or in an area (areal data); a lake or an air mass would have to be mapped as occurring over an area (areal data); and a volume of freight moved over a given railroad line could only be mapped as a line.

Ordinal Scales. Ordinal scales involve nominal classification, but they also differentiate within a class of data on the basis of rank according to some quantitative measure. Rank only is involved; that is, the order of the variables from lowest to highest is given, but not any definition of the numerical values. For example, we can differentiate major ports from minor ports, intensive from extensive agriculture, or among small, medium, and large cities, hot and cold temperatures, and so on. Such ordinal scales enable the map reader to tell that some place, linear, areal, or volumetric variables compared to others are larger or smaller, more or less important, younger or older, and so on, but they do not indicate any specific magnitude of difference.

Interval Scales. Interval scales add the information of distance between ranks to the description of kind and rank. To employ an interval scale, we must use some kind of standard unit (which may be quite arbitrary) and then express the amount of difference in terms of that unit. For example, we can differentiate among temperatures by using a standard unit, the degree (Celsius or Fahrenheit), among city sizes by using a person (as a unit of population) as a standard unit, and among differences in elevation by employing standard units of linear measure such as the meter or foot.

Although interval scales of place, linear, areal, and volumetric variables provide more information than nominal and ordinal scales, one must be careful not to infer more than is warranted by the nature of the unit and the scale to which it applies. For example, one should not say that 40°F. is twice as warm as 20°F.

The three scales, nominal, ordinal, and interval, are progressive in their descriptive efficiency. Everything is nominal; ordinal adds ranking; and interval assigns magnitudes to the ranks. There is no problem in recognizing to which class data belong, but the symbolization of data may make scaling less clear. For example, if a scale of values, such as 0–100, is divided into four ranges, 0–25, 25–50, 50–75, 75–100, and then any value in one range would be represented on a map by a symbol of one size called "range graded," then the symbolization fits logically between ordinal and interval scales. One cannot form ratios from the four range-graded symbols, but there is more information about the distance between ranks than there is in strictly ordinal ranking.

Ratio Scales. A ratio scale is a refinement of an interval scale. It provides magnitudes that are intrinsically meaningful by employing an interval scale which begins at a zero point that is not arbitrary as the zeros of the Fahrenheit and Celsius temperature scales are. Such a set is called a *ratio scale.* Elevation above a datum, barometric pressure, the Kelvin temperature scale, depth of snow or precipitation, populations of cities, and tons of freight are all measured on ratio scales. Most measures pertaining to length, area, and volume are developed on ratio scales. Ratio measures relating to the concept of geographical density (number of items per unit of area), although they are absolute magnitudes, have a different kind of geographical validity, since the quantities refer to areas instead of to points.*

From the point of view of mapmaking there is no difference between the symbolization of geographical data categorized on interval and ratio scales. In both instances a range is being displayed, and from the point of view of representation, it is immaterial whether or not the scale begins at an arbitrary zero. The map user, however, must be very careful when interpreting such maps, since the two scales are very different.

BASIC STATISTICAL CONCEPTS AND PROCESSES

A large share of mapping today (and probably even more in the future) is basically statistical, either because it is derived from published statistical data or because the manipulation of the data for mapping follows statistical principles. The first characteristic is obvious; the second is less so, but probably more important. For example, the mode is one kind of statistical average and the making of a soil map demands the mapping of the modal soil categories; the imagery created by satellite scanning systems requires analysis and processing by statistical methods; and national map accuracy standards for large-scale mapping are based upon statistical concepts. Furthermore, much of the data with which a cartographer works consist of samples obtained in various ways, and the realistic portrayal of these data cannot be accomplished unless the cartographer has a clear understanding of the reliability of the data and the relative appropriateness of the various statistical measures as descriptive devices.

In the use of nominal, ordinal, interval, and ratio scales, in the manipulation of data, and in the selection of categories, the cartographer must use statistical techniques. Every cartographer, therefore, must be familiar with some of the basic concepts of statistical method.

It is clearly not possible in an introductory textbook where the main theme is cartography and not statistics, to go into much detail concerning the variety of statistical techniques that the cartographer might find useful. Nevertheless, the preparation of most kinds of maps usually requires familiarity with the general concepts involved. A good book on statistical method is a useful reference; a number of titles are included among the references at the end of this chapter.

The first step in the cartographic process, whether the map is general or thematic, is to decide on a hierarchy of importance for the different classes of data to be mapped. Then one must select the basic conceptual forms in which they are to be displayed, for example, whether each data set is to be conceived as a discrete or a smooth distribution. After these basic decisions are made, the cartographer may then proceed to deal with a variety of problems concerning the data.

*See Phillip C. Muehrcke, "Concepts of Scaling from the Map Reader's Point of View," *The American Cartographer 3,* no. 2 (1976): 123–41, for a fuller discussion of the relation between scaling systems and the interpretation of mapped data.

When statistical data are obtained from different sources, it usually is necessary to equate them so that they provide comparable values. For example, different countries use different units of measure such as metric, long or short tons, U.S. gallons or Imperial gallons, hectares or acres, and so on. Frequently the units must be further equated to bring them into strict conformity. If, for example, we were preparing a map of energy reserves, it would not be sufficient to change only the volume or weight units to comparable values, but it would also be necessary to bring the figures into conformity on the basis of their BTU ratings. It is also frequently necessary to process the statistics so that unwanted aspects are removed. A simple example is the mechanics of preparing a map of rural population based on county data. The total populations of incorporated divisions as well as the totals for counties may be provided in census tables; if so, the populations of the incorporated divisions must be subtracted from the county totals.

Another illustration is provided by the well-known regional isothermal map. If, for example, the objective of the map is to reveal relationships among temperatures, latitudes, and air masses, the local effects of elevation must be removed from the reported figures. This involves ascertaining the altitude of each station and the conversion of each temperature value to its sea-level equivalent.

After the statistical data have been made comparable, the next step may be to convert them to mappable data. Such statistics as ratios, per hectare yields, densities, percentages, and other indices must be calculated before plotting.

A calculator or a personal computer is of constant use to a cartographer. If the calculations are many or difficult, it is well worth the initial tabulating time to prepare the data for subsequent machine processing. A large amount of data of all kinds is available in machine usable form, and the cartographer need only obtain copies of the necessary information. The entire processing can be done by programming, and the result can be printed tabulations or machine-produced plots. Effort spent in initial search for data in the form easiest to process will often save considerable time.

Absolute and Derived Data

All maps fall in one of two classes: they portray either (1) absolute, observable qualities or quantities, or (2) derived qualities or quantities. Examples of the first class are maps showing categories of land use or roads, the production or consumption of goods, or the elevations of the land surface above sea level. The qualities or quantities are those observed concerning a single class of data, and they are expressed on the map in absolute terms according to some measurement scale, for example, grassland, output of hydroelectric power by states, or numbers of persons by counties. In this group many combinations are possible, and several kinds of values may be presented at once. In no case, however, are the data expressed as a relationship, such as per capita consumption of services or percentage change in population density.

The second group of maps portray derived values which express either some kind of summarization or some sort of relationship between two or more sets of data. Examples of this second group include number of persons per square kilometer, average July temperatures, per capita income, or categories of similar ground slope. This group, showing derived rather than absolute values, includes four general classes of relationships: averages, ratios, densities, and potentials. Each class will be discussed in the following sections, but be alert to the possibility that derived quantities, especially averages and ratios, can be misleading if the characteristics of the data set and its processing are not clearly understood by the mapmaker and presented to the map user.

Averages. Averages are probably the commonest kinds of derived variable. They are often called *measures of central tendency* because a selected quality or quantity is used to characterize a series of qualities or quantities. There are many kinds of averages, but in general only three are commonly used in cartography.

The *arithmetic mean* is widely used. Most maps of climate, income, production, and the other elements dealt with in the study of the physical and cultural character of regions are based on arithmetic means derived by the reduction of large amounts of statistical data. It is usually symbolized by $\overline{X}$ in the following equation:

$$\overline{X} = \frac{\Sigma X}{N}$$

where ΣX is the summation of all the X values, and N is the number of occurrences of X.

The mean sometimes should be areally weighted. For example, if a map of the United States showing the value of farmland per acre by states were to be prepared from data reported by county averages, and if some counties in a state were much smaller than others, then to give equal weight to all values would seriously bias the average for that state. Therefore, whenever the values (X) in a distribution are in any way related to areal extent, they must be weighted for their areal frequency. This is mostly easily done by multiplying each X value by the area of the region to which it refers, summing these products, and dividing the sum by the total area. The general expression for any areally weighted mean is, therefore,

$$\overline{X} = \frac{\Sigma aX}{A}$$

in which ΣaX represents the sum of the products of each X value multiplied by its area, and A is the total area, that is, Σa. The areally weighted mean is also called the *geographic mean*.*

The *median* is another kind of derived measure of central tendency. If we rank (arrange in numerical order) all the values of a variable from the lowest to the highest, the median is the value in the middle; that is, half the values will be higher and half lower than the median value. Using the previous illustration of a map showing in this case the median value of farmland per acre by states obtained from data reported by counties, then for each state the county values would be ranked and the midpoint value would be taken as the representative value.

As before, if the counties within a state varied greatly in size, this could seriously bias the value for the state. In that case, the median should be areally weighted. When this is done, the geographic median is that value below which and above which half the total area occurs. The method of calculation requires that the values be ranked, but in addition the area associated with each value is placed alongside it. Then, beginning at either the top or bottom of the array, the areas associated with each observation are progressively summed. The geographic median value is the one whose associated area, when it is added to the accumulating sum, makes the sum equal to half the total area.

The *mode* is a third kind of average. It is the value or characteristic that occurs most frequently and is the basis for maps that portray the predominant occurrence of kinds of areal phenomena such as land use, soil character, vegetation cover, linguistic areas, and the like. Determining the mode is often quite simple, such as in the case of the predominant fuel used at generating plants or the party gaining the most votes in an election. Large-scale maps of nominally scaled areal distributions, such as land use, simply locate the categories where they are within boundaries visible on the ground. Each such mode is absolute; there is no other within the boundary. On the other hand, small-scale maps of such distributions are more difficult because individual categories are too small to be mapped. The modal class must then be derived by determining which occupies a greater proportion of an area than any other class. Actually, or at least in principle, the entire area to be mapped is divided into smaller unit areas and the modal class is then

*Two other kinds of means, the geometric and harmonic, are not often used in cartography; their special applications will be found in books on statistical methods.

derived for each unit. Often this decision is not easy. For example, note the grassland category in Figure 6.1.

The determination of the modal quality of an area can be very elusive unless the distribution being mapped is visible. Even then, it is possible that two or more categories equally comprise an areal unit or that the variation within a unit area is so great that less than 25 percent occurs in any single class. The smaller the unit areas, the less likely this is to be a problem. Determination of the modal class is a different problem in land-category (land use, vegetation, and so on) mapping from LANDSAT-derived data (Multispectral Scanner—MSS, and Thematic Mapper—TM), since the radiation from each ground-scene area scanned, called the instantaneous field of view (IFOV), results in only a single digital measurement, called a picture element or *pixel*. For all four bands of the MSS the IFOV is approximately 80 by 80 m, but because of the timing and geometry of the system, the pixels refer to somewhat smaller areas, about 0.5 hectares (1.25 acres). For most of the bands of the TM the IFOV is much smaller.

FIGURE 6.1 The actual vegetation cover distribution in one unit area. This is one of many serving as the source data for a vegetation cover map of a large region on a scale too small to show details less than unit area size. Since grassland covers the largest area of the unit, that is, occurs most frequently, it is the modal class, and this entire unit would be mapped as grassland.

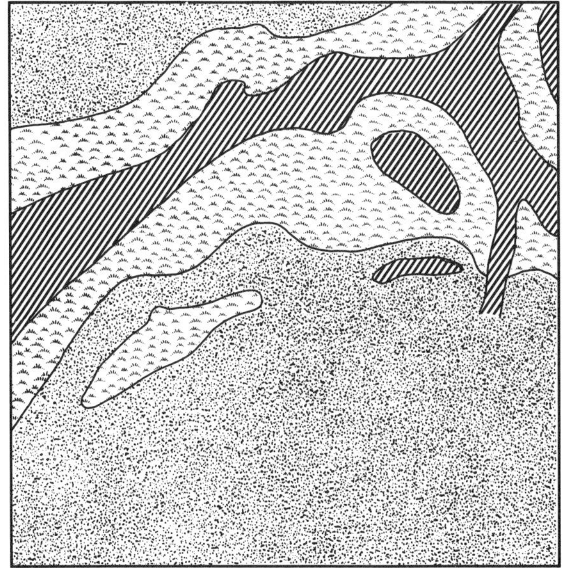

Each of the three kinds of measures of central tendency will in some instances provide mapping data that truly represent the character of the distribution. In other cases, the distribution may be such than an average is not very representative. There are indexes of variation that help to describe how well the arithmetic mean, the median, and the mode characterize a series or array. These will be described later.

Ratios. The second class of derived quantities consists of measures, such as *ratios* or *rates, proportions,* and *percentages,* in which something is measured per unit of something else, or in which some element of the data is singled out and compared to the whole. Maps that show the percentage of rainy days, the proportion of all cattle that are beef cattle, mortality rates, or the rate of growth or decline of some phenomenon are examples. In this group the numerical value mapped will ordinarily be the result of one of the following basic kinds of operations:

$$\text{a ratio or rate} = \frac{n_a}{n_b}$$

$$\text{a proportion} = \frac{n_a}{N}$$

$$\text{a percentage} = \frac{n_a}{N} \times 100$$

where n_a is the number in one category, n_b the number in another category, and N the total of all categories.

As used by the cartographer, such statistics sometimes take on the characteristics of a spatial average. For example, twelve persons per square kilometer is a ratio obtained by dividing the number of people by the total number of square kilometers. This kind of ratio is the basis of the density concept treated below.

Maps of these kinds of derived quantities are made to show variations from place to place in the relationship mapped, and they are usually prepared from summations of statistical data either over area or through time. The appropriateness of the quantity obviously depends on the specific use to which it is being put, but a few words of caution are in order. Percentages, ratios, and rates, when mapped on the basis of enumeration units, are usually assumed by the map reader to extend more or less uniformly throughout the enumeration unit. If the phenomenon does not in fact so occur, then the mapped data may be quite misleading, just as it may be if too few of the items occur. Thus, a value of 100 percent of farms with tractors could be the result of only one farm with one tractor in a large area.

Quantities that are not comparable should never be made the basis for a ratio. For example, we ought not even calculate let alone map the number of tractors per farm by dividing the total number of tractors by the number of farms in a county unless the farm sizes (or some such significant element) are relatively comparable. Common sense will usually dictate ways to insure comparability.

Densities. The third class of derived quantities consists of those that are commonly called *densities*. These kinds of measures are employed when the major concern is the relative geographical crowding or sparseness of discrete phenomena. Examples are maps of the number of persons (or trees, cows, or whatever) per square kilometer, or the average spacing between phenomena such as service stations or mail collection points. The density is computed by

$$D = \frac{N}{A}$$

where N is the total number of phenomena occurring in an enumeration unit, for example, a county, and A is the area of the unit. The *average spacing* between phenomena, another way of looking at density, is computed by

$$\bar{S} = 1.0746 \sqrt{\frac{A}{N}}$$

where $\bar{S}$ is the average spacing of the elements or the mean distance between them in linear values of the same units used for A when a hexagonal spacing is assumed, which is the arrangement when the elements of N are equidistant. If we assume a square arrangement, the interneighbor interval, as it has been called, is the square root of the reciprocal of the population density expressed in the linear units of the measure (for example, square kilometers) used to calculate the density.

Density is more closely related to the land than other averages and ratios, and the significant element in the relationship is area. Thus, for example, 5000 items in an area of 100 square kilometers is a density of 50 per square kilometer; if arranged hexagonally, each individual would be about 150 m from neighbors and, if arranged rectangularly, paired individuals would be slightly closer. In many instances, a density value derived from the (1) total number within, and (2) the total area of a statistical unit is not as significant as one which expresses the ratio between more closely related factors. For example, the relation of the number of people to productive area in predominately agricultural societies is frequently found to be more useful than is a simple population to total area ratio. If the data are available, we can easily relate population to cultivated land, to productive area defined in some other way, or to that spatial segment that is important to the objective of the analysis.

When working with densities and average spacings, the cartographer is limited in the detail that can be presented by the sizes of the statistical units (such as townships, counties, or states) for which the enumeration of the numbers of items has been made. Generally, the larger the units, the less will be the differences among the values. In many cases the initial data must be supplemented by other sources in order to present a distribution as close to reality as possible.

A relatively precise way to do this is to estimate the density of a portion of the enumeration district by using supplementary data. It is then necessary to adjust the value for the other portion so that the modifications are consistent with the original data. This may be accomplished as follows.

Assume, for example, an enumeration district with a known density, *D,* of 90 (Fig. 6.2). Assume further that the district is divided in two portions, *m* comprising 0.7 of the total area, and *n* making up 0.3 of the area. If the density of area *m* is estimated to be 15, then the density of area *n* must be calculated in order that the two regions together have a density of 90.

The basic equation is

$$D_n = \frac{D}{1 - a_m} - \frac{D_m a_m}{1 - a_m}$$

in which D_n = the density in area *n* (see Fig. 6.2)
D = average density of the area as a whole (number of units ÷ area)
D_m = estimated density in part *m* of area
a_m = the fraction (0.1 to 0.9) of the total area comprised in *m*
$1 - a_m$ = the fraction (0.1 to 0.9) of the total area comprised in *n*

D_m and a_m are estimated approximately. It is not necessary to measure a_m accurately, since the amount of error in a rough estimate is likely to be less than the amount of error in the best possible estimate of D_m.

Study of neighboring areas sometimes gives a clue to a value that may reasonably be assigned to D_m. For example, other maps may show what would appear to be similar types of distribution prevailing over D_m and over the whole of *E*, an adjacent unit. It would be reasonable, therefore, to assign to D_m a density comparable with the average density in *E*.

Having assigned estimated but consistent densities to the two parts of an area, we may then

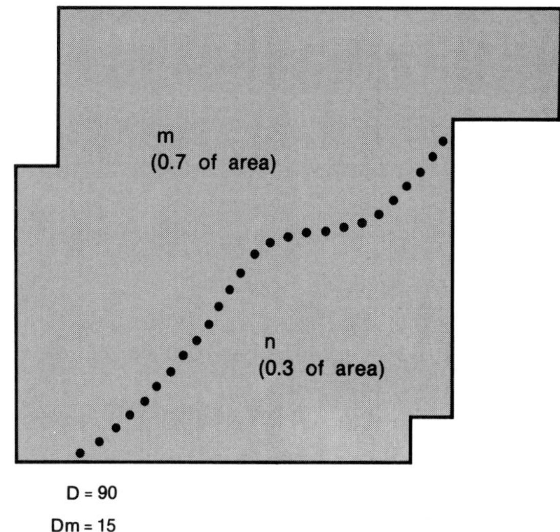

FIGURE 6.2 A hypothetical enumeration district divided in two portions, *m* and *n*, by the dotted line. See text.

again divide each or one of these parts into two subdivisions and work out densities for those two in the same manner. The process may be repeated within each subdivision.

The calculations merely ensure consistency in apportioning values within the limits of territorial units for whose subdivisions no statistical data are available. The method can be applied in the mapping of any phenomena for which statistics are available only by larger units, but the quality of the resulting map depends on the validity of the estimates of D_m. Figure 6.3 illustrates the refinement that can be made.

Potentials. The fourth kinds of derived quantities are employed in *potential* maps. These maps assume that the individuals comprising a distribution, such as people or prices, interact or influence one another, directly with the magnitudes of the phenomenon and inversely with the distance between their locations. Because this assumption is derived from the physical laws governing the gravitational attraction of inanimate masses, it is called a *gravity concept.* It has

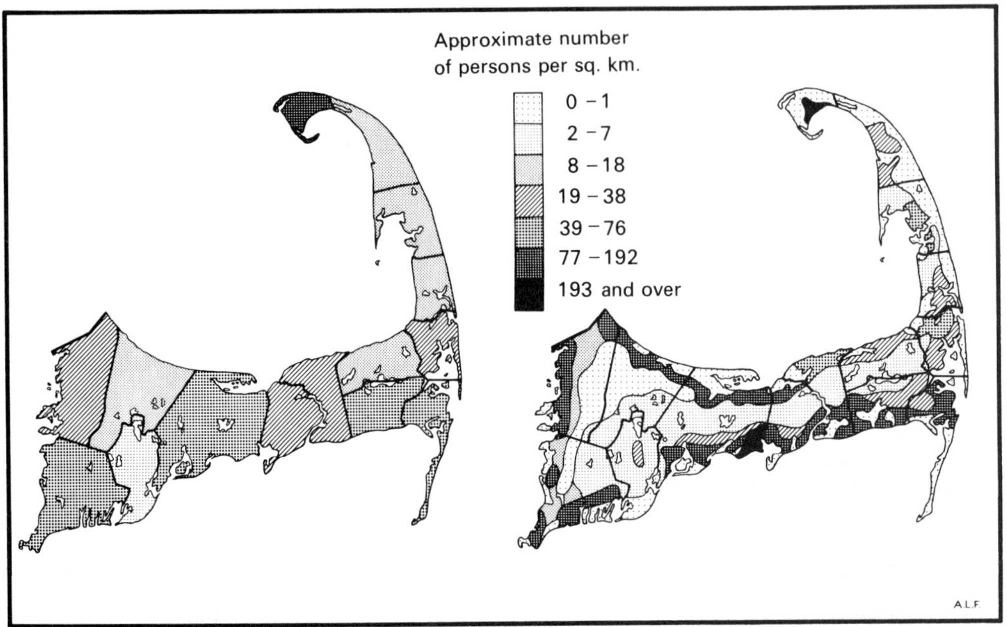

FIGURE 6.3 The map on the left shows the density according to whole census enumeration districts, while the map on the right shows the refinement that can be developed by applying the system described in the text for estimating densities of parts. (Redrawn from *The Geographical Review*, published by the American Geographical Society of New York.)

been applied to a variety of economic and cultural elements.

The value of the potential at any point is the sum at that point of the influence of all other points on it plus its influence on itself. The potential P of place i of phenomenon X will be

$$P_i = X_i + \sum \frac{X_j}{D_{ij}}$$

where X_j is the value of X at each place involved and D_{ij} is the distance between place i and j. The preparation of a potential map requires that this summation be repeated for each place, a task for a computer since if, for example, j is only 50, more than 2500 calculations will be required. The resulting map values will be in numbers of units of X at particular points. It is apparent that the relationship can easily be amended by inserting a constant in the equation or by defining in special ways the distance between the places.

Appropriate Averages and Indexes of Variation

As pointed out earlier, averages are probably the most common way of summarizing data in cartography. It is an unhappy fact, however, that an average is but a simplified statement, and one cannot "tell by looking" whether it is a good or poor summary of the data. Accordingly, it is the responsibility of the cartographer (1) to understand the character of the data set being mapped, and (2) to do whatever is possible to make the map user aware of the critical characteristics of the data so that the summarization will not mislead. In this section we will direct our attention to the first task by listing the appropriate averages to use for classes of data and describing the appropriate measures of indexes of variation, which suggest how valid the averages are. The second responsibility is much broader and involves such subjects as simplification and classification pro-

TABLE 6.1 Scaling Systems: Their Appropriate Averages and Indexes of Variation

Scale	Average	Index of Variation
Nominal	Mode	Variation Ratio
Ordinal	Median	Decile Range
Interval	Arithmetic Mean	Standard Deviation
Ratio	Arithmetic Mean	Standard Deviation

cedures, methods of symbolization, and so on, all of which will be dealt with in later chapters.

Table 6.1 lists the appropriate averages with which to summarize data scaled in each of the four ways and the index of variation suitable to evaluate the quality of the summary.

Mode—Variation Ratio. In a set of variables scaled nominally, each is distinguished qualitatively but not quantitatively. When unsummarized absolute values are mapped, such as actual land use, kinds of road surfaces (gravel, black top, concrete), or classes of coal produced at each mine, the actual facts are mapped. On the other hand, when the scale of the map requires it, or the objective calls for it, a derived average called the *mode,* the class that occurs most frequently, is mapped. The mode can be either the predominant class or occurrence of a set of variables, such as minorities in cities, or if it is an area phenomenon, it can be the class that occupies the largest proportion of the area.

The *variation ratio* (v) indicates how representative of the distribution the mode is. It provides an index of the proportion of nonmodal cases; v varies from a value of near 1, which indicates that nearly all units are nonmodal, to 0, which indicates the occurrence of only the modal class. Therefore, the nearer v is to zero, the better the quality of the mode as a summarizing statement. The variation ratio is calculated by

$$v = 1 - \frac{f_{\text{modal}}}{N}$$

where f_{modal} is the frequency or number of occurrences in the modal class and N is the total of all occurrences. For example, suppose we contemplated mapping the predominant source of farm income by states of the United States from such data as those in Table 6.2.

Two of the values of v in Table 6.2 are near 0.50, which indicates that those modes are not very representative.

A common kind of map of data scaled nominally is one in which the areal frequencies of the variables are mapped, such as land type (upland, lowland, hills, swamps, and so on) or vegetation. The mapped categories are determined on the basis of the predominant or modal class in a series of unit areas, such as was described earlier and illustrated by Figure 6.1. In that case the frequency of each variable is the proportion of the total area it occupies in each mapping unit. The modal class is, therefore, the category occupying more area than any other category, not necessarily more than one-half of the area. When we

TABLE 6.2 Farm Income—Cash Receipts, 1980 ($1,000,000)

State	Crops	Livestock	Gov't. Payments	Total	v
Alabama	696	1,140	23	1,859	0.39
Alaska	7	4	—	11	0.36
Arizona	937	783	5	1,725	0.46
Arkansas	1,531	1,457	35	3,023	0.49
California	9,390	4,149	14	13,553	0.31
Colorado	965	2,220	18	3,203	0.31
Etc.	—	—	—	—	—

are concerned with areal frequency, the calculation of the variation ratio must be weighted by area and becomes

$$v = 1 - \frac{a_{modal}}{A}$$

where a_{modal} is the area of the unit occupied by the modal category and A is the total area of the unit. The areally weighted variation ratio indicates the proportion of the unit occupied by the nonmodal category. The nearer it is to zero, the better that area mode represents the unit. In Figure 6.1 the variation ratio for the area mode, grassland, is 0.43, which indicates that nonmodal classes occur in nearly half the area.

Median—Decile Range. The *median* is the quantity or quality midway in an array of ranked variables, that is, the class that neither exceeds nor is exceeded by more than half the observations. It can be used as a descriptive measure for summarizing data ranked ordinally or for data ranked on interval and ratio scales.

To illustrate the use of the median with data scaled ordinally, we will use the data in Table 6.3, an example of three counties typical of the kind of ranked data needed to prepare a state map showing farm quality by counties. The data consist of farms which have been classified in four categories from I (good) to IV (poor) on the basis of a variety of characteristics, such as the value of buildings, management practices, soil productivity, and others. The categories are ordinal values, which cannot be added to provide an arithmetic mean for each county. To choose the modal class as representative might be unsatisfactory, since, in our sample of three counties, there are more cases of nonmodal classes than of the modal class in two of them. Accordingly, the median might be selected as the average to use.

For each county in Table 6.3 the ordinary median is easily derived by finding the rank (farm class) above which and below which half the total number of farms occur. When, as in this example, area is a significant element in the variable, then the median should be areally weighted. This is done by progressively summing the areas of the farm classes in each county from the top (or bottom) and finding the class associated with half the total area. It will be noted in Table 6.3 that of the three counties, two have different ordinary and geographic medians.

A median class may or may not be representative of the character of an area where the values (farms) are widely scattered among the classes. For example, in Washington County there is a significant proportion of cases other than the median. In that county there are 720 class III farms (the median class), but there are 870 nonmedian class farms. On the other hand, in Union County the median of class I is an excellent characterization of the county.

An appropriate descriptive statistic to evaluate dispersion among the ranks of an ordinal scale is a *quantile range,* such as the *decile range.** The decile range (d) is the number of classes between those in which the first (d_1) and ninth (d_9) deciles occur; in effect, $d = d_9 - d_1$. The first decile is the value below which 10 percent of the variables occur and the ninth decile is the value above which 10 percent occur. When used with ranked classes, the value of d can range from 0, when less than 10 percent of the variables occur in classes outside the median class, to one less than the total number of classes. The nearer d is to the total number of classes, the poorer the median is as a representative average. For example, in Table 6.3 the total number of farms in Washington County is 1590; the first decile (159 farms) is included in class IV and the ninth decile is included in class II. The span between II and IV is $2 = d$.

The decile range should be areally weighted when considerable differences occur in the value–areal extent relationships. In that case the weighted decile range is based on deciles deter-

*A quantile is obtained by dividing a distribution into equal segments; quartiles divide it into four categories, deciles into ten categories, and so on, to centile (or percentiles) which divide it into one hundred categories.

TABLE 6.3 Counties: Numbers and Areas of Farms by Class; Medians and Decile Ranges

County	Farm Class	Number of Farms	Area (km²)	Cumulated Area	Median	Geog. Med.	*d*	*d* (Geog.)
					Median		Decile Range	
Union	I	936	896	896	I	I	1	0
	II	41	41	937				
	III	30	26	963				
	IV	45	27	990				
	Total	1,052						
Washington	I	120	293	293				
	II	400	582	875		II		3
	III	720	601	1,476	III		2	
	IV	350	221	1,697				
	Total	1,590						
Wood	I	20	14	14				
	II	42	25	39				
	III	123	127	166	III		2	
	IV	171	236	402		IV		1
	Total	356						

mined for 10 percent of the total area involved, not 10 percent of the number of variables. For example, in Table 6.3 a considerable number of larger farms in Wood County is in class IV. Wood County has an ordinary median of III and a decile range of 2. When areally weighted values are determined, however, the geographic median becomes IV and the geographic decile range drops to 1, indicating that areal weighting improved the average. On the other hand, the opposite occurs in Washington County where farm size decreases with farm class. In that case the geographic median of II with a decile range of 3 is not as good an average as the ordinary median of III with a decile range of 2. The determination of whether the ordinary median or the geographic median will provide the more representative average is made after careful inspection of the data; the chosen procedure is then applied uniformly to all the data.

Arithmetic Mean—Standard Deviation. Although the concepts of the mode and median underlie much mapping of qualitative and ranked distributions, the great majority of maps based on averaging numerical data do so by employing the *arithmetic mean.* All kinds of occurrences through time, such as precipitation, passenger traffic, production data, and so on, are usually summarized by deriving the mean of the series using the formulas given earlier. Such data are measured on either interval or ratio scales. The calculation of a mean, either ordinary or areally weighted, is relatively simple, but the evaluation of how well the mean serves as a measure of the "central tendency" of the series is not quite so simple.

If a series of observations, such as low temperatures each day or the number of automobiles passing a point between 4:00 P.M. and 5:00 P.M. daily, fall within a narrow range, then the mean of the series is generally a good summary. Conversely, if the numbers vary greatly, that is, if they are widely dispersed, then the mean is not a good summary. The most useful measure or index of the dispersion is a statistic called the *standard deviation,* which reveals how widely the observations depart from the mean. It is based upon an important concept in statistical analysis known as a *normal distribution,* which describes the frequency with which various values occur in a set of observations. It represents a kind of "ideal"

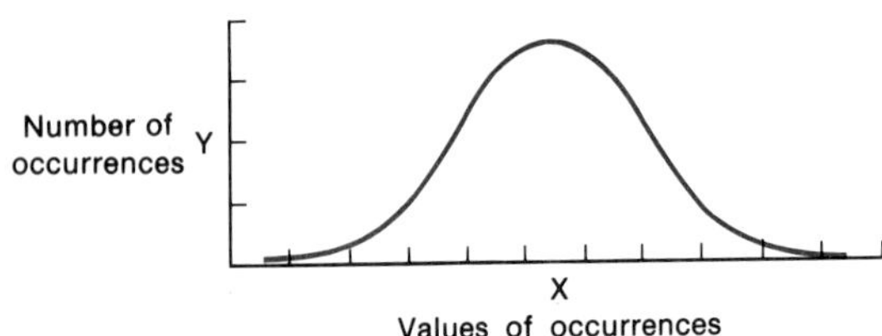

FIGURE 6.4 A normal curve results from plotting the values of a normal distribution on the X axis against the frequencies of their occurrences on the Y axis. The mean is the value at the peak of the curve.

series based upon the theory of probabilities, and the distributions of a great many variables come quite close to it.*

In a perfectly normal distribution, the values near the arithmetic mean of the series occur most often. The greater the deviation, that is, the departure of a value from the mean, the less often that value occurs. When this relationship is plotted on an ordinary frequency graph, as in Figure 6.4 where the frequency of the occurrence of the values is plotted on the Y axis against the values on the X axis, the result is a bell-shaped curve called the *normal curve* of a normal distribution. The standard deviation (s) is a way of describing the dispersion of the values around the mean of a normal distribution. It is a number, in units of the series, obtained by the following formula:

$$s = \sqrt{\frac{\Sigma(X - \overline{X})^2}{N}}$$

in which s is the standard deviation, and $(X - \overline{X})$ is the difference between each value and the mean. When working with data, if we do not wish to go to the trouble of determining the differences between each value and the mean, we may use the following formula:

$$s = \sqrt{\frac{\Sigma X^2}{N} - (\overline{X})^2}$$

As pointed out in connection with the determination of the mean, when data are in any way related to areal extent, it is necessary to take this into account. The easiest way to do this is to use the areal extent as the frequency. The expression for an areally weighted standard deviation then becomes

$$s = \sqrt{\frac{\Sigma aX^2}{A} - \left(\frac{\Sigma aX}{A}\right)^2}$$

in which ΣaX^2 is obtained by first squaring each X value, then multiplying it by the area it represents, and summing the products. The term $(\Sigma aX/A)^2$ will be recognized as the square of the geographic mean.

As is evident from the first of these formulas, the standard deviation is the square root of the mean of the squared deviations from the arithmetic mean of the distribution. It is sometimes referred to in other ways. For example, the mean of the squared deviations is called the *variance*, and the standard deviation is, therefore, the square root of the variance. The standard deviation is also often called the *root mean square*.

The importance of the standard deviation is that the arithmetic mean minus $1s$ and plus $1s$ of a normal distribution includes slightly more than two-thirds (68.27 percent) of all the values in the series regardless of the numerical difference in the high and low values. Consequently, the smaller the value of s, the closer the values of the series cluster around the mean and therefore the better the mean characterizes the series. Graphically, the area under the normal curve between plus $1s$ and minus $1s$ includes 68.27 percent of the total area under the curve (Fig. 6.5). Furthermore, regardless of how the values in a normal distribution occur, the range between $2s$ above and $2s$ below the mean will include at least three-quarters (75 percent) of the observations.

*Although many phenomena approximate a normal distribution of values, some do not. A description of other kinds of distributions is more appropriately a topic for study in statistics.

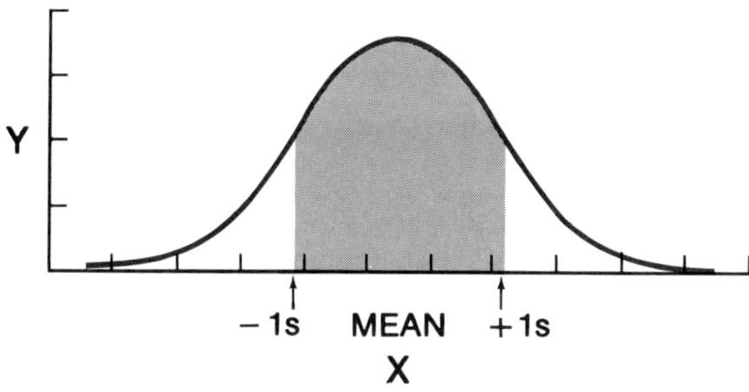

FIGURE 6.5 The shaded area under the curve in Figure 6.4 shows the proportion of the occurrences included within one standard deviation (plus and minus) from the mean.

The mean and standard deviation are widely used in both descriptive and inferential statistics. Maps of distributions such as rainfall variability, probability of temperature extremes, potential crop yields, and so on depend upon standard deviations. In addition to its descriptive usefulness, the standard deviation is used to set class limits for mapped distributions. It has been suggested that only by using the mean and standard deviation for class limits it is possible to obtain comparability of distributions through time for statistics about distributions like birth and death rates, diseases, or income.

One ordinarily thinks of obtaining means and standard deviations from numerical data such as published meteorological data, census reports, agricultural and industrial production summaries, and the like. On the other hand, it is not uncommon for cartographers to be called upon to obtain and work with data that must be derived from information that is already mapped. This requires some different approaches. To illustrate some of these, let us suppose we were called upon to determine the average elevation above sea level as well as the standard deviation of the elevations of a region for which we had contour maps. To rid ourselves of the detailed complications of an ordinarily highly irregular land surface, we will substitute a perfectly conically shaped island mapped with contours at a 30-m interval. The substitution is for convenience only; it does not change the procedures. The island map is shown in Figure 6.6, which shows the elevations (z) and labels the areas between successive contours as $a_1, a_2 \ldots a_9$. Inspection shows that different elevations occur more or less frequently. In this instance, the geographic frequencies are the relative areas of the occurrences; consequently, it is necessary to obtain the areas of the spaces between the contours. This can be done accurately by computer computation if the contours have been digitized or by measuring them with a planimeter. The results obtained are expressed in any convenient square units, such as square centimeters. The measurements obtained from the original drawing of Figure 6.6 (before reduction for printing) are shown in the first two columns of Table 6.4. The mean elevation may be determined in two ways: by calculation or by graphic analysis.

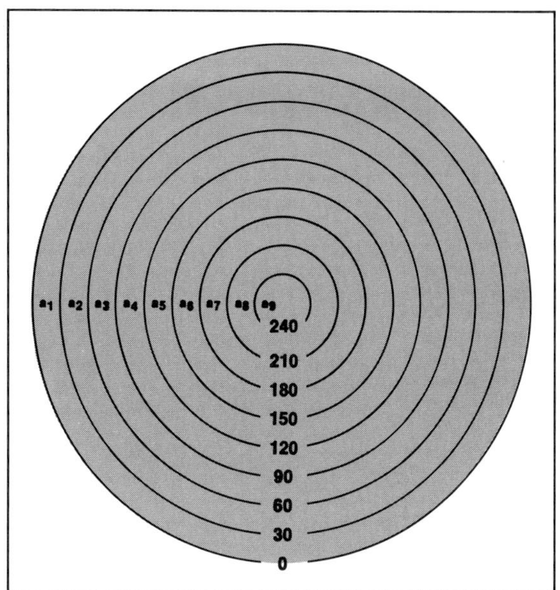

FIGURE 6.6 A conically shaped island delineated by contours at a 30-m interval. See text.

Using the areally weighted mean formula given above, the data in Table 6.4 yield

$$\bar{z} = \frac{9329.25}{102.65} = 90.9 \text{ m}$$

TABLE 6.4 Calculation of the Mean Elevation of the Island Shown in Figure 6.6

Elevation Classes (meters) (z)	Midvalue of Elevation Classes (z_m)	Map Areas of Elevation Classes (cm^2) (a)	(az_m)	Cumulated Areas (cm^2) ($a_1 + a_2 + a_3 \ldots$)
0–30	15	21.42	321.30	21.42
30–60	45	18.84	847.80	40.26
60–90	75	16.45	1233.75	56.71
90–120	105	14.06	1476.30	70.77
120–150	135	11.55	1559.25	82.32
150–180	165	8.84	1458.60	91.16
180–210	195	6.39	1246.05	97.55
210–240	225	3.81	857.25	101.36
240–270	255	1.29	328.95	102.65
Total		102.65	9329.25	

The same result can be obtained graphically by constructing a cumulative frequency graph. Additionally, a cumulative frequency graph can be used to obtain a value for a geographic median using interval or ratio data instead of ordinal data. A cumulative frequency graph is constructed by plotting the values of the distribution on the Y axis of an arithmetic graph in the order of their values against the extent of their progressively cumulated areas on the X axis. The paired data from the first and fifth columns of Table 6.4 may be graphed, as shown in Figure 6.7, by plotting the value of the upper-class limit of the first elevation category, z_1 (30), on the Y axis and the area, a_1, (21.42) on the X axis. The next pair, z_2 (60) and $a_1 + a_2$ (40.26), is similarly plotted. The addition of each area value to the sum of the preceding areas (after arranging them in the ascending order of z values) assures that the curve, which results from joining the plotted points with a smooth line, will rise to the right. The area under the curve (bounded by the curve, the baseline, and an ordinate erected at the value of the total area) may then be measured and expressed in square units. When this total is divided by the length of the baseline from the origin to the total area (measured on the graph in linear terms of the same units used to express the area under the curve), the quotient is the height of a rectangle with the same area as that under the curve, one side of which is the length of the baseline. The height of the rectangle when plotted on the Y axis of the graph is the mean height of the curve and is the geographic mean.

FIGURE 6.7 A cumulative frequency graph. The horizontal dashed line is the height of a rectangle which has the same area as that included between the curve and the baseline. The height of that rectangle on the Y axis is the geographic mean. The vertical dashed line is erected at the midpoint of the cumulated areas which are plotted along the base. The Y value of the curve at the point of intersection with the vertical dashed line is the geographic median.

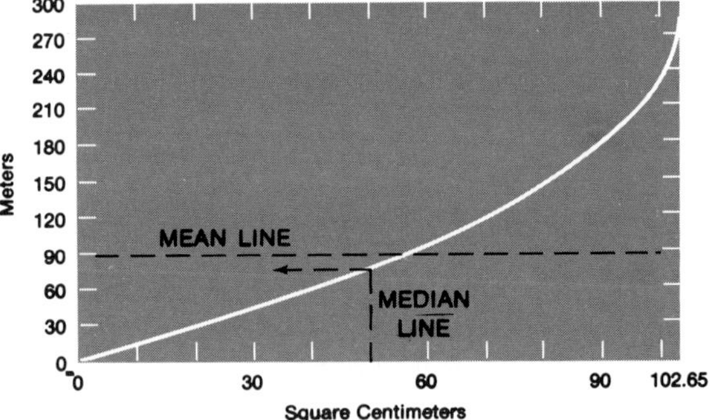

In Figure 6.7 the area under the curve (on the original drawing, before reduction) was 13.42 cm^2. The length of the base as plotted was 8.74 cm. Division of the area by the base (13.42/8.74) resulted in a quotient of 1.535 cm. This value, when measured on the Y axis from the base, gives a rectangle with a height having an elevational value of 90.65 m, which is the geographic mean. In Figure 6.7 this height is shown by the dashed horizontal line. The difference of approximately 0.25 m between the mean determined arithmetically (90.9 m) and the mean determined graphically (90.65 m) results from the graphic process and the consequent rounding of numbers, as well as the fact that midvalues of classes were used in one computation, whereas upper-class limits were used in the other. This kind of graph is also occasionally called a *hypsographic curve* because of its use to determine mean land heights (mean hypsometric values) exactly as illustrated here. It may, of course, be applied to any other kinds of data that are expressed on an interval or ratio scale and that have actual or assumed continuous areal extent.

The cumulative frequency graph may also be used to determine areally weighted percentiles, quartiles, and so on, as well as the areally weighted median, of a series scaled on interval or ratio bases. The median is determined by erecting an ordinate at the midpoint of the baseline and reading the Y value of its intersection with the curve. The areally weighted median elevation of the island in Figure 6.6, as derived from the graph in Figure 6.7, is close to 81 m; that is, half of the map area as shown is above and half below that elevation. Percentile and quartile values are similarly obtained by subdivision of the base and by determining the Y values of ordinates erected at such points. The curve can also be used to find the total area of all those sections lying above or below any given elevational value.

Many problems of cartographic analysis include data with open-end classes, that is, the top or bottom class may be only available on a "less than . . ." or a "more than . . ." basis. This is the case with the series of elevations shown by the contours on the island in Figure 6.6, wherein the contour of highest value is 240 m. By definition, the area inside that contour is above (that is, more than) 240 m, but by exactly how much is not shown. In other cases, the neat line or map edge may cut through a statistical unit area and produce a similar result. When the limits of all classes are not given, the mean value cannot be accurately ascertained, but judicious estimation and interpolation will usually make significant errors unlikely.

To illustrate the calculation of an areally weighted standard deviation, that statistic for the island shown in Figure 6.6 is calculated below. The necessary data are shown in Table 6.5, and it will be seen that midvalues are employed.

$$s = \sqrt{\frac{1{,}261{,}424.25}{102.65} - \left(\frac{9329.25}{102.65}\right)^2}$$
$$= 63.47 \text{ m}$$

This means that one could expect that about two-thirds of the island has an elevation within 63.5 m of the mean; or, about two-thirds of the island lies above approximately 27.5 m (90.9 − 63.47 m) and below aproximately 154.5 m (90.9 + 63.47 m); that is, about two-thirds lies between those two elevations.

The Standard Error of the Mean. Many maps are made from data that have been obtained by sampling or that may be assumed to be a sample. For example, one may compute a mean from every tenth observation in a very large series, which is clearly sampling. One may also compute a mean from a series of temperature observations, say for the last 50 years. This may also be considered a sample, since it contains only a portion of all possible values; all the temperatures that occurred earlier than 50 years ago are not represented. It is reasonable to assume that a mean calculated from a sample will not be the same as a mean derived from the total popula-

TABLE 6.5 Calculation of the Standard Deviation of Elevations on the Island Shown in Figure 6.6

Midvalue of Elevation Class (m) (z_m)	Map (or actual) Area of Class (cm^2) (a)	(az_m)	(z_m^2)	(az_m^2)
15	21.42	321.30	255	4819.50
45	18.84	847.80	2025	38151.00
75	16.45	1233.75	5625	92531.25
105	14.06	1476.30	11025	155011.50
135	11.55	1559.25	18225	210498.75
165	8.84	1458.60	27225	240669.00
195	6.39	1246.05	38025	242979.75
225	3.81	857.25	50625	192881.25
255	1.29	328.95	65025	83882.25
Total	102.65	9329.25		1,261,424.25

tion. An inference regarding the validity of a sample mean may be made by calculating the standard deviation of the mean. It is usually called the *standard error of the mean* and is symbolized by $s_{\bar{x}}$. It is obtained by dividing the standard deviation of the sample values by the square root of the number of the sample values that were used in calculating the mean. Its expression is

$$s_{\bar{x}} = \frac{s}{\sqrt{N}}$$

where s is the standard deviation of the values of X, and N is the number of X values that entered into the calculation of the mean of X.

Put in cautionary terms, two relationships evident from the standard error of the mean should be kept in mind:

1. The greater the standard deviation of the sample values, the less confidence one can put in a mean derived from them.
2. The smaller the number of sample values used to arrive at a mean, the less confidence one can have in it.

Although this is not the place to go into it in detail, the principle underlying the standard error of the mean is applicable to the confidence one can place in a distribution map based upon an areal spread of values, such as summary values for states or meteorological or other observations. In such a map each state or the area of each region is being represented by one value which is, in effect, a kind of "mean" for that whole region. The lower the density of the observations, or in other words the greater the area per value, the smaller the "sample size" and the less confidence the display warrants.

GENERALIZATION

In most aspects of our existence, we are surrounded by detail and complexity. In order not to become mired in confusion, we often focus on the more universal characteristics of things rather than on their individual, unique qualities. For example, one such operation is the derivation of averages of various sorts to characterize distributions, as we saw in the preceding section, but those are only a few of a great many ways we derive or induce general characteristics from particulars. Others include such actions as categorizing or classifying arrays and eliminating visual complexities by simplifying outlines. These are common, everyday ways of helping us to comprehend; collectively they are called *generalization*.

The various methods of generalization are, of course, applied in cartography, not only because it promotes understanding, but also because some processes of generalization are made necessary by the fact that reduction from reality to a smaller scale is a fundamental part of mapmaking. Only in recent years has it been possible for a person to see very much of the earth at once. Although the view from a space vehicle high above the planet may arouse awe, the scene is not likely to be very meaningful. The earth is simply too large and the phenomena too complex for anyone to grasp very much about any sizable area by direct observation from afar. The solution is to apply systematic reduction. The contraction of geographical space as an aid to science and understanding is analogous to the use of magnification to aid in the study of phenomena that are too small to be directly observed. In the senses used here, both processes, reduction and enlargement, are means to the same end—comprehension—but the consequences of these processes are quite different.

Unmodified magnification (for example, with a microscope) is accompanied by a general loss of clarity, a reduction in the intensity of coloring, and an increase in the relative visual importance of the specific compared to the general. Unmodified reduction, when applied to earth phenomena, is likewise accompanied by unavoidable changes. Distances separating features and widths and lengths of features are shortened; adjacent discrete items become more and more crowded; and intricate configurations may appear chaotic. Such consequences of reduction are a mixed blessing. On the one hand the shrinkage of space is what makes it possible to see the geographical arrangement of phenomena, but on the other hand, the increased complexity and crowding tend to compound confusion. To counteract these undesirable effects, two fundamental operations must be performed on the wealth of data. One is to limit our concern to those classes of information that will serve the purpose for which the map is being made, an operation we call *selection*. The other is to fit the delineation of the selected data both to the scale of the map and to the requirements of effective communication. This we call *simplification*.

As we will see, simplification is part of the general procedure called cartographic generalization, but selection is not. *Selection,* as the term is used here, is the intellectual process of deciding which classes of information will be necessary to serve the purpose of the map. No modification takes place; the choice is either to portray roads or not to portray roads, to include or not include major hydrographic features, or to name or not name by lettering all cities over 150,000 population. Once the cartographer has made this selection, the operations of generalization can be performed on the selected information.

The Elements and Controls of Generalization. There are several operations included in cartographic generalization; it is convenient to group them into four categories of processes, termed the *elements* of cartographic generalization, as follows:

Simplification: The determination of the important characteristics of the data, the elimination of unwanted detail, and the retention and possible exaggeration of the important characteristics.

Classification: The ordering or scaling and grouping of data.

Symbolization: The graphic coding of the scaled and/or grouped essential characteristics, comparative significances, and relative positions.

Induction: The application of the logical process of inference.

These fundamental and complex operations are all consciously accomplished by the cartographer in the practice of mapmaking, but each map provides a different set of requirements. The "mix" of the processes of generalization will, therefore, vary from map to map. The manner in which each will be performed depends on the dictates of the *controls* of cartographic generalization. These controls are:

Objective: The purpose of the map.

Scale: The ratio of map to earth.

Graphic limits: The capability of the systems employed for the communication, and the perceptual capabilities of the readers of the communication.

Quality of data: The reliability and precision of the various kinds of data being mapped.

It is only fair to observe that the separation of cartographic generalization into four categories of operations and four controls is in itself a generalization about generalization. The aim in doing so, like the aim of all generalizing, is to simplify to manageable proportions what is really a very compex intellectual-visual process. Generalizations are always needed—even about generalization.

The Elements of Generalization

The four elements of generalization are not clearly separable in many cartographic situations. Keep in mind that in the subsequent discussions of these, each is being treated individually without regard to the others. This is impossible for the cartographer to do in practice.

Simplification. Such things as roads, buildings, small streams, and so on can be delineated without being significantly enlarged only on very large-scale (such as 1:2,500) cartographic displays called plans. On all but the largest-scale maps, most features must be exaggerated to make them reasonably visible. For example, at the relatively large scale of 1:25,000 a street 20 m (66 ft.) wide, if shown true to scale, might normally be symbolized by two fine lines 0.8 mm (about 0.03 in.) apart; if reduced photographically to 1:100,000 they would only be 0.2 mm apart; while if reduced to 1:500,000, the lines probably would disappear to the unassisted eye.*

There are two primary objectives in the practice of cartographic simplification. One is to fit the information to the capability of the map to portray it at the chosen scale—far from a mechanical operation because an equally important aim is to maintain as far as possible the essential geographical characteristics of the mapped phenomena. Since the various symbols take up room on a map, it is readily apparent that as map scales get smaller, less and less data can be shown. Part of the solution to the problem can be taken care of initially by judiciously selecting only the necessary classes of data to be portrayed, thereby in a sense making room for it, but that is usually not enough. The information in the classes to be mapped normally must also be simplified by eliminating some of it and by reducing linear complexities, which can easily become overly intricate when reduced. These operations can be performed manually and with computer assistance. Those simplification processes consisting of particular statistical analyses and computer manipulations are treated separately in Chapter 11. In this section we are concerned with the principles involved in the decision about how much information to retain or discard and the maintenance of inherent geographical character.

The relative map space available for the portrayal of phenomena is obviously a function of scale, but it is important to keep in mind that the reduction of available space takes place as the square of the ratio of the difference in linear scales. For example, a region mapped at 1:25,000 will only occupy *one-sixteenth* as much map space when mapped at 1:100,000. It is quite obvious that the compression that results from scale reduction often allows only limited portrayal, and that sometimes a considerable proportion of each

*As a matter of fact, some phenomena commonly shown on maps, such as boundaries, are usually not even visible lines on the earth and in any case, they have no width. Yet they must often be shown by means of actual lines on maps.

selected class of data cannot be included. Figure 6.8 shows two examples of the kind of pressure exerted by great reduction and the consequent simplifications that resulted. What is to be discarded and what is to be retained depends on many factors, among the more important of which are: (1) the relative importance of an item, (2) the relation of that class of data to the objective of the map, and (3) the graphic consequences of retaining an item.

By analyzing what cartographers have actually done in the way of reducing the *number of items* from one compilation scale to another, F. Töpfer developed what he called the *radical law.** In its basic form, the relationship can be expressed by the equation

$$n_c = n_s \sqrt{S_c/S_s}$$

where

n_c = the number of items on a compiled map with a scale fraction of S_c

n_s = the number of items on a source map with a scale fraction of S_s**

The basic principle expressed by the radical law depends on the ratio of the scales of the compiled and source maps rather than on whether they are relatively large or small. For example, compiling at a scale of 1:192,000 from a 1:24,000 source generates the same simplification problems as compiling at a scale of 1:500,000 from a 1:62,500 source.

The problems of simplification are compounded by the nature of the data. For example, within a reasonable scale range, such things as islands or lakes are generally represented without exaggeration and a relatively large number can be retained, whereas items such as settlement symbols with their associated names take up a much larger proportion of space, and relatively fewer can be retained. For these reasons the authors of the radical law found it necessary to introduce two constants, C_e and C_f, in the right side of the basic equation. Constant C_e is called the *constant of symbolic exaggeration* and takes three forms.

C_{e1} = 1.0 for normal symbolization, that is, for elements appearing without exaggeration.

C_{e2} = $\sqrt{S_s/S_c}$ for features of areal extent shown in outline, without exaggeration, such as lakes and islands.

C_{e3} = $\sqrt{S_c/S_s}$ for symbolization involving great exaggeration of the area required on a compiled map, such as a settlement symbol with its associated name.

Constant C_f is called the *constant of symbolic form* and also takes three forms.

C_{f1} = 1.0 for symbols compiled without essential change.

C_{f2} = $(w_s/w_c)\sqrt{S_c/S_s}$ for linear symbols in which the widths of the lines on the source map (w_s) and the newly compiled map (w_c) are the important items in generalization.

C_{f3} = $(a_s/a_c)\sqrt{(S_c/S_s)^2}$ for area symbols in which the areas of the symbols on the source map (a_s) and the newly compiled map (a_c) are the important items in the generalization.

The basic equation, with or without modification by the introduction of the constants, is solely a statement of the number of items that can be expected on the newly compiled map. Although its primary value is theoretical rather than practical, in that it expresses in symbolic form the fundamental characteristics of simplification, it is useful in several important ways.

*F. Töpfer and W. Pilliwizer, "The Principles of Selection" (with introduction by D. H. Maling), *The Cartographic Journal* 3 (1966): 10–16. The original work was in German; the authors' original symbols have been changed here to reflect what they represent in English.

**When dividing one fraction by another, one inverts the divisor fraction and then multiplies. With map scales expressed as fractions (e.g., $\frac{1}{250{,}000}$), the division S_c/S_s is accomplished by dividing the denominator of the source map scale by the denominator of the compiled map scale.

THE ORKNEY ISLANDS

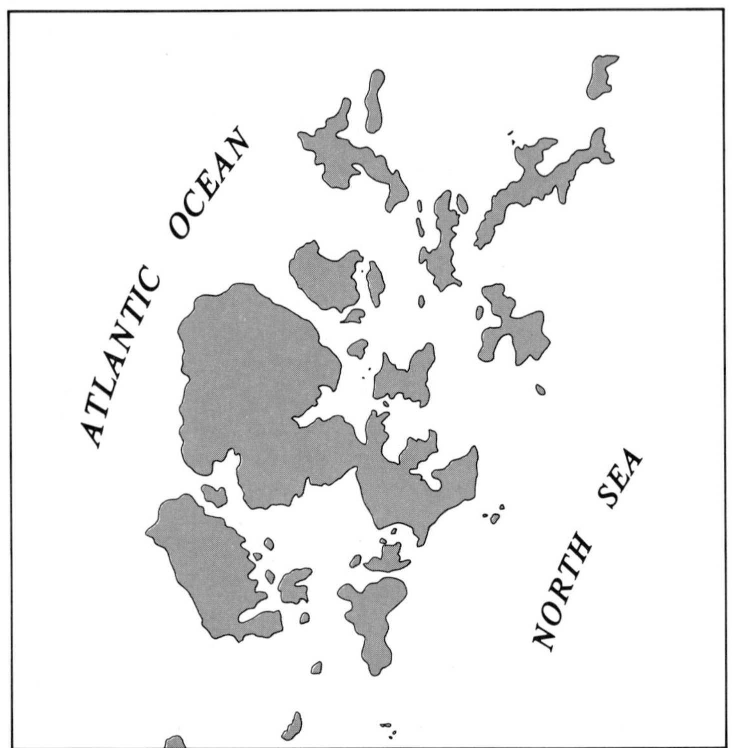

1:1,000,000

1:10,000,000

PART OF THE LAKE REGION OF WISCONSIN AND MICHIGAN

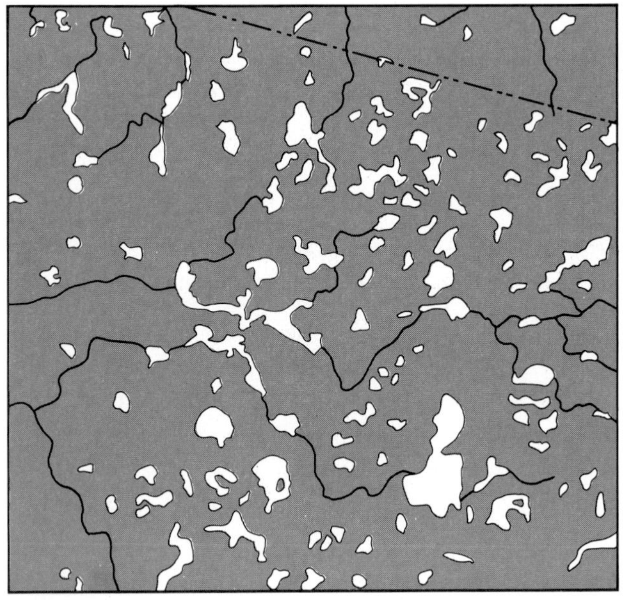

1:500,000

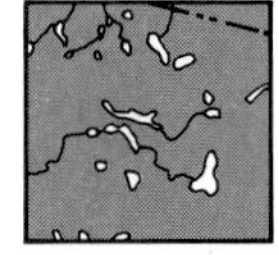

1:2,500,000

FIGURE 6.8 Reasonable delineations of two regions at two scales. Keep in mind that the larger scales are in themselves relatively small-scale, so that a great amount of simplification had already been introduced.

Straightforward applications of the principle of simplification are appropriate for: (1) point data sets (for example, towns on a road map), (2) linear data sets, such as roads or streams, (3) areal data sets that consist of numerous small similar items within a region, such as lakes or islands, and (4) including in computer algorithms to evaluate computer generalization schemes.

Although the law specifies, with high probability, how many items or how much detail can be retained when working from a larger to a smaller scale, it cannot, of course, specify which of the items should be included and which should be discarded. One cannot, for most smaller-scale maps, reduce the number of features within a class, such as rivers or cities, on purely objective grounds, such as size. Importance is a subjective quality; a simplification of cities on a map of the United States which included only those of more than 100,000 inhabitants would eliminate many in the western United States that are far more "important" in their regions than many of those that would have been included in the more populous eastern United States.

For many maps the question of which individual data element to retain and which to eliminate is a difficult task. The choice can be deduced from the purpose of the map and the place assigned the particular data distribution in the visual hierarchy planned for the map. This determination demands that the cartographer be knowledgeable about the data being mapped. For example, a recent study of the variations in the generalization of stream networks at different scales with computer assistance (which makes use of the radical law) suggests incorporating in the cartographic data base such criteria as discharge (mean flow, peak flow, and so on) and morphological (width, depth, profile, channel type, and so on) characteristics.* As more and more such maps are made, the ability to specify such criteria will become increasingly critical. The simplification problem of *which* specific data to retain remains one of the cartographer's foremost problems.

In the simplification operation the elimination of information regarding a feature or an area must be done in a way that maintains, as far as possible, its intrinsic geographical nature. Some things are unavoidable, however; the lengths of irregular lines (rivers, coasts) become shorter and the areas bounded by irregular lines (lakes, countries) become smaller as the lines themselves become simpler. Nevertheless, when the geographical character of a region or phenomenon is an important aspect of a map, then its distinctive characteristics should be portrayed to the degree the scale allows even if that requires exaggeration. For example, a highly meandering stream, such as the lower Mississippi River, shown at a small scale should have some meanders even though the loops may be far larger than reality at that scale. Similarly, a region studded with hundreds of lakes, all of which are too small to survive the reduction, should have some lakes; a winding road should be so shown; and a highly serrated coast should retain some of that character. Of course, when map scales are extremely small, such details cannot be shown.

When the objective demands it or the scale requires it, the simplification must become essentially diagrammatic. This means eliminating the elements in such a way that the simplification is obvious and the map user will be quite aware of it. Diagrammatic generalization is not a simple operation. It has some of the characteristics of a caricature in that it involves careful analysis of the basic form characteristics of outlines (continents, countries, lakes, islands, and so on) and the retention of those distinctive shapes in the delineation (Fig. 6.9). In many cases it is far better to show an area with clean, firm, smooth lines, which emphasize the main characteristics of its form, than to try to be highly "accurate." Rudolph Arnheim put it well:

> . . . *an image of reduced size is not obtained simply by leaving out details. Artists as well as cartogra-*

*V. Gardiner, "Stream Networks and Digital Cartography," *Cartographica 19* (1982): 38–44.

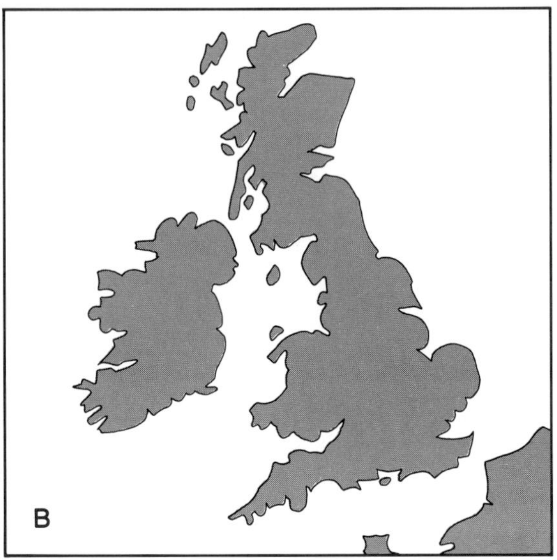

FIGURE 6.9 Two representations of Great Britain and Ireland at the scale of about 1:15,000,000. The one on the left (A) is basically simplified to fit the scale and is suitable for a reference map intended to give an impression of detailed precision. The one on the right (B) is a diagrammatic generalization suitable as a base on which to display some thematic data. Note that (B) captures basic shapes, which tend to be masked by the detail in (A).

*phers realize that they face the more positive task of creating a new pattern, which serves as an equivalent of the natural shape represented. Such a newly created pattern is by no means the same thing as a sandpapered copy of the original. The reduction of size provides the mapmaker with a degree of freedom, which he can use to make his visual images more readable and more pertinent.**

Classification. The same goal of expressing the salient character of a distribution is the objective of classification, but the methods employed are different. Instead of eliminating some of the data, classification modifies the data in its attempt to typify.

Classification is a standard intellectual process of generalization that seeks to group phenomena in order to bring relative simplicity out of the complexity of differences, or the unmanageable magnitudes of information. It is difficult to imagine any intellectual understanding, beyond the very elementary, that does not involve classification. We classify without thinking about it; we sort numerical data into averages, above and below average, extremes, and so on, or we sort all the varieties of phenomena into simple classes such as "roads," "rivers," or "coastlines."

Some of the more common processes of classification are: (1) the allocation of similar qualitative phenomena, such as land use or vegetation, into categories (cropland, forest) or quantitative data into numerically defined groups, and (2) the selection of location and the modification of the data element at that location to create a "typical" data element for portrayal on the map.

The manipulations that the cartographer performs in the process of classification are treated in Chapter 11. They include class interval selection and various agglomeration routines. One classification manipulation, called clustering, is

*R. Arnheim, "The Perception of Maps," *The American Cartographer 3* (1976): 5–10.

illustrated here, since it involves little statistical manipulation and clearly illustrates the distinction between simplification and classification.

Clustering is a cartographic necessity whenever numerous discrete items characterize a distribution and, at the reduced scale of the intended map, it would become undesirable or impossible to portray every individual item. Two options are available to the cartographer: one is a method of simplification; the other is a method of classification. Both seek to "typify" the distribution. Figure 6.10 illustrates the two options. In Figure 6.10B a predetermined number of the individual items are portrayed on the reduced-scale map, and the others are eliminated; this is simplification. In Figure 6.10C the items have been grouped and then an average or typical position is designated to represent each group; this is classification.

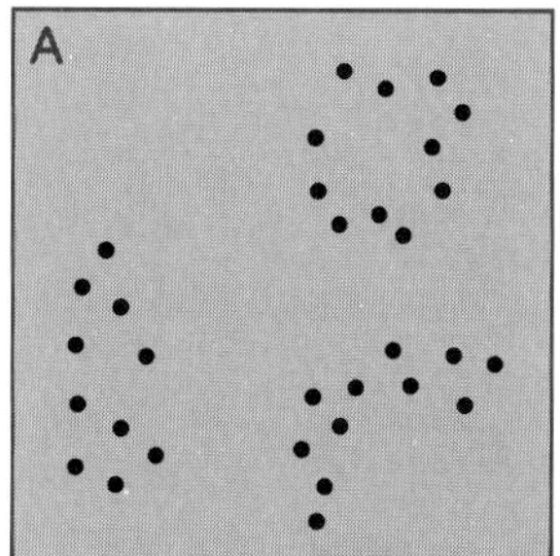

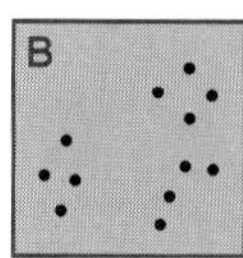

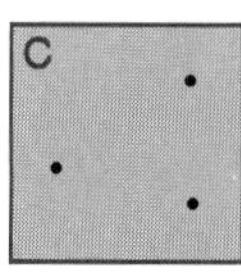

FIGURE 6.10 A given point distribution in (A), for example, thermal springs, is shown at a reduced scale in (B) and (C). In (A) each spring is shown. In (B) each of the three clusters is *simplified* by retaining only four of the springs; in (C) each cluster is *classified* by a single point symbol, since in (C) each dot typifies the cluster but is not placed where a spring is actually located.

Symbolization. After simplification and classification have been applied to the data selected for mapping, the cartographer turns to the process of symbolization. Symbolization is the assignment of various kinds of marks to the summarizations resulting from classification and to the essential characteristics, comparative significances, and relative positions resulting from simplification. This graphic coding makes the generalization visible; obviously this process is most critical to the success of any map. Judicious simplification and classification can be nullified by poor symbolization, while on the other hand, good symbolization can enhance their effectiveness. Unfortunately, good symbolization can also impart an unwarranted impression of precision and accuracy to poorly simplified or classified data.

All the marks on a map are symbols, from boundary lines to the dots representing cities and from the blue ocean to the hummocks showing marshlands; by using these marks, the cartographer is symbolizing a concept, a series of facts, or the character of a geographical distribution. The very fact of symbolization is generalization, sometimes to a rather small degree, as when we characterize the internal areal administrative hierarchy of townships and counties by a set of boundary symbols, or sometimes to a high degree, as when we symbolize an entire complex urban area with a single dot.

It is important to observe, furthermore, that the degree of this kind of generalization commonly varies widely within a map. Some symbols, such as the lines of the graticule, which indicate latitude and longitude, involve almost no generalization, while on the same map we may include a large, smooth arrow symbolizing the migration of population from one region to another, the arrow taking no note whatever of the many different routes taken by the peoples who moved.

In the total process of symbolization the cartographer must assign relative graphic prominence to the features included on the map. Ordinarily this aspect of symbolization is dealt with in the design of the graphic presentation (see Chapter 7), when we are concerned with levels of visual significance and the visual hierarchies of the communication. On the other hand, in practice, we cannot separate visual and intellectual elements so clearly, and generalization primarily by the processes of simplification and classification during compilation is made easier if some of the basic design decisions have been made before beginning the map. When making the initial selection of data, the cartographer should have at least a general idea of the intended visual hierarchy.

Induction. Induction or inductive generalization is the process of extending the information content of a map beyond the selected data by means of logical geographical inferences. For example, if we have average January temperatures for a series of stations, we can, by suitable "logical contouring," construct a set of isotherms. The set of isotherms conveys far more information and has resulted from making logical inferences about the probable occurrence of average January temperatures in the areas between the data points. Any logical extension of data founded on accepted associations is inductive, such as the mapping of a classification system (for example, climates or soils), where data are actually available only for a relatively few points.

The induction that can result from a mapmaking effort may extend beyond that added by the cartographer. A map user, with personal knowledge, may further amplify the map information already arrived at inductively. Through induction, map users often uncover hypotheses that cannot be generated in any other manner. This is one of the primary utilities of a map.

The Controls of Generalization

The elements of generalization are operations performed by the cartographer. The controls of generalization are the factors that influence how each process is performed. It is impossible for this book to do more than suggest the operations of the controls, because there is no limit to the number of possible combinations of objectives, scales, graphic limitations, and qualities of data. Nevertheless, a few observations can be made to suggest their functioning.

Objective. The reason for making a map clearly has a great deal to do with how it should be designed. The kind of audience to which it is aimed—geographically sophisticated or ignorant, children or adults, and so on—is one basic factor. How it is to be used is another. Is it to be studied with no time limit, as might be the case with a topographic map or a general atlas map, or is it to be used as an illustration to be shown briefly on a screen during a slide presentation? Is it primarily to provide a great deal of general geographical information, or is it to display the structure of a particular distribution? Before beginning, the cartographer must answer these kinds of questions; otherwise, there will be no basis for making the many decisions that are involved in the generalizing process.

Scale. Obviously the scale of the finished map-to-be is a fundamental factor in the kind and degree of generalization that will be employed. As a general rule the smaller the scale, the greater will be the degree of generalization required. At large scales, such as in plans and topographic maps, most of the generalization is classification and symbolization, since scale is not a serious constraint.* In thematic mapping, the situation is quite different. When the cartographer has complete freedom to select the scale of a thematic map, that decision may be greatly influenced by the generalization needed to fit the objective.

Probably the most important task of the cartographer when generalizing the various categories of geographical data within the context of scale is to keep the various degrees of generalization reasonably "in tune" with the objective and with the chosen scale. At each scale there is a range of generalization that will "fit" the scale, that is, the treatment will be neither too detailed nor too general: the difficulty comes in the attempt to maintain a balance among and consistency within categories such as coasts, boundaries, or roads. The combinations of scale, data, and objective are legion; in practice, it is nec-

*J. S. Keates, "Symbols and Meaning in Topographic Maps," *International Yearbook of Cartography* 12 (1972): 168–81.

essary to leave such matters to "experience" and "good judgment."

The cartographer should remember that when working at very small scales, there may be a great scale range within the map. For example, on the conventional Mercator's projection the values of S (see Chapter 5) given in Table 6.6 show the relative exaggeration of areas at the given latitudes. This means that at latitude 60° on Mercator's projection, there is four times as much map space available to represent the same size phenomena as there is at the equator. There is a natural tendency to use all the space available.* Consequently, in maps on Mercator's projection, there is usually far more simplification in the equatorial regions than in the high latitudes. Whether one *should* be consistent in the degree of simplification and other factors or not depends on the objective of the map. Whether one *can* be consistent is an equally pertinent question with respect to computer-assisted simplification and symbolization: Algorithms to enable computer-assisted simplification rarely compensate for the areal distortion or the angular compression that may result from projection.

TABLE 6.6 Values of S in Mercator's Projection

Latitude	S
0°	1.0
20°	1.1
40°	1.7
50°	2.4
60°	4.0

Graphic Limits. To distinguish among the symbols that cartographers use, we employ one or a combination of variations of the primary graphic elements: hue, value (tone), size, shape, spacing, orientation, and location (see Chapter 7). Our abilities to use these elements are subject to three types of limitations—physical, physiological, and psychological. These limits act as important controls on the entire generalization process. The physical limits are imposed on the graphic elements by the equipment, materials, and skill available to the mapmaker. The physiological and psychological limits are imposed by the map user's perceptions and reactions to the graphic elements and their modulations. All types of limitations are extremely important in controlling the amount and degree of generalization that a cartographer can successfully use.

The physical limits include factors such as the overall maximum format size that can be used, the available line width sizes, lettering styles and sizes, color screens, preprinted symbols, dimensionally stable film or plastics, specialized symbol templates and the individual's or machine's ability to use these factors. To date, in computer-assisted cartographic production the graphic limits in terms of the availability and variety of materials and options are still generally, on the average, more restrictive than those in manual cartography. Nevertheless, there are conspicuous exceptions, and the consistency obtained by computer assistance generally exceeds that of manual rendering. In some areas computers have extended the limits of accuracy and precision beyond what is humanly possible. For example, the repeatability and consistency of computer-assisted products can be measured in thousandths of a millimeter, a capability that manual production cannot attain. It is safe to say that in the areas of accuracy and precision, computer-assisted production has reduced the graphic limitations to such a degree that the physiological and psychological aspects are predominant, provided, of course, that the map reader is a human and not a machine!

The psychological aspects of graphic limits are a function of the perceptions of the map-reading audience, and these have been shown to vary consistently from one symbol type to another. For example, a line twice as wide as another will

*Perhaps the operation in cartography of Parkinson's Law that work expands so as to fill the time available for its completion!

usually look that way, but a circle with twice the area of another circle will look significantly less than twice as large. The physiological aspects refer to such things as our ability to distinguish among hues, sizes of type, or tones of gray. Our ability to judge differences in tonal values is considerably less than our abilities to produce them. These kinds of limits obviously affect the processes or elements of generalization. In many cases such limits specify the number of classes we may employ, or the minimum distance between points that a computer-assisted simplification algorithm can reasonably use.

Quality of Data. It is readily apparent that the more reliable and precise the data, the more detail is potentially available for presentation. In cartographic generalization the converse is unusually important; that is, the cartographer must sometimes go to considerable pains not to let the map give an impression of an accuracy greater than the source material warrants.

Just as basic as the quality of the data is to proper generalization is the scholarly competence and intellectual honesty of the cartographer. Intellectual honesty is particularly important in cartography, because a well-designed and well-produced map has about it an authoritative appearance of truth and exactness. The cartographer must, therefore, take unusual pains to ensure that the data are correct and that their presentation on the map does not convey a greater impression of completeness and reliability than is warranted. For example, there is no theoretical limit to the number of isolines which can be interpolated from a set of point-data, such as temperatures or densities of population; yet if the set is small and widely spaced, a large number of isolines would give an impression of detail and accuracy quite out of line with the quality of the data.

One of the most difficult tasks for the cartographer is to convey to the map reader a clear indication of the quality of the data employed in the map. When writing or speaking, words such as "almost," "nearly," and "approximately" can be included to indicate the desired degree of precision of the subject matter. It is not easy to do this with map data.

There are several ways in which the cartographer may proceed. One is to include in the legend, when appropriate, a statement concerning the accuracy of any item. Another and more common method, on larger-scale maps, is to include a reliability diagram, which shows the relative accuracy of various parts of the map. It is also good practice to include in the legend, if warranted, terms such as "position approximate," "generalized roads," or "selected railroads," so that an idea of the completeness and accuracy may be given to the reader.

Computer-assisted cartography in some ways compounds the problem of accuracy and completeness. As the number of machine-usable cartographic data bases increases, the chances for misuse also increase. Any data-base creator must include information indicating its quality to aid the cartographer in avoiding misuse. When using existing map sources for data, the cartographer abides by the rule "Always compile from larger scale to smaller scale." When a machine-readable data base is used, unless information regarding the input scale has been included, the cartographer has no way of knowing whether this rule was followed or whether a previous generalization was enlarged.

SELECTED REFERENCES

Blalock, H. M., *Social Statistics,* 2d ed., rev. New York: McGraw-Hill, 1979.

Chang, Kang-tsung, "Measurement Scales in Cartography," *The American Cartographer* 5 (1978): 57–64.

Court, A., "The Inter-Neighbor Interval," *Yearbook of the Association of Pacific Coast Geographers* 28 (1966): 180–82.

Davis, J. C., and M. J. McCullagh (eds.), *Display and Analysis of Spatial Data,* NATO Advanced

Study Institute. New York: John Wiley & Sons, 1975.

Keates, J. S., "Symbols and Meaning in Topographic Maps," *International Yearbook of Cartography* 12 (1972): 168–81.

Lewis, P., *Maps and Statistics* New York: John Wiley & Sons, 1977.

Muehrcke, P. C., "Concepts of Scaling from the Map Reader's Point of View," *The American Cartographer* 3 (1976): 123–41.

Swiss Society of Cartography, *Cartographic Generalization: Topographic Maps,* Cartographic Publication Series No. 2. N.p., 1977.

Taylor, P. J., *Quantitative Methods in Geography: An Introduction to Spatial Analysis* Boston: Houghton Mifflin Company, 1977.

Unwin, D. J., *Introductory Spatial Analysis* New York and London: Methuen, 1981.

Wright, J. K., " 'Crossbreeding' Geographical Quantities," *Geographical Review* 45 (1955): 52–65.

Wright, J. K., "Map Makers are Human: Comments on the Subjective in Maps," *Geographical Review* 32 (1943): 527–44.

7

Graphic Presentation and Design

We obtain information in a variety of ways, but for most of us probably by far the largest share of our knowledge of wide-area geographical relationships results from looking at prepared maps. Some of these are more or less mechanical images, such as air photographs and satellite views, while others are created by arranging marks to form a visual representation of selected spatial phenomena. In order to display the data, we employ an almost unlimited variety of signs. By relating identifiable graphic characteristics of the marks to chosen attributes of the data, we assign qualitative and quantitative meanings to the signs, and they then become designated symbols. By arranging the symbols in the horizontal plane, we endow them with geographical meaning, and the display becomes a map.

In any graphic system of communication the various signs obviously must be distinguishable, just as the letters of the alphabet used in a language must appear different, so that we do not get mixed up as to the sounds they represent. Furthermore, by the systematic use of graphic similarities and differences among the signs, we can express likenesses and distinctions among the data they symbolize. The various signs must be carefully chosen and arranged, so that they form an effective presentation, and this calls for careful attention to the principles of graphic communication. We shall begin the consideration of graphic presentation by surveying the design process.

Graphic Design

Graphic design is a vital part of cartography because effective communication requires that the various signs (lines, tones, colors, lettering, and so on) be carefully modulated and fitted together. Just as an author—a literary designer—must employ words with due regard for many important structural elements of the written language, such as grammar, syntax, and spelling, in order to produce a first-class written communication, in parallel fashion the cartographer—a map designer—must pay attention to the basic principles of graphic communication.

Maps are made with the fundamental objective of conveying geographical information, and the processes of data gathering, symbolization, choice of scale, and projection are all focused to that end. Although the substantive aspects of preparing a map can be quite complex, the designing of it as a graphic display is equally important and complicated. The manner of presentation of the many map components so that together they appear as an integrated whole, devised systematically to fit the objectives, includes a variety of elements. Regardless of the positional accuracy or essential appropriateness of the data, if the map has not been carefully designed it will be a poor map.

In this chapter we will survey the principles of graphic communication and some of the fundamental elements of map design. Three important components of cartographic design, color, pattern, and typography, are unusually complex and require individual treatment in the following two chapters.

The Design Process

The activities and procedures involved in arriving at design decisions in cartography are extremely varied. Maps are made by large organizations, such as a federal agency or a large company, and they are also prepared in small shops or by individuals. Design decisions in large organizations usually follow elaborate procedures, which attempt to ensure that all relevant aspects are considered; user surveys are often undertaken; prototypes may then be prepared and subjected to criticism; and proposals move from one department to another with the ultimate decisions being made at the highest administrative level. Changes do not come easily in such a system.* At the other end of the gamut is the individual cartographer who must make most of the design decisions alone and who is not greatly constrained by previous output.

---

*Clarence R. Gilman, "Map Design at USGS: A Memoir," *The American Cartographer 10,* no. 1 (1983): 31–49.

Automated and computer-assisted cartography has had a great impact on mapmaking but mostly in matters of accuracy, speed, cost, consistency, freedom from tedious operations, and flexibility. In both automated and computer-assisted cartography several aspects of graphic design have been affected, some positively and some negatively. On the positive side, the flexibility of procedures and the ease of making changes have made it relatively easy to develop prototypes and to try design options before becoming locked into a specific plan. In strictly manual cartography, it is likely to become too expensive and time consuming to make a change after one gets very far in the construction process. On the negative side, until now the capability of a large proportion of the computer-driven plotters and printers to modulate the signs has been rather limited. Because it has been difficult to do much graphic organization, such maps have been quite plain with a minimum of associated geographical information. Rapid developments are taking place, however, and much greater options can be expected.

Regardless of the system, the normal objective in map design is to evoke in the minds of viewers the desired image of the spatial environment appropriate to the intended purpose of the map, whether it be the composite details of a general map as in topographic and atlas maps, or the structural character of a distribution, as in a thematic map. Furthermore, the necessary critical information must be displayed if the map or chart is to be used for specific operations, such as navigation by automobile or water craft.

Because graphic structures involving spatial data are often relatively complex, any one design problem will likely have a considerable number of possible solutions. In this and the next two chapters we are concerned with the kinds of choices that must be made and the bases for the various principles and precepts that influence the decisions. Whether the ultimate choices are made by collective institutional activity or by individual judgment makes no difference. It is well to keep in mind, however, that familiarity and tradition are powerful forces, which are not lightly disregarded, especially by large organizations.

Decisions about graphic design in cartography are not easy. Most choices are compromises, since it is normal for the "intellectual" and the "visual" objectives to be in conflict. If, for example, we wish to develop a strong graphic contrast at a coastline, this means dark on one side and light on the other. If we wish to indicate greater depth of ocean by deeper shades, then the shallow water is light. A shoreline contrast then requires dark land, but that means type and symbols on the land will not be very visible. One can only proceed by making choices after weighing all the pros and cons.

Whatever the system, the designing process logically can be thought of as consisting of three successive stages or operations. Imagination is the primary element in the first stage of the designing process. One scans the various possibilities and considers all the ways that the problem can be approached and tries to visualize the different solutions. The result is a general idea of the approach, which involves making decisions such as the relation of the map to others, the format (its size and shape), the basic layout, the selection of the data to be incorporated, the graphic organization of the data to be displayed, and so on.

The second stage involves the development of a specific graphic plan. Here the various alternatives are weighed within the limits of the general plan. At this stage decisions are made regarding particular kinds of symbolism, color use, typographical relationships, general line weights, and the like, in terms of how they will fit together graphically. By the time the second stage is completed, all but minor decisions have been made.

The third stage is that of preparing the detailed specifications for the construction of the map, whether by automated or manual methods—all symbols, line weights, tones, colors, lettering sizes, and so on. Nothing is left unplanned. It is obvious that preparing detailed specifications requires a thorough understanding of all the processes involved in map construction.

Cartography and Creativity

Cartography does not qualify as an aesthetic art form like painting, music, fiction, or dance. The functionalism of cartography together with the limitations imposed by geographical reality put too many constraints on the cartographer to allow "full freedom of expression." On the other hand, the variety of graphic media and design possibilities, as well as the significant role of craftsmanship throughout the history of cartography, show that mapmaking indeed merits the characterization of being a mixture of art, science, and technology.

The preparation of a map is not a mechanical process like taking a photograph. Instead, it involves assembling, processing, and generalizing diverse data and then symbolically displaying them as a meaningful, functional portrayal. That is a highly creative operation, and an important part of it is the development of the graphic design. It is a complex task, since there is an almost unlimited number of options for organizing the visual character of the display, and the choices involve a combination of intuition and rational choice. Since the ultimate aim is excellence in communication, some combinations will come closer than others to meeting that criterion, but some inner subjective sense is not a requirement for acceptable design. Like the study of literary composition, the basic elements of graphic composition lend themselves to systematic analysis, and the principles can be learned. This is not meant to imply that every individual can become highly creative merely through study; we are not all equally endowed with creative intuition. As has been often repeated, one can be taught a craft (the mastery of the materials and techniques with which one works), but one cannot be taught the art.

A basic requirement in graphic design is a willingness to think in *visual* terms, uninhibited by prejudices resulting from previous experience. The range of imaginative innovation must, of course, be disciplined to some extent, since cartography, like many fields, has developed traditions and conventions; to disregard them completely would inconvenience the user of the maps, which would in itself be poor design.

Three propositions, which have been suggested by various students of aesthetics and visual expression, may be advanced as guides to the student of functional map design:

1. Beauty or elegance may occur in functional graphic design, but it will be a consequence of good design in a favorable context.
2. Something that is well designed will not look so; in other words, the design itself should not appear contrived.
3. Simplicity, in the sense of being clear and uncomplicated, is highly desirable and is a result of excellence. Because simplicity is relative in a context, it cannot be defined, but it can be recognized.

Objectives of Graphic Map Design

There are a great many ways to symbolize (that is, to encode) geographical data, concepts, and relationships, but assigning specific meaning to the various kinds of distinctive marks, to their variabilities, and to their combinations is only the first of two steps in cartographic design. The second step is to arrange the marks in a total composition that will make the viewer see the result the way the cartographer intended. The two aspects are not really separable, of course, since one must select the symbolism with the ultimate objective in mind.

The communication objectives of cartography range along a continuum from general maps, such as atlas or topographic maps, to thematic maps, like those of population or rainfall. An individual map usually involves some of each aim, but it is worthwhile to contrast the two as they affect map design. A "pure" general map, at one end of the continuum, tries to display a variety of kinds of geographical information, and in theory at least, neither any class of selected phenomena nor any place is more "important" than any other. A "pure" thematic map, at the other end of the continuum, is primarily concerned with portraying the overall form or structure of a given spatial distribution. It is the structural relationship of each part to the whole that is important. It is a kind of graphic essay dealing with the spatial variations and interrelationships of some geographical distribution. The fundamental design objectives (and problems) of general maps and thematic maps are therefore quite different.

In a general map each item or attribute to be mapped must be encoded in a unique fashion, so that the meaning will not be confused with some other; therefore, the graphic character of each symbol must be distinct. Unless emphasis is specifically desired, no mark should appear more important than another, and no group or class of marks should dominate. Obviously, that characterization refers to a "pure" general map, but such a map will be rare, because usually some geographic characteristics are thought to be more "important" than others and will consequently be given visual emphasis. In any case, individual legibility and graphic contrast among the symbols are primary aims.

In a thematic map, on the other hand, the possible modulations of marks and the chosen symbolic system must be made to work together graphically to evoke the overall form of a distribution. Although spatial images and graphic integration are primary aims, an adequate locational "geographical matrix" must also be provided.

It bears repeating: Although the fundamental objectives of general and thematic cartography are opposite and lead to distinct graphic treatments, in practice most maps actually combine the objectives to some degree. The cartographer must recognize the combination and proceed to design accordingly.

Classes of Symbols

There is obviously an unlimited variety of spatial data that can be mapped, all of which must be represented by symbols. In order to consider the ways in which signs can be employed to symbolize the variety of data, it is helpful to classify them. We can recognize three classes of symbols: *point, line,* and *area.*

1. *Point symbols.* Point symbols are individual signs, such as dots, triangles, and so on, used to represent place or positional data, such as a city, a spot height, the centroid of some distribution, or a conceptual volume at a place, such as the population of a city. Even though the mark may cover some map space, when conceptually it refers to a location, it is a point symbol.
2. *Line symbols.* Line symbols are individual linear signs used to represent a variety of geographical data. Just because a line symbol is used does not mean that the class of data being represented is linear; for example, contours are lines used to represent elevations and depths (point data) from which volumes may be determined.
3. *Area symbols.* Area symbols are some sort of marking extending throughout a map area to indicate that that region has some common attribute, such as water, administrative jurisdiction, or some measurable characteristic. When used this way, an area symbol is graphically uniform over the entire area it represents.

   Another kind of area symbol consists of nonuniform markings, such as tonal variations like those in a snapshot, to portray continuous ordinal variation of some phenomenon from place to place. This sort of areal symbolism, called shading, is quite complex and can be used to present both tangible things, such as the land surface, or far more abstract variations, such as ratios.

Figure 7.1 shows a few examples of the great variety of point, line, and area symbols used to portray some kinds of nominal, ordinal, and interval-ratio data. The three classes of symbols, along with lettering, comprise the basic building blocks of cartographic presentation; but in order

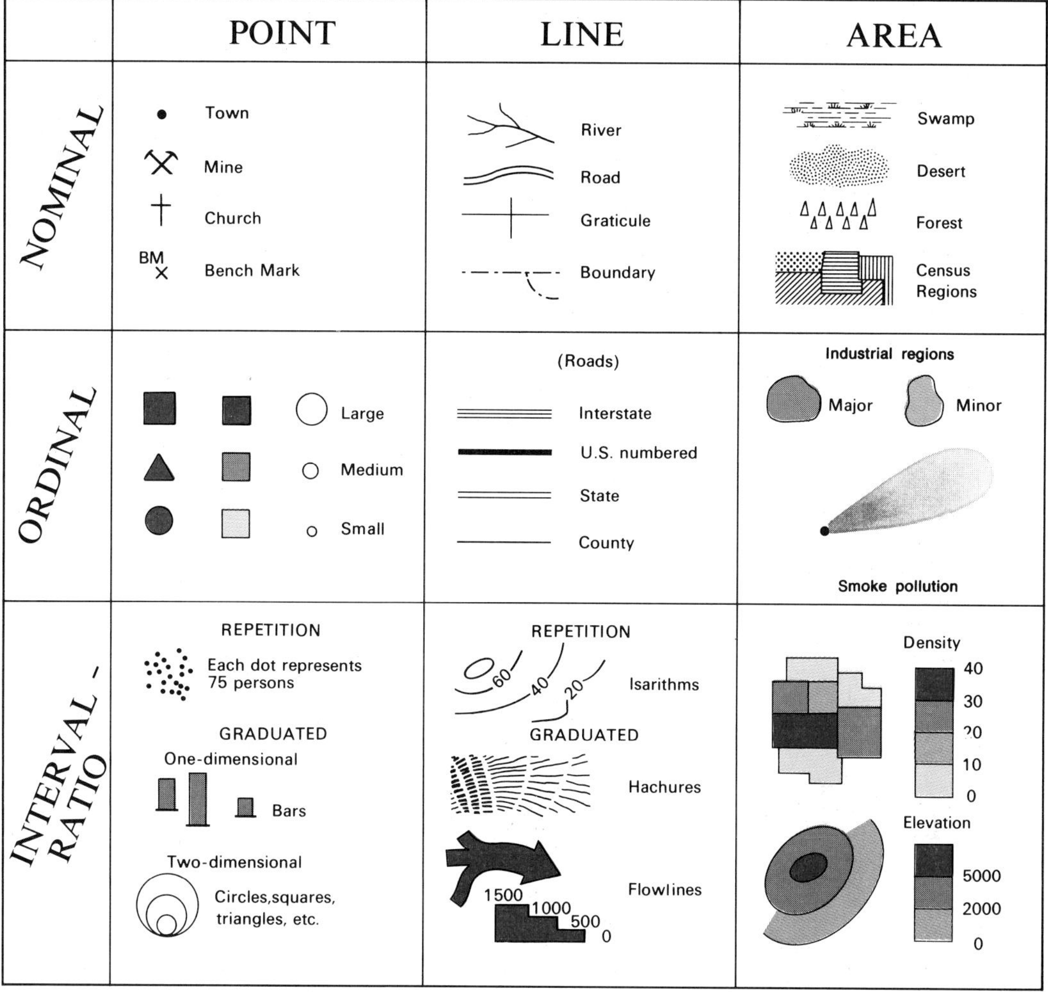

FIGURE 7.1 Some examples of the three classes of symbols (point, line, area) and how they might be used for a few of the kinds of nominal, ordinal, and interval-ratio data.

to endow the marks with meaning, to make them appear similar or different from one another, and to make them appear more or less prominent, their appearances must be modulated by making appropriate use of what we can call the primary graphic elements.

Primary Graphic Elements

Positional distinction among data is the primary purpose of a map, but if that were all a map could reveal, there would be little communication. In order to represent the different data in meaningful fashion, we must vary the appearance of the symbols, and we do so by systematically adjusting their graphic quality. Marks on the map plane may be made to appear more or less distinctive and prominent in an infinite number of ways, but when we reduce all the variations and combinations to their elemental characteristics, we find that differences in appearance result from making changes in *hue, value, size, shape, spacing, or-*

ientation, and *location*. These are the primary graphic elements and they, together with the classes of symbols (and typography), comprise the essential ingredients of all graphics. Just as there is an almost unlimited number of ways of combining the sounds of speech for audible communication, there are almost no bounds to the ways we can combine the graphic elements for visual communication. Their distinctive characters, as applied to point, line, and area symbols, are given in Figure 7.2. The primary graphic elements are:

1. *Hue*. Hue is a very important and complex visual perception, which will be discussed at greater length in the next chapter. Here suffice it to point out that the common use of the term "color" really refers to hue. When we say things are different colors, we are usually describing characteristics such as blue, green, red, and so on, most of which cannot be printed on these pages. Consequently, Figure 7.2 can only suggest the hue differences of red, blue, green, and yellow.

2. *Value*. As a graphic quality, value refers to the relative lightness or darkness of a mark, whether of black or any other hue. A surface that reflects a noticeable amount of light is said to have a *tone*. Perceptually, a given tonal surface (as physically measured) may appear different under varied viewing circumstances; consequently, when referring to the sensation of tone, it is better to use the term value, which refers to the perceptual scale of tones. In the perceptual scale of values light is referred to as high value and dark as low value. When maps are printed we tend to associate "more" with dark and "less" with light, but that may be reversed on a video screen.

3. *Size*. Marks vary in size when they have different apparent dimensions—diameter, area, width, height. Usually the larger a sign, the more important it is thought to be.

4. *Shape*. Shape is the graphic characteristic provided by the distinctive appearance of (1) a regular form, such as a circle or triangle; (2) the outline of an irregular area, such as a state or island; or (3) the contour of a linear feature, such as a river or coast.

5. *Spacing*. When a sign is made up of an arrangement of component marks, such as a series of dots or lines, their spacing may be varied. A fine spacing is one in which the marks are close together; it contrasts with a coarse spacing. Regular arrangements are often described as being composed of so many lines per inch (lpi) in common English use or lines per centimeter (l/cm) in the International System.

6. *Orientation*. Orientation refers to the directional arrangement of an elongated individual mark or the parallel lines of marks as they are positioned with respect to some frame of reference. Such a frame may be the map border, graticule, or the circumference of a circle.

7. *Location*. Location in the visual field, that is, on the map plane, is generally applicable only to those components that can be moved about, such as titles, legends, or some of the typography. The locations of most of the symbols are prescribed by the geographical ordering of the data and are only susceptible of such changes as can be introduced by using different projections or shifting the map frame.

The primary graphic elements are not easy to define with words, since words are a foreign language to graphic communication. Furthermore, the words that we must use do not have specific meanings out of context. It is worth reiterating that here we are using the terms for the graphic elements in the restricted sense of how graphic

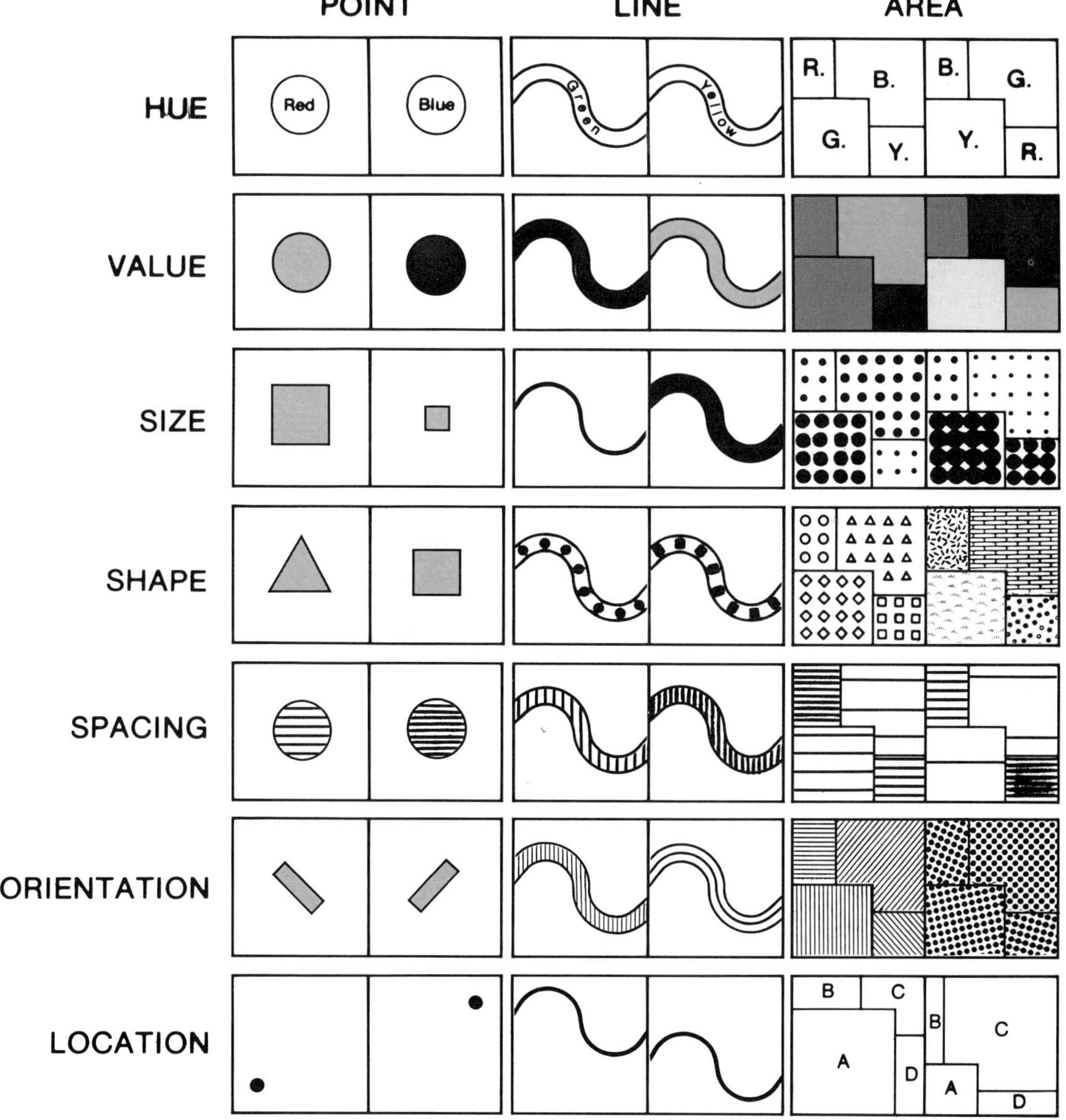

FIGURE 7.2 The primary graphic elements with some examples of their simple application to the classes of symbols.

marks may be made to appear different from one another.*

Although the foregoing list includes the primary intrinsic ways that marks may be varied, there are others that may assume importance in some instances. Examples are *focal quality,* which is the attribute of being sharply distinct or, conversely, fuzzy; *chroma,* the relative hue saturation with value held constant; and *position* in a series, which may give distinctiveness to a mark relative to other marks. More complex are those that combine variations in the graphic elements. For example, differences in the sizes and spacing of the marks comprising a pattern, such as an array of dots or crossed lines, results in what is called a difference in *texture;* and variation in the hue-value-chroma relationship results in a change in *intensity.*

Because cartography is a visual medium, it involves various principles of graphic communication that are critical to good visual composition, just as things like syntax, paragraphing, organization, phrasing, and so on are indispensible to good literary composition. But while the general objectives of communication by means of language and by means of graphic presentation may be similar—clear, unambiguous transfer of information—there are many differences between the two media. Among the more important of these differences is the fact that the constituents of a graphic display are visual stimuli and that people react to graphic presentation in quite different ways than they do to written and spoken communication.

Perception of Graphic Complexes

Through long familiarity, the sounds of language and the written words that encode the sounds tend to be "transparent" in that their meanings become known to us without our paying much attention to the actual sounds or the physical appearance of words. This is not true with graphic presentation. The marks are "opaque"; we pay a great deal of attention to the way they look and to their arrangement.

A person reading or being spoken to receives the information in serial fashion, that is, the words follow one another in sequence. The reader or listener is thus programmed to receive the thoughts in a definite order. Graphic commmunication is quite different in that a person receives the visual impressions synoptically (all at once) instead of in a sequence. Our perception of every mark on a map is simultaneously affected by its location and appearance relative to all the other marks. This means that instead of structuring our communication sequentially, we must always deal with it as a whole. Everything on a map is related visually to everything else; if we change one thing, we change everything.

When people view a display of any kind, they process it perceptually. Just as when people hear a series of words or other sounds, there is an automatic attempt to "make sense out of it," so do they—unconsciously—attempt to organize what they see in a visual field. Perceptually this involves assigning visual meaning and importance to the various marks and to the shapes, sizes, orientations, values, hues, and so on. Because we tend to reject visual monotony and ambiguity, we will "organize" the display so that it

*The term "visual variable" was introduced in 1967 by Jacques Bertin in *Sémiologie graphique,* 2d ed. (Paris: Gautier-Villars, and Paris-La Haye: Mouton & Cie, 1973). The essentials of that work were condensed in Jacques Bertin, *La graphique et le traitement graphique de l'information* (Paris: Flammarion, 1977), and published in English as *Graphics and Graphic Information Processing,* translated by William J. Berg and Paul Scott (Berlin and New York: Walter de Gruyter & Co., 1981). The second edition of the larger work has been published in English: Jacques Bertin, *The Semiology of Graphics,* translated by William J. Berg (Madison, Wisc.: University of Wisconsin Press, 1983).

Although there is considerable similarity among the entries in the lists of primary graphic elements herein and Bertin's list of visual variables, they are not all the same. For example, instead of "location" (one variable in cartography), Bertin treats the two dimensions of the plane (*x,y*) as individual variables (important for processing diagramed statistical information); and a systematic variation of both size and spacing comprise Bertin's visual variable *grain* (Fr.), which has been translated in English as "texture."

"makes sense" visually. It is inevitable, therefore, that viewers of graphic displays will see them structurally: some marks will seem more important than others, some shapes will "stand out," some things will appear crowded, some colors will dominate, and so on. If these visual relationships among the graphic stimuli coincide with the cartographer's intentions, the basis for effective communication has been established.

A useful way of analyzing graphic design is to think of it as consisting of a process in which a set of signs (symbols) is modulated (graphic elements) in order to attain desirable qualities, such as clarity and legibility, in the map. These qualities we can call the *components* of graphic design. The factors like objective and scale, which affect the treatment of the components, we can call the *controls* of graphic design.

COMPONENTS OF GRAPHIC MAP DESIGN

The graphic constituents of design are those attributes of the marks used for representation that either by themselves or in organized array are visually significant to the total graphic presentation. Only the most important can be considered in this book. They are clarity and legibility, visual contrast, figure-ground, balance, hierarchical structure, color and pattern, and typography. The important components, color, pattern, and typography, are relatively complex and will be treated separately in the next two chapters.

Although space allows only a summary treatment of each of the graphic components, it should be kept in mind that the abbreviated attention that can be paid to them here is not a proper measure of their importance. It is worth observing that cartography involves the combination of many ingredients, and that it is simply not possible to provide instructions "from the ground up" in all the topics that are fundamental to it, whether they be mathematics, geography, or graphic composition. Nevertheless, the basic principles involved can be treated.

Clarity and Legibility

The transmission of information by means of the coding built into signs of various sorts (lines, letters, tones, and the like) requires that they be clear and legible. Although the various graphic elements treated in this chapter have other functions to perform in a graphic composition, one of our basic aims is to see that they are clear and legible. These qualities can be obtained by the proper choice of lines, shapes, and colors and by their precise and correct delineation. Lines must be clean, sharp, and uniform; colors, patterns, and shading must be easily distinguishable and properly registered (fitted to one another); and the shapes and other characteristics of the various symbols must not be confusing.

One very important aspect of legibility is size. No matter how nicely a line or symbol may be produced, if it is too small to be seen, it is useless. There is a lower size limit below which an unfamiliar shape or point symbol cannot be identified. Although there is some disagreement as to the exact measure of this threshold, a practical application sets this limit as being a size that subtends an angle of about 1 min. at the eye. That is, no matter how far away the object may be, it must be at least that size to be identifiable. This limit sets an absolute minimum, since it assumes perfect vision and perfect conditions of viewing. Because of the unreasonableness of these assumptions, it is wise for the cartographer to establish the minimum size somewhat higher, and it may be assumed that 2 min. is more likely to be a realistic measure for average (not "normal") vision and average viewing conditions. Table 7.1 (based on 2 min.) is useful in setting bottom values of visibility.

It should be remembered that some map symbols have length as well as width; in such cases (for example, with lines) the width may be reduced considerably, since the length will enhance the visibility. Similarly, other characteristics, such as contrasting colors or shapes, may increase visibility and legibility, but even though the existence of a point symbol on the map is made visible by such devices, if it does not stand

TABLE 7.1 Approximate Minimum Sizes for Legibility of Point Symbols

International		U.S. Customary	
Viewing Distance	Size (width)	Viewing Distance	Size (width)
50 cm	0.3 mm	18 in.	0.01 in.
2 m	1.15 mm	5 ft.	0.03 in.
5 m	2.9 mm	10 ft.	0.07 in.
10 m	5.8 mm	20 ft.	0.14 in.
15 m	8.7 mm	40 ft.	0.28 in.
20 m	11.6 mm	60 ft.	0.42 in.
25 m	14.5 mm	80 ft.	0.56 in.
30 m	17.4 mm	100 ft.	0.70 in.

at or above the sizes given in Table 7.1 it is not likely to be legible. In other words, it might be seen (visible), but it might not be read or recognized (legible).

A second quality affecting legibility is also operative in cartography. Generally, it is easier to recognize something we are familiar with than something that is new to us. Thus, for example, we may see a name in a particular place on a map and, although it may be much too small to read, we can tell from its position and the general shape of the whole word what it is.

Visual Contrast

The fact that the symbols of a map are large enough to be seen does not in itself provide clarity and legibility. An additional graphic component, that of visual contrast, is necessary. No element is as important as contrast. Contrast is the basis of seeing and, assuming that each item on a map is large enough to be seen, the manner in which a sign differs from its background and the adjacent signs affects its visibility.*

Visibility is indispensable, of course, but it must not be assumed that maximum contrast is automatically desirable. In some cases, separate components of a map differ only to a limited degree, and it is desired that the symbols indicate this. As will be pointed out later, many degrees of contrast can be incorporated in hierarchic structures. Contrast is achieved by modulating the graphic elements. There are a few general principles of contrast that can be suggested here.

An observer resents visual ambiguity, and the differentiation among symbols should always be sufficient so that there is no doubt about it. Furthermore, the critical eye seems to accept moderate and weak graphic distinctions passively and without enthusiasm, so to speak, whereas it relishes greater contrasts. The desirable quality of a display, variously termed "crisp," "clean," and "sharp," seems to be obtained in large degree by the amount of contrast it contains. Figure 7.3, a simple matrix of line widths, illustrates one aspect of this. The most "interesting" sections are those where there is considerable contrast among the widths. On the other hand, Figure 7.3 also

FIGURE 7.3 Size contrast of lines. Uniformity produces unpleasant monotony. The general areas outlined in red appear to be more interesting or "crisp" than areas where there is less contrast.

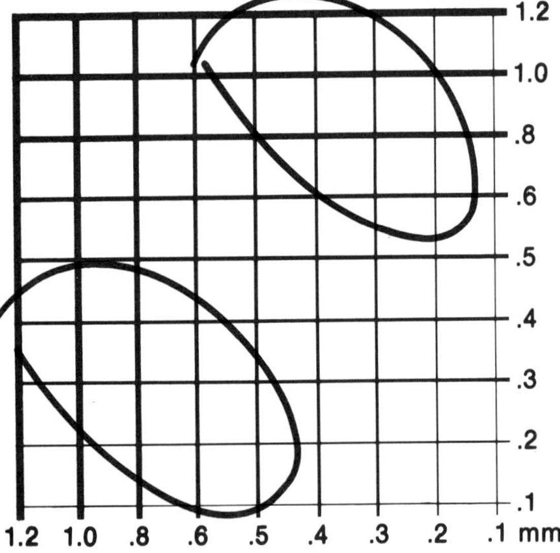

*A study of various factors affecting contrast is Michael W. Dobson, "Visual Information Processing During Cartographic Communication," *The Cartographic Journal* 16, no. 1 (1979): 14–20.

illustrates that too much contrast can be unpleasant.

The graphic elements are additive in that if two signs are varied in two ways (for example, in both shape and size), the contrast between them will be greater than if only one element had been modulated. Variation of only one may be sufficient to achieve the needed contrast if the sign is plain, but the more complex it is, the larger the number of elements that must be varied to obtain contrast.

Contrast is a subtle visual element in some ways, and in other ways it is blatant. The character of a line and the way its curves and vertices are formed may set it apart completely from another line of the same thickness. The thickness of one line in comparison to another may accomplish the same thing. The relative darkness of tonal areas or the differences in hue of colored sections may be made similar or contrasting. The shapes of letters may blend into the background complex of lines and other shapes, or the opposite may be true. If one element of the design is varied as to hue, value, size, or shape, then the relationship of all other components will likewise be changed. It requires careful juggling of the graphic variations among the symbols on a kind of trial-and-error basis to arrive at the right combination.

Visual Balance

Designing a graphic composition requires a number of preliminary decisions pertaining to balance. These decisions involve problems of layout concerning the general arrangement of the basic shapes of the display. The basic shapes may include land-water masses, titles, legend boxes, color areas, and so on.

Balance in graphic design is the positioning of the various visual components in such a way that their relationship appears logical or, in other words, so that nothing seems conspicuously out of place. In a well-balanced design nothing is too light or too dark, too long or too short, in the wrong place, too close to the edge, or too small or too large. Layout is the process of arriving at

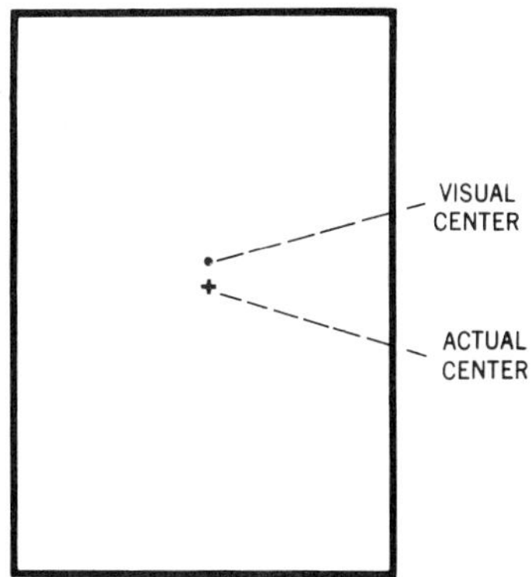

FIGURE 7.4 The visual as opposed to the actual center of a rectangle. Balancing is accomplished around the visual center.

proper balance by getting rid of what appear to be misfits.

Visual balance depends primarily on the relative position and visual importance of the basic parts of a map, and thus it depends on the relation of each item to the optical center of the map and to the other items, and on their visual weights.

The optical center of a map is a point slightly (about 5 percent of the height) above the center of the bounding shape or the map border (see Fig. 7.4). Size, value, brilliance, contrast, and to some extent a few other factors influence the weight of a shape. The balancing of the various items around the optical center is akin to the balance of a beam or lever on a fulcrum. This is illustrated in Figure 7.5, where it can be seen that a visually heavy shape near the fulcrum is balanced by a visually lighter but larger body farther from the balance point. Many other combinations will occur to the reader.

The aim of the cartographer is to balance the visual items so that they "look right" or appear natural for the purpose of the map. The easiest way to accomplish this is to arrange the main shapes in various ways with respect to the map frame until a combination is obtained that will present the items in the fashion desired. Figure 7.6 shows some rough preliminary sketches of a map trying out several ways of arranging the var-

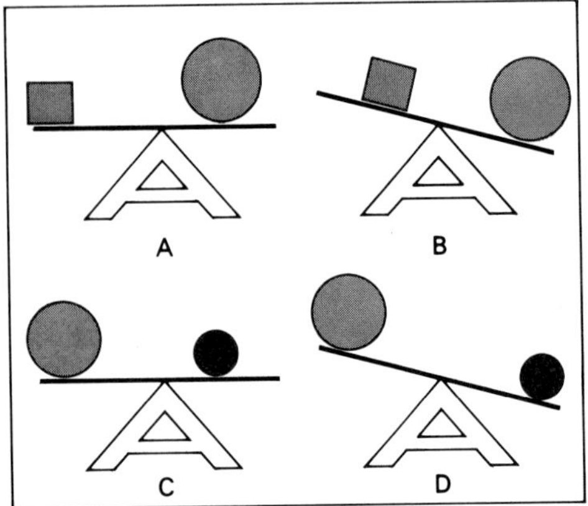

FIGURE 7.5 Visual balance. (A), (B), (C), and (D) show relationships of balance. (A) and (B) are analogous to a child and an adult on a seesaw; (C) and (D) introduce relative density or visual weight, darker masses being heavier.

ious shapes (land, water, title, legend, and shaded area). Many general reference map series use the margins for titles, legends, scales, acknowledgments, and the like. These masses must also be appropriately balanced.

The format, that is, the size and shape of the image area on which a small-scale map is to appear, is of considerable importance in the problem of balance and layout. Shapes of land areas are, of course, prescribed, and can be difficult to fit. They can also vary to a surprising degree on different projections; in many cases the desire for the greatest possible scale within a prescribed format may suggest a projection that produces an undesirable fit for the area involved. Likewise, the necessity for fitting various shapes (large legends, complex titles, captions, and so on) around the margins and within the shape sometimes makes the format a limiting factor of more than ordinary concern. Generally, a rectangle with sides having a proportion of about three to five seems to be the most pleasing shape (see Fig. 7.7).*

The possibilities of varying balance relationships to suit the objective of the map are legion. Try analyzing every visual presentation, from television displays to printed advertisements, in order to become more versatile and competent in working with this important factor of graphic design.

Figure-Ground

A complex occurrence in human visual perception is the *figure-ground* phenomenon. The eye and the mind, working together, react spontaneously to any visual array, whether or not it is familiar. They tend immediately to organize the display into two basically contrasting perceptual impressions: a figure on which the eye settles and which it sees clearly, and the amorphous, more

*For a large map to be viewed at normal reading distance, it is desirable to have the short dimension vertical. Not only does this reduce neck-stretching, it also allows easier use of bifocals.

FIGURE 7.6 Preliminary sketches of a map made in order to arrive at a desirable layout and balance.

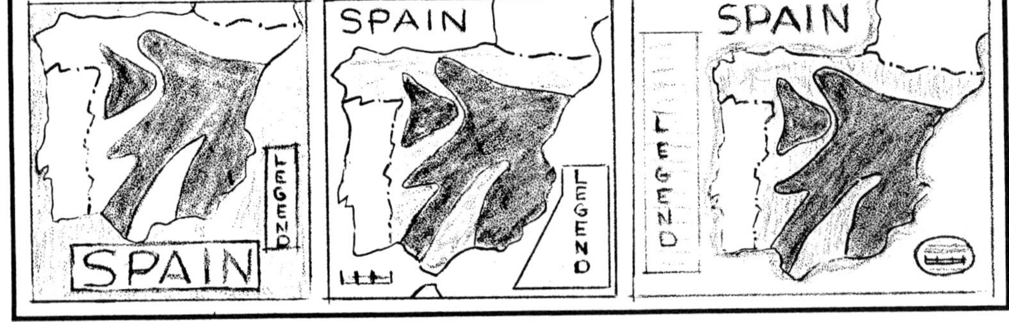

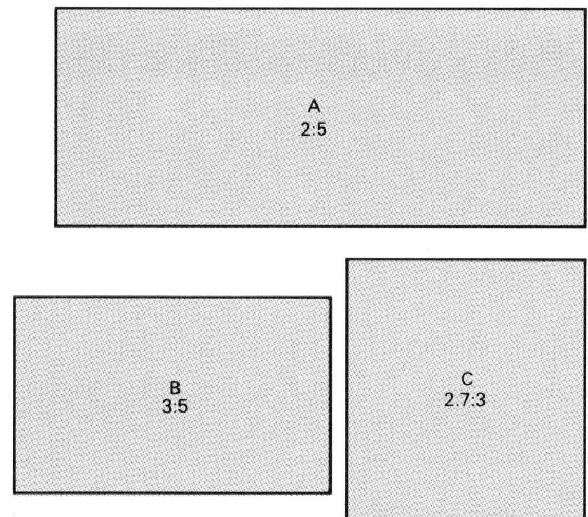

FIGURE 7.7 Various rectangles. (B), with the ratio of its sides 3:5, is generally considered to be more stable and pleasing than the others.

or less formless ground around it. The figure is perceived as a coherent shape or form, with clear outlines, which appears in front of or above its surroundings. The figure is not confused with the less distinct, somewhat formless ground. While this spontaneous visual organization is occurring, the previous experience of the observer is also brought into play by mentally processing the probabilities of what the displayed geographical form relationships might be.

The separation of the visual field into figure and ground is automatic; it is not a conscious operation. It is a natural and fundamental characteristic of visual perception and is therefore a primary component in graphic map design. For both general and thematic maps the user must be able to tell land from water, and recognize and separate the outline of the town, island, or harbor from its surroundings—in short, the user should be able to focus immediately on the cartographer's objectives without visually fumbling and groping to find what he or she is supposed to be looking at.

A simple illustration of this phenomenon of perception is provided by Figure 7.8A in which a simple line, wandering across the visual field, merely separates two areas and is quite ambigu-

FIGURE 7.8 Four simple sketch maps to illustrate various aspects of the figure-ground relationship. See the text for explanation.

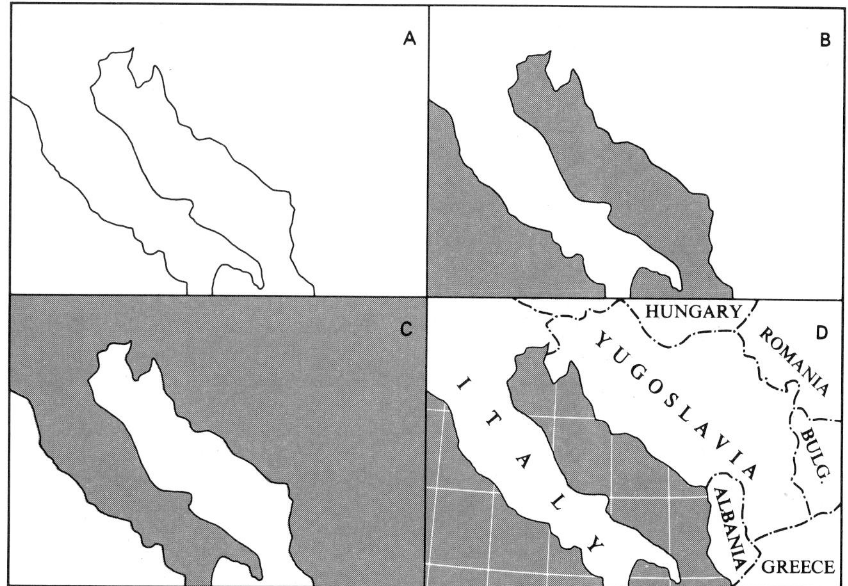

ous. Assuming that the viewer knows he or she is supposed to see an area separated into familiar shapes, the user is confused as to which is land and which is water; unless the reader is unusually familiar with the area, he or she will continue to be perceptually confused. In Figures 7.8B and 7.8C, areas are clearly defined from one another, but there is still the problem of which is land and which is water. In Figure 7.8D, by the use of land-based names, familiar boundary symbols that usually occur on land areas, the graticule to "lower" the water, and slight shading to separate the land from the water, the land clearly emerges as the figure. Of course, once we have clarified the display by observing Figure 7.8D and recognizing the area, recognizing land and water in A, B, and C of Figure 7.8 becomes much easier. It is important to observe, however, that the fundamental perceptual relationships are always there; regardless of how well acquainted a map user may be with an area, the kind of presentation illustrated in Figure 7.8A is subconsciously perceptually ambiguous and therefore confusing and irritating.

It is impossible to develop in depth in this textbook the rather large variety of visual stimuli that are involved in the promotion of desired figural perception. Like all components of graphic design, complex interrelations are involved. Only some of the basic principles can be suggested.

Differentiation must be present in order for one area to emerge as the figure. The desired figure area must be visually homogeneous, and the homogeneity of the whole visual field (the entire map) must not be stronger than the desired figure; otherwise the whole map will become the figure and its surroundings the ground. Such differentiation may be promoted in a variety of ways, such as by differences in color, value, and spacing.

Closed forms, such as islands, entire peninsulas or countries, and other complete entities are more likely to be seen as figures than if they are only partially shown. If all of the peninsula of Italy had been shown in Figure 7.8A, it would have appeared more easily as figure. In general, the smaller surrounded area tends toward figure.

Familiarity exercises considerable influence in the promotion of figure. We readily fix on a well-known shape. In terms of map design, for example, this suggests that the less well known an area is which is desired as figure, the greater must be the application to it of the other factors that promote figure.

Brightness (tonal value) difference promotes the emergence of figure, and all other things being equal, the darker tends toward figure. In graphic composition, all other things are not usually equal, and attention must always be focused on the particular set of visual circumstances. For example, in Figure 7.8D, the lighter area tends toward figure because of the overriding effect of other influences.

Good contour is the graphic equivalent of "logical" or "unambiguous." When something appears continuous, symmetrical, or sensible, it will lead toward figure-ground differentiation. Good contour also involves the tendency for the viewer to move toward the simplest "visual perceptual explanation" of graphic phenomena. For example, in D in Figure 7.8 the graticule appears to be continuous "beneath" the land, thus "raising" the land area above it and helping it to become figure.

The principle of good contour finds many applications in cartography in addition to the common practice in small-scale black and white sketch maps of placing the graticule only on the water in order to help the land become figure. The breaking of a line for lettering or other map components is possible because, given the requisite character of the line, visual logic—good contour—will "continue" the line where it is not actually shown.

Articulation of an area, in the broad sense of its being segmented by a complex of internal markings, tends to lead toward the emergence of figure. In Figure 7.8D there is considerable detail on the land, which is made up of complex lettering and boundaries in contrast to the simplicity of the graticule on the water. This helps to cause the land to emerge as figure. Things such as city symbols, names, rivers, transportation routes, and relief representation are all components that can contribute to figure.

Area is important in leading to figure-ground differentiation. Generally, the tendency is for smaller areas to emerge as figure in relation to larger areas. Some research has suggested that for thematic maps the figure-to-ground ratio (total map area minus area of figure divided by area of figure) should be between approximately 1:4 and 1:1.5. When ratios are larger than 1:4, the ground may overwhelm the figure; when ratios are smaller than 1:1.5, there may be confusion between figure and ground.

In composing a graphic communication, the cartographer must work with the characteristics influencing the figure-ground relationship with great care. Some geographical relationships simply have "bad contour," are large, cannot be closed, and so on; in these cases the cartographer must lean more heavily on other graphic elements. Each composition is a new problem, and standard specifications cannot be made any more than we can say that a well-designed paragraph will contain a given number of words or sentences.

Hierarchical Organization

The communication of spatial phenomena always involves substantive elements of differing significance, and the foregoing sections have dealt with components of graphic design that can be employed to portray this. In addition, the effective influences of hue, value, and intensity of color, as well as pattern, will be treated in the next chapter. Most communications about reality involve more complex relations, which require a kind of internal graphic organization or structuring. For example, the basic form of a distribution in a thematic map is more important than the base data against which it is displayed; the classes of roads shown on a general map may range in importance; or a map may show numerous areal categories, such as types of soil, vegetation, or crops, which may be separated into two groups in which the components of each group are more closely related than are the two groups. It is the cartographer's objective to separate the meaningful characteristics visually and to portray graphically the significant likenesses, differences, and interrelationships.

Because this usually involves "levels" of relative importance, we can call it *hierarchical organization,* and it is among the most important of the graphic components of map design. It calls for sophisticated application of variations of the graphic elements in interrelated fashion. Three kinds of hierarchical organization can be recognized, here called extensional, subdivisional, and stereogrammic.

Extensional organization is primarily concerned with the portrayal of networks of lines of varying significance, although it can also apply to point symbols. For example, road systems are comprised of several classes of roads; drainage systems involve several orders of streams; and railroad and airline systems have main lines or routes and feeder components. The central place concept in geography recognizes a hierarchy of settlements ranging from a metropolitan center downward to other cities, towns, villages, and hamlets. A producing center may be connected to consuming centers.

Extensional organization is usually basically ordinal. Even though interval or ratio scales may be employed in the individual symbolization, the overall graphic objective is to portray relative importance, and the graphic elements of size, value, and color are the means most often employed. Figure 7.9 is an example of extensional organization.

Subdivisional organization is employed to portray the internal relationships of a hierarchy. For example, cities are divided into wards and precincts; the land use categories of agricultural and nonagricultural may each be subdivided into several related uses; and different classes of bedrock or soil may each comprise several associated types. Figure 7.10 is an example of complex subdivisional organization.

Subdivisional presentation is primarily concerned with area symbols and generally employs

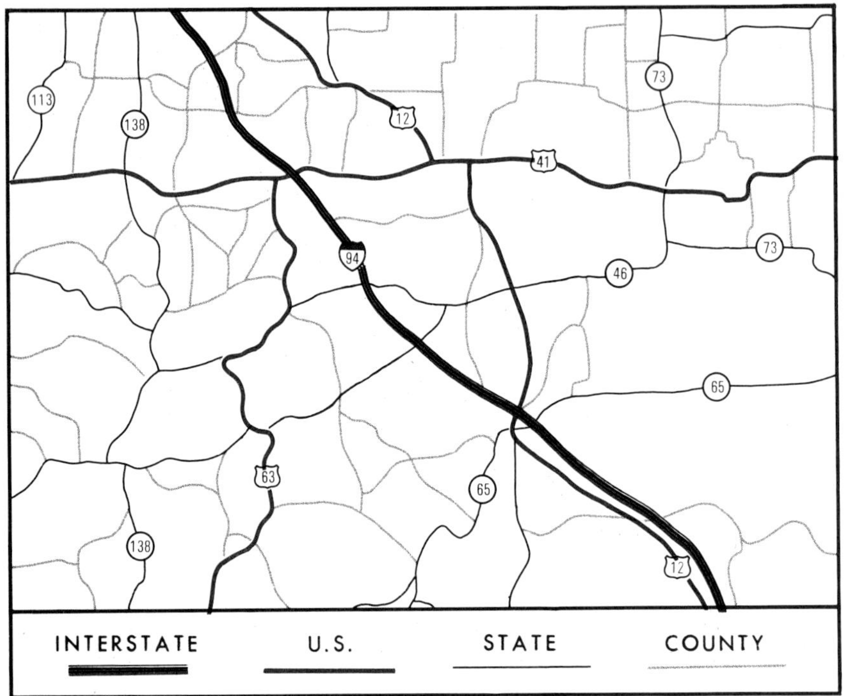

FIGURE 7.9 An example of extensional hierarchical graphic organization in which a set of roads is graded according to relative importance.

FIGURE 7.10 An example of subdivisional hierarchical organization in which the primary division is between humid and dry climates, with a secondary subdivision based on temperature, and a tertiary subdivision based on desert versus steppe.

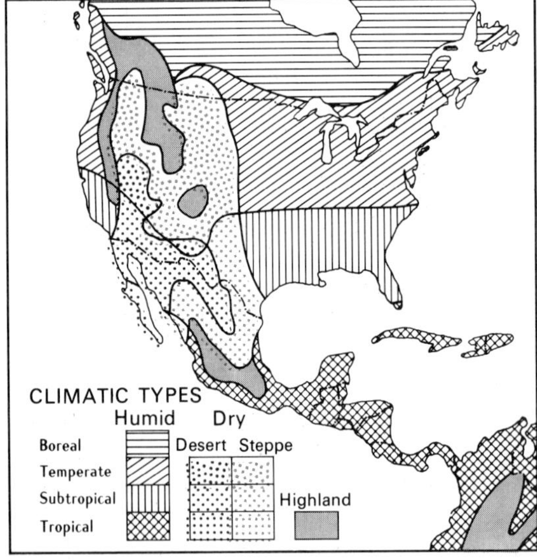

the differences in hue and pattern. The substantive differentiation may be solely nominal or may be made up of a combination of nominal and ordinal scaling. Since at least two levels or organization are involved, each of which incorporates subdivision, the elements of visual contrast and similarity must be judiciously employed. The primary categories must have distinctive visual homogeneity, and the visual variance between classes must be greater than the visual variance among the several subdivisions of each major category.

Stereogrammic organization differs from subdivisional structure in that it is concerned with giving the map user the impression that the components lie at different visual levels. Such an objective is fairly common in both general and thematic cartography. The purpose is to assist the reader to focus on one set of data and, while doing so, to subordinate the rest of the data.

General maps of all scales portray many categories of data; it is often desirable to prepare them in such a way that a user can visually separate one category (boundaries, the drainage system, the contours, or whatever) from the remainder in order to be able to concentrate on the single element.

Often in thematic cartography, the cartographer wishes to emphasize one portion of the map or a particular relationship on the map. For example, as illustrated in Figure 7.11, the objective may be to show territories that have changed hands in Europe since 1900. In order to accomplish this, it is helpful to make them appear above the background base data so that the eye will focus on these areas and will only incidentally look at the geographical matrix "beneath" them.

Stereogrammic organization must, of course, draw heavily on the principles involved in figure-ground differentiation, but many related cues to depth perception may be invoked, such as superimposition, progression of size and weight, value progression (all illustrated in Fig. 7.12), as well as differences in hue and saturation, which have some connotations of depth.

Hierarchical organization is sometimes made difficult by the constraints of reality. "Important" linear and areal phenomena may well be smaller than less important ones; a significant item may occur inconveniently in a faraway corner of the map; and a color convention, such as blue water, may prevent a desirable series of hues. In any given design problem any one of the graphic components of design may be of paramount importance, but solving problems of hierarchical organization are often likely to be the most important and demanding tasks of the map designer.

CONTROLS OF GRAPHIC DESIGN

Just as the applications of the elements of generalization are influenced by a set of controls, so are the components of graphic map design. Since they are fundamental to cartography, three of the controls—objective, reality, and scale—are controls in both generalization and graphic design: Maps are made for a reason; they are analogues of real areas; and they are all made to a scale. In graphic map design two other controls are especially important: audience and technical limits.

All the components of graphic map design will not be equally significant in all maps, but the full set of controls is applicable in every case. The controls interact, of course, and it is often useful to contemplate the set all together at the outset of a map project. If one can anticipate the kinds of problems likely to be faced, it is often possible to influence the initial conception of the project before it has progressed to the point where some unanticipated constraint unduly complicates the graphic design problem.

Brief comments on the character and operation of the individual controls are given below. Of necessity, they are general. You may find it useful to consider each one in relation to a specific map.

Technical Limits

The control called *technical limits* refers to the way a map will be constructed or produced. In computer-assisted cartography, it obviously has to do with the capabilities of the hardware; various pen sizes, line widths, and tonal or color variations may be included in the repertoire of the machines. Various limits may be prescribed for a printed map: one color, two colors, or four colors; hand-drafted in ink or scribed; with halftone or limited to line—the list is almost endless. The cartographer should know the full set of technical limits within which the map will be produced before considering the possibilities of graphic design.

Objective

It is obvious that the purpose for which the map is being made is the essential determinant of its final form. All aspects of symbolization and graphic design must be consciously fitted to the purpose as carefully as possible.

Of particular importance in the operation of objective as a control is the breadth of the pur-

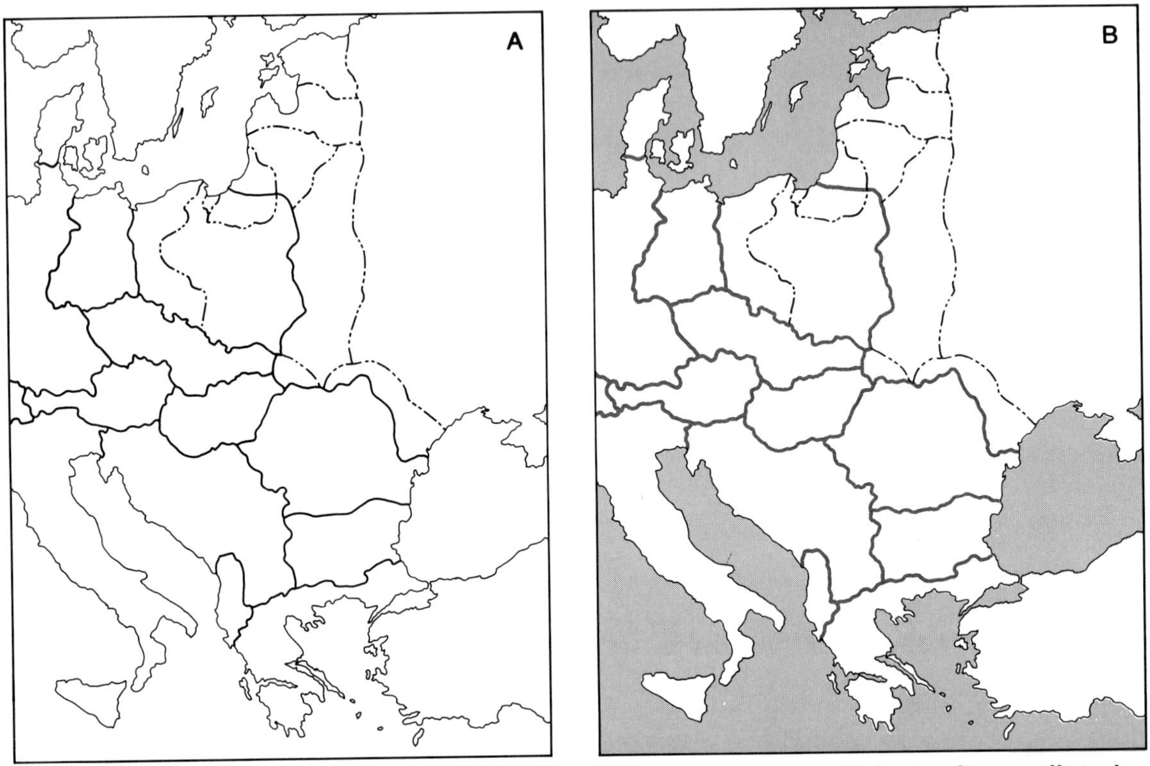

FIGURE 7.11 An example of stereogrammic hierarchical graphic organization. In (A), all elements lie generally in the same visual plane. In (B), the land has been made to appear above the water, and the more modern boundaries have been made to rise above the general visual plane of the land.

FIGURE 7.12 Some examples of depth cues that may be useful in stereogrammic organization. (A), (B), (C), and (D) illustrate various kinds of superimposition, (E) illustrates a progression of size, and (F) illustrates a progression of value. Like the graphic elements, depth cues may be used additively, as for example in (G).

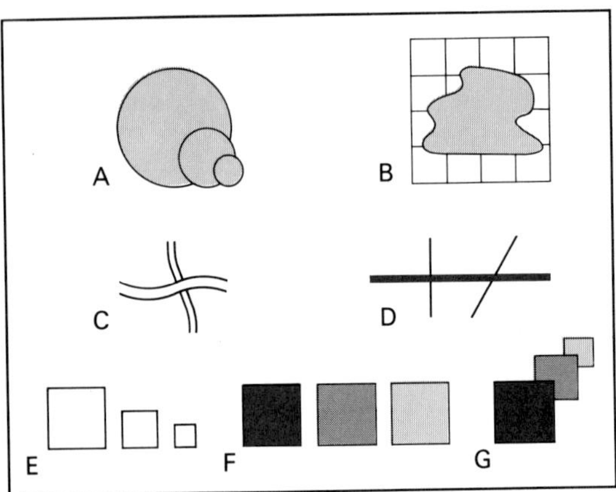

pose. General maps are inherently multipurpose. In thematic mapping, the greater the number of identifiable objectives, the more difficult it is to arrive at a graphic design that incorporates solutions to all of them. Because many printed maps are difficult, costly, and slow to make, there is a strong tendency to combine many objectives in one map. Furthermore, there seems to be a kind of fear among some cartographers that a map may be used for purposes other than those which prompted its preparation. Obviously, any knowledgeable map user will not measure distances improperly, or the like, but the handiest map is often the easiest place to look for something, such as a boundary, river, or city. It is a mistake to try to anticipate possible improper uses, and to make allowance for them complicates the graphic design and leads to visual complexity.

Another significant aspect of the objective is that of the total appearance or "look" of the map. Whether the map should appear dark or light, crowded or open, precise or approximate, and the general subjective feeing to be conveyed, such as graceful, bold, traditional, or modern, are important aspects of map design; such affective objectives may be as important as the substantive objectives in many cases.

Reality

The overall geographical dimensions and character of the regions or distributions being mapped are given and, except in unusual circumstances, they cannot be much modified by the cartographer; regardless of what is done, Chile will remain long and narrow; the Boston–New York–Philadelphia–Washington megalopolis area is heavily populated with many comparatively large cities; some land forms are extremely rugged; soil variations are often very complex; and areas of recent glaciation tend to have enormous numbers of lakes. Each of these kinds of reality, and many more like them, place serious constraints on the graphic design possibilities; they must be anticipated at the outset of the planning stage of a map or a map series.

Scale

Scale is prescribed by the format and by its relation to the area being mapped. From a conceptual point of view, scale operates in subtle ways. The smaller the scale, the "farther away" one is from the area being mapped, and this intuitive feeling should be matched with the graphic design. Although one can say, in general, that the smaller the scale, the smaller the line weights and type sizes should be, no specific prescription can be made because other controls are also relevant. Nevertheless, under ordinary circumstances, a magnifying glass should not be needed to use a map. In addition to type sizes, the employment of the combinations of signs and graphic elements may be affected in series maps at different scales. The relation among textures and colors and the various scales of related maps should appear to be logical.

Audience

Both general and thematic maps are made for a variety of audiences. These range from young schoolchildren to college students, and from the lover of the outdoors to the technical engineer. Each group may be different in terms of its familiarity with graphic and symbolic conventions, its geographical knowledge, and its perceptual limitations in general. For the geographical unsophisticate, the important basic shapes, important names, and the like should be emphasized in the graphic design. A map on a wall being viewed by an audience seated at desks will require a different treatment of graphic character than a map of the same sort in a textbook on the desk or the map on a television or monitor screen. A map on a television screen is likely to be seen only briefly and thus cannot have much lettering. The perceptual limits of the audience may sometimes be a significant factor in the design. For example, older people have more difficulty seeing small type and must hold visual material closer to them. A perfect example of this aspect of map design, albeit not at all graphic, is provided by the problems of making maps for the partially sighted and totally blind, wherein close attention must be paid to things such as tactile discriminability.

The functioning of the audience as a control may be extended to include the conditions under which a map may be used. Legibility requirements for road maps to be used in a moving automobile must be higher than for a map to be viewed in a stable condition. Maps and charts to be used at night in boats, ships, and aircraft must be usable in dim light so that bright illumination, which would greatly affect night vision will not be needed.

PLANNING GRAPHIC MAP DESIGN

Thoughtful and effective presentation of complex relationships demands careful planning and, when graphic means are employed, the display must be as carefully organized and planned as any other sort of communication. When we plan to write something, we first prepare an outline; it is also necessary to outline a graphic communication.

Each graphic ingredient of a map should be evaluated in combination with the others in terms of its probable effect on the reader. To do this requires a full and complete understanding of the objectives of the map to be made. We can scarcely imagine writing an article or planning an oral presentation without first arriving at a reasonably clear decision concerning (1) the audience to which it is to be presented, and (2) the scope of the subject matter. With these considerations in mind, the writer assesses the relative significance of each item to be included.

We use the term "outline" in its broad sense as a brief characterization of the essential features of a communication. Although notes can be jotted down to keep track of ideas, the outline for a graphic presentation is *graphic*. Some of it can be done by visualizing relationships, but often it requires preliminary sketches with the roughing in of the major elements or experimentation on the monitor. As planning progresses, a representative section of the map can be prepared to evaluate the graphic relationships. For printed maps final determinations of colors and values can often be decided at the proofing stage, but the basic decisions must be made at the outset.

The Graphic Outline

As a rule a great variety of geographical data is presented on printed general maps, and basic decisions focus on how numerous the various items in each class are and consequently how prominently they may be portrayed. All sorts of hierarchical problems arise involving road, stream, and boundary precedence and the visual sequence of symbols and type sizes.

An outline for a general map or map series usually begins with a document listing all the categories of data to be included. The next stage is a preliminary design plan including a prototype of a map, or section of one, normally prepared in proof. This stage lasts through several trials until the final specifications are decided and production begins.

Thematic maps may appear to be less complicated in that they usually involve fewer categories of data, but they pose a different kind of graphic problem in that they generally involve complex hierarchical decisions. This requires preparing visual "outlines." Figures 7.13A, B, C, and D illustrate how such a graphic outline may be prepared for a thematic map. The assumption is made that the planned thematic map is to show two related (hypothetical) distributions in Europe. The fundamental organizational elements of the communication are as follows.

1. The place—Europe.
2. The data—the two distributions to be shown.
3. The position of the data with respect to Europe.
4. The relative position of the two distributions.

Any one of these four elements may be placed at the head of the graphic outline, and the order of the others may be varied in any way the cartographer desires. In Figure 7.13A the design places the organizational elements in the general order of 1-2-3-4; in B, 2-3-4-1; in C, 3-1-4-2; and in D, 4-2-3-1. Other combinations are, of course, possible. It should not be inferred that the positioning of the elements in the graphic outline can be as exact and precise as in a written outline. In the latter, the sequence is reasonably assured, since the reader or listener will start at one point and proceed in one direction to the end, whereas in a graphic presentation, the viewer, at least potentially, sees the display all at once. It is up to the cartographic designer to attempt to lead the viewer by applying the various princi-

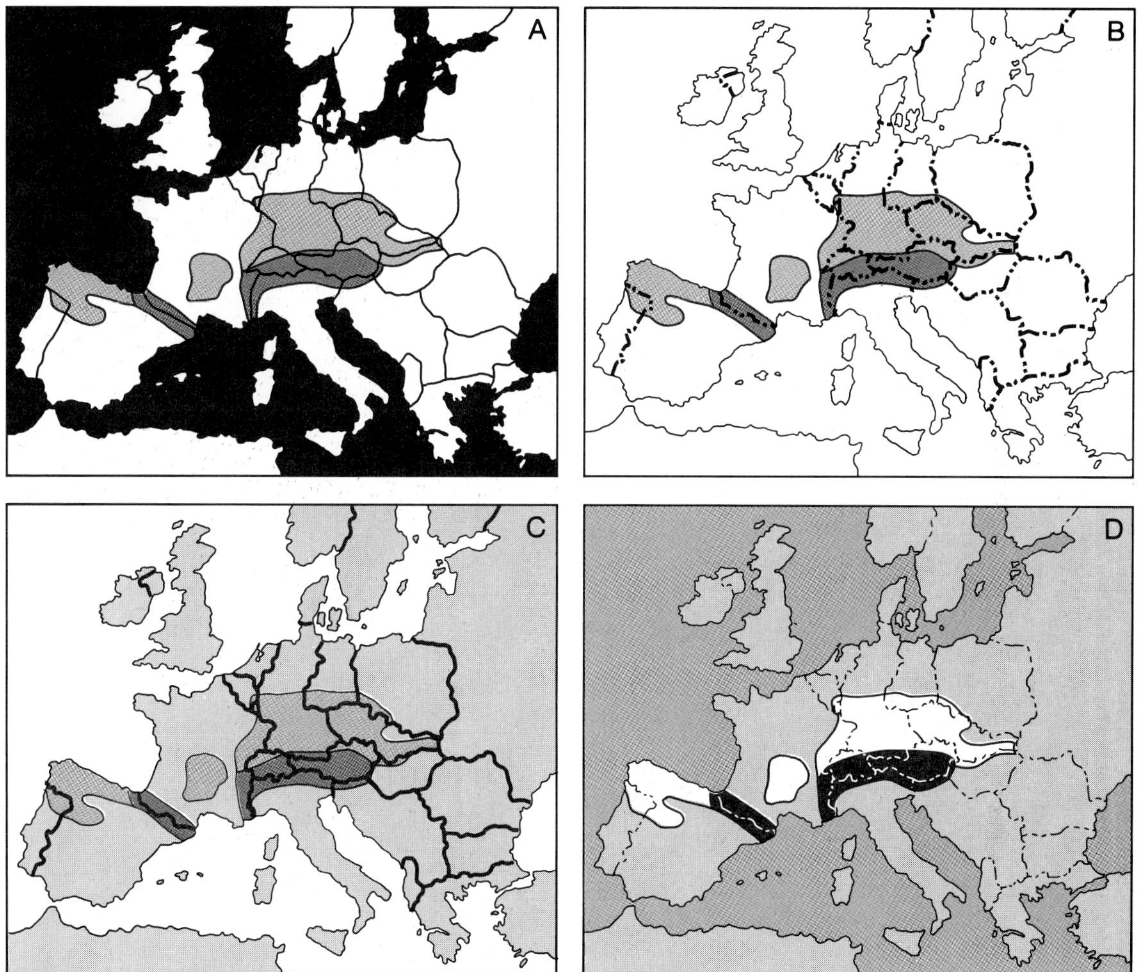

FIGURE 7.13 Examples of variations in the primary visual outline. See text for explanation.

ples discussed earlier in the section on the components of graphic map design.

It is appropriate at this point to digress slightly in order to emphasize one of the more difficult complicating factors that a cartographer must face in cartographic design. Any mark on a map has, of course, an intellectual connotation, but it also has a visual impact. It is difficult to remove the former in order to evaluate the latter, but many times it is not only necessary but definitely desirable. Considerable experimentation is necessary for an electronic display. Illustrators can turn their works upside down; advertising layout artists can "rough in" outlines and even basic blocks of type as design units without "spelling anything out"; and because the intellectual connotations cannot always be predicted, cartographers, for their own purposes, will often obtain better designs for thematic maps if they do likewise, except for obvious, well-known shapes, such as continents and countries.

After the structuring of the major organizational elements has been decided, attention must be shifted to the second stage of graphic outlining. Just as with a written outline, the major items (the primary outline) are first determined; then the position of the subject matter within each major topic is decided. In the case of a map, the graphic presentation of the "detail" is primarily a matter of clarity, legibility, and relative contrast of the detail items.

Titles, Legends, and Scales

These standard elements serve two major purposes in cartography. Naturally, they have a denotative function in identifying the place, subject matter, symbolization, and so on, but for many maps they also serve as visual masses that can be positioned to provide the graphic organization of the map (Fig. 7.14).

A title serves a variety of functions. Sometimes it informs the reader of the subject or area on the map and is therefore as important as a label on a medicine bottle. But this is not always the case, since some maps are obvious in their subject matter or area and really need no title. In some instances the title may be useful as a shape to be used to help balance the composition.

It is impossible to generalize about the form a title should take; it depends entirely on the map, its subject, and its purpose. Suppose, for example, that a map had been made showing the density of population per square kilometer in Canada according to the 1981 census. The following situations might apply.

1. If the map appeared in a textbook devoted to the general worldwide conditions at that time with respect to the subject matter, then just

   CANADA

   would be appropriate because the time and subject would be known.
2. If the map appeared in a study of the current worldwide food situation, and if it were an important piece of evidence for some thesis, then

   CANADA
   POPULATION PER SQUARE
   KILOMETER

   would be appropriate.
3. If the map appeared in a publication devoted to the changes in population in Canada, then

   POPULATION PER SQUARE
   KILOMETER
   1981

   would be appropriate, since the area would be known but the date would be significant.

Many other combinations might be appropriate, but there is no need to belabor the fact that the wording of the title must be tailored to the purpose. Similarly, the degree of prominence and visual interest given to the title, through the style, size, and the boldness of the lettering must be fitted into the whole design and objective of the map.

FIGURE 7.14 Titles, legends, scales, and insets may be arranged in various ways in the graphic organization of a map.

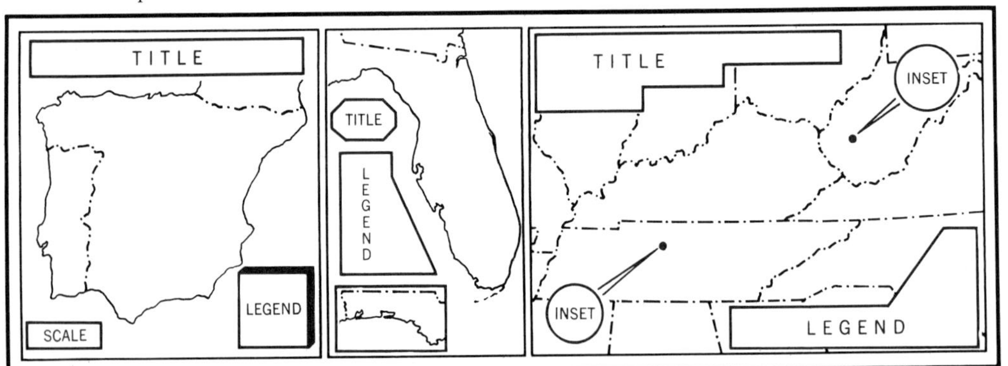

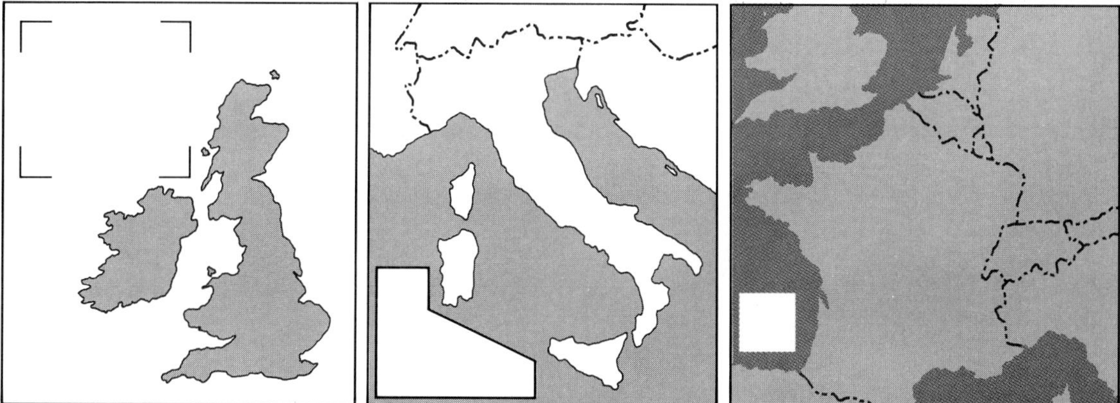

FIGURE 7.15 Examples of variations in the prominence of map legends. Note the operation of the principles of figure-ground relationships.

Legends or keys are naturally indispensable to most maps, since they provide the explanation of the various symbols used. It should be a cardinal rule of the cartographer that no symbol that is not self-explanatory should be used on a map unless it is explained in a legend. Furthermore, any symbol explained should appear in the legend *exactly* as it appears on the map, drawn in precisely the same size and manner. The arrangement of the component parts of a legend, such as the series of colors or patterns and point and line symbols being employed, is worthy of careful attention. As a general rule, a range of values is arrayed vertically with the lowest values at the bottom.*

Map legends can be emphasized or subordinated by varying the shape, size, or value relationship. Figure 7.15 illustrates several variations. In the past it was the custom to enclose titles and legends in fancy, ornate outlines, called *cartouches,* which, by their intricate scrollwork, called attention to their presence. Today it is generally conceded that the contents of the legend are more important than its outline, so the outline, if there is any, is usually kept simple, and the visual importance is regulated in other ways.

The statement of scale of a map also varies in importance from map to map. On maps showing road or rail lines, air routes, or any other phenomenon or relationship that involves distance concepts, the scale is an important factor in making the map useful. In such cases the statement of scale should be placed in a position of prominence, and it should be designed in such fashion that it can be easily used by the reader.

The method of presenting the scale may vary. For many maps, especially those of larger scale, the Representative Fraction (RF) is useful because it tells the experienced map reader a great deal about the amount of simplification and selection that probably went into the preparation of the map.* A graphic scale is much more common on small-scale maps, not only because it simplifies the user's employment of it, but because an RF in the smaller scales is not so meaningful.

*Alternative suggestions have been made. See Alan A. DeLucia and Donald W. Hiller, "Natural Legend Design for Thematic Maps," *The Cartographic Journal* 19, no. 1 (1982): 46–52.

*It should be remembered that if a map to be printed will be changed in size by reduction, this will not change the printed numbers of the RF. On a map designed for reduction, the RF must be that of the final scale, not the construction scale.

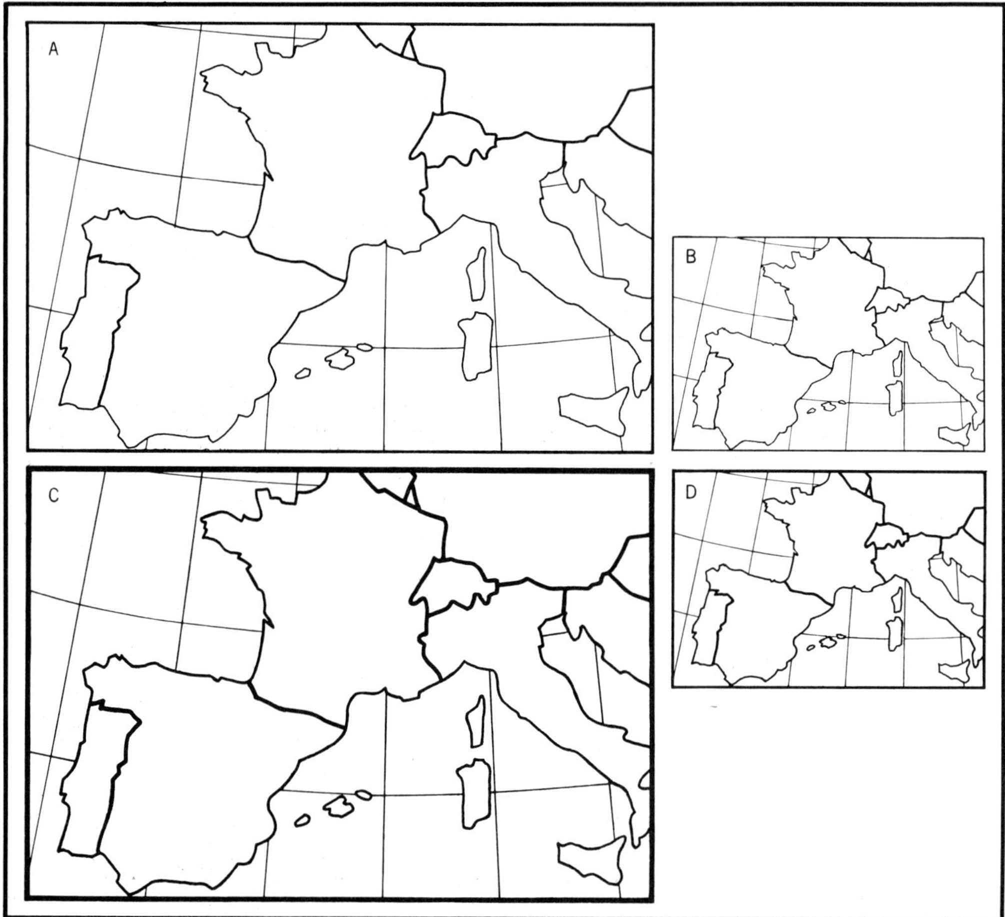

FIGURE 7.16 Effects of reduction. In (A) when manual artwork is designed for appropriate line contrasts at drawing scale, then reduction (B) will decrease the contrasts too much. In (C) when artwork is designed for reduction, then reduction (D) produces appropriate line contrast relationships.

Effects of Reduction

With scribing, most printed maps are now constructed at the reproduction scale. Nevertheless, a considerable number of maps are still prepared by pen and ink methods and are usually drafted at a scale larger than the reproduction scale, that is, for reduction. This is done for a variety of reasons, the most important of which is that it is often impossible to draft in ink with the precision and detail desired at the scale of the final map.

Drafting a map in ink for reduction does not mean merely drawing a map that is well designed at the drafting scale. On the contrary, it requires the anticipation of the finished map and the designing of each item so that when it is reduced and reproduced, it will be "right" for that scale. A map must be designed for reduction as much as for any other purpose.

The greatest problem in designing for reduction involves line widths. In general, a map on which the lines appear correct at the drafting scale will appear a little "light" if it is reduced. Consequently, the mapmaker must make the map

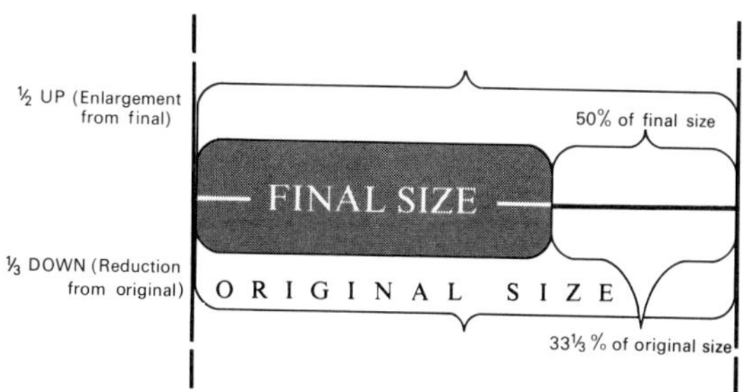

FIGURE 7.17 Relation of enlargement to reduction.

overly "heavy" in order to avoid its appearing too light after reduction. It is also necessary to "overdo" the type sizes somewhat, just as it is necessary to make lines and symbols a little too large on the drawing. Figure 7.16 illustrates some of these relationships.

Maps of a series should appear comparable and, if they are drawn for reduction, they should be drafted for the same amount so that similar lettering, lines, and the like will appear comparable. This may necessitate changing the scale of base maps, which is troublesome, but it will insure that the line treatment and lettering in the final maps will be uniform.

Specifications for drafting and for reduction are given in terms of linear change, not areal relationships. It is common to speak of a drawing as being "50 percent up," meaning that it is half again wider and longer than it will be when reproduced. The same map may be referred to as being drafted for one-third reduction; that is, one-third of the linear dimensions of the original will be lost in reduction. Figure 7.17 illustrates the relationships.

Since it is common practice in large printing plants to photograph many illustrations at once, it is also desirable, for economy, to make series drawings for a common reduction.

SELECTED REFERENCES

Arnheim, R., *Art and Visual Perception,* rev. ed. Berkeley, Calif.: University of California Press, 1974.

———, "The Perception of Maps," *The American Cartographer* 3 (1976): 5–10.

Arvetis, C., "The Cartographer-Designer Relationship, A Designer's View," *Surveying and Mapping* 33 (1973): 193–95.

Bertin, J., *Graphics and Graphic Information Processing,* trans. W. J. Berg and P. Scott. New York: Walter de Gruyter, 1981.

———, *The Semiology of Graphics,* trans. W. J. Berg. Madison, Wisc.: University of Wisconsin Press, 1983.

Cuff, D. J., and M. T. Mattson, *Thematic Maps: Their Design and Production.* New York: Methuen, 1982.

Dent, B. D., "Visual Organization and Thematic Map Communication," *Annals of the Association of American Geographers* 62 (1972): 25–38.

Gilman, C. R., "Map Design at USGS: A Memoir," *The American Cartographer* 10 (1983): 31–49.

Keates, J. S., *Cartographic Design and Production.* London: Longman Group Ltd. 1973

Keates, J. S., *Understanding Maps.* Harlow, Essex. England: Longman Group Ltd., and New York: Halstead Press, John Wiley & Sons, 1982.

Morrison, J. L., "Computer Technology and Cartographic Change," in *The Computer in Contemporary Cartography,* ed. D.R.F. Taylor. New York: John Wiley & Sons, 1980, pp. 5–23.

Muehrcke, P. C., "An Integrated Approach to Map Design and Production," *The American Cartographer* 9 (1982): 109–22.

Petchenik, B. B., "A Verbal Approach to Characterizing the Look of Maps," *The American Cartographer* 1 (1974): 63–71.

Robinson, A. H., "A Program of Research to Aid Map Design," *The American Cartographer* 9 (1982): 25–29.

Taylor, D.R.F., ed., *Graphic Communication and Design in Contemporary Cartography,* Progress in Contemporary Cartography, Vol. II. New York: John Wiley & Sons, 1983.

8

Color and Pattern

Perception of graphics is very complex, as it is a compound of responses to visual stimuli, recognition, curiosity, and a host of other reactions. Although we can resolve visual differences into their basic characteristics, which comprise the classes of signs and the primary graphic elements, some graphic components of maps are clearly more distinctive than others and seem to be unusually significant in creating the "look" of a map. These are: color, represented in the graphic elements by hue and value; pattern, such as parallel lines, stippling (dots), and cross-hatching, a combination of signs and several of the graphic elements; and typography, the common alphanumeric marks (letters and numbers). Throughout the history of cartography, mapmakers have paid special attention to these, and in this chapter we will consider the basic characteristics of color and pattern and their employment in cartography. Chapter 9 will deal with typography as a major element in mapmaking.

Color especially has played a major role, partly because of its usefulness as an aid to clarity. Even a small amount of color can make an enormous difference in the appearance of a map. A graphic complex of numerous different black lines, representing boundaries, coasts, rivers, roads, lakes, to say nothing of such things as canals, railroads, contours, and transmission lines, in an extreme combination can convey nothing but utter confusion. Yet, if the various phenomena are carefully color coded (for example, bounded areas tinted to show their extent, rivers and coastlines distinguished from roads and boundaries, and so on), visual order replaces chaos.

Pattern is not as versatile an element as color because it is largely used as an area symbol; yet, that has always been a major element of many maps. For some four hundred years before the development of lithography and color printing, the production of patterned area symbols on various kinds of printing plates called for great ingenuity. Today pattern is used alone (monochromatically) or, very often, in conjunction with color.

We will begin the consideration of these two distinctive components of graphic design with color; its characteristics are among the most complex of the primary graphic elements.

Color in Cartography

The desirability of color has been so great that considerable attention has been devoted to ways of applying it to maps. Before maps were printed, color was drawn or painted on manuscript maps. After the fifteenth century, when the printing of maps became common, most color still had to be applied by hand to each sheet because color printing was prohibitively expensive and technically difficult. Map tinting became a trade, and many people were so employed by mapmaking establishments. Often templates and other elaborate ways of ensuring that the colors were put in the right places were worked out. Such practices extended even to large-scale topographic series. With the development of lithography and then photography in the nineteenth century, techniques for printing color were developed, and since that time, a steadily increasing number of maps have used it.

Color on a map allows greater detail; it adds visual interest; it increases the design possibilities; and it adds greatly to the possibilities for hierarchical graphic structures. Because color may be used efficiently to code similarity and dissimilarity between and within a limited number of classes of phenomena, it is a great aid to clarity. All these advantages are not without their cost, of course. Printed color is relatively expensive, and it calls for special treatments in map construction and reproduction. Color in video displays (TV) is readily available.

Cartographers have employed color for so long that a good many conventions have developed that are important in considering the design of a map. Furthermore, color arouses aesthetic reactions and connotes concepts (red with warm, blue with cold, green with vegetation, and others) that may be important in creating an effective graphic communication. Equally important in cartographic design involving color are the physiolog-

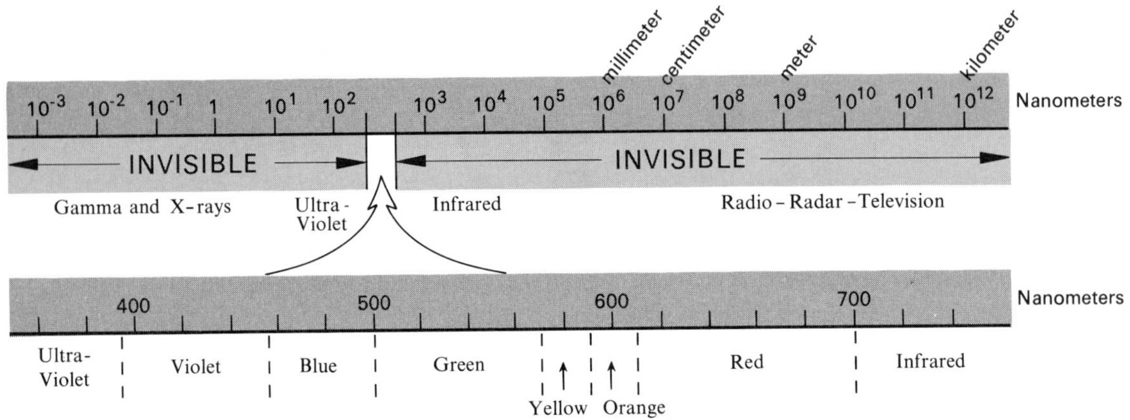

FIGURE 8.1 A portion of the electromagnetic spectrum with wavelengths shown in nanometers. Only a small portion of the entire spectrum is visible. One nanometer (nm) = one billionth of a meter (10^{-9} meter). See Color Figure 1.

ically based perceptual aspects, such as sensitivity, visual acuity, and contrast.

In order to discuss color as one of the more important visual reactions it is first necessary to appreciate how it occurs as a visual sensation.

THE SENSATION OF COLOR

Spectral Colors

The visual sensation of light occurs because of the stimulation of receptors in our eyes by a portion of the electrogmagnetic spectrum. That spectrum ranges from the very short wavelengths of X rays and gamma rays to the very long wavelengths used by radar and television (Fig. 8.1). Only a tiny portion of the spectrum stimulates the receptors in our eyes, namely, the wavelengths ranging from approximately 400 to 700 nm.*

When the source of illumination emits the full range of visible wavelengths in suitable portions, we call it white light. When white light (such as sunlight) is separated into its component wavelengths by the refraction of a prism or of raindrops, and these are reflected to the eye, the various wavelengths provide the attribute of color called hue (blue, yellow, red, and so on).* The shorter wavelengths are the violet-blues, near the 400-nm end of the spectrum, and the longer wavelengths are the reds, near the 700-nm end. The order of the spectral hues is that of the rainbow, and is shown in Color Figure 1.

Pure spectral colors are not often seen except when white light is refracted, as by a gem, a water droplet, and the like, but they provide most of the basic names we use to identify hues (violet, blue, green, yellow, orange, and red). The order of their wavelengths is significant in cartography because hues sometimes are employed in what is called a *spectral progression*.

Common Colors

The myriad colors we see in nature and in all fabricated things are rarely spectral hues but instead are almost always various combinations of wavelengths. This happens because illuminated surfaces absorb a proportion of some of the wavelengths of the light that falls on them (incident light) and reflect the remainder. For example, a living tree leaf absorbs much of the wave-

*A nanometer (nm), previously called a millimicron (mμ) is 1 meter times 10^{-9}, or one billionth of a meter.

*Defective color vision occurs in about 4 percent of men and 0.4 percent of women. The more common types affect the perception of reds and greens. Only a very small proportion are truly "color blind," that is, have no chromatic response.

lengths of the spectrum except for those in the 500- to 550-nm range, a large proportion of which it reflects. Because of this selective absorption and reflection, it appears green to our eyes. Paper that reflects all the wavelengths of sunlight in about the same proportion will appear white, and a surface that absorbs most of the wavelengths and reflects very little will appear black.

An ordinary surface will reflect at least a small proportion of all the wavelengths, but it will gain its distinctive character from those wavelengths it reflects the most. The concept of colors as combinations of wavelengths is most easily understood by displaying their character on a spectral reflectance curve. Data are obtained from a spectrophotometer, an instrument that measures the proportions of the various wavelengths of light in a color, and these data then may be plotted on a graph whose X axis is wavelength and whose Y axis is percent reflectance. Graphs of several colored surfaces are shown in Figure 8.2.

In addition to the general color indicated by the higher (more dominant) section of the curve on a graph, other characteristics are shown by the shape of the curve. A relatively flat curve indicates a lack of a dominant hue; if the flat curve is high on the graph it approaches white; if it is low, it is a dark gray. The higher the curve, the lighter the color; the more peaked the curve, the stronger or more intense the color will appear.

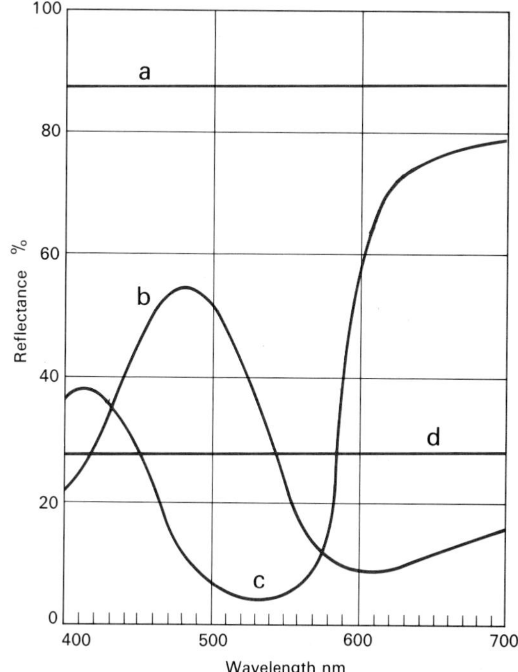

FIGURE 8.2 Four spectral reflectance curves: (A) a white surface; (B) a "greenish blue" ink; (C) a "red-purple" ink; and (D) a dark gray surface. (Curves B and C after Nimeroff.)

The Dimensions of a Color

The character of any color is a combination of three qualities. Depending on whether one is considering color from the point of view of the stimuli or the response, one arrives at somewhat different definitions of the major components, but a description of a color is incomplete unless all three are included. The cartographer is, of course, concerned with what the map reader sees; consequently, we will consider the dimensions primarily from the perceptual point of view. The three dimensions are: (1) hue, (2) value, and (3) intensity, or chroma. As we will see later, specific terms and definitions are employed in particular systems of color specification.

Hue. *Hue* is the attribute of color that we associate with differences in wavelength. When we identify something as red, green, or yellow, we are describing its hue. Hues occur in great variety because wavelengths can be combined in an infinite number of ways. The arrangement of hues into an orderly sequence or series requires the adoption of some point of view or purpose; each adoption will result in a different sequence. The most familiar sequence is that in the spectrum (rainbow), which occurs when the wavelengths in white light are refracted (see Color Fig. 1).*

*The spectral series is obviously "natural," since it occurs in nature, but perceptually it is not a logical or orderly sequence.

Some hues are described as *primary* in that all other colors may be created by a suitable mixture of them. A primary color is not a spectral hue (one wavelength) but, like all nonspectral hues, is a combination of wavelengths in which one portion of the range is dominant. Blue, green, and red light are called the *additive* primaries, since all other colors can be created on a "white" reflective surface or the phosphor dots of a picture tube by combining light of these three colors in the required proportions (Color Fig. 2). All color video displays, such as television, function by combining the additive primaries.

Colors on a paper result from pigments being applied to the surface, which, when illuminated by white light, absorb or subtract some of the wavelengths; what is reflected to the observer are the remaining wavelengths. The *subtractive* primaries are cyan (greenish blue), magenta (purplish red), and yellow, and most colors can be created by a suitable mixture of these pigments (Color Fig. 3). This is an ideal, of course, since the pigments are never "pure" and since the paper and the incident light will always have some special characteristics of their own.

Value. All colors may be ranked in terms of their lightness or darkness in a given situation. Various terms are used to describe this attribute, the more common being luminosity, brightness, reflectance, and value. *Luminosity* is the quality of seeming to emit light; yellow is inherently a more luminous hue than blue. *Brightness* is the attribute of sensation by which an observer is aware of differences in luminance. For example, we may say that one surface is brighter (or lighter) than another. *Reflectance* is the fraction of the incident light reflected from a surface; for example, we might say that the reflectance of surface A is 30 percent, meaning that 30 percent is reflected and 70 percent absorbed. Reflectance and transmission are measured by a photometer, often called a *densitometer.*

Value is the sensation of lightness or darkness of a smooth tone (of any hue) as rated on a gray scale. A gray scale is a progression of gray tones from black to white evenly spaced according to some definition (Fig. 8.3). Because the apparent lightness or darkness of a color is affected by its surroundings, the values of two surfaces may be different, even though their reflectances are the same (Fig. 8.4). Furthermore, perceptual judgments of equal differences in value do not parallel equal physical differences in reflectance. Consequently, as we will see later, an equal-contrast gray scale is not the same as one in which black-white ratios change linearly.

Intensity. A third attribute of any color has to do with its richness. Some colors are brilliant, such as a strong red, while another example of that same red hue may be quite weak, as would be the case were we to mix the red with a large amount of gray of the same value. Various terms, such as chroma, saturation, and purity, are employed to describe this dimension of color in different systems of color description or specification. In each case they are precisely defined, and this gives the terms somewhat different meanings.

The dimensions of hue and value are uncomplicated attributes. On the other hand, the third

FIGURE 8.3 A photographic gray scale in which the steps are equal differences in black/white ratios, that is, percent reflectance measures. Note the apparent "waviness" caused by (visual) induction. Note also that a middle gray does not appear to be halfway between black and white. (From a Kodak Gray Scale, courtesy Eastman Kodak Co.)

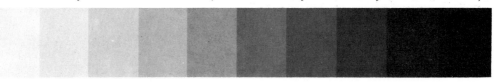

FIGURE 8.4 Visual value is a perception that cannot be measured instrumentally. The two gray spots have the same reflectance, but they are clearly different in value.

dimension—intensity, chroma, saturation, purity—whatever it is called, involves scaling the colors according to a combination of hues with whites and grays. All colors vary in their richness and brilliance, and in many circumstances it is convenient and more meaningful to refer to the relative "intensity" of a color, although the term has no precise definition in colorimetry.

SYSTEMS OF COLOR SPECIFICATION

Color has intrigued scientists, artists, and cartographers for centuries. Theories as to the nature of light and the physiologic basis for the occurrence of color in the eye-brain perceptual mechanism are very complex, and even today there is less than a full understanding of this mysterious phenomenon. On the other hand, the practice of employing colors either on a video screen or on a printed surface is an everyday occurrence. Furthermore, we can now describe and specify colors without ambiguity, so that they may be duplicated, detailed in plans and contracts, and studied with rigor. In general, the two areas, the production of colored maps on the one hand and the understanding of the graphic elements comprising color on the other, have been treated as quite separate subjects until recently. Ideally they are all part of the same subject, since selecting colors effectively requires an appreciation of color theory, perception, and identification, while preparing a display so that the chosen colors are obtained requires a basic understanding of color production technology.

In this section we will first examine briefly the tint screen system of specification, normally used in lithographic color printing, from the point of view of some of the perceptual consequences. A more detailed treatment of the actual technology will be found in Chapters 17 and 18. Then we will introduce the two systems of color specification used in rigorous research on color perception and widely employed in a great variety of industrial, commercial, and scientific activities.* One, known as the CIE system, is "objective" in the sense that it is based on instrumentation and the mathematical analysis of the physical characteristics of light. The other, known as the Munsell system, is "subjective," being based on human perceptual reactions to colors. This is not the place to go deeply into the theory and application of these two systems, but they will be briefly described, because the fundamental difference between their approaches to the phenomenon of color will help to impress on the mapmaker the complexity and importance of color in cartography.

The Tint Screen System

In lithography the printing surface either accepts or does not accept ink. Consequently, there are two ways of printing a gray (or a tint of some other color): (1) Use a gray ink, or (2) print a set of tiny, closely spaced dots. Generally the first alternative is uneconomical and the second is commonly used, with the cartographer or the printer using what are called *dot tint screens,* which serve as masks in the construction or printing stages of the process (see Chapter 18). These screens are precisely made films with closely spaced dots of a given size arranged in a rectangular pattern

*Ideally, treatment of the systems of specification should come first, but because the use of tint screens is helpful in understanding color, we begin with that.

on each one. The spacing of the dots in a set of screens is identified by the number of lines of dots per inch (lpi).* A set of 65-line screens is considered coarse because the dots can be seen, while a 150-line set is considered fine, since no pattern is visible and only a smooth tonal effect results. The rating of the tone produced by a dot tint screen is designated as the percentage of the tone produced by its use; therefore, a 10 percent screen will produce a light tone and an 80 percent screen, a dark tone.** The usual set of dot tint screens varies by 10 percent increments, although other percentages are available.† Figure 8.5 shows a series of 10 percent increments and the shapes of the dots produced.

Two-Color Combinations. Often the economics of map production and printing restrict the use of ink colors to one or two. When only one color is used, the choice of tint screens is usually limited to ten gradations plus the uninked paper color. Two inks provide a vastly increased number of possibilities, since any of the ten gradations of the one color can be combined with any of the ten gradations of the other color. The choice of a combination is made from a two-color chart. Many of the illustrations in this book use black and dark red printing inks, and Figure 8.6 is a two-color chart showing the possible combinations of those two colors.

Process Color Combinations. The reproduction by printing of paintings, color photographs, and other items with continuous tone color modulations is done by a technique called the four-color process (see Chapter 18). The four ink colors are the three subtractive primaries, cyan, magenta, and yellow, along with black. Inks match-

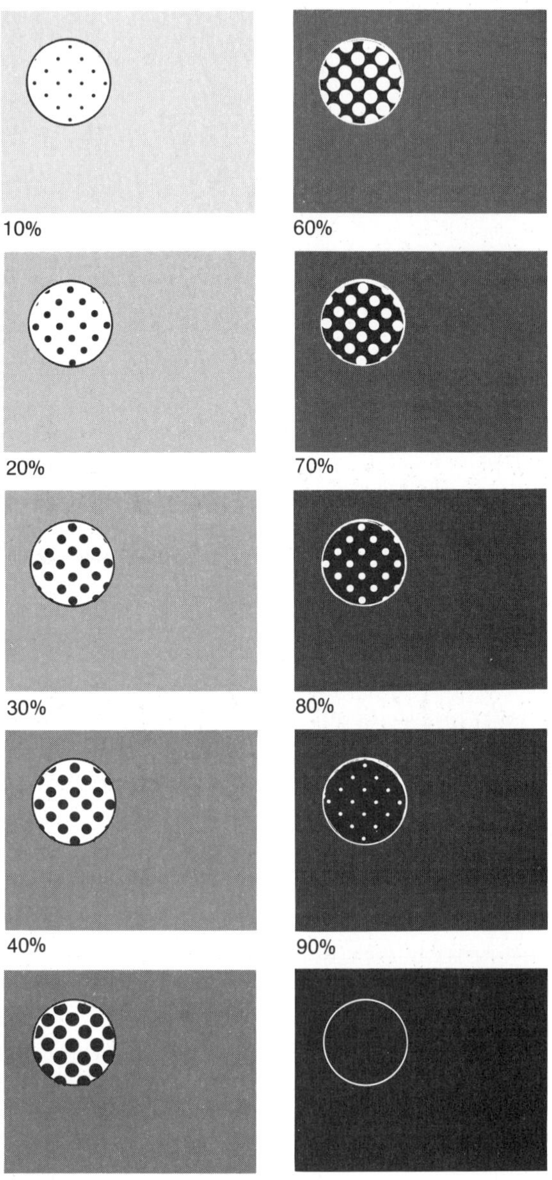

FIGURE 8.5 A series of dot tint screens with enlargements showing the shapes of the dots. Screens are negative and a screen that will produce a 30 percent tint will actually look like a 70 percent screen.

*In the International System of Units, the measure is lines per centimeter (l/cm).

**Screens are identified by their result, so that a 10 percent screen will be mostly opaque with small transparent dots; when exposed, only the areas of the transparent dots will print.

†Computer-driven laser platemaking techniques now make it possible to produce dots of any size, thus allowing any desired percentage of area inked, although most screening is still done in 10 percent increments. Various other kinds of screens are used in the graphic arts to produce special effects, such as straight line and mezzotint, but these have not been used extensively in cartography.

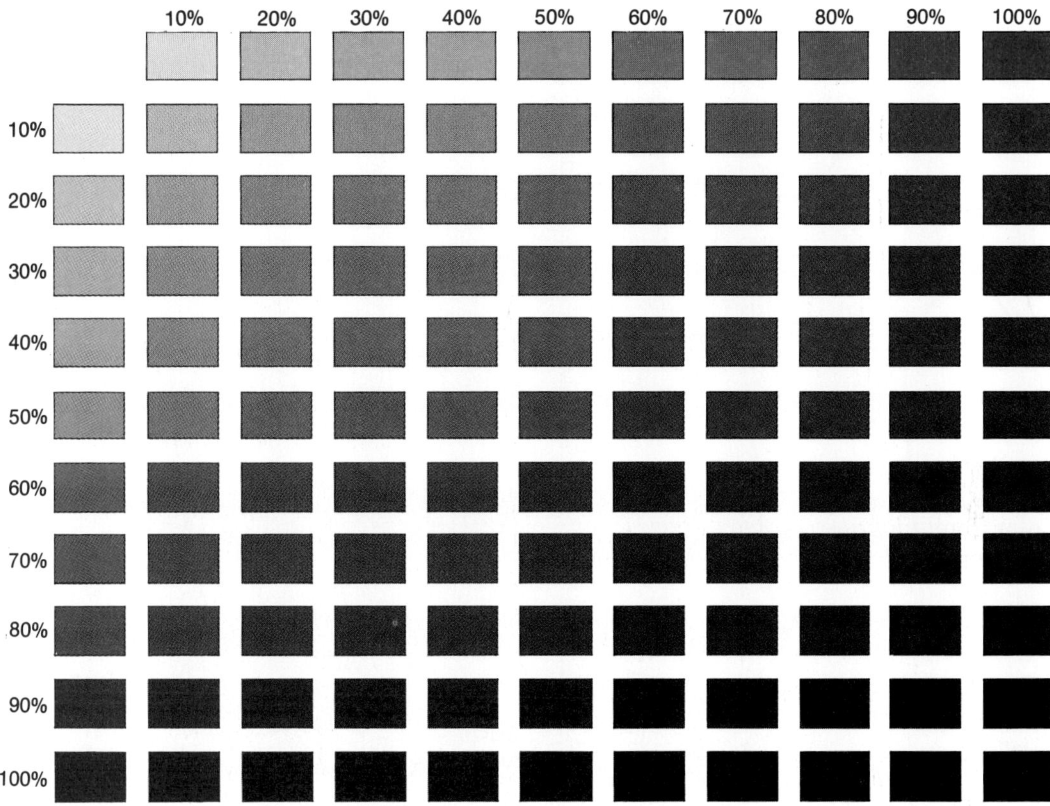

FIGURE 8.6 A two color tint screen chart showing the possible combinations of the two inks used in printing this book. It is apparent that relatively few of the combinations are usable, especially when both ink colors are dark.

ing the subtractive primaries are called "process" colors, and they can be used with the tint screen system to produce various hues. Figure 8.7 shows the spectral reflectance curves for solid printings of the three process inks used by the U.S. Defense Mapping Agency, and the four possible solid overprintings of the inks. Note the decrease in reflectance when two of the inks are overprinted and the fact that when all three are overprinted, a dark brown approaching black results.

When tint screens are combined, various intermediate hues are obtained. Prediction of the hue produced by overprinting screens is complicated by the facts that ink tends to spread, and that the screens are aligned at different angles causing some dots to be only partially overprinted. For example, an analysis of the consequences of overprinting 10 percent black, 20 percent yellow, 30 percent magenta and 40 percent cyan on white paper predicts the following percentages of colors in a unit area: white 20 percent, cyan 20 percent, black 14 percent, magenta 12 percent, blue (cyan plus magenta) 12 percent, yellow 7 percent, green (cyan plus yellow) 7 percent, red (magenta plus yellow) 4 percent, and dark brown (cyan plus yellow plus magenta) 4 percent. The colors are visually integrated, and in this case will give an impression of a particular blue with a value and intensity rating.*

*See A. Jon Kimerling, "Color Specification in Cartography," *The American Cartographer* 7, no. 2 (1980): 139–53, for a thorough analysis of this general problem. Example from p. 144.

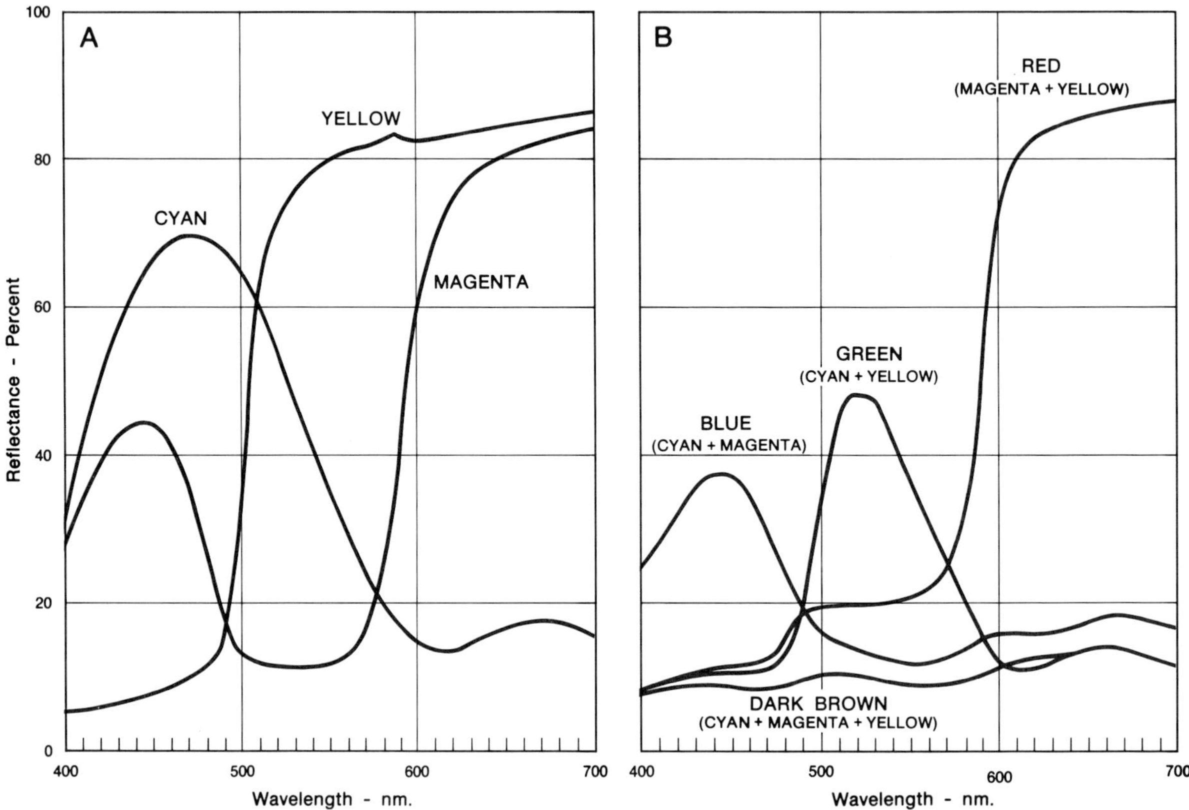

FIGURE 8.7 Smoothed reflectance curves showing: (A) the process inks used by the U.S. Defense Mapping Agency, and (B) the four solid (not screened) ink overprinted combinations (After Kimerling.)

The CIE System

The *Commission International de l'Éclairage* (CIE), also known as the International Commission on Illumination (ICI), has developed a widely used system of colorimetry. It allows the precise specification of any color in numerical terms. The procedure is employed by the U.S. National Bureau of Standards to define color, and it is the basis for a standardized color identification system employed in the U.S. Defense Mapping Agency.

There are three prerequisites for an objectively based scheme, namely, the definition of: (1) standard illuminants, (2) a standard observer, and (3) standard primaries.

CIE Standards. It is readily apparent that any surface can reflect only the wavelengths contained in the incident light. The CIE early defined several light sources: A, a tungsten incandescent lamp; B, noon sunlight; and C, average daylight from an overcast sky. Subsequently the CIE added several others. Each standard illuminant can be characterized by a spectral reflectance curve (Fig. 8.8).

Color is a perceptual phenomenon, and any useful system of specification must ultimately be based on that fact. In order to do this, the CIE system incorporates the color-matching abilities of a "standard observer," which is based on the responses of an average, normal human eye, which have been extensively studied.

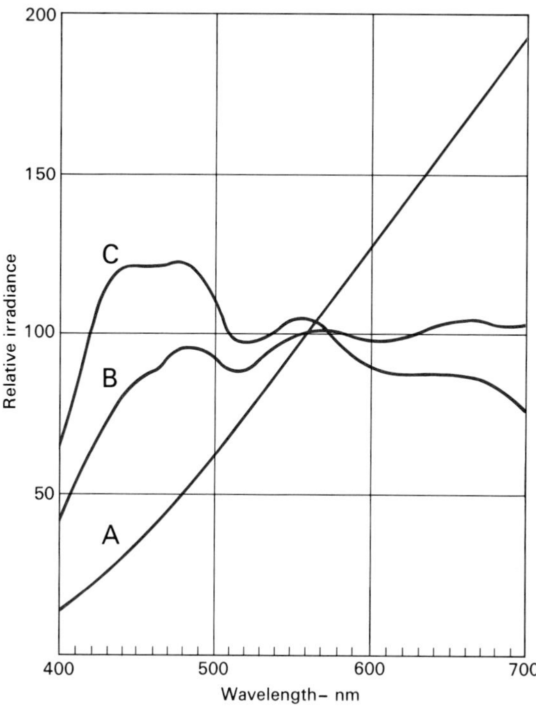

FIGURE 8.8 Relative spectral irradiance of CIE standard illumination sources: (A) incandescent lamp; (B) noon sunlight; and (C) average daylight from an overcast sky. (From Nimeroff.)

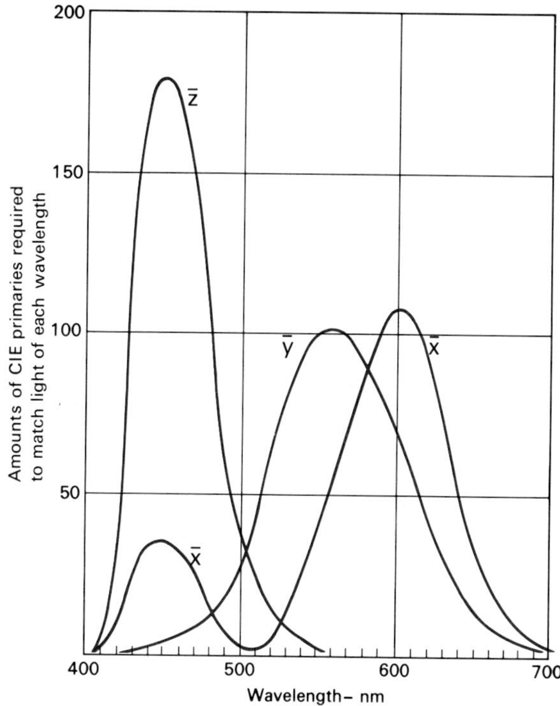

FIGURE 8.9 The CIE standard observer (1931) is defined by the amounts of the three selected primaries (lights) needed to be mixed to match the wavelengths of the spectrum.

As noted above, any color can be matched by mixing additively lights (not pigments) in suitable proportions of three primary colors—red, green, and blue—each of which will consist of a combination of wavelengths. The standard observer in the CIE system is defined by the amounts of the three primaries needed to match light of each spectral wavelength. These varying proportions can be plotted as spectral reflectance curves; they are designated $\bar{x}$, $\bar{y}$, and $\bar{z}$ and are shown in Figure 8.9.

Chromatic Coordinates. The required amounts of the three additive primaries needed to match any particular color are called the *tristimulus values* of that color. They are designated X, Y, Z. The tristimulus values are, in effect, the summation of the amounts of the red (X), green (Y), and blue (Z) primaries (in combination with each other, of course) needed to match each wavelength of the color in question. The tristimulus values are obtained by exposure to a device called a spectrophotometer, which gives the values by automatic computation. The tristimulus values are then converted to chromatic coordinates by obtaining the ratios $x = X/(X + Y + Z)$ and $y = Y/(X + Y + Z)$. The analogous ratio $z = Z/(X + Y + Z)$ is not needed; x and y carry all the information, since the sum of all three ratios is one. The values x and y are the fractional amounts of the red and green primaries needed to match the hue and saturation of the particular color. In the CIE system tristimulus value Y equals the total luminous reflectance ("lightness" or "darkness") of the color. The chromatic coordinates x and y are then entered on the CIE chromaticity diagram.

The chromaticity diagram is shown in Figure 8.10. The smooth red curve shows the gamut of points (locus) representing all the hues of the visible spectrum. The positions along this line of selected wavelengths (nanometers) are identified. The sloping straight line in the lower portion of the diagram represents the magentas and purples, which fall between red and blue but do not occur in the spectrum. The use of the chromaticity diagram is as follows:

1. The chromatic coordinates of the illuminant, for example (C), are plotted. This position on the diagram is achromatic (without color) for that illuminant. One can visualize the chromaticity diagram as showing all the spectral hues at full saturation around the inside of the locus, all of which fade in saturation toward C until there is no color at the point representing the illuminant.
2. The intersection of the chromatic coordinates of the color being analyzed is plotted. For example, a green ink might have chromatic coordinates of $x = 0.212$ and $y = 0.348$. These are plotted as F on Figure 8.10.
3. A straight line from C through F will intersect the locus of the spectrum at G. This intersection designates a wavelength which is called the *dominant wavelength*. In the illustration this is 495 nm.
4. The ratio CF to CG (percent) defines the *purity* of the color.

   The *luminosity* (luminous reflectance) of the color is given by tristimulus value Y.

FIGURE 8.10 The chromaticity diagram (also called a Maxwell diagram) shows the CIE primary (additive) colors at the apexes. The gamut of the spectral hues is shown by the smooth curve (locus), and the sloping line between 400 and 700 nm shows the chromaticities of the mixture of the extremes of the visible spectrum. These do not occur in the spectrum. All physically realizable chromaticities occur within these limits. See text for explanation of plotted points C, F, and G.

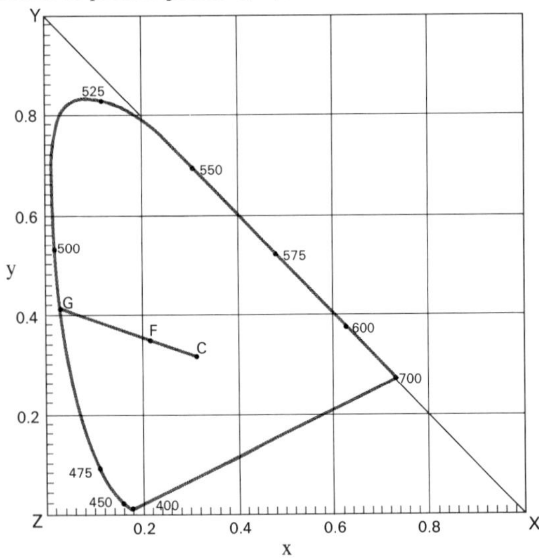

The three values, *dominant wavelength, purity,* and *luminosity,* specify the color in the CIE system. The *dominant wavelength* is the hue and may be defined as the monochromatic light which, when mixed in suitable proportion with the illuminant light, will match the color in question. The *purity* is the degree of saturation ranging from full color (100 percent) to the achromatic stimulus of the specified illuminant. The *luminosity* is the lightness or darkness of the color.

The U.S. Defense Mapping Agency has developed a color identification system based upon CIE color specifications. The relationship among overprintings of the process colors and the form of that color space in the CIE system is illustrated in Figure 8.11 and Color Figure 4. In Figure 8.11 the hexagon outlines in plan the range of (predicted) possible CIE coordinates for all possible combinations of solid cyan, magenta, and yellow inks with the range of 0 to 100 screen percentages of the other two. If the third dimension, Y (luminosity), were added, the position of the graphed hexagon in Figure 8.11 would be deformed with the yellow apex high, the blue apex low, and the other apexes at intermediate levels "above" the plane of the chromaticity diagram. The perspective diagram in Color Figure 4 shows a view, from the magenta-red side, of the color space or solid defined by the process inks. The

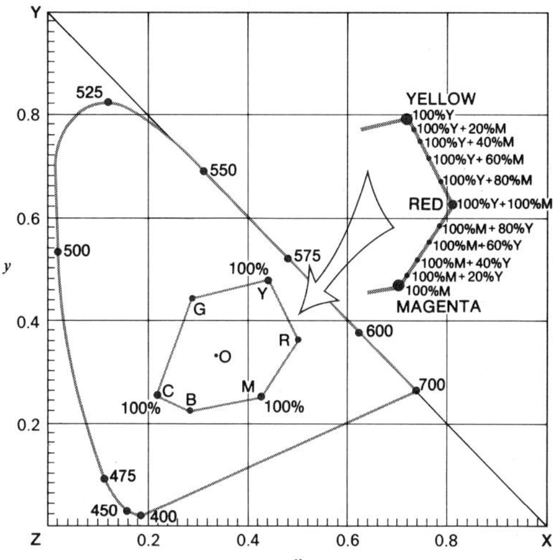

FIGURE 8.11 The plan representation on the chromaticity diagram of the (predicted) locations of all possible 20 percent interval dot screen combinations of each of the U.S. Defense Mapping Agency process colors with the other two. All combinations of all three colors and less than 100 percent screen values will lie within the plan hexagon. See also Color Figure 4. (After Kimerling.)

color solid should be visualized as an irregular hexagonal prism with facets converging to two points, very dark brown (theoretically, black) at the bottom and white at the top.

The great advantage of the CIE system is that any color may be precisely specified in physical terms. It must be emphasized, however, that it is based on the characteristics of additive light and not on the characteristics of subtractive pigments and our perception of them. Except for computer-generated map displays on color cathode ray tubes (still not a widespread practice), it is pigments that are put on printed maps and it is people (who are not spectrophotometers) who look at them. Consequently, although the CIE system provides a rigorous method for analyzing and specifying colors, which is extremely useful in production and research, it is apparent from the chromaticity diagram and Color Figure 4 that equal distances in the color solid do not represent equal visual differences. A mathematical transformation is possible, but at present, systems in which color differences are based entirely on human perceptual judgments are the closest we have to that ideal. Several such systems have been advanced; the best-known and the one most widely used in the United States is the Munsell system.

The Munsell System

This system, named for its originator, A. H. Munsell, an American painter and student of color, identifies a color in terms of three dimensions: hue, value, and chroma. Each dimension is divided into a sequence of steps arranged so that the steps are equally spaced from a *perceptual* point of view. Each step in each scale is assigned a designation, and the combinations of these constitute a notation system that refer to the more than 1500 actual samples of the Munsell colors.

The Munsell color system is used by industry and government for many purposes that range from matching soil colors and color photogrammetry in survey work to color coding of wires in electronics. It forms the basis for a standardized system of color names worked out by the Inter-Society Color Council and the National Bureau of Standards (ISCC-NBS). Munsell colors physically exist as painted chips.*

Munsell Dimensions. Each color is specified by reference to scales of hue, value, and chroma. Hues are equally spaced from one another in the sense of equal visual steps from one hue to the next. The one hundred hues are arranged in a circle so that each hue is perceptually related to the one next to it (in other words, adjacent hues are chromatically similar). There are ten major hues, arranged as shown in Figure 8.12 and illustrated in Color Figure 5. The order of the hues corresponds to the order of the hues in the spectrum with the addition of purples. A clockwise

*Information and catalogs of materials are available from Munsell Color, Macbeth Division of Kollmorgen Corp., 2441 No. Calvert St., Baltimore, MD 21218.

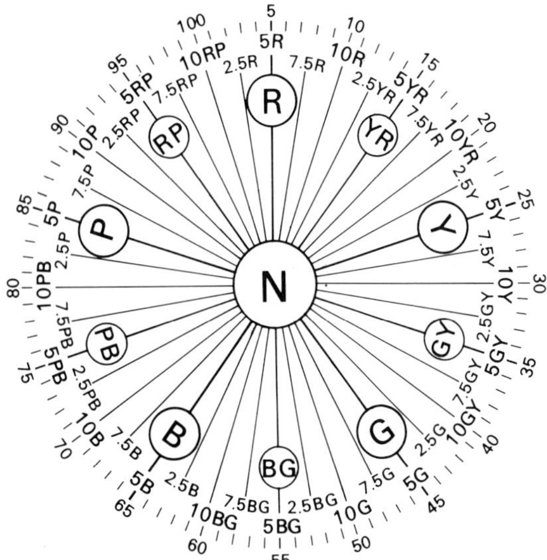

FIGURE 8.12 The Munsell circuit of 100 equally spaced hues. The ten major hues, ten steps apart, are named R (red), YR (yellow-red), Y (yellow), GY (green-yellow), G (green), BG (blue-green), B (blue), PB (purple-blue), P (purple), and RP (red-purple). They increase in chroma outward from an achromatic stimulus, N (neutral), at the center. Hues opposite one another on the circuit are called complementary colors. (Courtesy Munsell Color, Macbeth Division of Kollmorgen Corporation.)

progression of colors in the hue circle is similar in order to a counterclockwise progression around the spectral wavelength locus on the CIE chromaticity diagram.

The ten major hues, consisting of five principals and five intermediates, may be referred to by their initials (for example, R for red or BG for blue green). More commonly, each of the spaces between the major hues is divided equally (visually) into four divisions, as shown in Figure 8.12, and these are designated by a numerical/letter combination, such as 2.5G or 10YR. Numerals alone from 1 to 100 can be used.

All colors are ranked as to value in relation to a range of achromatic grays, from black to progressively lighter tones to white; this is called the *value scale.* Although the value scale is continuous from "pure" black (low value) to "pure" white, it is usually thought of as consisting of the steps shown in Figure 8.13. Black is designated as 0/ and white as 10/ on the scale, with equally spaced grays between. A tone midway between black and white would be designated as 5/. Fine distinctions can be made by employing decimals. The spacing on the Munsell value scale is perceptual and is not the same as a photographic gray scale (Fig. 8.3) in which the spacing is made by equal changes in the physical black-white ratios.

The dimension of chroma in the Munsell system is the degree to which a hue departs from a gray tone of the same value. It is analogous to the descriptive term intensity and to the concept of purity (saturation) in the CIE system, except that in a chroma range a value ranking is specified and is kept constant. The scales of chroma extend from /0 for a neutral gray in equal steps toward full saturation for that hue and value. Color Figure 6 shows the ranges of value and chroma for complementary Munsell hues 5.0 blue and 5.0 yellow-red.

Munsell Notation and Color Solid. The relationship among the three dimensions of color in the Munsell system can be visualized as forming a three-dimensional color space or solid. We may use the earth as an analogy (Fig. 8.14). The axis of the earth forms the scale of values from black at one pole to white at the other. Halfway between, at the plane of the equator, is value 5/. The hue circle is arranged longitudinally with the higher chroma values occurring at the "equator." In this view "latitude" from one pole to the other represents value and "longitude" represents hue. Chroma would vary along latitudinal planes from high chromas at the surface inward to the achromatic axis.

Because hues vary in their intrinsic lightness and because of the physical limitations of materials, the Munsell color solid, defined by the outer series of colors, is irregular, assuming the shape illustrated in Figure 8.15. When selected color

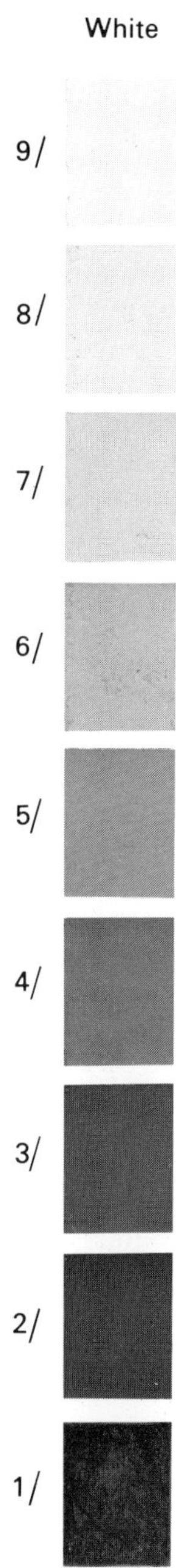

FIGURE 8.13 The Munsell value scale from 1/ to 9/ of equally spaced grays (0/ is black and 10/ is white). The range is continuous but is here shown in steps. (Courtesy Munsell Color, Macbeth Division of Killmorgen Corporation.)

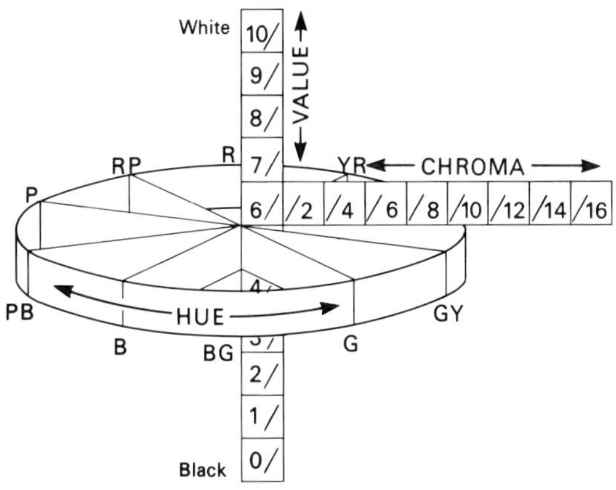

FIGURE 8.14 The arrangement of the Munsell hue, value, and chroma scales in color space. If the hue circuit at value 5/ and full chroma were made to correspond to the equator of the earth, then value would be latitude (0 to 180 degrees), hue would be longitude (0 to 360 degrees), and chroma would increase with perpendicular distance outward from the axis. See also Color Figure 7.

chips are arranged on vertical planes radiating out from an axis, they form the "color tree" shown as Color Figure 7.

In Munsell notation the designation of a particular color is given symbolically by H V/C; H is the hue designation, such as 5R; V/ is the value ranking; and /C is the chroma. A strong red might be 5R 5/14, a pale pink 5R 8/6, and a deep red 5R 3/10.

All the Munsell colors have been analyzed by the National Bureau of Standards according to the CIE system, and the relation between the two systems may be appreciated by viewing the plot on the chromaticity diagram (for illuminant C) of the ten major hues at value level 5/ and chroma level /6, the highest even chroma level at which all the hues exist (Fig. 8.16). Although the hues are equidistant visually in the Munsell system, they arrange irregularly on the chromaticity diagram. Furthermore, all the hues are the same chroma level but, as is shown by the relative distances from C to the points on the locus, in the CIE system they have marked differences in purity.

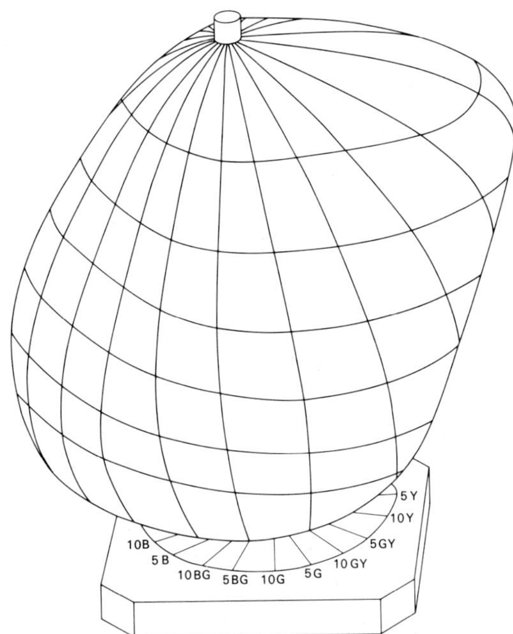

FIGURE 8.15 The three-dimensional shape of the Munsell color space as seen from the blue-green side.

FIGURE 8.16 The chromaticity coordinates (x and y) of the ten major Munsell hues (at value 5/ and chroma /6) plotted on the CIE chromaticity diagram. (Data from Kelly, Gibson, and Nickerson.)

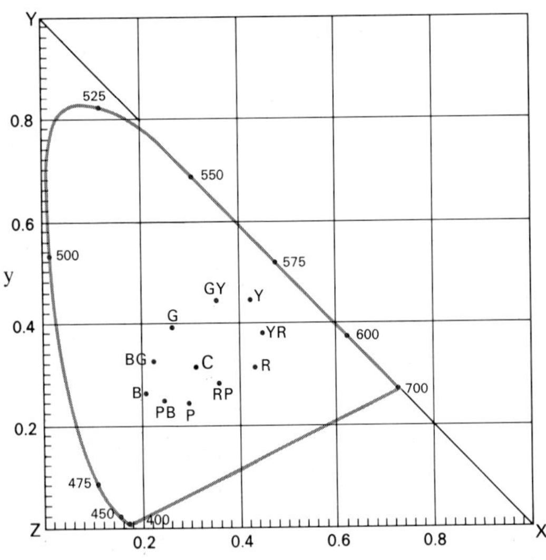

Systems for Computer Graphics

The availability of color in electronic displays has made it necessary to develop systems of color specification that can be used to direct the computer to produce colored lines, areas, and other symbols. As in color television, the displays employ light to create the effects, so the systems are based upon the three additive primaries of red, green, and blue. By modulating the relative output of the three primaries, the various hues, values, and saturations can be attained.

The manipulation of color for cartographic displays on the screen of a cathode ray tube is as yet relatively uncommon, although it is widely employed in motion pictures and television for backgrounds and especially for advertising and logos. Nevertheless, the student should be aware of the existence of such systems. They are comparable to the subtractive tint screen systems used to designate colors for printing. We can look forward to the day when the colors on the screen can be readily duplicated automatically in hardcopy output, either directly as a graphic or indirectly as film for printing. Several systems for hard-copy output with electronically controlled ink jets are now in use. Of course, they employ the subtractive primaries, and algorithms are available for the conversion of additive color specifications to the equivalent subtractive, and vice versa.

The basic requirement of such a system is that any color (hue, value, saturation) be specified by a unique position in a color space in which all the achievable combinations of the additive primaries can occur. Several such systems are now in use. One in wide use is called the RGB model, so named for the three additive primaries, red, green, and blue. Although the exact form of the system depends upon the software, the RGB model may be visualized as a cube in which position is specified by values for the *x*, *y*, and *z* coordinates, namely, red, green, and blue, respectively (Fig. 8.17). The maximum possible number of coordinate intervals is 255, which if they were

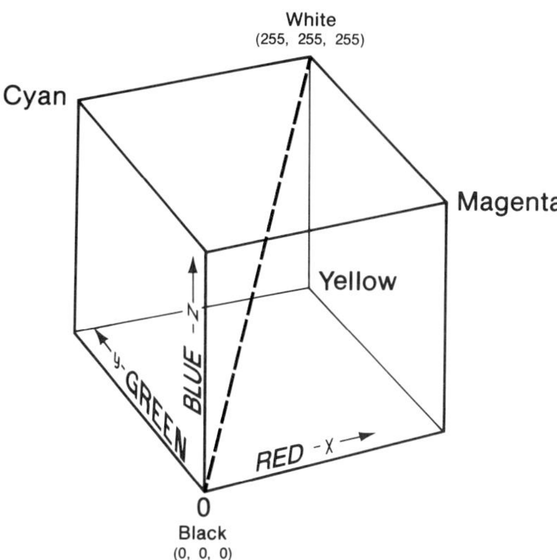

FIGURE 8.17 Diagram illustrating the RGB color model. The red diagonal is the locus of equal amounts of the three additive primaries and is thus the progression of the achromatic values (neutral grays) from black to white.

all made available would permit 255^3 (= 16,581,375) combinations of the three primaries. Naturally, that is far more than are needed, and the coordinate steps (not necessarily equal) are chosen to fit the needs. For example, if fifteen equal steps in each dimension were selected, it would allow 3,375 combinations of the primaries.

In the RGB cube a coordinate position of (0, 0, 0) designates black (without light) while the opposite corner with a coordinate location of (255, 255, 255), the maximum of the three primaries, denotes white. The diagonal between those two corners becomes the locus of all achromatic values. The other corners of the cube represent hues resulting from the mixtures of two of the additive primaries, blue + green = cyan, blue + red = magenta, and red + green = yellow (see Color Fig. 2).

There are other basic models, such as HSV (hue, saturation, value), HLS (hue, lightness, saturation), which employ other systems for specifying a position in color space. The exact form of each model depends upon the system and its software.*

CHOOSING COLORS FOR MAPS

Colors are used on maps in a great variety of ways. On some maps large areas, such as oceans or land areas, are all colored alike, while on others only a few line symbols, such as roads, streams, or contours, are colored. The range of categories on quantitative maps is often symbolized with colors to enhance the display, as on maps showing statistics by enumeration districts or relief maps showing various elevation levels. In all such cases, the choice of colors to be used is vital to the map's success in portraying the data effectively and, equally important, in not creating bothersome combinations and contrasts that are garish or draw unwanted attention.

As a general precept, the choice of how the graphic elements are combined and modulated with the signs should be made on the basis of how best to serve the objective of the map. That rather self-evident procedure is by no means as simple as it might appear. Designing a map, as with most other primarily functional things, involves making choices among conflicting principles and objectives, which is why design problems often lead to compromises. For example, we may need to differentiate a set of lines by hue, but other aspects require that the lines be very thin; unfortunately, it is difficult to distinguish the colors of thin lines. The only procedure is to recognize the conflict and assign priorities.

In discussing the choice of colors, it is necessary to recognize a third set of primary colors in addition to the additive and subtractive. This third set consists of the *perceptual* primaries, which comprise those hues we perceive as being dis-

*James D. Foley and Andries Van Dam, *Fundamentals of Interactive Computer Graphics* (Reading, Mass.: Addison-Wesley Publishing Company, 1982), pp. 593–623.

tinctive. These are blue, green, yellow, red, brown, black, and white. All the thousands of other colors appear to us as mixtures of two or more of the perceptual primaries as, for example: pink (red and white), orange (red and yellow), purple (red and blue), turquoise (blue and green), tan (brown and yellow), and so on. Particularly important is the fact that we see black and white as colors, even though we know that from the physical point of view, white is a combination of all wavelengths and black is the absence of any light.

Another aspect of perception is important. A map appears as a complex of symbols either luminous on a screen or printed on paper, both of which serve as a background somewhat in the sense of the figure-ground phenomenon described in the last chapter. Ordinarily we do not think of the background as being one of the colors of the map. Thus, we think of a map printed in one ink on contrasting paper, usually lighter, as being a one-color or monochromatic map, even if it were printed with green ink on tan paper. On the other hand, we can print some areas of a map in solid or dark screens of the ink and arrange for lines and lettering on those areas to be unprinted so that they appear as light on a dark background. When this is done, the marks that are lighter than their background are called "reverse." Luminous marks on a screen are not called reverse.

For printed maps the cartographer has available the solid colors of the inks to be used in the printing, the separate tint screen values of those inks, and the colors that result from their tint screen combinations. If only one color is available, it is usually black printed on a white background. With one color the graphic element value is available along with the others, such as size, shape, and so on. With two colors, the combination of hues and values allows some variation in intensity. With all three process colors, all three dimensions of color are available. The three dimensions of color are not equally significant from the perceptual point of view.

Value (relative darkness or lightness) is a primary graphic element, since it is the basic factor in the recognition of graphic differences. The interplay of light and dark, as for example in the type on this page and as in shadows and other such contrasts, is essential to clarity, legibility, and the recognition of three-dimensional form. Without value contrast, it would be difficult to see differences among map marks.

Value is the critical dimension from the point of view of perceptibility, and for that reason, it must logically rate as the most significant color dimension. On the other hand, value contrast is such a common aspect of seeing that we tend to take it for granted, and instead of value, hue is the dimension that draws by far the most interest. For that reason, it ranks as the primary dimension of color. People actively like or dislike hues and find them intriguing; when asked to comment on the appearance of something, they usually mention hue before value or intensity. Different cultures rank hues differently in desirability. We tend to associate hues with various reactions such as warm and cool, and hues are profoundly involved in aesthetic reactions.

Intensity (chroma, purity) seems to be by far the least significant of the three dimensions of color although on occasion, especially when misused, it can be extremely important.

It bears repeating that the three dimensions do not occur separately. Although we may react quite positively to the "redness" of a bright red ink, it also needs to be judged as to its value and intensity.

Principles in the Choice of Colors

Color is sometimes a decorative element in maps, but in most cases, it is used to symbolize or enhance the symbolism of data. Accordingly, we shall concern ourselves here with the essentially nonaesthetic aspects of its perception. Some of these are *physiologically* based; that is, the human eye-mind perceptual mechanism functions in certain ways, which set primary limitations on what we can do. Some of the perceptual patterns are *psychologically* based, resulting in a variety of connotative and subjective reactions to color. Because of the importance of color in mapping and because of its long use, a variety of *conventions* have developed. Some are old and some are new, but all are important, if for no other reason than if we violate a convention, we may raise questions in the reader's mind and diminish the effectiveness of the communication.

In the following sections we will examine briefly the perceptual characteristics of hue, value, and intensity (chroma) and, in some cases, we will divide the consideration into those aspects that are physiologically based, those that are subjective and connotative, and those that are conventional. It is important to remember two things. One is that there are many aspects of the employment of color that are still mysterious and have not been sufficiently investigated to allow one to formulate valid precepts. The other is that, as with many situations in graphic design, everything desirable often is not possible; geographical facts, perceptual precepts, and cartographic objectives are often in conflict.

The rapidly growing use of various classes of computers with screens on which maps can be displayed raises numerous questions about the choice of colors. In many instruments color is not available or is severely limited, but in any case the image is luminous on a dark gray background. This is the opposite of the usual printed map with its darker image on a lighter background. Consequently, we may assume that displays and reactions to hues, values, and intensities may be quite different.

Soundly based precepts for the use of hues, values, and intensities on luminous maps on the screens of cathode ray tubes are few compared to the guidelines for hard-copy maps, whether they are being prepared for printing or for production by a computer-driven printer. Most of the following discussion concerns hard-copy maps.

The Employment of Hue

The physiological reactions of one individual to hue are likely to be slightly different from those of anyone else; yet, people have much in common. A number of generalizations of significance in cartography can be made having to do with sensitivity to hue, the apparent advance and retreat of hues, their relative luminosity, the relation of hue to visual acuity, and the phenomenon of simultaneous contrast.

Sensitivity. Most people are quite sensitive to differences in hues when the colors are placed side by side against a neutral (medium gray) background. On the other hand, the ability to recollect hues and to retain an impression, such as from a legend to a map, appears to be severely restricted, especially when the surroundings of the colors are different. As a general rule, then, when hues are being employed to distinguish one thing from another, they should be made as different as possible within the logic of the graphic hierarchical organization of the map. It should also be remembered that between 4 and 5 percent of the population has some degree of deficiency in color vision, which should emphasize the need for considerable differences in hue.

An aspect of sensitivity that is important in cartography is the fact that when colored symbols (lines or dots) are very thin or small, it is difficult to distinguish hues, and additional contrasts of other graphic elements, such as shape, is often necessary.

Our sensitivity to hues varies; the eye is "attracted" more to some hues than others. No precise values of relationship are available, but most students rank the hues ordinally in the following order (most sensitive first): red, green, yellow, blue, and purple.

Advance and Retreat. One physiological phenomenon that has had a profound effect on cartography by influencing a convention is that light rays entering the eye are refracted in inverse relation to wavelength. This means that, theoretically, blue items focus in front of the retina, while red ones focus behind the retina. In fact, the eye accommodates with the result that a red object should appear slightly nearer than a blue. The recognition of this relationship led to the proposal in the late nineteenth century of arranging hues in spectral order to represent elevations on relief maps; the highest elevations, since they were red, would therefore appear closer to one looking "down" on the map. In fact, the effect is slight, and so many other factors are involved, such as the combinations of wavelengths that usually make up a color, that the phenomenon of advance and retreat can be ignored. Nevertheless, the spectral progression of blues for water, greens for lowlands, through yellows, buffs, and reddish colors at the highest elevations is a strong convention in cartography.

Visual Acuity. Many maps show a wealth of detail by fine lines and point signs on a colored background. Other things being equal, the more monochromatic the background, the easier it is for the eye to resolve the detail. A background color such as yellow, which combines high concentration in a narrow spectral range with high luminosity, will display fine black detail well. On the other hand, background colors that mix a variety of wavelengths, such as brown, would make it more difficult to distinguish fine complexity. It is unfortunate that a medium brown has become the conventional color for contours; they are difficult to follow in complex graphic situations.

Blue is relatively poor as a defining color; as a consequence, an involved blue coastline with a blue tint on the water areas makes it difficult to see the line clearly. Because blue is conventionally so employed, one must see that value relationships are used to assist as much as possible.

Simultaneous Contrast. Whenever two colors are adjacent, they modify one another, not only in hue, but also in value. When a hue is surrounded by another color, the surrounded color tends to shift in appearance toward the complementary color of the background. For example, a green on a yellow background will appear bluer than the same green on a blue background (Color Fig. 8). The phenomenon of simultaneous contrast is more significant in cartography in connection with value, but its existence in connection with hue makes it necessary to refrain from using hues that are very similar, since they may be diffucult to identify in different parts of a colored map.

There are three subjective aspects of the perception of hue that the cartographer must keep in mind when using color on maps: the phenomenon of individuality of hues, their symbolic connotations, and their affective value.

Individuality of Hues. As mentioned earlier, the perceptual primaries are hues that appear quite unique and distinct, while other colors look like combinations of these individual hues. For example, it is well known that a mixture of blue and yellow pigments results in green, but green does not look like a mixture, although orange does look like a mixture of red and yellow. Many such apparent mixtures are possible, such as reddish brown, greenish yellow and, of course all the tints (white plus hue) and the shades (gray plus hue).

The phenomenon of individual hues is important in cartography in two ways. Individual hues should be used to symbolize distinctly different phenomena, and apparent mixtures should be used to portray items that share some of the attributes that are symbolized separately by the individual hues. For example, an area of predominately ethnic group A might be blue, another area mostly ethnic group B might be red, and a third area of mixed A and B might be purple.

Symbolic Connotations. Colors are widely associated with sensibilities and moods. Green is cool, red is warm, buffs are dry, blue is wet, yellows are sunny, and so on. Different cultures associate hues with various meanings, and the diverse symbolisms (for example, white for purity, black for mourning in Western culture) and significances are very complex.* Most such attributes of color are not directly important to the cartographer, but when using hues as area symbols, the need to parallel portrayals of things such as temperature, wetness, and so on with generally associated hues is apparent. Furthermore, if a map is to be attractive and the cartographer wishes the general reaction of viewers to be one of confidence, the color associations would be different than if the objective were to suggest the inherent complexity of a distribution.**

Affective Value. Some colors are liked more than others. As is the case with symbolic connotations, this preference apparently varies from one culture to another. Generally, the cartographer is primarily concerned with designing an effective communication, and aesthetics plays a minor role. Nevertheless, other things being equal, it is better to employ colors that are liked. Studies of affective value in the United States—which probably reflect Western culture—suggest that blue, red, and green are generally considered "pleasant"; blue is at the top, while violet, orange, and yellow are rated significantly lower. The differences do not seem to be large and would depend greatly on the specific character of a color and its strength.

During the long history of cartography, many color conventions have developed. Some are based on the connotations of hue, some have been contrived in order to standardize symbolism, and some have grown up more or less by accident. An example of a connotative convention is blue water; an example of a devised convention is red for igneous rocks, pinks for metamorphic, yellows for the Tertiary period, and so on; and an example of an accidental convention would be brown contours. The subject of color conventions in cartography is very complex, and space does not allow a full treatment. Some conventions are generally employed, while other, very complex systems have been devised for particular kinds of mapping, such as hydrogeologic or vegetation mapping. An example of a generally followed convention is the spectral progression for relief maps referred to earlier and discussed in a later section. Some conventions are conventions only in a limited area, such as a country. Some are widespread. Only a few of the generally accepted, broad, conventional uses of hue can be listed here:

1. Blue—water, cool, positive numerical values.
2. Green—vegetation, lowlands, forests.
3. Yellow, tan—dryness, paucity of vegetation, intermediate elevations.
4. Brown—landforms (mountains, hills, and so on), contours.
5. Red—warm, important items (roads, cities, and the like).

The Employment of Value

Value, the impression of lightness or darkness, is basic to clarity and legibility because of the importance of value contrast in definition. The basic precept is that the greater the value contrast, the greater the definition, and the greater the clarity and legibility.

As with the graphic element of hue, the effective employment of value by the cartographer is based on a number of perceptual aspects. Some of these are physiologically based, some are founded on our subjective reactions, and some are established conventions.

Of great importance is the fact that physiologically we are not very sensitive to differences in value, and our ability to recall or recognize a particular value is limited. When a quantative geographical series is being represented by value differences of a single hue, such as on a precipitation map or a choropleth map of economic

*See Henry Dreyfuss, *Symbol Sourcebook* (New York: McGraw-Hill Book Co., 1972), pp. 231–46.

**The significance of the "look" of maps is discussed in Barbara Bartz Petchenik, "A Verbal Approach to Characterizing the Look of Maps," *The American Cartographer* 1, no. 1 (1974): 63–71.

data, it is wise to limit the symbolization to four or five value steps, not including solid and white. If the hue being used is inherently light, even fewer steps can be employed.

Our limited sensitivity is further complicated by the phenomenon of simultaneous contrast, illustrated earlier by Figures 8.3 and 8.4. When different tones are used on a map, the value of any one area is affected by the adjacent lighter or darker tones. This phenomenon, called *induction* (not to be confused with induction in cartographic generalization), may be lessened by edging the tones with black (or very dark) or white lines. Although the use of smooth tones on a map may be aesthetically pleasing, the cartographer must be wary of potential recognition problems. In many instances it is wise to add some pattern to the tones to aid their recognition.

Another physiologically based perceptual phenomenon is irradiation, the apparent spread of the light at a dark edge, which occurs with extreme values, especially black on white and white on black. Given the same width of line as, for example, in type, the dark on a light background will appear slightly smaller than the reverse (white on a dark background) (Fig. 8.18).

Perhaps the most important of the subjective and connotative aspects is the fact that variation in value conveys an implication of magnitude variation; in printed maps the darker indicates more of whatever is being symbolized. It makes no difference whether gray or some other hue is being employed; we naturally make the deeper water, the denser population area, and the greater per capita income darker than the areas of lesser magnitude. Some phenomena in which the range from low to high magnitudes is inversely related create problems. For example, a high rate of literacy is the same as a low rate of illiteracy. In such instances, the cartographer must judge which aspect is to receive the emphasis.

There is an additional subjective and conventional aspect that sometimes aids the cartographer in making a decision when faced with the kind of problem just mentioned. Where a range of value steps is used to portray a distribution that has qualitative implications, the darker is usually associated with the less favorable. In the example above, we would normally use the darker values for the higher rates of illiteracy (or the lower rates of literacy) on the ground that not being able to read is unfortunate.

One other subjective aspect of value in cartography is important in the overall graphic plan for the symbolism on a map. Extreme values tend to dominate a composition, and the larger the area with an extreme value, the greater the dominance. What constitutes an extreme value depends on the graphic context, so to speak. White and black are the extremes in the range of values, but maps are usually printed on white or light-colored paper. Ordinarily, then, black would be perceived as the extreme, and large areas of black, such as if a considerable area of ocean were made black, would draw the attention. This would be undesirable if, in fact, the land areas were the regions of main concern. The opposite would occur if a white area were the only high-value region on an otherwise dark map.

The Equal-Contrast Gray Scale. The quantitative variations in many geographical phenomena, such as amounts of precipitation, temperature, depth of water, intensity of land use, and so on, are usually depicted by some technique that depends for its effectiveness on a series of classes, which usually are graphically differentiated by value contrasts. It is desirable to make the value contrasts correspond to the differences between the data classes. For example, if we have a range of five equally spaced data classes (e.g., 0–20, 20–40, 40–60, 60–80, 80–100), which are to be symbolized with value contrasts obtained with black ink and dot tint screens, it would be appropriate to choose five steps equally spaced in terms of value contrast on the gray scale. There are four options: (1) white plus three screens plus solid black; (2) white plus four screens; (3) solid black plus four screens; and (4) five screens. How to obtain equal steps turns out to be a surprisingly

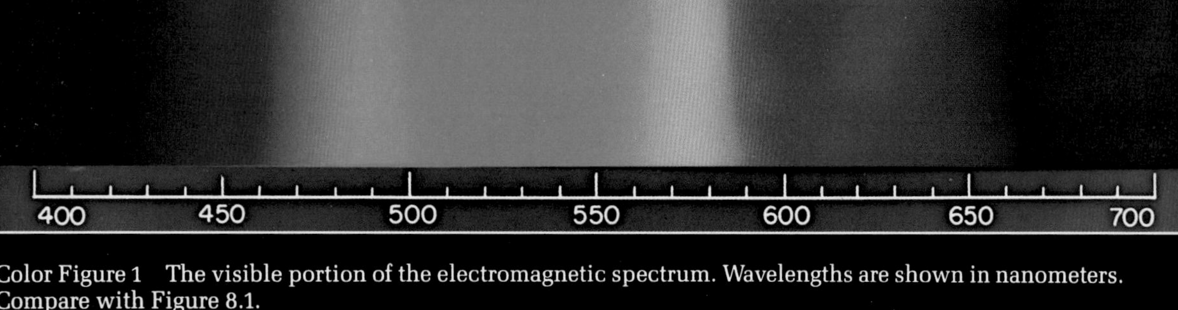

Color Figure 1 The visible portion of the electromagnetic spectrum. Wavelengths are shown in nanometers. Compare with Figure 8.1.

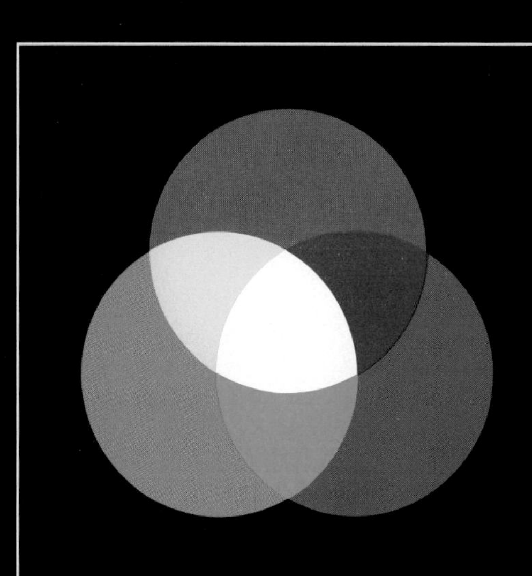

Color Figure 2 Additive mixing of light primaries as in electronic displays. Color cathode ray tubes (for example, television) use this form of color mixing.

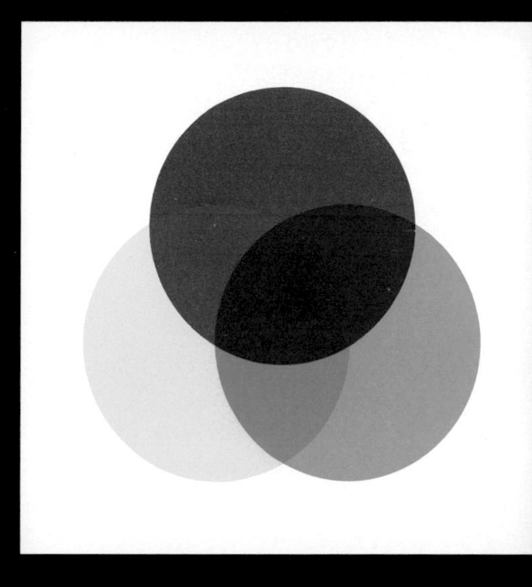

Color Figure 3 Subtractive mixing of pigment primaries as in painting and printing.

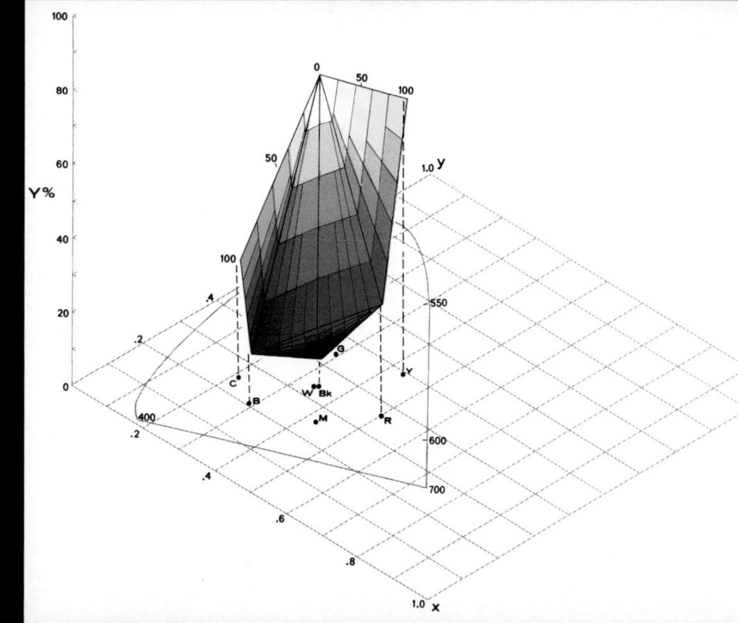

Color Figure 4 Color solid showing combinations of process inks in CIE chromaticity space as seen from the magenta-red side. The large number of possible combinations of the three inks are represented in the solid by only five screen percentages: 0, 25, 50, 75, and 100 percent. (Courtesy of A. Jon Kimerling and *The American Cartographer*).

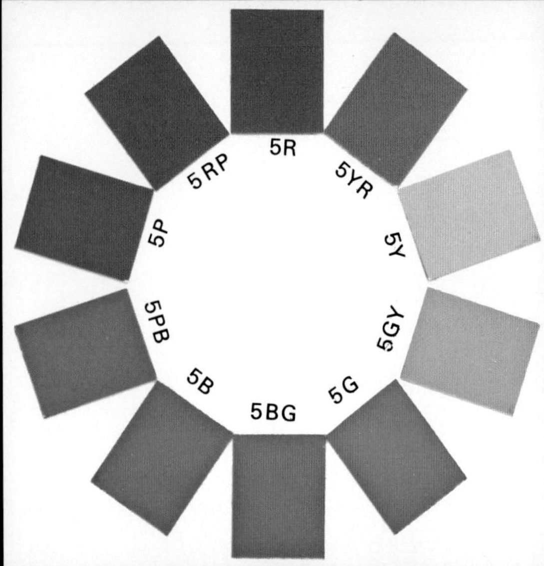

Color Figure 5 The Munsell color circle showing the five principal hues—red, yellow, green, blue, and purple—and the five intermediates at mid-value and strong chroma.

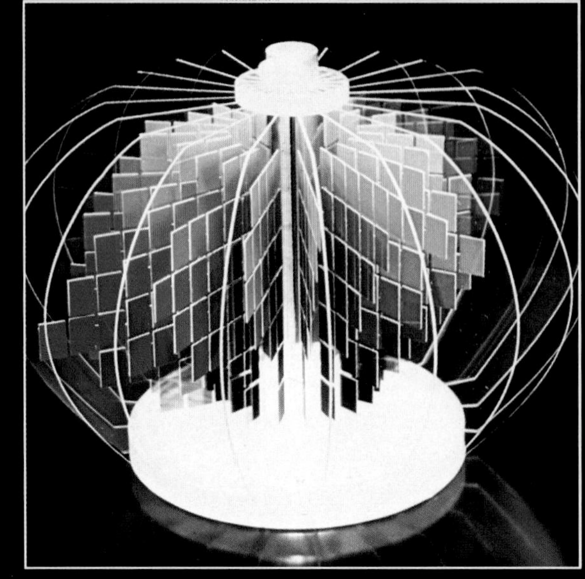

Color Figure 7 Munsell color chips, like those in Color Figure 6, arranged in three dimensions. Hue, value, and chroma variation is continuous, of course, and the complete array would take the form of the solid shown in Figure 8.15. (Courtesy Munsell Color, Macbeth Division of Kollmorgen Corporation).

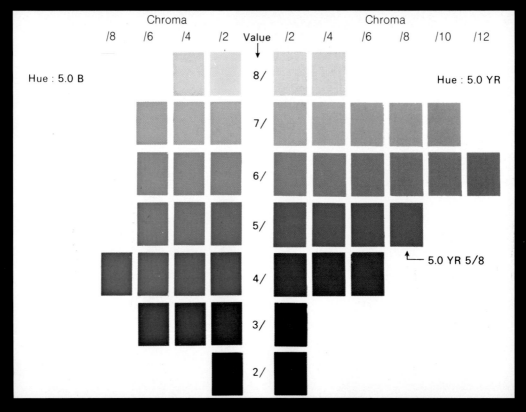

Chroma
/8 /6 /4 /2 Value /2 /4 /6 /8 /10 /12
Chroma
Hue : 5.0 B
Hue : 5.0 YR
8/
7/
6/
5/
4/
3/
2/
5.0 YR 5/8

Color Figure 6 Value and chroma variations of two complementary Munsell hues: 5.0 blue and 5.0 yellow-red. Note that extreme values are not shown and that chroma differences are shown by

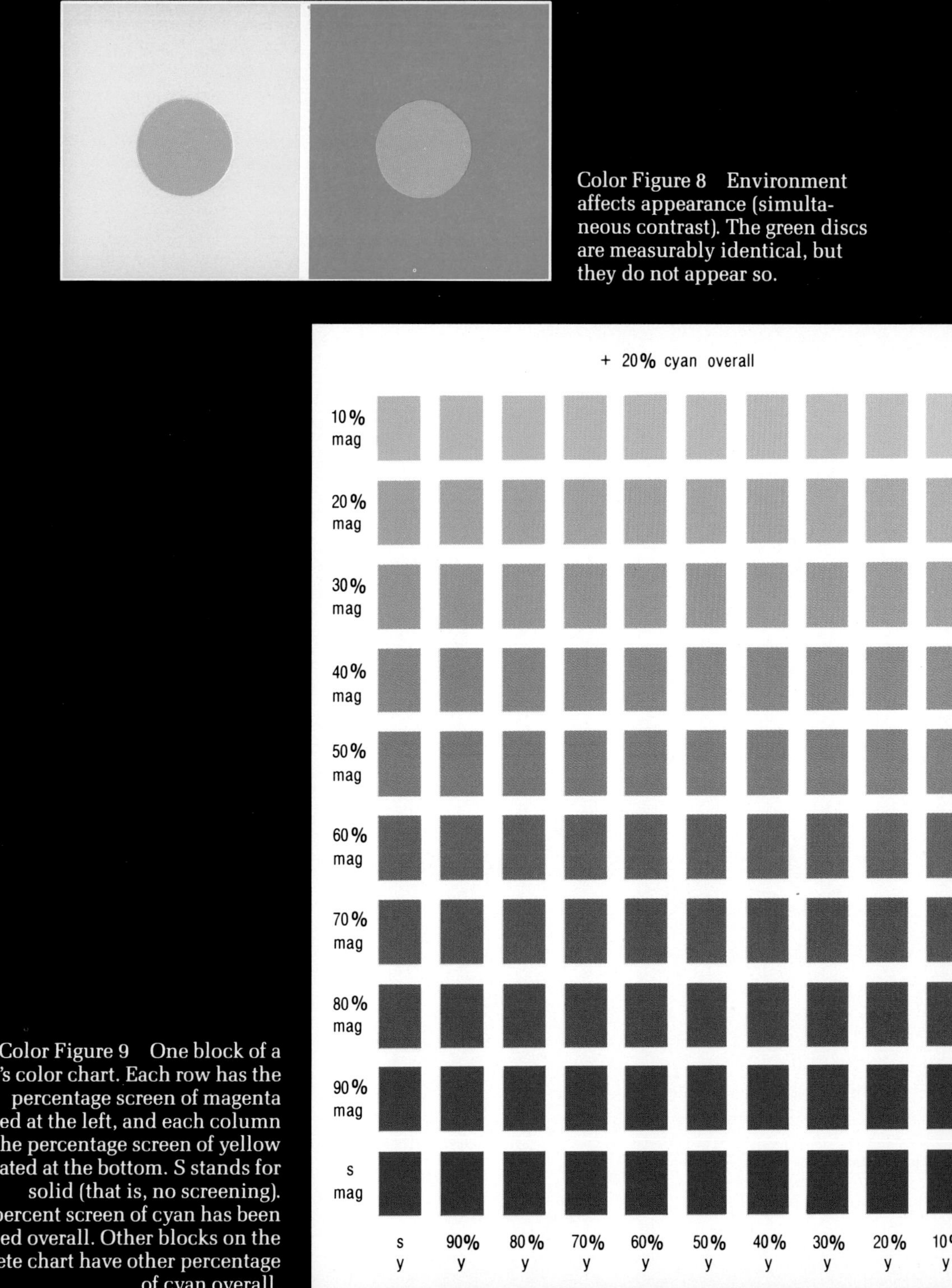

Color Figure 8 Environment affects appearance (simultaneous contrast). The green discs are measurably identical, but they do not appear so.

Color Figure 9 One block of a printer's color chart. Each row has the percentage screen of magenta indicated at the left, and each column has the percentage screen of yellow indicated at the bottom. S stands for solid (that is, no screening). A 20 percent screen of cyan has been added overall. Other blocks on the complete chart have other percentage of cyan overall.

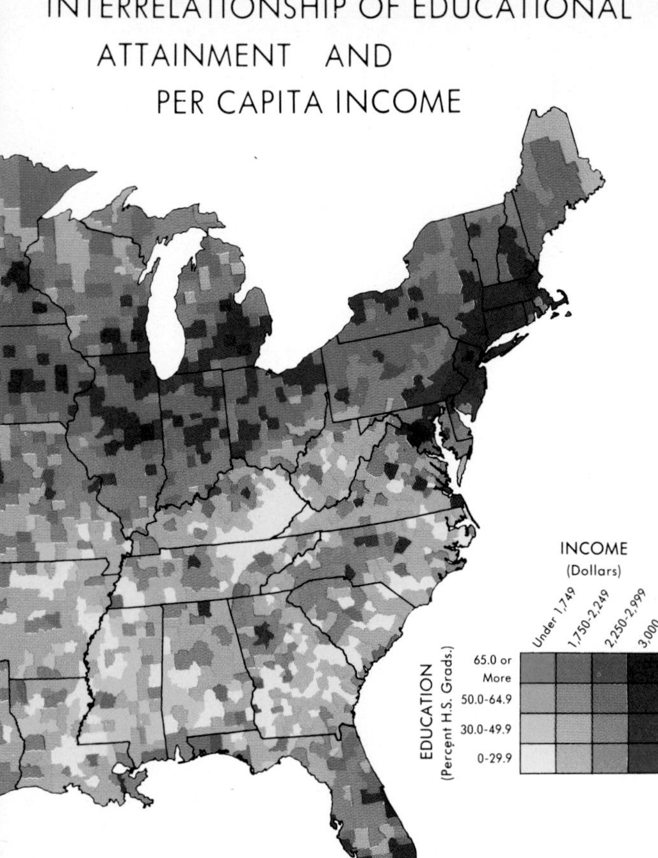

Color Figure 10 A portion of a complex color choropleth map produced, in this case by computer output on a microfilm plotter. This "two-variable" map was made by combining two "single-variable" maps. (Courtesy Geography Division, U.S. Bureau of the Census).

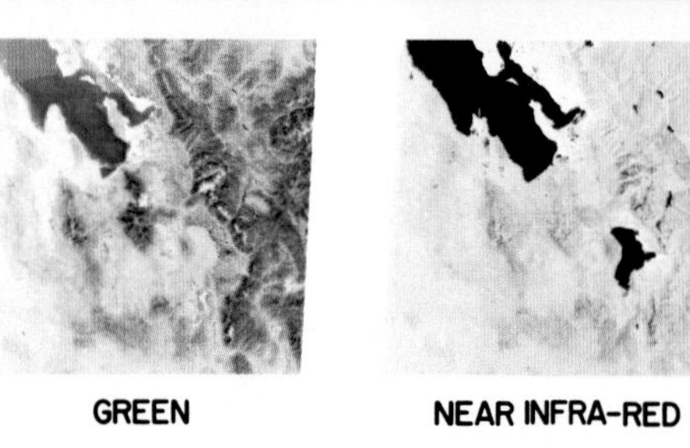

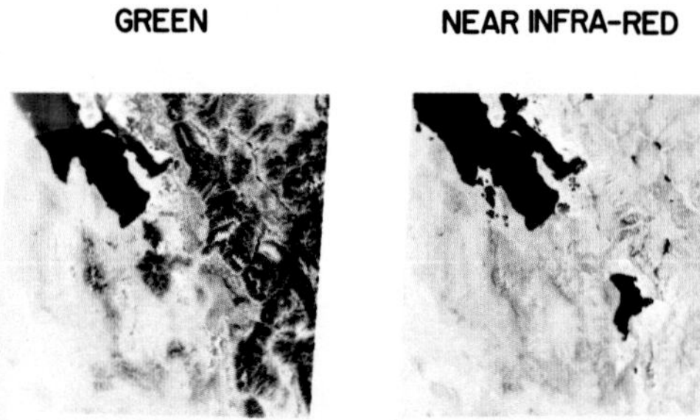

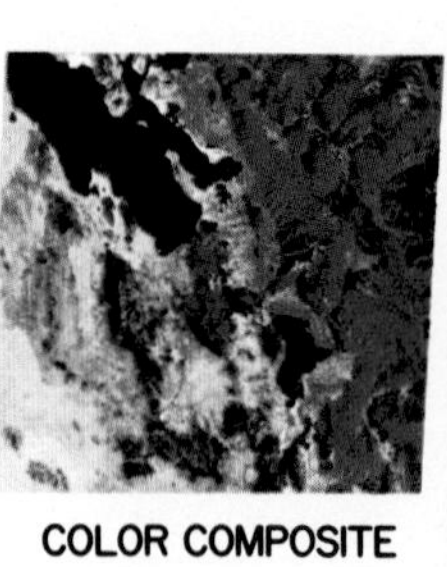

Color Figure 11 Images from the four bands in LANDSAT's multispectral scanner differ in appearance because the amount of energy reflected from features varies with the wavelength. Band 4, the green band, is useful for delineating areas of shallow water. Band 5, the red band, emphasizes cultural features. Bands 6 and 7 sense the near-infrared and infrared, respectively, and emphasize vegetation and land-water boundaries. Band 7 provides the best penetration of atmospheric haze. (Photos by NASA).

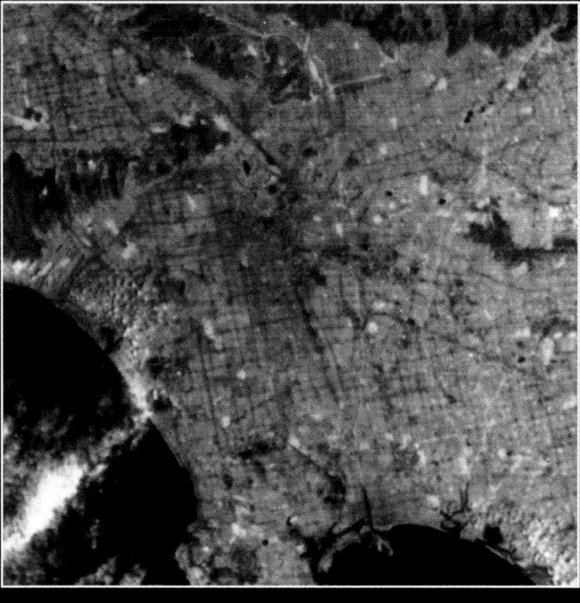

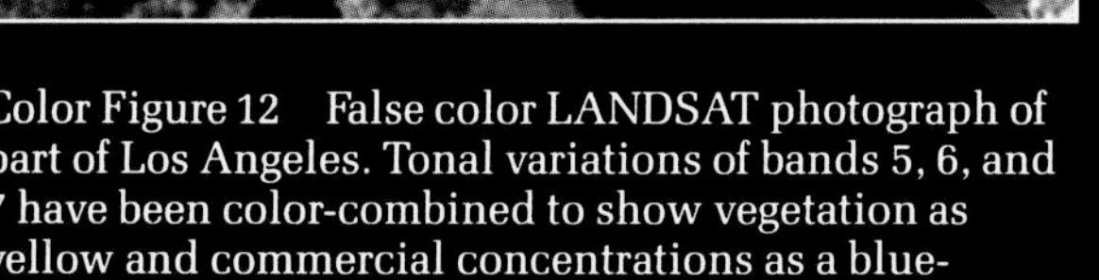

Color Figure 12 False color LANDSAT photograph of part of Los Angeles. Tonal variations of bands 5, 6, and 7 have been color-combined to show vegetation as yellow and commercial concentrations as a blue-purple gray. (Photos by NASA, Hughes Aircraft Company, and General Electric Company).

Color Figure 13 Electronic enhancement of the center section of Color Figure 12 processed from the LANDSAT computer-compatible digital tapes. Statistical techniques are used to correlate and assign colors to combinations of the tonal levels of the several spectral bands. Heavy commercial areas show as yellow. (Photos by NASA, Hughes Aircraft Company, and General Electric Company).

Color Figure 14 Multispectral scanner data from LANDSAT 3 has been used to create this illustration taken from a 1:250,000 sheet of Las Vegas, Nevada. (Courtesy of U.S.G.S.).

Color Figure 15 A composite color image of bands 2, 3, and 5 from the Thematic Mapper sensing system aboard LANDSAT 4. Dyersburg, Tennessee. (Courtesy of U.S.G.S.).

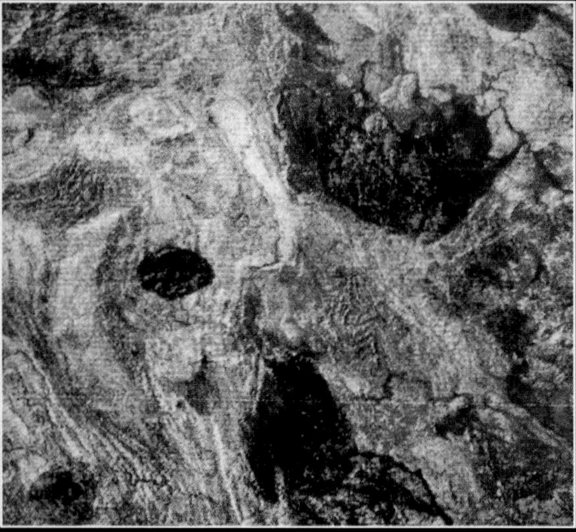

Color Figure 16 A combination of contrast stretching and band ratioing. After each band was contrast stretched, the ratios were printed; bands 4/5 in blue, bands 5/6 in green, and bands 6/7 in red. (From T. Lillesand and R. Kiefer, *Remote Sensing and Image Interpretation*, Copyright 1979, John Wiley & Sons, New York.)

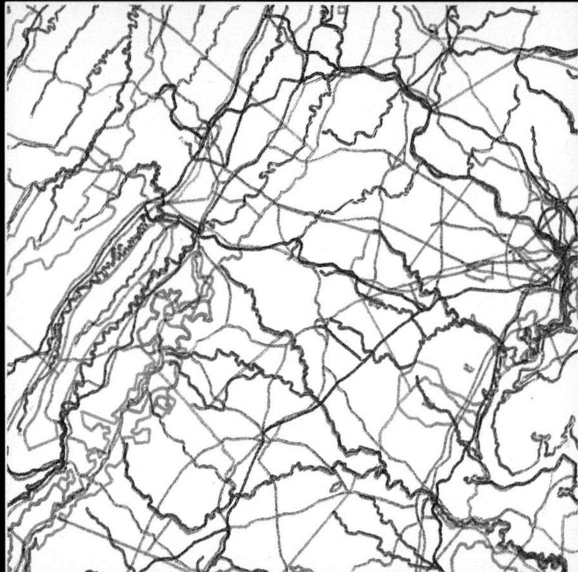

Color Figure 17 Portion of a color plot of the planimetric cartographic data taken from the Middle Atlantic (Region 3) 1/2,000,000-scale U.S. GeoData tape. Color codes on the original map were blue (streams and water bodies), red (roads), black (railroads), green (federally administered land), and brown (political boundaries). (Courtesy U. S. Geological Survey).

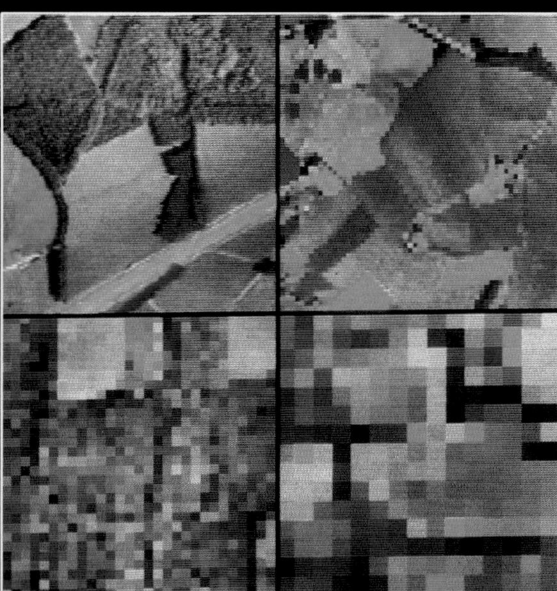

Color Figure 18 Images produced with the digital data of a 12 channel MSS scanner on an I²S—System. Pictorial degradation results from progressive doubling of pixel width from one image to the next. Data acquisition and enhancement: Belfotop-Eurosense. Image made available by courtesy of Prof. Ir. van Zuijlen—ITC, The Netherlands. Image produced on order of The Netherlands Remote Sensing Board.

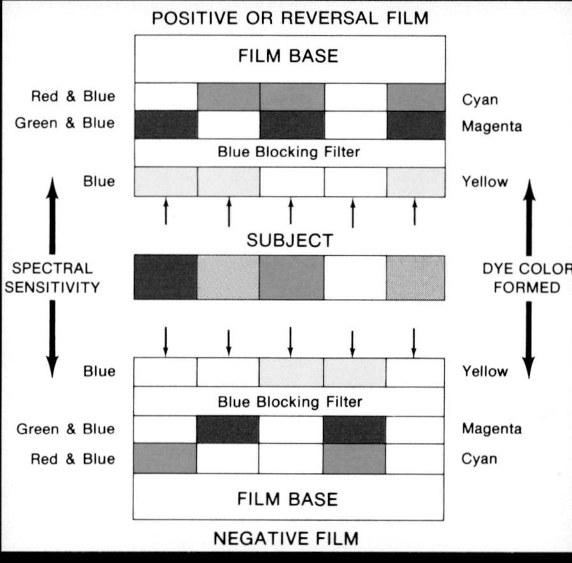

Color Figure 19 Color film consists of three emulsion layers, each sensitive to different wavelengths of light. After the film is processed, the blue-sensitive layer contains yellow dye, the green-sensitive layer contains magenta dye, and the red-sensitive layer contains cyan dye.

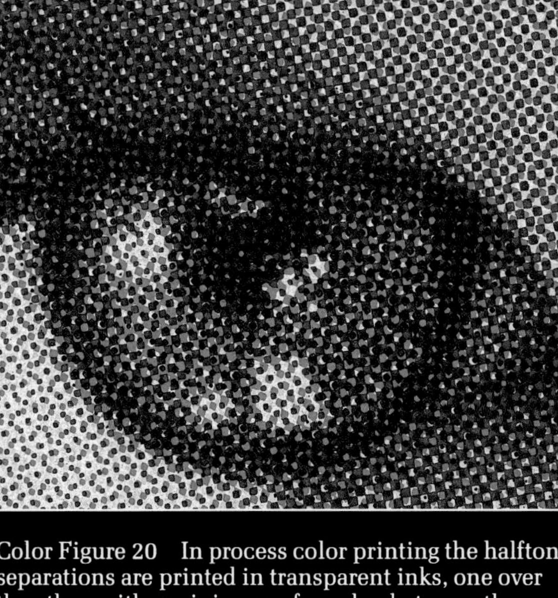

Color Figure 20 In process color printing the halftone separations are printed in transparent inks, one over the other, with a minimum of overlap between the cyan, magenta, and yellow dots. When the resulting dot pattern is viewed, the individual color dots combine and blend optically to produce the visual effect of full color. Magnified dot pattern is illustrated.

Color Figure 22 These high quality duplicates of several types of multiple color maps were made using a system of contact xerography. (Courtesy Kimoto USA, Inc.).

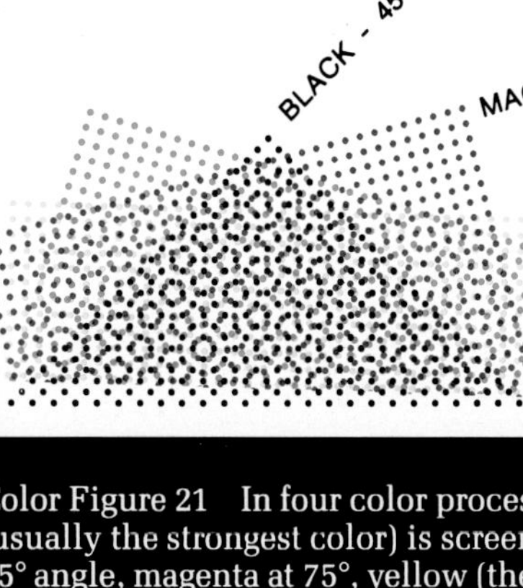

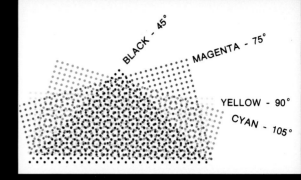

Color Figure 21 In four color process printing black (usually the strongest color) is screened at the standard 45° angle, magenta at 75°, yellow (the least noticeable color) at 90°, and cyan at 105°.

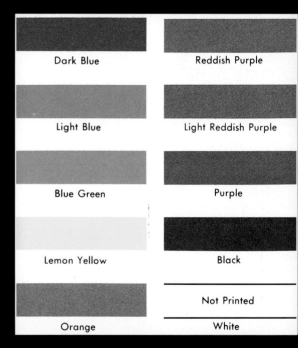

Color Figure 23 The U.S. Department of Defense had developed a system of Standard Printing Colors that ar derived primarily from these eight basic colors, plus black and transparent white. (Courtesy Department of Defense).

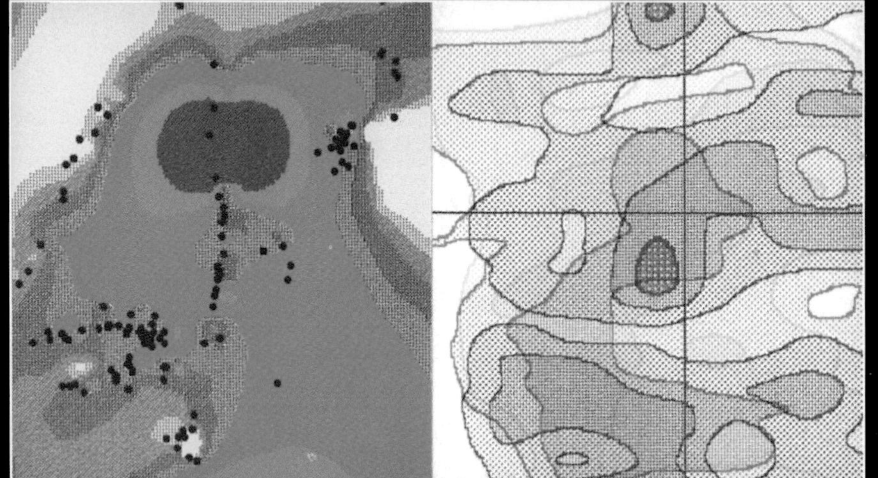

Color Figure 24 These multiple color line, pattern, and area fill plots demonstrate the flexibility of ink-jet color graphic printers and suggest a wide range of possible cartographic applications. (Courtesy PrintaColor Corporation).

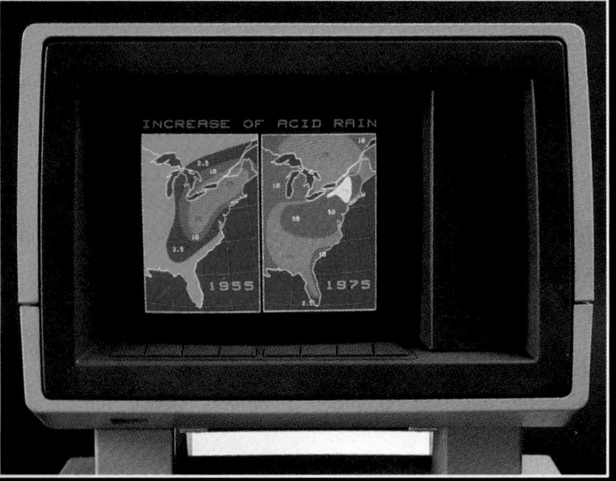

Color Figure 25 Color raster displays make it possible to depict areas in solid colors, which is an advantage when map making requires area fills, shading, or three-dimensional modeling. (Courtesy Hewlett Packard).

Color Figure 26 Portion of digital land cover map, scale 1:100,000 (Tacoma, Wash., 30 minute by 60 minute quadrangle). This map was prepared on a large-format laser plotter from satellite data in a digital format, with land use coded in color. Pixel value: 0.46 hectare (1.13 acres). (Courtesy of U.S. Geological Survey).

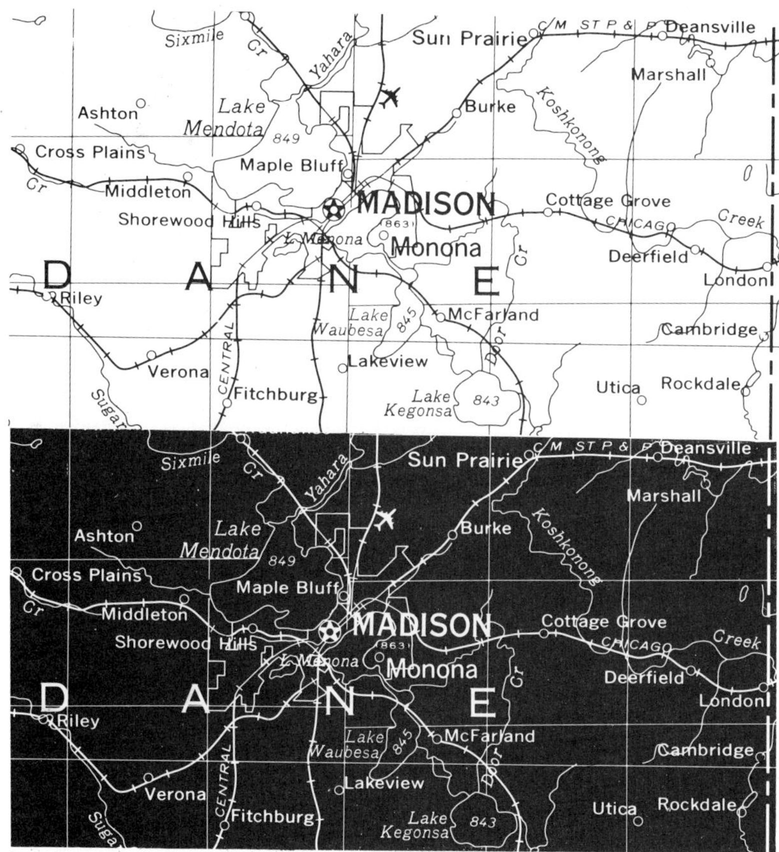

FIGURE 8.18 An example of the phenomenon of irradiation. The black on white and the reverse (white on black) lines and lettering are the same dimension, but the reverse appear slightly larger.

difficult problem from both the psychophysical and production points of view.*

The perceptual aspects have been extensively studied since the nineteenth century. For most other scales the series (continua) for which we might wish to establish equal contrasts, such as lengths, widths, or areas, are unbounded at one end. That is, a line can be progressively "more" (wider or longer) without limit. The gray scale is different in that it is bounded by black at one end and white at the other. To complicate things further, either direction on the scale can be considered "more," either progressively more white or more black. In most of the "equal-value" gray scales that have been derived, it has been assumed that white is at the high or "more" end of the scale. This has resulted in equal-value gray scales in which the value steps, toward the lighter end, are allotted greater differences in percent reflectance, such as the Munsell value scale, a graph of which is shown in Figure 8.19.

*Psychophysics is the study of the relations between the physical magnitudes of stimuli, such as lengths, sounds, gray tones, and so on, and the magnitudes of our sensory responses to them.

The complete Munsell scale of values extends from pure black (0) to pure white (10), and that range cannot be attained with ordinary white paper and black ink. Printed solid black ink on white paper reflects about 8 percent of the incident light, and the paper itself reflects only about 75 percent, which means that the ordinary range of perceived black to white extends only from about 3.3 to approximately 8.7 on the Munsell scale. That segment of the Munsell value range can be transformed into a perceived equal-contrast value range from 0 to 100 percent to provide guidance in determining tonal steps in map production. Such an equal-contrast gray scale is graphed in Figure 8.20.*

The graph may be used to derive dot tint screen percentages (black ink) to produce desired visual differences in a series of tones as follows:

1. Determine the number of classes to be symbolized, for example, five in an equal-interval series.
2. Choose the tones to be used for the first and last classes. Example: first will be white and last will be black; the range of perceived value contrast is therefore one hundred.
3. Divide the range by one less than the number of classes: 100/4 = 25; the number 25 is the wanted difference in value contrast between each of the five tones.
4. Find tint screens that correspond to 0, 25, 50, 75, and 100 perceived blackness percentages on the graph (Fig. 8.20) by extending horizontals from the perceived blackness values to the curve, and then reading downward to the corresponding tint screen percentage (nearest whole number). Example: 0 = 0 percent (white); 25 = 13 percent screen; 50 = 35 percent screen; 75 = 62 percent screen; 100 = 100 percent (solid black).

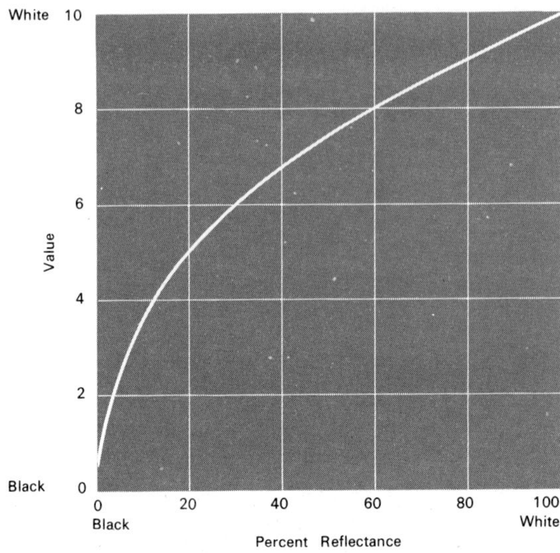

FIGURE 8.19 The relation between the Munsell value scale and percent reflectance. The theoretical limits are pure black (0 percent light) and pure white (100 percent reflectance). Compare Figure 8.13.

Suppose one does not wish to use either solid black or white as part of an equal-contrast series of four classes, and instead proposes to use a 20 percent screen for the first class and a 70 percent screen for the last. By reversing the above use of the graph, a 20 percent screen is found to have a relative blackness value of 33 and a 70 percent screen, a value of about 81. The range is 48; then 48/3 = 16, which is the wanted contrast between successive classes, giving a relative blackness series of 33, 49, 65, and 81. Proceeding as in the previous example, the corresponding tint screens are found to be 20 percent, 34 percent, 50 percent, and 70 percent.

The mapped classes in a quantitative series may not be equal. For example, we may wish to have a series of five classes such as 0–10, 10–25,

*Based upon (1) the Munsell value scale, and (2) measurements of reflectance of known percent area inked samples of Defense Mapping Agency process black ink and standard white paper. (A. Jon Kimerling, "The Comparison of Equal Value Gray Scales," *The American Cartographer* [in press].) The labels on the ordinate and abscissa scales of the graph are a necessary compromise with reality: Percentage reflectance does not translate linearly to percent area inked, which in turn does not necessarily translate linearly to tint screen percentage; and every printing situation (paper, ink, press, and so on) will be different. Consequently, for each particular set of circumstances, the curve might be somewhat different, but probably not by much.

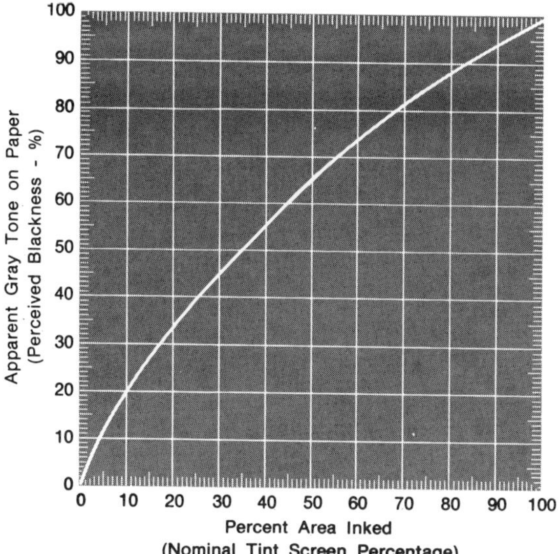

FIGURE 8.20 Graph showing the relation between apparent gray tone on paper (percentages of perceived blackness) and nominal tint screen percentages. See text for explanation and use.

25–45, 45–75, and over 75. If one were to choose white for the lowest class and class midpoints for the relative blackness values for the other four (17, 35, 60, 82), then the tint screen series would approximate 0, 8, 22, 45 and 70 percent.

The foregoing procedure is straightforward, but the production road is full of hazards and difficulties. Two of these are primary: (1) the availability of tint screens, and (2) modification of the screen values in the production process. These are dealt with in Chapter 18.

The magnitude and direction of possible variations in screen percentages that may occur in the production process are generally unpredictable.* In most cases the variations, although perhaps noticeable, are not likely to upset a series of gray tones completely, since the differences within one map will usually tend in one direction. Nevertheless, the cartographer is well advised to keep in mind the following strategies when symbolizing with dot tint screens:

1. Use as few classes as possible.
2. Space the tones as far apart on the perceived blackness scale as is consistent with the data.
3. Use very low and very high percentage screens only if absolutely necessary.
4. Combine patterns with smooth tones if there is *any* likelihood of the tones being difficult to distinguish from one another.

All the precepts and problems associated with choosing gray tones produced by black ink are applicable to a series of tones printed with an ink of any other color. There is one important principle to keep in mind, however, when using hues other than black: The lighter the value of the solid ink, the more difficult it is to obtain distinguishable tones.

The Employment of Intensity

Our sensitivity to differences in intensity (or chroma), with hue held constant, is not very great. Furthermore, it is difficult to obtain chroma (as precisely defined) differences in map production without also obtaining value differences. Furthermore, value differences of a hue result in differences in intensity. Therefore this dimension of color is less controllable and thus less useful to the cartographer than are hue and value. This is not to say that intensity is unimportant. Conventionally and subjectively, the greater the intensity, the greater the magnitude implication.

A significant physiologically based aspect of intensity is that, up to a point, the larger the area of a color, the more intense it will appear. This is important in two ways. Unless the cartographer is very careful in selecting colors from the very small samples on a printer's color chart (Color Fig. 9) by keeping in mind the relative map areas on which the colors are to be used, it may turn out that a color that seems modest on the chart will be overwhelming in its intensity on the map. Just as extreme values dominate a composition, so do extreme intensities. This principle is im-

*Mark S. Monmonier, "The Hopeless Pursuit of Purification in Cartographic Communication: A Comparison of Graphic-Arts and Perceptual Distortions of Graytone Symbols," *Cartographica* 17, no. 1 (1980): 24–39.

portant in a second way. If two similar hues or two values of the same hue are used in the legend and a primary distinction between them is their intensity, then larger map areas of the less intense color (in the legend) may be confused with smaller map areas of the more intense color.

Color Selection

In addition to the foregoing considerations of employing hue, value, and intensity, there are some important general considerations that one must keep in mind.

In mapping nominally scaled phenomena with colored area symbols, the main problem for the cartographer is to select colors that match the communication objectives. If there is no ranking involved (that is, if no area is more important than another), then colors should be selected that carry little or no magnitude implication. If some of the nominal categories logically can be thought to be "more important" than others as, for example, might be the case when mapping urban-nonurban or agricultural-forest distributions, then a magnitude implication may be desirable.

When many colors are required, it is difficult to select a set that does not include noticeable differences in value and intensity, and as we have seen, the darker and more intense colors inherently connote greater magnitude or importance.

When there are many areas, as for example on a detailed geological or soil map of a complex area, variations in value and intensity cannot be avoided. Often in such cases the more intense colors are used for the smaller map areas to give them more visibility, and conversely, to refrain from allowing a large area to overwhelm the portrayal. When legends include a large number of categories, the difficulty of carrying recognition from the legend to the map is lessened by adding patterns to the colors and sometimes by identifying each category by a number or a letter.

Graded Series. One of the common tasks of the cartographer is to select colors to be used to portray progressive classes on maps in which the phenomena have been ranked on interval or ratio scales. Two avenues are open: (1) to use several hues, and (2) to use a series of values of one hue.

The first is quite common, stemming in part from connotative associations of hue, such as warm with red and cool with blue, and in part from the common convention of portraying successive elevations on relief maps with a spectral progression of hues. On the latter the water is blue, lowlands are green, intermediate elevations are yellow-brown, and high areas are reddish. The spectral progression on relief maps is a convention of long standing and thus is well known through regular appearance on school maps and on atlas maps for the general public. Except for its familiarity, it is graphically illogical with little to recommend it both because the connotative associations of some of the hues detract from the map's effectiveness and because the value changes result in marked variations in the perceptibility of other data being shown. A complete spectral progression generally should not be used for data other than elevation.

Several tests have shown that a progression of values of one hue is the most efficient in conveying the magnitude message of a simple graded series; it is probably even as effective as a two-hue series with strong connotative aspects such as a red-to-blue series to portray warm to cool temperatures. On the other hand, it is quite likely that very distinctive hues used in the legend of a complex graded "two-variable" series, such as is shown in Color Figure 10, make it possible for the map user to carry categories more easily from the legend to the map and vice versa.

A major concern of the cartographer in color selection for a map of a graded series is to make sure that the colors selected for the map areas do not result in undesirable linear contrasts. A large area of a light color (such as yellow) next to one or more colors that are considerably darker will result in the boundary between the colors being very prominent visually when, in fact, it may be of no great consequence.

PATTERNS

The term "pattern," when used to refer to a graphic symbol, denotes any systematic repetition of *visible* marks covering a section of a display. Very little study has been directed toward the understanding of patterns, their perceptual consequences, and our ability to discriminate among them.

Some basic terminology is necessary in discussing them. The character of a pattern depends upon the way the component marks have been structured and modulated with the qualities of the primary graphic elements: (1) *size,* that is, whether the marks are small or large, thin or thick, and so on; (2) *shape* (of the marks), such as round dots, wavy lines, tiny tree symbols, crosses, or whatever; (3) *spacing,* that is, whether the marks are close together or far apart, often specified by the number of lines per in. or cm; and (4) *orientation,* that is, the directional characteristics of the marks in relation to the map frame and the viewer (horizontal, vertical, or diagonal). Of course, the marks may be dark on light or light on dark (reverse). As a class of graphic structures, patterns are very complex, and this complexity probably has had much to do with frustrating their systematic study.

Hundreds of distinctive patterns are possible, and although there is no systematic classification of them, it is convenient to recognize three general categories:

1. Line patterns composed of usually straight but sometimes wavy parallel lines. The lines may be of any width and spacing. When two sets of lines are crossed, the result is called *cross-hatching,* and the sets of lines may intersect perpendicularly or at an angle (Fig. 8.21).
2. Dot patterns composed of round dots in either a rectangular or triangular array (Fig. 8.21). A rather evenly spaced, irregular distribution of small marks giving a more or less smooth appearance is called *stippling.*
3. Miscellaneous patterns are of great variety, ranging from the tufted grass symbols commonly used to show swamps and marshes to arrays of crosses or other distinctively shaped marks (Fig. 8.21). Many such patterns have become conventional in specific subject-matter fields, such as geology or soil science.

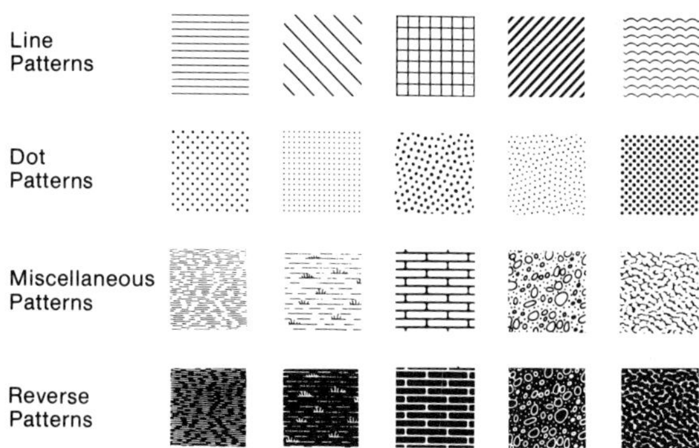

FIGURE 8.21 A variety of common patterns.

The several characteristics of pattern are perceived in integrated fashion in an impression impossible to describe by a single measure (just as it cannot be done for color); instead, we must rely on the individual graphic elements, such as shape, size, orientation, and so on. An important variation among some patterns is the impression of relative overall darkness (Fig. 8.21). This should not be confused with the concept of value, which refers to the lightness—darkness of smooth tones. The textures of most patterns are so coarse as to preclude a ranking on a value scale; yet, when one is put alongside another, it may appear "darker" and even more "intense" or "busy."*

Patterns are widely used in cartography for several purposes. Probably the most common use is as an area symbol to impart a unified quality to a geographical area, such as an area of a given kind of bedrock, climatic regime, or administrative jurisdiction. An associated use is to add some

*The phenomenon of relative darkness is analogous to the concept of "color" in typography where, because of the design of the type and the manner of its composition, a block set in one type style will appear darker or lighter overall than a block set in another.

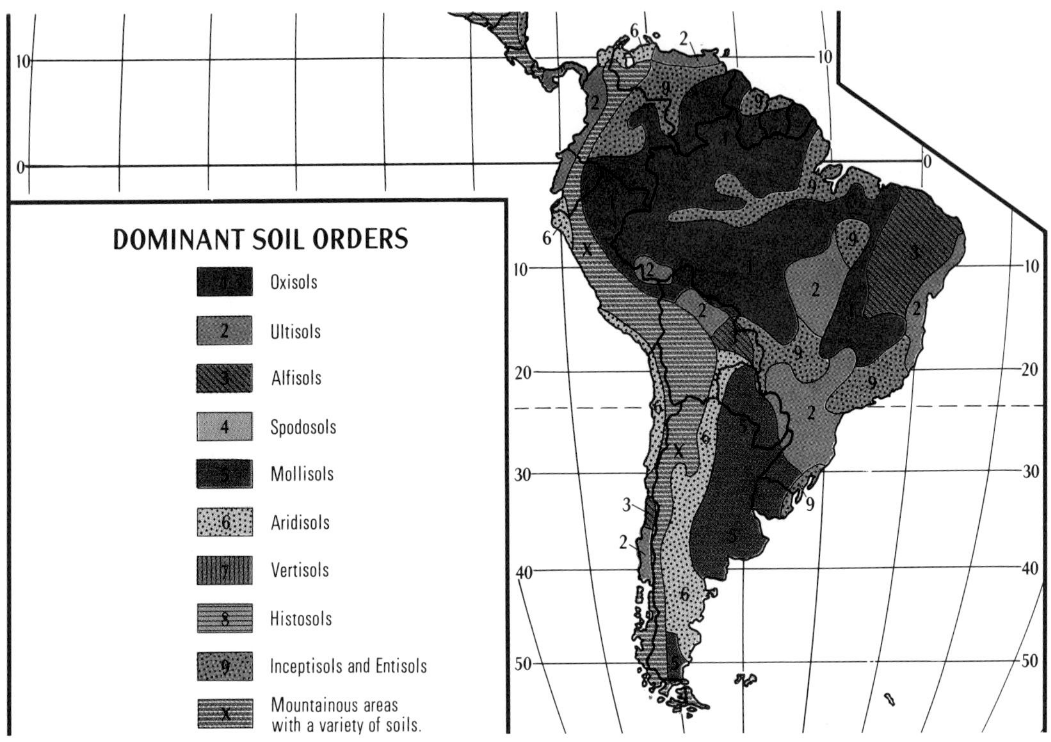

FIGURE 8.22 An example of a two-color map in which pattern has been employed to make areas appear more distinctive. Note that numerals in the classes provide an additional aid to recognition. (From *Fundamentals of Physical Geography*, 3d ed., by Trewartha, Robinson, Hammond, and Horn, Copyright 1977, McGraw-Hill Book Co., Inc.)

graphic distinctiveness to printings of uniform tones, especially on maps with a great many area symbols, so that each one may be more easily identified and matched with the legend.

The production of pattern on a map may be accomplished in several ways. For drawings to be photographed, a variety of preprinted patterns are available. Patterns may be produced by using negatives or positives in the same manner as dot tint screens. Combining tint screens at other than 30° angular separation will produce a (usually undesirable) effect called moiré, as explained in Chapter 17. Occasionally moiré can be used as a pattern.* Patterns can also be produced in CRT displays and output devices directed by a computer. To date, the computer-generated patterns have generally been limited in variety and have tended to be rather coarse, but that can be expected to change for the better.*

The Employment of Pattern

As pointed out, the primary reason for employing pattern as a symbol is to give individual areas the

*A. Jon Kimerling, "Cartographic Guidelines for the Use of Moiré Patterns Produced by Dot Tint Screens," *The Canadian Cartographer* 16, no. 2 (1979): 159–67.

*Computers can be programmed so that plotters can produce line and dot patterns with any given level of percent area inked. Occasionally, this capability has been used for a gradation of classes, but most patterns, especially the coarser, are not seen as impressions of value. For example, Figure 8.25 is approximately 50 percent area inked, but it is not seen as a 65 percent gray tone.

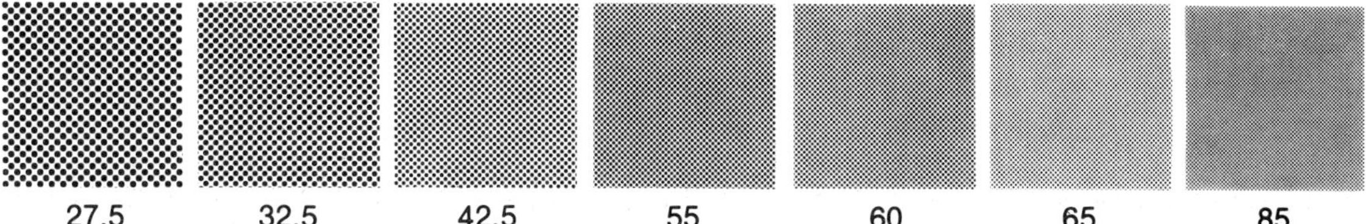

FIGURE 8.23 A series of dot area symbols in which the variables, orientation, and percent area inked are held essentially constant but spacing is varied. The numbers below the blocks show the numbers of lines per inch in each symbol. Note that textures coarser than about 40 are primarily seen as patterns without value, while a texture as fine as 85 appears primarily as a value area without providing much impression of pattern.

quality of homogeneity. For these purposes it is obviously necessary to make the patterns distinctive. On a color map, often a simple addition of dark or light dots or lines is sufficient to enable the user to recognize the symbolized areas easily, as illustrated in Figure 8.22.

Patterns used alone with a single color are often dot or line patterns. Because it is generally desirable to use fine-textured patterns rather than coarse ones, obtaining the desired contrast is sometimes a problem. One useful technique, which will result in a darkness contrast, is to use a pattern normally in one area and then use the same pattern in another area but lighten it by tint

FIGURE 8.24 A simple monochrome map contrasting the use of line and dot patterns. Line patterns are perceptually relatively unstable, and all but the finest-textured should be used with caution.

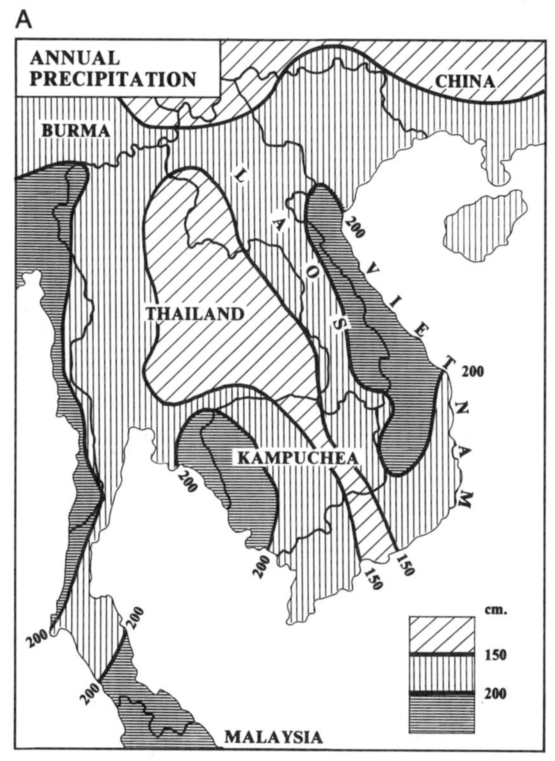

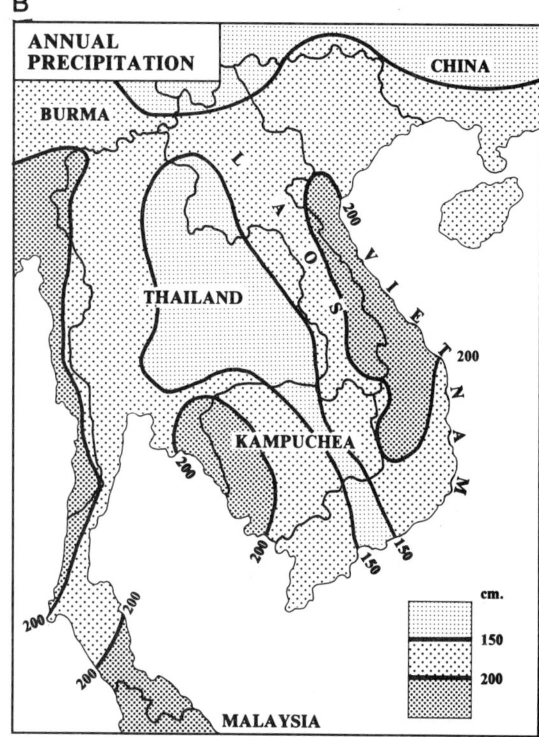

screening it. One must be careful not to suggest ordinal ranking of nominal data when doing so would not be appropriate.

Dot patterns having more than seventy-five lines per inch will usually be perceived as gray value areas without an impression of pattern, while those with fewer than forty lines per inch will probably not convey much impression of value to most readers (see Fig. 8.23). As long as the component marks are bold and the textures of the patterns relatively coarse, differences in both structure and orientation are generally quite noticeable. The important point in obtaining contrasts among patterns is to vary as many of the characteristics as possible instead of trying to hold some constant. The more elements that are held constant, the greater the difference must be in those that are varied.

Any line anywhere has, in the eye of the viewer, an orientation, and the viewer will tend to move his or her eyes in the direction of the line. Orientation differences among patterns provide a strong contrast. Furthermore, if irregular areas are differentiated by line patterns, as in Figure 8.24A, the reader's eyes will be forced to change direction frequently. Consequently, the reader will often experience considerable difficulty in focusing on the various data boundaries. If the line patterns of Figure 8.24A are replaced with dot patterns, as in Figure 8.24B, the map is seen to become much more stable, the eye no longer has a tendency to jump, and the positions of the boundaries are much easier to distinguish. Names are also easier to read against a dot background than against a line background. If, however, the parallel lines are fine enough and closely enough spaced, the perceptual effect is largely one of relative darkness, and it will have but a slight suggestion of direction.

Many parallel-line patterns are definitely irritating to the eye. Figure 8.25 is an exaggerated example of the irritation that can occur from using parallel lines. The probable cause is that the eyes cannot quite focus on one line. Whatever the reason, the effect is somewhat reduced if the lines, regardless of their width, are separated by white spaces greater than the thickness of the lines. Generally, a cartographer should be very wary of using any but the finer-textured line patterns.

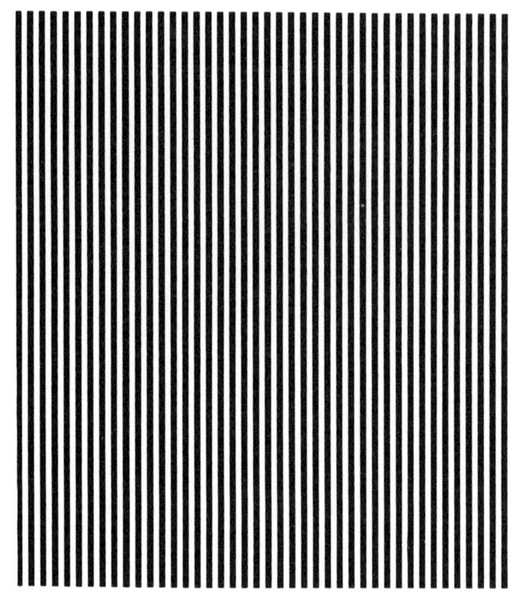

FIGURE 8.25 Parallel line patterns can be very disturbing to the eyes.

SELECTED REFERENCES

Beck, J., "The Perception of Surface Color," *Scientific American* 233 (1975): 62–75.

Birren, F., *The Story of Color.* Westport, Conn.: Crimson Press, 1941.

Castner, H. W., and A. H. Robinson, *Dot Area Symbols in Cartography: The Influence of Pattern on Their Perception,* Technical Monograph No. CA-4. Washington, D.C.: American Congress on Surveying and Mapping, Cartography Division, 1969.

Cox, Carleton W., "The Effects of Background on the Equal Value Gray Scale," *Cartographica* 17, no. 1 (1980): 53–71.

Cuff, D. J., and M. T. Mattson, *Thematic Maps:*

Their Design and Production. New York: Methuen, 1982.

Evans, R. M., *An Introduction to Color*. New York: John Wiley & Sons, 1948.

Kimerling, A. Jon, "The Comparison of Equal Value Gray Scales," *The American Cartographer*, (in press).

———, "Color Specification in Cartography," *The American Cartographer* 7, no. 2 (1980): 139–53.

———, "Process Color Diagrams," *The American Cartographer* 8, no. 2 (1981): 180–81.

———, "Cartographic Guidelines for the Use of Moiré Patterns Produced by Dot Tint Screens," *The Canadian Cartographer* 16, no. 2 (1979): 159–67.

———, "A Cartographic Study of Equal Gray Scales for Use with Screened Gray Areas," *The American Cartographer* 2 (1975): 119–27.

Meyer, M. A., F. R. Broome, and R. H. Schweitzer, Jr., "Color Statistical Mapping by the U.S. Bureau of the Census," *The American Cartographer* 2 (1975): 100–17.

Monmonier, Mark S., "The Hopeless Pursuit of Purification in Cartographic Communication: A Comparison of Graphic-Arts and Perceptual Distortions of Graytone Symbols," *Cartographica* 17, no. 1 (1980): 24–39.

Munsell, A. H., *A Color Notation*, 12th ed. Baltimore: Munsell Color, 1975.

Nimeroff, I., *Colorimetry*, National Bureau of Standards Monograph 104. Washington, D.C.: U.S. Government Printing Office, 1968.

Olson, J. M., "The Organization of Color on Two-Variable Maps," *Proceedings*, International Symposium on Computer-Assisted Cartography. Falls Church, Va.: American Congress on Surveying and Mapping, 1977, pp. 289–94 and color insert ff. p. 250.

Robinson, A. H., "Psychological Aspects of Color in Cartography," *International Yearbook of Cartography* 7 (1967): 50–59.

Sargent, W., *The Enjoyment and Use of Color*. New York: Dover Publications, Inc., 1964.

Shellswell, M. A., "Towards Objectivity in the Use of Colour," *The Cartographic Journal* 13 (1976): 72–84.

Standard Printing Color Identification System, Technical Report No. 72-2, St. Louis, Mo.: Aeronautical Chart and Information Center (ACIC), 1972.

Stoessel, O. C., "Standard Printing Color and Screen Tint Systems for Department of Defense Mapping, Charting and Geodesy Services," *Proceedings*, ACSM Technical Sessions, ACSM-ASP Fall Convention, 1972.

Williamson, Glen R., "The Equal Contrast Gray Scale," *The American Cartographer* 9, no. 2 (1982): 131–39.

9

Typography and Lettering the Map

Some cartographers have claimed that the names on maps are a "necessary evil" because they crowd and complicate the representation. They argue that a view of the earth from above is unencumbered by names. On the other hand, maps are made to show where things are, and to do this it is important to be able to tell what's what. Only a person thoroughly familiar with a given area would not need names to identify the mapped items.

Names are also important as part of the graphic display. In ordinary reading, such as you are now doing, most of the words are "transparent" in the sense that we pay little or no attention to their appearance. On the other hand, the lettering is a significant, "nontransparent" component of a visual display, and it is an important element in graphic design. As a matter of fact, the graphic quality of a map depends heavily on the style and placement of the names, and these often provide clues to how, where, and by whom the map was made.

The process of selecting the type, preparing the names, and placing them in position is collectively called "lettering the map." Where there are a considerable number or variety of names, it is among the more complex and time-consuming parts of the cartographic process. Although the computer is extremely useful in preparing type, placement of names on maps for good graphic quality has not yet been entirely automated.

To emphasize the close association between the process of lettering and the techniques for construction and production, we will first survey briefly the major changes which have taken place in this essential aspect of mapmaking.

The History of Map Lettering

In the manuscript period, all map lettering was, of course, done freehand, and each copy of a map had to be laboriously hand lettered by the calligrapher. The styles were varied and ranged from cramped and severe to free flowing (Fig. 9.1).

After the latter part of the fifteenth century, when maps began to be duplicated by woodcut printing and copper engraving, lettering became the problem of the craftsman. Small lettering in woodcuts was difficult, and various ways of in-

FIGURE 9.1 A portion of one of the maps in Sir Robert Dudley's atlas, *Dell' Arcano del Mare* (1646–1647). The engraver, A. F. Lucini, was fond of ornate calligraphy.

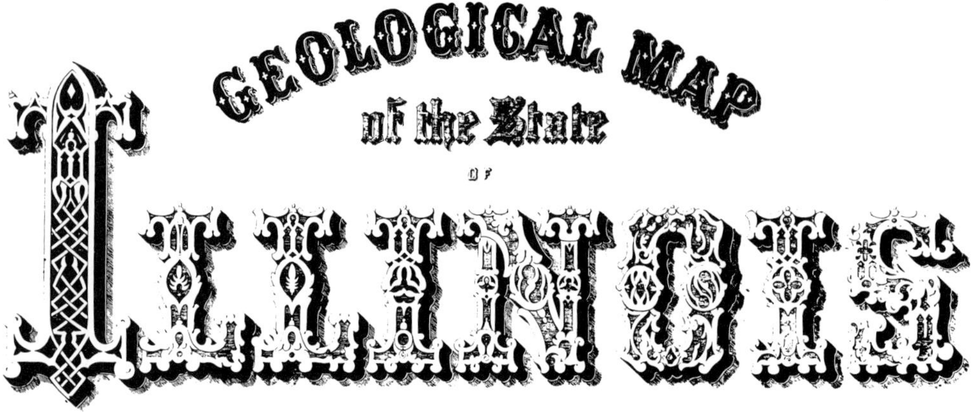

FROM ORIGINAL SURVEYS, MADE UNDER THE DIRECTION OF

A. H. WORTHEN, STATE GEOLOGIST.

FIGURE 9.2 Ornate lettering in the title of a lithographed map of 1875. The name "Illinois" is more than 1 ft.long on the original.

corporating movable type were devised, such as inserting type in holes cut in the blocks. The copper engraver cut letters with a burin or graver in reverse on the plate. The great Dutch atlas makers included many pictures of animals, ships, and wondrous other things for, as Hondius explained, "adornment and for entertainment," but their lettering was generally well planned in the classic style and well executed.

As might be expected, when hand lettering was done by those more interested in its execution than in its function, it tended to become excessively ornate. The trend toward poor lettering design seemed to accelerate after the development of lithography in the early 19th century and continued well into the Victorian era. Lettering and type styles in general became so bad by the close of the nineteenth century that there was a general revolt against them, which caused a return to the classic styles and greater simplicity. Figure 9.2 is an example of ornate lettering in the title of a nineteenth-century geological map. The fancy lettering of that and earlier periods provides good examples of manual dexterity, since they are intricate and difficult to execute, but they are examples of poor communications design because they are hard to read and call undue attention to themselves.

Until the mid-nineteenth century most names on a map were created by craftsmen, either freehand with a pen or brush on a manuscript, carved with a knife in a woodblock, or incised with a graver in a copperplate. Metal type as used to print the text in books was employed in mapmaking only sparingly, such as for titles or blocks of lettering. In the 1840s a process called cerography (wax engraving) allowed the mapmaker to press type into a wax surface to make a mold from which a printing plate could be made.* This direct use of type made it much easier to letter maps and led to some overcrowding, but maps of the highest quality were still lettered by hand, either by engraving or pen, well into the first half of the present century. The introduction of the stick-up process in the 1920s and 1930s, whereby names set in type were printed on thin, clear plastic which was then wax-backed for adhesion

*David Woodward, *The All-American Map, Wax Engraving and Its Influence on Cartography* (Chicago and London: The University of Chicago Press, 1977).

to the map, made freehand lettering all but a thing of the past in most map construction. The development of photographic and electronic typesetting devices, which are making the use of metal type unnecessary, has further expanded the role of typography in mapmaking. Today, scores of type styles in a complete range of sizes are available, and the cartographer must plan very carefully the typographic component of a map, so that it will function efficiently and fit the graphic design properly.

The Functions of Lettering

The utility of a general map depends to a great extent on the characteristics of the type and its positioning. General maps usually have a great many names, and the recognition of the feature to which a name applies, the "search time" necessary to find a name, and the ease with which they can be read are all important to the function of the map. The type on a thematic map generally does not run the gamut of functions that it does on general maps. Nevertheless, it must be made to fit into the display in such a way as to enhance the communication without drawing undue attention to itself.

Like all other marks on a map, the type is a symbol, but it is considerably more complicated in its function this way than most symbols. Its most straightforward service is as a literal symbol: The individual letters of the alphabet, when arrayed, encode sounds that are the names of the features shown on the map. Generally, this is the most important role of the lettering in the communication system that is the map.

In several ways the type on a map functions as a locative graphic symbol. By its position within the structural framework, it helps to indicate the location of points (such as cities); it more directly shows by its spacing and array things such as linear or areal extent (as of mountain ranges and national areas); and by its arrangement with respect to the graticule, it can clearly indicate orientation.

By systematically employing distinctions of style, form, and color (hue), the type on a map may be used as a means of showing nominal classes to which the labeled feature belongs. For example, we can identify all hydrographic features in blue type; within that general class, we can further show open water by all capital letters, and running water by capitals and lowercase letter forms. By variations of size, it can portray the ordinal characteristics of geographical phenomena, ranking them in terms of relative area, importance, and so on.

In a more subtle way, the type on a map serves as an indication of scale. Primarily by its size contrasts with respect to other factors such as line width and symbol size, we can often tell "by looking" that one map is a larger or smaller scale than another.

We must approach the problem of choosing the type for a map as we do other aspects of cartographic design, by settling clearly on the objectives of the map and then fitting the selection of type to them.

Planning for Lettering

The phrase "lettering a map" means the preparation of all aspects of this phase of mapmaking, which includes all the alphanumeric material. It involves the entire process from the beginning to the end and requires careful consideration of many factors.

The more elaborate and complex the map, of course, the more elements must be considered; but in general there are at least seven major headings to the planning checklist. The complexities of the map and its purposes will add subheads to the following major elements.

1. The style of the type.
2. The form of the type.
3. The size of the type.
4. The contrast between the lettering and its ground.
5. The method of lettering.
6. The positioning of the lettering.
7. The relation of the lettering to reproduction.

The style refers to the design character of the type; it includes elements such as thickness of line and serifs. The form refers to whether it is composed of capitals and/or lowercase, its stance (upright or slanted), roman or italic, or combinations of these and other similar elements.*

The methods of lettering include the mechanical means whereby the type or lettering is affixed to the map. The positioning of the lettering involves the placement of the type. Also important is a consideration of when and where on the map, and in the construction schedule, the lettering is done.

Regardless of the kind of map, the lettering is there to be seen and read. Consequently, the elements of visibility and recognition are among the major yardsticks against which the choices and possibilities are to be measured.

ELEMENTS OF TYPOGRAPHIC DESIGN

Type Style

The cartographer is faced with an imposing array of possible choices when planning the lettering for a map. The number of different alphabet designs from which to select may surprise you, but the cartographer must also settle on the wanted combinations of capital letters, lowercase letters, small capitals, roman, italic, slant, and upright forms. There is no other technique in cartography that provides such opportunity for individualistic treatment, especially with respect to the monochrome map. The cartographer who becomes well acquainted with type styles and their uses finds that every map or map series presents an interesting challenge.

Letter style exhibits a complex evolution since Roman times.* The immediate ancestors of our present-day alphabets include grandparents such as the capital letters the Romans carved in stone and the manuscript writing of the long period before the invention of printing. After the development of printing, the type styles were copies of the manuscript writing, but it was not long until designers went to work to improve them. Using the classic Roman capitals and the manuscript writing as models for the small letters, there evolved the alphabets of uppercase and lowercase letters that it is our custom to use today. There are three basic classes.

One class of designers kept much of the free-flowing, graceful appearance of some freehand calligraphy, so that their letters carry something of the impression of having been formed with a brush. The distinction between thick and thin lines is not great, and the serifs, short angular attachments to the ends of the lines or strokes forming a letter, are smooth and easily attached. Such letters are known as classic or old style. They appear "dignified" and have about them an air of quality and good taste that they tend to impart to the maps on which they are used. This style has a neat appearance, but at the same time it lacks any pretense of the geometric (see Fig. 9.3).

A radically different kind of face was devised later; for that reason it (unfortunately) is called modern. Actually, the modern faces were tried out more than two centuries ago, although we think of them as coming into frequent use around 1800. These typefaces look precise and geometric, as if they had been drawn with a straightedge and a compass—and so they were. The difference between thick and thin lines is great, sometimes excessive, in modern styles (see Fig. 9.4). Both old style and modern types have serifs. All upright forms of these type styles are sometimes loosely classed together as roman.

A third style class includes some varieties that

---

*In common typographic terminology, capital letters (e.g., A) are often called "uppercase" and small letters (e.g., a) are called "lowercase." The terms are a holdover from the time when all type was set, one letter at a time, by a compositor standing before a case holding the type. The capitals were in the upper part of the case and the small letters in the lower.

*A well-illustrated, interesting treatment of the development of lettering and type styles is Alexander Nesbitt, *The History and Technique of Lettering* (New York: Dover Publications, 1957).

CHELTENHAM WIDE

Cheltenham Wide

CHELTENHAM WIDE ITALIC

Cheltenham Wide Italic

GOUDY BOLD

Goudy Bold

GOUDY BOLD ITALIC

Goudy Bold Italic

CASLON OPEN

FIGURE 9.3 Some classic or old style type faces. (Courtesy Monsen Typographers, Inc.)

are definitely modern in time but not in name, as well as some of older origin. This class, which is more and more important in modern cartography, is called sans serif (without serifs) and has about it an up-to-date, clean-cut, new, nontraditional appearance. There is nothing subtle about most sans serif forms. There are many variants in this class, some of which include variations in the thickness of the strokes (see Fig. 9.5). Sans serif forms are sometimes called Gothic.

There are several other less common styles, which are occasionally used on maps, such as text and square serif (Fig. 9.6). Text, or black letter, is dark, heavy, and difficult to read.

FIGURE 9.4 Some modern style letter forms. (Courtesy Monsen Typographers, Inc.)

BODONI BOLD

Bodoni Bold

BODONI BOLD ITALIC

Bodoni Bold Italic

MONSEN MEDIUM GOTHIC

Monsen Medium Gothic Italic

COPPERPLATE GOTHIC ITALIC

FUTURA MEDIUM

LYDIAN BOLD

DRAFTSMANS ITALIC

FIGURE 9.5 Some sans serif letter forms. (Courtesy Monsen Typographers, Inc.)

This listing by no means exhausts the possibilities. There are literally hundreds of variations and modifications, such as the open letter, light or heavy face, expanded or condensed, and so on.

In the selection of type, the cartographer must be guided by certain general principles that have resulted from a considerable amount of research by the students of alphanumeric symbolism as well as from the evolved artistic principles of the typographer.

Recognition depends on the occurrence of familiar forms and on the distinctiveness of those forms from one another. For this reason, "fancy" lettering or ornate letter forms are hard to read, and text lettering is particularly difficult. Conversely, well-designed classic, modern, and sans serif forms stand at the top of the list, and apparently they rate about equally. Ease of recog-

FIGURE 9.6 Examples of text and square serif letter forms. (Courtesy Monsen Typographers, Inc.)

CLOISTER BLACK

Cloister Black

STYMIE MEDIUM

Stymie Medium

nition also depends, to some extent, on the thickness of the lines forming the lettering. The thinner the lines relative to the size of the lettering, the harder it is to read. The cartographer must therefore be careful in the selection of lettering, since, although the bold lettering may be more easily seen, the thicker lines may overshadow or mask other equally important data. The type on a map is not always the most important element in the visual outline; instead, it may even be desirable that the type recede into the background. If so, light-line type may be an effective choice.

The problem of the position of the lettering in the visual outline is significant. For example, the title may be of great importance, whereas the balance of the type may be of value only as a secondary reference. Size is usually much more significant than style in determining the relative prominence, but the general design of the type may also play an important part. For example, rounded lettering may be lost along a rounded, complex coastline, whereas in the same situation angular lettering of the same size may be sufficiently prominent.

Cartography conventionally uses different styles of lettering for different nominal classes of features, but this may be easily overdone. Although there is little evidence, there does seem to be some indication that the average map reader is not nearly so discriminating in reactions to type differences as cartographers have hitherto thought. The use of many subtle distinctions in type form for fine classificational distinctions is probably a waste of effort.

As a general rule, the fewer the styles, the better harmony there will be. Most common typefaces are available in several variants; it is better practice to use these as much as possible (see Fig. 9.7). If styles must be combined for emphasis or other reasons, good typographic practice allows sans serif to be used with either classic or modern. Classic and modern are normally not combined. Differences in style are best used for nominal differentiation; size and boldness are more appropriate for ordinal, interval, and ratio distinctions.

Sans serif seems to be assuming a more dominant position in cartography, while the old style and modern types are used more often in literary composition. It is not uncommon to find a sans serif used in the body of a map with a contrasting style, usually old style, employed for the title and legend materials.

Futura Light

Futura Light Italic

Futura Medium

Futura Medium Condensed

Futura Medium Italic

Futura Demibold

Futura Demibold Italic

Futura Bold

Futura Bold Italic

Futura Bold Condensed

Futura Bold Condensed Italic

FIGURE 9.7 Variants of a single type style. The Futura style has a larger number of variants than is usual, but the list is representative except that expanded (opposite of condensed) is missing. (Courtesy Monsen Typographers, Inc.)

The cartographer is usually less concerned with what has been called the "congeniality" of type style, that is, the correspondence between the subjective impression one gains from a type design in relation to its use. There is no question that readers respond to various styles with reactions such as "authoritative," "delicate," "strong," "arty," and "clean."

Type Form

Individual alphabets of any one style consist of two quite different letter forms, capitals and lowercase letters. These two forms are used together in a systematic fashion in writing, but conventions as to their use are not well established in cartography. In general, past practice has been to put more important names and titles in capitals and less important names and places in capitals and lowercase. Names requiring considerable separation of the letters are commonly limited to capitals.

The tendency in cartography is for hydrography, landform, and other natural features to be labeled in slant or italic, and for cultural features (manmade) to be identified in upright or roman forms. This can hardly be called a tradition, since departure from it is frequent, except in the case of water features where it is clearly a strong convention. The slant or italic form seems to suggest the fluidity of water.

There is a fundamental difference between slant and italic, although the terms are sometimes used synonymously (see Fig. 9.8). True italic in the classic or modern faces is a cursive form similar to script or handwriting. Gothic or sans serif italic and slant are simply like the upright letters tilted forward. Classic and modern italic forms, being much more cursive, are considerably harder to read than their upright counterparts; however, it is doubtful that there is much difference in ease of recognition between the upright and slant letters of sans serif.

Type Size

When we are concerned with perception and legibility, the subject of type size is quite complex, because different type styles at one specified size

FIGURE 9.8 Differences between italic and slant forms. (Courtesy Monsen Typographers, Inc.)

Kennerley—an upright Classic

Kennerley in the Italic form

Monsen Medium Gothic—upright

Monsen Medium Gothic Italic

may well appear to be different sizes, to say nothing of the light, medium, bold, and extra bold variations of one style. To appreciate how this can occur requires a brief description of the formation of the letters which make up our normal alphabet.

As already noted, there are two basically different kinds of letter forms, lowercase and capitals. The majority of lowercase letters are usually the same vertical height (called the x-height), namely a, c, e, m, n, o, r, s, u, v, w, x, and z. Forms of this same size make up part of other lowercase letters, and is also referred to as the "body." Some lowercase letters have an element called an *ascender* that extends upward from the body, as in b, d, f, h, k, l, and t, whereas others have a *descender* below the body, as in g, j, p, q, and y. The dot in an i and a j may or may not be placed as high as an ascender. The ratio between the x-height of the body and the length of the ascenders and descenders is entirely arbitrary; the body may be small and the ascenders and descenders long or the body may be large and the ascenders and descenders short. In contrast, capital letters (of a given size and style) are all the same height, usually about as tall as the distance from the base of the body of lowercase letters to the top of the ascenders in lowercase letters of the "same" size. These relationships are illustrated in Figure 9.9.

The size of type, which refers to its height as on a printed page, is commonly designated by *points,* 1 point being nearly equal to 0.35 mm (1/72 in.). With the development of photo-electronic means of producing type, size may also be expressed as a measure of its height in millimeters or inches. The system of designating by points originated when all type was metal, and the point size specifically refers to the distance between the upper and lower edges of the cast metal block on which the type face appears, which in most styles comes close to being the distance between the upper limit of an ascender and the lower limit of a descender (Fig. 9.9). Thus in 18-point type, that total distance will be approximately 6 mm. It is important to note, however, that the capital letters of 18-point type are likely

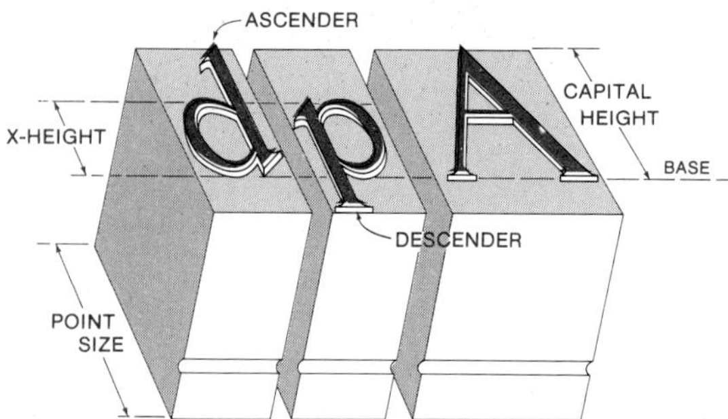

FIGURE 9.9 The point size of a type is a specific dimension of the block on which the type face is located. Among different type styles there is no consistent relation between lengths of ascenders/descenders and body height, but capitals usually are as tall as from the base of the body to the top of the ascender.

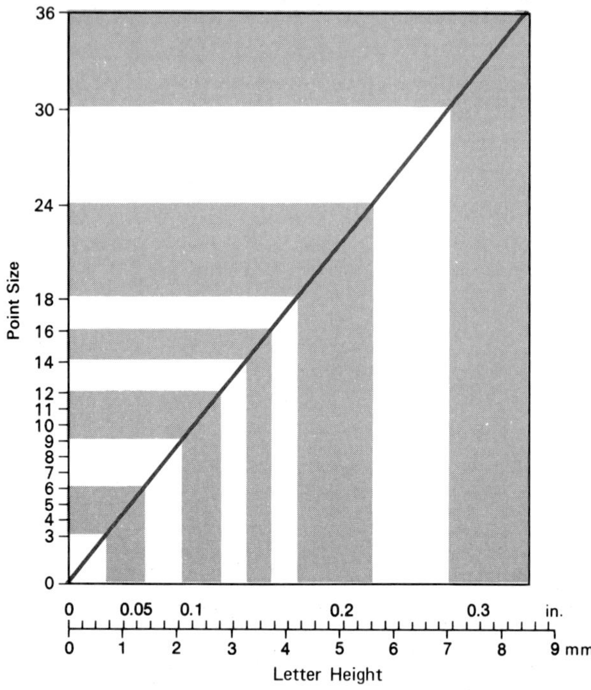

FIGURE 9.10 The relation between letter height and the corresponding approximate point sizes of capital letters.

to be only about 4 mm high while capital letters that are actually 6 mm high (approximately $\frac{1}{4}$ inch) will be those of a 24-point size type.

The lowercase letters of dissimilar styles of type, all of a given point size, may differ markedly in the sizes of the x-height or body, whereas there will be little variation in the heights of the capitals. Although a large share of composition (the production of type for printing) is now done photoelectronically, the system of designating sizes by points is still standard procedure.

Reading type when it is set in continuous, spaced lines, as on this page, is very different from the perception of type on maps. Through long experience in reading text we have learned to recognize quickly the shapes of thousands of words, based a good deal on the patterns of ascenders and descenders. On the other hand, type on maps is mostly used for unfamiliar names which must fit in given spaces. Furthermore, they may be angled, curved, or widely letter spaced, that is, there may be some map distance between each letter of a name. Except in occasional descriptive legends type does not occur set in lines or blocks. Consequently, since there is considerable variation among type styles in the apparent sizes of lowercase letters (x-height with ascenders and descenders) a more useful general index of type size in cartography is the height of the capital letters. Therefore, Figure 9.10, which equates point size with letter height, is based upon the normal height of the capital letters in the various point sizes, not the distance between the top of an ascender and the bottom of a descender.

Perhaps the most common kinds of typographic decisions facing the cartographer concern the various sizes to be used for the great variety of items that must be named on maps. Traditionally, specifications for type size are usually based on the magnitude of the object being named or the space to be filled, but the lettering must also be graded with respect to the total design and intellectual content of the map. Much of the criticism, however, conscious or unconscious, that is leveled at map type is aimed specifically at size—or lack of it.* There seems to be

*Particularly for maps to be projected on a screen before a large audience.

a widespread tendency among amateur (and even some professional) cartographers to overestimate the ability of the eye.

In Chapter 7 it was observed that, assuming no other complications (the assumption is a bit unrealistic), the eye reacts to size in terms of the angle the object subtends at the eye. Generally, with normal sight, an object that subtends an angle of 1 min. can just be recognized. Letter forms are complex, however, and it has been determined that about 3-point type is the smallest, just recognizable type at usual reading distance. Normal sight is, however, a misnomer; it certainly is not average. It is safer to generalize that probably 4-point or 5-point type comes closer to the lower limit of visibility for the average person.

A common employment of type size variations in cartography is for the purpose of ordinal, interval, or ratio ranking or categorizing of places, usually settlements, in terms of their relative population size. The ability of map readers to recognize differences in size has been carefully analyzed in several studies. In general, the results reveal that the great majority of map users are not very sensitive to small variations in type sizes when the different sizes are not adjacent to one another, which is usually the case on a map. One study, which tested discrimination among sizes of Futura Medium type (see Fig. 9.7) with names set in capitals and lowercase letters, obtained the following results:*

1. Size differences of less than 15 percent (for example, between $7\frac{1}{2}$ and $8\frac{1}{2}$ points) are consistently unrecognizable.
2. Differences of more than approximately 25 percent in letter height are highly desirable.
3. Within the range of $5\frac{1}{2}$ to 15 points, any paired combination differing by 2 to $2\frac{1}{2}$ points can be used safely, for example, $5\frac{1}{2}$ and $7\frac{1}{2}$ in the smaller sizes and $8\frac{1}{2}$ and 11 points in the larger.

*Barbara Gimla Shortridge, "Map Reader Discrimination of Lettering Size," *The American Cartographer* 6, no. 1 (1979): 13–20.

It is reasonable to assume that these recommendations may be extended to most other typefaces. It also appears that when we wish to use one style and form of type in several sizes to differentiate categories of data (for example, sizes of cities), we should limit the number to three. People seem to be able to tell "small," "medium," and "large" apart, but have difficulty with four or more. For more than three categories we probably should introduce significant differences in form. In the interest of promoting ease of recognition ("findability" when looking for a name), the cartographer should use the largest type sizes possible, consistent with good design.

Frequently, cartographers are called on to prepare presentations for groups. They must face questions such as, "What is the minimum type size that may be read, under normal conditions, on a wall map or chart from the back or middle of a 10-m (33-ft.) room?" Figure 9.11 is a nomograph, the construction of which is based on the assumption that if a particular size, at normal reading distance from the eye (45 cm, approximately 18 in.), subtends a certain angle at the eye, then any size lettering, if viewed at such a distance that it subtends the same angle is, for practical purposes, the same size. Thus, 144-point type at about 10 m (33 ft.) from the observer is the same as approximately 8-point type at normal reading distance, since each circumstance results in nearly the same angle subtended at the eye. It will be seen from the graph that legibility diminishes very rapidly with distance. For example, any type of 16-point size or smaller cannot be read even at 3 m (10 ft.) from the chart or map.

The preparation of maps and diagrams to be projected by means of 35-mm slides or overhead transparencies involves several complicating factors that should be investigated with care.*

*An excellent treatment of these problems with specific directions is given in H. W. Brockemuehl and Paul B. Wilson, "Minimum Letter Size for Visual Aids," *The Professional Geographer* 28, no. 2 (May 1976): 185–89.

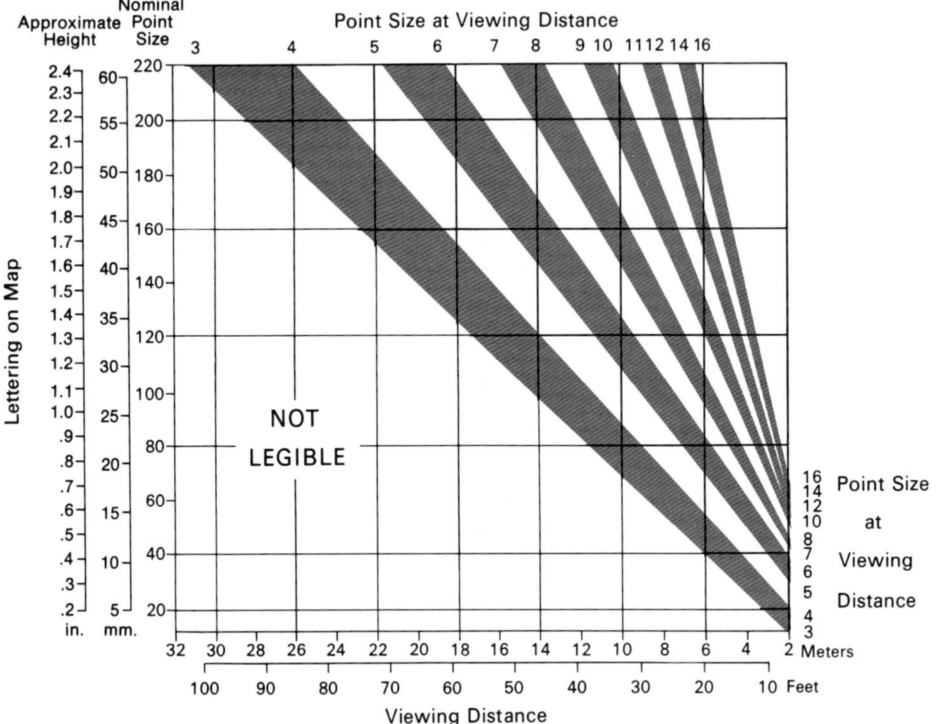

FIGURE 9.11 A nomograph showing the approximate apparent point sizes of lettering (capital letters) when viewed from various distances. A viewing distance may be extended vertically to a desired point-size band (in color) and then read horizontally to the left to the equivalent size. Or one may start at the left with a size on a map and reverse the process. Note that equivalent sizes less than 3 points are not legible.

Type Color

In typography, "color" of type refers to the relative overall tone of pages set with different faces. Here it is restricted to the actual hue (such as black, blue, gray, or white) of the letters and the relation between the hue and value of the type and that of the ground on which it appears. Commonly, lettering of equal intrinsic importance does not appear as equal in various parts of the map because of ground differences. Even when the same typeface must be used in various graphic situations because of other design requirements, the cartographer should be aware of these possible effects. Perhaps, within the limits of the design, the cartographer may be able to correct or at least alleviate a graphically inequitable situation.

Stated in general terms, the perceptibility of lettering on a map (other effects being equal) depends on the amount of visual contrast between the type and its ground. Putting aside effects that result from texture of ground, type size, and so on, the basic variable is the degree of value contrast between the type and the ground on which it stands (Fig. 9.12). Thus, black type on a white ground would stand near the top of the scale and, as the tonal value of the lettering approached the tonal value of the ground, visibility would diminish. This is of concern either when large regional names are "spread" over a considerable area composed of units that are colored or shaded differently or when names of equal rank must be placed on areas of different values.

Commonly, the lettering on maps is either dark (such as black) on a light ground or light (such as white) on a dark ground. The latter is called *reverse* (or reversed) lettering. Type is often added

CARTOGRAPHY

FIGURE 9.12 Perceptibility and legibility depend on lettering-background contrast.

to one of the color flaps of a multicolored map (such as blue for hydrography). Regardless of the color of the print and of the ground, if the value contrast is great, the lettering will be visible.

Screened lettering on a light or dark ground or reverse lettering on a relatively dark ground is an effective way to create contrast in the map lettering. The production of screened or reverse lettering is not difficult in the map construction process (see Chapter 18), and it provides a very effective way to categorize classes of lettering.

LETTERING THE MAP

Methods of Lettering

During the first quarter of this century much of the lettering for maps was done freehand, that is, by a person using pen and ink. Freehand lettering is slow and therefore costly. If it is to be done well, it requires much more calligraphic skill than most cartographers enjoy. The desire for lower costs, speed, and standardization led to the development of mechanical lettering systems. Today almost all mapmaking uses some method other than freehand for finished map construction. These may be grouped into three categories: (1) stick-up lettering, (2) computer-assisted lettering, and (3) mechanical systems for ink lettering. Freehand lettering is now largely restricted to the compilation phase and to the occasional special map produced by free-lance cartographers and illustrators.

No matter how the lettering on a map is produced, an important element is spacing, or the relative distance between letters. In type set by machine, the spacing is mechanical, and consequently, words often appear ragged and uneven. Visual spacing in which the appropriate amount of space appears between each letter and its neighbor is far better (Fig. 9.13). For this reason, many names and titles made by stick-up need to be positioned one letter at a time. The mapmaker will soon learn that there are different classes of letters according to their regular or irregular appearance and according to whether they are narrow, average, or wide. They must be separated differently, depending on the combination in the word. Mechanical spacing should always be avoided.

Stick-up Lettering. Stick-up lettering in which type is affixed to the map has a number of advantages over other methods of lettering, in addition to the fact that it is generally faster and requires less skill. Any of hundreds of type styles and sizes can be selected. If the position first selected for the name turns out to be unsuitable, the name may be relocated without difficulty. The letters may be used as composed in a straight

FIGURE 9.13 Mechanical spacing on the left compared to the much better visual spacing on the right. In the mechanical spacing there are visual gaps, such as between the P and the A, while other letter seem crowded together, such as in MIL.

SPACING SPACING

MILES MILES

unit, or they may be cut apart and applied separately to fit curves. Sans serif typefaces are easier to work with in this respect, since there are no serifs to interfere with cutting and curving.

The type used for stick-up lettering can be printed on opaque white paper but more often on thin plastic sheets, the backs of which are then given a light coating of colorless wax. The words, numbers, or letters may then be cut from the sheet, and burnished onto the artwork by rubbing. Rubbing is usually done with a smooth tool on a sheet laid over the film to protect it. Light rubbing will hold the lettering in place but will allow it to be repositioned. Final burnishing is postponed until all positioning decisions have been made.

It is possible, of course, to print or typewrite the letters on other materials and to affix them to the map in other ways. Gum-backed paper, thin tissue, and other materials have been employed, and liquid adhesives, self-adhesive materials, and so on have all been tried. The standard process, however, uses lettering on thin plastic film or cellophane with a wax backing.

There are two ways of producing such stick-up materials: by nonimpact (photographic, electronic) and impact (typewriters, printing presses) methods. A number of typesetting devices of each type are manufactured for the purpose, and each machine has its special characteristics (Fig. 9.14). They will not be detailed here. Suffice it to say that nonimpact methods provide several advantages, including the ability to vary the size of the lettering. This can be accomplished easily because in these devices the letters exist in an analog or digital template (matrix) and, by enlargement or reduction, the print can be selectively scaled to the desired size. With impact systems, on the other hand, a separate set of matrices or typewriter heads must be obtained for each size desired.

The cartographer must make a complete type list consisting of all names and other lettering to appear on the map or series of maps for which stick-up is to be used. Spellings must, of course, be reviewed carefully. It is desirable to obtain several copies of the composition. Then, it is only necessary to repeat names, such as "river" or "lake," as many times as the total number of map occurrences of that size divided by the number of copies of the composition to be obtained. All lettering must be categorized in the listing according to type style, size, and capital and lowercase requirements. The names will be set as listed and, since the probability of inadvertent omission is relatively high, it is good practice to obtain extra alphabets of the styles and sizes being used.*

Another type of stick-up lettering consists of alphabets (and symbols) on the "underside" of plastic. When burnished, the letters will come off the plastic and adhere to the map. This kind of lettering is primarily useful for titles and larger names.

Computer-Assisted Lettering. The increased use of computer-driven graphics output devices in recent years has encouraged cartographers to explore various methods of lettering under electronic control. Any automated device suitable for generating graphic artwork can also be used to form letters. This means that maps can be lettered with character printers, vector plotters, and raster printer/plotters (see Chapter 18). As can be expected, however, the results will vary widely in flexibility (type style, size, boldness, spacing) and graphic quality. At present, computer-assisted typography tends to appear somewhat rigid and mechanical, lacking the character supplied by subtle variations in line widths and curvatures. Although many machine options are already possible (see Fig. 9.15), each letter is rather stiffly drawn, and placement appears to be firm and mechanical (see the section on Automated Positioning, below).

To take advantage of computer assistance in lettering, a list of place-names and other typography must be available in computer compatible (digital) form. If this is done on a project-by-project

*Commercial firms that will supply stick-up type to specifications include: Supertype, 1300 W. Olympic Blvd., Suite 208, Los Angeles, CA 90015 (213/388-9573), and JCS, P.O. Box 225393, Dallas, TX 75265 (214/363-5600).

FIGURE 9.14 On the left is the Varityper Comp/Edit 6400 and the sample on the right illustrates the systems modifications capability. (Photo courtesy of Varityper).

FIGURE 9.15 Examples of the Hershey lettering fonts currently capable of being used in computer-assisted cartographic production.

FONTS

Simplex Script
Aa Bb Cc Dd Ee Ff Gg Hh Ii Jj Kk Ll Mm Nn
Simplex Roman
Aa Bb Cc Dd Ee Ff Gg Hh Ii Jj Kk Ll Mm Nn
Simplex Numbers And Special Symbols
@ [ ]) − + < = > * (% : ? ! , ' ; / . □
Complex Roman
Aa Bb Cc Dd Ee Ff Gg Hh Ii Jj Kk Ll Mm
Complex Italic
Aa Bb Cc Dd Ee Ff Gg Hh Ii Jj Kk Ll Mm
Complex Numbers
@ [ ]) − + < = > * (% : ? ! , ' ; / . □
Duplex Roman
Aa Bb Cc Dd Ee Ff Gg Hh Ii Jj Kk Ll Mm Nn
@ [ ]) − + < = > * (% : ? ! , ' ; / . □
Triplex Roman
Aa Bb Cc Dd Ee Ff Gg Hh Ii Jj Kk Ll Mm
@ [ ]) − + < = > * (% : ? ! , ' ; / . □

digital record to drive the selected lettering device. Alternatively, the cartographer may use an existing data base containing all place-names for a region that has been created by a central mapping authority (see the section on the Geographical Names Data Base, below). By working from this computer-generated list, the cartographer can select only those names that are needed for a given map.

Mechanical Systems for Ink Lettering. Mechanical devices are available to enable someone who is unskilled in freehand lettering to produce acceptable ink lettering. With them we can obtain neat lettering, but it should be emphasized that much of the lettering produced with these aids appears rather mechanical and looks a little like a mechanical drawing. The complexities of good quality map lettering require more versatility than can be easily obtained by such means. Nevertheless, mechanical lettering aids will continue to find a place in "do-it-yourself" cartography, especially for the produc-

basis, the cartographer will need to prepare the usual type list and then enter this information into a computer file. The computer can then use this tion of graphs, charts, and diagrammatic maps. Most of these devices require a special pen that is guided either mechanically or by hand with the aid of a template.

Leroy is the patented name of a lettering system involving templates, a scriber, and special pens. A different template is necessary for each size of lettering. The template is moved along a straightedge, and the scriber traces the depressed letters of the template and reproduces them with the pen beyond the template. A variety of letter weights and sizes in capitals and lowercase is possible by interchanging templates and pens.

Wrico is the patented name of a lettering system involving perforated templates or guides and special pens. The lettering guides are placed directly over the area to be lettered and are moved back and forth to form the various parts of a letter. The pen is held in the hand and is moved around the stencil cut in the guide. To prevent smearing the guide rests on small blocks, which hold it above the paper surface. A considerable variety of letter forms is possible, including condensed and extended. A different guide is necessary for each size, and pens may be varied.

Varigraph is the patented name of a lettering device which also involves a template with depressed letters and a stylus. The device is actually a sort of small, adjustable pantograph that fits over a template. The letters are traced from the template and are drawn by a pen at the other end of the pantograph-like assembly. Adjustments may be made to make large or small and extended or condensed lettering from a single template. Templates of a variety of letter styles are available.

Freehand Lettering. It is important that a cartographer become reasonably adept at freehand lettering, at least with pencil, in order to do the necessary lettering for compilation. Furthermore, the cartographer, in designing a graphic composition, will be called on to lay out titles, legends, and the like, and to work out the spaces and sizes needed prior to obtaining stick-up.

When laying out the lettering plan for titles and for names that are to be curved or spaced for stick-up positioning, one uses guidelines. They are simply what their name implies, lines to serve as guides for the lettering, whether stick-up, mechanical, or freehand. They may be drawn with a straightedge or a curve, but guidelines are more easily made with any of a number of patented plastic devices designed for this purpose. These devices have small holes in which a pencil point may be inserted. The device may then be moved along a straightedge or curve by moving the pencil. By placing the pencil in other holes, parallel lines may be drawn. Guidelines consisting of three parallel lines, as shown in Fig. 9.16 are drawn for freehand lettering. The bottom two determine the x-height of the lowercase letters, and the upper guideline indicates the height of the capital letters and ascenders.

Positioning the Lettering

Map reading is affected greatly by the positioning of the names. When properly placed, the lettering clearly identifies the phenomenon to which it refers, without ambiguity. Equally important is the fact that the positioning of the type has as much effect on the graphic quality of the map as does the selection of type styles, forms, and sizes. Incongruous, sloppy positioning of lettering is just as apparent to the reader as are garish colors or poor line contrast.

Large mapmaking establishments tend to develop policies regarding the positioning of type partly to obtain uniformity, partly because it is cheaper, and partly because we can assign parts of such activity to a machine. The excessive, systematic, overall result—*all* names parallel, for example—tends to set an unfortunate standard that should not be blindly followed. The cartographer, as in all other matters involving the structuring of graphic composition, should be guided by principles and precepts based on the function of the map as a medium of communication. Generally, the object to which a name applies should be easily recognized, the type should conflict with the other map material as little as possible, and the overall appearance should not be stiff and mechanical.

As previously observed, one of the important functions of type on a map is to serve as a locative device. The lettering can do this in three ways: (1) by referring to point locations, such as cities; (2) by indicating the orientation and length of linear phenomena, such as mountain ranges; and (3) by designating the form and extent of areas, such as regions or states. The first rule of positioning type is to position the lettering so that it enhances the locative function as much as possible.

It will be convenient in the list of the major principles of type positioning to organize them as they refer to place, linear, or areal phenomena. There are a few general rules, however, that are independent of such a locative organization.

1. Names should be either entirely on land or on water.
2. Lettering should generally be oriented to match the orientation structure of the map. In large-scale maps this means parallel with the upper and lower edges; in small-scale maps, it means parallel with the parallels.
3. Type should not be curved (that is, different from rule 2 above) unless it is necessary to do so.
4. Disoriented lettering (rule 2 above) should never be set in a straight line but should always have a slight curve.
5. Names should be letter spaced as little as possible.
6. Where the continuity of names and other map data, such as lines and tones, conflicts with the lettering, the data, not the names, should be interrupted.
7. Lettering should never be upside down.

Conflicts among precepts frequently occur because of opposing requirements. There is no general rule for deciding such issues; the cartographer must make a decision in light of all the special factors. Figures 9.17 and 9.18 have been prepared to illustrate the foregoing principles of type positioning as well as the precepts that follow.*

FIGURE 9.16 Three guidelines are used for freehand lettering. For positioning type, only one guideline is needed.

Place Features. Generally, lettering that refers to place locations should be placed either above or below the point in question, preferably above and to the right. The names of places located on one side of a river or boundary should be placed on that same side. Places on the shoreline of oceans and other large bodies of water should generally have their names entirely on the water. Where alternative names are shown (Köln, Cologne), the one should be symmetrically and indisputably arranged with the other so that no confusion can occur.

Linear Features. Lettering to identify linear features should always be placed alongside and "parallel" to the river, boundary, road, or whatever to which it refers, never separated from it

*For many illustrations of good and bad practice in positioning lettering, see Eduard Imhof, "Positioning Names on Maps," *The American Cartographer* 2, no. 2 (1975): 128–44.

by another symbol. Where there is curvature, the lettering should correspond. Ideally, we would place the type along an uncrowded stretch where the lettering could be read horizontally, but often this is not possible. Complicated curvatures should be avoided, and generally the names should be somewhat letter spaced but not much. Such designations along rivers need to be repeated occasionally. If we must position lettering nearly vertically along a linear feature, it is easier to read it upward on the left side of the map and downward on the right. Where possible, it is good practice to curve lettering so that the upper portions of the lowercase letters are closer together because there are more clues to letter form in the upper parts. Names along linear items are better placed above than below the features because there are fewer descenders than ascenders in lowercase lettering.

Areal Features. Normally, the lettering to identify areal phenomena will be placed within the boundaries of the region. The name should be letter spaced to extend across the area, but the letters should not be so far apart that they do not appear as parts of a name and thus be lost among the other detail. Letters should not crowd against boundaries. As observed in the general rules, where any tilting is necessary, there should be added a clearly noticeable bit of curvature so that the name will not look like a printed label simply cut out and pasted on the map. Curvatures should be very simple and constant.

Automated Positioning

In spite of the advances in computer-assisted cartography, the placement of names on maps has yet to be accomplished in a way that meets desired cartographic quality. The problems that were noted previously with respect to automated lettering are bothersome, but effective integration of lettering with the other elements of the map image by strictly automated means has proven to be extremely difficult. The reason, of course, is that good placement requires making numerous decisions that involve spatial synthesis and judgment. The current generation of mapping software lacks these human abilities.*

To date, most automated positioning has been restricted to lettering set in a straight line, usually horizontally. This may be sufficient for constructing titles and legends, but represents only a crude approximation to the results that can be achieved by hand/eye placement of type within the map image. Although progress is being made, letter spacing, such as for the names of regions and for curved names, is still next to impossible to do well without manual override.

In fact, current research trends imply that semiautomated type placement involving human-machine interaction may be the most practical approach at the present time. With this procedure, type can be selected from an existing data base of geographical names, and tentatively placed on the map automatically according to a set of good positioning principles. Then, by using this "first approximation," the cartographer can shuffle names around and change type parameters to avoid overlap and to make all names legible. Once an acceptable arrangement is achieved, the computer is directed to create the finished lettering flap on a printing/plotting device. The overall result thus approaches an efficient merging of the advantages exhibited by both the computer and the cartographer.

Some estimates show that lettering accounts for almost 40 percent of total map production costs. Manual override of a computer-assisted system will help to reduce these costs, but the resulting lettering will probably also be more stiff and less flowing in position for some time. Ultimately the cartographer may experience greater flexibility in map lettering, but for the near future the cost savings gained by computer assistance may only result in less flexibility in map lettering. It may turn out that the whole concept of map

*Ongoing research in the area of artificial intelligence suggests that future generations of mapping software may be more able to mimic some of the human thought processes.

lettering will change to incorporate various computer possibilities.*

Geographical Names

Difficult as the solutions to the various problems concerning the positioning of names may be, often an even more difficult question is their proper or appropriate spelling. For example, do we name an important river in Europe *Donau* (German and Austrian), *Duna* (Hungarian), *Dunav* (Yugoslavian and Bulgarian), *Dunarea* (Romanian), or do we spell it *Danube,* a form not used by any country through which it flows! Is it *Florence* or *Firenze, Rome* or *Roma, Wien* or *Vienna, Thessalonkie, Thessoloniki, Salonika,* or *Saloniki,* or any of a number of other variants? The problem is made even more difficult by the fact that names change because official languages are altered or because internal administrative changes occur. The problem of spelling is difficult indeed.

The difficulties are of sufficient moment that governments that produce many maps have established agencies whose responsibility is to formulate policy and to specify the spelling to be used for names on maps and in official documents. Examples are the U.S. Board on Geographical Names (BGN), an interdepartmental agency, and the British Permanent Committee on Geographical Names (PCGN). The majority of such governmental agencies concern themselves only with domestic problems, but these two include the spelling of all geographical names as part of their function.

One of the major concerns of such an agency (and of every cartographer) is how a name in a non-Latin alphabet will be rendered in the Latin alphabet. Various systems of *transliteration* from one alphabet to another have been devised, and the agencies have published the approved systems. The Board on Geographic Names has published numerous bulletins of place-name decisions and guides recommending treatment and sources of information for many foreign areas. These are available upon application.*

The general practice in the United States is to use the conventional English form whenever such exists. Thus, *Finland* (instead of *Suomi*) and *Danube River* would be preferred. Names of places and features in countries using the Latin alphabet may, of course, be used in their local official form if that is desirable. The problem is very complex, and it is difficult to be consistent. It is easy for English speaking readers to accept *Napoli* and *Roma* for *Naples* and *Rome,* but it is more difficult for them to accept *Dilli, Mumbai,* and *Kalikata* for *Delhi, Bombay,* and *Calcutta.*

The problem is much too complex to be treated in any detail here, but it is important for the student of cartography to be aware of it. Above all, cartogaphers must not fall into toponymic blunders like placing on maps such names as *Rio Grande River, Lake Windermere,* or *Sierra Nevada Mountains.*

Geographical Names Data Base. In the interest of standardizing geographical names usage and in automating the use of place-names in cartography and other applications, the U.S. Geological Survey (USGS) has developed its *Geographical Names Information System* (GNIS). As the core of the system, a machine-readable national geographical names data base is being constructed. The primary source for the place-names in these files is the standard $7\frac{1}{2}$-minute quadrangle topographic map series published by the survey. Names information derived from the records of the U.S. Board on Geographical Names,

* Using a display screen, for example, the type might not appear on the map at all. Then, if the map user were to touch a feature or spot on the map, the appropriate name could be verbally announced or displayed in flashing light on the side of the map. Clearly, a fully developed system of this sort could make far more information available to the map user than simple place-name data. If this were to happen, the nature of the map would be changed dramatically from what we have come to expect.

*To enquire about domestic names, address: U.S. Board on Geographic Names, National Center, Stop 523, Reston, VA 22092. For foreign names address: U.S. Board on Geographic Names, Defense Mapping Agency, Building 56, U.S. Naval Observatory, Washington, DC 20305.

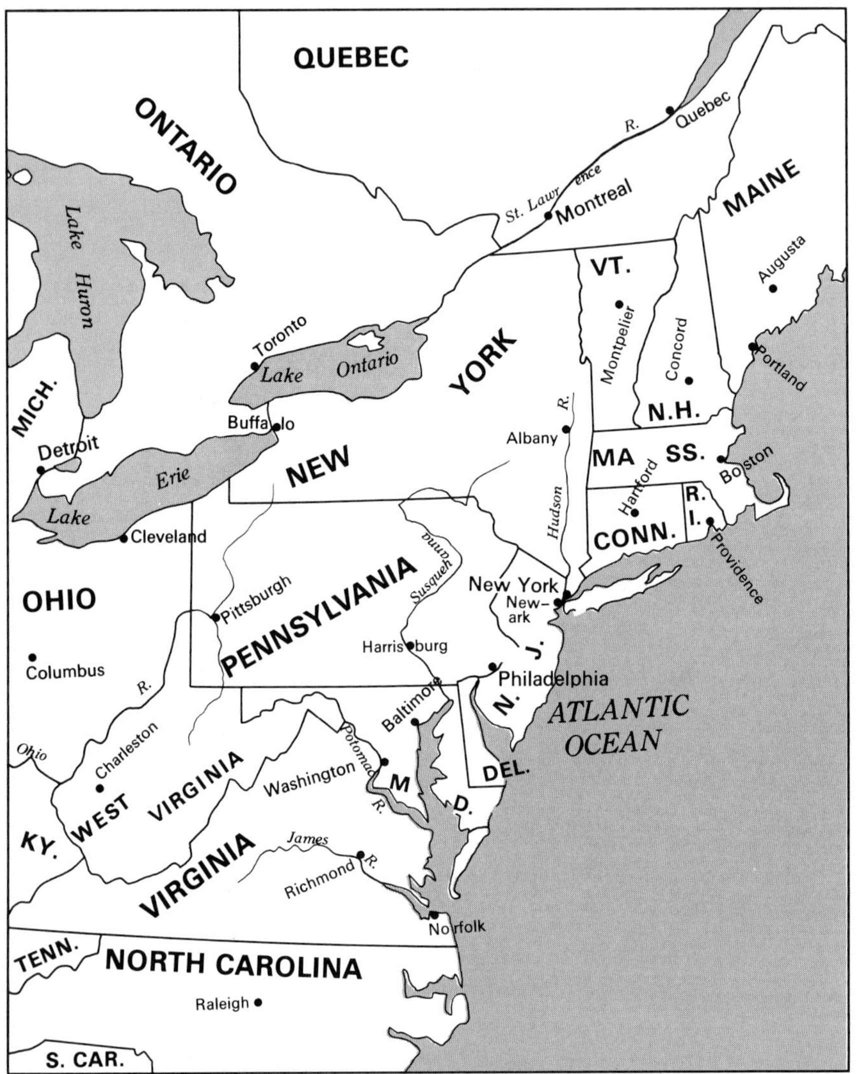

FIGURE 9.17 Most of the general rules about positioning lettering have been violated in almost every instance on this map. Compare Figure 9.18.

the U.S. National Ocean Survey charts, and other federal agency sources are also being integrated into the files. To date approximately 2 million names have been incorporated.

The comprehensive files include all named natural features (about 80 percent of the data base) and most major and minor civil divisions, dams, reservoirs, airports, and national and state parks. Named streets, roads, and highways are not included at this time. A number of data elements are coded with each feature included in the data base. In addition to the official name of the feature, these include one of seventy-two feature classes (school, cemetery, stream, summit, pop-

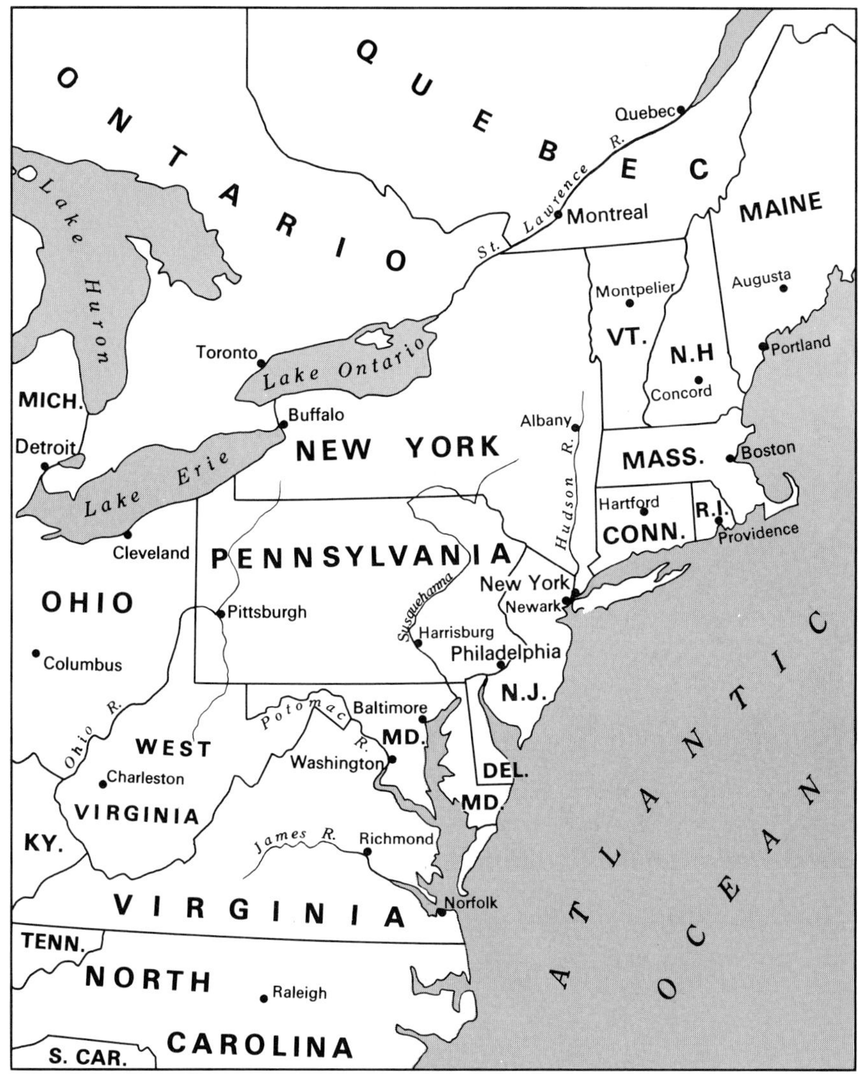

FIGURE 9.18 The same lettering as in Figure 9.17 has been positioned to better advantage on this map.

ulated place, and so on), location (state, county, and geographical coordinates), elevation, variant names (aliases), and the map sheet on which the feature is found.

Information from the GNIS is presently available in several forms. An outside computer terminal can be used for direct on-line access to the data base. Alternatively, USGS computer products such as magnetic tapes, computer printouts, and microfiche can be purchased through the National Cartographic Information Center (NCIC).*

*NCIC, U.S. Geological Survey, 507 National Center, Reston, VA 22092 (703/860-6039).

NATIONAL GAZETTEER--NEW JERSEY 1982 NJ33

FEATURE NAME	FEATURE CLASS	STATUS	COUNTY	COORDINATE	SOURCE COORDINATE	ELEV FT	MAP NAME
Central Park	park	ADMIN	Bergen	405352N0740034W			Hackensack
Central Park School	school	UNOFF	Salem	393941N0753023W			Wilmington South
Central Pier	locale	BGN	Atlantic	392122N0742529W		5	Atlantic City
Central Regional High School	school	UNOFF	Ocean	395325N0741220W			Toms River
Central School	school	UNOFF	Cumberland	392915N0750121W			Millville
Central School	school	UNOFF	Monmouth	401055N0740513W			Asbury Park
Central School	school	UNOFF	Monmouth	402406N0740601W			Sandy Hook
Central School	school	UNOFF	Middlesex	402611N0742357W			New Brunswick
Central School	school	UNOFF	Middlesex	403429N0743025W			Bound Brook
Central School	school	UNOFF	Somerset	403739N0742957W			Chatham
Central School	school	UNOFF	Essex	404626N0741356W			Orange
Central School	school	UNOFF	Essex	404722N0741847W			Caldwell
Central School	school	UNOFF	Warren	404808N0745943W			Washington
Central School	school	UNOFF	Bergen	405750N0740734W			Paterson
Centre		VARIANT					
See Centerville	ppl		Hunterdon	403218N0744516W			
Centre Bridge		VARIANT					
See Stockton	ppl		Hunterdon	402428N0745843W			
Centre City	ppl	BGN	Gloucester	394633N0751052W		80	Woodbury
Centre Grove	locale	BGN	Cumberland	392123N0750720W		77	Dividing Creek
Center Grove		VARIANT					
Centreton		VARIANT					
See Centerton	ppl		Salem	393131N0751005W			
Centreton		VARIANT					
See Centerton	ppl		Burlington	395944N0745224W			
Centreville		VARIANT					
See Knowlton	ppl		Warren	405549N0750142W			
Ceva Lake	reservoir	BGN	Mercer	401623N0744642W			Pennington
Ceva Lake Dam	dam	UNOFF	Mercer	401623N0744642W			Pennington
Chadwick	ppl	BGN	Ocean	395935N0740351W		8	Seaside Park
Chadwicks		VARIANT					
Chadwicks Beach		VARIANT					

FIGURE 9.19 A convenient product of the Geographical Names Information System is the Geographical Names Alphabetical Finding List, which consists of a set of spiral-bound volumes for each state. Information is presented as it is coded in the data base.

Printed and bound products, such as alphabetical and topical lists, can also be purchased (Fig. 9.19). Regardless of the form of output, the system offers accurate, ready reference information for automated name placement and map production. Because the data base is in machine-readable form, the information it contains can be correlated easily with the information in other computerized files.

SELECTED REFERENCES

Bartz, B. S., "An Analysis of the Typographic Legibility Literature," *The Cartographic Journal* 7 (1970): 10–16.

Brockemuehl, H. W., and P. B. Wilson, "Minimum Letter Size for Visual Aids," *The Professional Geographer* 28 (1976): 185–89.

Hirsch, S. A., "An Algorithm for Automatic Name Placement Around Point Data," *The American Cartographer* 9 (1982): 5–17.

Imhof, E., "Positioning Names on Maps," *The American Cartographer* 2 (1975): 128–44.

Keates, J. S., "Lettering," in *Cartographic Design and Production*. London: Longman Group Limited, 1973.

Nesbitt, A., *The History and Technique of Lettering*. New York: Dover Publications, 1957.

Ramano, F., "Typsetting's Brave New World: Struggling Toward the Year 2000," *American Printer* (1982): 25–28.

Shortridge, Barbara Gimla, "Map Reader Discrimination of Lettering Size," *The American Cartographer* 6 (1979): 13–20.

Updike, D. B., *Printing Types, Their History, Forms and Use: A Study in Survivals*. Cambridge, Mass.: Harvard University Press, 1922.

Yoeli, P., "The Logic of Automated Map Lettering," *The Cartographic Journal* 9 (1972): 99–108.

PART THREE

The Practice of Cartography: Data Manipulation and Generalization

10

Remote Sensing and Data Sources

REMOTE SENSING

Humans have built-in senses by which we can experience conditions of our environment. We "remotely sense" some of these conditions with hearing and sight. For example, the ear reacts to sound waves traveling through the atmosphere, enabling us to hear thunder, wind, and other sounds, and the normal eye is sensitive to some of the electromagnetic energy emitted and reflected by objects. In order to increase our range of perception, we have learned to make and use other kinds of sensors, which are sensitive to electromagnetic energy. Although remote sensing has long been an important part of mapping, especially in large-scale mapping, it is becoming increasingly so with modern developments in electronic data processing and satellites.

In the not-too-distant future when a person sits before a computer and begins to call forth data in map format, much of that data may come from data bases that were originally obtained by remote sensing techniques (for example see Color Fig. 11). The capabilities of remotely sensing the earth and converting that information into data of use to cartographers have never been more numerous. With each development in remote sensing the possibilities for cartographic use of the results increases several fold. Cartographers must be aware of and understand these developments in remote sensing to make the most use of them.

To understand modern remote sensing, we must first look at the characteristics of electromagnetic energy and the various ways it can be sensed. Then we will consider its usefulness to cartography.

Electromagnetic Energy

Incoming solar radiation from the sun strikes the earth or objects on the earth. A certain portion of that radiated energy is reflected from the earth or the object depending upon the surface character and material. Additional electromagnetic energy or radiation is continuously emitted from any object that has a temperature above 0°K (−273°C). All of this energy radiates in accordance with basic wave theory. The three components of the waves of radiant energy are the wavelength, wave frequency, and wave velocity.

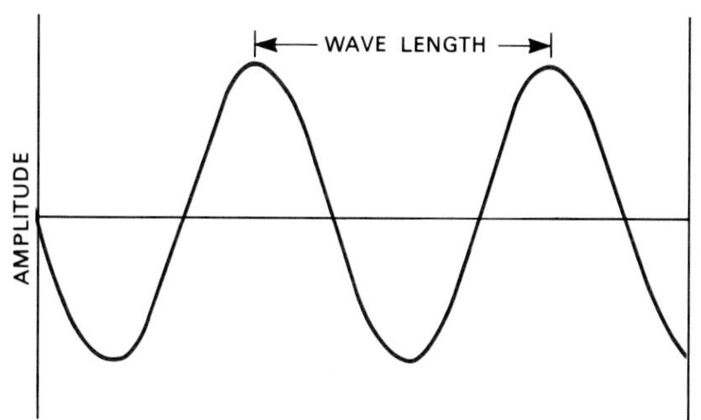

FIGURE 10.1 Wavelength is the distance between successive wave peaks.

Wavelength is the distance between the peak of one wave and the peak of the next (see Fig. 10.1). *Wave frequency* is the number of wave peaks that pass a given point in a specified length of time, and *wave velocity* equals the speed of light. Since the velocity is constant, the length of the wave and its frequency have a reciprocal relationship. In remote sensing we customarily refer to the wavelength to indicate a particular part of the electromagnetic spectrum. Because of the variety of substances on the earth and the great range of conditions, the waves of reflected or emitted electromagnetic energy can vary in length from the very short gamma rays and X rays, around 0.2 micrometers (μm)* in length, to radio waves, which can be many kilometers in length. A theoretical black body is used as a standard to compare emittance of radiation. There is a specific spectral distribution of emitted energy.

However, the amount of energy available is not the same at all wavelengths. Due to the selective absorption of electromagnetic energy by substances in the atmosphere like water vapor, carbon dioxide, and ozone, certain incoming specific wavelengths are more effectively blocked than others. This selected blocking affects the

*Micrometer (μm) = one millionth of a meter.

levels of electromagnetic energy that can be recorded at the surface of the earth (see Fig. 10.2). Further changes occur when the energy is reflected or emitted from the earth back through the atmosphere to a remote sensing system.

In addition to absorption, atmospheric scattering also affects the intensity and the wavelengths of radiation available to a remote sensing system. These combined effects—absorption and scattering—of the earth's atmosphere mean that the amount of available energy is not of the same magnitude for all wavelengths. The bands of the spectrum where atmospheric attenuation is slight, called *windows,* are the regions used for remote sensing (see Fig. 10.2).

For convenience, the continuum of electromagnetic energy is often divided into bands or regions such as ultraviolet, visible, infrared, and microwave. Boundaries of these bands are not precise and should be thought of only as zones of transition. In Figure 10.2 one can see within the largest window in the displayed portion of the spectrum that the visible band extends from about 0.4 to 0.7 μm, which means that our eyes can detect only a tiny portion of the entire spectrum. Photography can record wavelengths from about 0.3 to 1.2 μm or about three times the range the human eye can see. To sense wavelengths longer than 1.2 μm, instruments other than photographic cameras are used.

Although another principal window is a band from 8 to 14 μm, called the middle infrared, the window from 3.5 to 5.5 μm, is commonly used for sensing because the instruments designed for these wavelengths are less costly to construct. For passive microwave and radar sensing there are windows between 1 mm and 1 cm. Beyond the wavelength of 1 cm, the atmosphere is relatively "transparent" and accommodates various types of radio equipment.

For both very short and very long wavelengths the emittance is low; for some wavelengths in between, it reaches a maximum value, depending on the temperature of the black body (Fig. 10.2). This peak in the energy distribution curve is shifted to shorter and shorter wavelengths as the temperature of the black body increases. If, for example, a piece of iron is heated enough, its radiation curve would first intercept the long waves of the visible spectrum and appear to be a dull red. Upon further heating the peak of the curve would shift to shorter wavelengths and the color would change to orange, then yellow, and finally white.

Thus, we observe that remote sensing systems can detect and record energy in many parts of the spectrum including those that are invisible to the human eye. These new, unfamiliar views of our landscape can give analysts helpful new insights.

The numerous remote sensing devices that detect and record energy may be divided into active and passive systems. Passive sensors, like photography, detect natural energy, either reflected or emitted; active systems, like radar, generate the energy that is directed at and subsequently received from a target. Probably the most popular active system is radar, which operates in the microwave portion of the spectrum. An example of passive sensing is photography, which uses reflected solar energy, usually in the visible portion of the spectrum.

Electromagnetic energy can be detected and recorded by both photographic and nonphotographic means. An ordinary photograph is the result of a chemical emulsion being acted on directly by reflected energy, whereas the imagery produced by other nonphotographic remote sensing is the result of the detection of emitted or reflected energy and the conversion of the detected signals into a digital record and/or a picture-like format.

The Concept of Multispectral Sensing

Since the amount of energy reflected or emitted varies with wavelength, sensing simultaneously within separate narrow bands of the electromagnetic spectrum may provide data that show greater differentiation between an object and its background than do data from a single wide band.

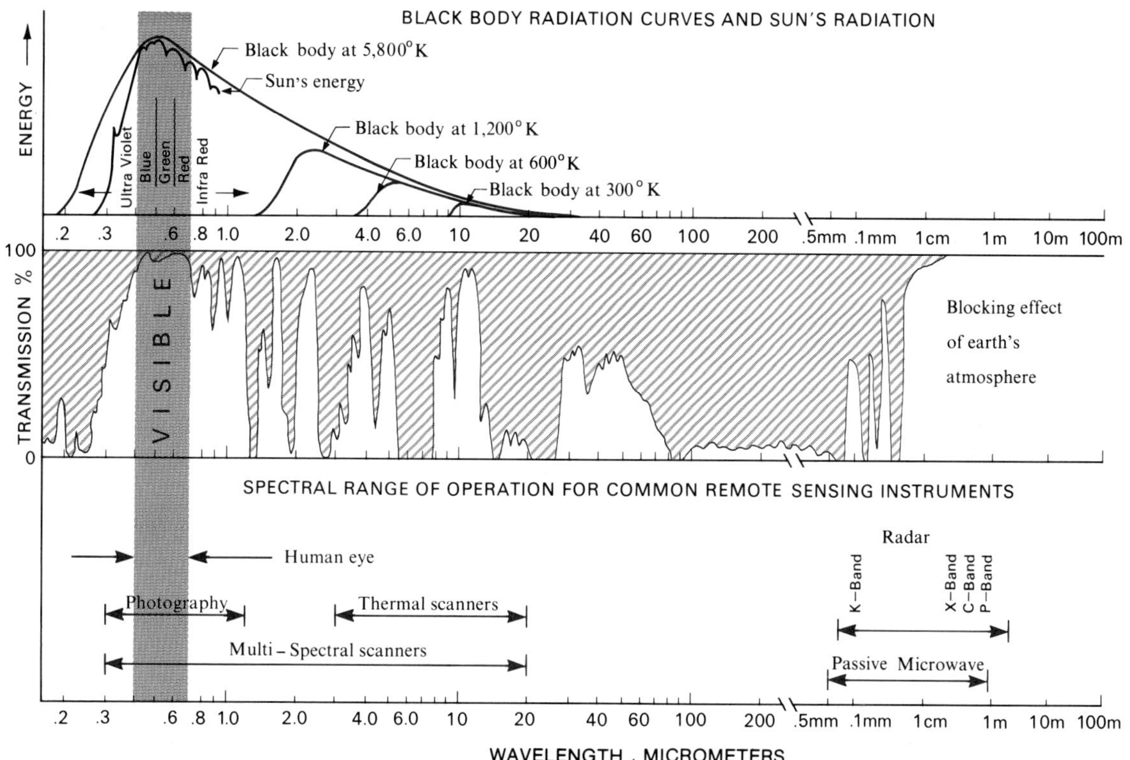

FIGURE 10.2 The electromagnetic spectrum showing the common bands of energy, energy curves for various black body temperatures, and windows where remote sensing can be accomplished. (After Scherz.)

This multispectral sensing can be accomplished by photographic as well as nonphotographic sensors, and separate images can be produced in black and white or they can be combined to produce color renditions. Transparencies from various wavelengths used in projectors equipped with color filters can be reconstituted into images in which the strengths of the colors can be varied. Often combinations can be found that make the desired information more evident.

Tonal variations on imagery indicate that changes are present; however, the eye is often unable to discern boundaries between tonal classes because of the subtle gradations. Instruments are available that can measure the optical densities of the film emulsion to establish these boundaries, or they can be established by electronic means from taped data. Bright colors can be assigned to the image classes, thereby enhancing the densities to make them strikingly visible (see Color Figs. 12 and 13).

Sensor Systems

For more than a hundred years remote sensing has been used to obtain geographical information. The aerial photograph was the first remote sensing device used to inventory and map characteristics of the earth systematically. Although some highly sophisticated systems have recently been developed to record earth data, the photograph, with its great resolving power, will undoubtedly continue to be widely used as a means of remote sensing.* Even though most of these newer sensors detect energy outside the visible portion of the spectrum and record that information in some form of imagery, many conventional photo-interpretation techniques and procedures are still applicable in its analysis.

In the past a major problem was how to obtain the needed number of photographs, but today the problem is how best to handle the tremendous volume of photographs and imagery that is available. Although statistical and computer techniques are used to convert some of the remotely sensed data to usable form, it appears that for the foreseeable future people will still play an important role in interpreting both photographs and other sorts of imagery.

As early as 1840 it was suggested that photographs be used for the purpose of mapping; during the following decades, a variety of schemes were devised to procure photography. Balloons, kites, rockets, and even carrier pigeons were used to send cameras aloft.

Pigeons carried breast-mounted cameras with automatic timing devices. When released several kilometers from their loft, they returned at a relatively constant speed and on a straight course, thus providing the means for obtaining a series of photographs along the entire route. With balloons and kites, cameras of considerable size could be sent aloft to provide photography for civilian and military uses.

The first photographs from an airplane were made by Wilbur Wright over Italy in 1909. At first, hand-held ground cameras were used but, by 1915, cameras specially designed for aerial photography were in use. Although facilities were rather primitive by today's standards, military forces were able to produce thousands of prints daily, and the usefulness of aerial photography was well established by the end of World War I.

In the civilian community advances continued to be made in the commercial and scientific uses of aerial photography, and during the 1930s government agencies made extensive use of aerial survey. The Agriculture Adjustment Administration systematically produced photography of agricultural activities over most of the United States, and photography was used by the U.S. Geological Survey for topographic mapping and by the Forest Service for timber inventories. State and local agencies recognized the value of aerial photography and began using it for planning purposes.

The greatest stimulus for photo interpretation probably came with World War II. Prior to the war many military leaders recognized that intelligence gathering by use of photo interpretation might well have great influence on the outcome of a future conflict. Germany made good use of this source of intelligence during its early offensive by photographing most military and transportation facilities across all of western Europe.

The English were forced to rely on photo interpretation for intelligence gathering after the retreat from Dunkirk in 1940. Procedures and methods they developed were invaluable to the United States when it entered the war.

The mass of photography that had to be handled required that many interpreters be trained; many of these people retained an interest in photo interpretation and contributed research to the field. Likewise, the urgent need for maps during the war resulted in greater use of photogrammetric methods, both because of the speed with which maps could be made and because the photographs provided a good source of information for inaccessible areas.

Improvements in aircraft, cameras, filters, and film emulsions have enabled aerial photography to maintain its role as an important source of information about the earth. Multilens cameras can record several simultaneous views of the same scene using different film-filter combinations, thereby obtaining views in various parts of the visible and near-visible spectrum. Panoramic cameras, by using a narrow slit that pans from side to side and exposes film held in the form of an arc, can produce high-resolution photography

*Resolving power is the degree to which a photographic emulsion or a detecting device can record and display small objects or small distances between objects.

of a large area in a single exposure. Strip cameras, by having rolled film move past a slit at a rate coordinated with the speed of the aircraft, can record a continuous image of the landscape along the flight line. Because of the great resolving power of black and white films and recent improvements in the fidelity of color films, one can usually expect much landscape detail for mapping from them.

Using rockets prior to 1950 and orbiting satellites since 1960, with increasingly complex sensor systems, images of the earth from increasing altitudes have become available. Photographic systems have given ground to nonphotographic systems where the electromagnetic energy is recorded as electrical signals, which are then relayed to the earth and converted to digits for processing. This processing often results in image (photograph-like) creation.

These developments appear to be continuing, and the detail possible from orbiting satellites increases. Today we are routinely sensing the earth from ground level to well over 900 kilometers (the SMS/GOES weather satellite orbits at 36,000 km) above the ground using a variety of photographic and nonphotographic sensing systems. The usefulness of these systems for monitoring human and natural activity on the earth is tremendous. Truly the past 100 years have seen a revolution in this field. The remainder of this chapter will discuss the analog photographic systems and the digital scanning systems most used today. A short section on the sources of remotely sensed data ends the chapter.

Analog Systems

Black and White Photography. Photography, which records energy between 0.3 and 1.2 μm, includes all of the visible spectrum and extends into the near infrared band. Panchromatic emulsions, which are used for most black and white photography, record wavelengths from about 0.3 to 0.7 μm, essentially the visible spectrum.

An object on the landscape is visible when its recorded tone is different from that of its background. If its tone is the same as the tone of the background, the object will not be visible and might be detected only if it casts a shadow.

In the visible and near-visible portion of the spectrum haze results from the scattering of the short wavelengths of blue light. To penetrate the haze, a filter that blocks the shorter wavelengths is used. If one wants to photograph only certain selected wavelengths, filters can be used that block all of the visible portion of the spectrum except the wanted wavelengths, which are allowed to pass through.

During manufacture, the emulsion of a film can be formulated so that it reacts only to particular wavelengths, and its sensitivity can be extended into the near infrared, energy that is not visible to the eye, but that can be recorded as an image on the film. The tone of an object resulting from infrared exposure may be quite different from that produced by reflected light. On panchromatic film an object may not be visible because its tone is the same as its background, whereas that same object and background might have contrasting tones on infrared film and be quite discernible. Since most of the shorter wavelengths are eliminated, black and white infrared photography is very effective in penetrating haze and can be used successfully on days when ordinary film would be unsatisfactory.

The greater contrast in images on infrared film makes it especially useful for some kinds of interpretation. For example, it is a favorite of foresters, since different types of forest cover can be more easily distinguished by variations in tone. Because water absorbs infrared radiation more effectively than it does visible light, it registers as a dark gray (or black) on infrared film, and so infrared photography is a convenient means of delineating hydrographic features.

Color Photography. During the last 40 years color photography has become more and more important. Initially it was much more expensive, and the resolution of its pictures was comparatively poor. But now color prints and transparencies are challenging black and white prints in quality and costs; they show landscape detail with ever-improving clarity.

Early experiments in color photography were conducted during the latter part of the nineteenth century, but it was not until 1935 that Kodachrome film was placed on the market. Slow film speeds and problems in processing initially discouraged extensive use of color film. In the past 30 years vast improvements have been made in film speed and resolution.

The human eye can distinguish only a limited number of gray tones but can recognize differences among a very large number of colors (hues). Since many elements of the landscape have special or unique colors associated with them, it follows that in some cases interpretation of a scene is made easier and more reliable when one is able to view it in color. Identification of objects involves consideration of the size, shape, tone, texture, and location of features. However, to determine the condition of objects, such as diseased or distressed vegetation, one may find that color difference is the only clue.

During World War II, false color or camouflage detection film was developed, allowing interpreters to distinguish easily dead vegetation or artificial camouflage materials from live vegetation. False color film differs from conventional color film in that the three layers of the emulsion are sensitive to infrared, green, and red instead of to blue, green, and red. An object that reflects infrared or longwave red appears red on the final print, those items that reflect visible red light appear green, and those reflecting green light are blue.

With conventional color photography the color of an object appears approximately the same as it does to the human eye; therefore if an item is painted the same color as its background, it will not be visible. On false color film objects painted with infrared-absorbing paints photograph blue or purple, whereas live, healthy deciduous vegetation with its high reflection in the infrared appears red (magenta). Military equipment such as guns and vehicles that have been painted the color of vegetation or sites that have been covered with dead vegetation can easily be distinguished from live, healthy vegetation on the basis of their color on the photography.

False color film is being used in a number of investigations including identification of land use, crops, geologic formations, forest species, vegetation disease, and water pollution. In urban land use mapping, for example, the film is useful because the use of longer wavelengths permits better penetration of haze. There is also sharp contrast between vegetation, which appears reddish yellow, and cultural features, which appear blue (see Color Fig. 12).

Multilens Photography. Although experiments in multispectral black and white photography were conducted as early as 1861, not until 100 years later was the technique used for aerial survey of terrain features. Since the late 1960s, missions of both space vehicles and aircraft have made extensive use of photography captured simultaneously in more than one wave band.

Cameras have been developed that have two or more lenses, each of which can be used with a different filter to expose a different kind of film. The result can be a series of photographs of the same landscape exposed at the same time, each recording reflected energy in a different portion of a visible or near-visible part of the spectrum (Fig. 10.3). Since the exposures are made simultaneously, variations in illumination are nonexistent, and the interpreter has pictures that differ only because of the particular wavelength used. Disadvantages of multilens photography are that the image format is smaller than with ordinary photography and there are many more images for an interpreter to attempt to comprehend.

Geometry of Photographs. The cartographer relies on photogrammetric methods to furnish planimetric and hypsometric positions for map preparation. Compared with field methods, pho-

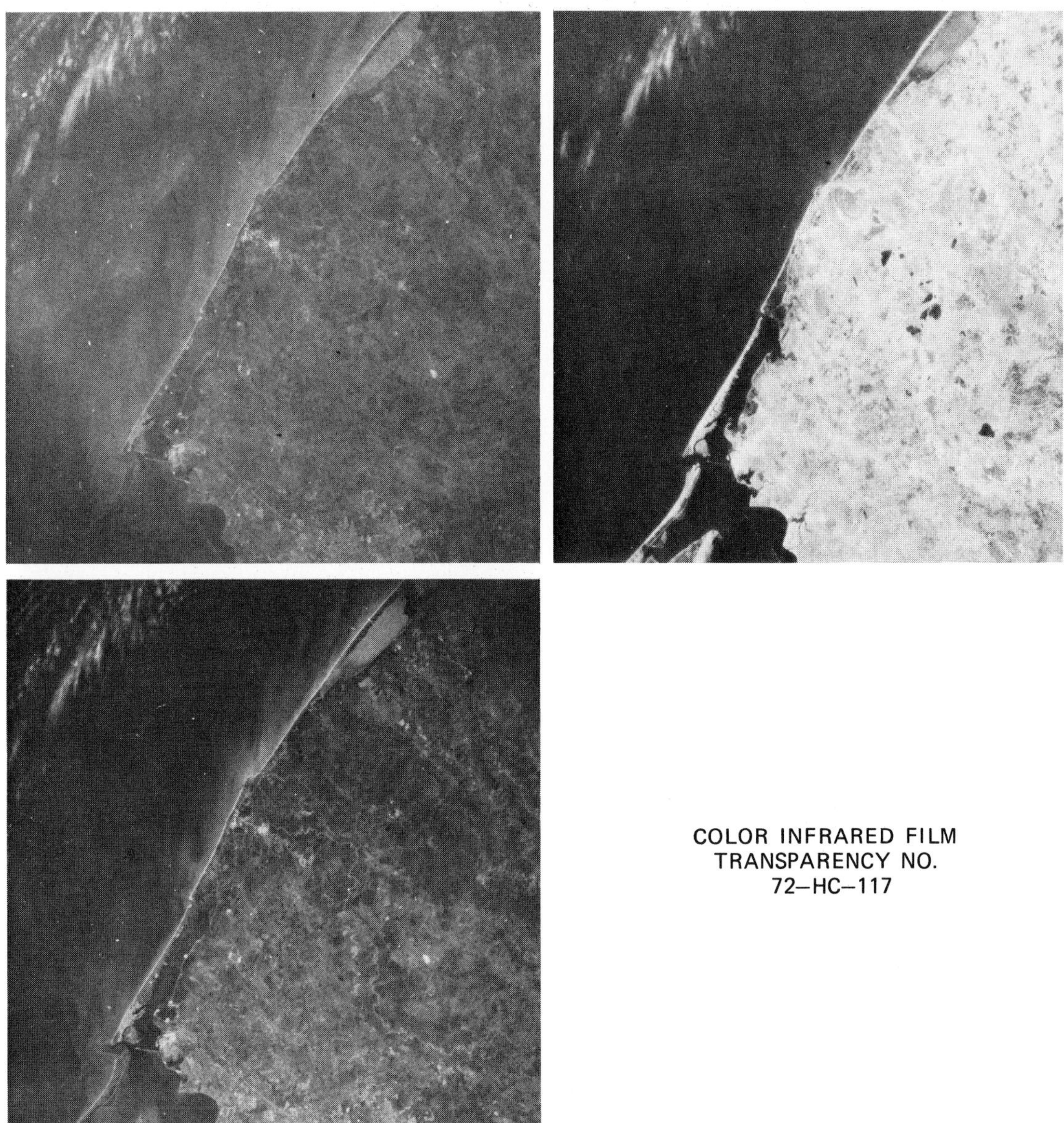

FIGURE 10.3 Multilens photography from space. Differences in tones on the three images are caused by the particular wavelengths recorded. (Images courtesy National Aeronautics and Space Administration.)

tographs can provide such information much more rapidly and at less cost. With limited ground control, photogrammetric methods can provide accurate large-scale maps for areas where it is difficult to conduct extensive field surveys because of terrain or climatic conditions. This advantage, of course, is especially important for military operations.

Most detailed large-scale maps made from aerial photographs are produced by government agencies or by commercial organizations working under contract. The high cost of precise stereoplotting instruments prohibits their use by individuals or small commercial firms. Some, however, can be found at educational institutions where they are used mainly for instruction and research.

Even without expensive equipment, cartographers can make use of photographs for extending control. Equipment that is not particularly costly can furnish control that is precise enough for the preparation of smaller-scale maps or for map revision.

The perspective projection of the image of the land surface on an aerial photograph causes the scale relationships to differ from those that would occur on an orthogonal projection of the land to a planimetric map. Disregarding things such as the distortion produced by the lens, paper, and film, the scale of a truly vertical photograph of perfectly flat terrain would be nearly the same as that of an accurate planimetric map. The occurrence of relief, however, causes greater variations in scale to appear because of the perspective view of the camera. Relative to one particular level of the terrain, higher points will be displaced away from the center of the photograph and lower points will be displaced toward the center. These differential variations in scale preclude our merely tracing information from photographs directly to large-scale line maps. The amount of displacement can be measured, and the disadvantage of not being able to trace information directly for large-scale map production is far outweighed by our ability to use displacement to determine distances above or below a chosen datum.

Our consideration here of the geometry of an aerial photograph will be limited to the aspects of vertical photographs with no tilt. Tilt refers to the deviation from the perpendicular of the lens of the camera in the airborne vehicle from the surface of the ground below. For most uses any aerial photograph with less than 3° of tilt is considered to be vertical (that is, the lens is perpendicular to the ground). A perfectly vertical photograph is rare. Many photogrammetric instruments allow the operator to correct for excessive tilt in a photograph when such a photograph is being used. The optical or lens distortion, which on a typical 9 by 9 in. photograph might be in the magnitude of a displacement of a fraction of a millimeter, can usually be ignored except for determination of elevations with stereoscopic plotting instruments. Distortion resulting from changes in the dimensions of the film base are usually very slight, but paper distortion of prints can be considerably more bothersome.

Scale. The precise scale ratio between two points on a vertical aerial photograph usually differs from that of the general or average scale. The general scale is the ratio of the focal length of the camera to the elevation of the camera with respect to some specific elevation on the landscape; it follows that this ratio will not be correct for any other elevation or datum.

$$\mathrm{RF} = \frac{C_f}{H}$$

where RF is the representative fraction, C_f is the focal length of the camera, and H is the height of the camera above the datum. Each of the infinite number of horizontal planes has its own specific ratio or scale. Figure 10.4 illustrates how scale varies with differences in the vertical positions of points on the landscape. The true map locations of two towers on the land surface at the same elevation are at points *A* and *B*. On the photograph at a scale of 1:20,000 these locations will be at *a* and *b*. Because of the perspective in the photograph, the tops of the towers will appear in the photograph at *a*′ and *b*′, which are clearly farther apart and therefore at a larger scale than *a* and *b*, for example, a scale of 1:18,000. In this particular case, we could plot the positions of the towers at a scale of 1:20,000 because the bases as well as the tops of the towers would be visible on the photograph. On the other hand, if

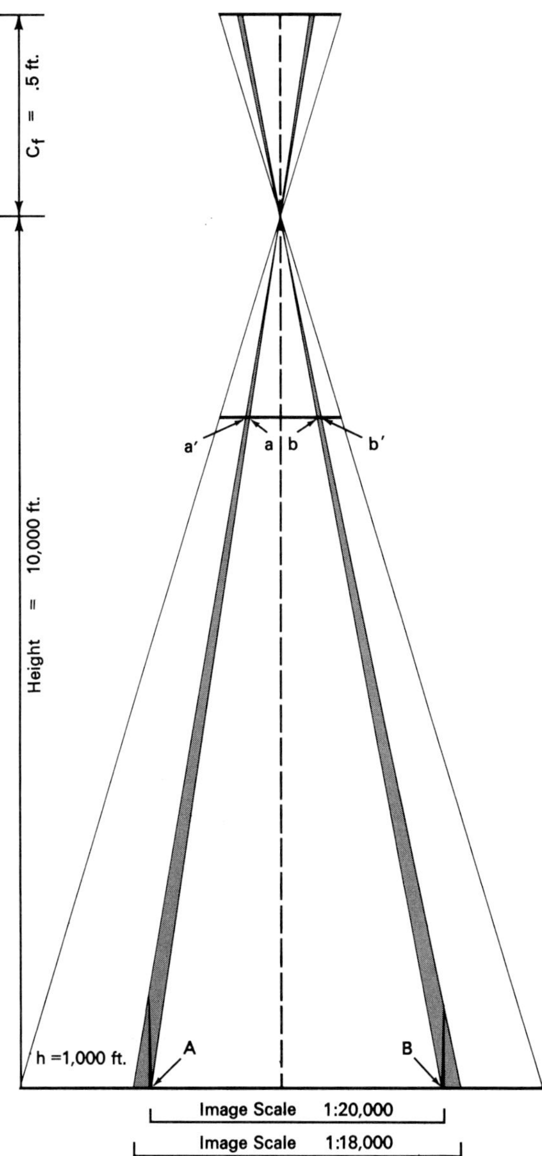

FIGURE 10.4 Because of the perspective projection of an aerial photograph, the locations of images on the film (and print) are determined by their vertical positions and their distances from the point directly beneath the camera.

the problem involved a hill, the base would not be visible and, therefore, we could not plot the position of a hill directly from the photograph at a scale of 1:20,000.

Displacement. Displacement because of relief occurs at a radial direction from the nadir, the point on the plane of the photograph located by the extension of a vertical line through the center of the camera lens which, on a truly vertical photograph, coincides with the principal point or geometric center. Straight lines across a photograph connecting opposite *fiducial* marks intersect at this principal point (Fig. 10.5). No displacement occurs at the principal point, then, when it is aligned with the perspective center of the camera lens, but image distance from the center of the photograph to any other point will depend on (1) the relative vertical location of that point, and (2) the distance of it from the principal point.

The amount of displacement changes directly with the vertical departure from a chosen datum and the distance from the principal point, and inversely with the height of the camera.

$$r_d = \frac{h \times d}{H}$$

FIGURE 10.5 Fiducial marks are located in the focal plane of the camera. They may be etched on the surface of the glass, which is in contact with the film or, in the case of the open-type focal plane, they are projections of metal that extend into the negative area.

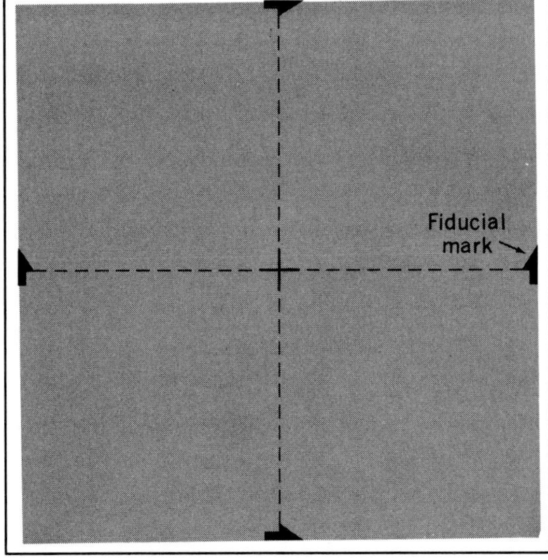

where h is the vertical departure from a chosen datum, (the height of the object), H is the height of the camera above the datum, d is the distance from the principal point of the photograph, and r_d is the radial displacement. For example, the radial displacement of the top of the tower at point A in Figure 10.4 is the length a'a on the photograph when the mapping scale is 1:20,000.

A comparison of similar triangles in Figure 10.6 also indicates this relationship; as either h or d increases, r_d (shown by red tints) also increases; and, as H increases, r_d decreases. Photographs made at low altitudes, then, show more displacement than those made at high altitudes. Although the latter have scale relationships that more closely approximate a map, the low-altitude photographs with the greater displacements turn out to be more useful for the precise determination of elevations.

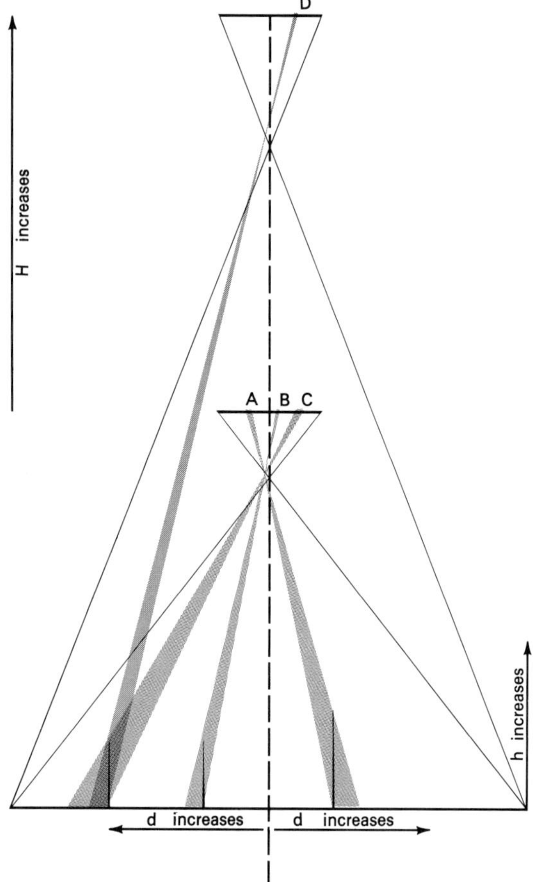

FIGURE 10.6 As the distance from the center, d, of the photograph increases, and/or as the height of the object, h, increases, the amount of displacement, r_d (shown in red tint and measured on the photograph) increases. As the altitude of the camera, H, increases, displacement decreases.

Parallax. We perceive depth in our vision in a variety of ways. With one eye we must rely on sizes of objects, clarity of detail, or whether one object appears in front of another. When using both eyes, each eye sends its own signal to the brain. Because our eyes are separated, each sees an object (if it is within approximately 600 m) from a different angle, and the subsequent signals to the brain cause the sensation of depth.

The displacement of objects on aerial photographs produces *parallax,* which is the apparent change in position of an object because of a change in the point of observation. This apparent change in position is the principal reason for our being able to view two photographs and produce an illusion of a third dimension. By viewing an object on one photograph with one eye and the same object on an overlapping photograph with the other, we are in effect viewing the object from two points that approximately represent the two camera stations. Each picture shows all objects in perspective from each camera station and our eyes each send a signal to our brain causing the image to take on an apparent third dimension.

On a photograph with no tilt, the parallax is a linear distance used for determination of elevation. It is a distance parallel with the axis of the camera stations and is dependent on the height of the object, the focal length of the camera, the distance between the camera stations, and the distances from the camera stations to the object. The algebraic difference of the parallax on two overlapping photographs is used to determine elevations using stereoscopic plotting instruments.

The parallax difference can be measured graphically; however, the error is likely to be somewhat larger than when using plotting instruments. Two overlapping photographs should be carefully aligned, as shown in Figure 10.7. The principal points and conjugate principal points (image of the principal point of the overlapping photograph) of both photographs must fall on a straight line. The average distance b between principal points and conjugate principal points is then computed. The parallax difference is the difference between the distances xx' and yy' and can be used in the following formula to determine the difference in elevation, Δe, between the two points.

$$\Delta e = \frac{H}{b} \times \text{parallax difference}$$

where H is the average height of the camera above the terrain.

On a truly vertical photograph, azimuths from the principal point are correct to any point on the photograph; this condition permits us to perform triangulation directly from the photographs. Since the direction from the principal point to its conjugate principal point is correct on each of two overlapping photographs, it follows that by superimposing these two lines we produce a baseline with ends from which all angles are correct. On pieces of overlay material we can mark the principal points of each photograph, and draw lines through the conjugate principal points and from the principal points through a third point c (Fig. 10.8). When the overlays are placed so the baselines are superimposed, we have produced a triangle in which the angles at a and a′ are correct. If these two angles are correct, the angle at c must, of course, also be correct (Fig. 10.9), the elevation of c having no effect on the horizontal angles. The three points have been located in a true planimetric relationship, and the scale of the three sides of the triangle is the same.

Sophisticated equipment has been invented to aid the cartographer in taking advantage of the geometric properties of photographs. Worthy of special note among this equipment is the orthophotoscope. By partially simulating the flight of

FIGURE 10.7 Graphically measuring parallax difference. The locations of one of the points on the two photographs is indicated by x and x′, and the second point indicated by y and y′. If the lines connecting the principal point to the conjugate principal point on each photograph are placed along a straight line, then the difference xx′–yy′ is the parallax difference.

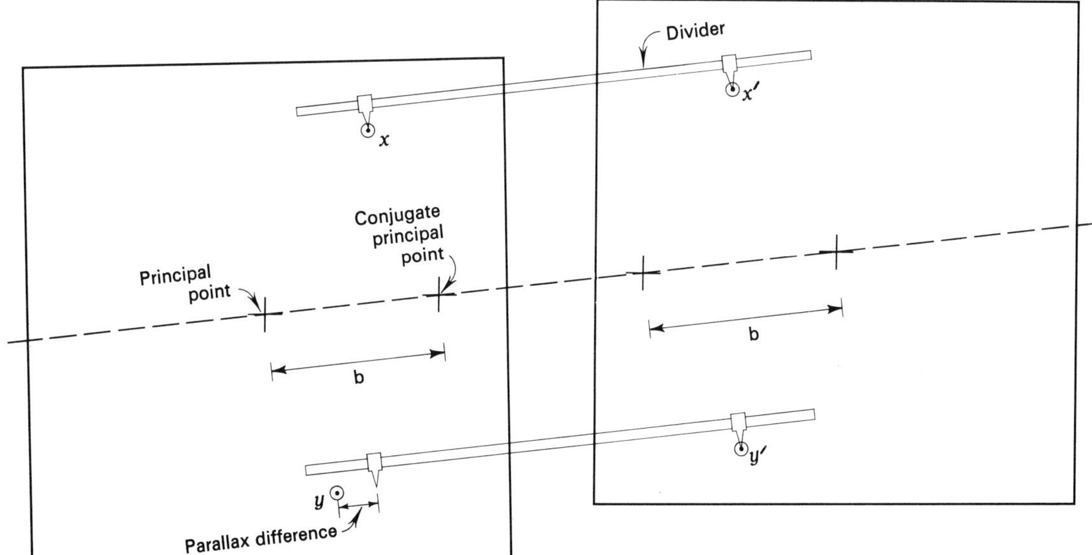

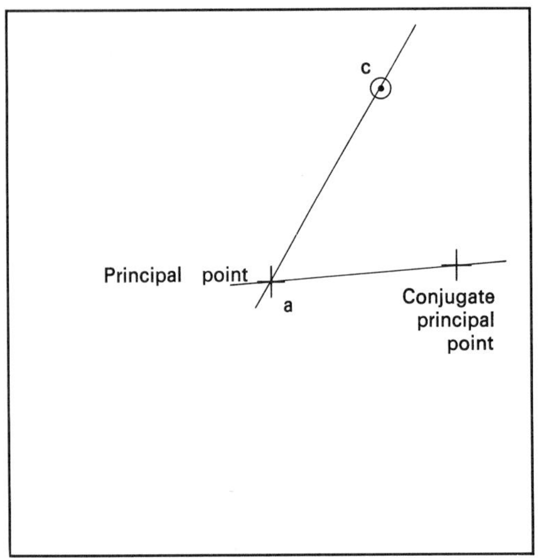

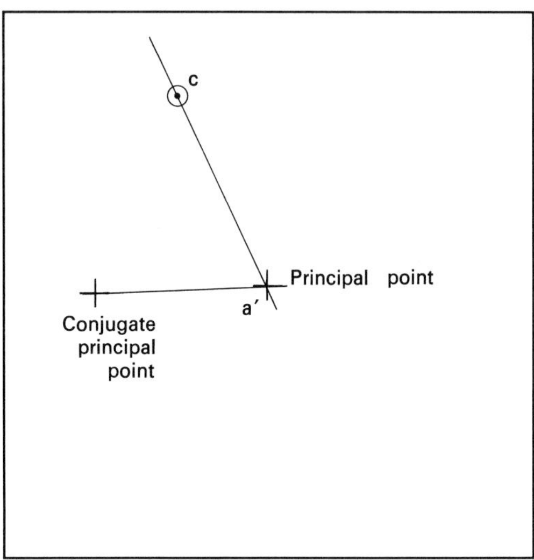

FIGURE 10.8 Due to the geometry of vertical aerial photographs, the angles at points a and a′ are correct.

the aircraft in the equipment and by the operator compensating for the effects of terrain relief on the ground, the geometric distortions of the photographs are essentially removed by the orthophotoscope and operator. The result is a planimetrically correct "image map."

Digital Systems

Photographic film emulsions can be made to be sensitive only in the range of 0.3 to 1.2 μm. In that range of the electromagnetic spectrum, and only in that range, can the photochemical detection and recording of reflected and emitted energy take place. For all other wavelengths of the electromagnetic spectrum one must use nonphotographic systems to detect data. Several nonphotographic systems will be discussed in the following sections. To avoid confusion, it must be mentioned that often the data that these nonphotographic systems detect are displayed or recorded as an image on photographic film to allow the cartographer to perform a visual analysis of the data. We refer to this data-collection activity as *digital,* since we are referring to nonphotographic sensors, which initially record the data as discrete electrical impulses on some recording device. Photographs, on the other hand, record a scene instantaneously on film and therefore are termed *analog sensors.*

FIGURE 10.9 the angles at points a and a ′ are correct; therefore, when the base lines are superimposed, the angle at c must also be correct.

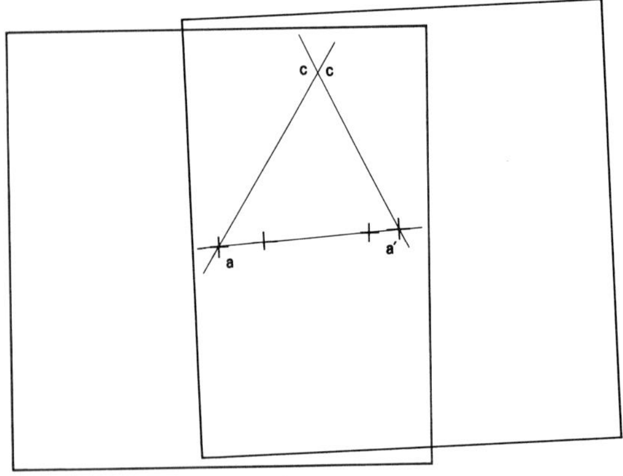

During World War II experiments were conducted in the use of nonphotographic sensors. The early work was to a great extent concerned with the detection of military targets by use of the infrared portion of the electromagnetic spectrum. Improvement in sensing systems has depended to a great extent on development of more efficient detectors. Nonphotographic sensors are now used in most portions of the spectrum, including the portion to which photographic films are sensitive.

Along with the improvement in sensing devices during the past 100 years we have seen a parallel development of new vehicles to carry the equipment aloft. The aircraft used for photographic missions during World War II were far superior to those used in World War I. Since the 1940s airplanes have been produced that can carry an arsenal of sensing equipment to extremely high altitudes. The 1957 launch of *Sputnik I* paved the way for the use of spacecraft to carry detecting devices far beyond the earth's atmosphere. Systematic use of orbital observations began in 1960 with the launch of *Tiros I,* and photgraphy produced by automatic cameras in an orbiting spacecraft became available in 1961. Although the pictures were made to monitor the attitude of the spacecraft, they served as a stimulus for future planned photographic missions. Manned space flights permitted selection of targets while in orbit, producing unique and valuable pictures.

Today we have had hundreds of space missions that have collected data and relayed it back to earth. For cartographic purposes the LANDSAT series of orbiting satellites has produced the most widely used data sets. The fifth LANDSAT was launched by the United States in 1984. Nations in addition to the United States and the U.S.S.R., the current leaders in this field, are soon to enter the space age collection of data for cartographic purposes. France and Japan are scheduled to orbit satellites within the nèxt 2 years.

Scanning Devices. Instead of recording in a wide-angle field the way a camera does, many nonphotographic sensing systems use scanning devices to detect the energy from one small element—ground cell—of the landscape at a time. These sensors have an instantaneous field of view (IFOV), which determines, along with the other parameters of the system, the resolution of the information that is recorded. As the vehicle (aircraft or satellite) moves along its line of flight, the scanner collects the energy from the landscape in a series of scan lines that are perpendicular to the line of flight. Rotating mirrors reflect the radiation received in the IFOV onto detectors. These spinning mirrors are synchronized with the speed of the sensor platform, so that as it moves along the flight line, a new swath, adjacent to the previous one, is scanned (see Fig. 10.10). The data that are collected in the form of electronic signals from narrow strips of the landscape are stored on magnetic tape and can later be processed to produce a picture-like format.

Geometry of Scanned Data. The geometry of scanned data is somewhat more complex than that of photographs, and it is impossible to overcome completely the effects of geometry for mapping purposes. This fact makes the use of imagery constructed from scanned data impractical for some precise cartographic work. These images usually yield thematic data that are displayed on a precise planimetric format obtained from another source.

FIGURE 10.10 Scanners detect energy along a scan swath that is perpendicular to the line of flight. The IFOV sweeps along each ground swath as the scanner platform moves along its prescribed track.

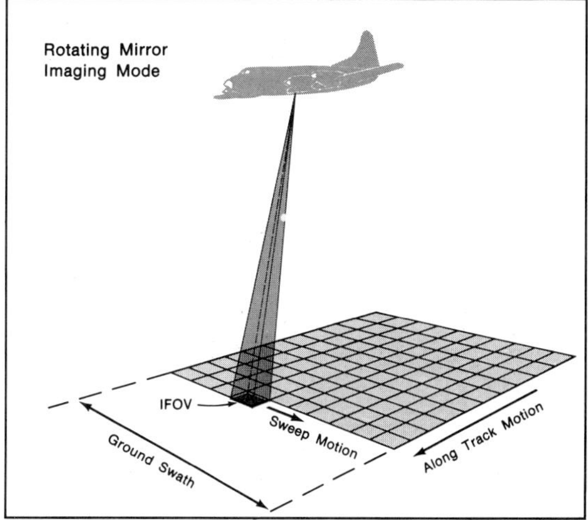

The distortion in scanned data results from both the geometry of the scanning device itself and from the pertubations that affect the vehicle that is carrying the device during its collection mission. Because of unwanted motions in an aircraft or spacecraft, a long series of perfectly parallel scan lines is practically impossible. Constant scale along a flight line, then, is unlikely, and because of the nature of scanning, scale along the scan line is also not uniform. The distortions fall into four categories:* tangential scale distortion, resolution cell size variations, one-dimensional relief displacement, and flight parameter distortions.

Tangential Scale Distortion. Because the rotating mirror in the scanner moves at a constant speed, the ground distance covered for any time increment increases toward the edges of the scan line. Although some CRT recorders can be programmed to compensate for this fact when displaying the data, most record at a constant speed, and the resulting image scale is therefore compressed at the edges of the images. This distortion is termed *tangential scale distortion*. It occurs only in the scan direction that is perpendicular to the line of flight (see Fig. 10.11). To correct the error in the planimetric positions resulting from this distortion, the following calculations are required. The planimetric y position for an image point p is found by first solving for V_p in:

$$V_P = \frac{Y_p \times V_{max}}{Y_{max}}$$

where Y_p = the image distance from the line of flight to the point p

Y_{max} = the distance from the line of flight to the edge of the image

V_{max} = $\frac{1}{2}$ the total field of view of the scanner

*T. Lillesand and R. Kiefer, *Remote Sensing and Image Interpretation* (New York: John Wiley & Sons, 1979).

Once V_p is computed, the correct ground distance for Y is:

$$V_p = H \tan V_p$$

where H = the height of the aircraft.

Resolution Cell Size Variations. As the IFOV moves out from the flight line, its ground cell size increases. As a result, the recorded value from a ground cell is the sum of the energy emitted/reflected from all the features in the ground cell being viewed. As the ground cell size increases, the recorded value becomes the value received from an increasingly larger zone of ground features. This decreases the value of the recorded signal for differentiating features on the earth. In cartography the result is that the scanned data at the edges of the image contain values received from a larger ground cell than the values received from ground cells near the center of the image.

One-Dimensional Relief Displacement. In contrast to photographs, relief displacement on scanned data is one-dimensional. The displacement is always perpendicular to the flight line, and images of features above the ground datum appear to lie on the ground away from the line of flight.

Flight Parameter Distortions. Scanned data are collected continuously or dynamically as the craft proceeds along its flight line. On the other hand, photography is an intermittent sampling at discrete instances along the flight line. Therefore, flight perturbations affect the relative positions of the resulting data more in scanned data than in photographed data. Figure 10.12 illustrates several distortions to the planimetry of a scanned image resulting from flight parameter distortions.

Scanning devices are used for the collection of data in the visible, near-infrared, and thermal ranges, and in concert as multispectral scanners in spacecraft.

FIGURE 10.11 Tangential scale distortion: The image (A) on the left is a conventional aerial photograph. The image (B) on the right is a scanned thermal image. The noticeable difference on image (B) is due to tangential scale distortion. (From T. Lillesand and R. Kiefer, *Remote Sensing and Image Interpretation*, Copyright 1979, John Wiley and Sons, New York.)

Thermal Sensing. The radiation distribution curve for the sun shows an energy peak in approximately the visible portion with most of the remaining radiation in the shorter wavelengths of the infrared (see Fig. 10.2). The earth, because of much lower temperatures, radiates energy at longer wavelengths. Curves for most objects on the earth peak near 10μm, although the curves vary with the temperature and the nature of the material of the object. Devices used to sense thermal radiation operate in wavelengths from about 3 to 14 μm and therefore are not de-

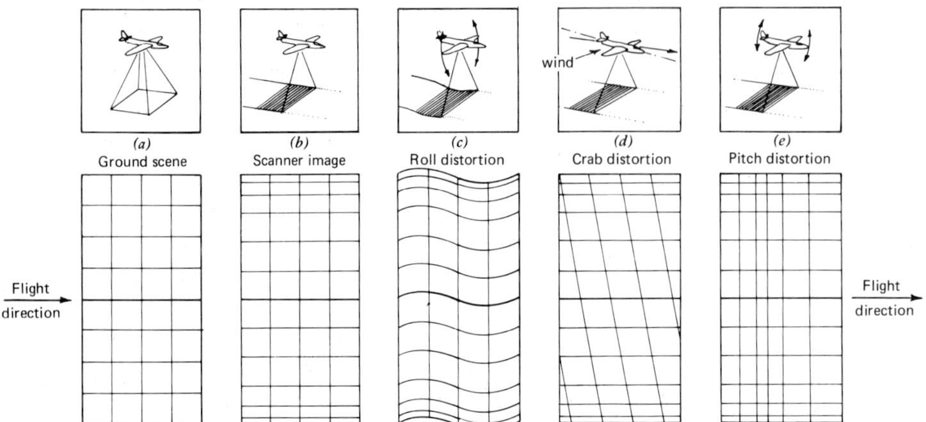

FIGURE 10.12 Flight parameter distortions: The five grids above illustrate the geometric distortion of normal photographs, scanned images, and distortions induced into scanned imagery by aircraft attitude deviations. (From T. Lillesand and R. Kiefer, *Remote Sensing and Image Interpretation*, Copyright 1979, John Wiley and Sons, New York.)

pendent on light. Sensors operating in the range of 3 to 5 μm, however, receive both thermal and reflected radiation and, during daylight hours, will receive roughly equal amounts of each.

The thermal sensor is a scanning device that uses a photon detector to receive the energy from rotating mirrors (see Fig. 10.13). This energy, which is received from along a scan line, is converted to electrical impulses proportional to the intensity of the energy received. The signal can be stored on tape and later used to produce a display on a television screen. Ordinary film can be exposed in synchronization with the display on the CRT to produce imagery of the thermal radiation.

Since the temperature of the scanning device itself causes it to emit infrared radiation in the same band in which it is operating, the scanner will affect the data. To overcome this problem, scanners are cooled to extremely low temperatures and enclosed in a heat-proof box. Scanners that operate in the band from 3 to 5.5 μm are less expensive, since they need to be cooled to only about −200°C, whereas those operating in the longer wavelengths must be cooled nearer to absolute zero.

The resulting variation along the thermal scan line is the result of temperature differences created by variations in the nature of the material or the form of the surface. These variations, when converted to tonal differences on a CRT or on a photograph, reveal these temperature differences in an image format (see Fig. 10.14). Tones can vary relative to each other depending on temperature changes through the day. Figure 10.15 shows a graph of the diurnal temperature variations for soil and rocks versus water.

The day-and-night capability of thermal sensing is an advantage over photography, but the resolution of landscape detail is much inferior. Thermal sensing is used successfully to detect and delineate the edges of forest fires; water pollution can easily be detected by locating warm-water discharges; and it can also be used for inventorying livestock and wild animals because of its day-and-night capability and its ability to discriminate between an animal and its background on the basis of temperature differences.

Multispectral Sensing. In one sense the concept of multispectral sensing combined with scanning device technology allows the cartographer the best of two worlds. Both reflected and emitted energy can be collected simultaneously

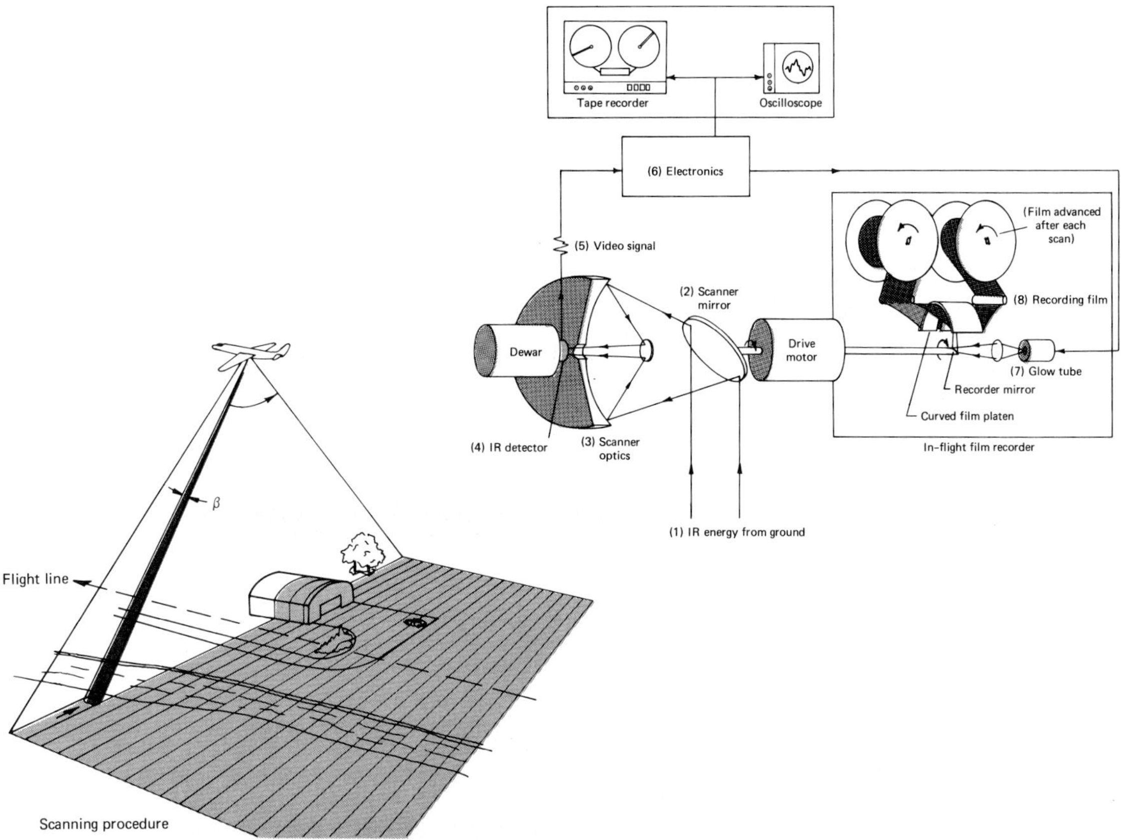

FIGURE 10.13 The operation of a thermal scanner. Below is the scanning procedure where β = IFOV. Above is the interior of the sensing system showing the scanning mirror and optics, the detector and possible outputs onto a tape recorder, an oscilloscope, or an inflight recorder. (From T. Lillesand and R. Kiefer, *Remote Sensing and Image Interpretation*, Copyright 1979, John Wiley and Sons, New York.)

by using a sensor system called a *multispectral scanner* (MSS). Figure 10.16 illustrates a MSS system.

The geometry of the MSS-scanned data is exactly like that of the thermal scanned data described above. The only difference is that the thermal scanner filters the incoming radiation and records just the thermal wavelengths on a single detector. The MSS scanner separates all the incoming radiation into various spectral components that are then independently recorded. As Figure 10.16 shows, a dichroic grating within the scanner system separates the reflected from the emitted incoming energy. Five channels are illustrated in Figure 10.16: devices with twenty-four channels have been tested. Final output from a MSS system may be either analog (images) or digital.

MSS systems can be carried aboard either aircraft or spacecraft. Each of the *LANDSAT* satellites has included a MSS system. *LANDSAT 1* and *2* had a four-channel MSS system with a ground resolution of about 80 m. Recording was in the 0.5 to 0.6 μm range (band 4), the 0.6 to 0.7 μm range (band 5), the 0.7 to 0.8 μm range (band 6), and 0.8 to 1.1 μm range (band 7) (see Color Fig. 11).

LANDSAT 3 included a five-channel MSS system with the same ground resolution but added a thermal band (10.4 to 12.6 μm). *LANDSAT 4*

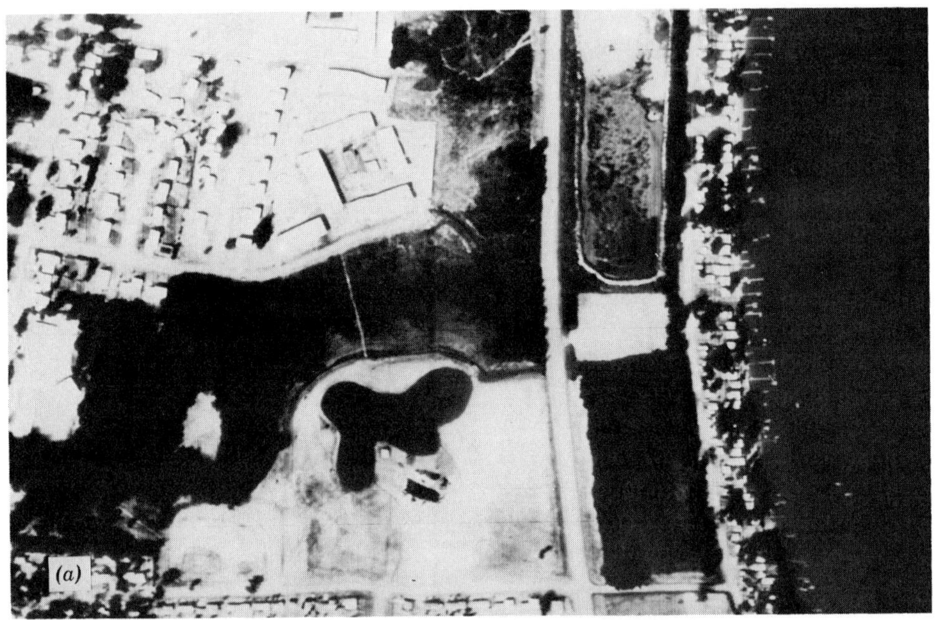

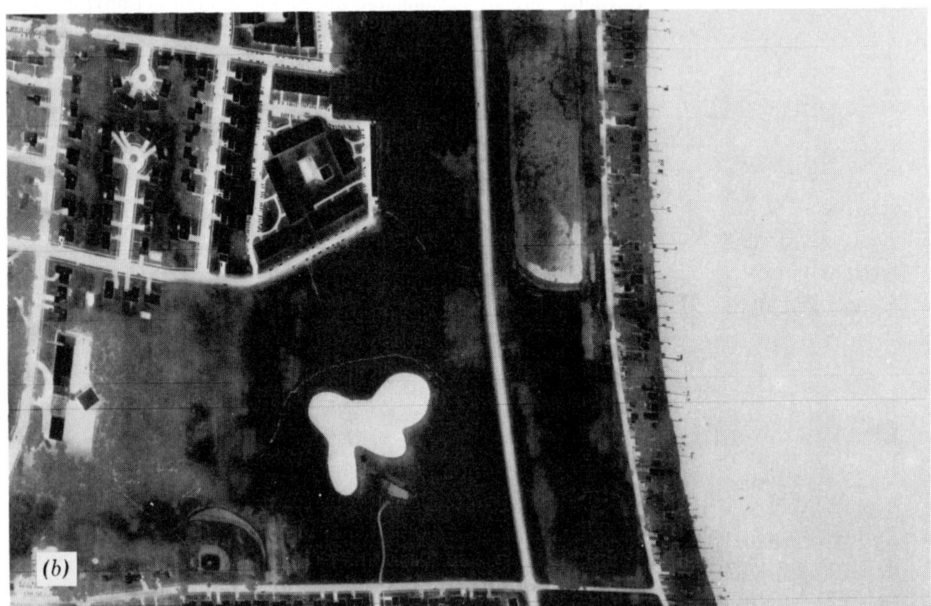

FIGURE 10.14 Two thermal images of the same area illustrating the difference between nighttime and daytime temperatures. Image (A) 2:40 P.M., image (B) 9:50 P.M. (From T. Lillesand and R. Kiefer, *Remote Sensing and Image Interpretation*, Copyright 1979, John Wiley and Sons, New York.)

also included a MSS system essentially like that described above minus the thermal band. Figure 10.17 and Color Figure 14 illustrate products made by using the type of data received from the MSS aboard *LANDSAT 3*.

The new addition to *LANDSAT 4* is the The-

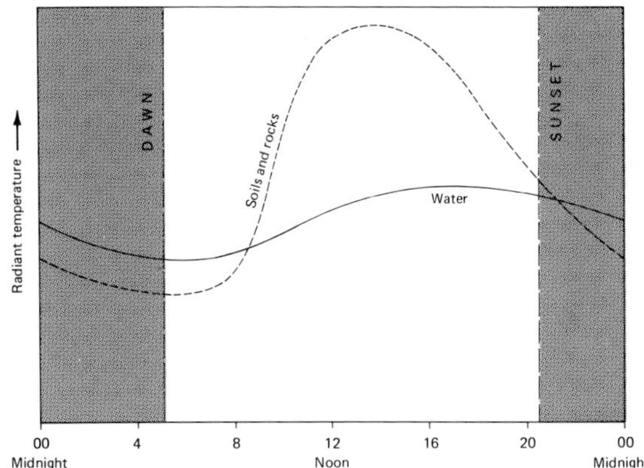

FIGURE 10.15 Graph of diurnal temperature variations for soil and rock versus water. (From T. Lillesand and R. Kiefer, *Remote Sensing and Image Interpretation*, Copyright 1979, John Wiley and Sons, New York.)

matic Mapper (TM) sensing system. The TM collects data in seven bands, six recording reflected energy in narrow spectral bands with ground resolution of 30 m and one thermal band (10.4 to 12.5 μm) with a ground resolution of 120 m (see Table 10.1). This scanning system results in imagery like that shown in Figure 10.18 and Color Figure 15. *LANDSAT 5* employs the same sensing systems as *LANDSAT 4*.

The usefulness of scanning devices, especially multichannel scanning devices, has been proven by many useful applications such as in crop and vegetation identification, soil moisture condition identification, and the distinguishing of rock formations. The principal advantage of multispectral imagery for applications purposes is that it enables the cartographer to employ spectral sig-

FIGURE 10.16 Diagram of a multispectral scanner system. Below is the scanning procedure where β = IFOV. Above is the scanner schematic. (From T. Lillesand and R. Kiefer, *Remote Sensing and Image Interpretation*, Copyright 1979, John Wiley and Sons, New York.)

(*b*) Scanner schematic

(*a*) Scanning procedure during flight

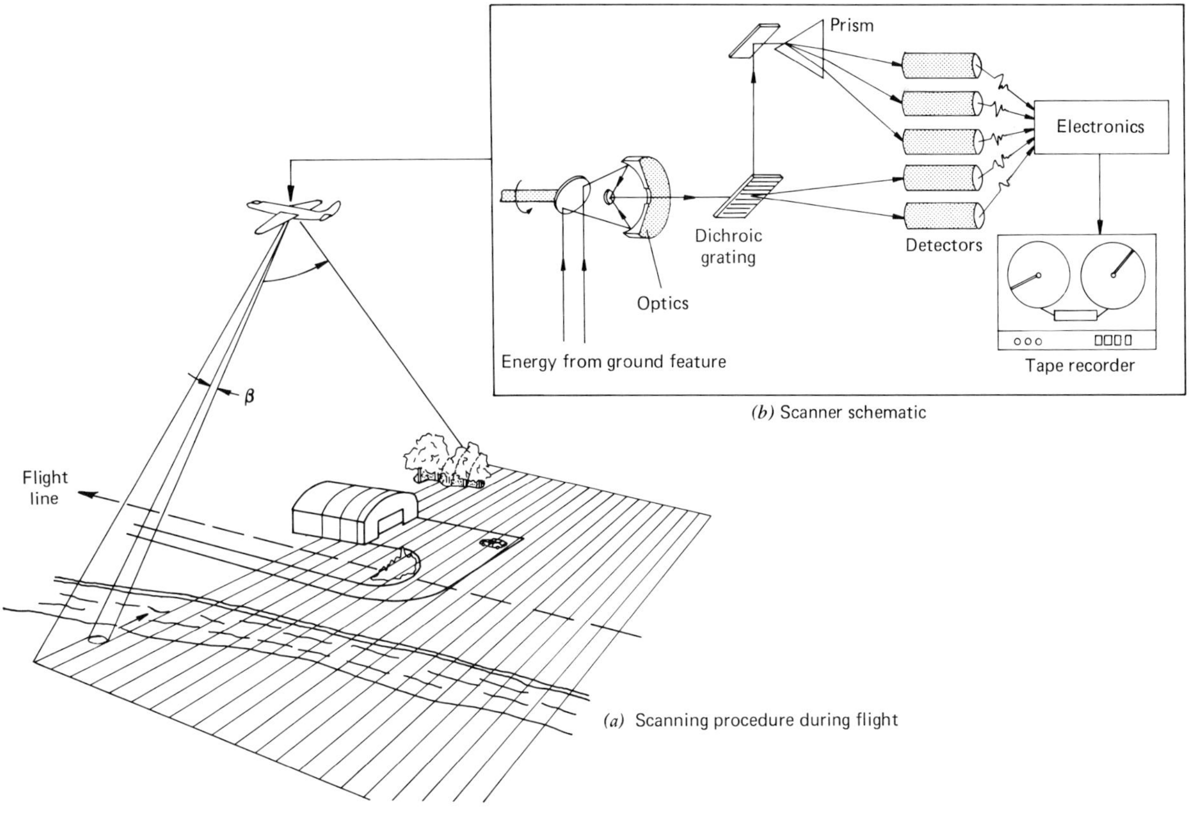

FIGURE 10.17 An example of MSS (band 4) imagery: the island of Spitzbergen. (Photo courtesy of USGS.)

natures to differentiate features on the imagery. A *spectral signature* is a record of the tonal appearance of a given object in the various bands of the electromagnetic spectrum. It enables the distinction between two objects whose separate identities are impossible to ascertain from any one image, but whose distinction is undeniable

TABLE 10.1 Thematic Mapper Bands: Sensing Ranges and Uses

Band 1:	0.45 to 0.52 micrometers Water body penetration, land use, soil, vegetation
Band 2:	0.52 to 0.60 micrometers Vegetation discrimination and vigor assessment
Band 3:	0.63 to 0.69 micrometers Vegetation and nonvegetation contrasts
Band 4:	0.76 to 0.90 micrometers Vegetation biomass, crop identification, crop-soil and land-water contrasts
Band 5:	1.55 to 1.75 micrometers Crop type, crop water content, and soil moisture
Band 6:	2.08 to 2.35 micrometers Rock formation discrimination
Band 7:	10.40 to 12.50 micrometers Thermal infrared for vegetation classification, vegetation stress, soil moisture

BAND 2

BAND 3

BAND 5

FIGURE 10.18 Representative of Thematic Mapper (TM) data are bands 2, 3, and 5 of the Dyersburg, Tennessee, quadrangle. These bands were used to create the map of Dyersburg in Color Figure 15. (Courtesy of USGS.)

given the various images resulting from multispectral sensing.

Microwave Sensing. Microwave sensing is concerned with wavelengths of the electromagnetic spectrum from about 0.1 cm to 100 cm where atmospheric attenuation is negligible. These electromagnetic waves are, of course, invisible to the eye and are detected by an antenna. At these longer wavelengths (compared to infrared and thermal infrared), there is very little energy available, and very sensitive equipment is necessary to detect the naturally radiated energy in this part of the spectrum.

One distinctive feature of microwave sensing is that the microwaves can penetrate haze, precipitation, clouds, and smoke. Another difference is that these waves give analysts a distinctive view of the environment compared to light or heat sensors, because of the length of the waves being recorded.

Passive Microwave Sensing. The microwave radiometer, a passive sensor, depends on energy that is emitted, transmitted, or reflected naturally from a surface. The emitted energy is related to temperature, and the transmitted energy has its origin in the subsurface of an object. During daylight hours, there is also a reflected component. The passive microwave sensors are comparable to the thermal sensors, except that they operate in longer wavelengths and sense with an antenna instead of with a heat-sensitive detecting device. Since somewhat better spatial resolution is possible in the shorter wavelengths (with a given antenna size), the passive sensors operate in that portion of the microwave spectrum from approximately 20 to 100 cm.

A microwave radiometer with a narrow-beam antenna can be attached to a scanning device that moves along a path transverse to the flight line. Signals received are amplified and stored on magnetic tape. Computers can assign colors to different levels of recorded energy to produce a false color image, or they can be displayed as numerical readouts. The weak signals would be no particular problem if the antenna were able to gather energy from a target for a long period of time. Because of the speed with which the scanning radiometer must operate, large areas must be sensed to gather enough energy for a satisfactory reading. Having to sense a relatively large area results in poor spatial resolution, which places some limitations on the uses of microwave imagery. Improvements in these sensors, however, promise much better imagery in the future.

Presently, data obtained by passive microwave can be used to determine soil moisture conditions, since one component of the energy originates in the subsurface. Other possible cartographic applications include mapping ocean surface conditions, inventory of the water content of snowfields, location of ice-water boundaries, and various kinds of geologic exploration.

Radar Sensing. Radar operates in the microwave portion of the spectrum with wavelengths from approximately 1 mm to 1 m and is an active system, furnishing its own source of energy. These wavelengths give radar the capability of penetrating clouds and haze. Since the system depends on no outside source of energy, it can operate during daylight or darkness. This day-or-night, all-weather capability can be successfully used for mapping in parts of the world where weather conditions have prevented the use of photography and some other types of sensing devices. The photolike format of the imagery permits the use of many of the photo-interpretation techniques for analysis.

Radar transmits a particular band of electromagnetic energy toward an object and detects a portion of the energy reflected from the object (backscattered) (see Fig. 10.19). Besides establishing the direction to the target, the system also determines the distance by measuring the elapsed time between transmitted pulse and the return of the reflected energy. Since each pulse yields such a small amount of information, thousands of transmissions per second are necessary to pro-

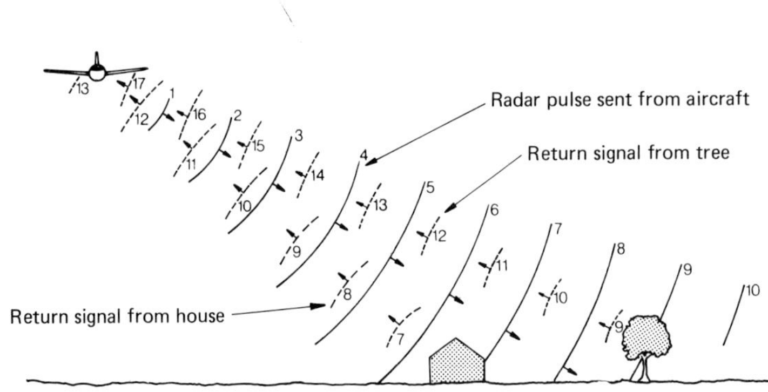

FIGURE 10.19 Diagram of active radar sensor showing wavefront location at seventeen consecutive time intervals. (From T. Lillesand and R. Kiefer, *Remote Sensing and Image Interpretation*, Copyright 1979, John Wiley and Sons, New York.)

duce a sample that can be converted to the photo-like image.

Radar that was developed during the early 1940s was of the PPI (Plan Position Indicator) type, which has a rotating antenna that provides a scan of a full 360°. The echo received is played back instantly in the form of a spot of light on a cathode ray tube. Reflecting objects such as storms, aircraft, or ships can be located in terms of direction and distance. For most remote sensing and cartographic applications, PPI lacks the necessary spatial resolution.

To produce photo like imagery, a system called side-looking airborne radar (SLAR) has been developed. Unlike the PPI antenna, which sweeps through a full circle, the SLAR antenna scans a swath of the terrain perpendicular to and to the side of the flight line (see Fig. 10.20). As the aircraft moves forward, a new strip of land is imaged. The relative intensity of the signal reflected from an object is converted to an image, and the location on the landscape is determined by the time elapsed between transmitting and receiving a given pulse (see Fig. 10.21). By recording the strengths of the signals from these narrow strips on a roll of film, an image of the terrain is produced.

Resolution of radar imagery is best at close range, and it can be improved by the use of longer antennae or by operating with shorter wavelengths. However, clouds and the atmosphere tend to absorb the shorter wavelengths. Also, there is a practical limit to the length of antennae that can be carried by aircraft. For good resolution conventional SLAR must operate in low-level, short-range situations.

To improve resolution at greater distances, *synthetic-aperture* radar has been developed. A relatively short antenna transmits and receives signals from a given target at regular intervals as the aircraft moves along the flight line. The backscattered signals are stored on film and later combined in a special way to produce the radar image. The effective length of the antenna can be the distance the aircraft travels along the flight line while a particular target is within view. An antenna only 1 m long can have an effective length of hundreds of meters.

Visual characteristics of size, shape, texture, tone, and shadows used in the interpretation of aerial photographs are also used in the analysis of radar imagery. Tonal differences are related to the variations in the signals returned to the antenna. The nature of this reflected energy is de-

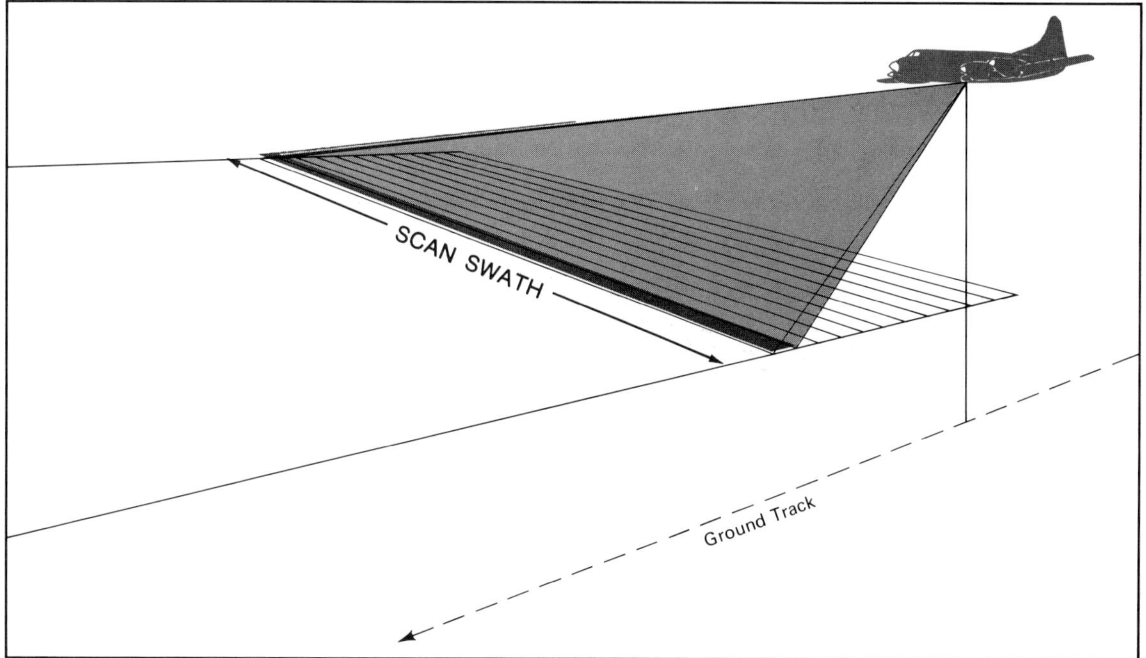

FIGURE 10.20 Side-looking radar scans a series of swaths only on one side of the aircraft.

pendent on the properties of the transmitted energy and the properties of the surfaces being sensed. The electromagnetic energy can vary in wavelengths, polarization, and direction. Properties of the surfaces that affect the reflected energy include roughness, slope, water content, dielectric and conducting properties, and attitude relative to the transmitter. Smooth surfaces and water, for example, appear dark; sand and rocks tend to be light tone.

Lack of a return signal causes a dark tone and, since the electromagnetic energy has little ability to penetrate solid objects, buildings, landforms, and vegetation can block the signal and cause shadows. Because of the angle at which the ground surface is being "illuminated," the shadow effects are similar to low-sun angle photography. This enhances the impression of relief and is useful in the analysis of the form of the land. To prevent the illusion of inverted relief, shadows should be oriented so they do not fall away from the observer. Usually surveys are made in the same "look" direction throughout the flight to produce consistency in shadow direction. To operate while flying in both directions of the traverse, the systems are arranged so they can view from either side of the aircraft as required.

Electromagnetic energy is composed of waves and vibrations that can be made to oscillate in a horizontal, vertical, or some other plane. When energy vibrates in a particular plane, it is said to be polarized. Radar can send signals horizontally or vertically, and the returning signal exhibits polarization. Images can be produced with the same polarization (HH) as was sent or with the opposite polarization (HV). Using different polarization modes produces images that can be quite different (Fig. 10.22). Shifts from band to band, then, can also produce images with different characteristics. Radar that operates in two bands and at two polarizations produces four different images simultaneously. Like the multispectral scanner or multilens camera, this provides a considerable amount of additional data for the analyst.

Geometry of SLAR Imagery. The geometry of SLAR imagery is different from that of either pho-

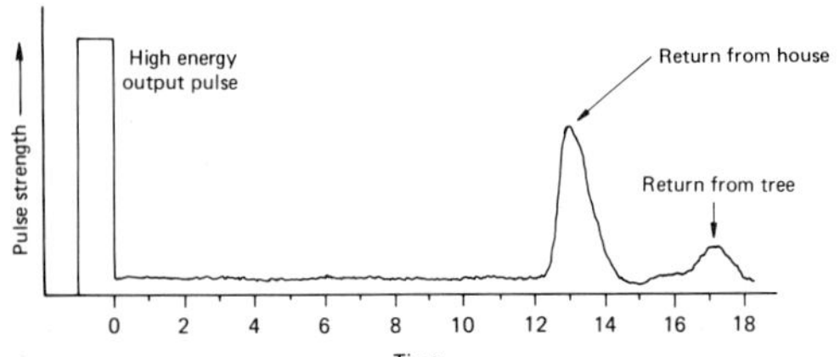

FIGURE 10.21 Plot of returned pulse strength against time. This illustrates a hypothetical return from the situation illustrated in Figure 10.19. (From T. Lillesand and R. Kiefer, *Remote Sensing and Image Interpretation*, Copyright 1979, John Wiley and Sons, New York.)

tographic imagery or scanned imagery. One might expect some similarities between the geometry of scanned imagery and of SLAR imagery, but fundamentally scanned imagery depends on an angle measurement, whereas radar imagery depends on distance. We will discuss three aspects of the geometry of radar imagery: scale distortion, the occurrence of relief displacement, and parallax.

FIGURE 10.22 Top image produced with same polarization. Bottom image produced with opposite polarization. (Courtesy of the Westinghouse Electric Company.)

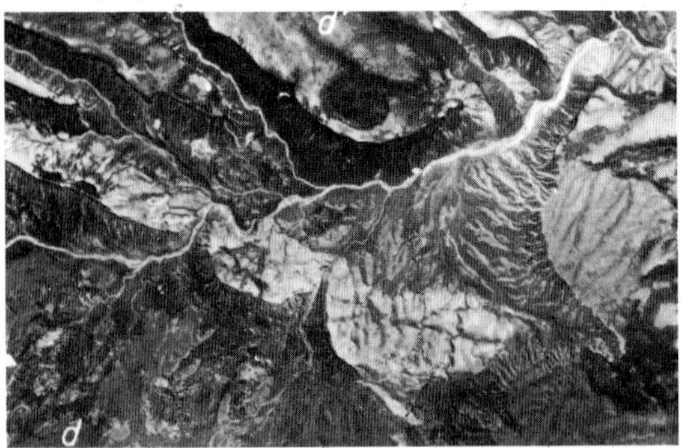

Scale Distortion. Radar uses two types of recording systems: slant range and ground range. Figure 10.23 illustrates the difference. Slant range records as unequal distances and sizes of features which are of equal size and spacing on the ground as demonstrated in Figure 10.23. Thus slant range data without correction is unusable for planimetric mapping. Ground range approximations from slant range data can be obtained if one assumes flat terrain and knows the flying height of the sensor by using the formula given below. From Figure 10.23 it follows that

$$SR^2 = H^2 + GR^2$$

and therefore

$$GR = (SR^2 - H^2)^{\frac{1}{2}}$$

The flight parameters also affect the accuracy of these calculations.

Relief Displacement. Relief displacement in SLAR imagery, as in scanned data, is one-dimensional and perpendicular to the line of flight. Displacement of features is inward rather than outward from the line of flight, however, since the radar pulse meets the top of the object before it meets its base. The signal from the top returns before the signal from the base and is recorded leaving the feature leaning toward the line of flight. This effect is called *layover*. The nearer the feature is to the sensor, the greater the displacement (see Fig. 10.24).

Slopes are also affected by this characteristic of radar imagery. There exists a slope where the radar pulse meets both the top and the bottom of the slope simultaneously (see hill C in Fig. 10.24). By this same reasoning hills can be foreshortened, like hill D in Figure 10.24, or affected by the layover effect as in hills A and B. Figure 10.24

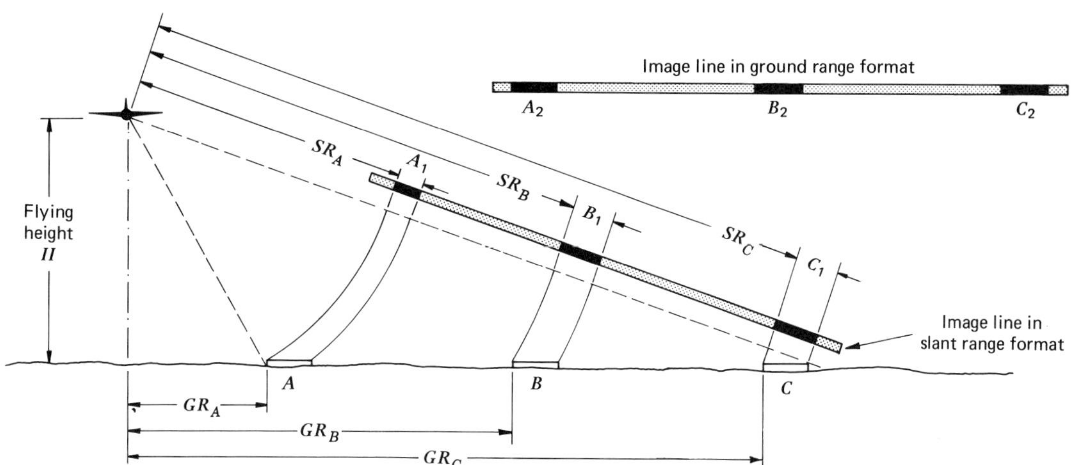

FIGURE 10.23 The difference between slant range (SR) and ground range (GR) geometry of a radar image. (From T. Lillesand and R. Kiefer, *Remote Sensing and Image Interpretation*, Copyright 1979, John Wiley and Sons, New York.)

FIGURE 10.24 Effects of terrain relief in causing displacement on SLAR images. (From T. Lillesand and R. Kiefer, *Remote Sensing and Image Interpretation*, Copyright 1979, John Wiley and Sons, New York.)

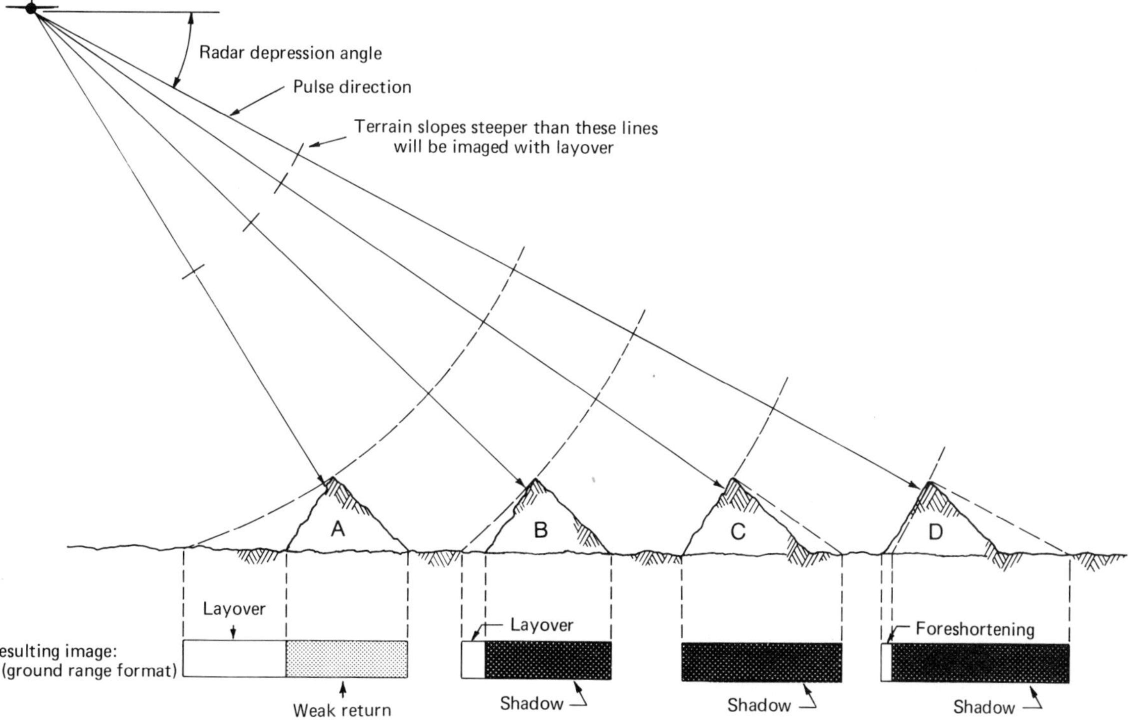

also shows that the shadow effect resulting from terrain increases with distance from the flight line. This becomes important in the calculation of parallax from radar images as described below.

Parallax. If an object is imaged from two different positions along a flight line, the relief displacement causes parallax in the imagery. This permits stereoscopic viewing. In radar imagery, however, because displacement is inward instead of outward, stereo radar imagery is taken along one flight line but from two differing altitudes. This results in altitude parallax, which avoids both the problem of the direction of illumination and the problem of sidelighting. If one reflects a moment on the images illustrated in Figures 10.23 and 10.24, it becomes evident why radar imagery from parallel flight lines represents problems whereas radar imagery from different altitudes along the same flight line avoids these same problems. As with photographic parallax, altitude parallax can be used to approximate feature heights.

Ultraviolet Sensing. Unless an external source of energy is used, sensing in the ultraviolet is limited to daylight hours, since it depends on reflected energy from the sun. Because of the blocking effect of the earth's atmosphere (mainly by the ozone layer) most of the energy in wavelengths shorter than 0.28 μm does not reach the surface of the earth. Consequently, developments in remote sensing in the ultraviolet have not progressed as rapidly as they have in the longer wavelengths.

Film emulsions are available that are sensitive to wavelengths down to 0.29 μm and that provide a high degree of resolution. Photography in these short wavelengths requires special lenses with a high quartz content, since most ordinary lenses are opaque to wavelengths shorter than 0.36 μm.

Scanning devices can also be used to detect and record in the ultraviolet. The impulses can be recorded on tape and used later to produce a photolike image. By electronically filtering out the "noise" in the data that is caused by atmospheric scattering of the short wavelengths, it is possible to produce better detail than is possible with ultraviolet photography.

To date little use of ultraviolet sensing has been made for traditional cartographic purposes. In other fields such as medicine or other uses where atmospheric scattering of energy can be minimized, ultraviolet sensing has been used and maplike products have been produced.

Multiple Linear Arrays. More recently an alternative to scanners for the detection and recording of data from aircraft or spacecraft has been developed. Linear arrays consist of a large series of very small single detectors mounted in a straight line and carried perpendicular to the line of flight (see Fig. 10.25). Several detectors can be joined to provide nearly 10,000 individual detectors in a single line.

The technique used is termed *pushbroom scanning*. It uses the forward movement of the spacecraft to sweep the linear array of detectors

FIGURE 10.25 Linear arrays detect energy along a strip perpendicular to the flight direction of the sensor platform. The line of detectors simultaneously records information for each strip. The rate of recording of the detectors is synchronized with the speed of the platform to create a continuous image.

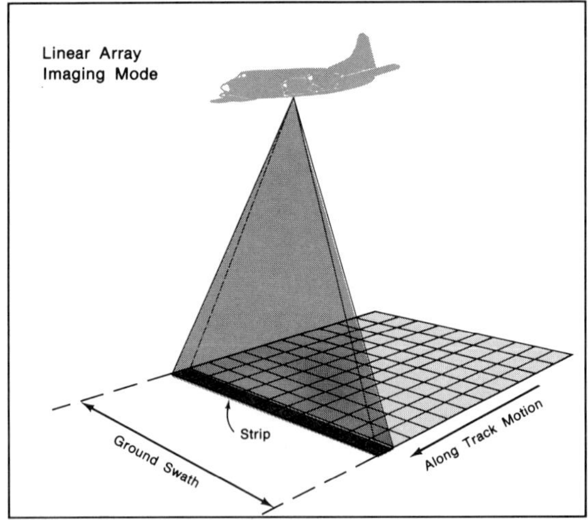

across the scene being imaged. The sampling rate of the detectors is set so that the forward motion of the spacecraft moves the array of detectors one ground resolution element forward between samples. Multiple arrays can be constructed where each line of detectors is set to sense in a finite range of the electromagnetic spectrum. Because solid state circuitry is used, there are no moving parts and the total weight is very little. Such sensors use little power, the geometric fidelity is high, and the life expectancy is long. All of these features are positive aspects for spacecraft instrumentation, which can not be easily repaired once in space.

The French SPOT satellite will be using linear arrays. Other experiments using linear arrays are scheduled for the United States' shuttle missions.

DATA SOURCES

Photo and Image Sources

Although a variety of federal agencies (as well as some local, regional, and state offices) hold aerial photographic coverage of parts of the United States, the best place to begin a search for coverage of any area is the nearest office of the National Cartographic Information Center (NCIC). The addresses of the NCIC offices are listed in Table 10.2. The National Cartographic Information Center is the office of the U.S. Geological Survey (USGS) in the Department of the Interior that helps individuals locate maps, photos, and other spatial data. NCIC's job is not necessarily to obtain copies of the information you seek, except in those cases where the USGS is the source, but to put you in contact with the source of the information you seek. Most of these NCIC services are free of charge.

NCIC maintains records of holdings on advance prints, maps, color separates, feature separates, out-of-print map reproductions, land-use and land-cover and associated maps, slope maps, digital terrain tapes, maps on microfilm, orthophotoquads, manned spacecraft images, LANDSAT images, computer-enhanced LANDSAT scenes, computer-compatible tapes of LANDSAT data, 35 mm viewing slides, photoindexes, aerial photography summary record system state-based graphics, microfiche indexes of aerial and space

TABLE 10.2 Addresses for Map Information

NCIC Headquarters
National Cartographic Information Center
U.S. Geological Survey
507 National Center
Reston, VA 22092
telephone 703/860-6045

NCIC Offices
Eastern Mapping Center-NCIC
U.S. Geological Survey
536 National Center
Reston, VA 22092
telephone 703/860-6336

Mid-Continent Mapping Center-NCIC
U.S. Geological Survey
1400 Independence Road
Rolla, MO 65401
telephone 314/341-0851

Rocky Mountain Mapping Center-NCIC
U.S. Geological Survey
Box 25046, Stop 504 Federal Center
Denver, CO 80225
telephone 303/234-2326

Western Mapping Center-NCIC
U.S. Geological Survey
345 Middlefield Road
Menlo Park, CA 94025
telephone 415/323-8111, ext. 2427

National Cartographic Information Center
U.S. Geological Survey
National Space Technology Laboratories
NSTL Station, MS 39529
telephone 601/688-3544

EROS Data Center
U.S. Geological Survey
EROS Data Center
User Services Center
Sioux Falls, SD 57198
telephone 605/594-6151

images, geographic computer searches, geodetic control data, reproductions of microfiche of state place-names, county maps, topographic maps on rolls of 35 mm microfilm, geographic coordinates of various U.S. and selected world names, microfiche of NCIC's map catalog, and numerous other map related items. The number of individual items, photos, maps, and so on about which NCIC has records numbers in the millions.

NCIC is fully aware of the holdings of the United States Forest Service, Bureau of Land Management, Bureau of Reclamation, U.S. Geological Survey, Bureau of the Census, Central Intelligence Agency, National Oceanic and Atmospheric Administration, National Ocean Service, Corps of Engineers, Federal Highway Administration, Federal Power Commission, Tennessee Valley Authority, Mississippi River Commission, International Boundary Commission, Library of Congress, Agricultural Stabilization and Conservation Service, Soil Conservation Service, National Archives and Records Service, National Aeronautics and Space Administration, and the Defense Mapping Agency.

The remainder of this discussion on sources only highlights some of the many sources. Again, our advice is to begin your search for sources by contacting NCIC.

Views from Space

LANDSAT Imagery. The most ambitious civilian undertaking to date in gathering earth inventory data by satellite is the LANDSAT series. As described above, five LANDSAT satellites have been placed in orbit. At an altitude of 900 kilometers each LANDSAT satellite could scan a particular earth scene once every 18 days. Each image contains 185 km by 185 km of earth area. One of the principal advantages of the LANDSAT coverage has been its continuity. Unfortunately a break in that continuity may occur as the plans to continue the series are in doubt.

LANDSAT is extremely popular among students, engineers, and scientists not only in the United States but in many other nations of the world. Imagery from LANDSAT has been used to investigate landform, land-use, geology, shallow seas, the environment, and vegetation. To order LANDSAT coverage through NCIC or the EROS Data Center in Sioux Falls, South Dakota, you must describe exactly the area of interest, either by geographic coordinates or by drawing an outline on a map. A phone call will appraise you of the current cost and available forms for your desired imagery.

Manned Spacecraft Imagery. During the manned Gemini, Apollo, and Skylab series of spacecraft missions, many photographs were taken by on-board mounted cameras and by cameras held by the astronauts. Some of these views are outstanding, and the availability of such imagery can be ascertained by filling out a geographical search inquiry form from either NCIC or the EROS Data Center.

Views from the Air

Tens of millions of aerial photographs of the United States exist. These photographs are a major tool in many planning and construction fields. The USGS has developed a computerized system, the Aerial Photography Summary Record System (APSRS), which helps users determine if aerial photographs cover a particular place, whether they meet other special conditions that the potential user specifies, and where copies can be obtained. NCIC offices have this APSRS system. The APSRS tells the user about cloud cover, types of photography, date, time, and the agency that has the film. NCIC then can explain how to obtain the aerial photograph from that agency.

People desire aerial photographs for uses ranging from a decorative wall hanging to determining precisely the extent of erosion on a particular slope. Scientists, planners, and environmentalists are interested in aerial photography to help study conditions on the earth's surface and changes in these conditions. Cartographers use aerial photography to produce accurate planimetric line maps. Since parts of the United States have been photographed at regular intervals since the late 1930s, a vast store of data and potential information is available. This imagery varies in scale and quality; enlargements may be purchased of most images. Again, the inquirer is first referred to NCIC for information.

For those people desiring historical information, there are photographs taken from the air dating from the nineteenth century. Copies of these images can also be obtained, but coverage is limited. NCIC can assist the user in determining both what coverage is available and how it can be obtained.

A special series covering over a hundred metropolitan areas in the United States has been created to provide images from NASA, Skylab manned space flights, and LANDSAT. They are usually available as color photographs or color infrared photographs. They have proved to be useful for regional planning and for general use. NCIC has complete information on this Major United States Metropolitan Area series.

Availability of foreign aerial photographic coverage is less certain. A citizen of a foreign country often encounters delays or refusals in attempts to obtain aerial photographic coverage of certain areas. Although such delays are sometimes understandable, the potential user in need of foreign aerial photographs is urged to start trying to obtain them well in advance. Satellite imagery, especially LANDSAT, is somewhat easier to obtain.

SELECTED REFERENCES

Avery, T. E., *Interpretation of Aerial Photographs*, 3rd ed., Minneapolis, Minn.: Burgess Publishing Co., 1983.

Chevrel, M., M. Courtois, and G. Weill, "The SPOT Satellite Remote Sensing Mission," *Photogrammetic Engineering and Remote Sensing* 47, no. 8, (1981): 1163–71.

Colwell, R. N. (Ed.), *Manual of Remote Sensing*, 2nd ed. Falls Church, Va.: American Society of Photogrammetry, 1983.

Estes, J. E., J. R. Jensen, and D. S. Simonett, "Impact of Remote Sensing on U.S. Geography," *Remote Sensing of Environment* 10 (1980).

Harger, R. O., *Synthetic Aperture Radar Systems, Theory and Design*. New York: Academic Press, 1980.

Holkenbrink, P. E., *Manual on Characteristics of Landsat Computer-Compatible Tapes*. Washington, DC: Department of Interior, U.S. Geological Survey, EROS Data Center Digital Image Processing System, 1978.

Kraus, K., "Rectification of Multispectral Scanner Imagery," *Photogrammetric Engineering and Remote Sensing* 44, no. 4 (1978): 453–57.

Lillesand, T. M., and R. W. Kiefer, *Remote Sensing and Image Interpretation*. New York: John Wiley & Sons, 1979.

Lintz, J., Jr., and D. S. Simonett (Eds.), *Remote Sensing of the Environment*. Reading, Mass.: Addison-Wesley, 1976.

Richason, F., Jr. (Ed.), "Remote Sensing: An Input to Geographic Information Systems in the 1980's," *Proceedings*, Pecora VII Symposium, Sioux Falls, S.D. Falls Church, Va.: American Society of Photogrammetry, 1982.

Shlien, S., "Geometric Correction, Registration and Resampling of LANDSAT Imagery," *Canadian Journal of Remote Sensing* 5, no. 1 (1979): 74.

Skolnick, M. I., *Introduction to Radar Systems*, 2nd ed., New York: McGraw-Hill, 1980.

Slama, C. C. (Ed.), *Manual of Photogrammetry*, 4th ed. Falls Church, Va.: American Society of Photogrammetry, 1980.

Slater, P. N., *Remote Sensing: Optics and Optical Systems*. Reading, Mass.: Addison-Wesley, 1980.

Townshend, I. R., *The Spatial Resolving Power of Earth Resources Satellites: A Review*. NASA Technical Memorandum 82020. Greenbelt, Md.: Goddard Space Flight Center, 1980.

Ulaby, F. T., R. K. Moore, and A. K. Fung, *Microwave Remote Sensing: Active and Passive*. Vol. 1. Reading, Mass.: Addison-Wesley, 1981.

U.S. Department of Interior, Geological Survey,

LANDSAT Data User's Handbook, rev. ed. Arlington, Va.: USGS Branch of Distribution, 1979.

Williams, D. L., and V. V. Salomonson, "Data Acquisition and Projected Applications of the Observations from Landsat-D," in *Joint Proceedings of the American Society of Photogrammetry and the American Congress on Surveying and Mapping,* 1979 Fall Meeting, Sioux Falls S.D. Falls Church, Va.: American Congress on Surveying and Mapping and American Society of Photogrammetry, 1979.

Wolfe, W. L., and G. J. Zissis (Eds.), *The Infrared Handbook.* Washington, D.C.: U.S. Government Printing Office, 1978.

11

Simplification and Classification Processes

Almost everything on a map is generalized. As we saw in Chapter 6, cartographic generalization can be defined in terms of four interrelated sets of processes: simplification, classification, symbolization, and induction. Before specifying the design of a map, or before selecting symbolization for the data to be used on the map, the cartographer must manipulate the data. Data manipulation requires the cartographer to apply one or more of the classification and/or simplification processes.

Simplification processes consist of eliminating unwanted data or modifying existing data. Classification processes consist of typifying distributions of points, networks of lines, or the agglomeration of areas. In this and the next three chapters, we will explore in depth three of the elements of cartographic generalization: (1) simplification, (2) classification, and (3) symbolization. We also will introduce some statistical methods, specifically linear regression and correlation analysis, that are useful to cartographers.

Cartographers working without benefit of machines must complete each generalization independently. This work is then either photgraphically reproduced or traced, and subsequently may be completely regeneralized to make additional maps. It is fairly safe to say that no two generalizations done by humans are identical. Moreover, it is difficult, if not impossible, for the cartographer working manually to provide a detailed verbal description of how a given generalization is achieved. Strict definitions for the elements of cartographic generalization for manual cartography are unnecessary, therefore, except in a strictly pedagogical sense. There is truly a subjective element of "art" in manually produced cartographic generalization.

The coming of computer assistance in cartography has had a profound effect on cartographic generalization. Cartographers have been forced to rethink and to define more explicitly the processes used in making a map, but in the process, cartographers have gained additional, alternative processes to aid in developing generalizations. Computers demand unambiguous instructions. Initially this led to the development of explicit, and sometimes limiting, algorithms to achieve cartographic generalization. But after some of the early routines were written, it became evident that the power of computer assistance had opened the door to a vast array of sophisticated, often statistical, processes which a cartographer could employ to aid with generalization.

Most of the generalization processes now routinely used in computer-assisted cartography were impossible with manual cartographic generalization. As an example, consider interpolation models used in isarithmic mapping (see Chapter 13). In manual processing, the cartographer has a practical choice between two methods: strict linear interpolation or "eyeball" sketching interpolation. Today, using machine processing, the number of interpolation methods available for use is infinite. Another example is the determination of class intervals in choroplethic mapping (see Chapter 14). There are many methods for the manual computation of class limits, but only with computer-assisted iterative solutions have theoretical statistical criteria been used in classifying the data. This has enormously increased the number of exceptable class interval determination methods available to cartographers.

Of course, the changes brought about by computer-assisted cartographic technology can be viewed as both blessing and curse. The student training for a career in cartography must now be aware of and conversant about the generalization processes that are available with computer-assisted cartography. The cartographer of today and tomorrow must be able to select intelligently the most appropriate generalization routine rather than be able to perform cartographic generalization manually. The switch to computer technology frees the cartographer from the drafting table, but it requires more knowledge and the ability to reason with that knowledge. As a result, cartography has become less a manual activity and more a mental activity.

COMMON SIMPLIFICATION AND CLASSIFICATION DATA MANIPULATIONS

In theory, the data manipulations involved in simplification and classification are easily distinguished. In practice, however, a total manipulation routine, regardless of whether it is manual or computer-assisted, may contain elements of both. By viewing the results of computer-assisted production, the order in which simplification or classification is performed cannot be determined without first seeing the details of the algorithm being used. In manual production it is impossible to discern which processes were used and their order of application.

Once selected, data may exist in many forms: (1) tabular values, (2) maps or photographically recorded images, (3) text (verbal) accounts, (4) computer-readable strings of coordinates (vectors), or (5) machine-stored arrays of picture elements (rasters). The actual manipulations that the cartographer employs will vary, depending on the initial data form. We will examine below some of the more common manipulations for both simplification and classification applicable to each data form.

The purpose of simplification manipulations is to assist the cartographer in conveying a message via the map. Today the amount of data available relative to every aspect of our lives is constantly increasing. This is useful if finer and finer detail about a prescribed topic is needed. But too much data can be as much of a problem for the cartographer as too little data. When mapping an area, the cartographer must convey the characteristics of the distribution in a manner suitable to the scale of the map. Knowing that intricate detail or complexity attracts attention, the cartographer must avoid detailed presentations on portions of the map that will detract the reader's attention from the overall trends in the distribution. To do this, the cartographer must simplify. Otherwise, the map will include what can be termed "noise" that will hinder its effectiveness for the user. Too often this noise, because it represents either more known data or more precisely recorded data, is highly valued by the cartographer, and there is a reluctance to simplify such "good" data. In data that have been remotely sensed, small variations, often the result of the mechanics of the sensor system, are also termed "noise." Simplification techniques are used to remove this noise to enable the user to interpret the image more clearly. In this instance, cartographers are eager to remove this "bad" noise. The two situations are analogous, however, with respect to conveying information to a user, and theoretically the noise should be removed from the data in both cases to enable the reader to see the trends of the distribution easily.

Simplification is required in another situation that has recently arisen because of the use of computer assistance in cartography. Data are converted to digital records; the accuracy depends on the scale of the source data and the resolution of the computer hardware used for the conversion. Most often these data are then used for output at scales smaller than that of the source data. In addition the output device may have a coarser resolution than the input device. In such cases more data exist in the digital record than can be plotted by the output device. These data cause a loss in efficiency to the entire mapping process. Simplification procedures provide the cartographer with the means to eliminate these unneeded and unusable data.

As an assist to the cartographer, simplification procedures are therefore critically necessary. Without them the cartographer may inadvertently—and easily, we might add—create a map that fails to convey its message to the user because the message is cluttered by noise or fails to be efficient in terms of data storage and manipulation.

Simplification Manipulations

Simplification is defined as the determination of the important characteristics of the data, the elimination of unwanted detail, and the retention and possible exaggeration of the important characteristics. Simplification of data can consist of either manual, statistical, or computer-assisted manipulations. Manual manipulations tend to be the most subjective. Data taken from maps, photographically recorded images, and text materials tend to be easily simplified subjectively by cartographers. Such simplification mainly takes the form of omissions or deletions, and without strict rules probably no two cartographers would perform exactly the same manual simplification manipulations for a given map. For example, in manually compiling a map of North America on a Lambert equal-area base at a scale of 1:20,000,000, some islands in the Caribbean and in the Canadian Arctic and some lakes in the interior of the continent are omitted. Likewise, some of the bays or inlets along the coastlines are deleted. The cartographer-compiler employs great subjectiveness, and consequently the compiler's skill and knowledge about the data being simplified vitally affect the usefulness and quality of the resulting map.

In recent years cartographers and earth scientists have increasingly applied statistical manipulations to tabular values or to data read manually from maps. The quantitative revolution in the field of geography led to the use of common statistical methods like moving averages, surface-fitting techniques, and regression and correlation analysis, and to the analysis of the results of these manipulations by cartographers. (Since regression and correlation are among the most basic statistical manipulations a cartographer must learn to use today, a review, with cartographic examples, is presented at the end of this chapter.)

With the introduction of computer-assisted cartography, the simplification manipulations available to a cartographer have exploded in both number and sophistication. Data exist in machine-usable form in either a vector or a raster format. The statistical techniques available for application become much more complex than simple regression and correlation analyses and now include multiple regression and correlation, factor analysis, discriminant analysis, curvilinear interpolations, and others. In dealing with raster data, routines such as ratioing, principal component analyses, density slicing, and others are routinely performed. These manipulations are more objective in the sense that the outcome is repeatable given one specific algorithm applied to a set of data. Subjectivity remains in that the cartographer must select an algorithm. In this sense generalization is like map projections. Cartographers must be trained to select and use correctly the generalization or projection most appropriate for a given purpose. The cartographer no longer has to perform the actual manipulations; the machine does that with a degree of precision that cartographers could never attain manually.

Simplification routines can be classed into two categories: (1) elimination routines, and (2) modification routines.

Elimination Routines. Two classes of operations can be identified in simplification by elimination of data: those that eliminate points and those that eliminate features. Point elimination simplifies a string of coordinate points that defines a linear feature or outlines an area by eliminating all but a chosen few points that are deemed most important for retaining the character of the line being mapped. Feature simplification retains some of a large number of small items and omits others to maintain the essential character of the distribution.

There are two reasons for the application of elimination routines: (1) to accompany a reduction in scale of the map being made; that is, the existing data are too detailed for portrayal at the desired scale of the map, or (2) to de-emphasize a data set that is to play a minor role in the cartographic presentation.

Point Elimination. The map in Figure 11.1 illustrates point elimination to simplify the outline of Portugal. The points that are emphasized by black dots on the left map have been retained and connected by straight line segments to make

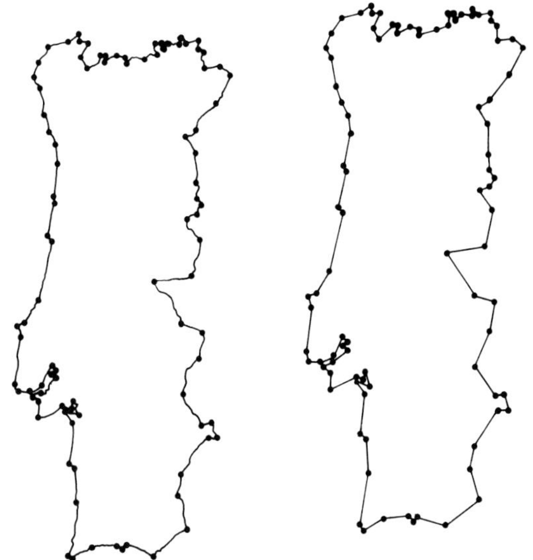

FIGURE 11.1 The simplification of the outline of Portugal by point elimination. The points indicated on the map to the left were retained on the map to the right where they were connected with straight line segments. All points not selected on the map to the left were eliminated in the production of the map on the right. (Courtesy of the American Congress on Surveying and Mapping.)

the map at the right. Certainly another cartographer might choose to retain other points, but the points denoted by dots on the left map of Figure 11.1 must be retained on the right map for the process to be one of simplification. A machine algorithm could also have selected points from the left map to be connected by straight line segments to form the map on the right. Therefore, Figure 11.1 could have been produced either manually or by machine by employing a simplification routine.

Figures 11.2 and 11.3 illustrate two computer-assisted simplification manipulations: (1) systematic retention of every *n*th point in the data file, and (2) random retention of 1/*n*th of the points. In the systematic retention of every *n*th point the cartographer creates a new data file by taking the first coordinate point in a data string and every *n*th point thereafter. The new file is then plotted. The value of *n* is rather subjectively determined, but the radical law (see Chapter 6) may be of some use. The larger the value of *n*, the greater will be the simplification.

In the random retention of 1/*n*th of the points, as in Figure 11.3, the new file is created by randomly selecting 1/*n*th of the points. Again the larger the value of *n*, the greater will be the simplification (see Fig. 11.4). The systematic retention and random retention of points are just two of many available computer-assisted simplification procedures.

In the manual elimination of points defining a feature, the manipulation consists of deleting the points that are visually unimportant, and a "feel"

FIGURE 11.2 A computer-simplified map of Sardinia produced by systematically retaining every fifteenth point in the original data file. (Data and algorithm courtesy of D. Brophy.)

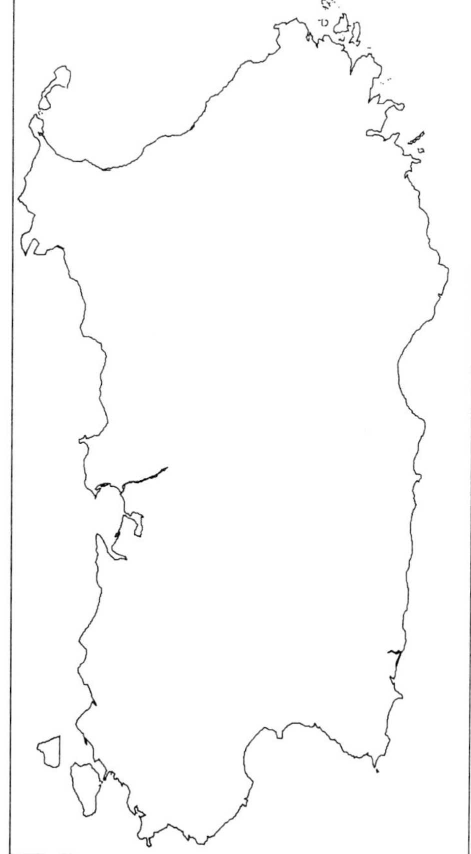

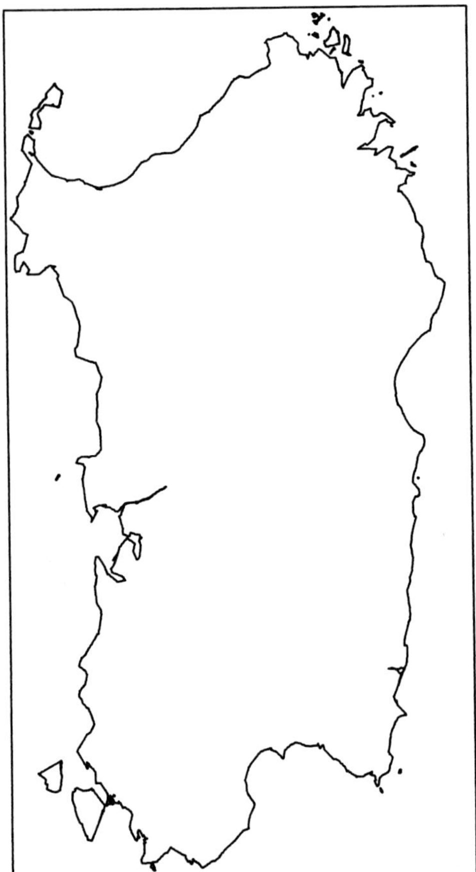

FIGURE 11.3 A computer-simplified map of Sardinia produced by randomly selecting one-fifteenth of the points in the original data file. (Data and algorithm courtesy of D. Brophy.)

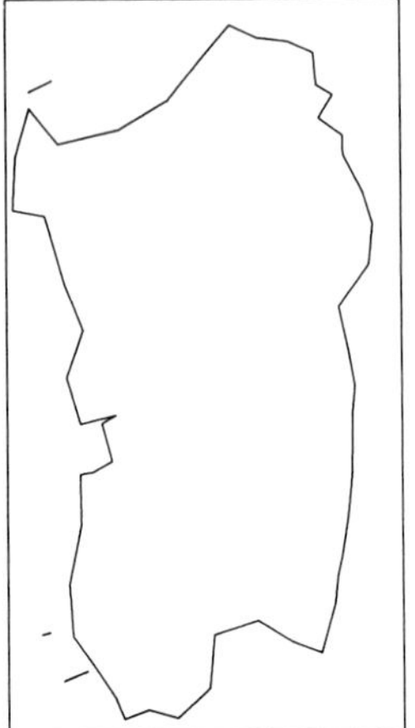

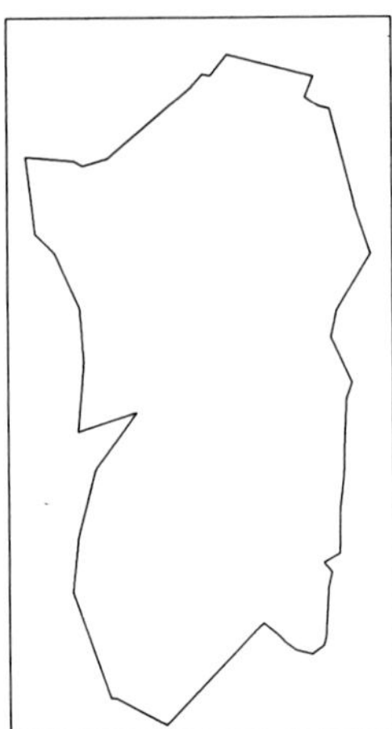

FIGURE 11.4 The map of Sardinia on the left results from the systematic retention of every 200th point, and the map on the right results from the random selection of 1/200 of the points in the original data file. (Data and algorithms courtesy of D. Brophy.)

for this elimination process, particularly when it is manually applied, probably comes only after considerable experience in cartographic compilation. For manual applications, the radical law has only marginal usefulness for this subjective process except in the case of place-names or of discrete point objects to be mapped, as in Figure 11.5. For linear features and for areas (for example, distributary streams or groups of islands) the actual features usually vary considerably in size. Therefore, size is often more important than the use of the radical law in the decision to retain or eliminate a feature.

In computer-assisted cartography the simplification manipulations illustrated in Figures 11.2 and 11.3 become less overtly subjective. Subjectivity is involved in the creation of the initial computer file, but once the data are in machine-readable form, repeatable simplification manip-

FIGURE 11.5 Simplification by point elimination. Application of the radical law might suggest a reduction to one-fifth of the original number of points. The point elimination process illustrated clusters the original points into groups of five and selects one (designated by red overprint) of the five to represent the group on the generalized map.

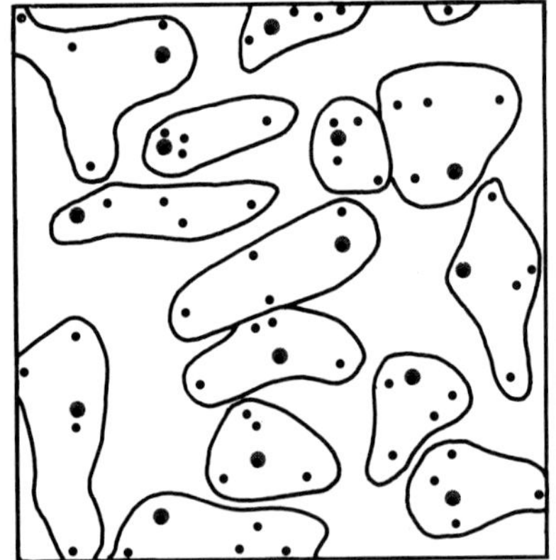

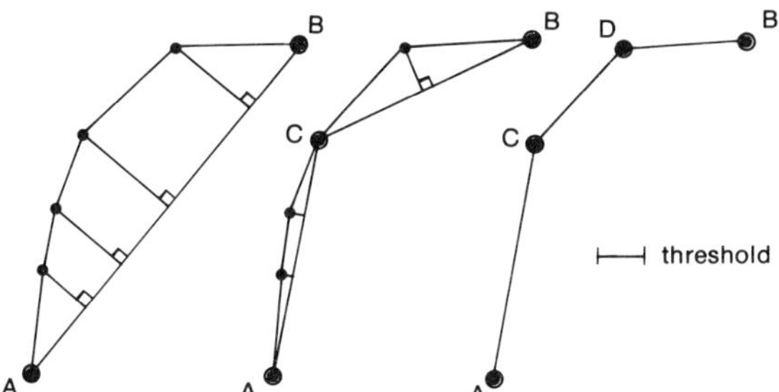

FIGURE 11.6 The successive stages in the application of the Douglas-Peucker algorithm for line simplification. (1) The initial line whose endpoints are connected by a straight line, AB. (2) Point C having the greatest perpendicular distance to line AB in (1) is selected for retention. Lines AC and CB are drawn. (3) The elimination of points between points A and C, because no perpendicular exceeds the threshold and the retention of point D because its perpendicular distance to line CB does exceed the threshold. (Courtesy of Mark S. Monmonier.)

ulations can be performed. Because of their lack of consideration for the characteristics of the lines being simplified, these two rather simplistic elimination routines are most often used to reduce the large amounts of data that have been digitized prior to initial plotting at a reduced scale. For example, when data are digitized on equipment with resolution of .0001 inch and then reduced by a factor of ten and plotted by equipment with equivalent resolution, the amount of data potentially available in the digitized file is ten times what can be physically plotted. In such cases one of these simple elimination schemes is appropriate as an initial simplification procedure.

Other criteria can be established to retain or eliminate points in a data file. The algorithm that is perhaps most common today was developed by Douglas and Peucker (see Fig. 11.6). The Douglas-Peucker algorithm allows the cartographer to specify a threshold that controls the amount of simplification. For a specified line segment the two end points are connected by a straight line and the perpendicular distances from all the intervening points to that line are calculated. If a perpendicular distance exceeds the specified threshold, the point with the greatest perpendicular distance is used as the new end point for the subdivision of the original line. The perpendicular distances from all the intervening points to the two new lines are calculated and compared to the specified threshold. If at any time none of the perpendiculars exceed the threshold, all of the intervening points are eliminated. The routine continues until all possible points have been eliminated.

Since the amount of data in machine-readable form is increasing rapidly, the aspiring cartographer must pay close attention to the simplification manipulations that can be applied to machine-readable data files, while at the same time realizing that elements of subjectivity and limits of hardware capabilities have determined the contents of those data files.

Feature Elimination. Simplification by feature elimination is illustrated in Figure 11.7, which shows the areas of forest cover of Sheboygan County, Wisconsin. In feature elimination simplification a feature is either shown in its entirety or omitted. In Figure 11.7 the smaller areas on the left map have been eliminated from the map on the right. Obviously, it is possible to combine point and feature elimination simplification by eliminating features and point-simplifying the outlines of the retained features.

Feature elimination, if done manually, will be inconsistent. In machine processing the criterion may be size of feature, proximity to neighboring features, or a combination. If either strictly size or proximity is used, the simplification routine requires only one numerical instruction. The cartographer may specify either the minimum size of feature to be retained or the minimum distance between features based on output scale and line width.

Feature elimination routines for data in vector form can be done by machine if relative importance rankings for each feature are specified in the data file. Simplification is then accomplished by specifying output containing all streams of rank 5 or above, or all roads of rank 2 or above, or whatever. If the numbers of features have been tabulated by rank, the radical law can be used

FIGURE 11.7 Simplification by feature elimination. Areas on the left map are either shown in their entirety or completely eliminated in the feature-simplified map on the right. (Courtesy of American Congress on Surveying and Mapping.)

to help the cartographer decide on the correct rank of features for display in accordance with the desired level of simplification.

Feature elimination can also be applied to raster data. The process sometimes called stenciling is used to delete or eliminate data contained for

FIGURE 11.8 An example of stenciling. The image on the right shows the results of stenciling the satellite image on the left. On the right image all nonwater pixels have been printed in black. This enables the user to concentrate only on the water areas. (Courtesy of P. Chavez, U.S. Geological Survey, Flagstaff, AZ.)

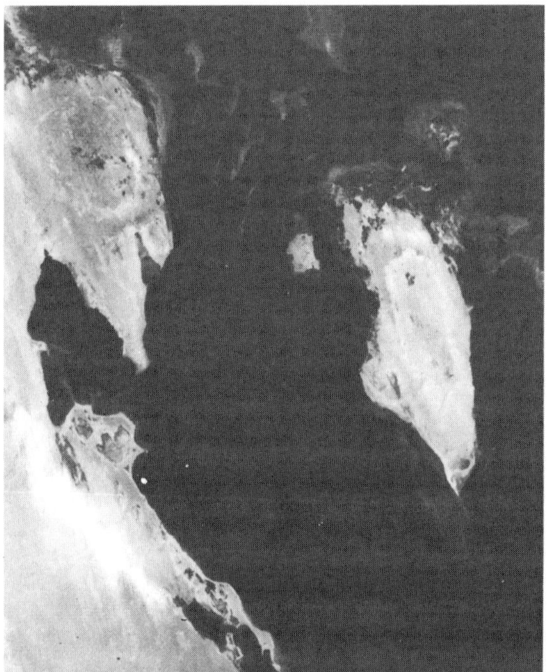

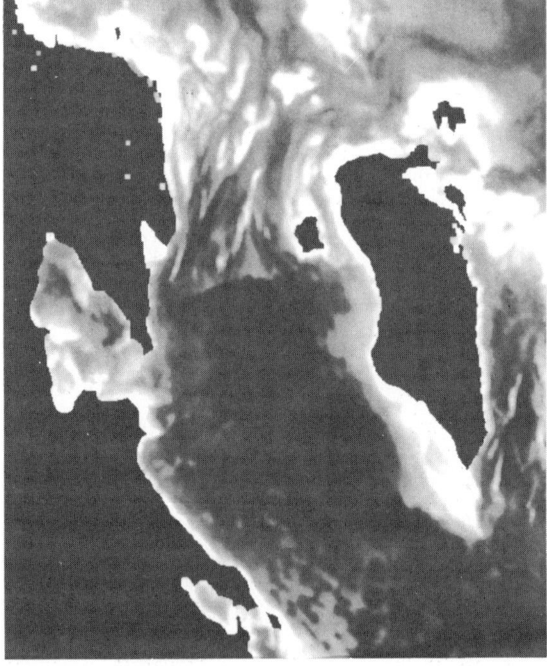

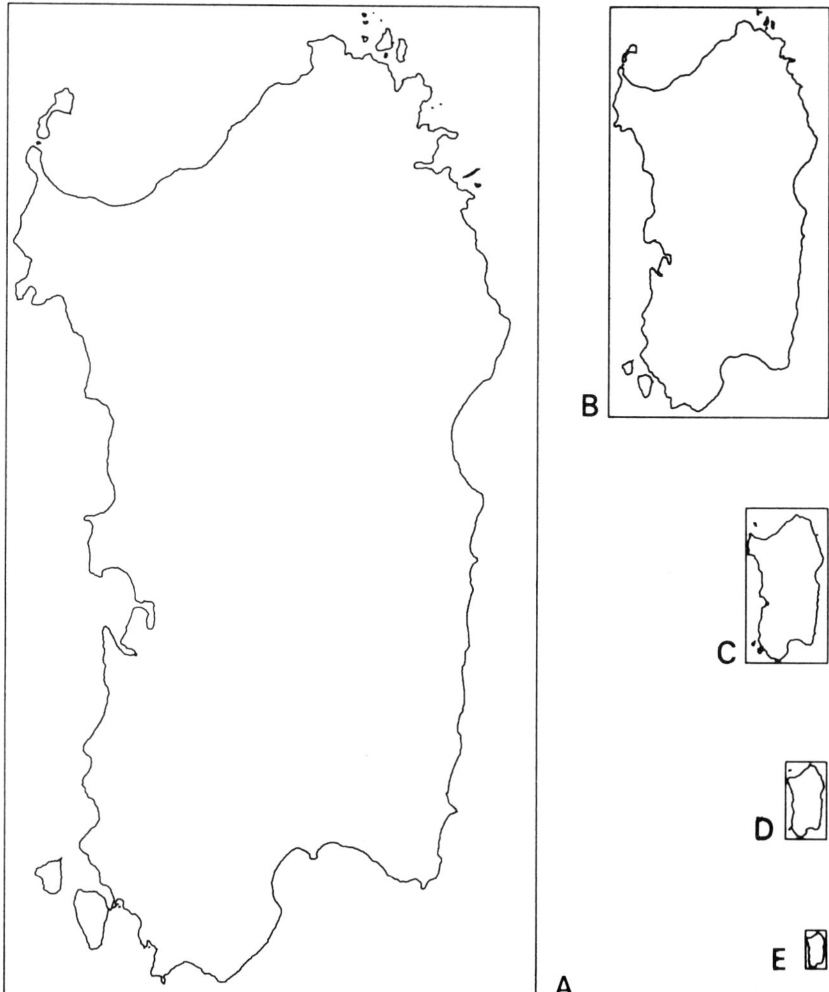

FIGURE 11.9 Simplification accompanied by scale reduction. Since the scale is successively reduced from (A) to (E), an increasing number of points in the outline of Sardinia must be eliminated. (Courtesy of American Congress on Surveying and Mapping.)

an image in raster format. For example, the right image of Figure 11.8 illustrates the removal of land areas from the image on the left in Figure 11.8, so that the investigator can more easily consider only the areas on the image covered by water. It is only required to know the different reflective values of land and water and to set all pixels of the image having reflective values indicating land to black. This emphasizes the remaining pixel data. Other simplification manipulations applicable to raster data will be discussed later.

Simplification can be applied along either of two dimensions: (1) a reduction in the scale dimension (see Fig. 11.9), that is, a change from some original scale to a smaller scale, or (2) at some constant scale dimension (see Fig. 11.10), that is, a detailed representation versus a simplified representation at the same scale. Along either of these two dimensions, the process of simpli-

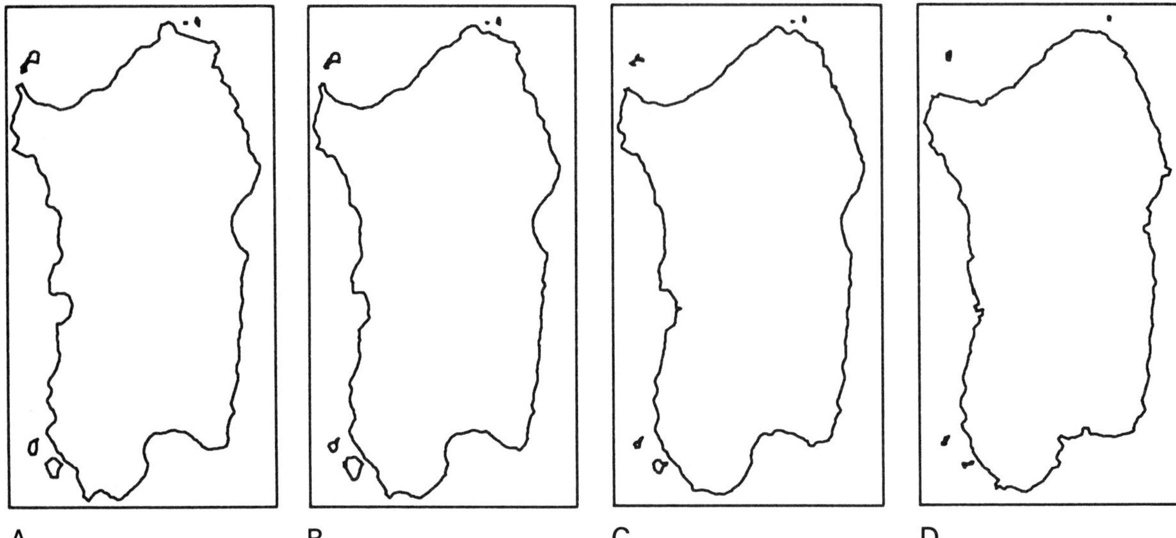

FIGURE 11.10 Simplification applied at a constant scale. The four maps of Sardinia (A through D) represent increasing simplifications of the coastline. (Courtesy of American Congress on Surveying and Mapping.)

fication can be applied to any data category, in both point and feature form. In practice, generalization usually calls for an application of the simplification process primarily along the dimension of scale reduction and secondarily along the dimension of constant scale. For stored data strings, simplification also usually involves both point and feature elimination.

Simplification accompanied by scale reduction is a rather straightforward concept. As the scale of the map is reduced, the physical space available for detail is reduced; therefore, simplification of the details is mandatory. Simplification applied at a constant scale is more subtle and perhaps more important in establishing a good map design. For example, if the outline of Sardinia as shown in the figures has been selected merely to tell the reader that Sardinia is the area of interest, a rather simplified outline, such as Figure 11.10D, is appropriate. Presumably the thematic data to be portrayed about Sardinia are the primary message of such a map. On the other hand, suppose the primary purpose of the map deals with ports or coastal shipping. In this case the detail on the delineation of the coastline is important, and an outline like that shown in Figure 11.10A is more appropriate. Usually the cartographer simplifies along both dimensions simultaneously; that is, the cartographer compiles from a larger-scale map or data base and selects the detail appropriate for the map design.

Modification Routines. Simplification by modification can also be categorized into two categories: (1) smoothing operators including line or surface-fitting routines, and (2) enhancement routines, including ratioing. These routines have generally come into use by cartographers rather recently and are applied almost exclusively to vector or raster data with computer assistance. Prior to computer assistance, these processes were too complex to apply manually.

Smoothing Operators. The class of simplification manipulations referred to as *smoothing operators,* including surface-fitting techniques, are appropriate for vector data representing linear features or areal boundaries, for regular matrices of data referring to points, or for raster data stored as picture elements.

Figure 11.11 illustrates the simplification of area/volume (raster) data using a smoothing operator. Each picture element or position is compared separately to its neighbors. After evaluation the picture element may be modified to conform more closely to its neighbors. It is important that the modification does not destroy the individual picture element or combine it with another picture element; it must merely modify its value.

Moving Averages. Figure 11.12 represents the commonly used smoothing operator of a *moving average*. It can be applied to any vector string of data, linear data, or matrix of point values (most often evenly spaced, but not necessarily). The weights and the number of neighboring points considered determine the effect of the application of a moving average. For example, as the number of neighboring values included in the moving average increases, the amount of simplification taking place is increased (compare line C in Fig. 11.13 to line B from Fig. 11.12). Furthermore, the more evenly apportioned the various weights assigned to the neighboring values by the moving average, the greater will be the simplification (compare line D to line B in Fig. 11.14). If large weights are assigned by the moving average only to close neighbors, the amount of simplification is reduced. The lower portions of Figures 11.12, 11.13, and 11.14 illustrate the effects of two-dimensional smoothing operators applied to a matrix of elevation data representing a statistical surface. For data in raster format, smoothing operators, almost identical mathematically to the one used in Figure 11.15, are termed *enhancement procedures*. They are discussed below.

Theoretically the selection of weights should be based on the amount of autocorrelation (see discussion at end of this chapter) suspected in the distribution being mapped. If autocorrelation is high, large weights assigned to nearby points result in little simplification, and greater simplification can be obtained by assigning even weights to the neighboring points considered. Alternatively high autocorrelation requires that the number of nearby points considered must be increased (in other words, consideration must be given to points farther from the point in question) in order to increase simplification. On the other hand, to capture the major features of a distribution having low autocorrelation, high weights should be assigned to nearby values, and consideration of few neighboring values is recommended. Since points at greater distances have little correlation with the point under consideration when low autocorrelation occurs, there is no justification for considering many neighboring points. In such cases it is best to use few points and let the assignment of weights determine the

FIGURE 11.11 A smoothing operator has been applied to the previously classed picture elements in (A) to obtain (B). (Courtesy of S. Friedman.)

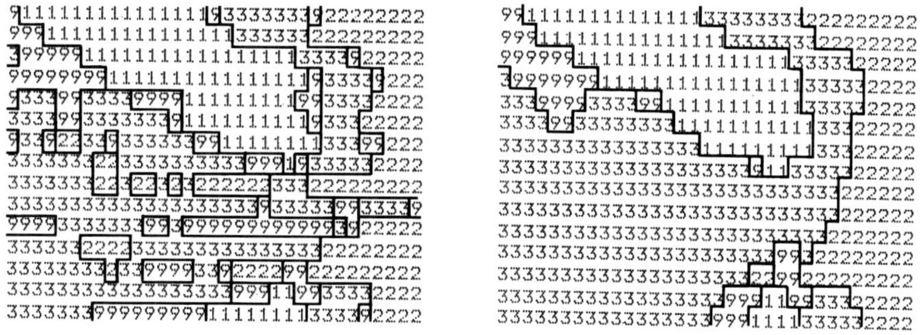

```
911111111111111193333333922222222
999111111111111111333333222222222
399999111111111111111113333922222
999999991111111111111111933339222
933399333399991111111119933332222
333399333333391111111111933322222
933922339333333991111111133399222
333333322333333333399919333332222
333333322322323222222333222222222
333333333333333333339333339933339
999933333339939999999999993922222
333333222233333333333333322222222
333333332339999339222299222222222
333333333333333333999119933332222
333333399999999911111111333392222
```

A

```
991111111111111113333333222222222
999111111111111111333333322222222
999999111111111111111113333322222
399999991111111111111111333332222
333999933339911111111111333332222
333399333333331111111111133322222
333333333333333311111111133322222
333333333333333333339113333322222
333333333333333333333333333222222
333333333333333333333333333222222
333333333333333333333333332222222
333333333333333333333399322222222
333333333333333333332299222222222
333333333333333333999119933322222
333333333333333339991111333332222
```

B

LINEAR DATA

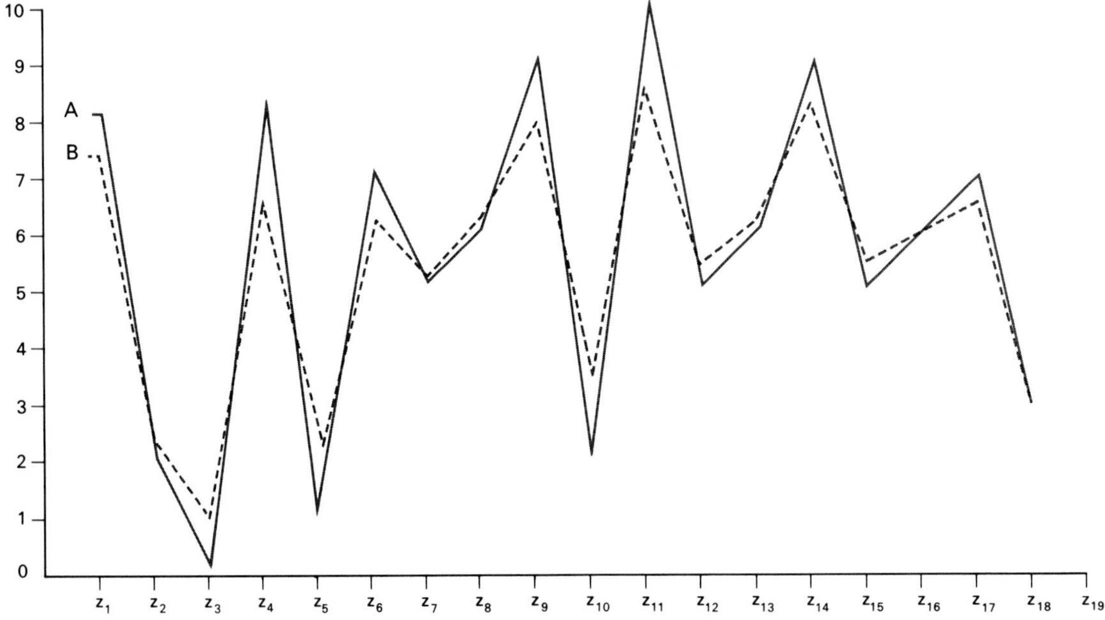

Smoothing operator: for B

$$\hat{z}_1 = .9z_1 + .1z_2$$
$$\hat{z}_i = .1z_{i-1} + .8z_i + .1z_{i+1}$$
$$\hat{z}_n = .1z_{n-1} + .9z_n$$

AREA DATA

A – Original data array

0	6	7	0	9	5	3
0	0	10	2	9	1	6
6	3	6	5	1	10	3
8	1	10	7	0	10	2
9	3	4	4	6	1	10
8	6	0	2	7	1	7
7	6	5	5	4	1	1

B– Smoothed data array

0.36	5.46	6.25	1.59	7.77	5.15	3.24
0.63	1.14	8.47	2.93	7.83	2.14	5.58
5.34	3.51	5.70	5.15	2.08	8.56	3.48
7.43	2.23	8.59	6.40	1.32	8.59	2.99
8.37	3.66	4.03	4.12	5.52	2.05	8.80
7.73	5.82	1.05	2.57	6.04	1.87	6.28
6.97	5.88	4.85	4.76	3.82	1.54	1.36

Smoothing operator: for B

for Z_{ij}

$$\hat{z}_{ij} = .03z_{i-1,j-1} + .03z_{i,j-1} + .03z_{i+1,j-1} + .03z_{i-1,j} + .76z_{ij} + .03z_{i+1,j} + .03z_{i-1,j+1} + .03z_{i,j+1} + .03z_{i+1,j+1}$$

for corners (appropriately rotated)

$$\hat{z}_{11} = .85z_{11} + .06z_{21} + .06z_{12} + .03z_{22}$$

for edges (appropriately rotated)

$$Z_{12} = .06z_{11} + .79z_{12} + .06z_{13} + .03z_{21} + .03z_{22} + .03z_{23}$$

FIGURE 11.12 Linear data: The results (line B) of using a three-term moving average with uneven weights to smooth linear data (line A). Area data: The smoothed down array (B) obtained by using a nine-term, two-dimensional, unequally weighted moving average to smooth the original data (A).

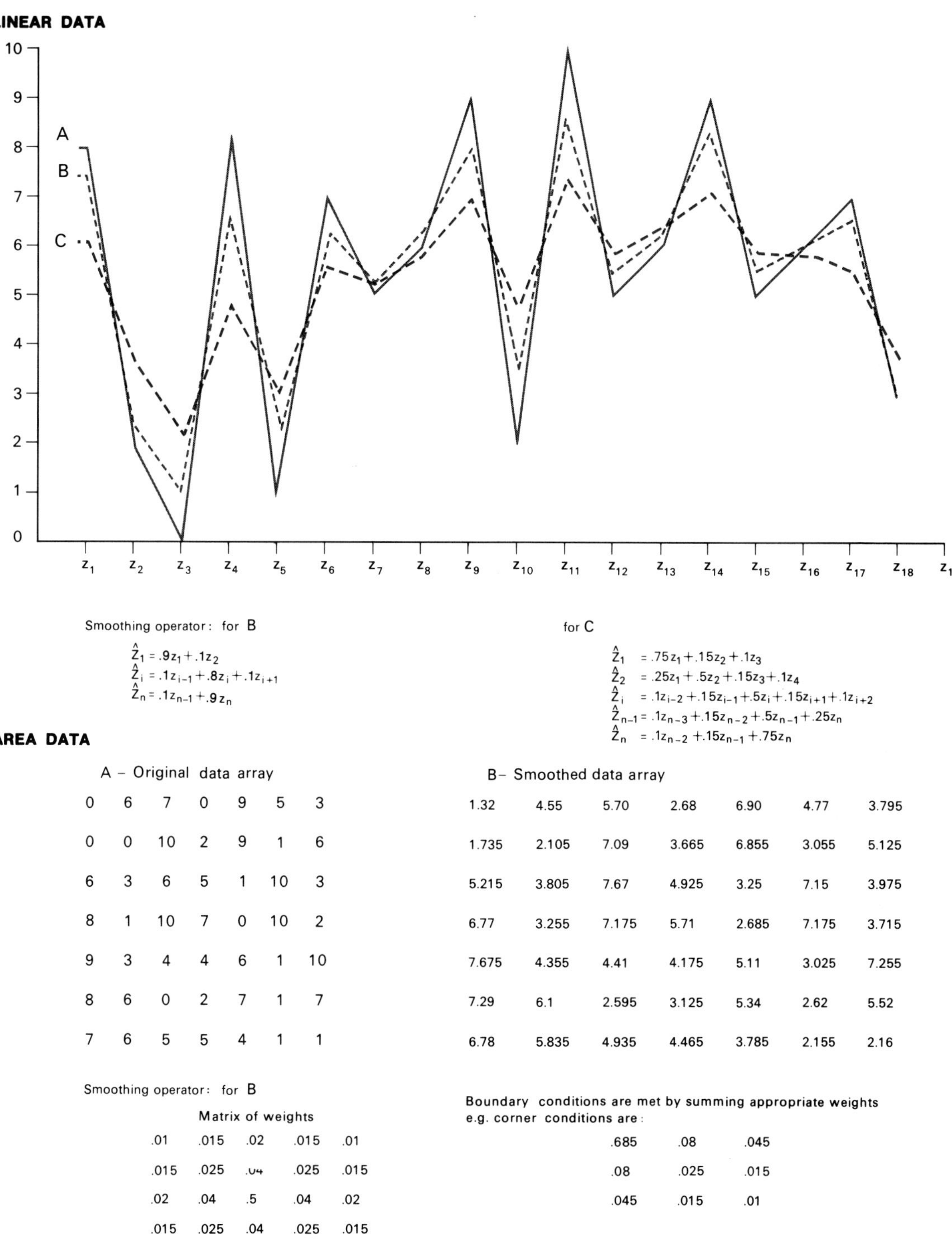

LINEAR DATA
10
9
8
7
6
5
4
3
2
1
0
A
B
C
z_1 z_2 z_3 z_4 z_5 z_6 z_7 z_8 z_9 z_{10} z_{11} z_{12} z_{13} z_{14} z_{15} z_{16} z_{17} z_{18} z_{19}
Smoothing operator: for B
$\hat{z}_1 = .9z_1 + .1z_2$
$\hat{z}_i = .1z_{i-1} + .8z_i + .1z_{i+1}$
$\hat{z}_n = .1z_{n-1} + .9z_n$
for C
$\hat{z}_1 = .75z_1 + .15z_2 + .1z_3$
$\hat{z}_2 = .25z_1 + .5z_2 + .15z_3 + .1z_4$
$\hat{z}_i = .1z_{i-2} + .15z_{i-1} + .5z_i + .15z_{i+1} + .1z_{i+2}$
$\hat{z}_{n-1} = .1z_{n-3} + .15z_{n-2} + .5z_{n-1} + .25z_n$
$\hat{z}_n = .1z_{n-2} + .15z_{n-1} + .75z_n$
AREA DATA
A – Original data array
0 6 7 0 9 5 3
0 0 10 2 9 1 6
6 3 6 5 1 10 3
8 1 10 7 0 10 2
9 3 4 4 6 1 10
8 6 0 2 7 1 7
7 6 5 5 4 1 1
B– Smoothed data array
1.32 4.55 5.70 2.68 6.90 4.77 3.795
1.735 2.105 7.09 3.665 6.855 3.055 5.125
5.215 3.805 7.67 4.925 3.25 7.15 3.975
6.77 3.255 7.175 5.71 2.685 7.175 3.715
7.675 4.355 4.41 4.175 5.11 3.025 7.255
7.29 6.1 2.595 3.125 5.34 2.62 5.52
6.78 5.835 4.935 4.465 3.785 2.155 2.16
Smoothing operator: for B
Matrix of weights
.01 .015 .02 .015 .01
.015 .025 .04 .025 .015
.02 .04 .5 .04 .02
.015 .025 .04 .025 .015
.01 .015 .02 .015 .01
Boundary conditions are met by summing appropriate weights
e.g. corner conditions are:
.685 .08 .045
.08 .025 .015
.045 .015 .01

FIGURE 11.13 Linear data: The results (line C) of using a five-term moving average with uneven weights to smooth linear data (line A). Area data: The smoothed data array (B) obtained by using a twenty-five-term, two-dimensional, unequally weighted moving average to smooth the original data (A).

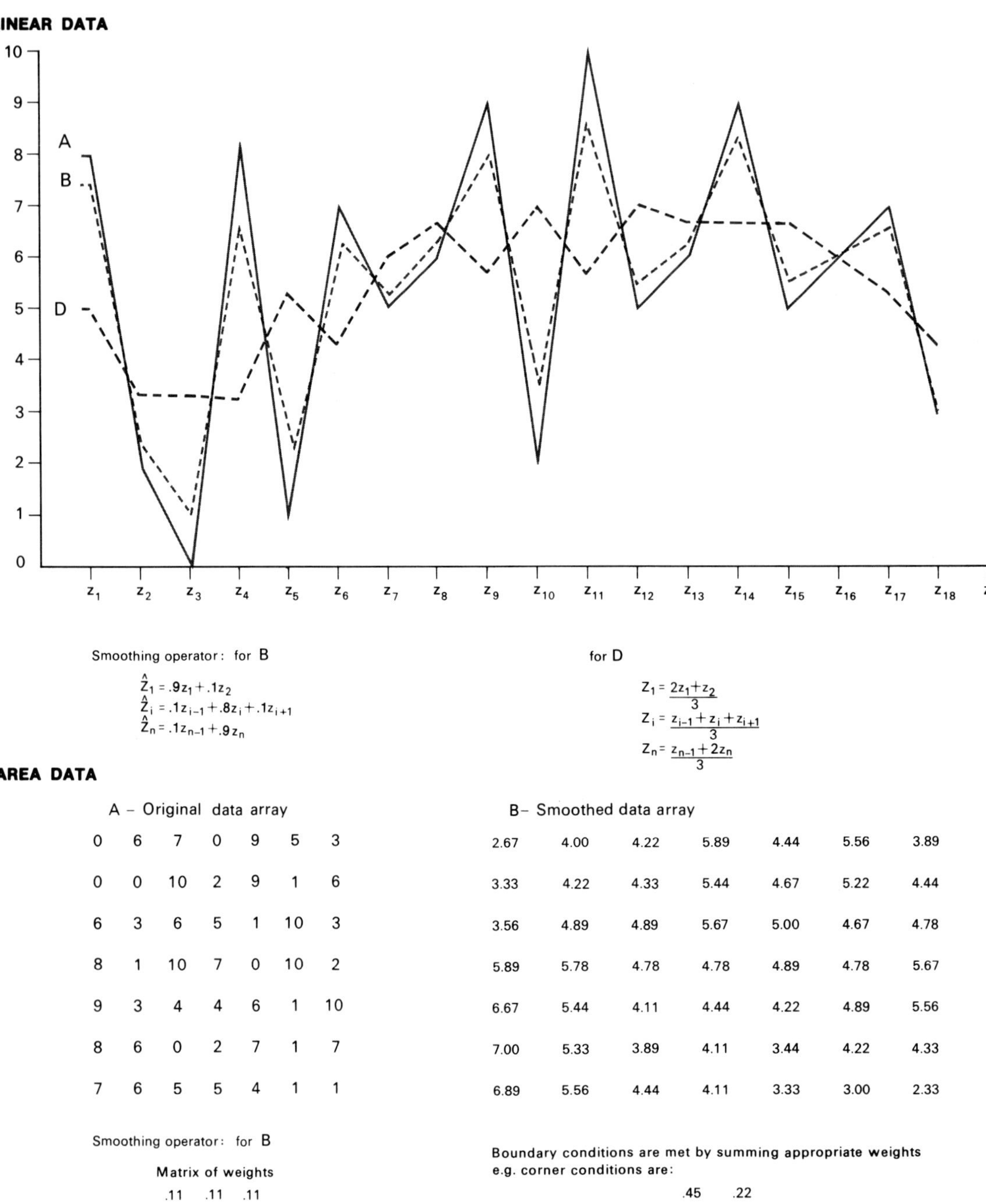

A – Original data array

0	6	7	0	9	5	3
0	0	10	2	9	1	6
6	3	6	5	1	10	3
8	1	10	7	0	10	2
9	3	4	4	6	1	10
8	6	0	2	7	1	7
7	6	5	5	4	1	1

B– Smoothed data array

2.67	4.00	4.22	5.89	4.44	5.56	3.89
3.33	4.22	4.33	5.44	4.67	5.22	4.44
3.56	4.89	4.89	5.67	5.00	4.67	4.78
5.89	5.78	4.78	4.78	4.89	4.78	5.67
6.67	5.44	4.11	4.44	4.22	4.89	5.56
7.00	5.33	3.89	4.11	3.44	4.22	4.33
6.89	5.56	4.44	4.11	3.33	3.00	2.33

Matrix of weights

.11	.11	.11
.11	.12	.11
.11	.11	.11

Corner conditions

.45	.22
.22	.11

FIGURE 11.14 Linear data: The results (line D) of using a three-term, equally weighted moving average to smooth linear data (line A). Area data: The smoothed data array (B) obtained by using a nine-term, two-dimensional, equally weighted moving average to smooth the original data (A). Equal weighting tends to increase the amount of simplification. Compare this figure to Figures 11.12 and 11.13.

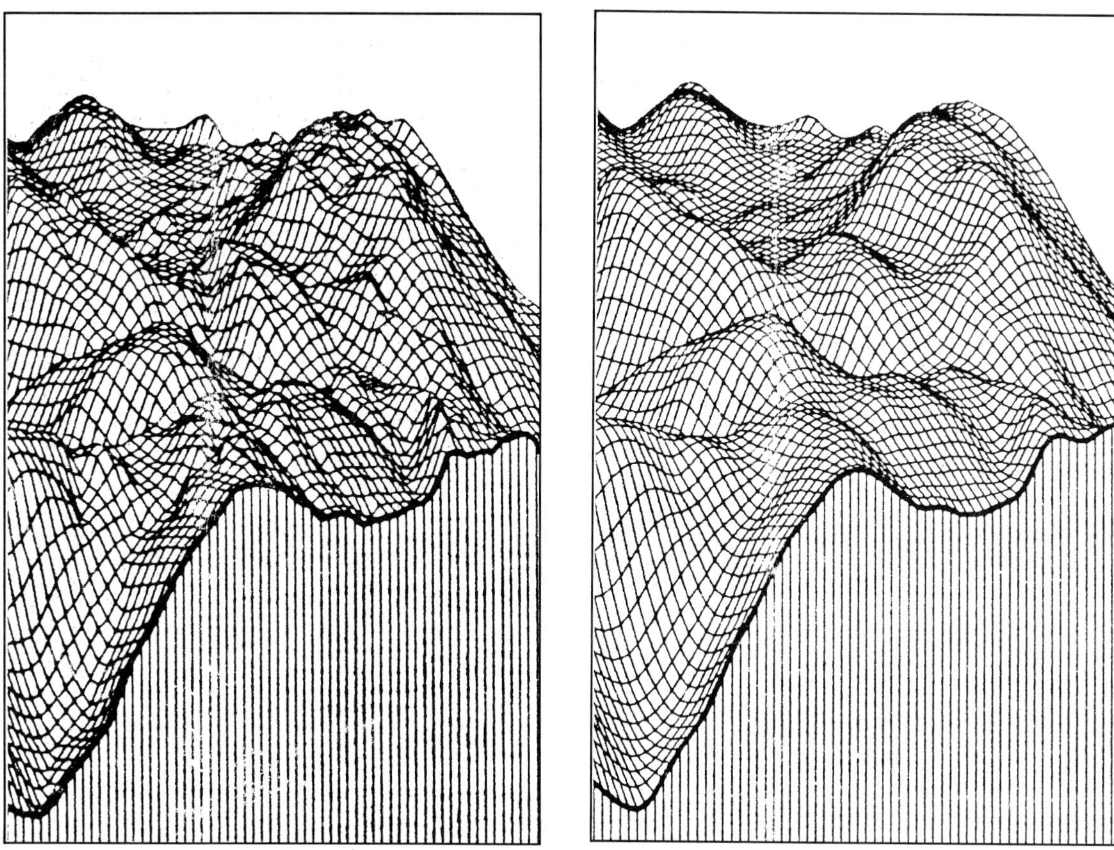

FIGURE 11.15 The right illustration is a plot of the original data matrix. The left illustration is a plot of the same data after the application of a two-dimensional smoothing operator.

amount of simplification that is accomplished. The cartographer is making the tacit assumption that the distribution is highly autocorrelated when simplification is performed using many neighboring points and even weights. The effect of such a selection is simply to smooth the surface greatly, and the cartographer should only make such a selection if that is the appropriate intent. Greatly smoothing a distribution with low autocorrelation results in a meaningless simplification.

Surface Fitting. Another group of simplification manipulations is *surface-fitting techniques.* In this category are several techniques that the cartographer can use to approximate lines and surfaces or to estimate values for logical contouring (leading to induction). Examples of the line-type fitting techniques are the regression line analyses mentioned at the end of this chapter. These constitute a kind of linear "surface fitting" in two dimensions only. For area/volume data the procedure can be likened to fitting a regression equation in three dimensions instead of in two dimensions.

The simplest surface-fitting technique is the "plane of best fit" (see Fig. 11.16). At least three noncollinear points (points not lying in a straight line) located in a three-dimensional space are necessary to fit such a plane, and preferably more than three points are used. One coordinate of each point, say the z value, is selected as the dependent variable, as in regression, and the other two coordinates are independent variables. An equation $z = ax + by + c$ is fitted to the points by the following matrix operation.*

$$\begin{bmatrix} N & \Sigma x & \Sigma y \\ \Sigma x & \Sigma x^2 & \Sigma xy \\ \Sigma y & \Sigma xy & \Sigma y^2 \end{bmatrix} \begin{bmatrix} c \\ a \\ b \end{bmatrix} = \begin{bmatrix} \Sigma z \\ \Sigma xz \\ \Sigma yz \end{bmatrix}$$

Most computers have software routines to compute the values for "c," "a," and "b." The resulting equation represents a simplification of the original data. The process can be easily extended to higher-order equations by using machine computation, and widespread use is being made of these simplification techniques in subsurface mapping. Figure 11.16 illustrates a "plane of best fit" as a simplification of a data distribution. Each fit gives a percentage explanation, in the regression sense, that provides an indication of the reliability of the fit.

A second group of similar techniques employs trigonometric series expansions to approximate a linear profile in two dimensions or a surface in three dimensions.* The mathematics are more involved, but the concepts are similar and the computer routines are readily available.

The cartographer uses surface-fitting techniques to simplify a data set. Often we are interested in showing trends through time or over areas. Simplification by surface-fitting techniques allows the cartographer to estimate these trends by computing equations for them. The equations in turn allow the interpolation of predicted values for points or areas for which no data are available. The calculations also allow an indication of the reliability of the trend. It is often more visually meaningful to map the trend than to map the complexities of the actual distribution. The intricacy of the map of the actual data may prove to be confusing to the map reader. One could almost conclude that mapping the results of surface-fitting can "enhance" the readability of a map. Surface-fitting techniques are not usually thought of as enhancement routines, however.

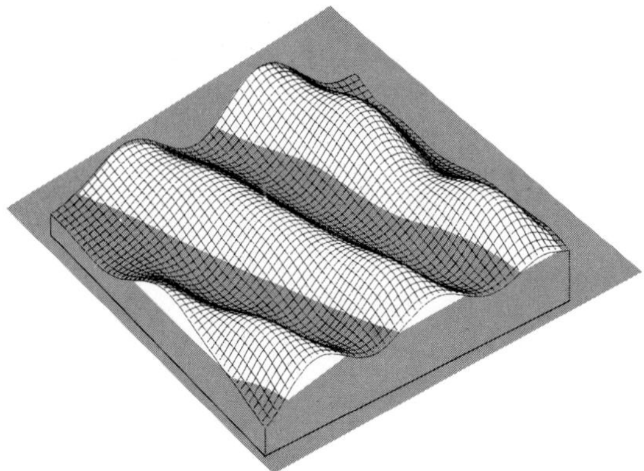

FIGURE 11.16 The shaded plane, with respect to the surface shown, is situated so as to minimize the sum of the squares of the deviations between points on the surface and corresponding points on the plane.

Enhancement Routines. Enhancement routines are applied to raster data, usually before image construction, and result in a modification of the individual elements (pixels) of data. Enhancement procedures using remotely sensed machine-readable data are similar to smoothing operators in that each individual data element is evaluated separately. The purpose is not necessarily smoothing, however; in some cases the purpose is to *increase* the difference (enhancement) between neighboring values, thus making the image resulting from the raster data easier to interpret. Regardless of the purpose, such a procedure is classed as one type of simplification, since it treats each data element individually, manipulating it with respect to neighboring values, and retaining it as an individual data element after modification (see Fig. 11.17).

Contrast Stretch. Raster data representing an image are often composed of pixel brightness

*For students unfamiliar with matrix manipulations, a good reference text is Frank Ayres, Jr., *Theory and Problems of Matrices*, Schaum's Outline Series. New York: Schaum Publishing Company, 1968.

*John W. Harbaugh and D. F. Merriam, *Computer Applications in Stratigraphic Analysis*. New York: John Wiley & Sons, Inc., 1968, 113–55.

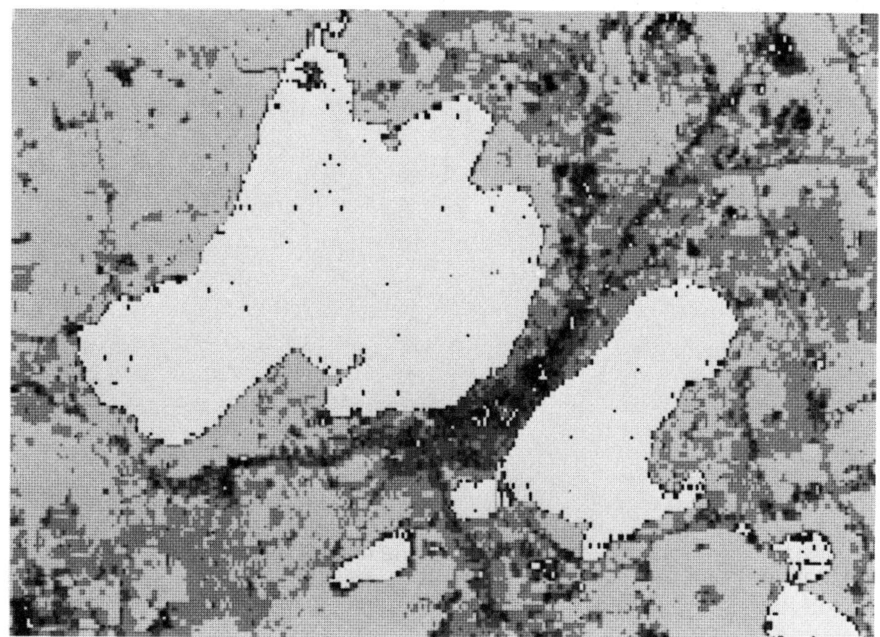

FIGURE 11.17 Map of Madison, Wisconsin, obtained by examining the changes in corresponding pixels in images taken in different bands of the electromagnetic spectrum. The procedure is one of simplification, since the individual data elements are separately evaluated and possibly modified. (Courtesy of S. Friedman, Earth Resources Applications: Image Processing Laboratory, Jet Propulsion Laboratory.)

values that occupy only a small portion of the possible brightness values. By the use of contrast stretch a cartographer can extend the range of pixel brightness values in an image over the entire range of possible values. This procedure increases the contrasts between pixels and aids in interpreting the image.

For example, Figure 11.18 illustrates a hypothetical histogram of raster brightness values obtained from an image. Note that the recorded pixels from the image have brightness values ranging from 60 to 158. The possible range of brightness values is 0 to 255. By using contrast stretch, the range 60 to 158 can be stretched to the range 0 to 255 as illustrated in Figure 11.19. Although Figure 11.19 illustrates a linear contrast stretch, other, nonlinear contrast stretches are also possible.

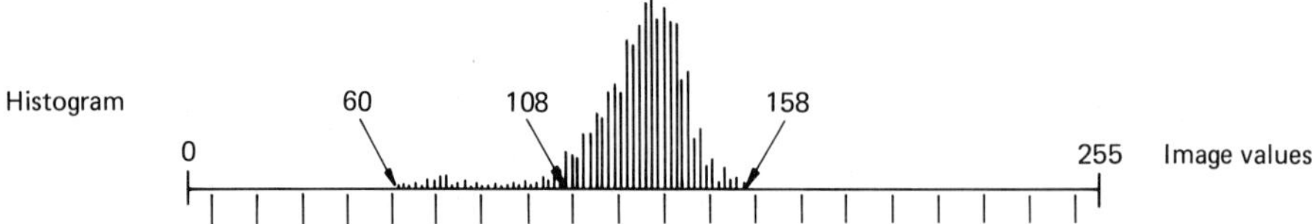

FIGURE 11.18 A hypothetical histogram of brightness values of individual pixels from an image. (From T. Lillesand and R. Kiefer, *Remote Sensing and Image Interpretation*, Copyright 1979, John Wiley and Sons, New York.)

Ratioing. Today it is quite common for a cartographer to have raster data representing images of a single scene captured in different parts of the electromagnetic spectrum (see Chapter 10). In such cases *ratioing,* another form of enhancement, is possible. A ratioed image is generated by calculating for each pixel of an image the ratio of the brightness values recorded for that pixel from two different bands of the electromagnetic spectrum. The ratioed brightness values may then be used to construct the image. These ratioed values are often contrast stretched, and several ratios may be combined to form a color image. Color Figure 16 illustrates the results of this procedure.

Although ratioing is an aid to interpretation, it does force the cartographer to make a number of decisions. When four or more images from different spectral bands have been taken, the number of possible ratios and combinations of ratios becomes staggering. Which ratios will best aid the interpreter? Statistical techniques like principal component analysis have been demonstrated to be helpful in forming the ratios that contain the most information without redundancies.

Other Enhancements. As we saw in Figure 11.11, smoothing operators can be applied to raster data. To enhance raster data, a smoothing operator consisting of a two-dimensional moving average can be applied to the brightness values of the individual pixels of an image. Two enhancement procedures are commonly used. First, if the data contain "noise," it may be useful to attempt to remove these minor fluctuations due to error from each of the pixels. To do so, an average of the adjacent pixel brightness values may be taken and used in place of the recorded pixel brightness value. This procedure is identical to the smoothing operator applied to Figure 11.15 and is called *low pass filtering.* It results in a statistically more reliable data set.

Alternatively a *high pass filter* is sometimes applied to increase or exaggerate differences that may be significant. The high pass filter is useful with noise-free data. It may be accomplished by calculating the average pixel brightness value in an area and then doubling the difference between an individual pixel and its neighborhood average value. This is sometimes called *edge enhancement,* since it tends to lighten the light values and darken the darker values in an area.

All of these techniques are simplification techniques. As you will note, many of them depend on computer assistance. We can only reiterate that before the availability and use of computers in cartography, simplification techniques were fewer and more subjectively applied.

FIGURE 11.19 A diagram showing a linear contrast stretch of the brightness values in the histogram in Figure 11.18. (From T. Lillesand and R. Kiefer, *Remote Sensing and Image Interpretation*, Copyright 1979, John Wiley and Sons, New York.)

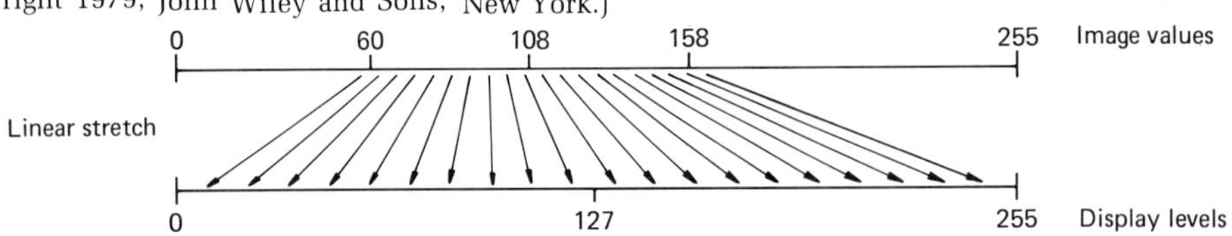

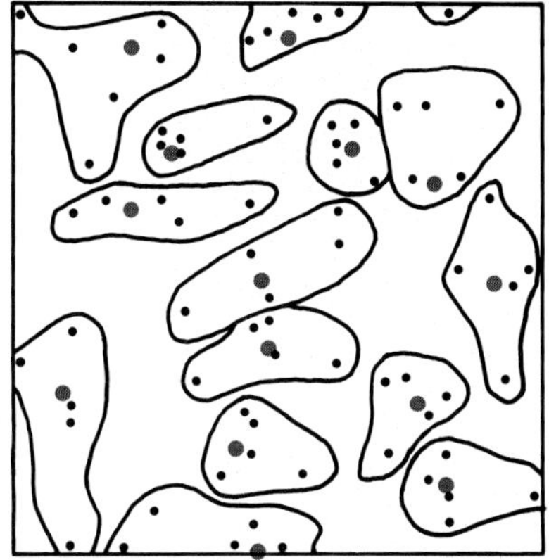

FIGURE 11.20 Classification of point patterns. After clustering the points, the cartographer selects a position and places a dot (shown in red) to "typify" the cluster. The "typical" position need not coincide with the position of any of the original data points.

Classification Manipulations

The primary purpose of classification is to assist the cartographer in processing data in order to convey the intended message efficiently to the map reader. This is identical to the purpose for simplification, but the methodology used to achieve classification is different.

Classification is defined as "the ordering or scaling and grouping of data." The major effort in classification is to "typify" the data set. In the process of typifying the data, none of the original data may actually be retained on the resulting map. Rather, a "typical" element replaces the true data element (see Fig. 11.20). Because of this emphasis on typifying, the most convenient way to discuss classification manipulations is by using the categories of place data, linear data, and areal or volumetric data.

Place Data Methods. Point clustering techniques comprise the manipulations of classification in this kind of data. They are used in two conditions: (1) where individual elements must be grouped and a typical location specified for the group, as in the construction of a dot map where, for example, one dot on the map represents ten houses in reality, and (2) where a complex pattern consisting of individual points must be typified and the "essence" of the point pattern retained on a map reduced in scale.

These manipulations remain rather ill defined and subject to the individual performing the manipulations. The agglomeration of points for simplification has already been illustrated in Figure 11.6. In classification (see Fig. 11.20) such agglomerations can also be done manually or with the aid of computers. In every case the cartographer must subjectively list some criteria that allow the agglomeration procedure to be applied. One such criterion is the specification of a starting point for the agglomeration; a second criterion is the direction of movement to be followed during the agglomerating procedure. Different results are obtained by starting at different points or by specifying different directions of movement.*

The selection of those criteria is slightly more critical in computer-assisted cartography than in manual cartography, since the cartographer retains more subjective control in the latter case. Although it has been satisfactorily demonstrated that the choice of these parameters significantly affects the outcome of the clustering algorithm, the relative advantages of one set of parameters over another is not known. As data become more readily available, research into clustering algorithms will be even more critical.

Even less is known about capturing the "essence" of a pattern of points. As of this time, we know of no computer algorithm that can generalize a pattern of points while retaining its "essence." Certain point patterns can be characterized by a statistical index referred to as the nearest neighbor statistic.* Computers can generate point patterns with the appropriate statistical indices, thereby characterizing or typifying a point distri-

*Barbara Gimla Shortridge, "Some Aspects of Error in Dot Mapping," unpublished M.A. thesis, University of Kansas-Lawrence, 1965.

*Joel L. Morrison, "A Link Between Cartographic Theory and Mapping Practice: The Nearest Neighbor Statistic," *The Geographical Review* 60 (1970): 494–510.

bution. We are unaware, however, of any cartographic use of these capabilities in attempts to generalize point patterns on maps.

Line Typification. The agglomeration of lines is rather uncommon. One example would be the combining of all air passengers traveling all airline routes used by all airlines between any pair of cities. The flow of passengers from Chicago to New York can be individually shown by airline X, airline Y, and airline Z, or these may be agglomerated into one flow line. Likewise, the distributary streams in a river delta may be agglomerated into a few lines to convey the essence of the distribution. Depending on the data, this classification manipulation of agglomeration may be performed either manually or with computer assistance.

These procedures are, as yet, not rigorously defined and require expert knowledge on the part of the cartographer of the distribution being portrayed. Eventually one could expect that geographers will develop a model for the outline of a fjorded coast or a meandering stream. When they do, these models will allow cartographers to "typify" meandering streams or fjorded coasts for generalizations on maps. Computer assistance will be required in this effort and will also enable ready use of the results.

In manual work cartographers have relied on the expertise of hydrographers or coastal geomorphologists to pass judgment on the coasts or river systems they portray on maps. The processes used have generally paid close attention to actual positions and have therefore been simplification processes. On occasion, especially where exaggeration is required, classification methods have been used. These attempts, however, are usually situation specific if not map specific.

Agglomeration of Areas. The agglomeration of areas is a very important manipulation in cartography. It can take place on nominally scaled data. Areas have certain characteristics: and Figure 11.21 illustrates an agglomeration of areas based on a specified set of criteria. The numbers

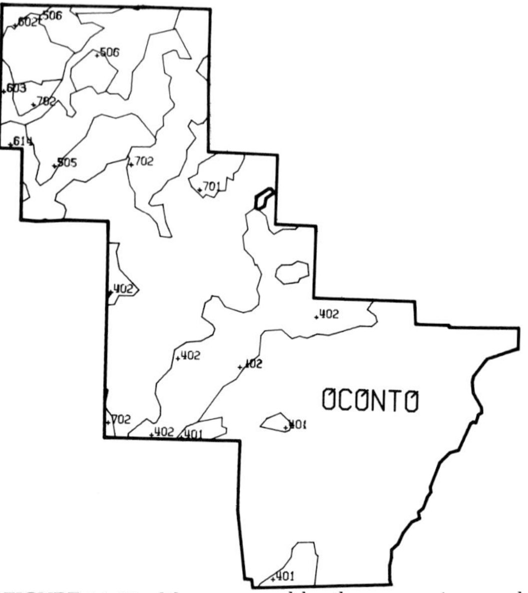

FIGURE 11.21 Map prepared by the successive overlay of nominal area regions. This computer-produced map of Oconto County, Wisconsin, consists of the intersection of the following digital data files: (A) county boundaries, (B) major water bodies, (C) soil regions with very severe erosion hazard, and (D) soil regions and soil associations. The numbers refer to soil region and soil association combinations occurring in regions with very severe erosion hazard. (Courtesy of P. Van Demark.)

associated with the areas designated in Figure 11.21 indicate regions where particular soils occur with various erosion hazard potentials. This computer-produced map uses the common manually applied overlay technique to form the regions. The agglomeration of nominally scaled areal data depends primarily on the size of the area unit to be agglomerated in relation to map scale. For example, there might be two areas of cropland separated by a tree-lined stream, as shown in Figure 11.22. Consider the area broken into fifteen small unit areas; at a comparatively large scale, one could map all areas, depending on the percentage of cropland in each small unit area (see Fig. 11.22A), whereas at a comparatively small scale, the area might be mapped as a single unit area consisting entirely of cropland (see Fig. 11.22B).

The problems of agglomerating ordinal, inter-

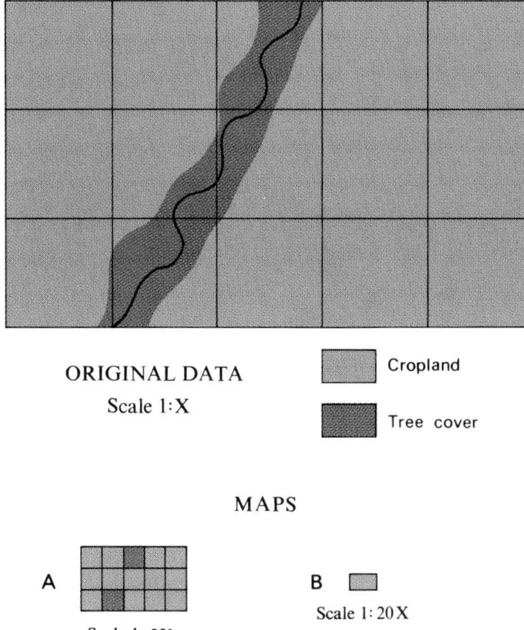

FIGURE 11.22 The *original data* are mapped at a scale of 1:X. (A) represents a smaller-scale agglomeration of the original data. (B) represents the further agglomeration of these units for an even smaller-scale representation.

val, or ratio data as opposed to nominal area data are similar in some respects. The agglomeration of areas using ordinal, interval, or ratio scaled data is a form of *dasymetric mapping* (see Chapter 14). The agglomeration of these volume data referring to an area is subject to more sophisticated classification manipulations. Foremost among these is class limit selection.

Class Intervals. The agglomeration of volumes at points or lines or over areas usually involves the grouping of like data elements into classes. This is appropriately called range grading. When a cartographer range grades values that refer to areas on an ordinal, interval, or ratio scale of measurement, the cartographer is mapping according to the choroplethic mapping method (see Chapter 14).

The problem of the selection of the optimal class intervals for a data distribution has been of concern to cartographers for many years, and computer assistance has not lessened the problem. When volumes existing as points or lines or over areas are mapped by range grading, the cartographer must select and apply a system of generalizing the array of data. Class intervals are the numerical categories of such a system, usually thought of as being bounded by class limits such as 0–2, 2–4, 4–8, and 8–16.

Prior to the specification of the class limits, the cartographer must decide on the number of classes for the map. The number of classes is also a generalizing process. Obviously, fewer classes means greater generalization, and minimal generalization usually implies more classes. However, much detail is not necessarily good, since excessive detail can easily draw the reader's attention away from more important aspects of the communication.

The controls on cartographic generalization effectively operate to limit the number of classes that can be shown on a map. For example, the graphic limits usually specify the maximum number of classes that can be used because the human eye has limited capabilities. The cartographer can show more classes if color is used than if the map is to be black and white, because the perceptual capabilities of the human eye can distinguish more colors than values.

On the other hand, the data quality control on generalization, along with the desired simplification, establishes a minimum number of classes that can be effectively used. If the data are of good quality, there can be more classes than if the data are of lesser quality.

A final factor to be considered is the significance of the various parts of the range of the data being portrayed. For example, we may be very concerned with the relative changes from place to place and would thus use a smaller interval in the lower portion of the range, since a change from 2 to 4 is the same relative change (100 percent) as a change from 50 to 100. On the other hand, we may be more interested in the absolute values of the higher part of the range; then we would make an opposite choice of intervals.

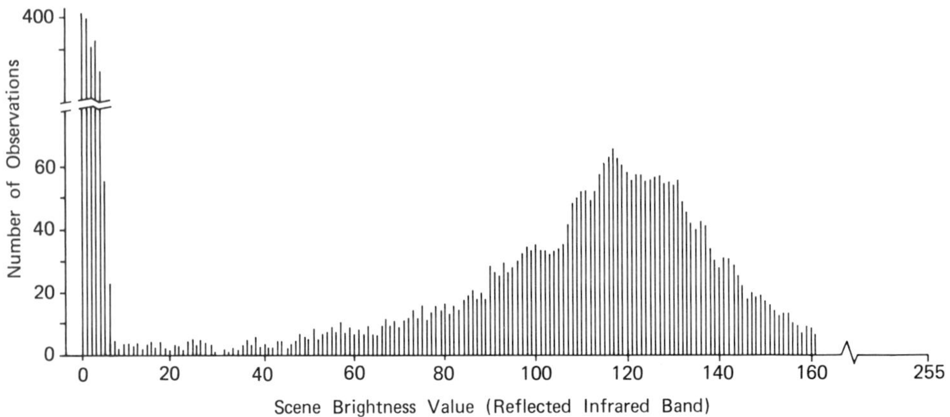

FIGURE 11.23 A histogram of the brightness values on a scanned image. (From T. Lillesand and R. Kiefer, *Remote Sensing and Image Interpretation*, Copyright 1979, John Wiley and Sons, New York.)

Since the controls of generalization place stringent limits on the number of classes a cartographer may use, the determination of the number of classes is not as much a problem for the cartographer as the determination of the class limits. The detailed treatment of class limit determination is given in Chapter 14.

Density Slicing. A procedure called *density slicing,* used in conjunction with the raster data representation of images, is an agglomeration manipulation. It is analogous to the specification of class limits and the determination of number of classes just discussed. The brightness values shown in the histogram in Figure 11.23 were obtained by scanning a film image with a scanning microdensitometer using a red filter. The values were classed into nine classes, and each class was symbolized (see Fig. 11.24). The number of classes and the class limits are the critical generalization decisions in this procedure and are identical in importance to the same decisions made in choroplethic or dasymetric mapping.

The classification manipulations that a cartographer may use are only beginning to take advantage of computer assistance. The potential for development in this area is great, and the impact of these developments on the future of cartography may be profound.

COMMON STATISTICAL CONCEPTS AND MEASURES

Since the introduction of computers into cartographic technology the sophistication of the statistical techniques that a cartographer is called upon to use in processing data prior to mapping has increased substantially. The cartographer must understand these analytical techniques and be able to use them wisely in a cartographic sense.

The concepts of regression and correlation form the basis for many of the procedures already discussed. Regression can serve to aid in point or feature elimination algorithms and as the basis for surface-fitting routines in cartographic simplification. The correlation coefficient is often used as an indicator of the adequacy of a surface fit or of a class interval scheme. Correlations are also used in factor analytical schemes which result in classification of data.

The extensions of linear regression to multiple regression and of correlation to autocorrelation have important consequences in cartography. Surface fitting is really a multiple regression technique, and the autocorrelation of a distribution is an important parameter for the cartographer to know in performing interpolation.

In the remainder of this chapter we review two very commonly used statistical techniques: linear

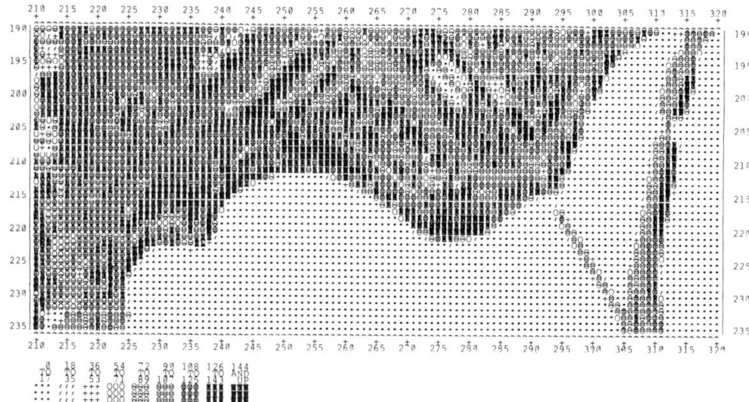

FIGURE 11.24 A plot of the density sliced image using the data in Figure 11.23 and grouping the brightness values into nine classes. (From T. Lillesand and R. Kiefer, *Remote Sensing and Image Interpretation*, Copyright 1979, John Wiley and Sons, New York.)

regression and correlation. These techniques are rudimentary and form the basis for several more advanced statistical data-processing techniques. What follows is a condensed introduction and reference, not a substitute for a book on statistics.

Regression Analysis

Regression can be used to provide solutions to numerous prediction problems and to summarize relationships between quantitative variables. A retail store chain may wish to predict sales volume for a given store based on its location, or a farmer may wish to predict harvest yields based on climatic factors. In its simplest form, regression analysis includes at least one independent variable or distribution (such as a measure of climate) and one dependent variable or distribution (in this case, yield). A simple model can then be established of the type

$$y = a + bx$$

where x is the independent variable, y is the dependent variable, and a and b are constants. One may readily recognize this as the equation for a straight line in a plane coordinate system, in which a is the intercept of the line with the y axis and b is the slope of the line. Figures 11.25 and 11.26 each illustrate the calculated linear regression equation for the observations of one independent variable x (per capita personal income) and one dependent variable y. In Figure 11.25 y represents per capita educational expenditures, while in Figure 11.26 y represents number of first-degree graduates. The areal units are shown in Figure 11.27.

The solution of a linear regression equation is

FIGURE 11.25 A scatter diagram with fitted linear regression line. Letters refer to areal units designated in Figure 11.27. $\hat{Y}$ refers to the predicted values for the dependent variable, and r refers to the coefficient of correlation. The data values are given in Table 11.1.

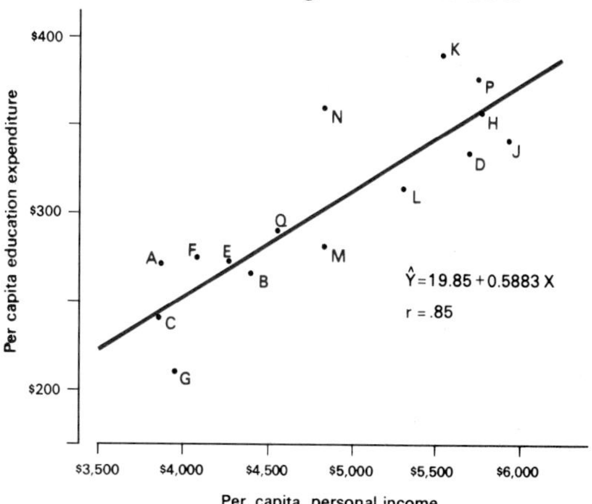

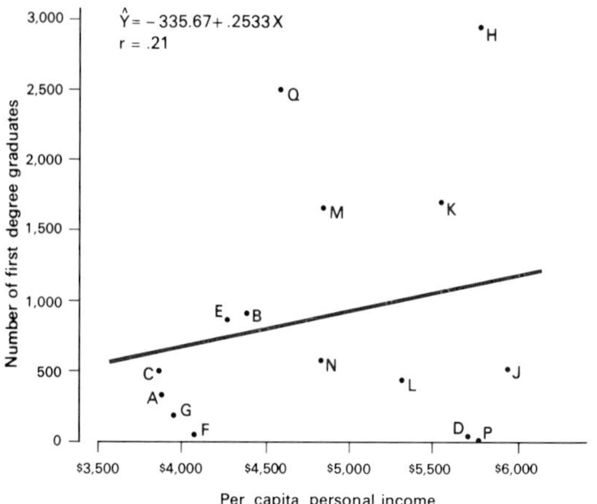

FIGURE 11.26 A scatter diagram with fitted linear regression line. Letters refer to areal units designated in Figure 11.27. $\hat{Y}$ refers to the predicted values for the dependent variable, and r refers to the coefficient of correlation. The data values are given in Table 11.1.

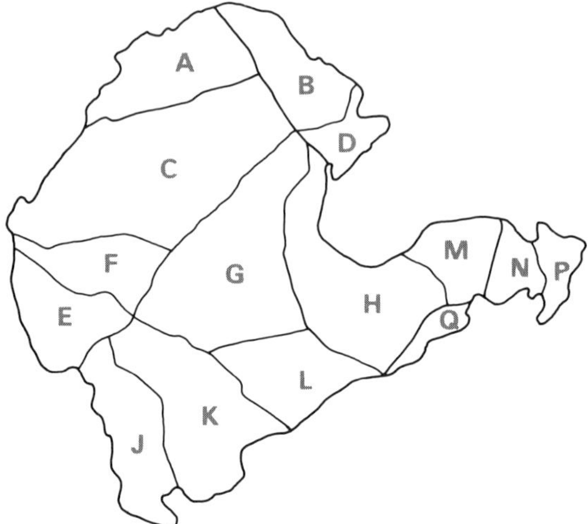

FIGURE 11.27 A map of a fictitious island showing the areal units to which the data in Table 11.1 refer.

usually based on the least squares criterion; this means that the regression line is so situated with respect to the graphed observations that the sum of the squares of the differences between observed y values and those predicted is minimized. (The observed y values are represented in Figs. 11.25 and 11.26 as points; the predicted values are read on the y axis for any given x value by moving vertically from the selected x value to the plotted regression line and then horizontally to the y axis to obtain the predicted value.) When this least squares criterion is employed, the correct values for the constants a and b are given by the following formula:

$$b = \frac{\sum_{i=1}^{n} (x_i - \overline{x})(y_i - \overline{y})}{\sum_{i=1}^{n} (x_i - \overline{x})^2}$$

$$a = \overline{y} - b\overline{x}$$

where

$$\overline{y} = \frac{\sum_{i=1}^{n} y_i}{N} \quad \text{and} \quad \overline{x} = \frac{\sum_{i=1}^{n} x_i}{N}$$

Table 11.1 and Figures. 11.25 and 11.26 are given as illustrations of this technique.

Maps are often created of either the observed values, the predicted values, or the residual values (the difference between the observed and the predicted), as shown in Figures 11.28 and 11.29. To prepare such maps, one should be aware of the assumption behind this type of analysis, and the selection of appropriate class intervals for mapping is critical (see Chapter 14).

From a geographic point of view the computed relationship of x and y does not take spatial position into account unless x or y are position-related variables (that is, variables whose measurement inherently contains some element of spatial position, such as latitude or longitude). In practice, most cases of x and y pairs of observations are not locational in that sense but are simply attached to places. For example, a regression analysis using hay production as an independent variable to predict numbers of cattle could be computed for a state using counties as the data collection units. In this case a pair of values of the dependent (number of cattle) and the independent (hay production) variables are associated with each county (location); however, location itself is not involved in the computations.

TABLE 11.1 Data for a Given Year for a Fictitious Island Having Fifteen Civil Districts (data used for Figs. 11.25 to 11.29)

Area	Per Capita Personal Income	Per Capita Educational Expenditure	Number of First-Degree Graduates
A	$3882	$273	330
B	4395	266	910
C	3870	240	500
D	5695	333	40
E	4282	273	870
F	4082	276	70
G	3952	210	240
H	5770	357	2920
J	5938	340	530
K	5550	390	1760
L	5304	314	460
M	4840	280	1670
N	4830	360	580
P	5745	376	0
Q	4570	287	2500

Another assumption underlying regression analysis is that the data are normally distributed. Only on the basis of this assumption can inferences about the strength of the relationship be made for the variables in general as opposed to the specific sample (set of points) used to derive the *a* and *b* values of the equation. The reliability of any prediction made on the basis of the data can then be assessed. The cartographer must be vitally interested in this aspect, since relative reliabilities should generally be indicated when mapping predicted values or residuals. For example, compare the maps in Figures 11.28 and 11.29. The predicted maps in each figure alone indicate nothing about their reliability; however, compare the corresponding scatter diagrams in Figures 11.25 and 11.26. The quality of the prediction mapped in Figure 11.28, as revealed by Figure 11.25 is much greater than the predicted map in Figure 11.29. One way to indicate reliability is to map residuals as well as predicted values as shown in Figures 11.28 and 11.29. Only the residual map tells us whether the predicted map is of much or little value.

We have discussed only the simplest case of regression analysis. Cartographers must often map the results of multiple linear regression (that is, regressions that include more than one independent variable) or curvilinear regressions (regressions that involve equations of curved instead of straight lines). Occasionally ordinal or interval information is incorporated into regression analyses. Although the computations become more involved or modified, the mapping problem and requirement remain relatively the same: map the predicted, observed, and residual values, and indicate their reliability.

Correlation Analysis

One important indicator of the strength of the linear relationship between two sets of data, y and x, is the correlation, r, between y and x. The value of r can vary between -1 and $+1$, where $r = 1$ indicates that an increase in x is associated with a corresponding increase in y, $r = -1$ indicates that an increase in x is associated with a corresponding decrease in y, and $r = 0$ indicates the absence of a predictable relationship, that is knowledge of the x value gives no predic-

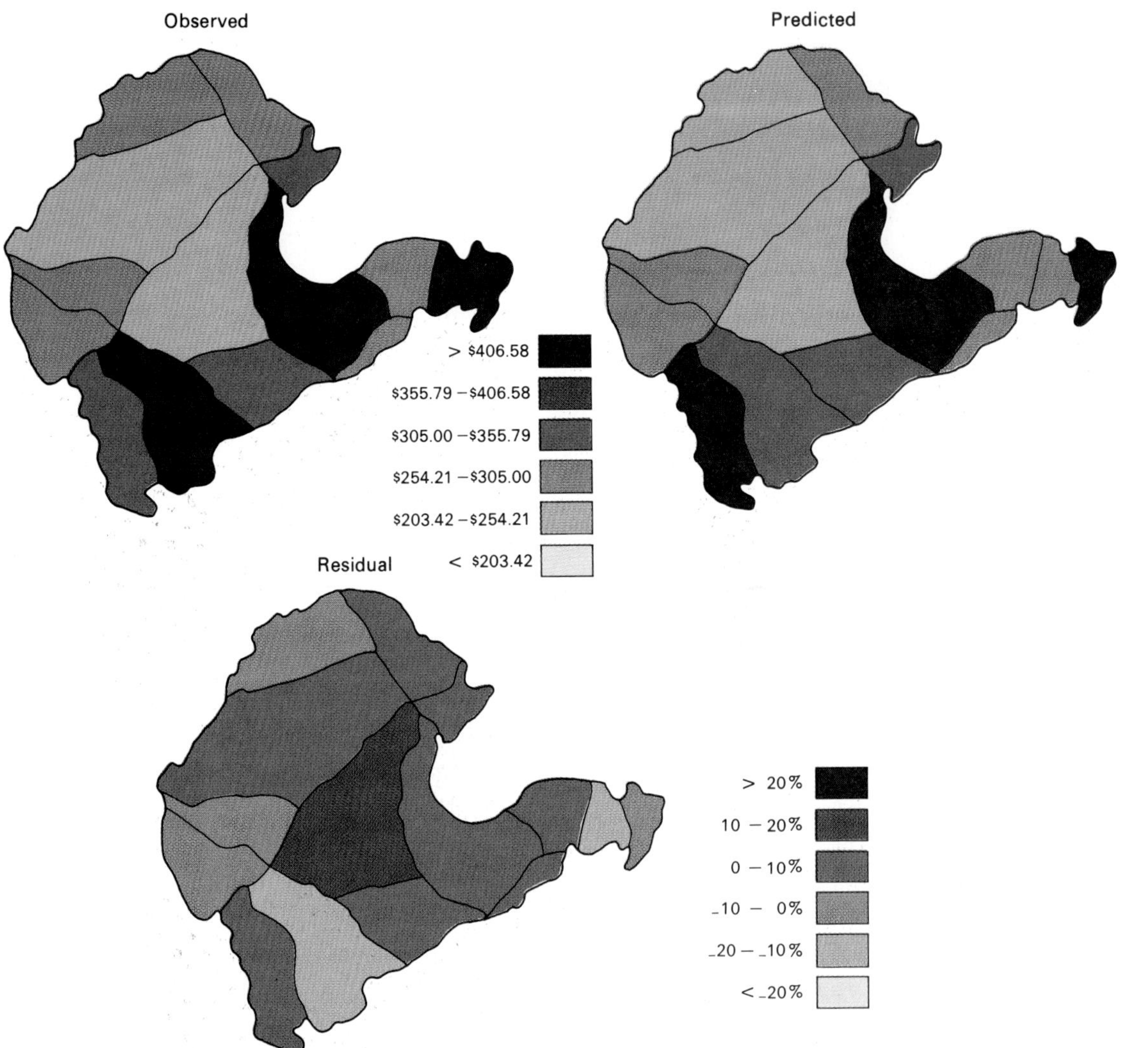

FIGURE 11.28 Maps of the fictitious island showing *observed* per capita educational expenditure, *predicted* per capita educational expenditure based on per capita income, and the *residuals* from the regression shown in Figure 11.25.

tive information about y. The formula for the coefficient of correlation is:

$$r = \frac{\sum_{i=1}^{n} (x_i - \bar{x})(y_i - \bar{y})}{\sqrt{\sum_{i=1}^{n} (x_i - \bar{x})^2 \cdot \sum_{i=1}^{n} (y_i - \bar{y})^2}}$$

The value r itself is a summary measure relating to an entire set of paired observations; therefore, the cartographer cannot map an $r = 0.5$, for example. It is possible that a separate value of r might be calculated for sets of data (for example, number of calls to the automobile club for aid in starting the car, and temperature) for a given location. If similar values for r were computed at a series of locations, this set of r values could be mapped.

Perhaps of more importance to the cartogra-

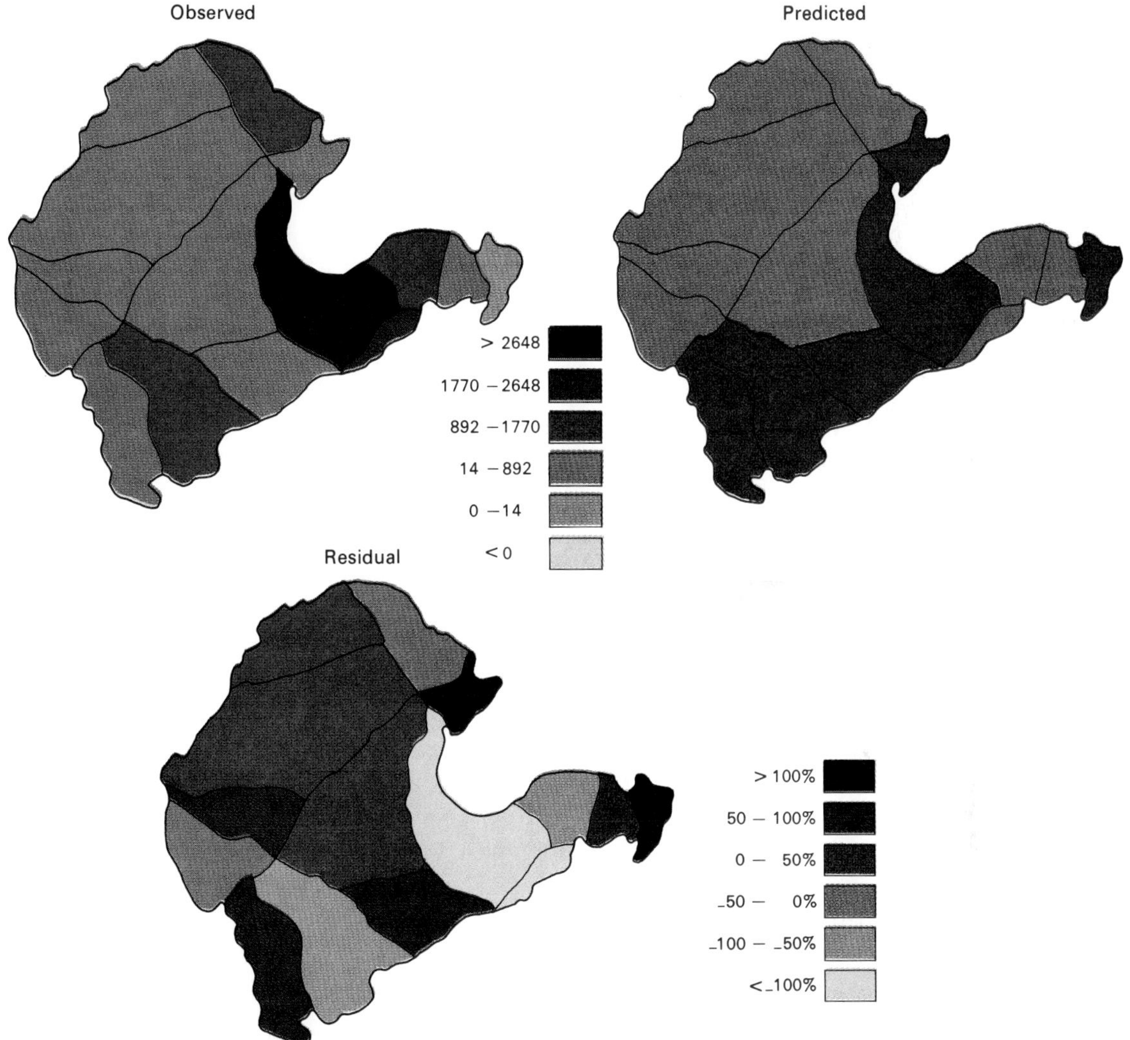

FIGURE 11.29 Maps of fictitious island showing *observed* numbers of first-degree graduates, *predicted* numbers of first-degree graduates based on per capita income, and the *residuals* from the regression shown in Figure 11.26.

pher is the concept of correlation when it is applied to neighboring values in one data set. The appropriate name is *autocorrelation,* and a series of coefficients of autocorrelation can be computed for a single spatially distributed set of data values.

The position of the data in the spatial array is used to determine which observations are paired for calculation of the autocorrelation coefficient (see Fig. 11.30). The formula for calculation can be identical to the above formula for correlation. Each point value is paired with another point value a certain disance (lag) and direction from it (see Fig. 11.31). The number of possible lags is numerous, and a value for *r* can be calculated for each lag. If the lags are systematically scaled in distance and direction, a plot of *r* values against lags can be obtained (Fig. 11.32). This is the autocorrelation function for the distribution being analyzed. The importance of this concept for car-

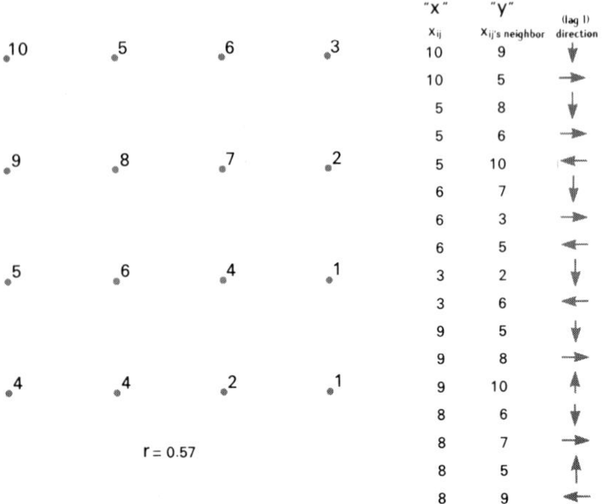

FIGURE 11.30 The calculation of an autocorrelation coefficient. The value for r, the lag one autocorrelation coefficient for the data array at the left, is obtained by defining x and y as shown in the right-hand column above. The equation for r is given in the text.

tographers will become more apparent in the following chapters, when cartographic generalization and portrayal of surface volumes are discussed.

FIGURE 11.31 For the matrix of data values indicated by points above, various lags can be defined as illustrated.

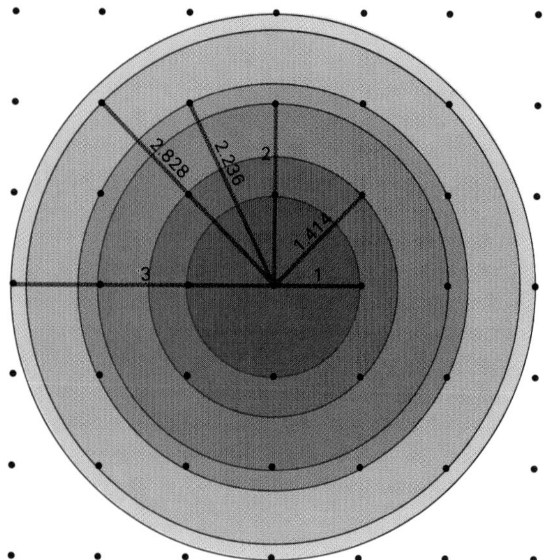

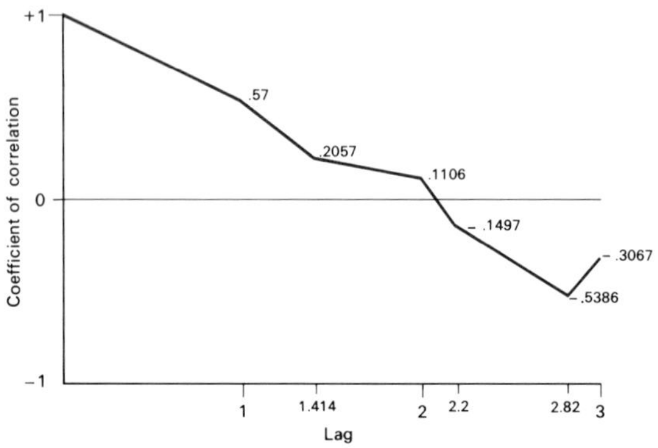

FIGURE 11.32 The plot of a graph, approximating the autocorrelation function for the data array shown in Figure 11.30, using six different autocorrelation coefficients calculated for increasing lags.

Finally the coefficient of correlation described and defined above is perhaps the most common correlation measure for interval or ratio data. Again, however, there are other measures of correlation more appropriate for nominal or ordinal scale data and for some instances of interval or ratio data. The addition to a map, showing the results for a regression analysis of the value of the coefficient of correlation, is one important way that a cartographer can indicate the reliability of the predicted results being portrayed.

Selected References

Baures, P. J., and J. Duveroy, "Statistical and Spatial Filtering: Application to Aerial Photographs," *Applied Optics* 17 (1978): 3395–3401.

Bernstein, R. (Ed.), *Digital Image Processing for Remote Sensing*. New York: IEEE Press, 1978.

Blalock, H. M., *Social Statistics*. 2nd ed., New York: McGraw-Hill, 1972.

Castleman, K. R., *Digital Image Processing*. Englewood Cliffs, NJ: Prentice-Hall, 1979.

Claire, R. W., and S. C. Guptill, "Spatial Operators for Selected Data Structures," *Proceedings of Auto-Carto V: Environmental Assessment and Resource Management*, Falls Church, VA: American Society of Photogrammetry and American Congress on Surveying and Mapping (1982): 189–200.

Davis, J. C., and M. J. McCullagh (Eds.), *Display and Analysis of Spatial Data*. NATO Advanced Study Institute. New York: John Wiley & Sons, 1975.

Douglas, D. H., and T. K. Peucker, "Algorithms for the Reduction of the Number of Points Required to Represent a Digitized Line or Its Caricature," *The Canadian Cartographer* 10 (1973): 112–22.

Draper, N. R., and H. Smith, *Applied Regression Analysis*. 2nd ed., New York: John Wiley & Sons, 1975.

Dutton, G. H., "Fractal Enhancement of Cartographic Line Detail," *The American Cartographer* 8 (1981): 23–40.

Freeman, H. and G. G. Pieroni (Eds.), *Map Data Processing*. New York: Academic Press, 1980.

Harnett, P. R., G. D. Mountain, and M. E. Barnett, "Spatial Filtering Applied to Remote Sensing Imagery," *Optica Acta* 25 (1978): 801–9.

Jenks, G. F., "Lines, Computers and Human Frailties," *Annals of the Association of American Geographers* 71 (1981): 1–10.

Krumbein, W. C., and F. A. Graybill, *Introduction to Statistical Models in Geology*. New York: McGraw-Hill, 1965.

Mandelbrot, B. B., *Fractals: Form, Chance and Dimension*. San Francisco: W. H. Freeman, 1977.

Marble, D. F. (Ed.), *Computer Software for Spatial Data Handling,* Volumes I, II, III. International Geographical Union Commission on Geographical Data Sensing and Processing and the U. S. Geological Survey, 1980.

Marino, J. S., "Identification of Characteristic Points Along Naturally Occurring Lines: An Empirical Study," *The Canadian Cartographer* 16 (1979): 70–80.

Peuquet, D. J., "An Examination of Techniques for Reformatting Digital Cartographic Data, Part 1: The Raster-to-Vector Process," *Cartographica* 18 (1981): 34–48.

Peuquet D. J., "An Examination of Techniques for Reformatting Digital Cartographic Data, Part 2: The Vector-to-Raster Process," *Cartographica* 18 (1981): 21–33.

Salichtchev, K. A., "History and Contemporary Development of Cartographic Generalization," *International Yearbook of Cartography* 16 (1976): 158–72.

Taylor, P. J., *Quantitative Methods in Geography: An Introduction to Spatial Analysis*. Boston: Houghton Mifflin, 1977.

12

Cartographic Generalization: Symbolization

SYMBOLIZATION

This chapter and Chapters 13 and 14 deal with the fundamentals of the symbolization of the vast array of data that can be communicated cartographically. A large number of maps, mostly thematic, employ graphic symbols in a surprising variety of ways to show the general or detailed characteristics of a single aspect of a class of geographical phenomena. In contrast, general maps portray location of a variety of past or present geographical data all at once.

Maps showing distributions in numerical form are one of the cartographer's stocks-in-trade. They are capable of surprising variety and can be used to present almost any kind of data. Few maps can be made that do not in some way present quantitative information, even if the amounts involved result from so simple an operation as different symbols for cities of different size for an atlas map. Column after column of numbers in tabular form frequently look forbidding, and the map usually can present the important geographical characteristics of the material in a far more understandable, interesting, and efficient manner. Tabular quantitative materials of various kinds, from governmental censuses, to the reports of industrial concerns, to the results of our own surveys, exist in staggering variety. With such a wealth of material, the cartographer would expect to find that a large percentage of effort is devoted to preparing this type of map. Many of the distributions mapped are derived parameters, that is, generalizations of numerical arrays of the sort suggested in Chapter 6, and the cartographer must proceed warily lest he or she stumble intellectually. The student must appreciate that figures *can* easily lie cartographically if they are improperly presented to the map reader. As one author has observed, "One of the trickiest ways to misrepresent statistical data is by means of a map."*

Hasty evaluation of data, or the selection of data to support conclusions unwarranted in the first place, is thoroughly unscholarly and results in intellectually questionable maps. Their production is reprehensible, since they may be, and unfortunately sometimes are, used by their authors or others to support the very conclusions from which they were drawn in the first place. The exhortation, frequently implied or stated in this book, "to prepare the map so that it communicates what is intended," is not contradictory to the preceding; it is merely a matter of integrity.

Another source of dangerous error in cartography is the map that provides an impression of greater precision than is justified by the quality and quantity of the data used in its preparation. To keep from making this serious mistake, the cartographer is sometimes forced to invent special symbols such as dashed lines or patterns of question marks; such devices may detract from the aesthetic quality of the map, but they will serve the more important purpose of preventing a map reader unfamiliar with the subject matter from falling into the common trap of "believing everything he or she sees."

Maps employ a great variety of symbolism and to try to classify precisely all the variations in the geographical concepts that can be symbolized is all but impossible; the multiplicity of categories would reach astronomic numbers. Consequently, we cannot illustrate in a practical manner the variety of ways a given symbol may be used. Instead, here and in Chapters 13 and 14 we will illustrate the means by which the various fundamental classes of geographical data (place, linear, areal, and volumetric) can be and usually are symbolized.

Since there is no limit to the variety of phenomena that a cartographer might be called on to map, and each phenomenon exists and can be measured on some measurement scale, the cartographer must decide which scale of measurement to use.

The cartographer can generalize ratio data to interval, ordinal, or nominal scales because these scales of measurement are nested. Similarly, volumetric data can occur at points, along lines, or over areas. However, a cartographer sometimes maps place, linear, or areal data by something other than point, line, or area symbols, respectively. Two examples illustrate this. A road

*Darrell Huff, *How to Lie with Statistics*. New York: W. W. Norton and Company, 1954.

is usually thought of as a linear phenomenon, but at a very large scale and for some purposes, the fact that its surface character is different from its surroundings may be the important matter as, for example, in the study of variations in albedo. Also, a city is frequently thought of as a geographical point phenomenon, but a given array of cities in a region can be considered as distinguishing that area from another with a different frequency of city incidence. The fundamental point is that we must first determine the attribute that is to be given the emphasis before the symbolization system can be chosen. When approached from this point of view, the cartographic representation of information is not difficult.

Symbolization as an Element of Generalization

The two aspects scale of measurement (nominal, ordinal, interval, and ratio) and type of data (place, linear, areal, and volumetric) are of primary importance to the cartographer as he or she approaches symbolization. All data sets exist in one combination of these two aspects. Generalization by symbolization is a determination of how these two aspects of a given data set will be portrayed on a map.

Cartographic generalization by symbolization is, therefore, of two levels: (1) a change in scale of measurement from the original data set, or (2) a change in the data type. Fortunately there are some rules that aid in symbolization. For the well-trained cartographer it is also fortunate that there are many choices in the process, but for the untrained, such a large selection is also the source of untold problems.

The specific discussion of symbolization in this text is given in this chapter and in Chapters 13 and 14. After an in-depth discussion of the symbolization problem and an introduction to the statistical surface, the use of point symbols for mapping data sets will be given in this chapter. Chapter 13 will treat mapping with line symbols, and Chapter 14 will treat mapping with area symbols.

Scale of Measurement Aspect. All data sets can be measured on one of the scales of measurement defined in Chapter 6. For cartographic purposes it is important to know the scale of measurement of a given data set, for the choice of simplification and classification procedures depends on it, and that choice provides the input for the start of the symbolization process. As we have seen, the four scales of measurement are nested: nominal < ordinal < interval < ratio. We can therefore generalize the scale of measurement from ratio to interval, ordinal, or nominal; from interval to ordinal or nominal; from ordinal to nominal; but that is the limit of permissible changes. We can not change from nominal to ratio, for example, and we must map nominal data as nominal data. There is no choice.

In terms of the cartographic portrayal of the scale-of-measurement aspect of a data set most symbols by themselves only connote nominal or ordinal data. The symbolization can be enhanced to portray interval or ratio scales by the use of text or numbers. It makes more sense to define the portrayal scales for cartographic purposes as nominal, ordinal, range-graded, and ratio. (In a range-graded portrayal the relative size of each individual quantity is not shown; rather the observations are classed and the range of each class is shown by a symbol of a constant size.) While in measurement theory the distinction between interval and ratio data is important, it makes no difference to the cartographic symbolization

FIGURE 12.1 Examples of possible legends on (A) ordinal, (B) range-graded, and (C) ratio scales of portrayal for a map indicating population of cities. In legend (C) only representative symbol sizes are portrayed. In legends (A) and (B) only the symbol sizes shown in the legend appear on the map.

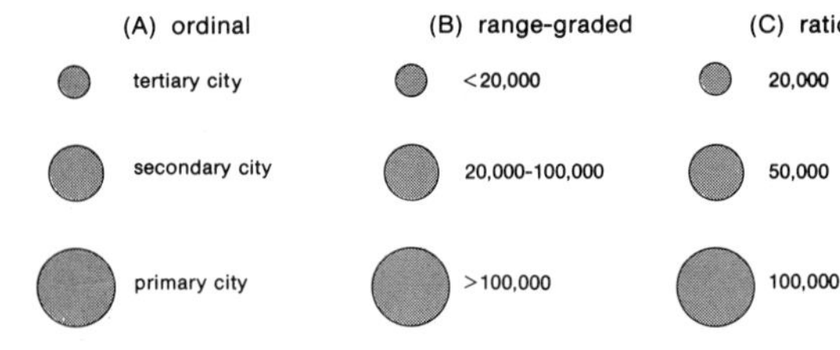

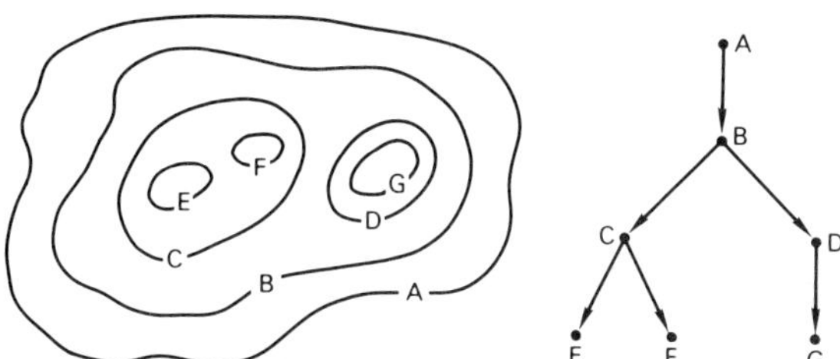

FIGURE 12.2 An example of the generalization of a contour plot to a network. (From *Computer-Assisted Cartography: Principles and Prospects* by M.S. Monmonier, Copyright 1982, Prentice-Hall, Englewood Cliffs, N.J. Reprinted by permission of Prentice-Hall, Inc.)

of data for mapping purposes. On the other hand, a distinction between ordinal, range-graded, or ratio portrayal does make sense cartographically, and we will use these terms in the following pages when referring to the scale of measurement for the portrayal of data (see Fig. 12.1). We will continue to use the term interval to refer to the existence or collection of the data prior to their mapping.

Data-Type Aspect. Data sets also have a data-type aspect of existence: they are place, linear, areal, or volumetric. On a two-dimensional map* the three dimensions of a volume can not be shown except by generalization by symbolization. Place data may be shown by point symbols, linear data may be shown by line symbols or point symbols, areal data may be shown by point, line, or area symbols, and volumetric data can also be shown by point, line, or area symbols. All can be portrayed by alphanumeric characters as well. Obviously, it would be nice if there were unique symbols for each class of occurrence. Since there aren't, the problem is more interesting and the cartographer has the choice of portraying a given aspect of a data set as it exists or by generalizing it.

Place, linear, and areal data are rather straightforward. Volumetric data are more complex in that volumes can occur at points (dollars in a bank), along lines (barrels of oil imported from Bahrain), or over areas (precipitation). Whenever a volume occurs over an area, a *statistical surface* exists (see discussion below). The cartographer has limited choices in generalizing by symbolization in this arena. Some of the possibilities, such as generalizing an area to a point or line in the construction of a network, actually remove the data set from a geographic format (see Fig. 12.2). Somewhat curiously, cartographers have traditionally not considered the loss of one or both of the two dimensions of an area as a cartographic problem. We do, however, consider foremost the reduction in the number of dimensions from three (globe) to two (flat surface) as a cartographic problem.

Therefore, in generalizing by symbolization, cartographers usually change the scale of measurement or the type of data. They do so by using the primary graphic elements discussed in Chapter 7. Somewhat more rarely, cartographers generalize by changing dimensions as in Figure 12.2. The reason why cartographers have not used a change in dimension as an integral part of their generalization by symbolization is probably related to their overriding concern with displaying "reality" on their maps. Abstractions derived from a change of dimensions can be very insightful in understanding a distribution's characteristics or in comparing distributions. Perhaps in the future cartographers will do more in this area.

The primary graphic elements themselves connote certain scales of measurement and vary in their usefulness for representing different data types. The correct use of symbolization, therefore, requires the cartographer to decide on the level of generalization by symbolization for each of these two aspects and then to select the primary graphic elements that will be used to encode the data set. This relating of the selected aspects to the primary graphic elements is the crux of cartographic symbolization. The success or failure of a map depends fundamentally on the cartographer's success or failure in this process.

*We are not concerned here with three-dimensional physical relief models or with tactual maps.

Of secondary concern in this case is good design. Unquestionably good design and execution of the chosen symbolization will enhance any map, and cartographers should strive for this goal. Conversely, poor design can block the message of the map from reaching the user. However, if the symbolization is incorrectly done, design, good or bad, can not make the map useful.

Graphic Elements in Symbolization

In this book we recognize the primary graphic elements as described in Chapter 7. Each of these graphic elements connotes a scale of measurement. In some cases they can be forced to connote any of the scales of measurement by the addition of text or numbers. Likewise, each of these graphic elements has greater usefulness for portraying some types of data than for others. A cartographer must learn the scale of measurement connoted by each graphic element and the efficiency with which each can symbolize a type of data.

The primary graphic elements of size, value, spacing, hue, orientation, and shape vary tremendously in their usefulness for cartographers. The position element will be dismissed here because the cartographer cannot vary the geographical ordering of the data sets (see Chapter 7). Thus our discussion is limited to the remaining six graphic elements: size, value, spacing, hue, orientation, and shape.*

Size. Size is one of the most useful graphic elements for symbolization. It connotes quantities and can be used to portray data sets at the ordinal, interval, or ratio scale of measurement. It is useful for place, linear, or volumetric data sets. Therefore, only nominally scaled and areal data sets cannot be symbolized by a variation in size.

Value. Like size, value is a much-used graphic element in symbolization. Value connotes order and can be used to display ordinal aspects of data sets. By the addition of legend numbers, value can be used to portray interval or ratio aspects of data sets. Value can show place, linear, or volumetric data. Because it is ordered, it should not be used for the nominal and areal aspects.

The graphic elements of size and value "order" the data in some way. In contrast to these "ordering" elements, the remaining four primary graphic elements are used to differentiate characteristics. These "differential" variables do not order; they merely connote a difference.

Spacing. Spacing refers to the spacing of a series of dots or lines that are the component marks of a symbol. In some cartographic literature spacing is used synonymonsly with texture. Others refer to a combination of variation in spacing and size as texture. When the spacing becomes very fine (close together) it cannot be distinguished from value. We will confine our discussion of spacing as used in symbolization to those cases where it is distinguishable from value. Spacing, therefore, connotes a nominal aspect of a data set. It can be used for place, linear, or areal aspects of data type, but not for volumetric data. It is, therefore, almost the exact opposite of value in the aspects that it can portray.

Hue. Hue is used as a graphic element as explained in Chapter 8. Hue connotes a nominal scale of measurement and can be used for place, linear, or areal data types.

Orientation. Orientation is used to differentiate nominal aspects of data sets and for place, linear, or areal data types.

Shape. Shape connotes a nominal aspect of a data set and can be used to portray place, linear, or areal data types.

All four differential variables of the graphic elements—spacing, hue, orientation, and shape—therefore are useful to portray the nominal scale of measurement and the place, linear, or areal aspects of data type. They should not be used for higher scales of measurement or for volumetric

*The previous edition of this text referred to shape, size, color, value, pattern, and direction as the visual variables.

data. Admittedly the cartographic convention of using the colors in the spectral progression to represent heights of land and depths of seas uses color for a non-nominally scaled phenomenon. Theoretically it is a poor choice for symbolization; in practice it is widely accepted and used.

However, some of these differential variables enhance map communication better than others. Texture is more important for areal data sets than for place or linear sets. Orientation is more useful for place than linear or areal data sets. Shape is best used for place or linear aspects, and hue is useful for all three data types.

Table 12.1 summarizes the symbolization problem. Generalization by symbolization can be done in two ways: (1) by a change in the scale of measurement aspect of a data set, or (2) by a change in the data type aspect of a data set. Once the specific aspects of the data set to be portrayed have been determined, the cartographer must select from among the appropriate graphic elements which will be used to construct the symbols needed to encode the data set. Certain combinations are not possible; among the possible some symbolizations are preferable to others (Table 12.1). If the cartographer pays attention to these ground rules and practices good design, the resulting map should be successful.

The Statistical Surface

The statistical surface, mentioned above, is one of the most important concepts in cartography, and cartographers must know the methods for mapping a statistical surface thoroughly. Simply stated, a statistical surface can be assumed to exist for any distribution that can be conceived of as being mathematically continuous over an area and measured on an ordinal, interval, or ratio scale of measurement.

It is possible to assume a statistical surface from volumetric data recorded at positions or for areas and to portray that surface by line or area symbols. The concept of the resulting statistical surface must therefore be clearly understood. The portrayal of a statistical surface will be discussed in both Chapters 13 and 14 under line and area symbolization. Because of its usefulness cartographers have even attempted portrayals of a statistical surface by point symbols, but rarely and

TABLE 12.1 The Symbolization Problem

	Scale of Measurement			
Data Type	Nominal	Ordinal	Interval	Ratio
Place	*hue* *shape* *orientation* spacing	*value* *size*	*value* *size*	*value* *size*
Linear	*hue* *shape* spacing orientation	*value* *size*	*value* *size*	*value* *size*
Areal	*hue* *spacing* shape orientation	X	X	X
Volumetric	X	*value* *size*	*value* *size*	*value* *size*

Those graphic elements shown in *italics* are of primary utility. X indicates that no graphic element is appropriate.

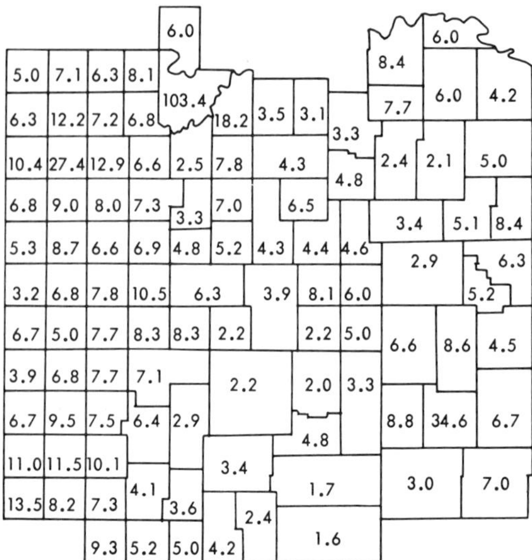

FIGURE 12.3 An array of statistical data for unit areas. The numbers are rural population densities for minor civil divisions in a part of Kansas. (Courtesy of G. F. Jenks and *Annals of the Association of American Geographers.*)

FIGURE 12.4 Elevated points whose relative height for each unit area is proportional to the number in that unit area in Figure 12.3. (Courtesy of G. F. Jenks and *Annals of the Association of American Geographers.*)

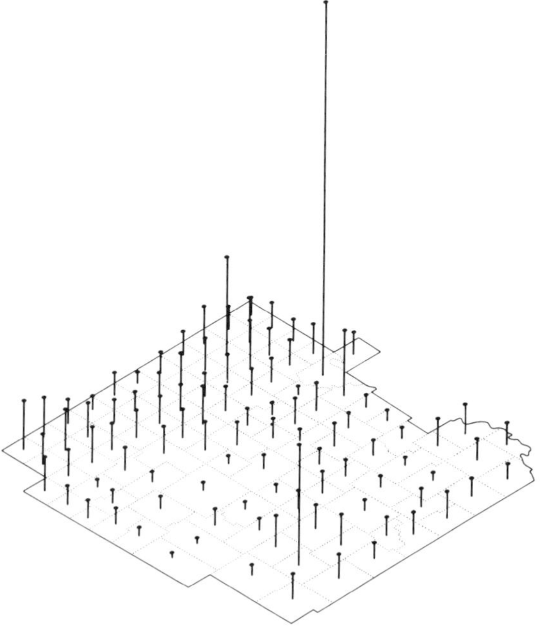

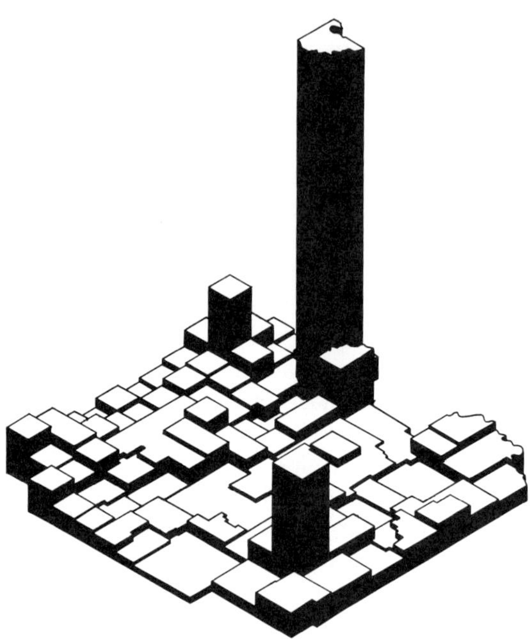

FIGURE 12.5 A perspective view of the statistical surface produced by erecting prisms over each unit area proportional in height to the numbers shown in Figure 12.3. (Courtesy of G. F. Jenks and *Annals of the Association of American Geographers.*)

with limited success; therefore we will not discuss them in this context.

If you think of the magnitudes of ratios assigned to unit areas or the sample values of continuous distributions (for example, temperatures at weather stations) as having a relative vertical dimension, you can visualize the sample values or the unit area values as forming a three-dimensional surface. For example, Figure 12.3 shows an array of numerical data that could either be a set of ratios, such as population densities, or a set of observations such as temperatures at weather stations, each arbitrarily assigned a location at the center of a unit area.

Figure 12.4 shows the erection of a column in each unit area; the length of the column is proportional to the number in that unit area in Figure 12.3. In Figure 12.5 the relative magnitudes of the values in the unit areas are emphasized by erecting over each area a prism that has a base shaped like the unit area and a height that is

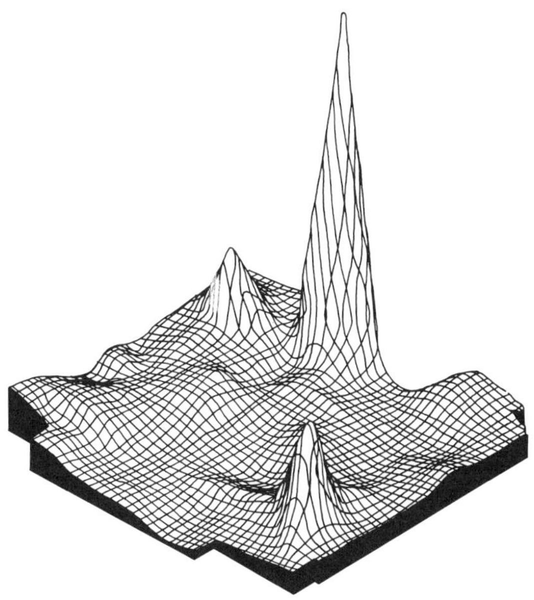

FIGURE 12.6 A perspective view of a smoothed statistical surface produced by assuming gradients among all the numbers shown in Figure 12.3. The elevated points in Figure 12.4 would all touch the surface shown here. (Courtesy of G. F. Jenks and *Annals of the Association of American Geographers*.)

scaled the same way as the length of the line in Figure 12.4. In a very real sense Figure 12.5 is a three-dimensional form of the familiar two-dimensional histogram. Figure 12.6, on the other hand, illustrates another way of showing the same statistical surface, but here the emphasis is on the magnitudes and directions of the gradients. All Figures 12.3 through 12.6, can be thought of as representations of a statistical surface consisting of a series of points, each with an *x*, *y*, and *z* characteristic. The *x* and *y* values refer to the horizontal, or planimetric, locations, and the *z* values refer to the assumed relative heights above some horizontal datum, such as the plane of the map.

Figure 12.4 is a symbolization of volumetric place data by line symbols, while Figure 12.6 is an alternative symbolization by line symbols of the same data. Figure 12.6 gives a vivid portrayal of the statistical surface in the opinion of many map users.

The concept of a statistical surface is important because its visualization allows the derivation of a great deal of additional information. For example, gradient is an important concept; there are numerous kinds of gradients, such as the slope of the land and variations from place to place of air temperature or atmospheric pressure, and many other more complex gradients derive from concepts such as population or economic potentials. Such quantities may be thought of as spatial vector quantities in that at each point of the distribution on the map there is a direction, a magnitude, and a sense. In contrast Figure 12.5 is the symbolization of a statistical surface by varying only heights of area units.

Figure 12.7 shows the same statistical surface using the graphic element value to portray the data in Figure 12.3. Figure 12.7 illustrates the

FIGURE 12.7 A choropleth (see Chapter 14) representation of the data given in Figure 12.3 using five range-graded categories of the graphic element value to portray the data.

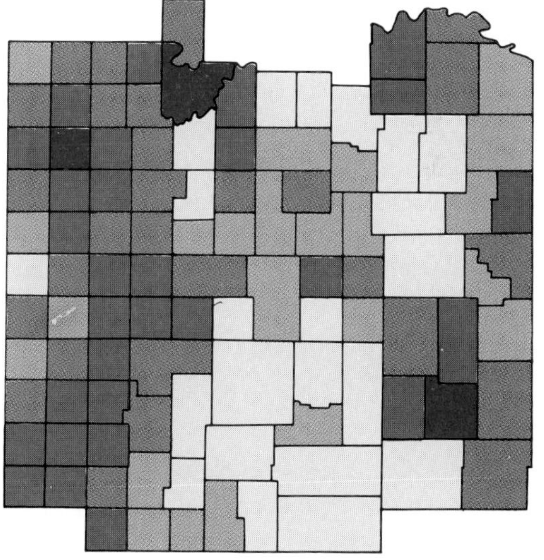

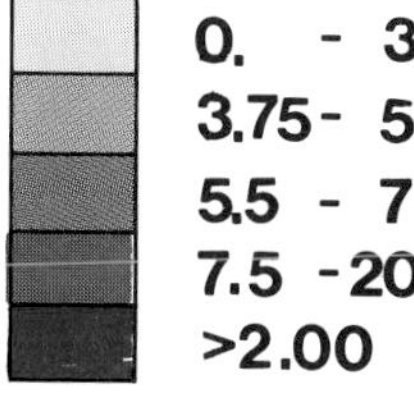

symbolization of a statistical surface by area symbols. Compare the visual impression from Figure 12.7 to the other portrayals of the same data set, but especially to Figures 12.3 and 12.5. With practice the portrayal in Figure 12.7 can become meaningful to a map reader, conveying patterns of areas of high data values or alternatively patterns of areas of low data values. Figures 12.4, 12.5, and 12.6 all give the visual impression of three dimensions. Figures 12.3 and 12.7 do not. Planimetric position is more easily ascertained in Figures 12.3 and 12.7, whereas the other three figures concentrate on giving an overall visual impression of the statistical surface.

Finally one measure of the importance of the statistical surface in cartography is the amount of space devoted to its symbolization in Chapters 13 and 14. It is because of its overriding importance that the concept is introduced early in our discussion of symbolization.

MAPPING PLACE DATA

Many phenomena that cartographers portray on maps exist in the real world at places. Examples abound: geodetic control points, telephone poles, and other features of concern to large-scale mappers; and villages, centers of areas such as counties, or political wards of concern to small-scale mappers. Regardless of the scale of portrayal the cartographer must obtain the data set from some source. Sets of data that exist at places are common, and they can occur at all four scales of measurement: nominal, ordinal, interval, and ratio.

Place data sets that a cartographer may wish to map are usually multifaceted. That is, the data have more than one attribute or characteristic. For example, one can conceive of a geodetic control point with both a nominal facet—it exists—and with a ratio facet—so many degrees of latitude and longitude. Similarly place data sets for large-scale urban mapping might include the set of telephone poles, electricity poles, fire hydrants, manholes, and so on. Each of these data sets has a nominal attribute, namely existence at some position. Most have additional attributes such as height, material from which they are made (wood or steel or aluminum), and so forth. In one sense all data that we map have the nominal attribute of existence, and we must portray that attribute. We normally imply the nominal attribute of existence by mapping a higher-scaled attribute of the same data set.

Examples of place data sets that a cartographer may wish to map include such things as first-, second-, and third-order control monuments and ordered facets of other place data sets. For example, one can consider an ordinal ranking of the trees in a timber stand as they relate to com-

FIGURE 12.8 Framed, nominally scaled pictorial symbols on a map promoting winter activities in a portion of the state of Wisconsin. The legend of the map lists fourteen symbols. (From a class project, University of Wisconsin-Madison.)

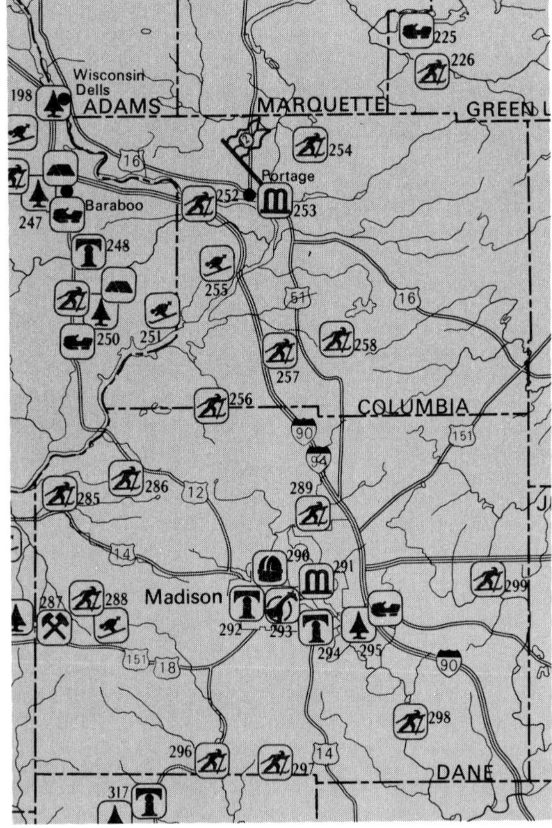

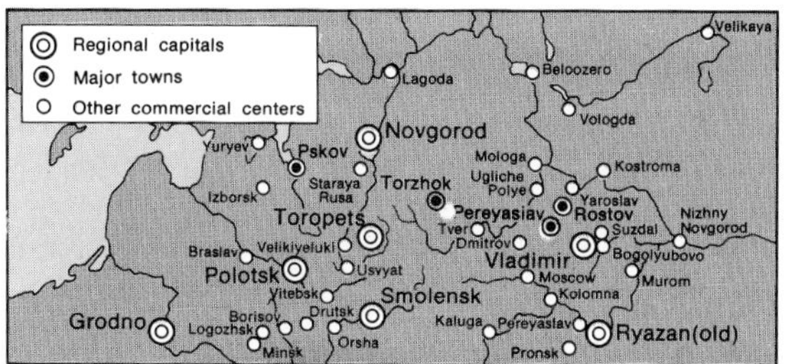

FIGURE 12.9 Ordinal place data symbolized by varying the graphic element size. The symbol sizes are enhanced by articulation within the symbol itself and reinforced by the size of the lettering.

mercial forestry. Some trees are ready to be cut, others nearly so, and finally some are young and need certain care such as fertilizer and pruning.

Interval-measured place data sets are less numerous than others. They are usually contrived from ratio point sets. Ratio point data sets include all sorts of economic data pertaining to households or individuals, including household income, size of family, and age of head of household. All of these exist as ratio data sets; yet the U.S. Census reports most of them as range-graded. In fact some of the data that the census collects from individuals actually are collected as range-graded data. The same holds for the Census of Agriculture and others. There is no shortage of place data measured on a ratio scale that can be mapped.

Mapping Place Data Sets with Point Symbols

Figure 12.8 illustrates mapping a nominal attribute of place data. The pictorial symbols on this map promoting winter activities illustrate differences in kind only. Figure 12.9 illustrates the mapping of an ordinal characteristic of place data, namely, the importance of various cities in a portion of the U.S.S.R. Figure 12.10 shows the range-graded mapping of a characteristic of place data.

Finally, Figure 12.11 illustrates the mapping of a ratio characteristic of place data. By comparing this map closely with Figure 12.10, it is easy to see that the cartographer ultimately determines how these data are mapped. Although the data exist as ratio data, the cartographer has chosen to map them as ratio in Figure 12.11 and as range-graded in Figure 12.10. Thus it is appropriate to say the Figure 12.10 is more generalized than Figure 12.11, and the generalization element used is symbolization.

Unfortunately not all place data are mapped with point symbols. If they were, the cartographer's symbolization problem would be easier.

FIGURE 12.10 The population of some cities in northeastern Ohio. Symbols are range-graded to denote the population of the cities (see Table 12.2).

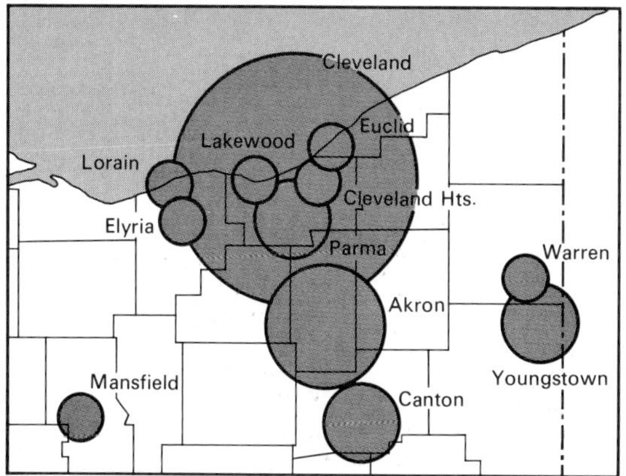

FIGURE 12.11 The population of some cities in northeastern Ohio. Symbols are proportionally scaled so that the areas of the symbols are in the same ratio as the population numbers that they represent (see Table 12.2).

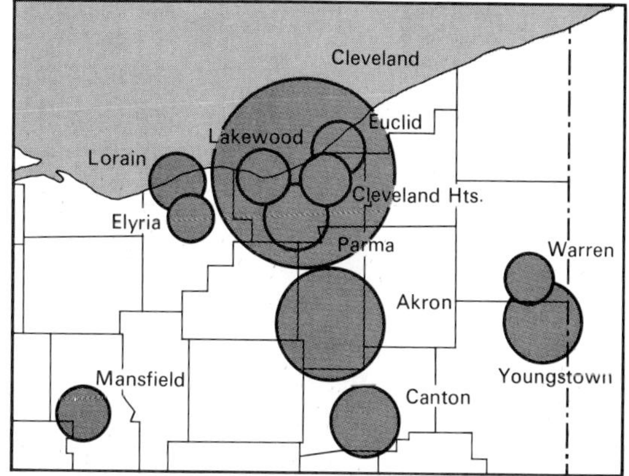

Likewise point symbols can map other than place data. Many place data sets can be mapped with point symbols as our examples will illustrate. Rarely would a linear data set be mapped by point symbols, but one common technique, dot maps, does use point symbols to map volumetric data sets by repetition of point symbols. Therefore, after a brief discussion of the graphic elements important in point symbolization, we will discuss in turn the use of point symbols to portray nominal, ordinal, range-graded, and ratio place data sets. The chapter ends with a discussion of the use of point symbols to portray volumes.

Graphic Elements Used in Mapping with Point Symbols

As seen in Table 12.1, the graphic element spacing is not used primarily in the mapping of place data sets. It may be used as a secondary variable, but it is rather difficult to vary spacing within a point symbol. For the most part relying solely on the graphic element spacing for the mapping of point symbols is not recommended.

It is also important to disassociate the graphic element spacing from the use of the word pattern to characterize the distribution of point symbols on a map. The repetition of point symbols on a map can create a visual pattern characteristic of the distribution being mapped—a very useful result of the mapping process. However, the pattern of point symbols on a map and the pattern of marks that creates the graphic element spacing are two distinctly different things. Cartographers must take care in using the terms, so that they are clearly understood.

Nominal Place Data Portrayal with Point Symbols. Mapping nominally scaled place data employs point symbols that may be classed as pictorial, associative, or geometric. Several graphic elements suggest nominal differentiation and are therefore commonly used in mapping nominal place data, including shape, hue, orientation, and to a lesser degree, spacing.

FIGURE 12.12 Stylized nominally scaled pictorial point symbols dealing with transportation. (From R. Modley, *Handbook of Pictorial Symbols*, Copyright 1976. New York: Dover Publications, Inc.)

Pictorial Symbols. Figure 12.12 illustrates the use of pictorial symbols for the representation of nominally scaled place data. These symbols could also employ different hues to aid in the communication. A pictorial symbol may be intricate, as shown in Figure 12.13, or stylized, as in Figure 12.12. For maximum effectiveness in communicating the map "message," pictorial symbols should communicate without the necessity for a legend but, in practice, a legend is usually provided. Maps of nominally scaled place data that employ pictorial symbols often err in two important respects: (1) the symbols are not easily distinguished one from another, and (2) too many symbols of roughly equivalent size are used. In an effort to illustrate that the data are nominally scaled, a cartographer may not vary the size of the various symbols, since doing so might connote a higher scale of measurement. By not differentiating sufficiently the different symbols, the cartographer may fail to communicate significant distributional patterns in a nominal distribution. The trend toward the use on maps of "framed" pictorial symbols of equivalent size especially promotes this error (see Fig. 12.8). The situation is easily avoided by using drastically varying shapes or using differing hues where possible. Different orientations of the same symbol are sometimes useful as well (see Fig. 12.14).

There should be no excuse for a cartographer to err by using pictorial symbols that are so similar that they are hard to distinguish. Yet exam-

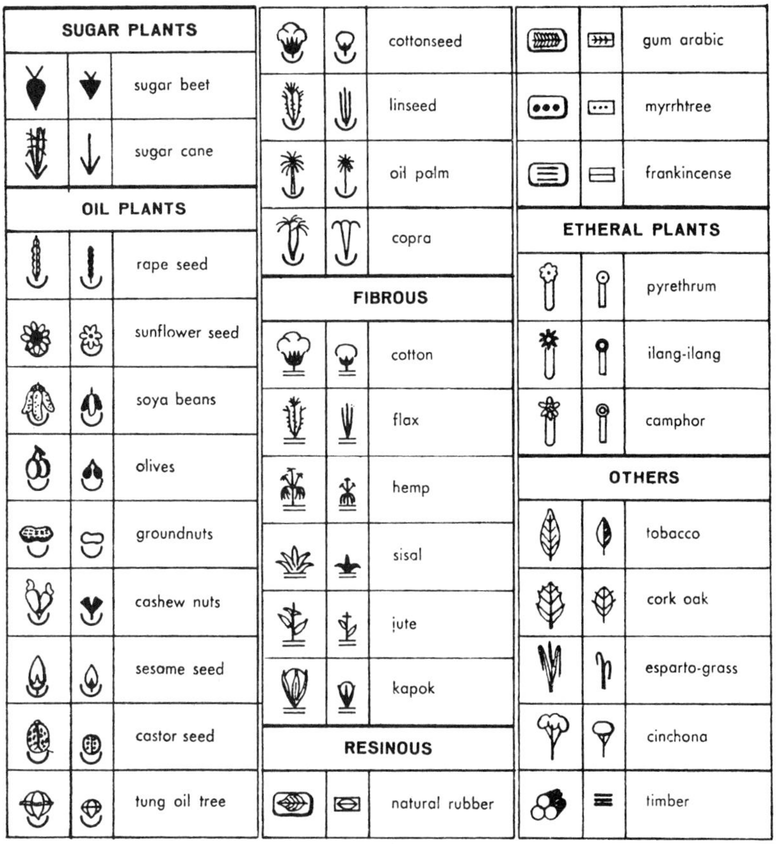

SUGAR PLANTS
sugar beet
sugar cane
OIL PLANTS
rape seed
sunflower seed
soya beans
olives
groundnuts
cashew nuts
sesame seed
castor seed
tung oil tree

cottonseed
linseed
oil palm
copra
FIBROUS
cotton
flax
hemp
sisal
jute
kapok
RESINOUS
natural rubber

gum arabic
myrrhtree
frankincense
ETHERAL PLANTS
pyrethrum
ilang-ilang
camphor
OTHERS
tobacco
cork oak
esparto-grass
cinchona
timber

FIGURE 12.13 Intricate nominally scaled pictorial point symbols representing crops. (From L. Ratajski, "The Methodological Basis of the Standardization of Signs on Economic Maps," *International Yearbook of Cartography* 11, no. 151 [1971].)

ples do occur. Note the pictorial symbols shown in Figure 12.15. Such pictorial symbolization schemes could ultimately lead to a dictionary of standardized symbols that would "handcuff" the cartographer's freedom for design and thus, in many instances, greatly retard the speed of communication of the map information.

The principal concern when employing pictorial representation for nominally scaled place data is to use shape, and, to a lesser extent, hue and orientation to make the symbols easily distinctive and identifiable.

Associative Symbols. Figure 12.16 illustrates some associative symbols for nominally scaled place data. Symbols in this class employ a combination of geometric and pictorial characteristics to produce easily identifiable symbols. As with pictorial symbols, the most powerful dimensions are shape and hue. A legend is highly recommended for a map that employs associative symbols, since they may be quite diagrammatic compared to pictorial symbols. Again, for nominally portrayed associative symbols as with pictorial symbols, the graphic elements of shape and hue are most useful. In Figure 12.16, for example, the variation of shape of the top part of the symbols for school and church assures accurate differentiation.

Geometric Symbols. The most common geometric symbols used to show nominally scaled

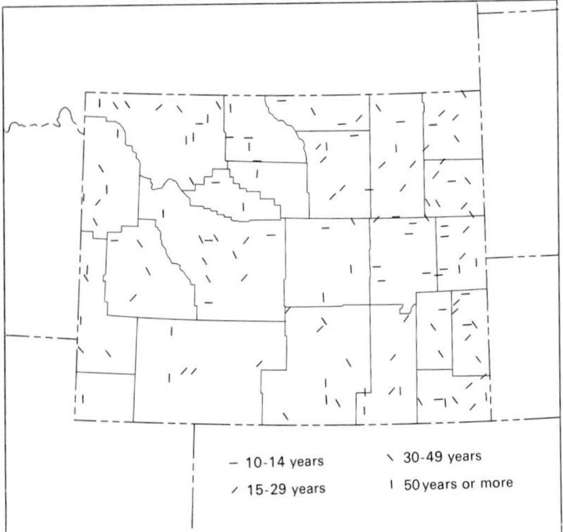

FIGURE 12.14 Nominally scaled symbols are used to indicate four classes of climatic stations with nonrecording gauges in Wyoming. The classes are delimited on a higher scale of measurement, but the depiction itself is nominal until the legend information is used.

BEAVER

MUSKRAT

MINK

OTTER

HERON

CRANE

SWAN

LOON

EAGLE

OLD FORT

BATTLEFIELD

BARRACKS

MILL POND

OLD MILL

BLACKSMITH SHOP

PIONEER CHURCH

FIGURE 12.15 Detailed pictorial point symbols. (From the legend of a map, "Legend Resource Inventory," Department of Resource Development, State of Wisconsin, May 1964.)

place data are circles, triangles, squares, diamonds, and stars; the graphic element of shape obviously is a most important factor. Hue can be used as a secondary element to differentiate the data, especially when some of the data classes are related. A legend is required on a map that employs geometric symbols. The major problems to be overcome in using geometric symbols are to ensure that each symbol is sufficiently distinctive and to make sure that the symbols do not connote higher than nominally scaled information.

Figure 12.14 illustrates the use of one geometric symbol, but this portrayal of the nominal differentiation by the graphic element orientation is only marginally acceptable. When the more visually prominent dimension of shape is used, as in Figure 12.17, the four categories are more apparent, and the distributional patterns are more obvious.

The categories (pictorial, associative, and geometric) can be thought of as occupying successive positions on a continuum of symbolization ranging from the analogic or mimetic at one end to the purely arbitrary at the other. The cartographer must be aware of where the selected symbolization lies on the continuum, its associated problems, and the appropriate corresponding need for providing a legend.

Therefore, using the graphic elements of shape, hue, orientation, and spacing for nominal point data sets, cartographers could theoretically display up to a maximum of four nominally scaled attributes of a given distribution. But doing so is

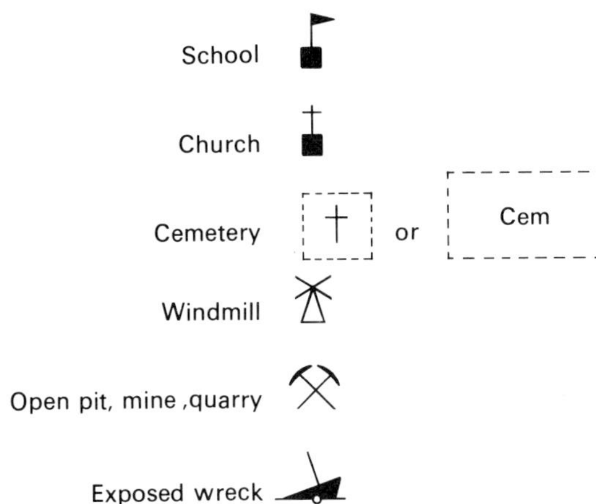

FIGURE 12.16 Associative symbols for nominally scaled positional data used by the U.S. Geological Survey on topographic maps.

not recommended. An example of the portrayal of two nominal attributes for one point data set is given in Figure 12.18.

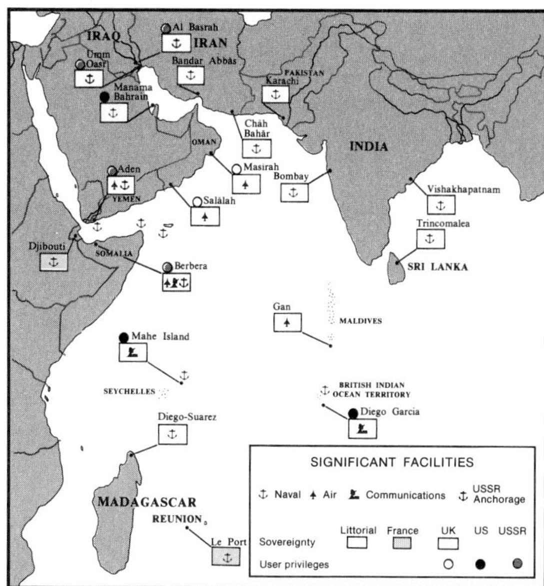

FIGURE 12.18 Map showing nominally scaled attributes of facilities in the Indian Ocean. Symbol shape and hue are used to differentiate the data. (Redrawn from *Indian Ocean Atlas*, p. 25, Central Intelligence Agency, August 1976.)

Ordinal Place Data Portrayal. As with nominal place data portrayal in cartography, the continuum from mimetic to arbitrary symbolization applies. However, a cartographer can employ the graphic elements size and value, allowing theoretically for the possibility of two ordinal characteristics or attributes plus the depiction of up to four nominal attributes of the data in one symbolization scheme using point symbols. The cartographer must be careful, however, not to overload the map with so much information that one attribute interferes with the reader's ability to understand another's information.

FIGURE 12.17 The use of the graphic element shape to portray the data given in Figure 12.14.

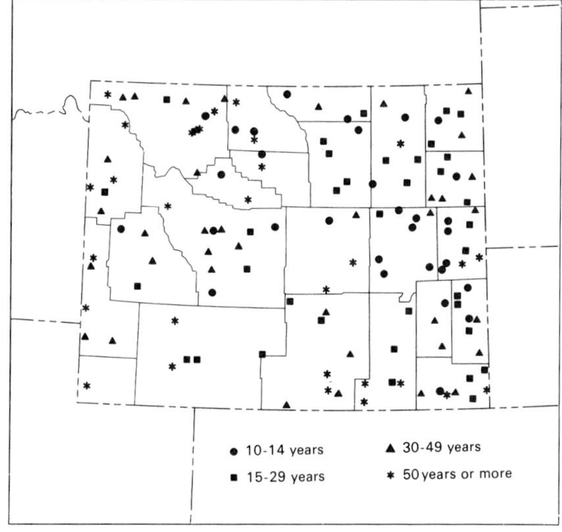

Figure 12.19 depicts the ordinal portrayal of one attribute of a data set. Oil reserves are depicted ordinally using the graphic element size to convey three classes of the data. The original data set may well have had ratio-scaled information, but the cartographer has chosen to show it ordinally. This is indicative of one way in which generalization is accomplished by symbolization in cartography.

Figure 12.20 uses the graphic element value to depict an ordinally scaled data set. The square

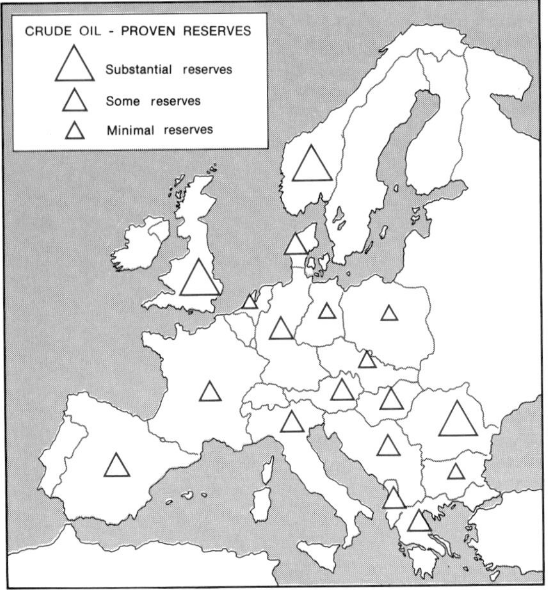

FIGURE 12.19 Ordinal symbolization by point symbols using size as the graphic element.

symbol is geometric and thus lies on the arbitrary end of the mimetic-arbitrary continuum. The two graphic elements in combination could also be used to reinforce the information being encoded into the map. Figure 12.21 is an example of this type of reinforcement, which, when feasible, is a useful tool for the map maker.

The cartographer who can work with color can also use the intensity or chroma of the color to portray a point data set on an ordinal scale of measurement. The use of color can work in much the same way as the use of value in Figure 12.20, or it can be used in conjunction with the graphic element size as in Figure 12.21. The use of color intensity is a very subtle differentiation and probably is risky to use alone if there is any question as to the map-reading audience's sophistication. On the other hand the use of this subtle property of color is an excellent way to reinforce the information being portrayed. Therefore, although it may not be wise to rely on a combination of either value or size along with color intensity to portray two different ordinally scaled attributes of a point data set, it would be quite acceptable

FIGURE 12.20 Ordinal symbolization by point symbols using value as the graphic element.

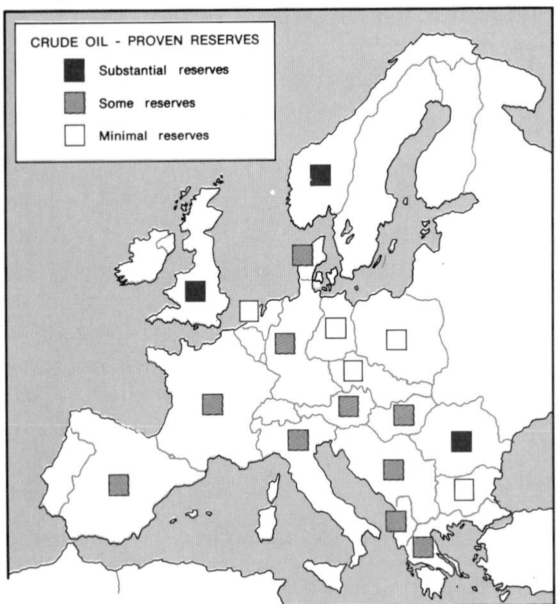

FIGURE 12.21 Ordinal symbolization by point symbols using both size and value as mutually reinforcing graphic elements.

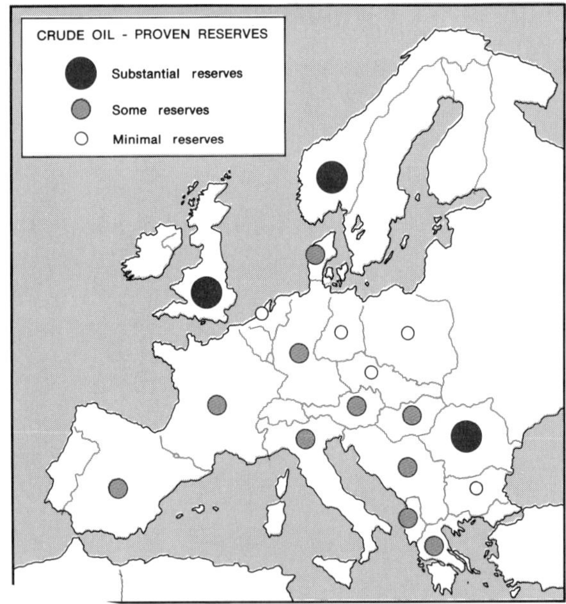

to use color intensity along with either size or value to reinforce the portrayal of a single attribute of an ordinally scaled place data set.

Range-Graded and Ratio Place Data Portrayal. For range-graded and ratio place data portrayal the continuum from mimetic to arbitrary symbols is also valid. The graphic elements appropriate in ordinal data portrayal are also appropriate in range-graded or ratio data display. Most often used and most important are the graphic elements of size and value. Symbolization attains range-graded and ratio scales of measurement primarily through the use of a legend. The symbols on the map, if there is no further explanation, really connote only an ordinal scale.

The French cartographer Bertin makes the significant observation that size is different from value in connoting range-graded and ratio data.* He maintains that size is far more efficient than value. Map readers will more readily attempt to set a ratio between the sizes of two point symbols, whereas they are much more likely to recognize that one value is greater than another. Hence, Bertin thinks that map readers can more easily get ratio information from a map that uses size variation in its symbolization yet only get ordinal information from a map employing value variation. Because we happen to agree with this hypothesis of Bertin, the use of the graphic element size in range-graded and ratio point data portrayal will be discussed below in some detail. Another reason for doing so is that variation in size is perhaps the most often-used variation of a graphic element in range-graded and ratio place data portrayal.

The Graduated Symbol Map

Graduated symbols (that is, symbols that are differently sized) are widely used; the most common is the graduated circle. The variation in size is used to symbolize either amounts at specific locations or totals that refer to enumeration units. Thus, they are useful (1) when point data exist in close proximity but are large in aggregate number, such as the population of a city, (2) for symbolizing totals of quantities such as tonnage, costs, and traffic counts, or (3) for representing the aggregate amounts that refer to relatively large territories. In the last instance, the territory is considered merely as a location, even though it obviously has areal extent. The area of a statistical unit is naturally a two-dimensional geographical quantity; however, when data referring to it are aggregated and symbolized by a point symbol (in combination with other statistical units), the areal unit has, in a sense, been reduced to only a place value for mapping purposes.*

The graduated circle is one of the oldest quantitative point symbols used for statistical representation. Near the beginning of the nineteenth century, it was used in graphs that illustrated the then-new census materials, and its first appearance on maps was in the 1830s. Since that time, it has been near the top of any list of quantitative point symbols in the frequency of its use; its ease of construction makes it likely that it will continue to be popular.

The Scaling of Graduated Symbols

When the graduated circle was first used to symbolize data measured on an interval or ratio scale, the variations in the actual areas of the symbols were made uniformly proportional to the numbers they represented. This has become known as the square root method of symbol scaling. Other methods of symbol scaling are range grading and psychological scaling.

Proportional Area Scaling or the Square Root Method. If two statistics have values to be rep-

*J. Bertin, *Graphics and Graphic Information Processing,* trans. William J. Berg and Paul Scott. Berlin and New York: Walter de Gruyter & Co., 1981, 196–7.

*This is an example of the inconsistency that results when we try to match up point data and point symbols, and linear data and line symbols.

resented that are in a 1:2 ratio (that is, the magnitude of the second is twice that of the first), the second circle is constructed so that its area is twice that of the first. Since the area of a circle is πr^2, and since π is constant, the method of construction is to extract the square roots of the data and then construct the circles with radii or diameters proportional to the square roots. The unit radius value, by which the square roots are made into appropriate plotting dimensions, may be any desirable unit, such as so many millimeters or hundredths of an inch; it is selected so that the largest circle will not be "too large" and the smallest circle will not be "too small." As long as the square roots are all divided by the same unit radius value, the areas of all the circles will be in linear proportion to the sizes of the numbers they represent (see Fig. 12.11).

Range-Graded Scaling Method. If the cartographer does not wish to show relative sizes of each individual statistical quantity, the data can be classified or range-graded. The cartographer would show all individuals in one class by a "standardized" symbol constructed to the size of the midpoint of that class (see Fig. 12.10). This process is referred to as the range grading of graduated symbols. In effect, the data are divided into class intervals, and one symbol size is assigned to each class.

Psychological Scaling Method. Extensive research in the psychophysical aspects of cartographic symbols has demonstrated that the perceptual response to differences between symbol areas is not a linear function; instead, the ordinary observer will underestimate the sizes of the larger symbols in relation to the smaller ones. The evidence is particularly strong that underestimation occurs between circle symbol areas. For example, to use the previous illustration, if the magnitudes of two quantities were in a 1:2 ratio and the actual areas of two circles representing them were in the same ratio, then a map reader would think the second was significantly less than twice the size of the first. When we

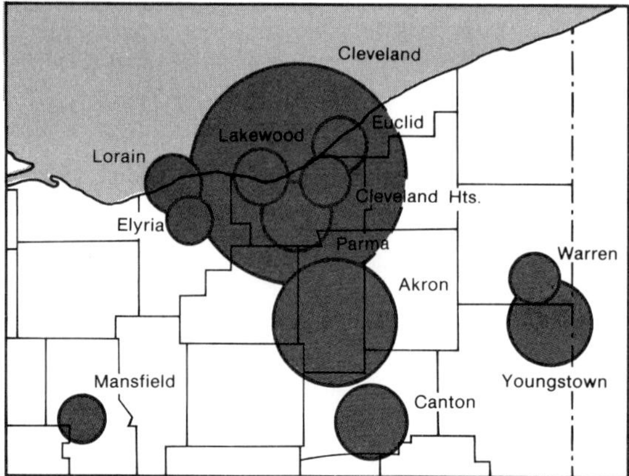

FIGURE 12.22 The population of some cities in northeastern Ohio. Symbols are psychologically scaled so that the areas of the symbols visually connote the correct ratios between the population numbers that they represent (see Table 12.2).

make the areas of the circles strictly proportional to the numbers they represent, we have, therefore, in effect reduced the visually apparent sizes of the larger circles in relation to the smaller, or to look at it another way, we have increased the visual significance of the smaller circles relative to the larger. This is unfortunate, since the primary purpose of making such maps is to symbolize quantities so that the map reader will obtain a realistic impression of the distribution being mapped.

We may compensate for underestimation by scaling the symbols to reflect the amount of expected underestimation (see Fig. 12.22). Instead of extracting the square roots of the data and making the radii of the circles relative to the square roots, the procedure is to (1) determine the logarithms of the data, (2) multiply the logarithms by 0.57 or some other value appropriate for the symbol being used, (3) determine the antilogarithms, and (4) divide the antilogarithms by the chosen unit value for the radii of the circles.*

*This procedure was devised by Professor James J. Flannery for graduated circles. Note that the square root of log n = log $n/2$ or log $n \times 0.5$. Multiplying the logarithms by 0.57 instead of 0.5 serves proportionately to increase somewhat the relative sizes of the larger circles so that they appear in proper relation to the smaller ones.

Figure 12.23 shows the difference in the sizes of two sets of circles constructed in the two ways.

An example will demonstrate the procedure. Table 12.2 gives the data for the circle maps in Figures 12.10, 12.11, and 12.22. The numbered columns of Table 12.2 show the calculations needed to determine the numbers that, when divided by the chosen unit radius value, will give the plotting radii of the circles. The antilogs provide directly the plotting values and need only be converted to map dimensions by dividing them by some convenient unit value. The maps shown in Figures 12.10, 12.11, and 12.22 have been constructed from the data in Table 12.2 using a unit radius value chosen so that the smallest circles have the same unit radius value.

Another aid in compensating for the underestimation is to use strategically selected "anchoring stimuli" in the legend that accompanies the map. If several circles are of a similar size, it is appropriate to use a legend circle in their general range. A solitary legend symbol is not often suitable, since the perceptual problem is one of underestimating differences between circles. Therefore, a legend design of at least three non-nested circles (see Fig. 12.24) is advisable for every graduated circle map.

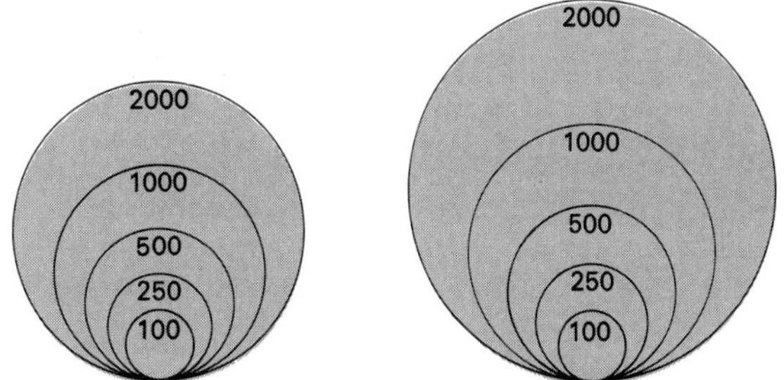

FIGURE 12.23 Two nested sets of graduated circles prepared from the same data show the difference between left, circles scaled, in proportional relation to area (according to square roots), and right, psychologically scaled to compensate for underestimation.

The problem of underestimation is less severe with range-graded circles as long as each chosen circle symbol is distinguishable from all of the others. In the case of range-graded symbols, all

TABLE 12.2 1970 Populations of Cities in Northeastern Ohio with More Than 50,000 Inhabitants.

	(1) 1970 Population (n)	(2) log *n*	(3) Antilog of (log *n* × 0.5)	(4) Antilog of (log *n* × 0.57)	Drafting Radii in cm (before reduction): Fig 12.10, 462.28 = 1.0 cm, Column 3 ÷ 462.28	Fig. 12.11, Range Graded	Fig. 12.22, 1000 = 1.0 cm, Column 4 ÷ 1000
Akron	275,425	5.44000	524.81	1261.23	1.135	1.25	1.26
Canton	110,053	5.04160	331.74	747.68	0.718	0.80	0.75
Cleveland	750,879	5.87557	866.53	2233.96	1.874	2.50	2.23
Cleveland Heights	60,767	4.78367	246.51	532.96	0.533	0.50	0.53
Elyria	53,427	4.72776	231.14	495.25	0.500	0.50	0.50
Euclid	71,552	4.85462	267.49	584.97	0.579	0.50	0.58
Lakewood	70,173	4.84617	264.90	578.52	0.573	0.50	0.58
Lorain	78,185	4.89312	279.62	615.29	0.605	0.50	0.62
Mansfield	55,047	4.74073	234.62	503.75	0.508	0.50	0.50
Parma	100,216	5.00094	316.57	708.82	0.685	0.80	0.71
Warren	63,494	4.80273	251.98	546.46	0.545	0.50	0.55
Youngstown	140,909	5.14894	375.38	860.79	0.812	0.80	0.86

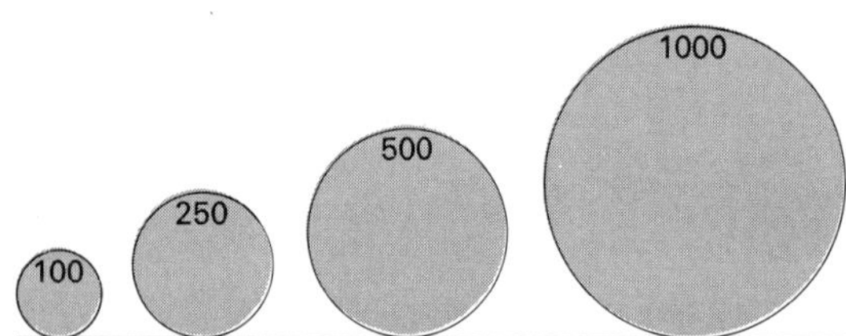

FIGURE 12.24 A legend design of nonnested symbols. This design is presumed to assist the reader in gaining the correct visual impression.

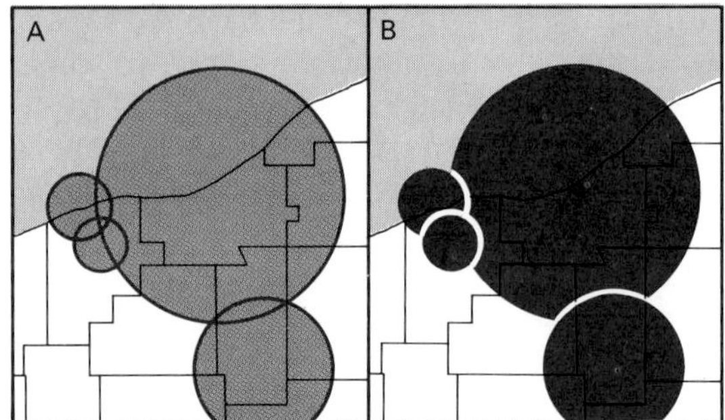

FIGURE 12.25 Two ways of solving the problem of overlapping circles: (A) The circles are allowed to overlap, since they are "transparent"; (B) Smaller circles are drawn "above" larger circles.

symbols should appear in the legend, and they may be nested to conserve space.

The difference between range-graded point symbols and square root or psychologically scaled symbols has its counterpart in the symbolization of volumes by line or area symbols. For example, choroplethic maps (see Chapter 14) are in essence range-graded; isarithmic maps are not. The difference is one of uniquely symbolizing every value on a continuum, as in square root or psychologically scaled point symbols and the isarithmic method, compared to the collapsing of values along the continuum into classes, as in range-graded point symbols and the choroplethic method.

Specification of Symbol Size. Another problem to be solved when using graduated symbols, especially graduated circles, is the selection of the unit radius. The cartographer may find that a desirable unit radius value for the map as a whole may cause the symbols in one part to fall largely on top of one another if their centers are placed approximately at the locations of the data they are to symbolize. Two of several ways to approach this problem are shown in Figure 12.25. The two small maps are taken from the data in Table 12.2.

The selection of the unit radius value with which to scale the circles is important and should be done with the aid of some preliminary experimentation. The ideal value is one that will provide a map that is neither "too full" nor "too empty." For example, Figures 12.26 and 12.27 show the area of land available for crops in some counties. When a small unit radius is chosen, the circles are too small to show much differentiation, as illustrated in Figure 12.26, and the impression is that there is practically no cultivated land in those counties. When a large unit radius is chosen, the representation again does not reveal much, and the impression is that practically all the land is cultivated, as in Figure 12.27.

Squares, Cubes, Spheres, and Other Point Symbols. Instead of proportional circles, we may employ almost any other geometric or pictorial figure. The circle is the easiest to construct and scale, but the requirements of design may make some other symbol desirable.

The square is relatively simple to use as a symbol to represent magnitudes, since all that is necessary is to obtain the square roots of the data and then scale the sides accordingly. This will grade the areas of the squares in linear proportion to the numbers they represent. There is clear evidence that the phenomenon of underestimation affects the map reader's impression of circles so scaled, but the evidence is less impressive with respect to squares. Range-graded and ratio-scaled squares are an alternative, but they appear to be used much less frequently than other point symbolization schemes.

Apparent Volume Symbols. It is not uncommon for a cartographer to be faced with a range of data so large that with range-graded or ratio scaling, both ends of the range cannot be effectively shown by graduated circles or squares. If the symbols are large enough to be differentiated clearly in the lower end of the scale, then those at the upper end will overshadow everything else and ruin the communication. Many attempts have been made to surmount this apparent difficulty by employing a symbol form drawn to look like a volume instead of an area, such as pictorial cubes or spheres, as shown in Figures 12.28 and 12.29.* This has been done by scaling the symbols according to the cube roots of the data, that is, by making the sides of the front of a perspective cube or the circle bounding a sphere proportional to the roots. If the symbols were actually three dimensional, their volumes would be in strict linear proportion to the numbers they represented. The cube roots of a series cover a much smaller range than the square roots, and this, of course, is what makes it *ostensibly* possible to portray the larger range of data graphically.

Although spherelike symbols may be very graphic and visually pleasing, especially when enhanced by good execution and design, they do not serve the purpose of effectively portraying the numbers that they are supposed to represent. Several studies have clearly shown that map readers evaluate the spheres not on their volume comparison, but on the basis of the map areas they cover, that is, as if they were graduated circles, including the characteristic underestimation. To what extent cubelike symbols suffer a similar perceptual fate is not yet known.

When graduated pictorial symbols such as soldiers, stars, and animals are employed, care must be taken to scale them properly. Their heights or diameters could be made proportional (as outlined above) to the data to be represented by a system that ensures that the *areas* covered by the symbols are visually related to the magnitudes they represent.* A safer course of action would be to employ these symbols as range-graded symbols, especially for pictorial symbols whose areas are almost impossible to evaluate visually.

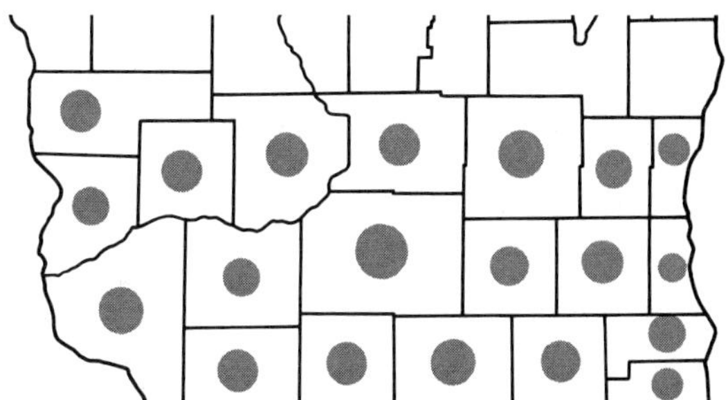

FIGURE 12.26 Area available for crops, by counties. The unit radius is too small.

Segmented Graduated Symbols. Cartographers have attempted to use parts of circles such as pie wedges or other symbols when precise locations are desired but large amounts of overlap would occur if an entire symbol were used (see Fig. 12.30). Such point symbols may be range-

FIGURE 12.27 Some data as in Figure 12.26, but the unit radius is too large.

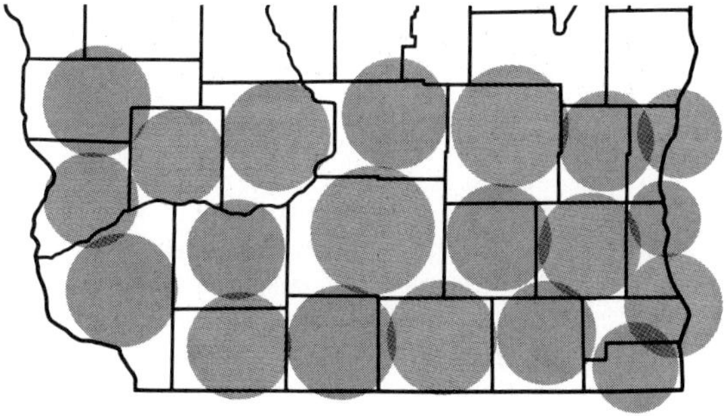

*The difficulty is more apparent than real. If one segment of a distribution does actually make all else pale by comparison, then that is the fact and it should be so portrayed. The primary responsibility of the cartographer is to prepare the map so that the visual impressions are correct.

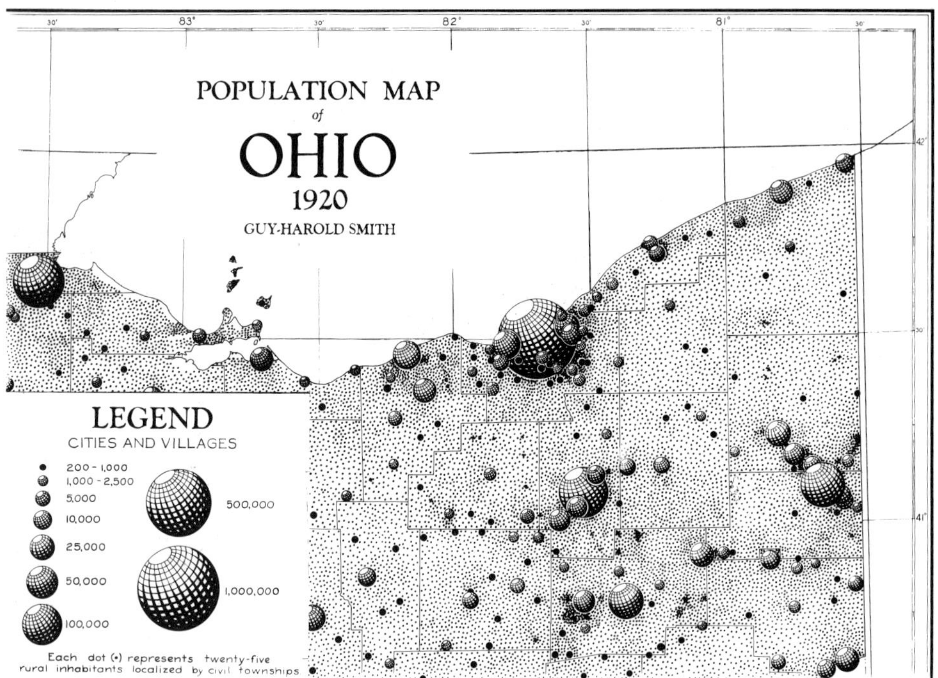

FIGURE 12.28 A portion of a population map of Ohio using spheres to depict volume data. (Courtesy of G. H. Smith and *The Geographical Review*, published by the American Geographical Society of New York, 1920.)

graded or they may be scaled on ratio basis. With this type of symbol several categories of data, including nominally scaled data, can often be presented simultaneously by showing several symbols at one location and differentiating them by hue or spacing (see Fig. 12.31).

Another way of symbolizing quantitative data is by segmenting the point symbol. Instead of using part of the symbol, the entire symbol is used, but it is segmented to illustrate the proportional parts that constitute the data. For example, the ethnic makeup of an urban population, or the value added by several categories of manufacturing, can be portrayed this way. The most common method is to use graduated circles and subdivide them as a round pie is cut. There is no generic term for this kind of symbol, but the subdivided circles are often called "pie charts."

Any relation of one or more parts to the whole can be shown visually by a segmented symbol. For example, Figure 12.32 shows the total amount of farmland in each county and, at the same time, what percentage of that total is available for crops. The procedure requires that the percentage be determined and that, by using a "percentage protractor," the various values be marked off on each circle.

It is important that the subdivision of each circle begin at the same point; otherwise, the reader will have difficulty in comparing the values. Also important is the decision of which portion to be shaded or colored.

Many other more complex kinds of segmentation can be employed. For example, a kind of

*There has not been enough research yet to determine to what extent underestimation affects the perception of these more complex symbol types. Determination of relative sizes of some geometric symbols has been studied by Robert L. Williams. See *Statistical Symbols for Maps: Their Design and Relative Values*. New Haven, Conn.: Yale University Map Laboratory, 1956.

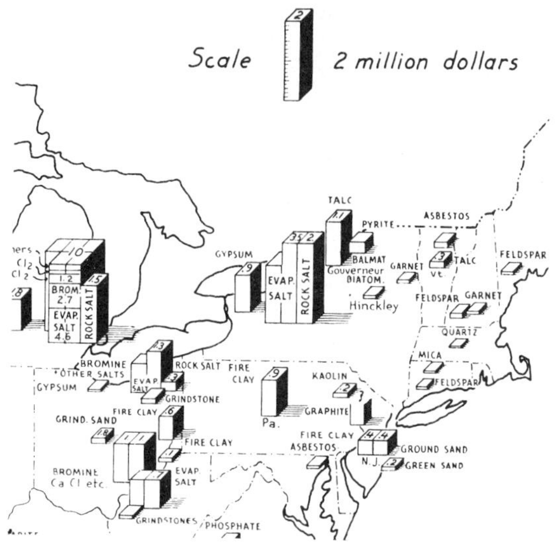

FIGURE 12.29 A map on which cubelike symbols, called block piles, are used to depict volume data. (From a map by E. Raisz in *Mining and Metallurgy*, American Institute of Mining Engineers.)

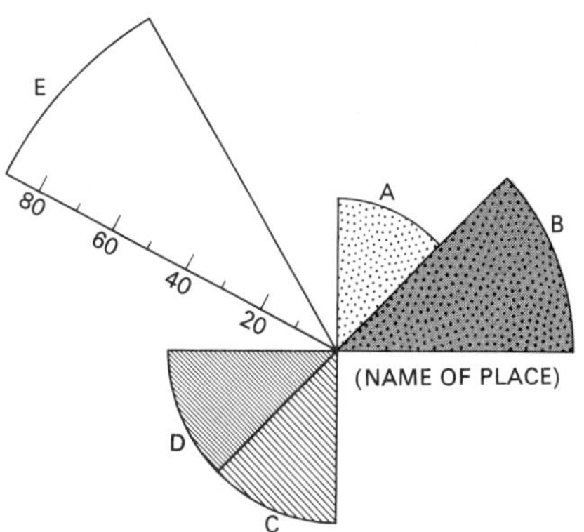

FIGURE 12.31 Examples of several ways to employ circle segments. (A) and (B) in the upper right quadrant can indicate nominal (hue) and ratio (size) characteristics (for example, amount of passenger versus freight traffic), (C) and (D) in the lower left quadrant can simply indicate different nominal (spacing) categories (for example, long grain versus short grain rice), and (E), with ratio (size) scaling, can indicate yet another category.

map conceptually similar to the pie chart can be constructed by employing one or more concentric circles to represent a portion of the total. The diameters of these interior circles are scaled according to the original data in the same manner as the circle representing the total. The visual

FIGURE 12.30 The legend and a small portion of a map illustrating the amount of information that can be coded in segmented, graduated point symbols. (From J. I. Clarke, *Sierra Leone in Maps*, Copyright 1966, London: University of London Press, Ltd.)

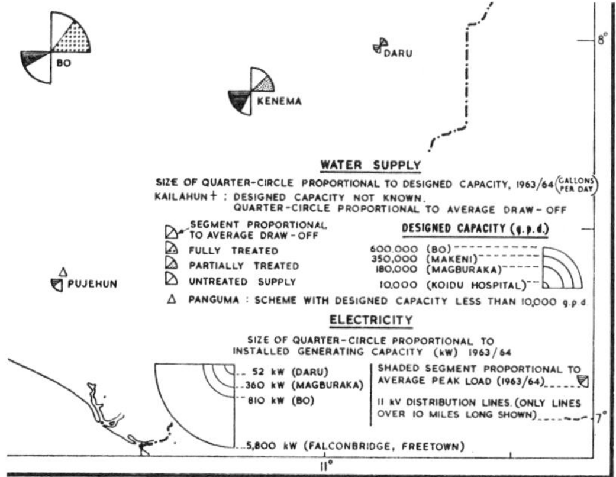

impression is not very efficient, however, since percentages are not easily read. Naturally, graduated squares, rectangles, and other figures may also be segmented in various ways to show parts of a whole, (see Fig. 12.33).

Directional and Time Series Point Symbols. A variety of more complex cartographic presentations can be accomplished by combining graphs and maps so that variations in directional components or changes through time can be given the added quality of relative geographical location. For example, the upper diagram in Figure 12.34 shows, for each of the stations indicated, the amount of darkness, daytime cloud, and sunshine throughout the year. The lower diagram shows, among other things, the directional component and the percent frequency of winds for August in the same area.

There is almost no limit to the amount of information that we can somehow encode in these kinds of point symbols. Of paramount importance, however, is the question of how much of the information will actually be available to the reader. As long as the reader can obtain a sig-

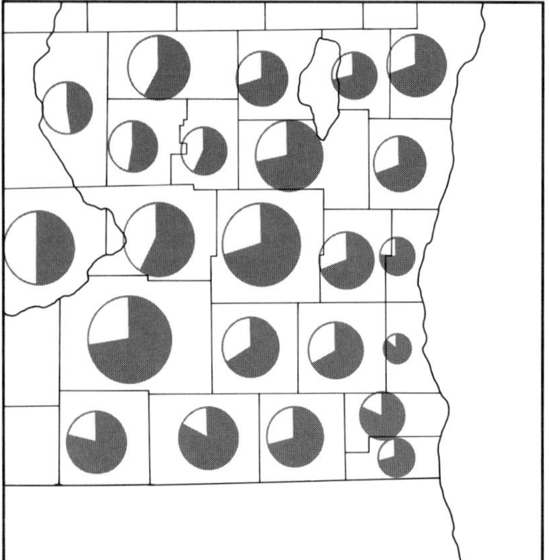

FIGURE 12.32 Land in farms and the percentage of that land available for crops by county in a portion of Wisconsin. The circles represent the land in farms. The percentage available for crops has been shaded on each circle.

FIGURE 12.33 The use of a complexly divided graduated point symbol (also with overlap). The reader may find it difficult to correctly estimate the various proportions of the symbols.

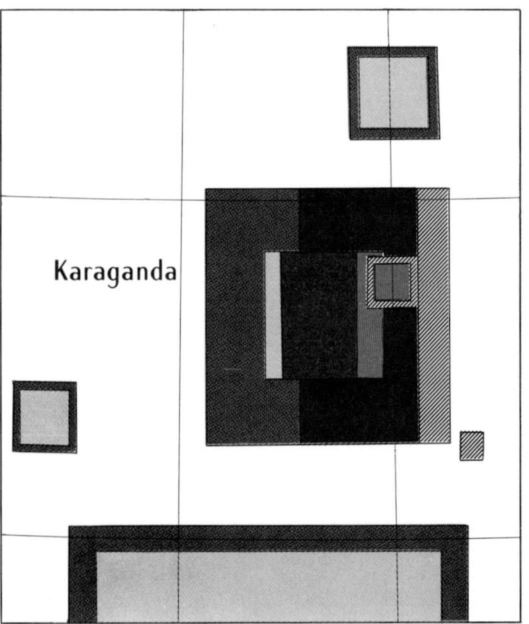

nificant geographical element (that is, an appreciation of similarities and differences from place to place) from the map, then the map is serving a purpose. If that part of the information is not readily obtainable, then the map is definitely a failure, and the data would probably be more easily used in tabular form.

Readability of Quantitative Data Symbolized by Point Symbols. The detailed readability of most point symbols has not been extensively studied. Therefore the precise degree to which a given symbolization scheme may be successful in communicating information to a map-reading audience is unknown. Most such research suggests that significant portions of even simple symbolization schemes may sometimes not be successfully communicated. One may assume that the more complex the symbolization scheme, the smaller the percentage of the information actually communicated to the map reader is likely to be, and therefore, cartographers should avoid complex, symbolization schemes. Users would have to be experts at scrutinizing symbols to obtain from the map information encoded with intricate schemes, and cartometric measurements from maps are time-consuming and exacting. Such information is much better provided in tabular form.

The use of detailed symbolization schemes should also be avoided for another reason. Data in machine-readable form make maps less necessary as storage devices for the spatial tabulation of data. Computer processing of such information is quicker and more exact than manual cartometric procedures. Many mechanical machine-driven plotters have the capability to execute any of the point symbolization schemes discussed in this chapter. Alternative scaling methods which require additional calculations are more accessible using machine-driven plotters than when the calculations must be performed manually. Therefore, there is no need for encoding large amounts of data into complex symbolization schemes. Machines can produce, in less time, several informative maps that employ simple schemes.

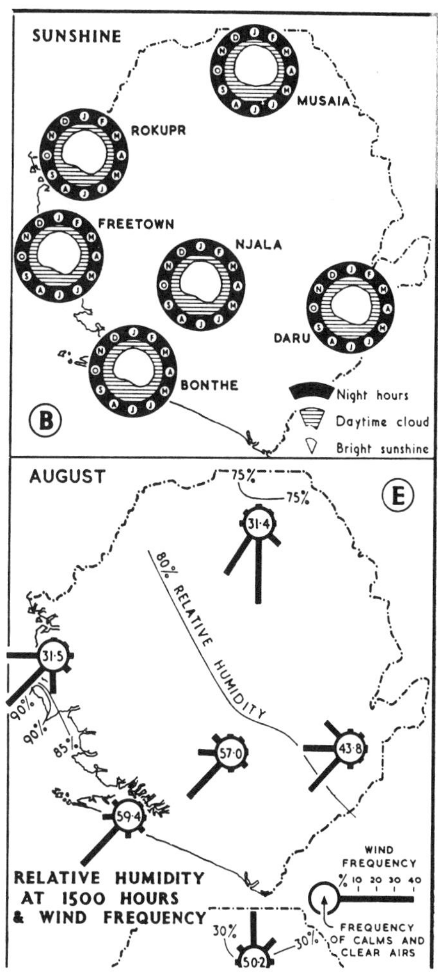

FIGURE 12.34 Cartographic representation of proportions of directional and sky condition data. (From J. I. Clarke, *Sierra Leone in Maps*, Copyright 1966, London: University of London Press, Ltd.)

Summary

It is easy to see that point symbols used to map place data can range from the mimetic or pictorial to the arbitrary or geometric regardless of the scale of measurement of the data set or the scale of measurement chosen for portrayal. All of the primary graphic elements can be easily employed (with the possible exception of texture) with point symbols. Spacing is limited to filling in ordinal, range-graded, or ratio point symbols. As such it can be used to reinforce other graphic elements when using point symbols.

For nominal portrayal of place data shape is probably the most useful graphic element. Hue, when it is available to the cartographer, is probably the second most useful. The graphic element direction or orientation is a distant third. (Review carefully the figures in this chapter.) For ordinal, range-graded, or ratio portrayal of place data size is probably the most useful graphic element. Value (the impression of tone) is also very important. Color intensity, when available, may also be used.

Point symbols can show several attributes of place data simultaneously, both nominally scaled (up to four attributes) and ordinally, range-graded, or ratio scaled (up to two attributes). The graduated circle is perhaps the most often used point symbol. However, it still communicates rather poorly to most map readers. Over the years cartographers have developed several different methods of scaling proportional point symbols to help map readers improve their ability to read data from a map. At all times, therefore, cartographers should take care to regard the map reader's capabilities, and a map should never be overloaded with complex point symbolization. Therefore, to attempt to show four or five attributes of place data with one complex symbolization scheme is probably doomed to failure from the start, not because it is impossible but because it will not effectively communicate to the map reader the information that the cartographer seeks to communicate.

VOLUME PORTRAYAL BY POINT SYMBOLS

Geographic volumes are of recurring concern to those who use maps. A geographic volume results from ordinal, interval, or ratio data collected over a geographic area, for example, census data. Enumerations are made, which refer to geographic areas, and the cartographer wishes to show the areal distribution of the data set, that is, its pattern or geographical arrangement, and to give some impression of the relative densities in different parts of the region.

Any real or conceptual quantity that can be thought of as existing in variable amounts from place to place can be symbolized by assigning a unit value to some point symbol and then putting the right number of these same-sized symbols in the right places on the map. This technique results in what is generally called a dot map, because simple round dots are the most frequently used symbol; however, any point symbol can be used. The kinds of quantities that can be portrayed in this way can range from slope values to people and even to percentages. Different colored or shaped dots (or other marks) can be used to show geographical mixtures, but fundamentally all we are doing is repeating a chosen symbol to show geographical frequency.

The Dot Map

To this point in our discussion we have been discussing the representation of data at a single point and its symbolization. The dot mapping method used for many years by cartographers uses a series of uniform point symbols to represent a quantity of data by the repetition of a point symbol. Each point symbol is equated to so much of the distribution being mapped, and its placement makes an attempt to show the location of the distribution. Conceptually the dot map is more complex than the more simplified point symbol map.

The dot map can show the details of the locational character of a distribution more clearly than any other type of map. Variations in pattern or arrangement, such as linearity and clustering, become apparent. The dot map provides an easily understood visual impression of relative density, accepted by the reader and easily interpreted on an ordinal scale, but it does not provide any absolute figures.*

Another advantage of dot maps is the relative ease with which they may be made manually. No computation is ordinarily necessary beyond determining the number of dots required by dividing the totals for each enumeration unit by the number selected as the unit value of each dot. The automation of the dot mapping method makes it even easier for the cartographer, since all that is necessary is for the cartographer to specify the unit value of the dot. In the automated production of dot maps, however, there still remains the lack of the accumulated knowledge that a cartographer may bring to the positioning of the dots on a manually produced map.

Although the map reader does not usually believe that he or she experiences particular difficulty in interpreting a dot map, research indicates that it is not as straightforward a symbolization as many have thought. It seems especially difficult for the untrained reader to estimate relative visual densities on an interval scale with much success, although ordinal decisions are apparently not difficult. The latter is probably the most important function to be performed by the dot map.

An important part of the cartographer's objective may be to have the reader gain considerable information regarding the various densities, relative as well as absolute, from the map. If so, the reader probably will need help. Most current dot maps have only a legend that tells the reader the magnitude of the unit value of the dot. It has been suggested that both design changes and reader training are needed to ensure that the reader gains the necessary information from the dot map.*

Dot maps ordinarily show only one kind of fact, for example, population or hectares of cultivated land but, by using different colored dots or different shaped point symbols, it is sometimes

*Theoretically, it would be possible to count the dots and then multiply the number by the unit value of each dot to arrive at a total but, in practice, this would be done only under duress!

*Judy M. Olson, "Experience and the Improvement of Cartographic Communication," *The Cartographic Journal* 15 (2) (December 1975): 94–108. One suggested design change is adding additional dots in dense areas to aid the reader. Suggested reader training consists of practice in the estimation of dot densities and an explanation of the tendency for underestimation of dot densities.

possible to include several different distributions on the same map.*

Making the Conventional Dot Map. Three important considerations affect the basic usefulness of a dot map: (1) the size of the dots, (2) the value assigned to a dot, and (3) the location of the dots.

The data needed for a conventional dot map consist of the enumeration of the number of items to be mapped in unit areas, usually civil divisions used as census statistical units. Rarely is it possible to employ so large a map scale that each single item may be shown by a dot, although theoretically the farmhouses on a topographic map might be termed a kind of dot map. Instead, it is usually necessary to assign a number of the phenomena to each dot. This is called the *unit value* of the dot. The number of dots is obtained by dividing the total number of items in each statistical division by the chosen unit value. For example, if a county had a total of 6000 hectares in corn, and a unit value of 25 hectares per dot had been chosen, then 240 dots would be placed in the county to symbolize the area devoted to corn production.

The Size and Value of the Dot. If the visual impression conveyed by a dot map is to be realistic, the size of the dot and the unit value assigned must be carefully chosen. The five dot maps shown here have been prepared from the same data; only the size or number of dots used has been changed. The maps show areas of potato production in Wisconsin.

If the dots are too small, as in Figure 12.35, the distribution will appear sparse and insignificant, and patterns will not be visible. If the dots are too large, they will coalesce too much in the darker areas, as in Figure 12.36, and give an

FIGURE 12.35 A dot map in which the dots are so small that an unrevealing map is produced. Each dot represents 16.2 hectares in potatoes in 1947.

FIGURE 12.36 A dot map in which the dots are so large that an excessively "heavy" map is produced, giving an erroneous impression of abundant potato production. The same data and number of dots are used as in Figure 12.35.

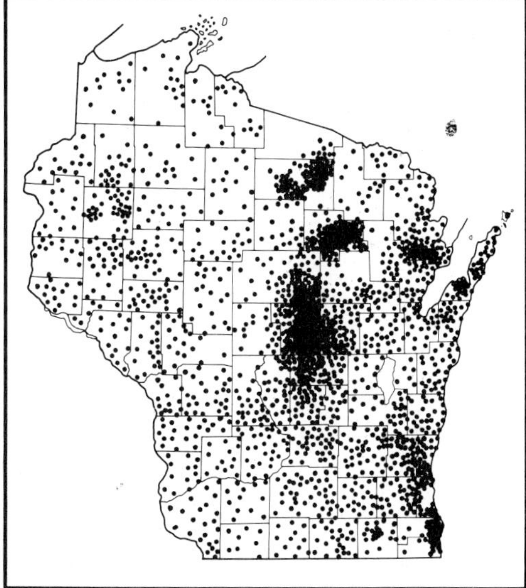

*A good analysis of the conceptual variations in dot mapping is given by Richard E. Dahlberg, "Towards the Improvement of the Dot Map," *International Yearbook of Cartography* 7 (1967): 157–67.

overall impression of excessive density that is equally erroneous. It appears in Figure 12.36 that there is little room for anything else in the region. Furthermore, when dots are so gross, they dominate the base data and generally result in an ugly map.

Equally important is the selection of the unit value of the dot; naturally, the two problems (size and unit value of the dot) are inseparable. The total number of dots should neither be so large that the map gives a greater impression of accuracy than is warranted, nor should the total be so small that the distribution lacks any pattern or character. These unfortunate possibilities are illustrated in Figures 12.37 and 12.38.

The selection of unit value and size of dots should be made so that in the denser areas (of a dense distribution) the dots will just coalesce to form a dark area. Figure 12.39 is constructed from the same data as the preceding examples but with a more wisely chosen dot size and unit value. Of course, if the distribution is sparse everywhere, then even the relatively dense areas should not appear dark.

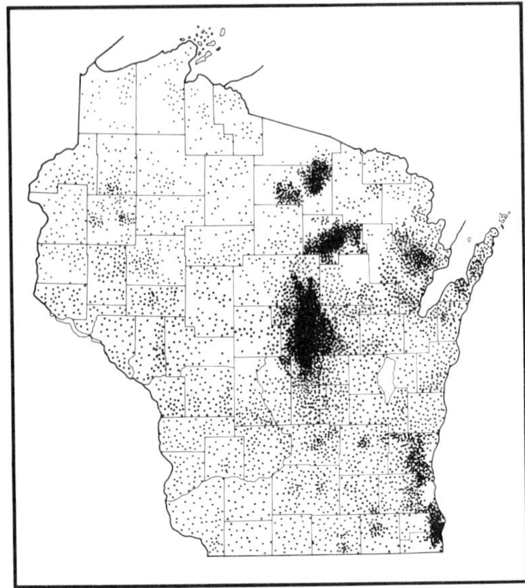

FIGURE 12.38 A dot map in which the unit value assigned to one dot is too small. Many dots must be drawn, causing the production of an excessively detailed map. The dots are the same size as those in Figure 12.37. Each dot in this example represents 67.0 hectares.

FIGURE 12.37 A dot map in which the unit value assigned to one dot is too large. Few dots can be drawn resulting in a barren map that reveals little pattern. Each dot in this example represents 60.7 hectares.

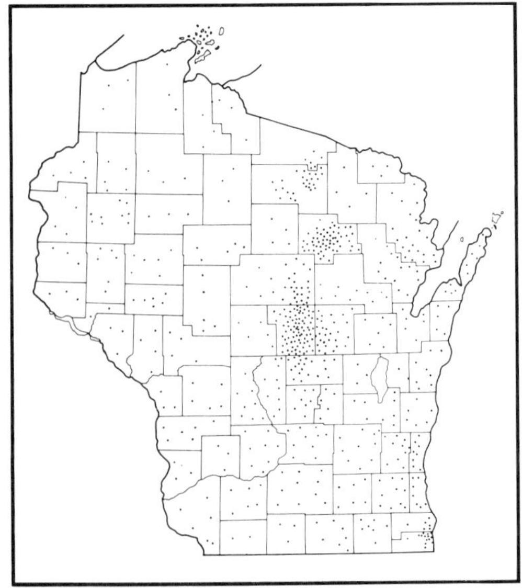

On relatively large-scale dot maps showing land use phenomena, it is possible to relate the dot size to the scale of the map by making the dot cover the scaled actual area. For example, we might make a map showing the area planted in wheat with a unit value of 500 hectares and a dot size that covers exactly 500 hectares at the scale of the map. We should not follow such a procedure blindly, however, since the elusive quality, "relative importance," of areal phenomena is often dependent on factors other than strict area relationship. By varying the relationship between dot diameter and unit value, the cartographer can settle on a compromise between the two that will best present the characteristics of the distribution.

Professor J. Ross Mackay developed an ingenious nomograph to assist in determining the de-

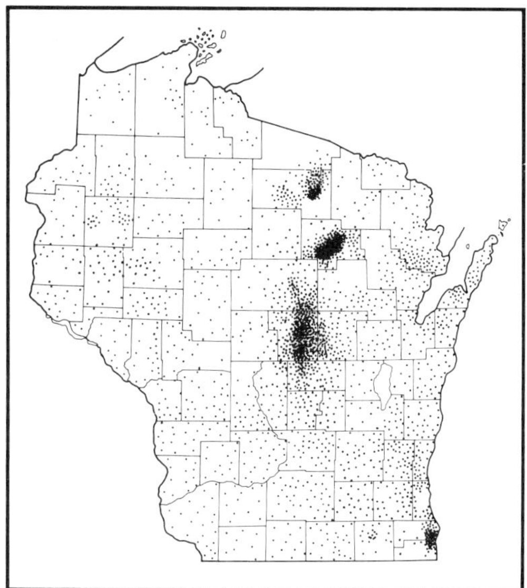

FIGURE 12.39 A dot map in which the dot size and dot unit value have been more wisely chosen than in the preceding examples. Each dot in this example represents 16.2 hectares.

sirable dot size and unit value.* This graph, with metric additions shown in Figure 12.40, requires a knowledge of the sizes of dots that can be made by various kinds of pens.

The Use of the Nomograph. The nomograph may be used in several ways, but perhaps the easiest is first to select three unit areas on the proposed dot map that are representative of a dense area, an area of average density, and an area of sparse density. A tentative unit value can then be selected and divided into the totals for each of the three statistical divisions. The map area of one of the divisions can be estimated in terms of square centimeters, and the number of dots per square centimeter can be deduced.

As in the previous example, assume that a county contains 6000 hectares of corn; a unit value of 25 hectares per dot is chosen, resulting in 240 dots to be placed within the boundaries of the division; assume further that the statistical division on the map covers 4.0 cm^2. This would mean that the dots would be placed on the map with a density of 60 dots per square centimeter. An ordinate at 60 is erected from the X axis of the nomograph. A radial line from the origin of the nomograph to a given dot diameter on the upper right-circular scale will intersect the ordinate. The location of the intersection on the interior scale will show the average distance between the dots if they were evenly spaced. The height of the intersect on the Y axis will indicate what proportion of the area will be black if that dot diameter and the number of dots per square centimeter were used. Also shown is the "zone of coalescing dots," at or beyond which dots will fall on one another.

If the initial trial seems unsatisfactory for each of the three type areas selected for experimentation, either the unit value or the dot size or both may be changed. We can enter the graph with any of several assumptions and determine the derivatives. It is good practice while using the nomograph actually to dot the areas (on a piece of plastic or tracing paper) in order to see the results and to help visualize the consequence of other combinations.

The cartographer should remember that the visual relationships of black to white ratios and the complications introduced by the pattern of dots make it difficult, it not impossible, for a dot map to be visually perfect. The best approach is by experiment after narrowing the choices by use of the nomograph.

Locating the Dots. Theoretically, the ideal dot map would be one with a large enough scale so that each single unit of the point data could be precisely located. Ordinarily, dot maps are small scale, but sometimes if the data are sparse enough, a unit value of one can still be used. It is usually necessary, however, to make the unit value of the dot greater than one, and the problem then

*J. Ross Mackay, "Dotting the Dot Map," *Surveying and Mapping* 9 (1949): 3–10.

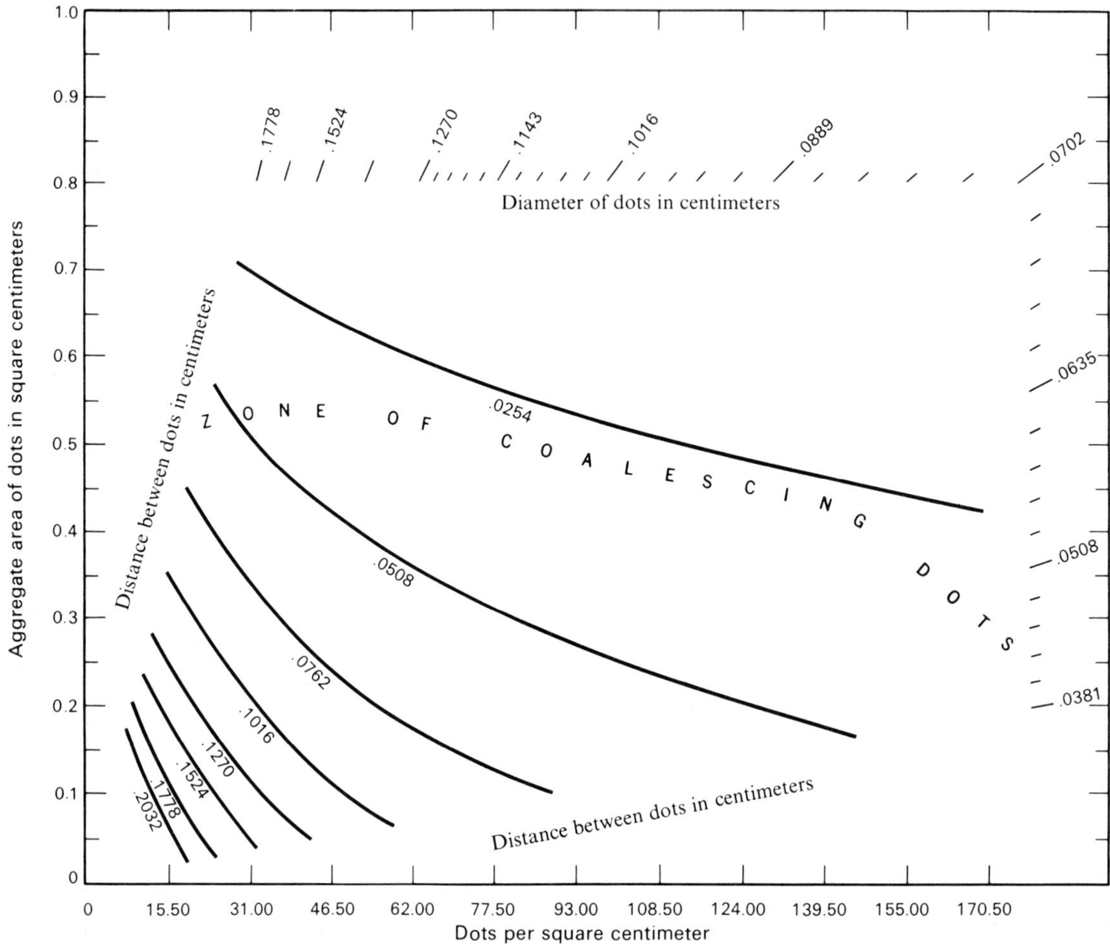

FIGURE 12.40 A nomograph showing the relationship between dot size and dot density. (Nomograph courtesy of J. R. Mackay and *Surveying and Mapping*, with metric notations added by author.)

arises of locating the one point symbol that represents several differently located units (see classification Fig. 11.20).

When generalizing, it is often helpful to consider the group of several items as having a kind of center of gravity and to place the symbol at that point. Usually, all the cartographer knows from the original data is the total number of dots to place within a unit area. Consequently, every available source of information must be used to assist the placement of the dots as reasonably as possible; such sources may be topographic maps and other distribution maps that correlate highly with the map being prepared. The quality of the finished map depends largely on the ability of the cartographer to generalize the distribution, that is, to bring all the pertinent evidence to bear on the problem of where to put the dots.

Considerable detail can be introduced into the map if it is prepared on the basis of the smallest civil divisions and is then greatly reduced. Placing the dots with reference to minor civil divisions can be done easily by using translucent material for the map and putting under it a map that shows the minor civil divisions to serve as a guide. Only the larger administrative units need ordinarily be shown as base data on the finished map. Special care must be exercised not to leave the guiding boundary areas relatively free of dots; if this is done they will show up markedly in the final map as white zones. We should also refrain from inadvertently producing wormlike lines of dots, unwanted clusters, or "tweed" patterns when placing the dots. Such regularity can easily occur and is quite noticeable by contrast with its amorphous surroundings. Machine-produced dot maps usually avoid these problems, but they suffer equally from a different set of possible pitfalls.

A dot map in which the requisite numbers of dots are evenly spread over the unit areas, like a patterned area symbol (although numerically correct), could better use some symbol system other than repetitive dots. Simple computer routines often err by spacing the dots too evenly. The dot map is best for distributions that have distinct internal arrangements, such as linearity, or are tightly clustered and dense in one area and sparse in other areas. Phenomena that exhibit uniform distribution characteristics are not well mapped by the dot map technique.

With increasing research more and more experimental dot maps can be expected to appear. The usefulness of the conventional dot map has assured that research devoted to its improvement will continue. The cartographer is well advised to warn the map-reading audience by appropriate statements in the map legend if the conventional dot map technique has been altered in any manner.

Machine Production of Dot Maps. Machine production of dot maps is possible. Figure 12.41 was prepared by one of the first dot mapping algorithms to be developed. The resulting map

FIGURE 12.41 A machine-produced dot map. (from Census of Agriculture, 1969, Vol. V, *Special Reports*, Part 15, Graphic Summary, p. 55. Washington, D.C.: U.S. Government Printing Office, 1973.

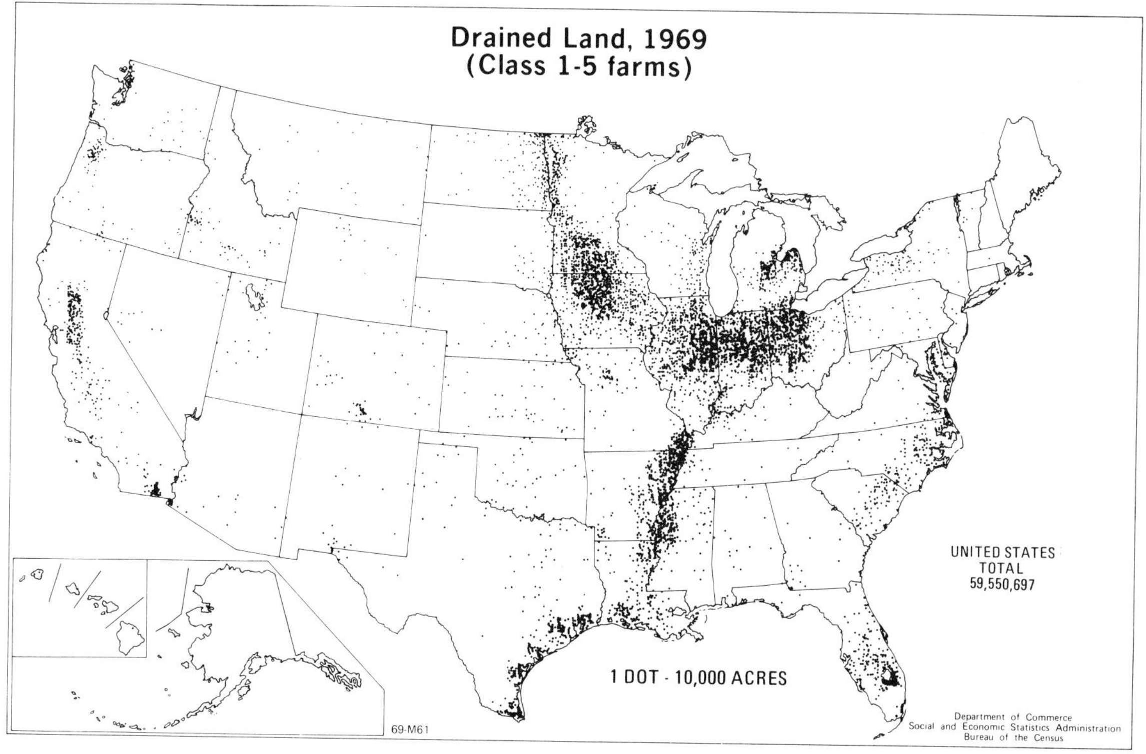

varies little from manually produced dot maps, but the savings in time and cost are large. Undoubtedly dot maps of the future will be predominantly done by machine. Whether machine or manually produced, the same factors requiring caution by the cartographer still apply.

The advantages of the machine production of dot maps are speed and ease of execution. Several maps can be tried using different levels of input parameters, such as one dot represents ten, fifty, or one hundred units. The cartographer can assess the resulting maps and use the one that most closely communicates the information that the cartographer wishes to communicate.

The disadvantage of machine production is that the accumulated knowledge that a cartographer brings to the placement of the dots on a conventional dot map is lost in machine processing. Algorithms that include statistical routines that contain varying probabilities for the placement of a dot at all of the possible positions are available, and the cartographer can use other machine-readable distributions to effect the dot placement. Undoubtedly further experiments will be made in this area.

Finally, perceptual research indicates a slight tendency for map readers to underestimate both the numbers of dots in an area and the difference between dot densities in two areas. For cartographers to be aware of this problem is to forewarn them. The usefulness and popularity of the dot mapping method for portraying volumetric data by repetitive point symbols insures that many dot maps will be produced in the future.

SELECTED REFERENCES

Arnberger, E., "Problems of an International Standardization of a Means of Communication Through Cartographic Symbols," *International Yearbook of Cartography* 14 (1974): 19–35.

Chang, K. T., "Measurement Scales in Cartography," *The American Cartographer* 5 (1978): 57–64.

Dahlberg, R. E., "Towards the Improvement of the Dot Map," *International Yearbook of Cartography* 7 (1967): 157–66.

Flannery, J. J., "The Relative Effectiveness of Some Common Graduated Point Symbols in the Presentation of Quantitative Data," *The Canadian Cartographer* 8 (1971): 96–109.

Groop, R. E., and P. Smith, "A Dot Matrix Method of Portraying Continuous Statistical Surfaces," *The American Cartographer* 9 (1982): 123–30.

Hsu, M. L., "The Cartographer's Conceptual Process and Thematic Symbolization," *The American Cartographer* 6 (1979): 117–27.

Jenks, G. F., "Contemporary Statistical Maps—Evidence of Spatial and Graphic Ignorance," *The American Cartographer* 3 (1976): 11–18.

Morrison, J. L., "A Theoretical Framework for Cartographic Generalization with Emphasis on the Process of Symbolization," *International Yearbook of Cartography* 14 (1974): 115–27.

Provin, R. W., "The Perception of Numerousness on Dot Maps," *The American Cartographer* 4 (1977): 111–25.

Robinson, A. H., "An International Standard Symbolism for Thematic Maps: Approaches and Problems," *International Yearbook of Cartography* 13 (1973): 19–26.

Rogers, J. E., and R. E. Groop, "Regional Portrayal with Multi-pattern Color Dot Maps," *Cartographica* 18 (1981): 51–64.

Slocum, T. A., "Analyzing the Communicative Efficiency of Two-sectored Pie Graphs," *Cartographica* 18 (1981): 53–65.

13

Mapping with Line Symbols

MAPPING LINEAR DATA

A surprising amount of linear information appears on most maps, but often neither the cartographer nor the map reader is fully aware of it. This information includes the outline itself, which defines the area of concern, the graticule definition, coastlines, political boundaries, rivers, roads, and much more. For the majority of these data, line symbols portray the linear features. Often the cartographer's design ingenuity is taxed to the limit by the range of linear features that are to be differentiated on the map. Providing sufficient contrast between these diverse data sets to enable efficient communication is often difficult, and many maps are less than satisfactory because of shortcomings in the line symbolization.

These linear data sets exist in the real world on nominal scales of measurement (for example, coastlines), and on ordinal scales (for example, the hierarchy of international boundaries, state boundaries, and county boundaries). Interval linear data sets include some kinds of flow lines (again, in reality the data may be on a ratio scale but mapped on a range-graded scale). Many flows are measured on a ratio scale, for example, gallons of water passing a stream gauging station.

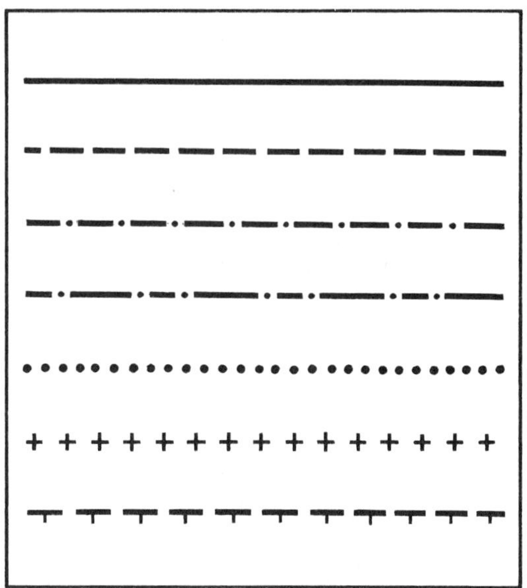

FIGURE 13.1 Examples of lines of differing character which are useful for the symbolization of nominal linear data.

Mapping Linear Data Sets with Line Symbols

Examples of line symbols used to portray nominal attributes of linear data sets are easy to find (for example, coastlines, graticules, and currents). Examples of the line symbols that can be used to portray these nominal data characteristics are shown in Figure 13.1. The use of line symbols to show ordinal data is illustrated in Figure 13.2. Figure 13.3 illustrates the range-graded portrayal of flow data, and Figure 13.4 illustrates the ratio portrayal of the same data given in Figure 13.3. Figures 13.3 and 13.4 illustrate generalization by the use of symbolization, Figure 13.3 being more generalized than Figure 13.4.

Linear phenomena can be placed on the same mimetic-to-arbitrary continuum as were place phenomena, but the distinction is often less useful, since mimetic lines are difficult to distinguish from arbitrary ones in most cases. A line in reality may simply be the abutment of two zones of differing character. It has no characteristics of its own, and thus the continuum from mimetic to

FIGURE 13.2 The use of line width (size) enhanced by the use of line character to denote the ordinal portrayal of civil administrative boundaries.

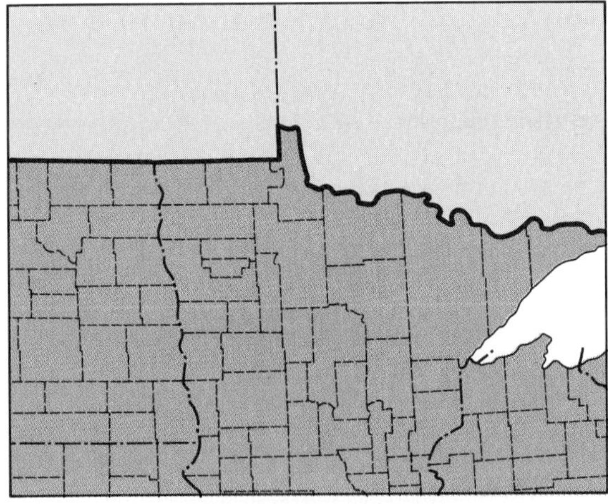

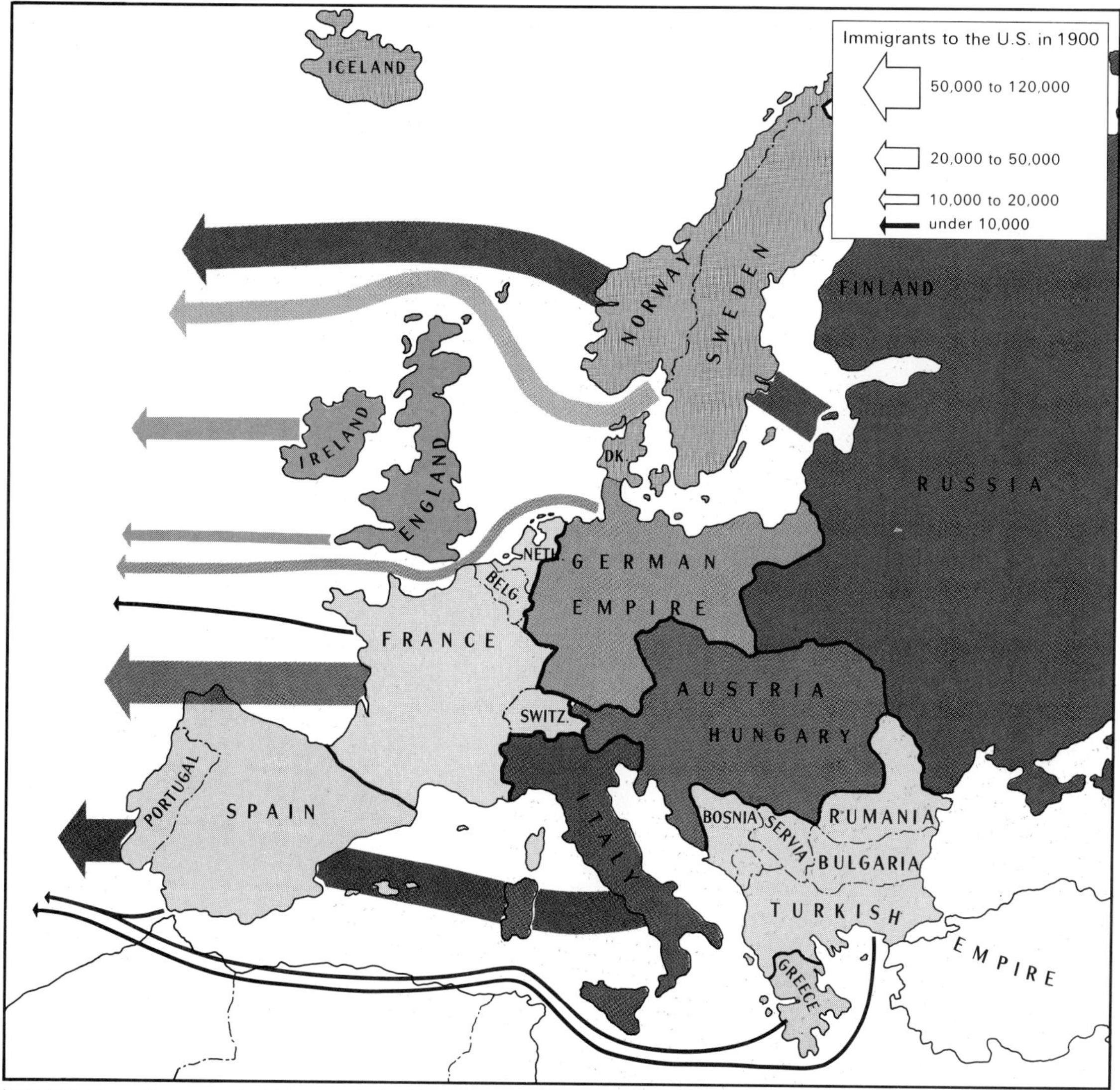

FIGURE 13.3 Range-graded line symbols. On this map of immigrants from Europe in 1900, lines of standardized width are used to represent a specified range of numbers of immigrants.

arbitrary does not exist. One of the most mimetic line symbols is the conventional line with cross ticks used to symbolize a railroad.

Mapping Using Line Symbols

Line symbols can be used to portray either place or volumetric ratio data in addition to linear data. The portrayal of ratio place data by line symbols can be accomplished either by their portrayal as flows or as gradients. For example, both the flow of emigrants from the rural southern United States to Chicago and the number of passengers flying from Chicago to New York represent flows measured at positions on the earth. Gradients may be portrayed by assuming a statistical surface such

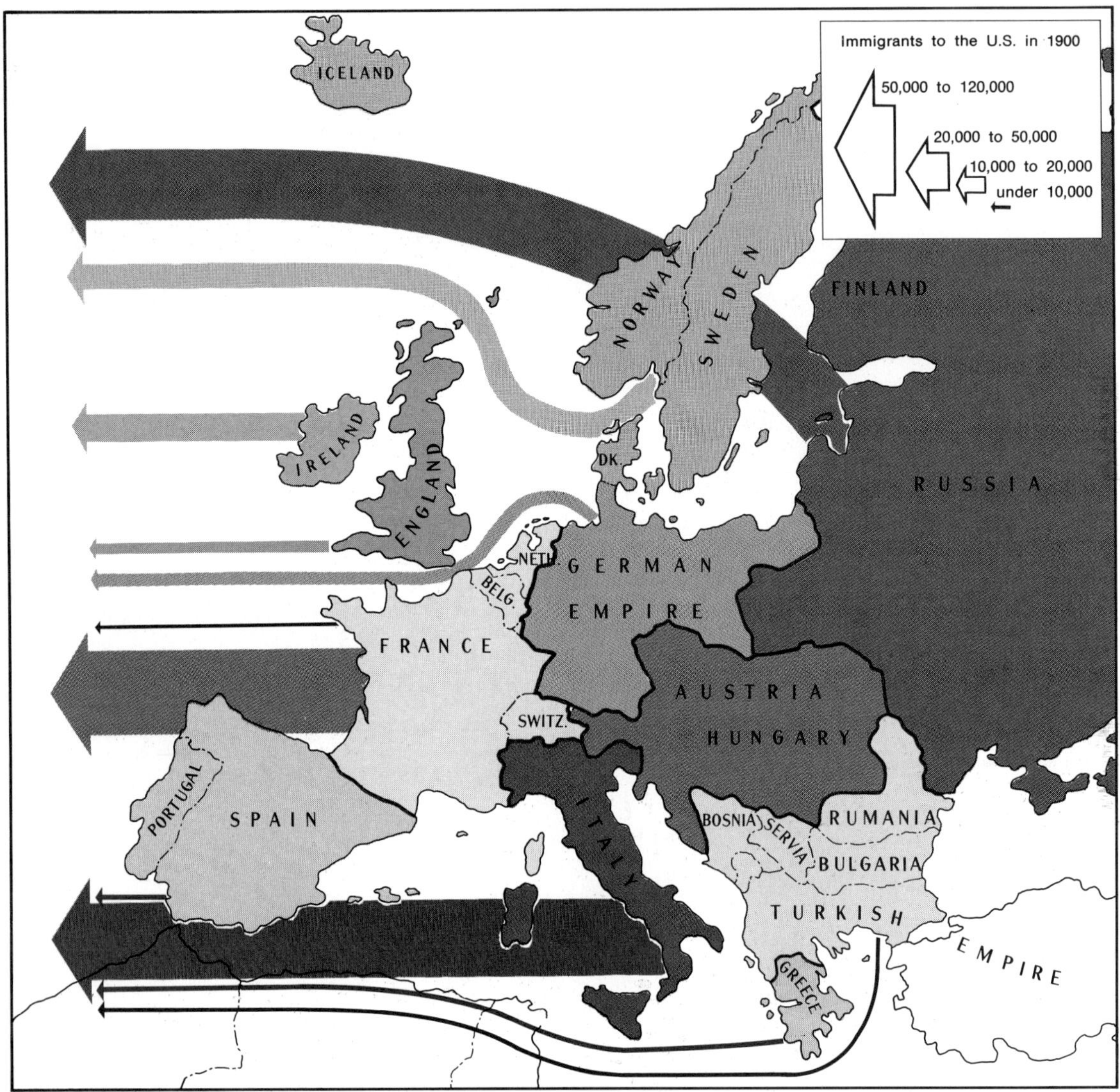

FIGURE 13.4 Ratio-scaled line symbols. On this map, which shows the same data as Figure 13.3, the lines are scaled proportional in width to the actual number of immigrants.

as might be formed by elevations above sea level or atmospheric pressures, both measured at positions.

Interval or ratio volumetric data referring to areas can also be portrayed by the use of line symbols. For example, the land surface form and population density can be portrayed by isarithms, or exports from one nation to another can be shown by flow lines. Tonnage shipped along rail lines or waterways can be shown by graduated flow lines, and profiles of a statistical surface can be shown by line vectors.

Three primary breakdowns, therefore, exist for mapping using line symbols: (1) nominal portrayal, using primarily variations in the shape of the line symbols, and also hue, if color is available; (2) portrayal of flows by varying the sizes of lines and secondarily employing differences in value or spacing, and (3) the assumption of a statistical surface (ratio-scaled volume) and its portrayal by line symbols.

The primary graphic elements also apply to mapping line symbols, but as was the case with point symbols, the use of spacing is somewhat restricted. Only where the lines are sufficiently wide for spacing to be discernible can we use this graphic element in line symbolization. Since the use of spacing is dependent on the width of the line symbol, spacing ordinarily can be used only in conjunction with ordinal, range-graded, or ratio portrayals, but since spacing connotes nominal scaling, it usually is used to map a nominal characteristic of a linear data set where the principal characteristic being mapped is ordinal, range-graded, or ratio-scaled.

The graphic element orientation also presents a problem in the symbolization of linear phenomena. Lines on the earth tend to have directions of their own, independent of their symbolization. Thus, it is very difficult to vary the orientation of most line symbolizations. However, directions measured at places, for example, ocean currents or wind directions, can be symbolized by arrows, and the orientation of these arrows is usually significant. Such symbols may be considered point symbols, and the distinctions between whether a symbol is a point or line symbol become rather subjective.

This chapter will first discuss how the primary graphic elements may be used in mapping data with line symbols and then will look at the portrayal of nominal data sets. Ordinal, interval, and ratio data sets will be treated under the portrayal of flows. Finally the portrayal of a statistical surface by line symbols will be discussed.

Graphic Elements Used in Mapping with Line Symbols

The graphic elements of shape, size, hue, and value are all useful in mapping data sets by line symbols. Problems come in trying to use orientation or spacing. The definitions of the primary graphic elements as they relate to linear symbols do vary somewhat from their use in point symbols. Size, value, and hue remain the same, but shape is somewhat different. When we speak of the shape of a line symbol, we are really talking about different line characters. Figure 13.1 illustrates some different lines that cartographers routinely use. In the following discussion we will refer to the variations among the lines in Figure 13.1 as differences in line character.

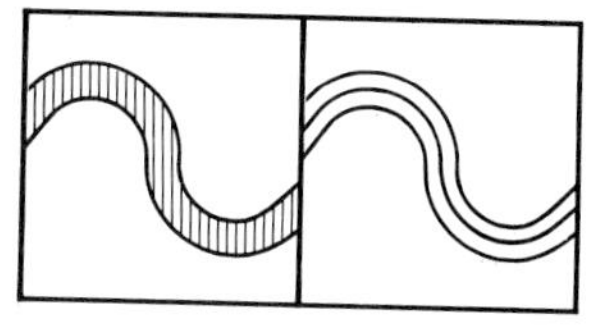

FIGURE 13.5 The graphic element orientation as applied to line symbols.

Orientation presents a different problem. Bertin defines line orientation as in Figure 13.5. Accepting Bertin's definition of line orientation, we must conclude that the usefulness of the graphic element orientation when applied to line symbols must be very limited. The graphic element spacing is likewise less important in mapping with line symbols. As we said, we will consider spacing as merely a secondary variable that cartographers can use to reenforce visual differentiation in line symbols that are graduated by width (size).

Although it is true that value and the dimension of color that we called intensity or chroma in Chapter 8 can be used for linear data symbolization, the effectiveness of these variables depends upon the width of the line symbol that uses them. Very narrow lines cannot be symbolized using value or color intensity to show ordinal, range-graded, or ratio-scaled attributes of a data set because the variations in value and color intensity are too subtle to allow easy visual differentiation. Even hue is less effective as the line symbol becomes thinner.

The primary graphic elements, therefore, for mapping linear data are shape (as defined above), size, and hue.

Mapping Nominally Scaled Data Using Line Symbols. When symbolizing nominally scaled linear data, such as coastlines or the routes of explorers, the cartographer must work primarily with the shape or character of the line symbol. The width of the line may also be varied, but only to the extent that this variation in size does not connote ordinal ranking. Consequently, varying line widths of the same character can be used only for widely divergent features. For example, a thin solid line could be used for the coastline and a thick solid line for the neat line without connoting ordinal ranking. However, a thin solid line for roads and a thick solid line for county boundaries could result in confusion, since some roads could follow county boundaries, and the varying widths could therefore connote an ordinal relationship among the roads.

The character of line symbols can be varied in many ways as seen in Figure 13.1. Depending on the cartographic situation at hand, the cartographer may wish either to select lines of varying character that will appear essentially equal in importance, or to use the character of line symbolization to establish relative visual importance of the different classes of nominally scaled data.

Several factors can affect the appearance of line symbols. These factors are: (1) size, (2) continuity, (3) brightness contrast, (4) closure, (5) complexity, and (6) compactness.*

Size variation in nominal symbolization has at least two consequences. First, the larger of two lines of the same color will appear more prominent, and without elaboration connote ordinal ranking. Therefore, the cartographer must be wary of using size variation in this way when symbolizing nominally scaled data. Second, size can be used to imply direction in two ways by varying size along a single line. For example, in the dotted lines in Figure 13.6, the larger dots in the foreground trailing to the smaller dots cause the smaller dots to recede into the background; this connotes direction. Varying the width of the solid line in Figure 13.7, as is often done in symbolizing rivers, also creates the effect of the direction of flow of a stream (it being presumed that in humid areas a river becomes larger as it flows from its source to its mouth).

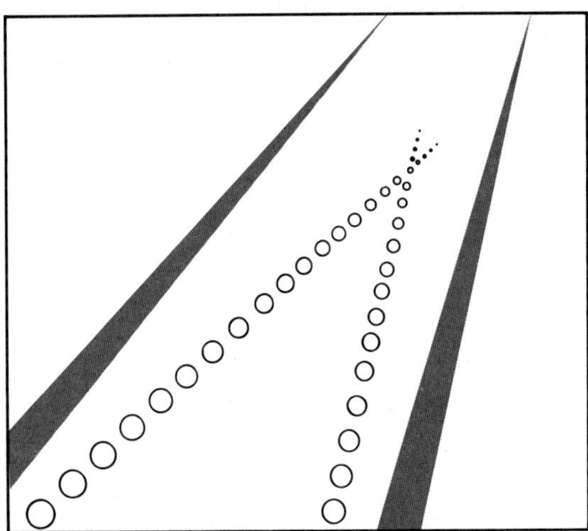

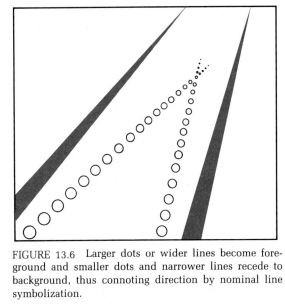

FIGURE 13.6 Larger dots or wider lines become foreground and smaller dots and narrower lines recede to background, thus connoting direction by nominal line symbolization.

Continuity is an important quality for the line symbolization of nominal data. It is illustrated in Figure 13.1. Continuity can vary from a solid line to a series of essentially point symbols that depend on proximity to establish closure and create a line or figure. Linear features occurring on the earth (such as roads, railroads, rivers, coastlines) tend to be shown with unbroken or solid lines. Imaginary features (political boundaries, the graticule, or coordinate system lines) may be shown with broken lines. This is only convention, and the convention is not strong except for perhaps roads and railroads. Often lines are broken to make room for other data. It would be relatively easy to find a broken coastline or power transmission line on a map although both are contin-

*Karen Pearson, "The Relative Visual Importance of Selected Line Symbols," unpublished master's thesis, Department of Geography, University of Wisconsin-Madison, 1971.

uous linear features as they occur on the earth. Likewise, the graticule may be shown by a solid line, especially over water; the division noted above is probably only true for a majority of maps. The relative perceptual merits of continuity are less well known, but there is some evidence to support the statement that a dotted line is judged less important than a solid or dashed line of equivalent width.

Brightness contrast affects the perception of line symbols by the use of the graphic element value. Lines of equivalent width but of contrasting brightness may be perceived as varying in importance. The darker tones usually connote "more" importance. Since this connotes an ordinal scale, if a nominal portrayal is sought, brightness contrast must be avoided.

FIGURE 13.7 Example of a stream network where direction of flow is connoted by increased width of lines. (From *Fundamentals of Physical Geography*, 3d ed., by Trewartha, Robinson, Hammond and Horn, Copyright 1977, McGraw-Hill, New York. Used with permission of McGraw-Hill Book Company.)

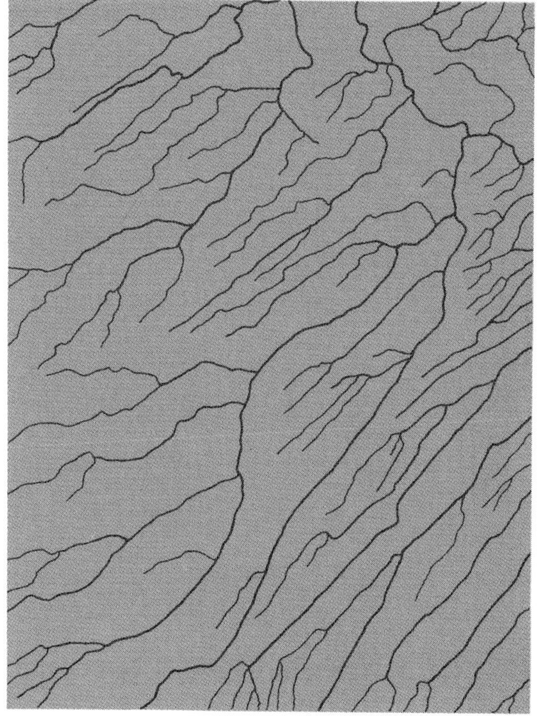

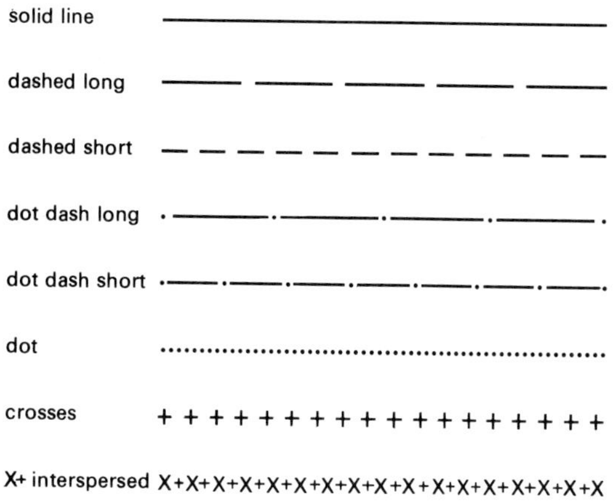

solid line
dashed long
dashed short
dot dash long
dot dash short
dot
crosses + + + + + + + + + + + + + + + + + + +
X+ interspersed X+X+X+X+X+X+X+X+X+X+X+X+X+X+X+X+X

FIGURE 13.8 Increasing complexity in the marks that compose a line symbol. The solid line is the simplest, and the x+ interspersed line is the most complex.

Visual closure occurs among groups of similar marks; therefore a series of dashes or dots will be viewed as a whole. The more closure a symbol has, the more visually prominent it is likely to be (see Fig. 13.1); hence, a solid line should be the most prominent.

Complexity is a term with many meanings. In this context it refers to the sequence of marks that compose the line symbol. All things being equal, the more complex a line, the more visually significant it will be. Figure 13.8 provides one example of increasing line complexity.

Compactness, another factor in nominal line symbolization, refers to the regularity of the symbol. The more compact linear symbol will possess greater visual weight, and a solid line is the most compact. It may only be possible to grade other levels of compactness by referring each line independently to a solid line.

These six factors allow the cartographer to symbolize nominal data by employing lines of differing types that either have equal visual importance or form a hierarchy of visual importance consistent with the objective of the map. Practice is the best teacher in the selection of which characteristic to vary in nominal scale line symbolization. The cartographer who is aware of the visual effects resulting from the above characteristics of line symbols should be in a position to employ lines suitably for any cartographic situation.

Mapping Flows by Line Symbols. The principal method of symbolizing ordinal, range-graded, or ratio data by line symbols portrays "connections" or "movements" along routes. For example, the flow of petroleum from the Middle East to Europe, of vacationers from the urban northeast of the United States to Florida, or the draft capacity of a canal system all occur along specific routes, either in the actual sense with the canals or in the more general sense with the commodity or population flows.

The techniques used in the construction of flow maps are easily observed from well-made examples. Range-graded or ratio scaling is most often accomplished by varying the thickness of the lines in proportion to the values, using a convenient unit width. The unit width is selected so that division of the data by the unit width results in a number that represents a map dimension in inches or millimeters, in the same fashion that a unit value is used to determine the drafting sizes of graduated circles. For example, if we were representing the flow of freight traffic between two cities, we could use a unit value of 1 mm for each 1 million metric tons; therefore, if the traffic total were 4.5 million metric tons, a line 4.5 mm wide would be shown between the two cities. There appears to be no tendency toward underestimation by map readers who view flow lines of variable width. Therefore, psychological scaling does not differ from numerical scaling. Lines may be shown as smooth curves, as in Figure 13.9, or as angular lines, as in Figure 13.10.

Movement along an actual route may be represented, or as in "origin and destination" maps the terminal points may simply be connected by straight lines. Arrowheads at the ends of lines are often used to show the directions of movement, although the varying thickness and angle with which "tributaries" enter frequently show flow adequately. Arrows or patterns may be placed along the lines, as in Figure 13.11, to show the direction or even to show relative movement in opposite directions. Lines can increase or decrease in width as the values change, but between tributaries a line should maintain a uniform width. Tributaries should enter smoothly in order to enhance the visual concept of movement.

In some instances, the range of the data is so large that a unit width value capable of allowing ratio differentiation among the small lines would render the large ones much too large. Conse-

FIGURE 13.9 Smoothly curving lines show movement on this flow map. (Map by G. B. Lewis from G. Manners, "Transport Costs, Freight Rates and the Changing Economic Geography of Iron Ore," *Geography* 52 (1967): 260–79.)

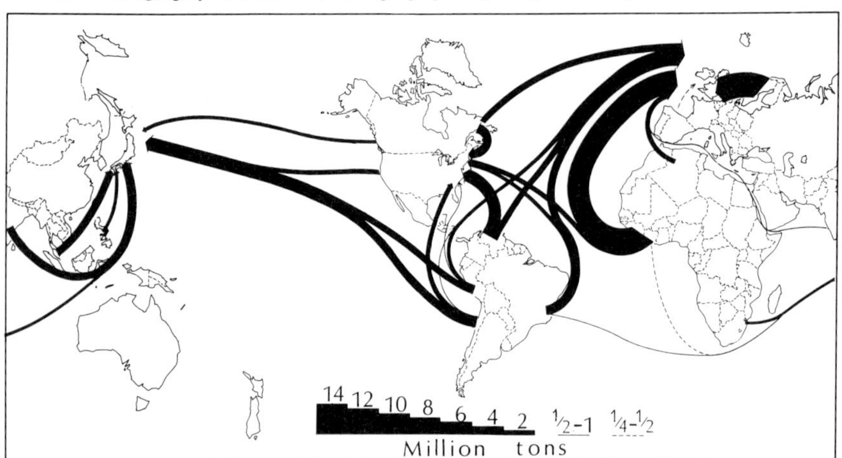

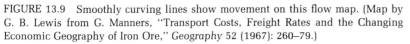

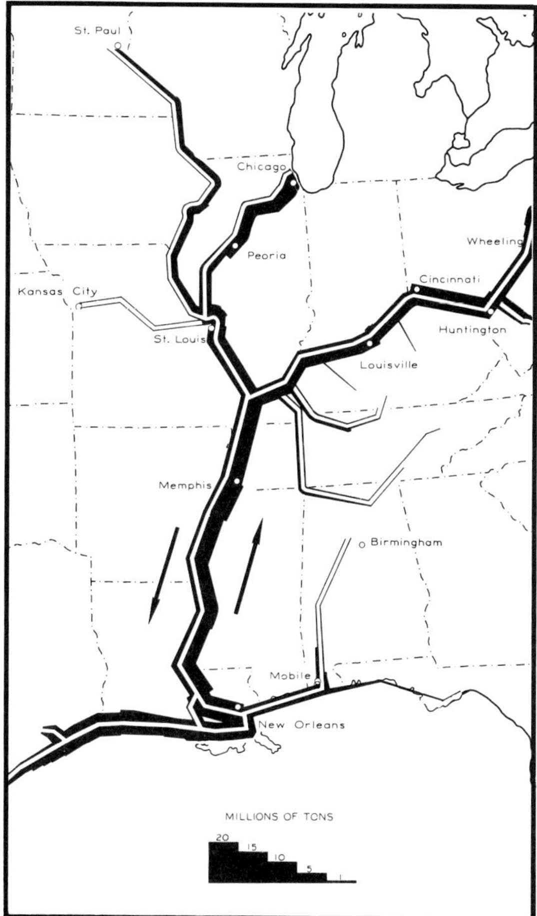

FIGURE 13.10 Angular lines show movement on this flow map. (From E. L. Ullman, *American Commodity Flow*, Copyright 1957, Seattle: University of Washington Press.)

quently, it is sometimes necessary to symbolize the smaller lines in some way, such as by dots or dashes (see Fig. 13.11).

Instead of using a system of symbolization that depends on variations in width, we can employ alternative methods. For example, we may replace the single line with several parallel lines, letting each of the lines represent a unit value, as in dot mapping. Another alternative is to categorize the data and symbolize the various classes with symbols that vary in pattern and value.

Range-graded line symbolization is accomplished, like all range-graded symbolization, by a finite number of symbol sizes, each representing an interval part of the range of the data. Figure 13.3 illustrates range-graded flow lines.

In general, the construction of proportional flow lines is a relatively simple task that is made even easier because human visual response to the varying widths of flow lines seems to be quite accurate and thus does not require scaling adjustments. Computer assistance is also available to aid the cartographer in the preparation of maps. Because of their excellent perceptual qualities, proportional flow lines are highly recommended where appropriate.

MAPPING THE STATISTICAL SURFACE BY LINE SYMBOLS

A statistical surface implies a base datum and a distribution of z values on an ordinal, interval, or ratio scale measured at right angles to that datum. By connecting the z values, a smoothly undulating *statistical surface* is formed, and the character of its surface is displayed on the map. The character of the statistical surface can be portrayed by at least four different categories of line symbols.

FIGURE 13.11 Arrows are used to show direction of movement on this flow map. (Map drawn by R. P. Hinkle. From E. L. Ullman, *American Commodity Flow*, Copyright 1957, Seattle: University of Washington Press.)

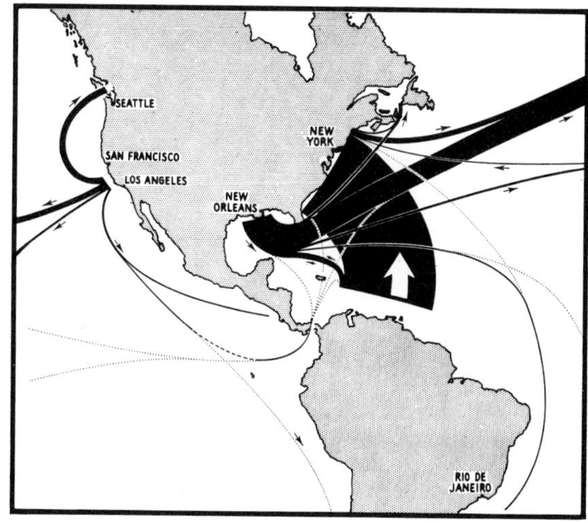

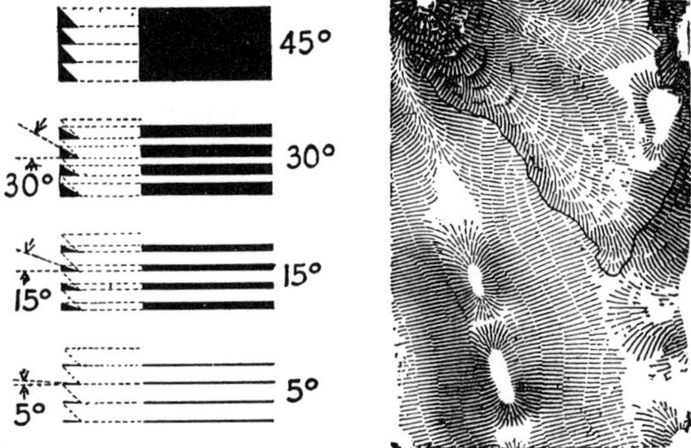

FIGURE 13.12 The line symbolization of slopes or gradients by the use of hachures. In this diagram hachures are equally spaced but vary in thickness depending on the value of the gradient being portrayed (Lehman's system.) (From *General Cartography* by E. Raisz, Copyright 1948, McGraw-Hill, New York. Used with permission of McGraw-Hill Book Company.)

One category uses flow lines. When the slopes of a statistical surface are considered as values existing at points, the surface may be portrayed by a line symbol called a *hachure*. A series of short lines are drawn parallel to the slopes on the statistical surface. Two options are available. In one the short parallel lines are equally spaced, and their thickness is varied to represent the steepness of the slope of the surface (Fig. 13.12). This system was specified in detail by the noted German cartographer, Lehman, around 1800. In the other variation the short flow lines are of even thickness, but their spacing is varied to represent the steepness of the slope, closer spacing representing steeper slope (Fig. 13.13). These forms of line symbolization of a surface are commonly called hachuring (see Chapter 15) and have been traditionally used to depict land surface.

FIGURE 13.13 The line symbolization of slopes or gradients by the use of hachures. The hachures are of equal thickness but vary in spacing depending on the value of the gradient being portrayed. Three systems are shown: Benoit, Bonne, and Hossard. The lines represent 5°, 15°, 25°, and 40° slopes. In these particular systems the length of the hachure is also significant. (From *Cartographie Generale:* Tome 1 by R. Cuenin, Copyright 1972, Editions Eyrolles, Paris, France.)

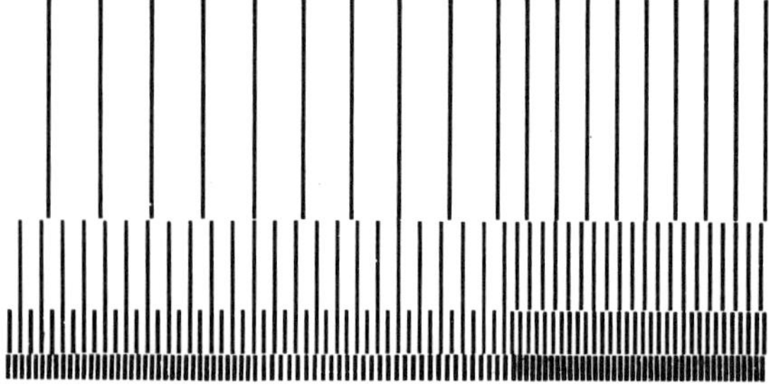

The major difficulty with hachures is that, although slope is their basis, they cannot practically be measured from the map, regardless of the precision underlying the representation. Flat areas, whatever their location, appear the same, and only rivers and streams or spot heights strategically placed give the reader clues to tell valleys from uplands. Another difficulty with hachuring is that its effectiveness, when printed in one color, depends greatly on the darkness of the ink. Thus, as darker inks are used to make the portrayal more effective, it creates a considerable problem because the other map detail becomes correspondingly obscured.

Naturally the hachure-like symbol need not be restricted to showing the gradients on the land surface; it can also be employed to show the directions and magnitudes of any sort of gradients, from temperature to population potential. It has not been used very often in these ways, however, since cartographers and geographers generally have thought that the gradients were adequately portrayed by other symbolization schemes, some of which provide considerably more information than gradient alone. Probably this attitude will continue. Nevertheless, if emphasis on the gradients of a distribution is paramount, the hachure-like symbol is an option.

In the other three categories of line symbols used to portray a statistical surface, one assumes a series of parallel (usually but not always equally spaced) planes intersecting a statistical surface, the intersections defining the lines that can be portrayed on a map. (1) If the series of parallel planes is parallel to the datum and the intersection lines are orthogonally projected onto one of the planes, a series of *isarithmic lines* result. (2) If the series of parallel planes is at a right angle to the datum, the lines of intersection show *profiles* across the surface of the smoothly undulating surface. (3) Finally, if the series of parallel planes is so situated that the angle between the planes and the datum lies between 0° and 90°, the lines of intersection represent *oblique traces*. Each method is important in cartography, and every cartographer should be aware of the assumptions, merits, and demerits of each method.

Isarithmic Mapping

When the interest in a geographical distribution is focused (1) primarily on the form of the distribution, that is, the organization or arrangement of the magnitudes, and directions (sense) of the myriad gradients that together constitute it, or (2) on the values at points of a truly continuous distribution, such as the land elevation, air temperature, or pressure, the isarithmic method is usually used.

In isarithmic mapping, the distribution is clearly conceived of as a volume, and in order to comprehend a volume visually, it is obviously necessary to see the shape of the outside surface enclosing it. Therefore, the character of a three-dimensional distribution or volume is most clearly mapped by delineating its surface.

The symbolization of real or abstract three-dimensional surfaces is difficult, and more time and effort probably have been devoted to it than to all other problems of symbolization put together. The principles involved in delineating such a surface are best illustrated by beginning with the familiar example of the surface of the land.

If the irregular land surface has been mapped in terms of *planimetry*, that is, the relative horizontal position of all points on the land, it is evident that there exists an infinity of points, each of which has, by reason of its location, an x, y, and z coordinate position with respect to the *datum* surface (spheroid) to which the earth's surface has been orthogonally transferred (see Fig. 13.14). By definition, the land surface is at all points either above or below the smooth assumed reference elevation, called a *datum*. If an imaginary surface, parallel to the horizontal datum and a given z distance from it, is assumed to intersect the irregular land surface, it must do so at all points having that z value. The *trace*, or the line of intersection, of these two surfaces will be a closed line. When this line, an isarithm, is orthogonally viewed, that is, perpendicularly projected to the map, it shows by its position the

FIGURE 13.14 The structural basis of an isarithmic map. The positions of elevations are first projected orthogonally to a spheroid. The "map" on the spheroid is then reduced and transformed to a plane, which the reader in turn views orthogonally. Horizontal planes (see Fig. 13.15) are then passed through the surface configuration. Curvature is exaggerated in this illustration.

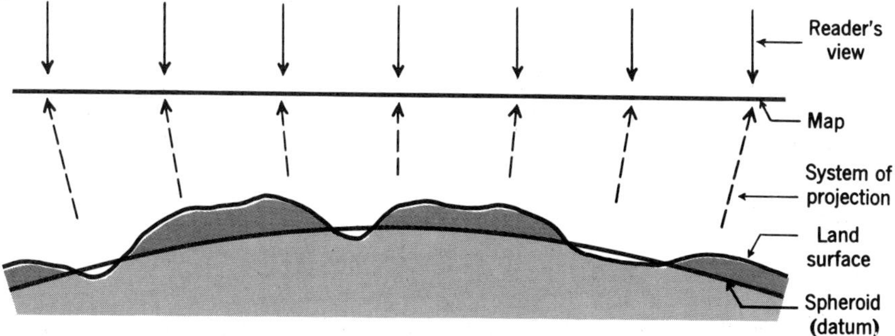

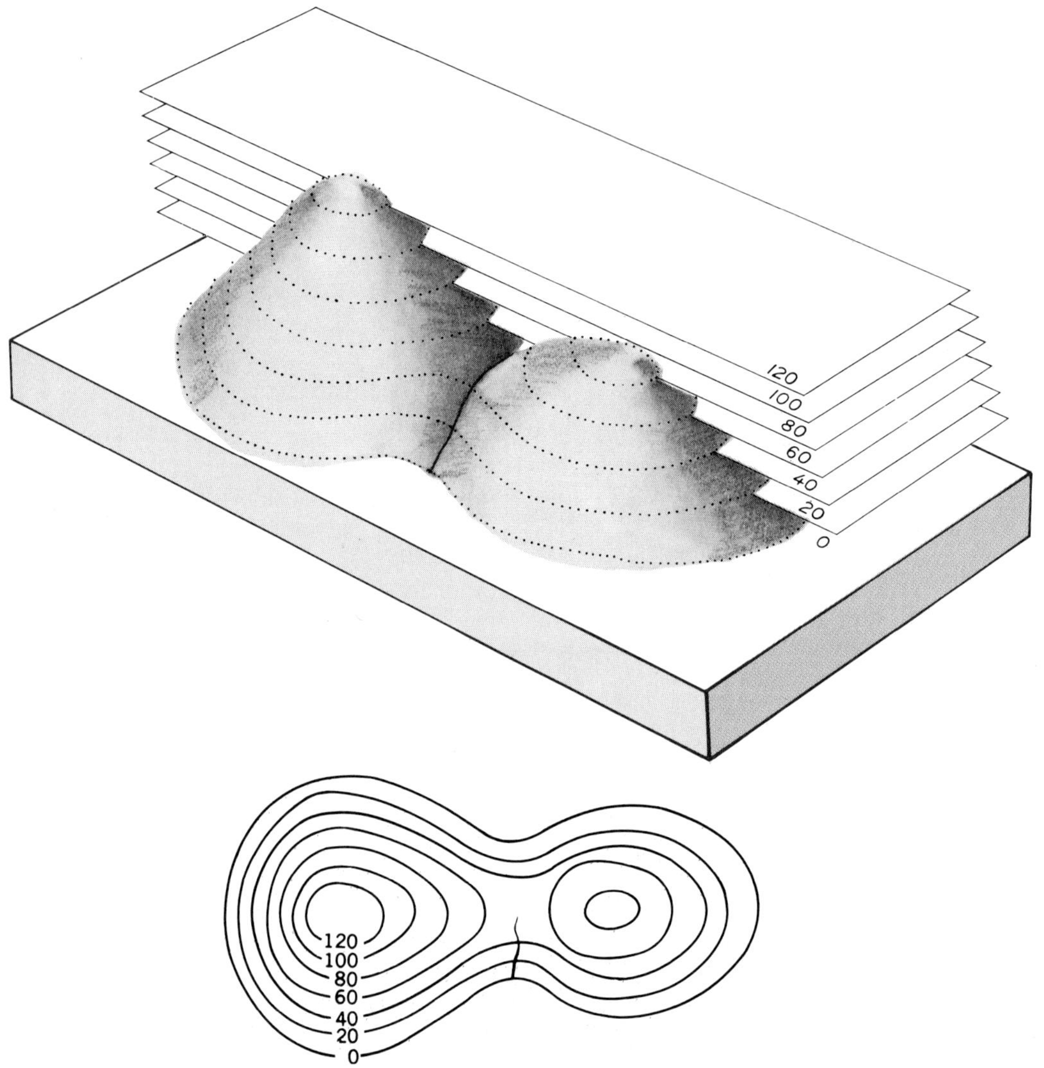

FIGURE 13.15 In the upper diagram, horizontal levels of given z values are seen passing partway through a hypothetical island. The traces of the intersections of the planes with the island surface are indicated by dotted lines. In the lower drawing, the traces have been mapped orthogonally on the map plane and constitute the representation of the island by means of isarithms.

x and y locations of all the points that have the particular z value it represents.

Figure 13.15 shows in perspective a hypothetical island, evenly spaced z levels, and below, an isarithmic map of the distribution of the z values on the surface of the island. In this case, z is elevation above the average level of the sea, which is defined as 0; it is therefore a contour map. The lowest or outer isarithm represents the average position of the shoreline. The next isarithm in Figure 13.15 is the trace of the plane spaced 20 z units above 0; it is in the same location as the average shoreline would be were the sea 20 z units higher.

The configuration of a three-dimensional surface is symbolized by the characteristic shapes and patterns of spacings of a set of isarithms, especially when the intersecting *z* planes are equidistant from one another. Smooth, steep, gentle, concave, convex, and other simple kinds of gradients and combining forms may be readily visualized from isarithmic maps, as indicated in Figure 13.16. For example, the bends of contours always point upstream when they cross a valley; they always point down slope when crossing a spur. The angles of slope, that is, the gradients or rates of change of elevation, of the land are shown by the relative spacings of the set of contours provided by equally spaced *z* planes. Profiles of the land surface along a traverse, or along a road or railroad, can easily be constructed from a contour map by working backward from the map to the profile. Detailed topographic forms and often even genetic structural details are readily revealed by the patterns of the contours on topographic maps.

It is not necessary that the *z* surface to be represented by isarithms be an actual visible, tangible surface, such as the land; the configuration of any three-dimensional surface may be mapped

FIGURE 13.16 The isarithmic spacings above and the profiles below illustrate in each diagram the manner in which the spacings of the isarithms show the nature of the configurations from which they result. All the forms may be described as variations of the general rule that if the isarithms are the traces of equally spaced z levels, the closer the isarithms are, the greater the gradient will be.

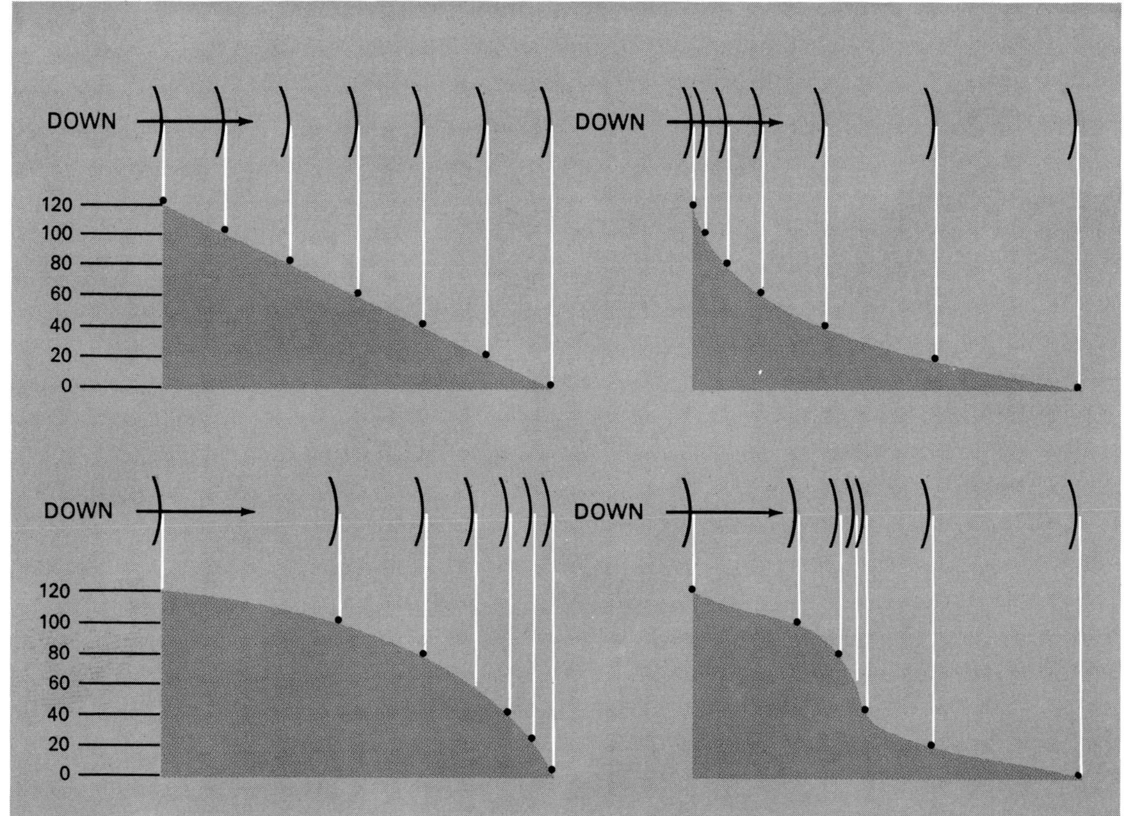

in the same way. For example, the form of a defined pressure surface in the atmosphere, such as the 500 millibar surface, may be mapped by passing *z* levels through the atmosphere. The shapes in nature are not visible, but the patterns of the isarithms show the gradients, the troughs, and the ridges of that surface in the same way that contours show the ups and downs of the land. The three-dimensional surface may even be an abstraction. For example, the *z* values may consist of some sort of ratio or proportion, such as persons per square kilometer. Anything that varies in magnitude and either actually exists or can be assumed to exist in continuous fashion over area constitutes a statistical surface. Its configuration can be mapped isarithmically.

Inferring the Statistical Surface. In order to map the traces of the intersections of horizontal levels with a statistical surface, it is necessary, as in following the proverbial recipe for rabbit stew, first to "catch" the statistical surface. This is easier said than done.

Only since the development of stereoviewing equipment for air photographs has it been possible to specify in its entirety the infinity of points on the land, each of which has its *z* value of relative elevation. In *all* other cases the statistical surface must be obtained from a limited number of *z* values, from which the totality of the surface must be *inferred*. In a very real sense, the infinity of *z* values constitutes a statistical population or universe but, because of practical limitations, only a sample of these points is usually available. Since we cannot know precisely the characteristics of the universe from which the sample has been taken, any extension of the characteristics of the sample to the universe or to a particular part of it constitutes an inference. Its validity can only be determined provided we have certain basic knowledge about the characteristics of the universe.

An example will clarify this. If air temperature is simultaneously observed at a series of weather stations, the temperatures at the given station positions constitute the sample of *z* values. If we want to make a synoptic (instantaneous view) map of the total distribution of temperatures (by means of *isotherms,* that is, isarithms of temperature), we can do so inductively only by making assumptions and inferences as to the nature of the temperatures that existed at the infinity of points that lie between the stations, for which, of course, actual *z* values are not available. It is apparent that the accuracy and representativeness of the given sample values, which the cartographer uses to locate the traces, are of considerable significance to the inference drawn of the total statistical surface. Various aspects of the probable validity of sample values are considered in subsequent sections.

Kinds of Isarithmic Mapping. A common practice in the past has been to give names to the isarithms employed for a particular kind of phenomenon. Thus we have isotherms (temperature), isobars (barometric pressure), isohyets (precipitation), and so on; the number of such terms is overwhelming. Such proliferation of the technical terminology serves no very useful purpose (except when necessary to distinguish between two sets of isarithms on one map).

There are two classes of *z* values that differ in terms of the precision with which they can specify a statistical surface: (1) actual or derived values that can occur at points, and (2) actual or derived values that cannot occur at points.

Actual values that can exist at points are exemplified by data such as elevation above or depth below sea level, a given actual temperature, the actual depth of precipitation, and thickness of a rock stratum. These kinds of values do exist at points. Only errors in observation or in the specification of the *xy* positions of the observation points can affect the validity of the sample values.

Derived values that can exist at points are of two kinds. One kind consists of averages or measures of dispersion, such as means, medians, standard deviations, and other sorts of statistics derived from a time series of observations made at a point. We can calculate a mean monthly temperature, an average retail sales figure, or an average land value for some particular place; the resulting numbers, although representative of magnitudes at the point in question, cannot, by their very nature, actually exist at any moment. A second kind of derived value that can occur at points consists of ratios and percentages of point values. Examples of these are the ratio of dry to rainy days at a particular place or the percentage of total precipitation that fell as snow. Such ratios are also incapable of existing at any instant, but they do represent quantities that apply to the point for which they are derived. Like measures of central tendency or dispersion, they are generally subject to more error than simple actual values. If they are rigorously defined and uniformly derived, they approach the validity of actual point values.

Quite different in concept is the derived value that cannot occur at a point. Representative of this class are percentages and other kinds of ratios that include area in their definition directly or by implication, such as persons per square kilometer, the ratio of beef cattle to total cattle, or the ratio of cropland to total land in farms. With such a quantity, only an average value for a unit area can be derived. Consequently, although it is perfectly legitimate to assume a statistical surface specified by these kinds of quantities, such a surface is dependent only on a series of average values for unit areas. Since each unit area represents a larger or smaller aggregate of *xy* points, no single point can have such a value. Nevertheless, in order to symbolize the undulations of the statistical surface by isarithms, it is necessary to assume the existence of such *z* values at specific points.

Absolute values that cannot occur at points but instead refer to areas, such as the populations of minor civil divisions, number of cows in counties, or potato production in states should not be assumed to constitute a statistical surface because their magnitude is affected by area; for example, 1000 persons in 20 km^2 is the same as 500 persons in 10 km^2. Unless such total values have been related in some manner to the area of the unit of collection, conceptually no statistical surface should be assumed to exist.

Because of the fundamental differences in the concepts of the two classes of statistical data, it is conventional to make a distinction among the isarithms employed to display their form. Different names are applied to them, and considerable confusion in definition and spelling exists in the literature. The following terminology seems to be consistent with modern usage.

An *isarithm* is any trace of the intersection of a horizontal plane with a statistical surface. It is thus the generic term; it may also be called an *isoline* or *isogram*. Isarithms showing the distribution of actual or derived quantities that can occur at points are called *isometric lines*. In contradistinction isarithms that display the configurations of statistical surfaces that are based on quantities that cannot exist at points, and that are therefore likely to be subject to a somewhat larger inherent error of position, are commonly called *isopleths*.

An observation is in order here. Much less confusion will result if you transfer the distinctions just made from the symbols to the surfaces they delineate. An isarithm is defined as the trace made by the intersection of a horizontal *z*-level plane with *any* three-dimensional surface. Whatever the nature of the surface may be, the function of the trace as a cartographic symbol is the same. Consequently, the "difference" between an isopleth and an isometric line is really an attempt to distinguish between the types of data and, as a consequence, the precision of the symbolization schemes; it is not a distinction between kinds of cartographic symbols.

Isarithmic maps serve two purposes: (1) they may provide a total view of the configuration of the statistical surface, for example, the form of the land, and (2) they may serve to portray the location of a series of quantities, for example, elevations at points. The ability with which they perform these functions depends on the validity and reality of the surface. For example, we can safely obtain values of elevations by interpolation from a contour map, within a certain margin of error. On many isarithmic maps, especially large-scale isoplethic maps, this cannot be done so readily. In such maps the isopleths serve more as "form lines," to delineate the general character of the surface, and less as a method of portraying specific values at points.

Elements of Isarithmic Mapping. When we are called on to prepare an isarithmic map, whether we are working with an isarithmic or isoplethic surface, we follow much the same procedure. In each case the numbers on our map, constituting the series of *z* values, are spaced some distance apart; whether they can or can not actually occur at points, we must, by assumption, inference, and estimate, produce isarithms that represent a continuous statistical surface. There are three elements of isarithmic mapping: (1) the location of the control points, (2) the gradient assumed, and (3) the number of control points.

Location of the Control Points. The location of each *z* value of the assumed statistical surface is called a *control point*. The *xy* positions and *z* values of these points constitute the statistical evidence (called "the control" in cartography) from which the character of the statistical surface is assumed and from which the locations of the isarithms on that surface are inferred.

The problem of choosing a location for the control point is not difficult when the statistical surface being mapped is based on actual or derived quantities that occur at points, as in isometric mapping. The location of each observation is then the location of the control point. This is not so, however, when the mapping distributions are derived from ratios or percentages that involve area in their definitions, as in isoplethic mapping, for example, the density of population (persons per unit of area). These kinds of quantities are derived from two sets of data based on unit areas. The resulting numbers, on which the ultimate locations of the isopleths depend, refer to the whole areal unit employed, and each is "spread" over the entire area of the unit. Therefore, there can be no points at which the values used in plotting the isopleths really exist. Nevertheless, in order to use the isarithmic method, we must assume positions for control points so that the isarithmic lines may be positioned.

When the distribution is uniform over an area of regular shape, the control point may be chosen as the center. If the distribution within the unit area is known to be uneven, the control point is normally shifted toward the concentration. Center of area may be considered as the balance point of an area having an even distribution of values without any unevenness of the distribution taken into account. The center of gravity takes into account any variation of the distribution. Figure 13.17 illustrates the concept. The four diagrams of statistical divisions show possible locations of the center of gravity and the center of area for uniform and variable distributions. They also illustrate the problem of locating the control point in regularly and irregularly shaped divi-

FIGURE 13.17 Problems in positioning the control points arise because either the center of area or the center of gravity may be used. These centers may not always coincide, or be representative of the distribution being mapped. (Courtesy of J. R. Mackay and *Economic Geography*.)

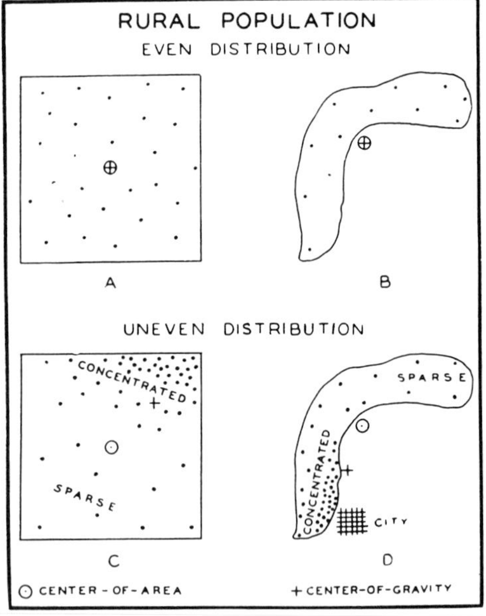

sions. In Figure 13.17A the center of gravity and center of area would, of course, coincide at the intersection of the diagonals of the square. Because the distribution in Figure 13.17B is even, the two centers also coincide, but they lie outside the irregularly shaped area at some point that is more representative of the whole division than any point within it. The distributions are uneven in Figures 13.17C and 13.17D, so the center of gravity is displaced away from the center of area. In each case, however, the center of gravity is probably the more reasonable location for the control point. In isoplethic mapping the cartographer has a great responsibility, since the control point location is assigned by the map author. The effects of poor control point location will be seen in the section on error in isarithmic mapping.

Interpolation. The selection of the isarithmic method implies the assumption that the distribution being mapped is continuous and smoothly undulating (that is, is composed of slopes). The basic characteristics of the configuration of most statistical surfaces are known only imperfectly, and an assumption is necessary as to the kinds of gradients that exist between the *z* values at control points.

As is illustrated in Figure 13.18, if two *z* values at different *xy* positions are represented, then the gradient between them may be represented by the straight line *a* or by some other gradient, such as the dashed lines *b* or *c*. It is entirely possible that *c* may represent the true slope; but unless evidence is available to indicate that *z* varies in a curvilinear relation with change in *xy*, such an assumption is obviously more complex than the gradient shown by *a*. In science, whenever several hypotheses can fit a set of data, the simplest is chosen. For most statistical data distributed over the earth the gradient is not known. Consequently, in the majority of manually drafted isarithmic maps, the linear gradient, *a*, is assumed in the construction of the isarithms. There are theories in some cases, such as that population density tends to be curvilinear (for example, *b* in

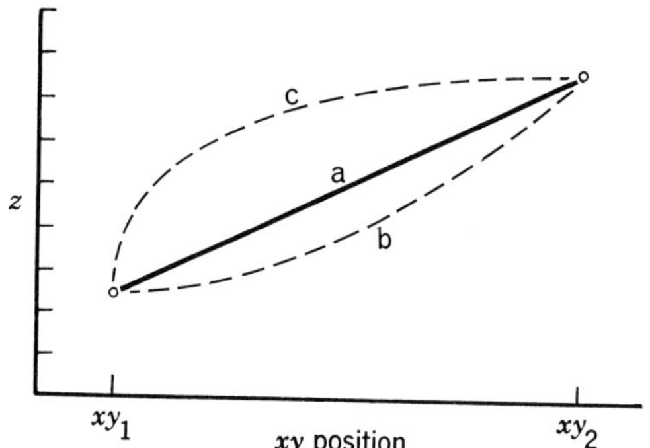

FIGURE 13.18 Three possible gradients which can be assumed between two control points.

Fig. 13.18). When such theories are known, they should be used.

Interpolation is the process of estimating the magnitude of intermediate values in a series, such as the *z* values along the lines shown in Figure 13.18. The control points of a statistical surface to be mapped by isarithms constitute the series; when no evidence exists to indicate a nonlinear gradient between control points, interpolation becomes a matter of estimating linear distances on the map in proportion to the difference between the control point values. For example, Figure 13.19 is a map on which are located *z* values at *xy* positions *a*, *b*, *c*, and *d*. If the position of the isarithm with a value of 20 is desired, it will lie $\frac{3}{10}$ of the distance from *a* to *b*, $\frac{3}{7}$ of the distance from *a* to *c*, and $\frac{3}{4}$ of the distance from *a* to *d*. Lacking any other data, the dotted line repre-

FIGURE 13.19 Linear interpolation between control points.

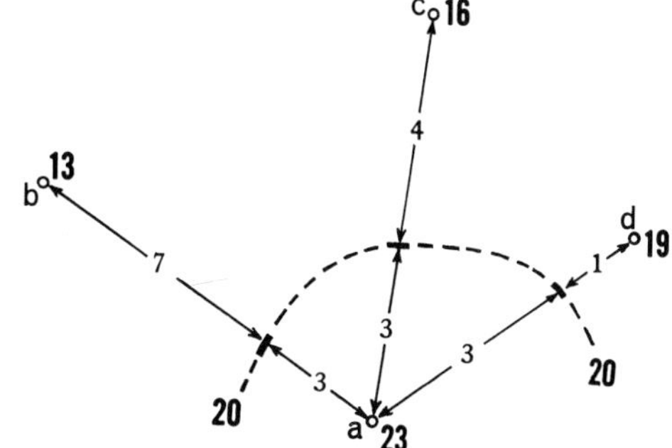

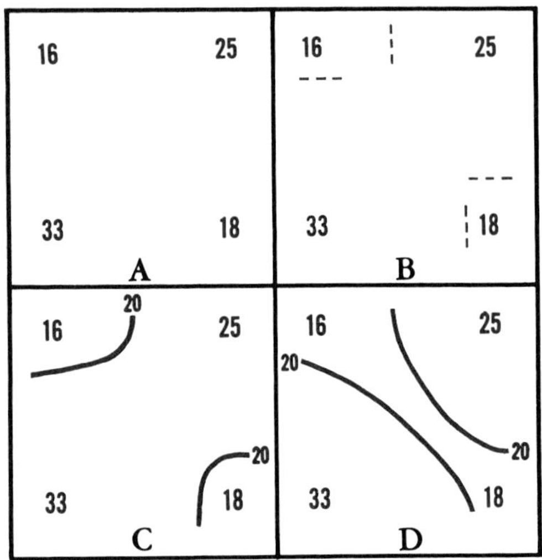

FIGURE 13.20 (A) The z values are arranged rectangularly. (B) The positions of the isarithms of value 20 between adjacent pairs have been interpolated. (C, D) The two ways the isarithm of 20 could be drawn through these points. (Redrawn, courtesy of J. R. Mackay and *The Professional Geographer.*)

senting the isarithm of 20 would be drawn as a smooth line through these three interpolated positions.

Figure 13.20 illustrates one of the common problems that arise when linearly interpolating among control points located in a rectangular pattern. Many census data are based on the rectangular minor civil divisions of the United States and Canada, and many other surveys use more or less rectangular subdivisions. When control points are arranged rectangularly, alternative choices will often arise concerning the location of an isarithm when one pair of diagonally opposite z values forming the corners of the rectangle is above and the other pair below the value of the isarithm to be drawn.

A careful examination of other relevant information may help to indicate which choice of the two alternatives is better. If this is not possible, averaging of the interpolated values at the intersection of diagonals will usually provide a value that will remove the element of choice (see Fig. 13.21). We are forced, in the absence of other data, to assume the validity of the average. This is often used in computer contouring routines. In isoplethic mapping, if the cartographer has control over the shapes of the unit areas, the problem of alternative choice can be prevented by designing the pattern of unit areas so that the control points have a triangular pattern.

In recent years several programs for computer interpolation of z at xy positions in isarithmic mapping have been developed. All common output devices can produce computer-interpolated and devised isarithms. As the use of computer interpolation increases, the complexion of the interpolation problem also changes.

Most often in today's computer routines, interpolation is done in two stages: (1) a primary interpolation, not necessarily linear, from the irregular original data points to a square or rectangular matrix of interpolated values, and (2) a secondary, usually linear, interpolation from the interpolated matrix of values to the positioning of the isarithmic lines. This two-stage interpolation can only be avoided when the original data set exists as a square or rectangular matrix of z values or when the complete set of original data is approximated by a surface approximation equation. As a result, most data are collected, where possible, in a rectangular matrix of values.

Because of cost (time) constraints, manual interpolation is almost solely linear. Machine interpolation is not necessarily linear. Many dif-

FIGURE 13.21 If the average of the interpolated values at the center is assumed to be correct, the isarithm of 20 would be drawn as in (B). (Redrawn, courtesy of J. R. Mackay and *The Professional Geographer.*)

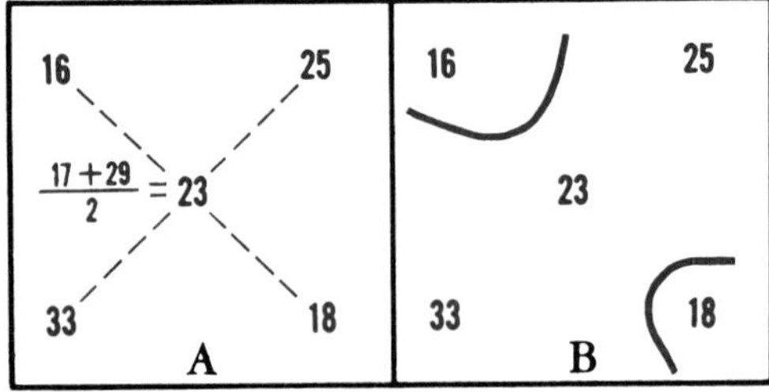

FIGURE 13.22 Four first-stage interpolation models applied to the same data set. The models are: W, weighted by distance squared; P, least squares plane of best fit; Q, least squares quadratic of best fit; and C, least squares cubic of best fit.

ferent mathematical procedures can be used for first-stage machine interpolation. For example, Figure 13.22 shows the same initial data set using four common first-stage machine interpolation methods. The choice among outputs can only be made with regard to the variation assumed by the cartographer to be present on the surface. Since the variations of only a few phenomena are actually known, cartographers face a major problem in selecting a first-stage interpolation method. For example, which map in Figure 13.22 most adequately portrays the data? In an actual mapping situation, there is presently no way to answer that question.

Number of Control Points. The final element of isarithmic mapping is the number of control points. There are several important concerns regarding the number of control points of which the cartographer must be aware. Initially, one of two conditions presents itself to the cartographer: (1) the data to be mapped must be collected by the cartographer, or (2) the cartographer is presented with a fixed set of data, and there is no way to enlarge the number of observations and no way to improve their accuracy. Unfortunately, the second case predominates.

In general, the larger the number of control points, the greater can be the detail of the surface portrayal. Since the three elements of isarithmic mapping do interact, however, a large increase in the number of control points is not always a blessing. There is a point of diminishing returns beyond which the increased improvement in surface portrayal is not worth the extra effort to process the additional control points. When this point of diminishing returns is reached depends on all the controls of generalization and the other two elements of isarithmic mapping. Certainly more control points that tend only to increase the clustering of the control points are not necessarily beneficial for surface portrayal, since this is more likely to lead to uneven generalization. Likewise, the purpose of the map may dictate a generalization level that requires only a minimum number of points. In any case, a cartographer should always avoid using too few points.

Control can be maintained over the three elements of isarithmic mapping when the cartographer gathers the data. A supplied set of data presents a different problem for the cartographer, and there are some guidelines for isarithmic mapping that will permit assessment of a fixed data scatter and number of control points.*

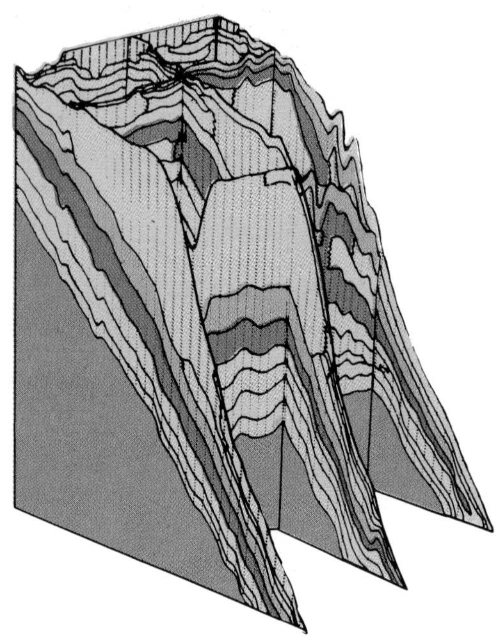

FIGURE 13.23 A series of profiles situated at right angles to each other gives a clear picture of the surface and its underlying strata. (Courtesy of John Wiley & Sons and J. W. Harbaugh and D. F. Merriam.)

Profiles

Another method of portraying the surface shell of an assumed statistical surface is to use profiles. A profile trace results from the intersection of a plane perpendicular to the *xy* datum and the statistical surface. Strictly speaking, a single profile trace does not constitute a map. A series of profile traces placed in relative position to one another, however, can enable the practiced observer to visualize a surface quite well (see Fig. 13.23). Cartographers have paid scant attention to using series of profile traces for displaying the statistical surface, and there has been little investigation of the perception of profiles.

The construction of a profile begins with an isarithmic map. The important steps are illustrated in Figure 13.24. The line along which a profile is desired is marked on the isarithmic map. The end points of the profile and the intersection of each isarithmic line with the profile line are transferred to a blank sheet of paper. A set of parallel ruled lines, spaced at a distance that reflects a suitable vertical exaggeration, is numbered to include the isarithmic values of each line crossing the profile on the map. The appropriate altitudes on the parallel lines are marked, depending on the isarithmic line value, and these altitudes are connected by a smooth line on the ruled parallel line set.

Figure 13.23 represents a series of profiles constructed in this manner along perpendicular directions. These surface profiles can also form the basis for the construction of an isometric block diagram. This technique can also be very instructive in helping a reader to learn to "visualize" a statistical surface.

*Joel L. Morrison, *Method-Produced Error in Isarithmic Mapping,* Technical Monograph No. CA-5, Cartography Division, American Congress on Surveying and Mapping, March 1971.

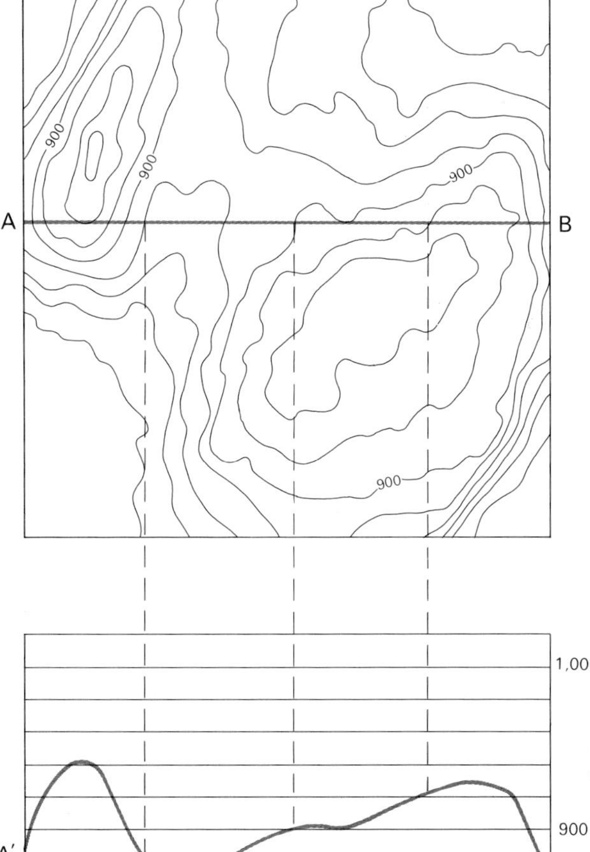

FIGURE 13.24 The construction of a profile from an isarithmic map. Line AB is drawn on the map. Intersections of the isarithms with line AB are perpendicularly projected to the appropriate line of a set of parallel ruled lines below. The resulting points are connected by a smooth line resulting in a profile A′B′.

Oblique Traces

Oblique traces result from the intersection of a series of planes with the base datum at some angle $\theta°$, where $0° < \theta° < 90°$. These traces may be graphically portrayed either in planimetrically correct position or in one- or two-point perspective.

Planimetrically Correct Oblique Traces. A vertical profile across the statistical surface, when viewed from the side and placed on a map, uses planimetric position in two dimensions; yet, the top and base of the profile are actually in the same planimetric position on the datum. It is impossible, therefore, to use planimetric position to express vertical dimension without producing planimetric displacement. In order to overcome this fundamental difficulty, it is sometimes desirable to substitute for the usual vertical or perspective profile across the surface a line that gives a similar appearance but is not out of place planimetrically. This kind of cartographic legerdemain may be accomplished by mapping the traces of the intersections of the surface with a series of parallel inclined planes (Fig. 13.25).

FIGURE 13.25 A reduced portion of a planimetrically correct terrain drawing of the Isle of Yell (Shetland Group). (Map drawn by P. Nicklin. Courtesy of Dr. N.J.W. Thrower.)

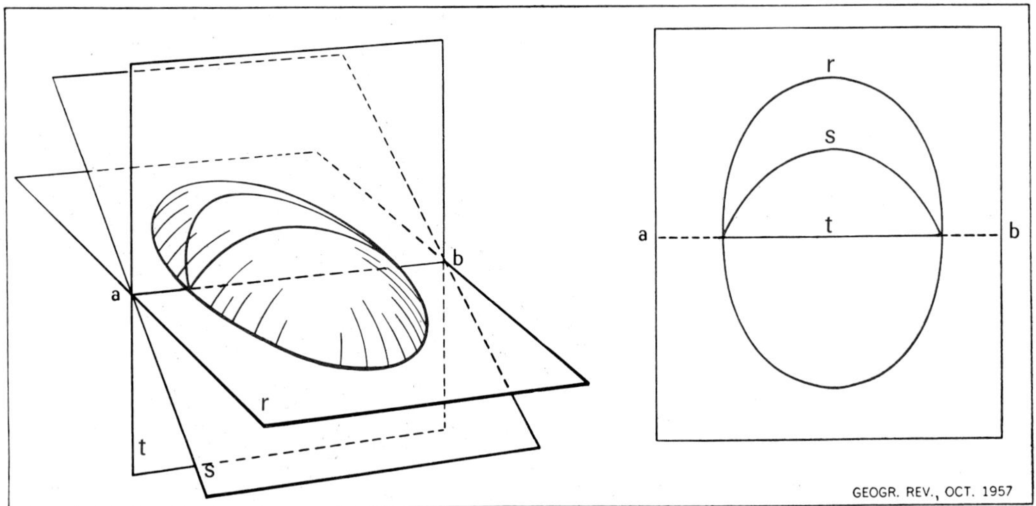

FIGURE 13.26 A comparison of the appearance on a map (right) of three traces produced by passing three differently oriented planes (r, horizontal, contour; s, inclined, oblique trace; and t, vertical, profile) through a single landform. The drawing at left shows the relationships in perspective. (Courtesy of *The Geographical Review.*)

FIGURE 13.27 Three principal variables that the cartographer must specify when using computer assistance to plot a series of perspective traces. θ is the angle through which the block is rotated in the *xy* plane. ϕ is the viewing elevation, and *d* is the viewing distance.

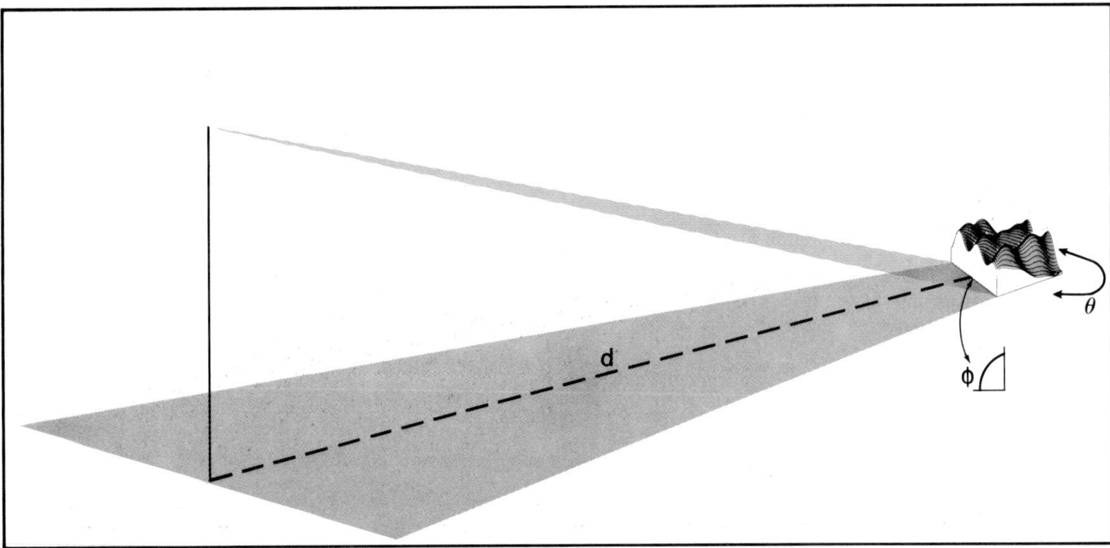

When a single plane parallel to the base datum is passed through the surface, the mapped trace of the intersection is an isarithm. If the plane is rotated on a horizontal axis, *ab* (see Fig. 13.26) and the successive traces plotted, and the result is still viewed from above, the traces produced run the gamut from the contour trace, *r*, to a straight line, *t*. The trace produced by the inclined plane, *s*, when viewed from directly above, has much the same appearance as would a conventional vertical profile if it were being viewed in perspective at an oblique angle as in a landform diagram. If the angle of inclination from horizontal is designated as θ, then when θ is zero, the trace of the ground with the plane orthogonally represented on a map is the conventional contour. When θ is 90°, the trace is a vertical profile, and its orthogonal representation is a straight line. When θ is greater than zero but less than 90° (*s* in Fig. 13.26), the trace will be that of the intersection of the ground with an inclined plane. A series of such traces on parallel inclined planes produces an *appearance* of the third dimension while still retaining correct planimetry.

The construction of an inclined trace is not difficult. It can be done either manually or with computer assistance.*

Perspective Traces. One of the first cartographic methods to be programmed to take advantage of computer assistance was the calculation and automatic plotting of perspective traces. Commonly used computer programs allow (1) the rotation, θ, about the vertical axis of the statistical surface, (2) changes in the viewing disance, *d* (the distance of the viewer's eye from the corner of the map), and (3) changes in the viewing elevation, ϕ (the angle at which the viewer is above the horizon) to be specified by the cartographer (see Fig. 13.27). Computer-drawn perspective traces can be drawn along two directions perpendicular to each other to form "fishnets" that tend to give a realistic impression of surface form (see Fig. 13.28) or along either the *x* or *y* direction only. Either one-point or two-point perspective can be employed (see Fig. 13.29). The ease of production and the resulting realistic impression insure this method a secure place in the cartographer's "bag of tricks" for the future.

FIGURE 13.28 Three computer-drawn sets of perspective traces: (A) "Fishnet" where traces are drawn in both the x and y directions. (B) Traces drawn only parallel to the x direction. (C) Traces drawn only parallel to the y direction.

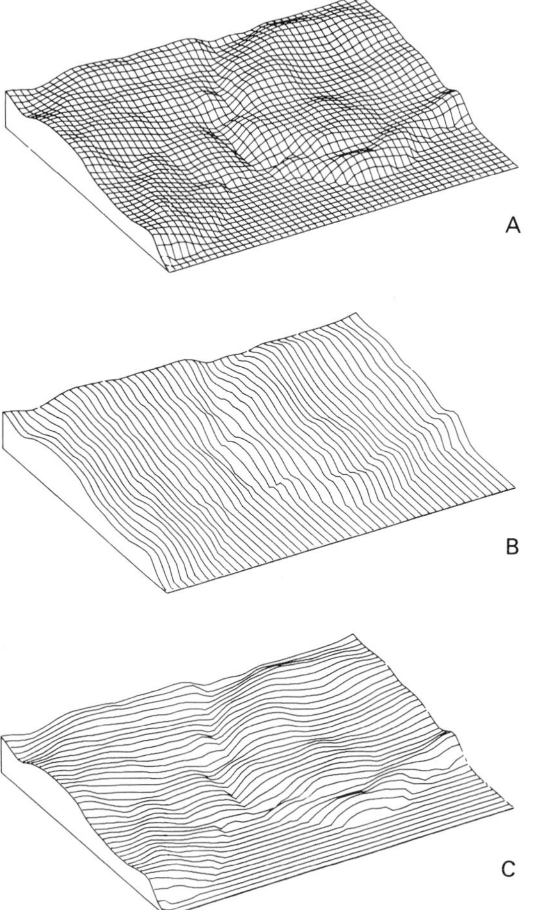

*The interested student is referred to the appropriate references for the detailed construction procedures. Arthur H. Robinson and Norman J. W. Thrower, "A New Method of Terrain Representation," *The Geographical Review* 47 (1957):507–20; Pinhas Yoeli, "Computer-Aided Relief Presentation By Traces of Inclined Planes," *The American Cartographer* 3 (1976):75–85.

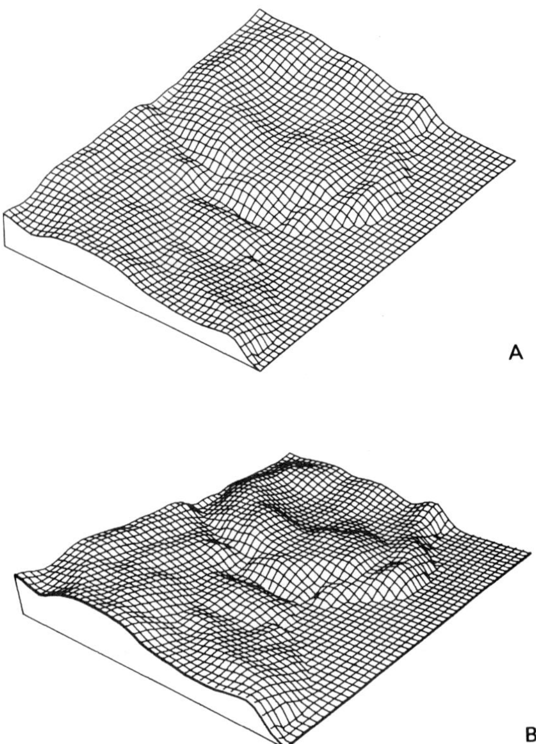

FIGURE 13.29 Two computer-drawn sets of perspective traces: (A) Using one-point perspective. (B) Using two-point perspective.

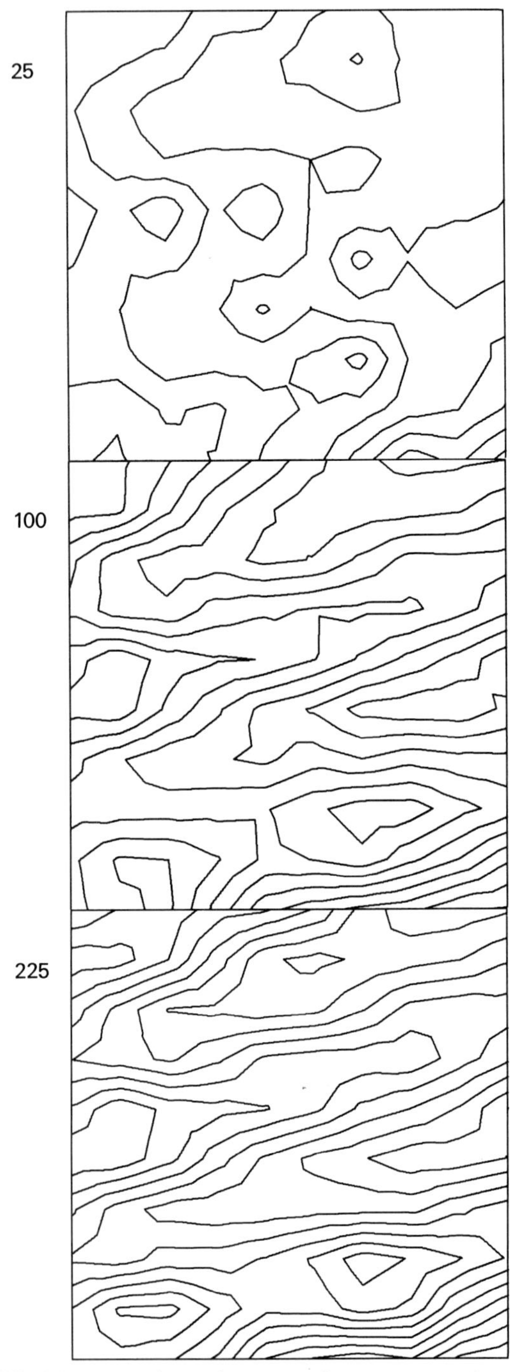

FIGURE 13.30 Three plots using the same interpolation model but with successively more initial data. The top plot uses 25 points, the middle plot uses 100 points, and the bottom plot uses 225 points.

Error in Isarithmic Maps

Because of the popularity of the isarithmic technique and because of its ease of execution given automated production, it is important that the cartographer know the error characteristics of an isarithmic map. Error in maps is an elusive thing and depends a great deal on the various definitions we employ. It ranges from the generalized "error" in a coastline to the "error" involved in ascribing a mean value to an entire unit area. Such a complex subject can be touched on only lightly in a textbook of this sort and, in many cases, only by implication. The process of isarithmic mapping, however, is subject to a considerable variety of errors stemming from the three elements of isarithmic mapping and from the

FIGURE 13.31 Six plots illustrating the effect of sample scatter. Plots 1, 2, and 3 are produced from random or more clustered than random scatters, especially plot 1, which is highly clustered. Plots 4, 5, and 6 are produced from scatters that are more systematic than random, especially plot 5, which is a square lattice of points.

quality of the data and the selection of the isarithmic interval.

For isometric mapping the errors arising from the three elements in isarithmic mapping are straightforward, but for isoplethic mapping the location of the control points, since the cartographer establishes their position during the mapping process, is subject to more error than in isometric mapping. It has been suggested that the most accurate way to map isoplethically may be to begin the process by making a dot map of the distribution.* The quality of the data obviously affects both isometric and isoplethic mapping, as does the selection of the isarithmic interval. The selection of an interval is fundamentally a generalization (classification) process, and the quality of data serves as a control on generalization. Each of these factors will be discussed briefly.

Method-Produced Error. Errors rising from the elements of the isarithmic mapping procedure are collectively called method-produced errors. The number of control points, their locations, and the selected interpolation model interact to affect the accuracy of any isarithmic map. Figure 13.30 illustrates the effect of increasing the number of control points. Figure 13.31 illustrates the effect of improving the location of the control points; Figure 13.22 illustrated the results obtained from employing different interpolation methods. Obviously, when three elements are used to the best of a cartographer's ability, the reliability of the isarithmic portrayal will be enhanced.

*J.A. Barnes, "Control Area and Control Points in Isopleth Mapping," *The American Cartographer* 5 (1978):65–9.

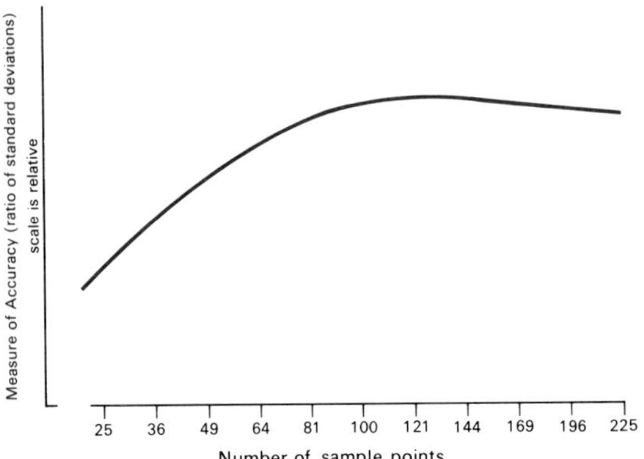

FIGURE 13.32 The characteristic curve of the hypothetical relationship between number of points and accuracy of the isarithmic map. The y axis scale is relative.

Figure 13.32 illustrates the hypothetical relationship between number of points and accuracy. Accuracy is usually very slight with few control points; then a rather dramatic increase in accuracy occurs as the number of control points increases. The diminishing returns mentioned earlier result in the leveling off of the curve as more and more control points are added. The cartographer should strive to balance the accuracy level desired against the cost of obtaining or using more control points.

Figure 13.33 illustrates the hypothetical relationship between location of control points and accuracy. From this figure it is obvious that clustered control point scatters (near the left hand edge of Figure 13.33) are to be avoided and that equispacing (2.0 on abscissa) is not as desirable as is "not quite equispacing" (to the immediate left of 2.0 on the abscissa). The best results are usually obtained by a scatter of control point locations that is more evenly spaced than random.*

In isoplethic mapping the cartographer has more control over the number and location of control points, because the data exist or are collected for areas. The sizes and shapes of these areas affect the locations of the isopleths. Elongated unit areas can create, in some instances, strong gradients across the orientation of the elongation; this in turn will induce the isopleths to lie in the direction of elongation. Figure 13.34 illustrates this phenomenon. In Figure 13.34A a hypothetical series of unit areas is shown together with isopleths located with an interval of two. In Figure 13.34B a smaller series of elongated minor civil divisions has been formed by aggregation from exactly the same distribution of data as shown in Figure 13.34A. The plotted isarithms show the same pattern. But when this process is followed with a different orientation of the elongated unit areas in Figure 13.34C, the isopleths are straightened and "pulled" into a different pattern.

There is not much the cartographer can do to compensate for this sort of effect, except to look for additional data that will either reinforce the pattern in an area of elongated unit areas or, conversely, lead to a modification of the possible artificial effect created by the elongation. Sometimes, by estimating densities of parts (discussed in Chapter 6), the cartographer can counterbal-

FIGURE 13.33 The characteristic curve of the hypothetical relationship between sample point scatter and accuracy of the isarithmic map. The y axis is relative. (The sample scatters are: AR, aligned random; UR, unaligned random; USR, unaligned, stratified random; US, unaligned systematic; and AS, aligned systematic.)

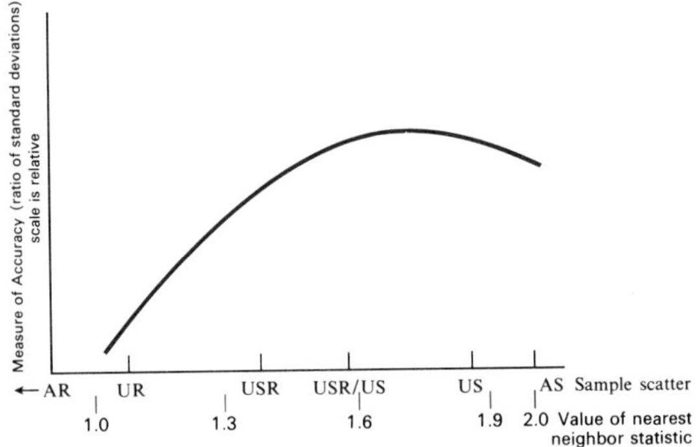

*Joel L. Morrison, "Observed Statistical Trends in Various Interpolation Algorithms Useful for First-Stage Interpolation," *The Canadian Cartographer* 11 (1974):142–59.

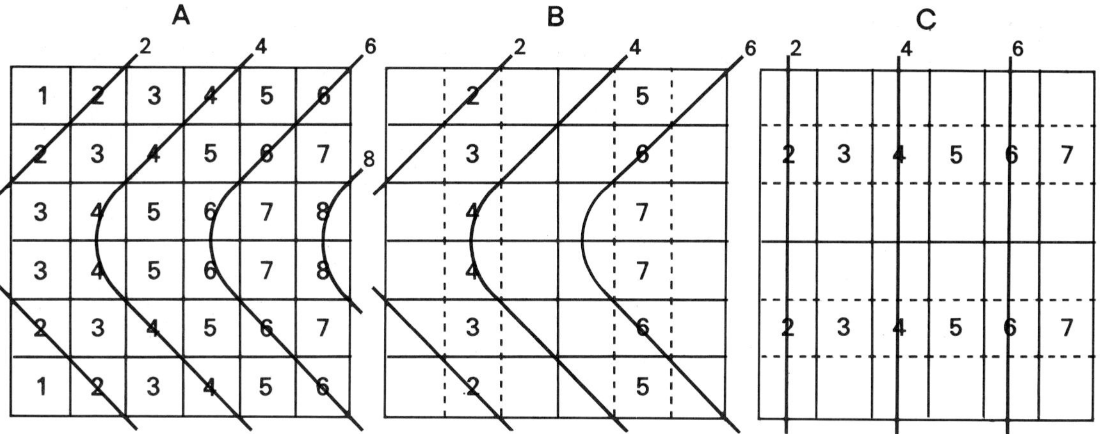

FIGURE 13.34 Diagram illustrating how elongated unit areas can sometimes affect the positions of isopleths.

ance this trend. Barnes has suggested that for isoplethic maps the following procedure might be optimum in reducing the method-produced error.

1. Prepare a dot map of the distribution.
2. Select a size for the control area. At map scale, on a transparent medium, draw a circle to represent this control area.
3. Arrange a triangular network of control points over the map, so that there will be complete coverage by control areas.
4. Calculate actual values at control points by centering the circular control area over each point and count the dots within the area.
5. Record the values to be mapped at the control points.
6. Locate the positions of selected isopleths by linear interpolation.
7. Draw the isopleths.

This procedure is logical and could be helpful in ensuring as accurate an isoplethic representation as possible given the data. The procedure has not been rigorously tested, however. The procedure does ensure that control areas are consistent in size, it allows the cartographer to choose the arrangement and number of control points, and it allows some control of the degree of generalization, or smoothness, of the resulting isoplethic map, depending on the size of the control area.

We do know that when the control points are unevenly arranged so that there is a greater density of points in one area than in another, an unevenness in the consistency of the treatment will result. This may provide more detail than we would want in the area of dense control, or it may lead to the converse, less detail in the area of sparse control than is wanted; hence, clustered control points tend to reduce overall accuracy. It is always good practice to show the control points on an isarithmic map to help the reader judge the quality of the map. This may be done by showing either the data points for isometric line maps or the unit areas (or control points) for isopleth maps.

With the advent of computer-assisted cartography the possibilities for using various nonlinear interpolation methods became practical for the first time. As illustrated in Figure 13.22, cartographers can only specify with confidence an interpolation method when a theory exists about the variation present in the phenomenon being portrayed. Where there is no such theory, the selection of the interpolation method must be done in conjunction with the number and location of control points and the generalization level desired for the given map. A total assessment of the method-produced error is difficult.

Quality of Data. As illustrated in Figure 13.35, any error in the z value at a control point can have as much effect on the location of an isarithm as changing the location of the control point. If the z value of a control point is correct but the xy position is incorrect, a displacement of the isarithm will result. If the xy position is correct but the z value is in error, the same thing will happen if a linear gradient is assumed. Clearly both the positions of the control points and the validity of the value at the control point have considerable effect on the accuracy of the statistical surface.

Several kinds of factors affect the validity of the control point value and hence the certainty with which the cartographer can locate the isarithm. Whenever there is a question concerning the accuracy of the z value, this doubt is automatically projected to the map; it there becomes transformed to a quesion as to the xy position of the isarithm. There is, therefore, always a zone on the map within which any isarithm may be located, depending on the certainty of the z values, and the width of this zone depends on the amount of error in the z value, assuming that the control point (the xy position) is correct.

There are three kinds of errors that commonly affect the realiability of the z values from which an isarithmic map is made: (1) *observational error,* (2) *sampling error, and* (3) *bias or persistent error.* A fourth, *conceptual error,* is related not so much to error as it is to the validity of the concept being presented by the map.

FIGURE 13.35 If a' is the xy location of a control point with z value a, and if the isarithmic plane has the value a, then a' will be the orthogonal position on the map of the isarithm. If the z value a is incorrect and really should be d, then the a isarithmic plane would intersect the surface at c and the isarithm on the map would be located at c'. If the a value were correct but the xy position of a should be at b, then b' would be the map position of the isarithm.

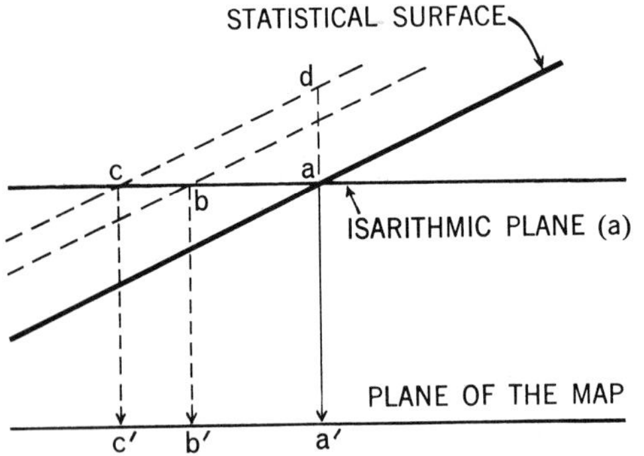

Observational error refers to the method used to obtain the z values. If they are derived by means of instruments operated or read by human beings, there is usually some inaccuracy, both in the instruments and in their reading by the observers. Observational error is not limited to instruments, however. A considerable amount of statistical data is based on various kinds of estimates, such as a farmer's estimate of the extent of cropland, the yield per acre, and the extent of soil erosion. Most statistical data are subject to observational error of one sort or another.

Sampling error is of several kinds. The most obvious is that associated with any map that purports to represent the distribution of an entire class of data, only a sample from which is actually known. Any map of mean climatological values, which is not specifically for a particular period, is, in effect, attempting to describe a total average situation (or statistical population) from only one relatively small time sample. Many kinds of data in censuses are collected by a sampling procedure. In this case, the statistical population exists at one time, but the cost of ascertaining it in its entirety is too great; thus, a sample is taken.

Bias or persistent error may be of many kinds. Instruments may consistently record too low or too high; the majority of weather stations may be in valleys or on hilltops; people show preferences for certain numbers when estimating or counting; and so on. Bias is difficult to ascertain but may have considerable effect on z values.

Conceptual error in the z value may be illustrated by the use of mean values computed from a series of observations with which to make a map of mean monthly temperatures. The mean may not describe the actual fact very well, since the dispersion of the values around the mean may be large. For example, the standard deviation of mean December temperatures in the central United States is around 4°F. There is therefore a high probability that approximately one-third of the time an individual December mean will be more than 4° above or below the average of the December means. A cartographer constructing a map that purports to show by isarithms the distributions of mean December temperatures will, in effect, be forced to locate each isarithm near the center of a map zone or band represented horizontally by several degrees of temperature gradient. Furthermore, it is necessary to realize that if individual year-to-year maps had been made, approximately one-third of the time the isarithm would lie outside the zone delineated by an average of 8° of temperature gradient. On even a small-scale map, such a zone would be of noticeable dimensions. It follows that minute isarithmic wiggles and sharp curves would be an affectation of accuracy not supported by the nature of the data.

It is important to take into account the various kinds of possible error in the validity of the *z* values. In many cases, some of the kinds of sampling error have already been ascertained and need only be obtained from the sources, for example, the U.S. Census of Agriculture. Standard deviations and standard errors of the mean are often available or are not difficult to compute. Simple logic and common sense will often provide enough of an answer for highly generalized maps. The total effect of all the possible sources of error and inconsistency from one part of an isarithmic map to another can function toward only one end, that is, to *smooth out the isarithms*.

Class Intervals in Isarithmic Mapping. Relative gradients on isarithmic maps are easily judged only when there is a constant interval; the general rule that the closer the isarithms, the steeper the gradient, holds only under such circumstances. Therefore, if primary interest is in the configuration of the statistical surface, the interval should be constant. By far, most isarithmic maps employ equally spaced intervals.

The recognition of form on the statistical surface from the patterns made by the isarithms, such as concave or convex slopes, hills or valleys, escarpments, and the like is difficult, if not impossible, when irregular intervals are used. The nature of slopes is easily revealed by the spacing of lines from an equal steps interval in the one case and almost completely hidden by the use of an unequal interval in the other. In some cases, when interest is concentrated in one portion of the range of values, an increasing or decreasing interval can be justified. This requires much more mental effort on the part of the map reader when trying to infer the form of the surface, but it does allow the cartographer to concentrate detail in one portion of the range. This should only be applied rarely and with sufficient notice to the map reader.

An equal steps or constant interval is ordinarily employed on contour maps of the land and isometric maps of other phenomena. On the other hand, there is no necessity for using only constant intervals, and often irregular intervals are used in isoplethic mapping.

Summary of the Isarithmic Method

Perhaps no cartographic method is as useful or often requested for surface portrayal as the isarithmic method. First, for the practiced reader, the configuration of the surface can be very graphic when equal-step intervals are used. Second, for extracting data from maps, detailed cartometric methods can be used (especially where accuracy permits, for example, on topographic maps). Third, many other cartographic methods are derived from isarithmic portrayals, among them profiles and oblique traces, hachuring, shading or shaded relief, and some dasymetric maps.

The variation between the two forms of isarithmic map, the isometric and the isoplethic, is not visually evident. The distinction only refers to the form of the original data used for the map: isometric maps result from place data, and isoplethic maps result from areal data. Since the error characteristics of isarithmic maps have received more attention than the comparable characteristics of other cartographic methods, the cartographer has more information to enable the production of effective communication with isarithmic maps than with most other methods.

SELECTED REFERENCES

Blumenstock, D. I., "The Reliability Factor in the Drawing of Isarithms," *Annals of the Association of American Geographers* 43 (1953): 289–304.

Downing, J. A., and S. Zoraster, "An Adaptive Grid Contouring Algorithm," *Proceedings of Auto-Carto V: Environmental Assessment and Resource Management.* Falls Church, VA: American Society of Photogrammetry and American Congress on Surveying and Mapping (1982): 249–56.

Elfick, M. H., "Contouring by Use of a Triangular Mesh," *The Cartographic Journal* 16 (1979): 24–8.

Griffin, T. L. C., and B. F. Lock, "The Perceptual Problem in Contour Interpretation," *The Cartographic Journal* 16 (1979): 61–71.

Hsu, M. L., and A. H. Robinson, *The Fidelity of Isopleth Maps: An Experimental Study.* Minneapolis, MN: University of Minnesota Press, 1970.

Lam, N. S., "An Evaluation of Areal Interpolation Methods," *Proceedings of Auto-Carto V: Environmental Assessment and Resource Management.* Falls Church, VA: American Society of Photogrammetry and American Congress on Surveying and Mapping (1982): 471–9.

Lam, N. S., "Spatial Interpolation Methods: A Review," *The American Cartographer* 10 (1983): 129–49.

McCullagh, M. J., and C. G. Ross, "Delaunay Triangulation of A Random Data Set for Isarithmic Mapping," *The Cartographic Journal* 17 (1980): 93–9.

Peucker, T. K., "A Theory of the Cartographic Line," *International Yearbook of Cartography* 16 (1976): 134–43.

Underwood, J. D. M., "Influencing the Perception of Contour Lines," *The Cartographic Journal* 18 (1981): 116–9.

14

Mapping with Area Symbols

MAPPING QUALITATIVE DATA RELATED TO AN AREA

In the introduction to Chapter 12 we explained that data related to areal units on the earth's surface exist on one of the four scales of measurement. When the aspect of the data set of interest is measured on a nominal scale, we refer to it as *areal data;* when measured on an ordinal, interval, or ratio scale of measurement, we refer to it as *volumetric data*. Since the cartographer can choose to "generalize" the scale of measurement, existing volumetric data sets can be mapped as areal data. The reverse is not true.

As we have seen, volumetric data can also be related to positions or to flows, and thus the cartographer may symbolize volumetric data by point or line symbols in addition to areal symbols. There is no separate class of volume symbols that a cartographer may use, although some point symbols can be constructed to give the visual impression of volumes, such as spheres or cubes.

Much data of interest are collected so that they refer to areas or can easily be related to areas on the earth's surface. On a nominal scale, the surface material of the earth is one such prominent distribution. Soil types, geologic structure, land use, land cover, and even the division between land and water are further examples of nominally measured aspects of distributions related to area. Zoning regions, landownership, police and fire jurisdictions, and school districts are examples of important nominal areal data sets in local areas.

There are also numerous data sets that a cartographer might map as ordinal, range-graded, or ratio data relating to areas. More data are collected for portrayal at range-graded or ratio scales than at ordinal scales. However, the map often portrays these data sets most efficiently on an ordinal scale. Almost all census data relate to areas. Some census data are published as range-graded, such as annual household income, age of head of household, number of years of education, and so on. Some summary census tables are "generalized" because the data are actually collected as ratio data sets but are combined to produce range-graded statistical tables by the census (sometimes to avoid disclosure of data related to an individual or single business). Many types of data are collected as ratio data: number of manufacturing establishments per county, number of dairy cows per county, and of course, number of inhabitants per county; or at a more detailed level, inhabitants per census enumeration district. The amount of data available is enormous.

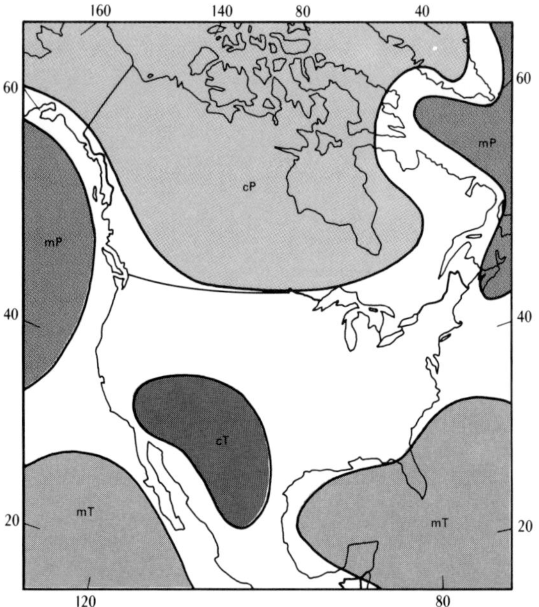

FIGURE 14.1 Portrayal of North American air masses and their source regions. Although data have quantitative characteristics, the intent of this illustration is simply to portray location of air masses. This can be accomplished by using nominal areal symbolization. (From *Fundamentals of Physical Geography* by Trewartha et al. Copyright © 1977, McGraw-Hill Book Co. Used with permission of McGraw-Hill Book Company.)

Mapping Areal or Volumetric Data Using Area Symbols

Figure 14.1 is an example of the mapping of one attribute of a data set on a nominal scale using

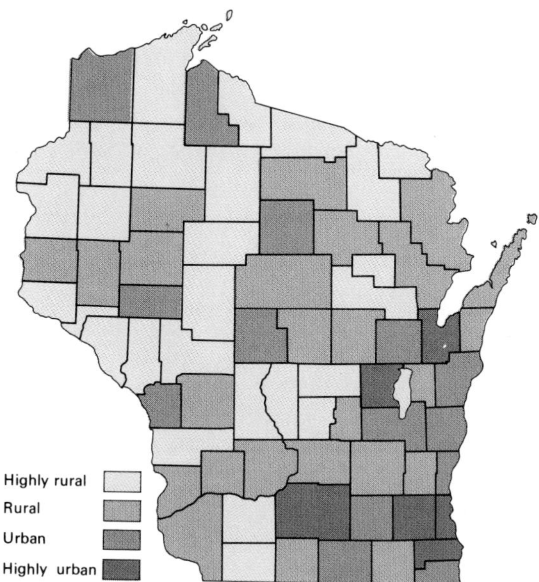

FIGURE 14.2 Map showing an ordinal ranking of the counties of Wisconsin, depending on degree of urbanization. Data from Table 14.1, column labeled "category."

area symbols. The data contain quantitative characteristics, but Figure 14.1 generalizes the data to show only a nominal attribute. Figure 14.2 is a common type of map for portraying volumetric data on an ordinal scale. The data are generalized and arranged in ranked classes based initially on a ratio scale. Figure 14.3 is an example of the range-graded portrayal of volumetric data, perhaps the most common type of map in this category. Data are arranged in finite classes, and each class is uniquely symbolized. Finally, Figure 14.4, a classless choropleth map (see below), is an example of the portrayal of ratio volumetric data.

Primary Graphic Elements Used in Mapping with Area Symbols

For mapping a set of data that is measured on a nominal scale and that refers to areas, the cartographer may use either pattern (arrangement or orientation) or the graphic element hue for differentiation. Most often we use "pattern" in this context as it is defined in Chapter 7. The visual variable texture, as defined by Bertin (see Chapter 12), is also associated with pattern. For convenience we will use the term "pattern" in discussing areal symbols used to portray data on a nominal scale. Later in the discussion we will refer to the spacing of patterns as a graphic element for area representation of volumetric data at ordinal or higher scales of measurement.

Forest, grasslands, national areas, culture regions—all the ways in which the character of one geographical region can be distinguished from another—are shown by using distinctive area symbols. The resulting maps are commonly called *qualitative distribution maps*. We have to be careful not to suggest ordinal relationships on qualitative (nominal) distribution maps. It is well known that darker and lighter values and, to a lesser extent, color intensities do suggest ordinal ranking. Darker values usually mean "more" and lighter values "less" of a given phenomenon. Likewise higher intensities usually suggest "more" and weaker intensities "less." As a consequence, only the primary graphic element hue is adequate in portraying nominal data, since hue does not

FIGURE 14.3 Map illustrating the range-graded classification of the counties of Florida. (Courtesy, J. M. Olson.)

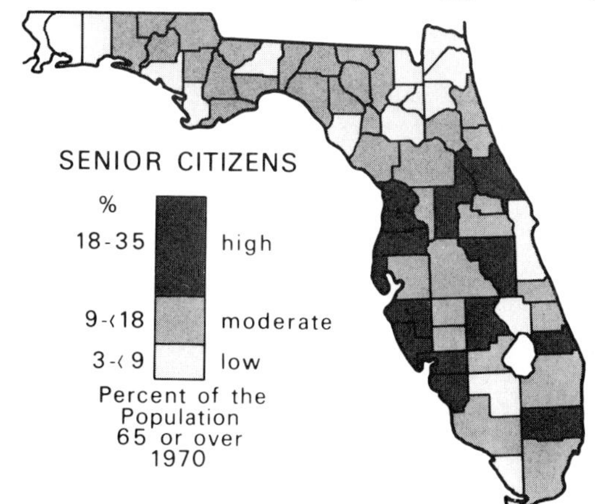

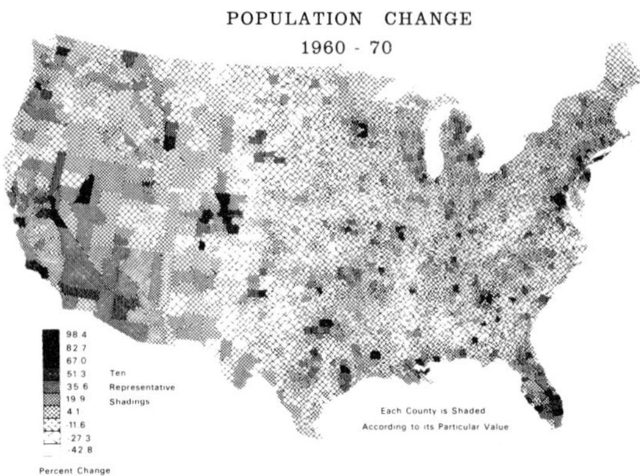

FIGURE 14.4 A classless choropleth map. The data are presented on a ratio scale of measurement. Only representative sections of the range are given in the legend. (Courtesy M. P. Peterson and the American Congress on Surveying and Mapping.)

connote ordinal ranking. Blue is not considered "more" or "less" than red or green, and so on.

When using patterns, the cartographer may differentiate by varying both orientation and arrangement. Figure 14.5 shows differentiation in pattern orientation and pattern arrangement for nominal area data. Care must be exercised in the use of patterns not to create a display that is so visually complex that excessive eye movement for the map reader results (see Fig. 8.23).

The use of areal symbols to portray quantitative data in cartography is also relatively straightforward. The most useful graphic elements are value and pattern spacing. Color-intensity may also be used. These graphic elements are applied to designated areas on the map. The procedures for arriving at meaningful lines that separate the areas on the map (that is, the areal extent of the symbols) is one of the cartographer's major problems.

The data may be place or areal, but the use of areal symbolizations in conjunction with any quantitative data implies a statistical surface. The cartographer can map place data by clustering the data into homogeneous groups and symbolizing the areas thus created by areal symbols. The result is a form of dasymetric mapping. The cartographer can also map areal data by applying areal symbols to the various map areas, resulting in either a simple choropleth or a dasymetric map. Finally, the cartographer may wish to display place or areal data constituting a statistical surface subjectively by continuous shadings.

Mapping Qualitative Distributions Using Area Symbols

Mapping nominal data is relatively straightforward, but some minor problems can be involved. Frequently, nominal categories are not geographically exclusive; that is, two or more categories often occur in the same area, such as botanical species, ethnic groups, or land use characteristics. Accordingly, if the cartographer does not want to generalize sufficiently to remove the mixture, symbolization that somehow represents the mixture or overlap must be employed. There are a number of methods, as suggested in Figure 14.6, none of which is suitable in all circumstances. If the cartographer is not careful, this practice can lead to incredibly complicated maps.

If color is being used, it is possible to choose hues that give the impression of mixture. For example, a red and a blue appear purple when superimposed, and that hue looks like a mixture. On the other hand, a hue produced by mixing blue and yellow produces green, and that hue does not look like a mixture of its components.

Certain disciplines have created nominal area data pattern conventions for mapping purposes. Geology has specified the use of particular patterns for certain rock types. The cartographer has no right to violate these conventions. Figure 14.7 illustrates some of the patterns used by geologists for nominal area differentiation.

Sometimes it is necessary to map nominally a distribution through two levels of its classification. This is called *subdivisional organization* (see Chapter 7). Preference could be to use hue to distinguish the major categories and to differentiate subclasses within categories by the use of

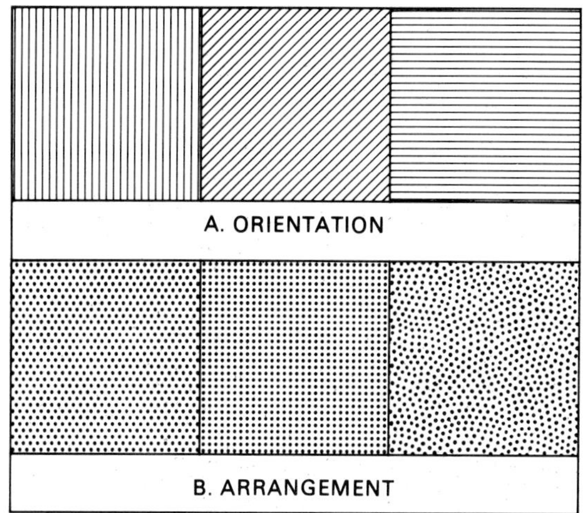

FIGURE 14.5 (A) a single area pattern has been oriented at three different angles to produce three visually different nominal area patterns. (B) The arrangement of the dots that compose the area pattern is varied, again producing three visually different nominal area patterns.

pattern. Hue tends to show easily the major breakdown of the distribution into its first-order classes. Pattern is more subtle and thus lends itself more readily to differentiation at the second order of classification, where the map reader is searching for more detailed information. It is easier to juxtapose patterns that are similar but that still retain their separate identities than it is hues.

MAPPING THE STATISTICAL SURFACE BY AREA SYMBOLS

When confronted with ordinal, interval, or ratio data relating to areas, that is, volumetric data, the cartographer must first decide whether point, line, or areal symbols will be used. If point symbols are selected, then the distribution may be shown either by using the dot mapping technique or by employing a graduated point symbol. If line symbols are selected, either flow lines or the isarithmic technique are possibilities. In the latter case a statistical surface is assumed. If the cartographer elects to use area symbols, then not only is a statistical surface assumed; it is implied. Essentially two techniques are available: the simple choroplethic or the dasymetric technique, both of which employ area symbols.

Therefore, to map a statistical surface composed of volumetric data collected over areas, the cartographer has three choices: one using line symbols, the isoplethic technique; and two using areal symbols, the choroplethic technique or the dasymetric technique. The choice among these methods depends on whether the interest is (1) as a distribution where the focus of interest is on the specific quantitative values at particular places, or (2) as a distribution where the focus of interest

FIGURE 14.6 Several methods of showing geographical mixture or overlap with area symbols.

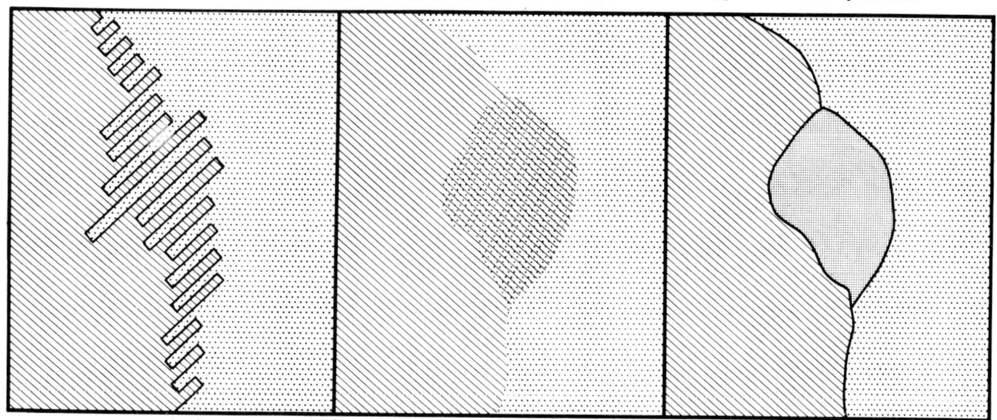

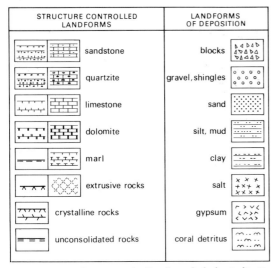

STRUCTURE CONTROLLED LANDFORMS	LANDFORMS OF DEPOSITION
sandstone	blocks
quartzite	gravel, shingles
limestone	sand
dolomite	silt, mud
marl	clay
extrusive rocks	salt
crystalline rocks	gypsum
unconsolidated rocks	coral detritus

FIGURE 14.7 Some standardized symbols for indicating lithologic data as suggested by the International Geographical Union Commission on Applied Geomorphology.

is on the geographical organization of the magnitudes and directions of the gradients (that is, in its surface configuration). When the emphasis in the mapping is on the locational representation of the z values, one of the two choroplethic techniques is appropriate. As noted earlier, when the emphasis in the mapping is on the gradients among the z values, their magnitudes, and their directions, the isoplethic technique is appropriate.

The simple choropleth, the dasymetric, and the isoplethic techniques are shown comparatively in Figure 14.8. The diagnostic characteristics (that is, the basic differences in appearance) among them are clearly evident. In the simple choropleth map the locations of the class limits employed coincide with the boundaries of the data-collection unit areas. In the dasymetric map the map positions of the class limits are independent of the boundaries of the enumeration districts, but the lines separating one class from another have no numerical value, as is also true in the simple choropleth map. In the isarithmic map the positions of the lines showing the isarithmic class interval employed are independent of the unit areas; they are lines along which a value is assumed to be constant.

In contrast, the cartographer must employ a form of dasymetric mapping, as explained below, when confronted with quantitative point data to be symbolized by areal symbols. By definition, every set of quantitative data, referring either to areas or to points that is to be symbolized by

FIGURE 14.8 Examples of three ways a set of z values that refer to enumeration districts or unit areas can be mapped: (A) a simple choropleth map, (B) a dasymetric map, (C) an isoplethic map.

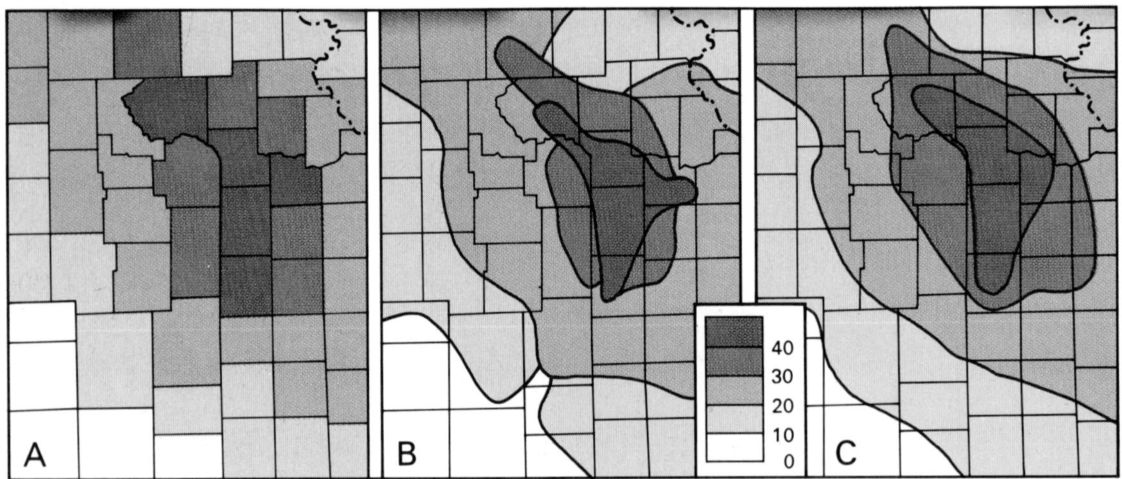

areal symbols constitutes a statistical surface. Therefore, there is only one *basic* method for representing a surface by areal symbols in a highly commensurable manner (having a common measure). It is called the *choroplethic method.*

Choroplethic Mapping

The choroplethic method consists of two highly commensurable techniques for the portrayal of a statistical surface by areal symbols. In one, called simple choropleth mapping (from the Greek words *choros,* place, and *plethos,* magnitude), the primary objective is to symbolize the magnitudes of the statistics as they occur within the boundaries of the unit areas, such as counties, states, or other kinds of enumeration districts.

In the other subclass, called *dasymetric* mapping (from the Greek words *dasys,* thick or dense, and *metron,* measure) the primary objectives are to focus interest on (1) the location and z magnitudes of areas having relative z uniformity, regardless of the unit area boundaries, and (2) the zones between which there occur more or less abrupt changes in these magnitudes. Both kinds of maps employ area symbols and, since they are quantitative, the darker visual value (gray tone) or the more intense color is assigned to the greater magnitudes. The simple choropleth map depends entirely on data collected by areal unit, but the dasymetric map can be created from data that initially refer either to areas or to points.

The Simple Choropleth Map. The simple choropleth map is, in effect, a spatially arranged presentation of statistics that are tied to enumeration districts on the ground. Tabular statistics are very convenient for many purposes, but frequently we want to refer to the array geographically instead of alphabetically and to be able to compare magnitudes in various places easily. This kind of map is made by symbolizing in some manner the amount that applies to the enumeration district or some other unit area. The symbolization scheme is usually range-graded.

The simple choropleth map requires the least analysis on the part of the cartographer. This technique may be used for many kinds of ratios, and when the statistical units are small, it provides sufficient variability in the map (if there is variability among the data) to take on some of the visual character of a continuous distribution. Ordinarily, however, it is employed when any other sort of map poses too many problems. For example, when mapping at a large scale, the cartographer may not know enough about the details of the distribution, or at a small scale the regional variations may be too inconsistent to attempt an isoplethic or dasymetric map. In effect, the simple choropleth map presents only the spatial organization of the statistical data, with no effort being made to insert any inferences into the presentation.

Absolute numbers alone are not ordinarily presented in a simple choropleth map. For many classes of data, the absolute quantities of a phenomenon (whether persons, automobiles, farms, or vacationers) are naturally functions of the areal size of the enumeration district. When we are concerned with the geographical distribution, we are normally concerned with comparisons of ratios involving area (density) or ratios that are independent of area (percentages or proportions), that is, when the effect of variations in the size of the enumeration districts has been removed.

The quantity mapped is therefore almost always some kind of average that is assumed to refer to the whole of a unit area, and the unit areas are range-graded in representation. Where class limits call for a change of symbolization, it occurs only at the unit area boundaries in a simple choropleth map. Since the boundaries are usually quite unrelated to the variations in the phenomenon being mapped, this adds a further measure of geographical obscurity to that of dealing entirely with averages.

Choropleth mapping is "safe" in that we symbolize the quantities where they are, but it is the least informative of the commensurable methods of mapping quantitative areal data. The production of relatively coarse simple choropleth maps is easily automated. Because of the speed and relatively low cost with which the computer can handle the data and direct the line printer, this kind of map is important as a first "glance" at a distribution. Figure 14.9 is an example of a simple choropleth map produced on a line printer.

The difficulties in producing choropleth maps of high visual quality completely automatically relate to computer hardware and software developments. Choropleth maps of high visual quality can now be obtained by photomechanical enhancement of computer line drawing output (see Fig. 14.10) or computer output on microfilm plotters (see Color Fig. 10). As use of the latter becomes more widespread, the cartographer should finally be able to obtain good quality automatic production of simple choropleth maps.

The Classless Choropleth Map. The earliest choropleth maps were classless, since their makers tried to give each unit area a tonal value matched to its place on a continuous scale of tones from the lowest to the highest. The attempts were not successful because tonal gradations could not be controlled in the production process. Recently the concept of the classless choropleth map has been revived because a computer can control output, which so far has mostly been line output by a plotter (Fig. 14.11).

A classless choropleth map is a ratio-scale portrayal similar in concept to proportionally scaled graduated point symbols. The term choropleth map or simple choropleth map has traditionally referred to a range-graded portrayal of the data set; it is thus more generalized than a classless representation.

A classless portrayal using a scale of tones can

FIGURE 14.9 A machine-produced simple choropleth map of Connecticut showing the average sale price per house by towns. (Map prepared by the Laboratory for Computer Graphics, Harvard University.)

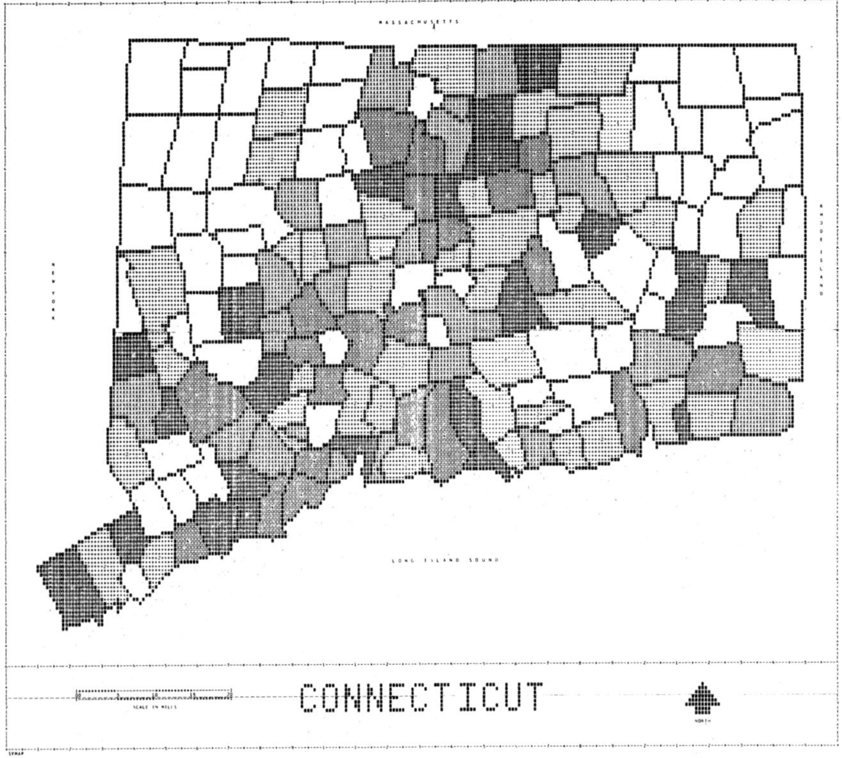

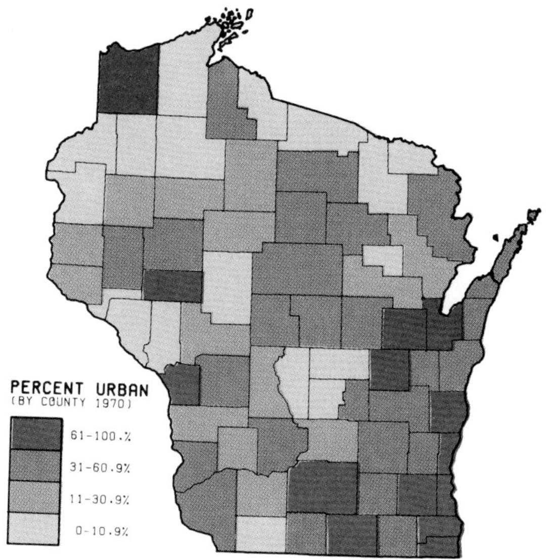

FIGURE 14.10 A map produced by photomechanically enhancing a computer line drawing obtained from a CALCOMP 1163 drum plotter. (Courtesy, University of Wisconsin Cartographic Laboratory.)

fairly represent the variations among the data values. Unfortunately, when only line output is available, a cross-hatch pattern has generally been used, and although it is reasonably effective in the higher ranges, it is less so in the low ranges where the pattern spacing becomes so large that it does not convey any effect of a gradation of tones.

The effectiveness with which a classless choropleth map portrays information compared to the effectiveness of a simple choropleth map is not known. On the one hand one could speculate that the classless map may appear more complex than its classed counterpart simply because the former is less generalized. On the other hand, the increased significance given to classed areas may suggest more uniformity than is desirable. There seems no question, however, that a classless choropleth representation may sometimes be the logical choice.

The Dasymetric Map. The nature of and knowledge about the phenomenon being mapped, or the inefficiency and variation of the unit areas employed to derive the data values, may make it desirable to employ the dasymetric system of representation for volumetric quantitative data. Some kinds of phenomena do not extend, even when abstractly conceived, across certain kinds of areas. Neither do they have distributional characteristics like those related to the sloping-surface concept associated with a continuously changing set of values. For example, agricultural population density in a region of extreme variation in the character of agricultural land is likely to show relatively uniform high density in favorable areas and sparse occupance in unfavorable areas. Changes would be rapid in zones that are not likely to coincide with the boundaries of the data-collection unit areas.

The dasymetric map, although often made from exactly the same initial data used for the simple choropleth map, assumes the existence of areas of relative homogeneity, which are also assumed to be separated from one another by zones of rapid change. By subdivision of the original statistical unit areas, additional detail may be added

FIGURE 14.11 A classless choropleth map. Each unit area is symbolized by a different area symbol in an attempt to reflect its actual value. (Courtesy M. B. Peterson and the American Congress on Surveying and Mapping.)

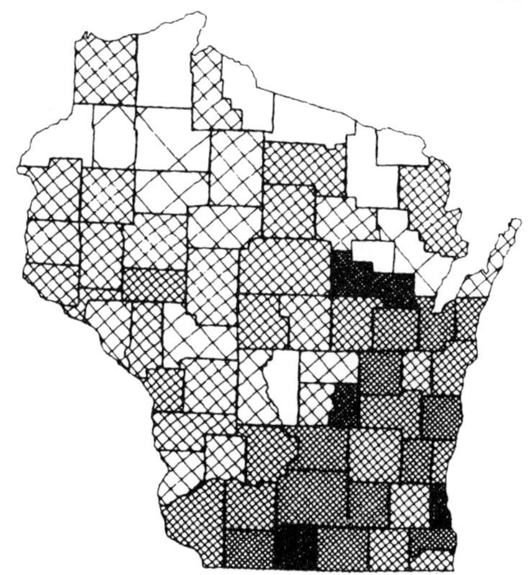

to the presentation on the basis of whatever knowledge the cartographer has about the data from field knowledge or any other kinds of available information.

The arithmetic involved in reconciling the numerical assumptions with the original unit area data has been explained and illustrated in Chapter 6. It should be emphasized that nothing in the data themselves will indicate to the cartographer the kinds of subdivision that should be made or where the zones of rapid change occur. Instead, this kind of information must come from other knowledge about the spatial relationships among the data and other geographical distributions. It may be observed in this connection that it is much more difficult to program the dasymetric process for automation than it is the simple choropleth, since it must always involve additional information, including associated distributions. The phenomena found in association with the distribution being mapped and on which the cartographer must base the rearrangement of the basic unit area data are of two kinds, (1) limiting variables, and (2) related variables.

Limiting Variables. These are variables that set an absolute upper limit on the quantity of the mapped phenomenon that can occur in an area. For example, suppose we were mapping percentage of cropland and the data showed that county A had 60 percent of its total area as cropland. If 15 percent of the area of the county were devoted to urban land uses, obviously that area could not be cropland. Exclusion of the urban area allocates all the cropland to the remaining sections of the county.

Suppose, furthermore, that a second limiting variable in county A were woodland, the areal percentage of which was known for minor civil divisions such as townships. If a township had 60 percent of its area in woodland, then the maximum cropland it could have would be 40 percent. Other areas of the county would, therefore, need to have a much higher percentage of cropland to make possible the original statistic of 60 percent for the county as a whole.

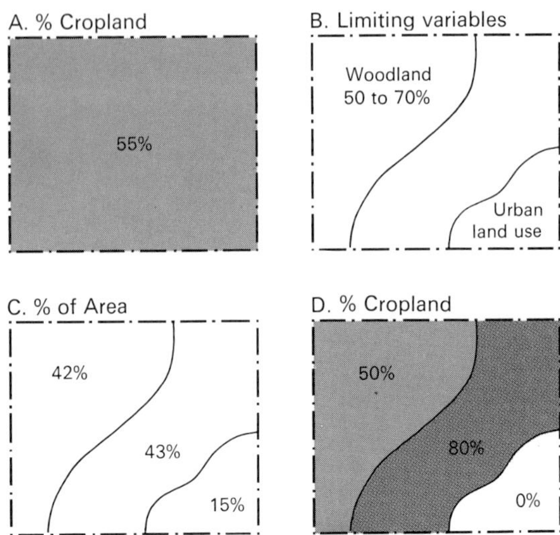

FIGURE 14.12 A knowledge of limiting variables can assist in dasymetric mapping. See text for explanation.

The diagrams in Figure 14.12 show how knowledge of the geographical occurrence of limiting variables enables the cartographer to add considerable geographic detail in dasymetric mapping to the simple choropleth representation. In Figure 14.12A all that is known is that the county in question consists of 55 percent cropland. In Figures 14.12B and 14.12C the limiting variables have been mapped: they consist of urban land use, consisting of 15 percent of the area of the county and an area of 50 to 70 percent woodland (occupying 42 percent of the area of the county). Using the formula for estimating densities of fractional parts and starting with urban land use, if we assume there is no cropland in the urban area, we may determine that the county outside the urban area consists of 65 percent cropland. That section, as shown in Figure 14.12B is almost half occupied by an area of 50 to 70 percent woodland. Assuming that the most cropland that could occur in the woodland area would be 50 percent, we may next determine that the area between the woodland zone and the urban area is approximately 80 percent cropland. A choropleth map would have found the county in only one class, mapped as Figure

14.12A; approached dasymetrically, there are three classes represented, mapped as Figure 14.12D.

Related Variables. The functioning of related variables in the dasymetric mapping process is much more complex than that of limiting variables. Related variables are those geographical phenomena that show predictable variations in spatial association with the phenomenon being mapped, but that cannot be employed in a limiting sense. To continue with the example of percentage of cropland, we may have available a map of surface configuration that shows that the land form varies from level on the one hand to very hilly, steep country on the other. There would normally be a high positive correlation between level land and percentage of cropland. Similarly, we could think of several other variables that would assist in predicting variations in the geographical occurrence of percent of cropland, such as types of farming regions and soil characteristics.

Related variables are more difficult to use properly because they cannot be employed in the strict manner possible with limiting variables. They are, however, very useful in helping the cartographer put some geographical "sense" into the bare statistics that are commonly gathered on the basis of enumeration districts.

Place Data Dasymetric Maps. The dasymetric technique can also be used to portray areas of uniform place data. The cartographer creates these areas of uniformity by clustering or grouping similar quantities into regions. Many different clustering techniques can be used. The equivalent of related and limiting variables may also be used to assist the cartographer in this form of dasymetric mapping.

Hammond* used a highly complex form of the dasymetric technique to portray from place data uniform areas of land surface form in the conterminous United States. Figure 14.13 gives some of the details of Hammond's classification scheme and a small portion of the resulting map. This type of dasymetric map represents the results of a complex level of geographic analysis classified for map presentation.

Elements of Choroplethic Mapping. The elements involved in choroplethic mapping include the size and shape of the unit areas, the number of classes, and the method of class limit determination. Only the latter two are important for dasymetric maps created from place data.

Size and Shape of Unit Areas. The unit areas of a region taken together form a pattern, and this pattern acts as a generalization filter. If the unit areas are large, the spatial variation of the data tends to be reduced or averaged out; if the unit areas are small, variation is preserved. If the unit areas vary greatly in size from one another, variation is preserved in one part of the region and lost in another, resulting in uneven generalization, an undesirable effect for a map. Ideally, for the best use of the choropleth method, unit areas should be of relatively equal sizes (preferably small) and of similar shape (see Fig. 14.14).

Since the data-collection-unit area boundaries must coincide with the discontinuities on the assumed statistical surface when the simple choroplethic technique is used, irregular shapes often cause distracting patterns on the map. Unit areas of similar shapes avoid this problem. In dasymetric mapping, the shapes of the data-collection unit areas are of less concern, since they are much less prominent in the mapping procedure.

The Number of Classes. The most common choroplethic method results in a range-graded symbolization. The number of classes symbolized therefore determines the detail that can be read from the map. The cartographic ideal is to present the maximum number of classes that can easily be read, but this depends on whether the map produced is monochromatic or color and

*E. H. Hammond, "Analysis of Properties in Land Form Geography: An Application to Broad-Scale Land Form Mapping," *Annals of the Association of American Geographers* 54 (1964):11–9.

SCHEME OF CLASSIFICATION

SLOPE (1st letter)

- A > 80% of area gently sloping
- B 50-80% of area gently sloping
- C 20-50% of area gently sloping
- D < 20% of area gently sloping

LOCAL RELIEF (2nd letter)

- 1 0-100 feet
- 2 100-300 feet
- 3 300-500 feet
- 4 500-1000 feet
- 5 1000-3000 feet
- 6 3000-5000 feet

PROFILE TYPE (3rd letter)

- a > 75% of gentle slope is in lowland
- b 50-75% of gentle slope is in lowland
- c 50-75% of gentle slope is on upland
- d > 75% of gentle slope is on upland

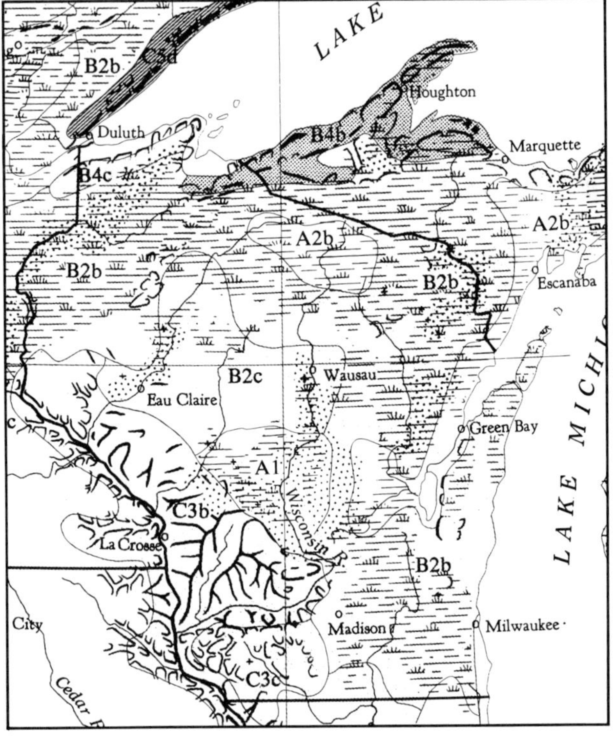

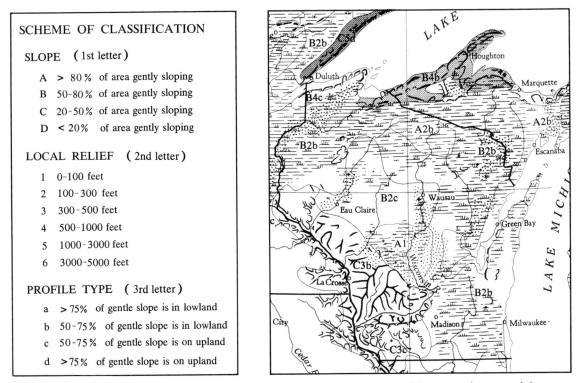

FIGURE 14.13 A portion of Hammond's land surface form map. The numbers and letters on the areas of the map are explained in the scheme of classification at the left. This map uses a highly complex dasymetric mapping methodology including initial place data. (Courtesy, E. Hammond and the Association of American Geographers.)

the nature of the distribution (see Fig. 14.15). Perceptually, a reader is limited to relatively few classes on a monochromatic map. Even when both pattern and value are employed, five to eight classes approach the maximum. When color is used, the number of classes can be increased. Normally the cartographer, by a process of trial and error, arrives at a convenient balance between the number of readable classes and the complexity of the distribution.

The classless choropleth map, of course, uses as many categories as there are unique values in the distribution shown on the map. Perceptually a classless choropleth map will probably not enable the map reader to determine an exact value for a unit area. On the other hand, the classless choropleth map sometimes presents a good impression of the overall distribution to the map reader. Especially if there are many small unit areas, the effect may be similar to that created by shaded relief.

Class Limit Determination. Perhaps no aspect of choroplethic mapping has received more space in the cartographic literature than methods for determining class limits. After a cartographer has decided on the number of classes for a given portrayal, the limits of these classes must be set. The methods outlined below are some of the many that cartographers have used. They are equally applicable to the establishment of isoplethic intervals as they are for establishing limits for range-

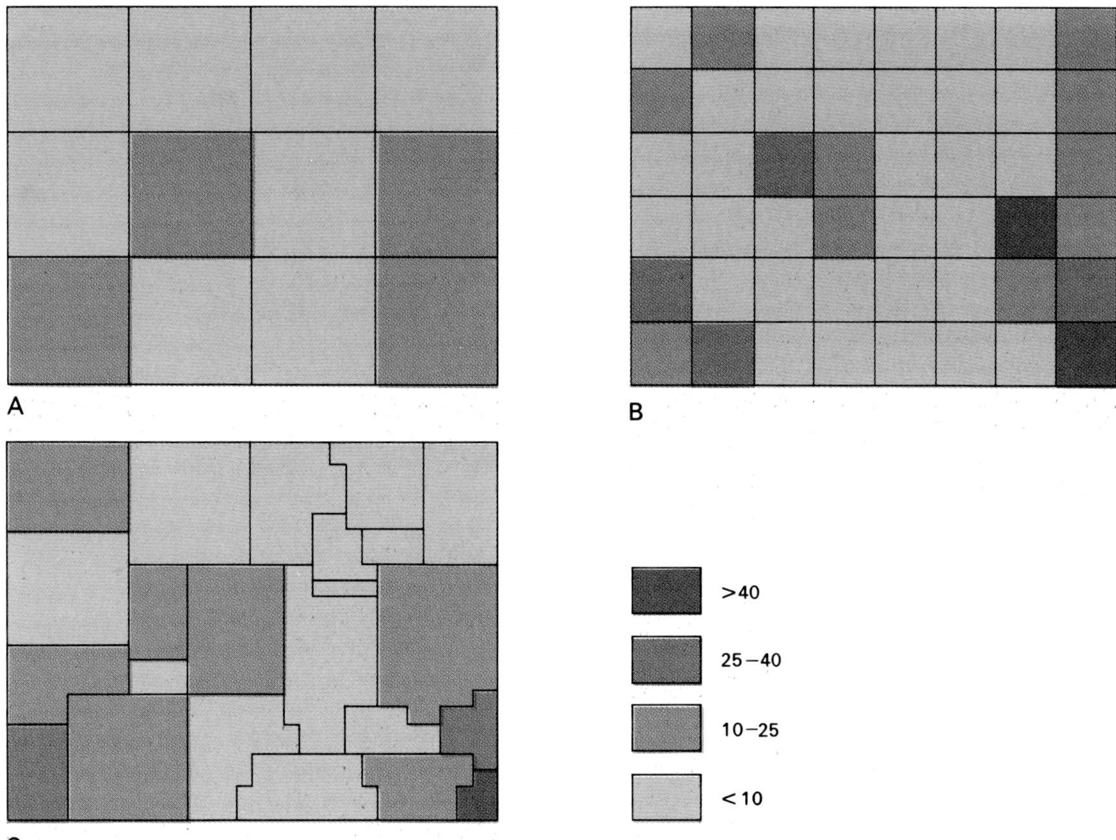

FIGURE 14.14 The same distribution mapped with three different sets of unit areas employing the same class interval scheme: (A) uses large units of equal area, (B) uses small units of equal area, and (C) uses irregularly sized unit areas.

graded choroplethic representations or range-graded proportional point or line symbols. (Our examples and the following discussion will concentrate on range-graded choroplethic mapping.)

Because they are so widely applicable in cartography, a considerable number of methods will be presented. Finally, as was the case with interpolation models in isarithmic mapping (Chapter 13), computer assistance has increased both the ease of use and the number of methods readily available. Because so many methods are available for the cartographer to use, it is relatively easy to select class-interval determination schemes that will range-grade the data in a way that presents a biased view or to mask the basic character of a distribution. Obviously the potential to enhance a portrayal by the judicious selection of class limits also exists.

Theoretically the cartographer should seek to highlight critical values in the distribution. For example, if a minimum of 40 percent of a county's population must be below the poverty level to qualify for a certain federal program, then this value might be chosen as one limit. If, however, there are no critical values, the cartographer should seek to maximize both the homogeneity within

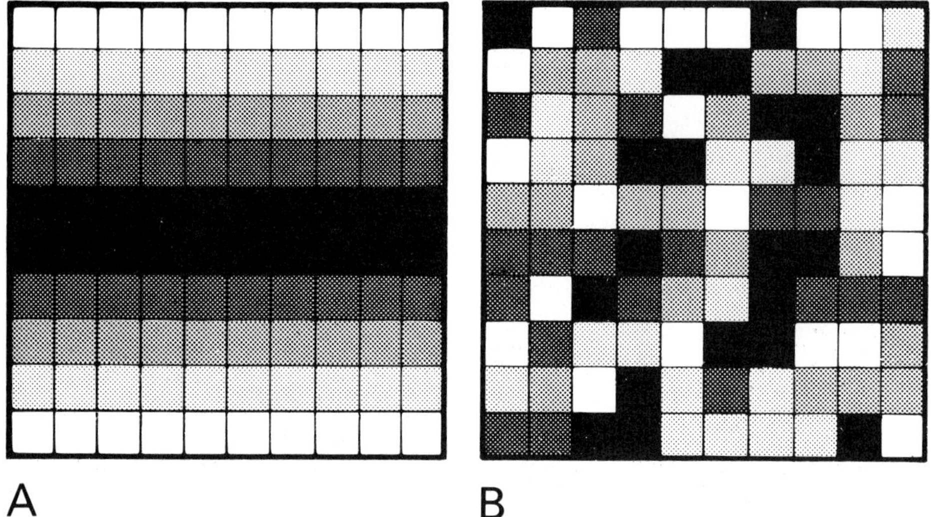

FIGURE 14.15 Both (A) and (B) employ five classes. Because of the rather simple distribution pattern shown on (A), it would be possible to increase the number of classes used. For (B) the lack of pattern in the distribution would make it even more confusing to the map reader if more classes were used. (Maps courtesy of J. M. Olson.)

each class and the differences between classes. Fortunately, with computer assistance, these criteria stated statistically can be approximated.

Kinds of Class Interval Series. There are many different kinds of class intervals, and the cartographer needs to analyze the immediate communication problem carefully when choosing one.

As Mackay has pointed out, class intervals are the mesh sizes of cartography, with the chosen limits forming the screen wires. We must choose the mesh size wisely, so that the "size sorting" of the distributional data will be most effectively accomplished. Among the more important of the cartographer's concerns in this connection is an analysis of the problem as related to generalization. As mentioned above, there is a direct relation between the number of classes and the amount of information that will be portrayed.

An infinite number of kinds of series exist for use by the cartographer in setting class limits. It is almost impossible logically to classify the various series according to their mathematical characteristics or the procedures used to derive them, but we can separate them into three major groups.

One is the equal-steps or constant series, which employs some kind of equal division of the data or the geographical area. There are several possibilities in this class.

A second class of series includes those in which the interval becomes systematically smaller toward either the upper or lower end of the scale. Generally, as has been pointed out, the tendency is to put the greater detail at the lower end because, in that section of the range, a small absolute difference is a large relative difference. There seems to be a considerably greater fascination for relative differences than for absolute differences; many distributions are highly skewed; and furthermore, in cartography there is usually much more room for such detail in the regions of sparse occurrences! A great many series increase or decrease toward one end of the range.

A third class is the irregular or variable series. This kind of series is used when the cartographer wishes (1) to call attention to various internal characteristics of the distribution (such as some values that may be significant in relation to other analyses), (2) to minimize certain error aspects, or (3) to highlight certain elements of the data range that would not be properly dealt with were a constant or a regular ascending or descending series being employed. These kinds of series are often chosen with the aid of graphic devices, such as the clinographic curve, the frequency curve, or the cumulative curve.*

Before the general availability of computers, the analysis of the consequences of using various kinds of class interval series was inordinately time-consuming. Now, however, we can quickly obtain measures that state the degree of total error (deviation from class means), its distribution among classes, and the ratio of between-class variation to the total variation for any number of classes desired.

Constant Series or Equal Steps. There are four major types of constant series that cartographers frequently use for range grading data (see Tables 14.1 and 14.2 and Fig. 14.16). The best known and perhaps the simplest is the series of *equal steps based on the data range* (see Fig. 14.16A). We obtain this kind of constant series by dividing the range between the high and low values by the desired number of classes to obtain the common difference. For convenience, the common difference is usually a round number. This common difference is successively added to each class limit, beginning with the lowest value, to obtain the next higher class limit. Major advantages of this series are its simplicity of execution and the comparability of the classes, especially in mapping with isarithms (such as contours). If we are mapping a distribution made up of a perfectly rectangular distribution of data observations, there is no difference between this series and the third and fourth series discussed below. A major drawback to this scheme is that the distribution of data observations is not taken into account. Thus a normal distribution or a highly skewed distribution could result in many data observations falling in a few classes and few or no observations in other classes. In fact, with highly skewed distributions, classes with no observations often result from this method.

A second type of constant series can be formed by use of the *parameters of a normal distribution* (see Fig. 14.16B). If we derive the mean and standard deviation, we may use them to set class limits such as the mean plus and minus one standard deviation, from one standard deviation to two standard deviations (above or below the mean), and so on (see Table 14.2). The more nearly normal the distribution, the more useful is this type for graphically portraying the areas and magnitudes of departures from the average.

If the data are normally distributed, the cartographer may employ fractional parts of standard deviations in an effort to create equal areas on the maps in each class or equal numbers of observations in each class. In essence the mean and standard deviations allow the cartographer to specify equal steps based on the data range, or equal numbers of observations per class or equal geographical areas (provided all observations refer to areas of equal size) for normal distributions. If applied to data that are not normally distributed, the use of the mean and standard deviation tends to result in classes with no data observations and to combine many of the observations into one or two classes.

A variation of the use of the mean for class limit determination is the method of "nested means" (see Fig. 14.16C).* In this method the overall mean of the data distribution initially divides the distribution into two parts. A mean is again calculated for each part, and each part is divided into two additional parts by this "second-

*Jenks and Coulson have devised a method of testing various class intervals series (as applied to a given distribution) against one another in a fashion analogous to the determination of the coefficient of variability. See George F. Jenks and Michael R. Coulson, "Class Intervals for Statistical Maps," *International Yearbook of Cartography* 3 (1963):119–34.

*Morton W. Scripter, "Nested-Means Map Classes for Statistical Maps," *Annals of the Association of American Geographers* 60 (1970): 385–93.

TABLE 14.1 Wisconsin Counties Ranked from Most Urbanized to Least Urbanized

Rank	Percent Urban	Observation (County)	Category	Area (km^2)
1	100.0	Milwaukee	Highly urban	619
2	81.6	Brown	Highly urban	1355
3	80.2	Waukesha	Highly urban	1440
4	78.2	Winnebago	Highly urban	1176
5	77.2	Dane	Highly urban	3100
6	76.1	Racine	Highly urban	873
7	74.9	La Crosse	Urban	1215
8	74.9	Rock	Urban	1867
9	73.3	Douglas	Urban	3393
10	71.5	Kenosha	Urban	707
11	69.2	Eau Claire	Urban	1681
12	68.6	Outagamie	Urban	1642
13	67.5	Ozaukee	Urban	609
14	61.1	Sheboygan	Urban	1311
15	60.2	Manitowoc	Urban	1526
16	57.4	Ashland	Urban	2686
17	57.1	Fond du Lac	Urban	1875
18	55.0	Lincoln	Urban	2331
19	52.2	Jefferson	Urban	1461
20	52.2	Wood	Urban	2103
21	49.6	Marathon	Rural	4103
22	49.4	Portage	Rural	2098
23	47.0	Washington	Rural	1109
24	46.9	Langlade	Rural	2222
25	45.8	Dodge	Rural	2310
26	44.7	Calumet	Rural	816
27	43.4	Marinette	Rural	3595
28	41.8	Green	Rural	1518
29	38.7	Dunn	Rural	2222
30	38.7	Walworth	Rural	1450
31	37.7	Monroe	Rural	2370
32	36.5	Kewaunee	Rural	857
33	36.3	Crawford	Rural	1518
34	35.4	Waupaca	Rural	1945
35	34.5	Chippewa	Rural	2655
36	33.7	Door	Rural	1272
37	33.6	Oneida	Rural	2885
38	32.8	Grant	Rural	3025
39	32.0	Sauk	Rural	2176
40	31.4	Green Lake	Rural	919
41	29.8	Richland	Rural	1513
42	28.9	Columbia	Rural	2015
43	28.4	St. Croix	Rural	1906
44	28.1	Oconto	Rural	2606

TABLE 14.1 continued

45	25.8	Rusk	Rural	2357
46	23.4	Pierce	Highly rural	1531
47	21.4	Barron	Highly rural	2243
48	21.4	Jackson	Highly rural	2590
49	20.4	Taylor	Highly rural	2536
50	20.3	Price	Highly rural	3284
51	19.9	Shawano	Highly rural	2328
52	18.8	Juneau	Highly rural	2059
53	16.9	Iowa	Highly rural	1971
54	15.2	Vernon	Highly rural	2085
55	9.1	Clark	Highly rural	3165
56	0.3	Waushara	Highly rural	1627
57	0.0	Adams	Highly rural	1753
58	0.0	Bayfield	Highly rural	3818
59	0.0	Buffalo	Highly rural	1844
60	0.0	Burnett	Highly rural	2176
61	0.0	Florence	Highly rural	1267
62	0.0	Forest	Highly rural	2616
63	0.0	Iron	Highly rural	1932
64	0.0	LaFayette	Highly rural	1665
65	0.0	Marquette	Highly rural	1184
66	0.0	Menominee	Highly rural	938
67	0.0	Pepin	Highly rural	614
68	0.0	Polk	Highly rural	2419
69	0.0	Sawyer	Highly rural	3297
70	0.0	Trempealeau	Highly rural	1914
71	0.0	Vilas	Highly rural	2246
72	0.0	Washburn	Highly rural	2113

TABLE 14.2 Five Class Limits Determined by Five Differing Methods Utilizing the Data Given in Table 14.1

A Equal steps based on the data range
Range 0 to 100%
Class 1 0 to 25
Class 2 25 to 50
Class 3 50 to 75
Class 4 75 to 100

B Parameters of a normal distribution

$\bar{z}$ = Mean = 34.8		s. = Standard deviation = 26.6
Class 1	$< -1s.$	0 to 8.2
Class 2	$-1s.$ to $\bar{z}$	8.2 to 34.8
Class 3	$\bar{z}$ to $+1s.$	34.8 to 61.4
Class 4	$> +1s.$	61.4 to 100

TABLE 14.2 continued

C	Nested means		
	First-order mean = 34.8		
	Second-order means = 13.8 and 58.2		
C	Class 1	0 to 13.8	
	Class 2	13.8 to 34.8	
	Class 3	34.8 to 58.2	
	Class 4	58.2 to 100	
D	Quartiles		
		rank	data values
	Class 1	72 to 55	0 to 9.1
	Class 2	54 to 37	15.2 to 33.6
	Class 3	36 to 19	33.7 to 52.2
	Class 4	18 to 1	55.0 to 100
E	Areal equal steps		
	Each class covers approximately $\frac{1}{4}$ of area of state.		
		data values	
	Class 1	0 to 15.2	
	Class 2	16.9 to 32.8	
	Class 3	33.6 to 49.6	
	Class 4	52.2 to 100	

order'' mean. Major advantages of this method are that no classes without data observations can be constructed and that at any level the distribution of the data can be said to be in equilibrium, since the absolute sum of the deviations of the class members from a mean dividing a distribution or distribution part into two classes is always equal for both classes divided by that mean. A major disadvantage is that only 2^n classes can be calculated, where n is the order of mean being calculated. Thus 3, 5, 6, 7, or 9, and so forth, classes are impossible with this method.

A third kind of constant series involves the employment of quantiles; a *quantile* is a division of the number of observations in the data array into equal parts. There may be any division, but the commonly used ones are quartiles (four), quintiles (five), sextiles (six), and so on, up to deciles (ten) or even centiles (one hundred). Quantiles are determined by arranging the observed values in the order of their magnitude, from the lowest to the highest. If we wished to obtain quartile values for interval or ratio data, for example, we proceed by counting one-fourth of the number of observations from the bottom of the ordered array to determine the first quartile value, and so on (see Table 14.2 and Fig. 14.16D). Quantiles are useful when the map is based on unit areas, especially in choropleth mapping, but if there are major differences in the sizes of the unit areas, much of the value of the quantile series is lost. If all the areas are of equal size, this method gives results equivalent to the type of constant series discussed next. This method can be useful for ordinally scaled data, since in ordinal scaling the data are ranked, and since this method depends only on the number of observations and not their value, it can be applied easily to ordinally scaled data.

A final type of constant series is what might be called *equal-area steps* or, in a sense, ''geographical quantiles'' (see Fig. 14.16E). In this kind of series the area of the map is divided into equal regions, the number depending on the cartographer's choice. The determination of the class limits for this kind of series can be accomplished by employing a cumulative frequency graph, as de-

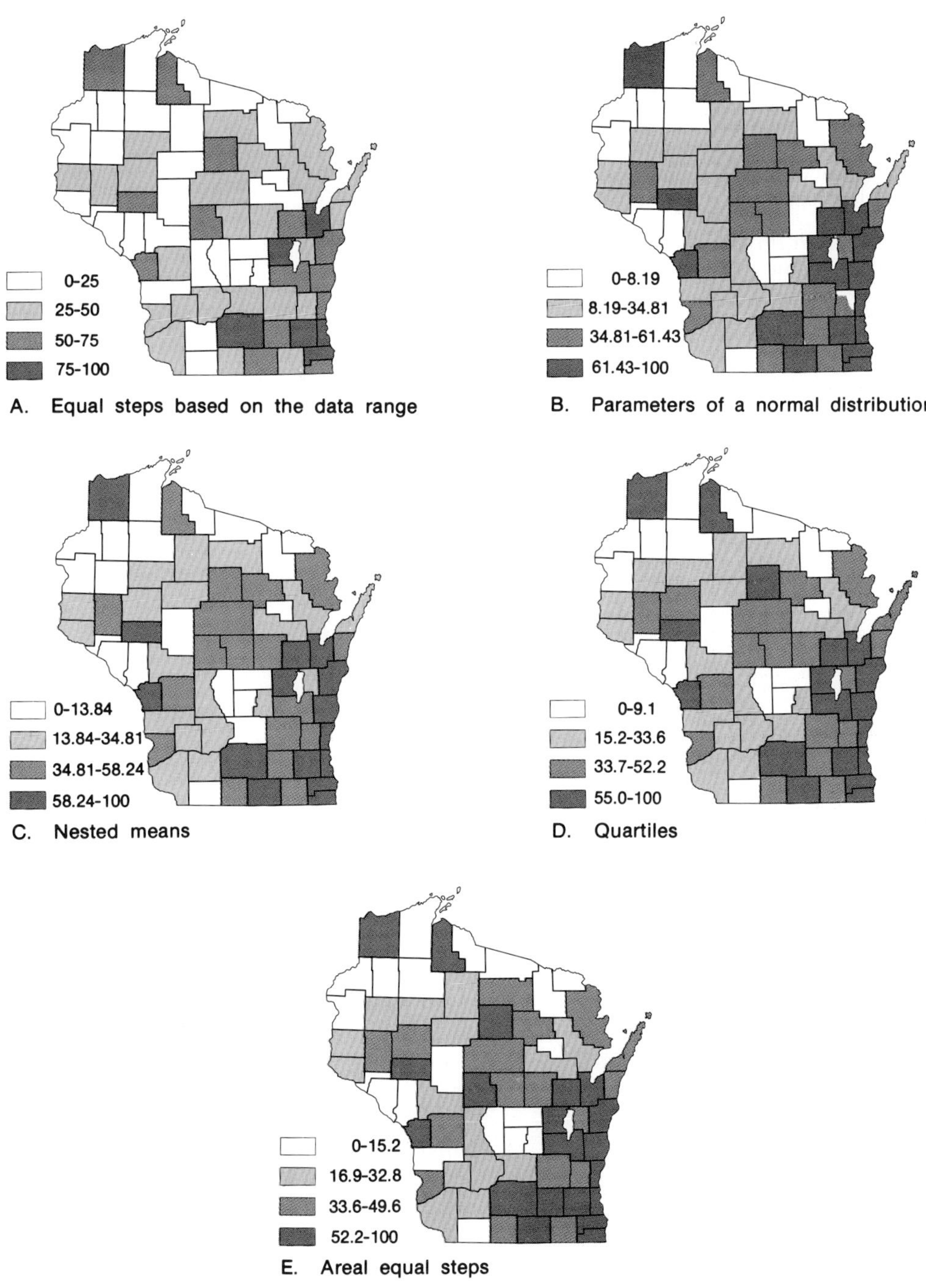

FIGURE 14.16 Five examples of constant series class intervals applied to the data in Table 14.1. The class limits are defined in Table 14.2.

scribed below. It cannot be done unless the areas of the unit regions are known (see Table 14.1). Furthermore, it has no applicability to range-graded point or line volumes. Finally, if the data are from a rectangular distribution and all unit areas have equal areas, there is no difference between this method and the equal-step method based on the data range or the quantile method. An added advantage of this method for cartographic purposes is that it specifies the class limits in such a way that each class occupies an equal portion of the map area (see Table 14.2).

Systematically Unequal Stepped Class Limits. This type of class limit determination scheme applies only to data measured on an interval or ratio scale.

Two groups of series with unequal steps are (1) arithmetic series in which each class is separated from the next by a stated numerical difference (not constant), and (2) geometric series in which each class is separated by a stated numerical ratio. The general form of the equation to produce class limits of either type of series is

$$L + B_1X + B_2X + \ldots + B_nX = H$$

where L = the lowest value
H = the highest value
B_n = the value of the nth term in the progression.

It is only necessary then to obtain B_n by some method and then solve the equation for X for any given values of L and H to establish the class limits. For *arithmetic progressions* the quantity B_n is obtained by

$$B_n = a + [(n - 1)d]$$

where a = the value of the first term
n = the number of the term being determined (the first, second, etc.)
d = the stated difference.

In *geometric progressions* the quantity B_n is obtained by

$$B_n = gr^{n-1}$$

where g = the value of the first nonzero term
n = the number of the term being determined (the first, second, etc.)
r = the stated ratio.

Both arithmetic and geometric series can take any of the following six forms, depending on the values of d and r.

1. Increasing at a constant rate.
2. Increasing at an increasing rate.
3. Increasing at a decreasing rate.
4. Decreasing at a constant rate.
5. Decreasing at an increasing rate.
6. Decreasing at a decreasing rate.

To illustrate systematically unequal stepped class intervals, twelve examples are calculated below and displayed in map format in Figure 14.17. The maps use the data in Table 14.1.

Note. In calculating many such series, fractions are common. In the following, all such fractions have been rounded to the most convenient whole number.

I. ARITHMETIC PROGRESSION EXAMPLES

A. Increasing at a Constant Rate

$$a = 1$$
$$d = 1$$
$$n = 1,2,3,4$$
$$B_1 = 1 + (0)1 = 1$$
$$B_2 = 1 + (1)1 = 2$$
$$B_3 = 1 + (2)1 = 3$$
$$B_4 = 1 + (3)1 = 4$$
$$L + \sum_i B_iX = H$$
$$0 + 10X = 100$$
$$X = 10$$

Class limits are 0 to 10
10 to 30
30 to 60
60 to 100

B. Increasing at an Increasing Rate

$$a = 1$$
$$d = n - 1$$
$$n = 1,2,3,4$$
$$B_1 = 1 + (0)(0) = 1$$
$$B_2 = 1 + (1)(1) = 2$$
$$B_3 = 1 + (2)(2) = 5$$
$$B_4 = 1 + (3)(3) = 10$$
$$L + \sum_i B_i X = H$$
$$0 + 18X = 100$$
$$X = 5.556$$

Class limits are 0 to 6
6 to 17
17 to 44
44 to 100

C. Increasing at a Decreasing Rate

$$a = 1$$
$$d = 1/n$$
$$n = 1,2,3,4$$
$$B_1 = 1 + (0)(1) = 1$$
$$B_2 = 1 + (1)(\tfrac{1}{2}) = 1.5$$
$$B_3 = 1 + (2)(\tfrac{1}{3}) = 1.67$$
$$B_4 = 1 + (3)(\tfrac{1}{4}) = 1.75$$
$$L + \sum_i B_i X = H$$
$$0 + 5.92X = 100$$
$$X = 16.89$$

Class limits are 0 to 17
17 to 42
42 to 70
70 to 100

D. Decreasing at a Constant Rate

$$a = 4$$
$$d = 1$$
$$n = 1,2,3,4$$
$$B_1 = 4 + (0)(-1) = 4$$
$$B_2 = 4 + (1)(-1) = 3$$
$$B_3 = 4 + (2)(-1) = 2$$
$$B_4 = 4 + (3)(-1) = 1$$
$$L + \sum_i B_i X = H$$
$$0 + 10X = 100$$
$$X = 10$$

Class limits are 0 to 40
40 to 70
70 to 90
90 to 100

E. Decreasing at an Increasing Rate

$$a = 10$$
$$d = -(n - 1)$$
$$n = 1,2,3,4$$
$$B_1 = 10 + (0)(-0) = 10$$
$$B_2 = 10 + (1)(-1) = 9$$
$$B_3 = 10 + (2)(-2) = 6$$
$$B_4 = 10 + (3)(-3) = 1$$
$$L + \sum_i B_i X = H$$
$$0 + 26X = 100$$
$$X = 3.846$$

Class limits are 0 to 38
38 to 73
73 to 96
96 to 100

F. Decreasing at a Decreasing Rate

$$a = 1$$
$$d = -1/n$$
$$n = 1,2,3,4$$
$$B_1 = 1 + (0)(-1) = 1 = 1.0$$
$$B_2 = 1 + (1)(-\tfrac{1}{2}) = \tfrac{1}{2} = 0.5$$
$$B_3 = 1 + (2)(-\tfrac{1}{3}) = \tfrac{1}{3} = 0.333$$
$$B_4 = 1 + (3)(-\tfrac{1}{4}) = \tfrac{1}{4} = 0.25$$
$$L + \sum_i B_i X = H$$
$$0 + 2.083X = 100$$
$$X = 48$$

Class limits are 0 to 48
48 to 72
72 to 88
88 to 100

II. GEOMETRIC PROGRESSION EXAMPLES

G. Increasing at a Constant Ratio

$$a = 1$$
$$r = 2$$
$$n = 1,2,3,4$$
$$B_1 = (1)(2)^{(0)} = 1$$
$$B_2 = (1)(2)^{(1)} = 2$$
$$B_3 = (1)(2)^{(2)} = 4$$
$$B_4 = (1)(2)^{(3)} = 8$$
$$L + \sum_i B_i X = H$$
$$0 + 15X = 100$$
$$X = 6.6667$$

Class limits are 0 to 7
7 to 21
21 to 47
47 to 100

H. Increasing at an Increasing Ratio

$$a = 1$$
$$r = n$$
$$n = 1,2,3,4$$
$$B_1 = (1)(1)^{(0)} = 1$$
$$B_2 = (1)(2)^{(1)} = 2$$
$$B_3 = (1)(3)^{(2)} = 9$$
$$B_4 = (1)(4)^{(3)} = 64$$
$$L + \sum_i B_i X = H$$
$$0 + 76X = 100$$
$$X = 1.316$$

Class limits are 0 to 1
1 to 4
4 to 16
16 to 100

I. Increasing at a Decreasing Ratio

$$a = 1$$
$$r = n/n - 1$$
$$n = 1,2,3,4$$
$$B_1 = (1)(0)^{(0)} = 1$$
$$B_2 = (1)(2)^{(1)} = 2$$
$$B_3 = (1)(\tfrac{3}{2})^{(2)} = 2.25$$
$$B_4 = (1)(\tfrac{4}{3})^{(3)} = 2.37$$
$$L + \sum_i B_i X = H$$
$$0 + 7.62X = 100$$
$$X = 13.1233$$

Class limits are 0 to 13
13 to 39
39 to 69
69 to 100

J. Decreasing at a Constant Ratio

$$a = 1$$
$$r = \tfrac{1}{2}$$
$$n = 1,2,3,4$$
$$B_1 = (1)(\tfrac{1}{2})^{(0)} = 1 = 1.0$$
$$B_2 = (1)(\tfrac{1}{2})^{(1)} = \tfrac{1}{2} = 0.5$$
$$B_3 = (1)(\tfrac{1}{2})^{(2)} = \tfrac{1}{4} = 0.25$$
$$B_4 = (1)(\tfrac{1}{2})^{(3)} = \tfrac{1}{8} = 0.125$$
$$L + \sum_i B_i X = H$$
$$0 + 1.875X = 100$$
$$X = 53.3333$$

Class limits are 0 to 53
53 to 80
80 to 93
93 to 100

K. Decreasing at an Increasing Ratio

$$a = 1$$
$$r = 1/n$$
$$n = 1,2,3,4$$

$$B_1 = (1)(1)^{(0)} = 1 = 1.0$$
$$B_2 = (1)(\tfrac{1}{2})^{(1)} = \tfrac{1}{2} = 0.50$$
$$B_3 = (1)(\tfrac{1}{3})^{(2)} = \tfrac{1}{9} = 0.111111$$
$$B_4 = (1)(\tfrac{1}{4})^{(3)} = \tfrac{1}{64} = 0.015625$$
$$L + \sum_i B_i X = H$$
$$0 + 1.627X = 100$$
$$X = 61.463$$

Class limits are 0 to 61
61 to 92
92 to 99
99 to 100

L. Decreasing at a Decreasing Ratio

$$a = 1$$
$$r = \frac{n - 1}{n}$$
$$n = 1,2,3,4$$
$$B_1 = (1)(0)^{(0)} = 1.0$$
$$B_2 = (1)(\tfrac{1}{2})^{(1)} = 0.5$$
$$B_3 = (1)(\tfrac{2}{3})^{(2)} = 0.44444$$
$$B_4 = (1)(\tfrac{3}{4})^{(3)} = 0.421875$$
$$L + \sum_i B_i X = H$$
$$0 + 2.3663X = 100$$
$$X = 42.26$$

Class limits are 0 to 42
42 to 63
63 to 82
82 to 100

There is obviously an infinity of possibilities using arithmetic and geometric progressions. The difficulty lies in assessing when it is appropriate to apply which particular progression. Because of this difficulty, such progressions have seen somewhat diminished use in class limit determination schemes. No strictly theoretical grounds have been advanced that demand the use of a progression, in spite of the fact that machine computation makes the calculations easier than ever before. Computer capabilities have made it possible for more complex statistical criteria to be used in class limit determination in place of progressions (see next section). Figure 14.17 provides examples of all twelve types of arithmetic and geometric progressions.

Irregular Stepped Class Limits. The irregular stepped interval techniques can be subdivided into two classes: graphic techniques and iterative techniques.

It must be emphasized at the outset that graphic approaches to the selection of class intervals may not be suitable for two reasons. On the one hand, the limits may be either ill defined or so oddly arranged in the series that it would be difficult for a reader to use them; and on the other hand, the arrangement of the values in numerical order (as is necessary in the preparation of the graphs) clearly destroys their geographical order or association. It is quite possible that some startling characteristic of a curve on a graph, such as a marked peak, depression, or flexure, may have no geographical significance whatever, since the data from which it derives may be widely and unsystematically dispersed. Three common graphic techniques employed by cartographers to aid them in selecting class limits are the frequency curve, the clinographic curve, and the cumulative frequency curve.

A *frequency graph* is prepared by arranging the z values of the data on the x axis of a graph plotted against the frequency of their occurrence of the y axis. Frequency in this sense means the number of times they occur. In the case of volumes distributed over areas, if the areas of the units are not widely different, the data collection units may simply be counted. If they are greatly different, we should employ a cumulative frequency graph. As an illustration of a frequency graph, Figure 14.18 has been prepared from the data in Table 14.1. The arrows indicate low points in the histogram which were selected as class limits. The data are mapped in Figure 14.19.

Generally, the low points on a frequency diagram are thought to be the most useful as class

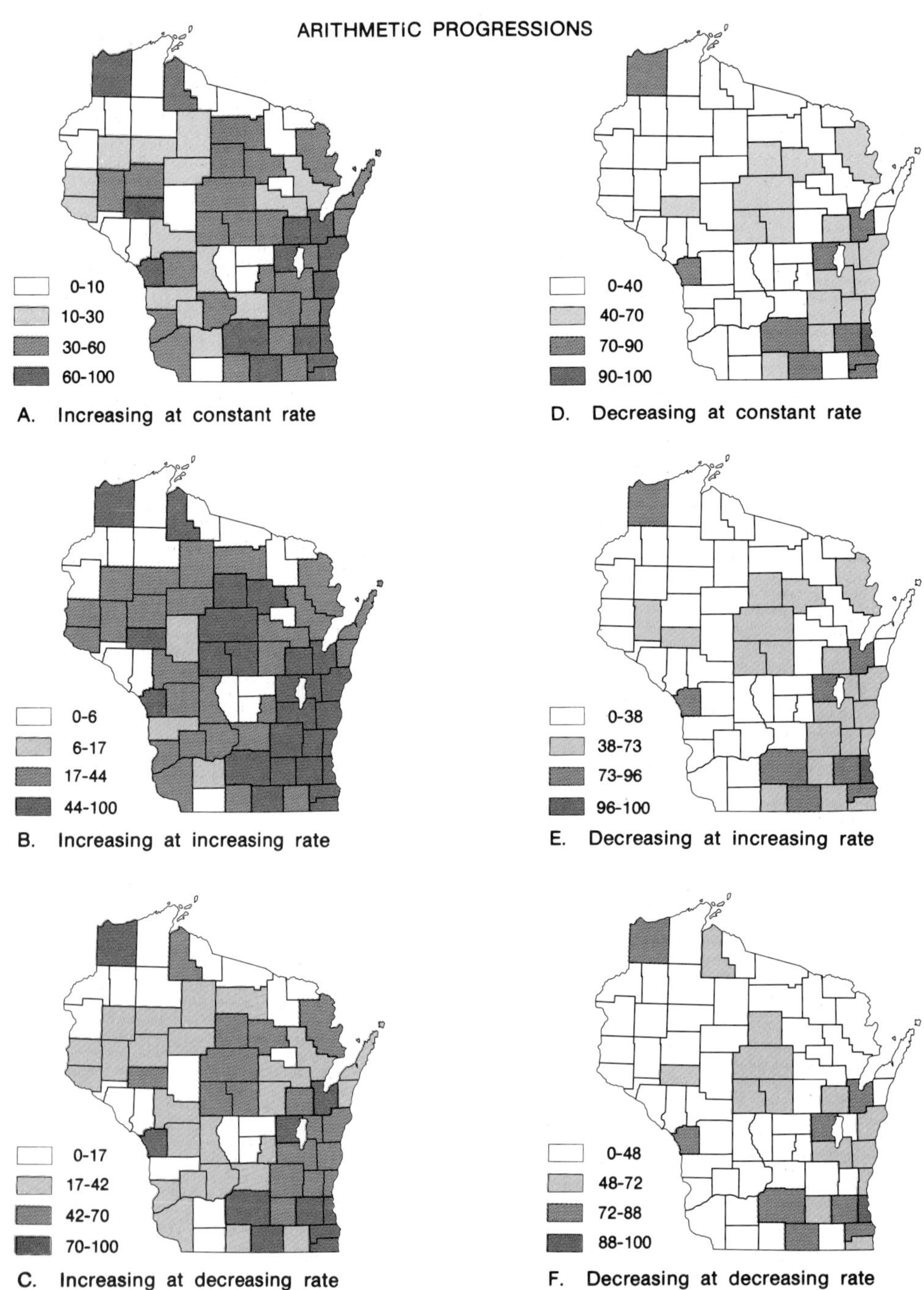

FIGURE 14.17 These twelve maps illustrate the use of arithmetic and geometric progressions in the calculation of class limits. See text for explanation of the calculation methods.

GEOMETRIC PROGRESSIONS

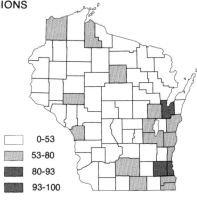

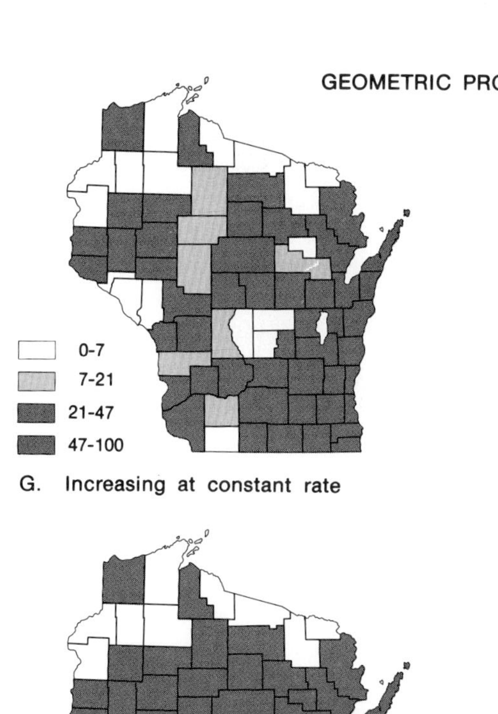

G. Increasing at constant rate

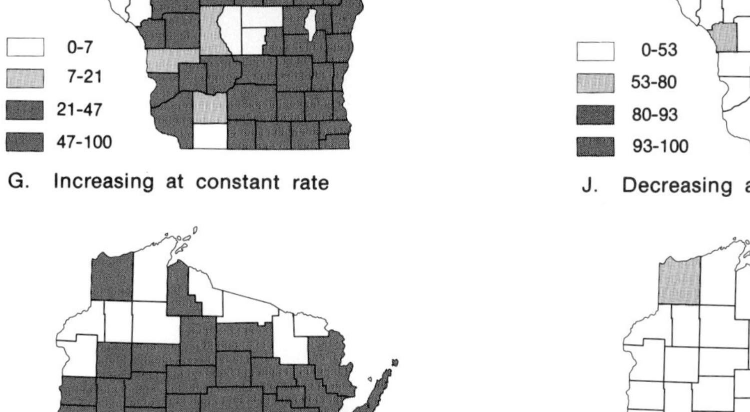

J. Decreasing at constant rate

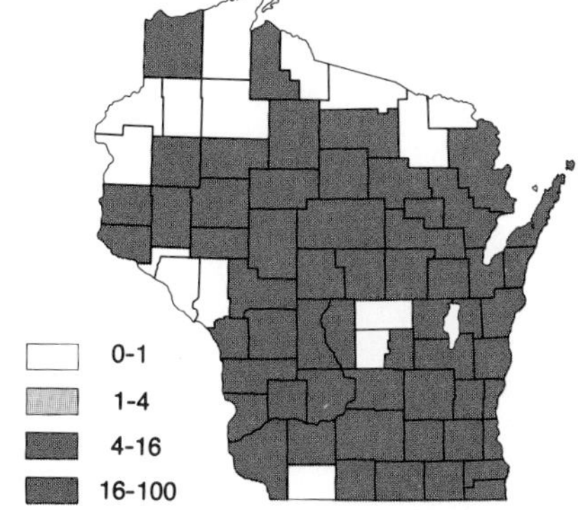

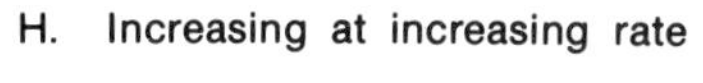

H. Increasing at increasing rate

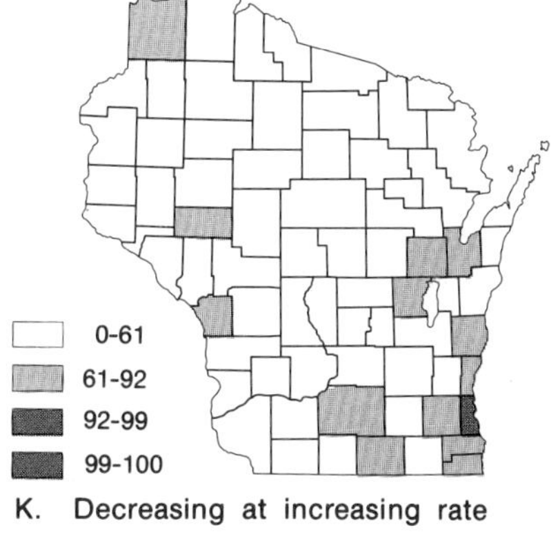

K. Decreasing at increasing rate

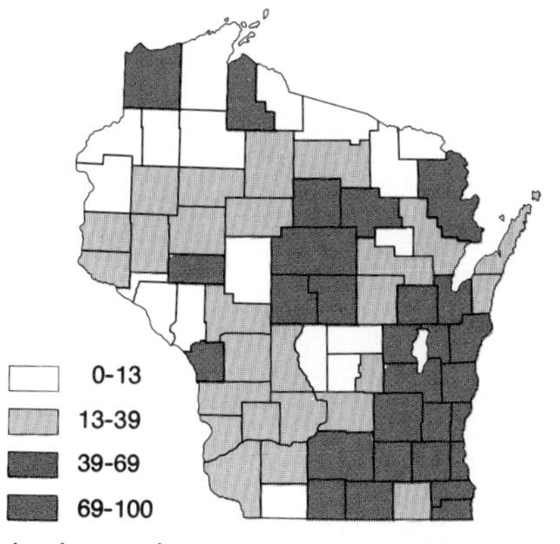

I. Increasing at decreasing rate

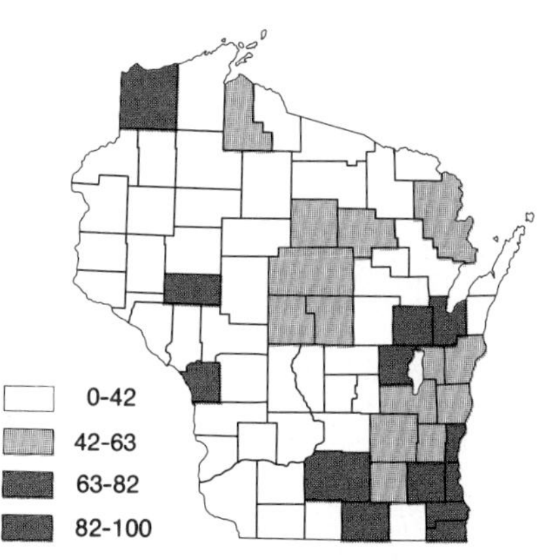

L. Decreasing at decreasing rate

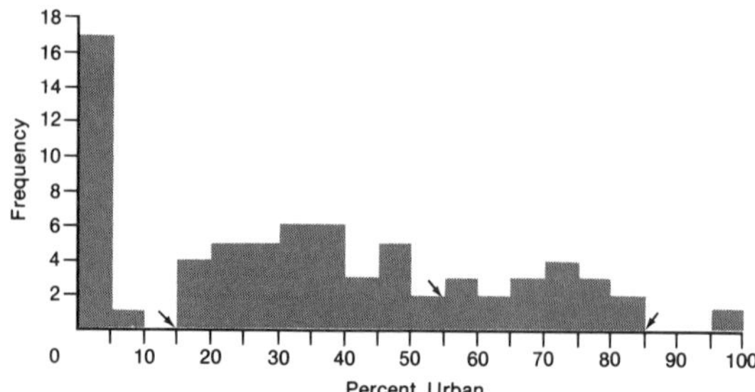

FIGURE 14.18 A frequency graph of the data in Table 14.1. Arrows denote the breaks in the graph that the cartographer has selected for class limits. (See Fig. 14.19.)

limits because they tend to enclose larger groups of similar values. Often, however, there may be few such low points. Frequency graphs are ordinarily quite useful to illustrate the numerical characteristics of the distribution, but they do not always provide clearly desirable class limits. They are relatively simple to construct, and most large computer installations have software that will quickly prepare a frequency curve or histogram of your data set. If there are low points on the frequency curve, they may be most useful for range grading point and line volume data. This method can be used with ordinal, interval, or ratio data.

FIGURE 14.19 A map of the data in Table 14.1 using class limits determined from the frequency graph in Figure 14.18.

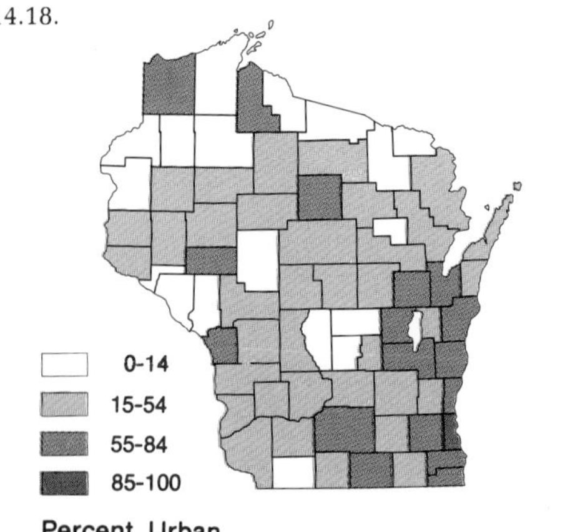

A *clinographic curve* is prepared by arranging the *z* values of the data in numerical order on the *y* axis plotted against the cumulative areas to which they refer on the *x* axis. The *y* axis is scaled arithmetically (evenly spaced steps), while the *x* axis is scaled, in percent of total area, with a square root scale from 0 to 100 percent. If data for enumeration units are being used, we may use the numbers of unit areas if they are of nearly equal sizes; otherwise we should employ the areas of the units. Figure 14.20 is a clinographic curve prepared from the data in Table 14.1. If point or line volumes are being used, each occurrence receives equal weight on the *x* axis. The data are mapped in Figure 14.21.

The critical points in a clinographic curve are the points where the slope of the curve changes. Ordinates at these flexures indicate the *z* values separating regions of different gradient, provided, of course, that the data adjacently plotted on the graph are, in fact, reasonably adjacent geographically. It is possible to obtain plots of clinographic

FIGURE 14.20 A clinographic curve of the data in Table 14.1. The arrows denote the points on the curve that the cartographer has selected to use for class limits. (See Fig. 14.21.)

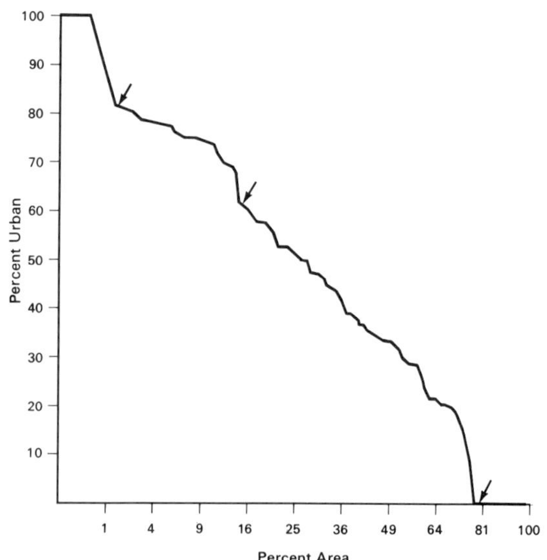

curves computed and drawn with machine assistance at most computer installations. The clinographic curve would appear to have significance for volumes distributed over areas, since it is possible to take area into consideration.

A *cumulative frequency curve* is prepared by arranging the z values of the data in numerical order on the y axis plotted against the cumulative area on the x axis. Both axes are scaled arithmetically, in contrast to the clinographic curve. Properly, the z value of each successive unit area value is plotted against the sum of its area and all preceding unit areas with lower z values. Again, for equal-size enumeration units or for point or line volumes, each numerically higher z value is equally spaced on the x axis. The graph will end on the right side with an x value equal to the total area included in the geographical area being mapped or the total number of observations for point and line volumes. A curve so plotted must rise continuously until it reaches the last plotted point. Figure 14.22 is a cumulative frequency graph of the data in Table 14.1.

For class limits that are likely to be significant in terms of a surface volume, the critical points are the tops and bottoms of "escarpments" on the curve, since these tend to segregate areas of different z values. The data are mapped in Figure 14.23, and the arrows on the graph in Figure 14.22 arbitrarily set the class limits. Obviously, the cartographer could have selected other points on the cumulative frequency curve. Probably more important is the fact that the cumulative frequency graph may be used to determine other values having cartographic significance. As described in Chapter 6, we may determine the geographical mean, the geographical median, or other functions of z values over areas such as "geographical quantiles" or equal area steps from a cumulative frequency curve.

Iterative techniques are too complex to calculate by hand; they require machine assistance. Normally, some logical statistical criterion is specified, and the computer iterates a solution so that the data are grouped in the way that most closely meets the specified criterion. Jenks first introduced class limit determination schemes employing iterative techniques that have statistical criteria based firmly in cartographic theory.*

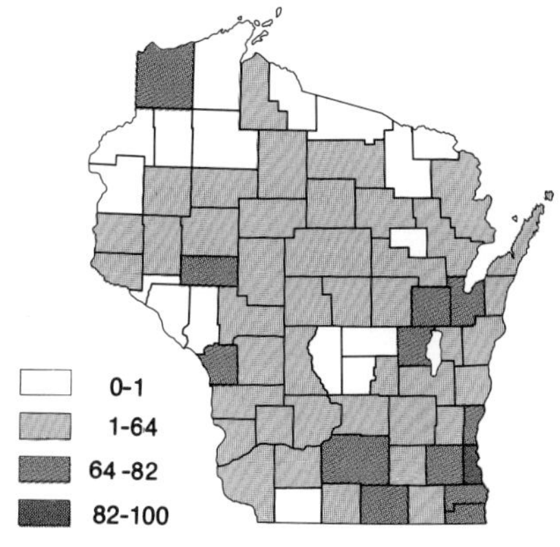

FIGURE 14.21 A map of the data in Table 14.1 using class limits determined from the clinographic curve in Figure 14.20.

More recently Jenks has specified two statistical criteria that can be met through machine iteration.**One, the *goodness of variance fit* (GVF), is useful when the cartographer wishes to minimize the squared deviations about the class means. The criteria to be satisfied is to maximize the quantity GVF.

$$\text{GVF} = \frac{\text{sum of squared deviations between classes}}{\text{total sum of squared deviations from the array mean}}$$

The total sum of squared deviations from the array mean, SDAM, minus the sum of squared deviations between classes is equal to the sum of squared deviations from the classs means, SDCM.

$$\text{SDAM} = \sum_{i=1}^{n} (z_i - \bar{z})^2$$

*George F. Jenks, "The Data Model Concept in Statistical Mapping," *International Yearbook of Cartography* 7 (1967): 186–90.

**George F. Jenks, personal communication.

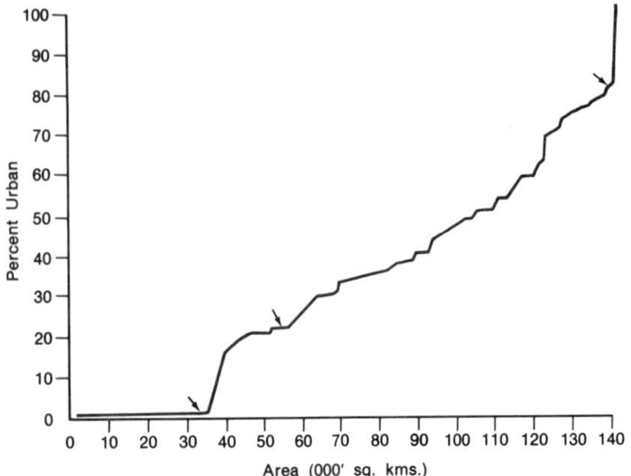

FIGURE 14.22 A cumulative frequency curve of the data in Table 14.1. The arrows denote the selected points on the curve for use as class limits. (See Fig. 14.23.)

where $\bar{z}$ = the array mean

$$\text{SDCM} = \sum_{j=1}^{k} \sum_{ij=1j}^{nj} (z_{ij} - \bar{z}_j)^2$$

where $\bar{z}_j$ = the mean of the jth class
where j = class number
i = observation number within class j

Using this criterion, the cartographer must first specify an arbitrary grouping of the numerically arranged data. The mean of each designated class is computed and the sum of the squared deviations between each observation and its class mean (SDCM) is calculated. The next step consists of moving observations from one class to another in an effort to reduce the sum of SDCM and thereby increase the GVF statistic. After observations have been moved between classes, the new class means, SDCM, and GVF are computed. This process is repeated until the GVF can no longer be increased.

FIGURE 14.23 A map of the data in Table 14.1 using class limits determined from the cumulative frequency curve in Figure 14.22.

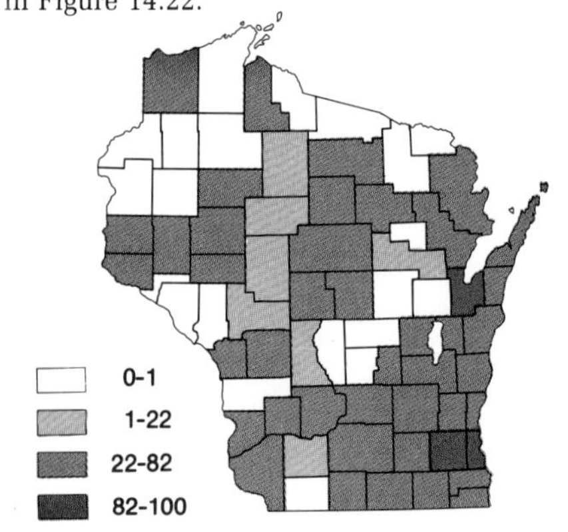

A second statistical criterion that can be utilized in conjunction with the class medians is to maximize the *goodness of absolute deviation fit*, GADF.

$$\text{GADF} = 1 - \frac{\text{sum of absolute deviations about class medians}}{\text{sum of absolute deviations about array median}}$$

The sum of the absolute deviations about the class medians, ADCM, and the sum of the absolute deviations about the array median, ADAM, are given by the following formulas.

$$\text{ADAM} = \sum_{i=1}^{n} |z_i - z_m|$$

where z_m = array median z-value

$$\text{ADCM} = \sum_{j=1}^{k} \sum_{ij=1j}^{nj} |z_{ij} - z_{mj}|$$

where z_{mj} = median z-value of class j

The procedure is similar to the one stated earlier for the GVF criterion. The cartographer must first specify an arbitrary grouping of the numerically arrayed data; then the median for each class and the ADCM value is computed. Observations are moved from class to class in an attempt to minimize ADCM and thereby increase GADF. The process is repeated until the GADF can no longer be increased.

The major advantage of these techniques is that the cartographer can attempt to maximize the homogeneity of each class and to maximize the heterogeneity between classes as defined by the statistical criterion employed. This is a worthy goal in range grading. It is only inappropriate when logic specifies otherwise.

These iterative techniques have come into widespread use by cartographers only since machine computation has been available. Additional iterative techniques could be specified by stating different statistical criteria. (See references to Evans and Monmonier at the end of the chapter.) Statistical criteria allow the cartographer to approach a theoretical goal in the range grading of data. For this reason they should receive the serious attention of all cartographers.

We have included quite a lengthy discussion on the selection of class limits. There still remains much research that could be conducted in this area, especially in the inclusion of these methods into computer algorithms. In spite of all the methods, cartographers still cannot specify with certainty which is the "best" method to use in any given situation. The potential exists, therefore, for misinformation to be transferred to the map reader under the impression that an objective class limit determination has been made.

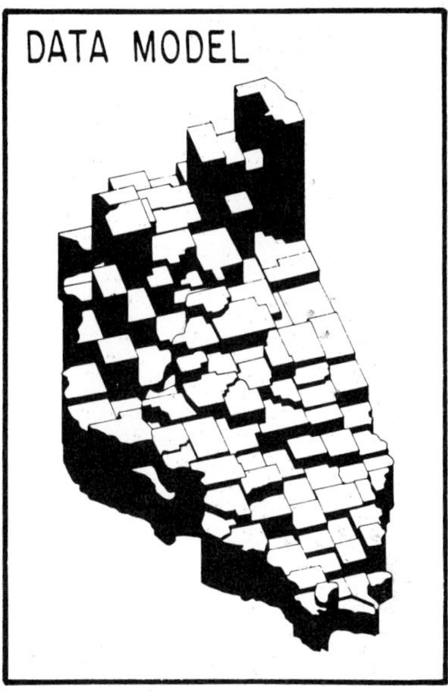

FIGURE 14.24 An unclassed three-dimensional map of a statistical distribution. Each unit area is shown as a prism whose height is proportional to the value of the mapped distribution, and whose base is a scaled representation of the unit area. (Courtesy of G. Jenks and F. Caspall and the *Annals of the Association of American Geographers*, 1971.)

Error in Choropleth Mapping. Jenks and Caspall have studied the error characteristics of choropleth maps that result from the number of classes and the class limit selection.* They have proposed that error index numbers be placed in the legends of all choropleth maps. Errors can be of three types, depending on whether the map is to portray (1) an overview of the distribution, (2) the geographical positions of tabular values from the distribution, or (3) significant boundary lines for the distribution. According to Jenks and Caspall, the overview error, which some believe to be the most important, can be calculated as the absolute volumes of prisms lying between the unclassed data in Figure 14.24 and a classed map model (Fig. 14.25). Similarly, they present statistics for calculating tabular and boundary value indices.

The errors resulting from the size and shape of the unit areas mainly affect the visual qualities of the map; these qualities also depend on the graphic design of the map. The choropleth method assumes that each unit area is homogeneous to the extent that the size and shape of a unit area masks internal variations, and this, of course, adds error to any choropleth map.

Each step of the dasymetric technique can also add error to the map. However, the error in producing the dasymetric map can be partially offset by the additional information in the form of related and limiting variables that the technique requires. Almost every step requires some subjective, but knowledgeable, decision to be made by the cartographer. As a consequence, the automation of the dasymetric technique has lagged somewhat behind that of other cartographic

*George F. Jenks and Fred C. Caspall, "Error on Choroplethic Maps: Definition, Measurement, Reduction," *Annals of the Association of American Geographers* 61 (1971): 217–44.

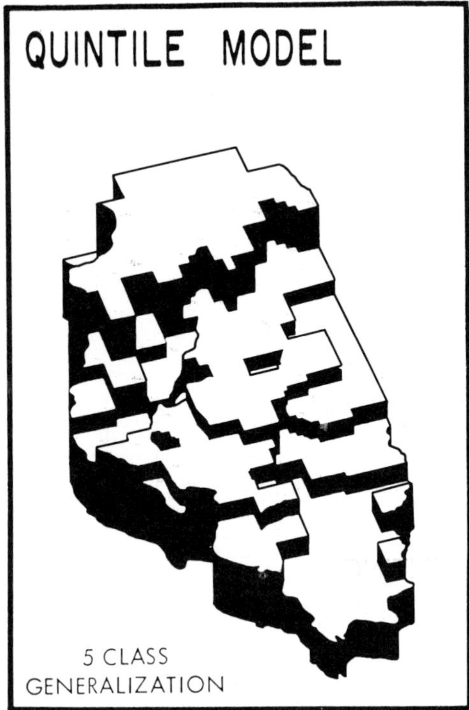

FIGURE 14.25 A five-classed, three-dimensional map of the statistical distribution shown in Figure 14.24. (Courtesy of G. Jenks and F. Caspall and the *Annals of the Association of American Geographers*, 1971.)

methods. No comprehensive study of the error characteristics of the dasymetric technique has been undertaken. The technique includes several different methodologies, and therefore the explication of the error characteristics of dasymetric maps may have to await the further definition of the technique itself. The same requirement is also necessary for the complete and successful automation of the dasymetric technique.

SELECTED REFERENCES

Armstrong, R. W., "Standardized Class Intervals and Rate Computation in Statistical Maps of Mortality," *Annals of the Association of American Geographers* 59 (1969):382–90.

Brassel, K. E., and J. J. Utano, "Design Strategies for Continuous-tone Area Mapping," *The American Cartographer* 6 (1979):39–50.

Chang, K. T., "Visual Aspects of Class Intervals in Choroplethic Mapping," *The Cartographic Journal* 15 (1978):42–48.

Cuff, D. S., and K. R. Bieri, "Ratios and Absolute Amounts Conveyed by a Stepped Statistical Surface," *The American Cartographer* 6 (1979):157–68.

Jenks, G. F., and M. R. Coulson, "Class Intervals for Statistical Maps," *International Yearbook of Cartography* 3 (1963):119–34.

Jensen, J. R., "Three-dimensional Choropleth Maps/Development and Aspects of Cartographic Communication," *The Canadian Cartographer* 15 (1978):123–41.

MacEachren, A. M., "Choropleth Map Accuracy: Characteristics of the Data," *Proceedings of Auto-Carto V: Environmental Assessment and Resource Management*. Falls Church, Va.: American Society of Photogrammetry and American Congress on Surveying and Mapping (1982):499–507.

———, "Map Complexity: Comparison and Measurement," *The American Cartographer* 9 (1982):31–46.

Monmonier, M. S., "Class Intervals to Enhance the Visual Correlation of Choroplethic Maps," *The Canadian Cartographer* 12 (1975): 161–78.

———, "Flat Laxity, Optimization, and Rounding in the Selection of Class Intervals," *Cartographica* 19 (1982):16–27.

Muller, J. C., and J. L. Honsaker, "Choropleth Map Production by Facsimile," *The Cartographic Journal* 15 (1978):14–9.

Peterson, M. P., "An Evaluation of Unclassed Crossed-line Choropleth Mapping," *The American Cartographer* 6 (1979):21–37.

Rowles, R. A., "Perception of Perspective Block Diagrams," *The American Cartographer* 5 (1978):31–44.

Tobler, W. R., "Choropleth Maps Without Class Intervals," *Geographical Analysis* 5 (1973): 262–5. See also: Dobson, M. W., "Choropleth Maps Without Class Intervals? A Comment," *Geographical Analysis* 5 (1973):358–60.

15

Portraying the Land-Surface Form

The portrayal of the land-surface form has always been of special concern to cartographers, but with few exceptions, such as the Han topographic map (Chapter 2), even those of the Middle Ages showed little of it, probably in part because there was so little metrical knowledge about it. To be sure, mountains were sometimes portrayed as piles of crags, and ranges were occasionally shown by a series of "fish scales" or similar conical symbols, or simply as bands of color to show where they were. Until reasonably accurate elevational and positional data became available on which the cartographer could base the representation, the land surface could not be well portrayed. In the Western world such measurement has come only in the last two centuries. As late as 1807 the heights of only about sixty mountains had actually been measured, and many earlier estimates had been ridiculously high or low.*

There is something about the three-dimensional (3-D) land surface that intrigues cartographers and sets it a little apart from other cartographic symbolization. First, it often requires more complex techniques as well as a balancing of alternatives. Moreover, the land surface is a continuous phenomenon; that is, all portions of the solid earth necessarily have a 3-D form; therefore, as soon as any of it is portrayed, all of it is, at least by implication. Also, the land surface is the one major phenomenon with which the cartographer works that exists as a graphic impression in the minds of most map users; they are, therefore, likely to be consciously or unconsciously critical of its portrayal.

Because of the relative importance to human beings of minor landforms, their representation together with other data has been a great problem in large-scale mapping. If the forms are shown in sufficient detail to satisfy the local significance of individual features, then the problem arises of how to present the other map data. On the other hand, if the cartographer shows with relative thoroughness the nonlandform data, which may be more important to the specific objectives of the map, then perhaps only a mere suggestion of the land-surface form can be shown, an expedient not likely to please either the mapmaker or the map reader.

An equally difficult task has been depicting the land surface on smaller-scale maps. Small-scale portrayal of the land surface is a major problem for atlas maps, wall maps, and other general reference maps, as well as for those thematic maps for which regional terrain is an important element in association with other distributions, such as climate, population density, and so on. The smaller scale requires considerable generalization, which is no simple task, as well as the balancing of the surface representation with the other map data, so that neither overshadows the other. A no less creative task is the portrayal, in bolder strokes, of the primary landforms on wall maps, so that important elements such as major regional slopes, elevations, or degrees of dissection are clearly visible from a distance. The specialized techniques for scientific terrain appreciation and analysis are reserved for the geographer and the geomorphologist.

In this chapter we will be concerned primarily with the portrayal of the 3-D land-surface form, not the land-surface character. The character of the surface not covered with vegetation ranges from sand to ice, and from smooth rock to scree and talus. Such techniques as rock drawing to characterize appropriate areas in mountainous regions have been raised to a high art but will not be treated here.

It should also be noted that, although this chapter is entitled "Portraying the Land-Surface Form," the principles apply equally to the solid surface beneath the waters of the earth. There are special problems of data gathering in connection with mapping underwater surfaces, of course, but otherwise there is no basic difference. The discussion here will focus on the land, since its appearance is relatively familiar and the graphic potentialities of the various methods can be easily grasped.

*For example, the Caucasus mountains (in the southern U.S.S.R.) were estimated by some in the seventeenth century as having an elevation of more than 50 *miles!*

The Historical Background

The story of the development of the portrayal of terrain is a recital of the search for methods suitable to a variety of purposes and scales. On large-scale maps the symbolization should both appear natural and be capable of exact measurement of elements of the land surface such as slope, altitude, volume, and shape. The major problem arises from the fact that, generally, the most effective visual technique is the least metrical, whereas the most metrical is the least effective visually. One of the major decisions of every mapping organization has been how to balance these opposing conditions. Modern developments in color printing have enabled the cartographer to obtain a relatively effective combination of techniques, without undue sacrifice of either desirable end.

The earlier representations of terrain on smaller-scale maps were concerned mainly with undulations of considerable magnitude and consisted of crude stylized drawings of hills and mountains as they might be seen from the side, such as those depicted in Figure 15.1. The perspective-like, oblique, or bird's-eye view became more sophisticated in the period from the fifteenth to the eighteenth centuries, when the delineation of the land surface developed along with the landscape painting of the period (Fig. 15.2). The eighteenth century was the period when the great topographic surveys of Europe were initiated, and for the first time, mapmakers had factual data with which to work, even though the first attempts were crude indeed. Delineation of the form of the land shifted from use of the oblique view to the plan view, perhaps because of the availability for the first time of extensive planimetric data. Symbols with which to present the form without planimetric displacement were developed.

In 1799 Johann Georg Lehmann, an Austrian army officer, systematized the use of short linear symbols called *Schraffen* in German and *hachures* in English and French. As proposed by Lehmann, each individual hachure, rather like a vector symbol, is a line positioned in the direc-

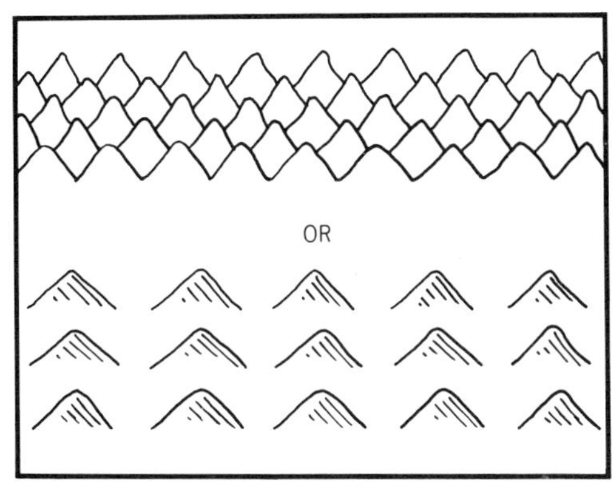

FIGURE 15.1 Some of the ways hills and mountains were portrayed in stylized form on early maps.

FIGURE 15.2 A small section of the 1712–1713 map of Switzerland, "Nova Helvetiae Tabula Geographica," by Johann Jacob Scheuchzer. Original scale ca. 1:230,000. (From facsimile published by De Clivo Press, 1971.)

tion of greatest slope; that is, its orientation on the map is normal (at right angles) to the contours. Lehmann proposed making the widths of the hachures proportional to the steepness of the slopes on which they lie: the steeper the slope, the thicker the line. The sense, or the direction down or up, is, however, not shown, and in some cases, without spot heights it is difficult to tell whether a blank area is a flat upland or lowland. When many such hachures are drawn close together, they collectively portray various aspects of the surface configuration. This feature turned out to be particularly useful on the then-recently initiated, large-scale, topographic military maps, and for much of the nineteenth century hachuring was widely employed.

Many varieties of hachuring were developed. Instead of varying the width of the lines according to the slopes, some used strokes of uniform width and then made their spacing proportional to the slopes. Hachuring provides an illusion of viewing an illuminated 3-D surface, and the light source can be assumed to be orthogonal to the map (Fig. 15.3) or from one side (Fig. 15.4). On small-scale maps or maps of poorly known areas, hachuring easily degenerated into "hairy caterpillars," an example of which is shown in Figure 15.5. Fortunately, these worms are now nearly extinct.

FIGURE 15.3 A section of a topographic map that portrays the land-surface form with vertically illuminated hachuring. (From Sheet 5473 [1894], Austria-Hungary, 1:75,000).

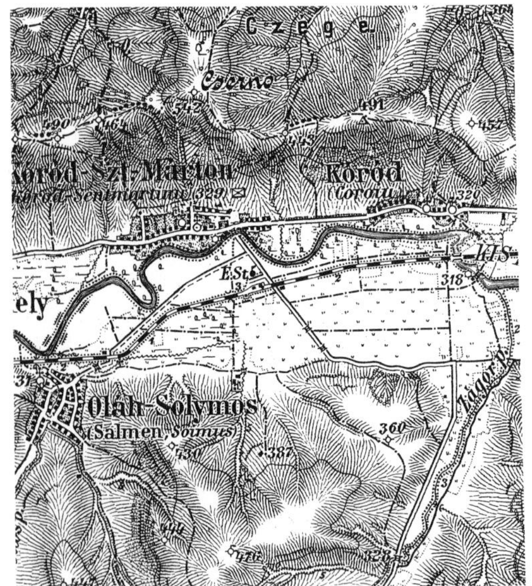

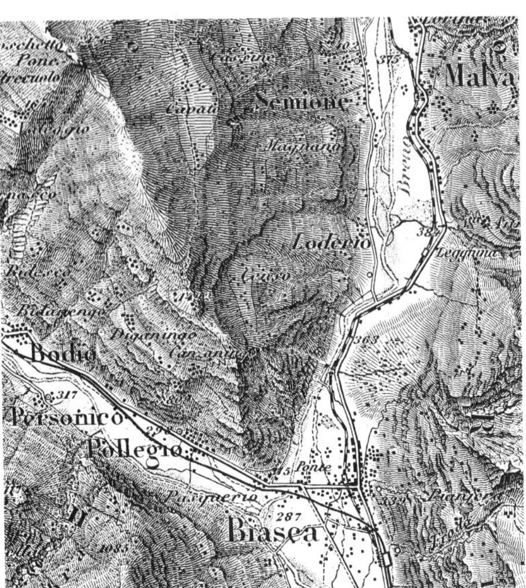

FIGURE 15.4 A section of a topographic map that portrays the land-surface form with obliquely illuminated hachuring. (From Sheet 19 [1858], Switzerland, 1:100,000, the "Dufour map.")

From the earliest use in the sixteenth century of lines of equal depth of water (isobaths), Dutch, and later French, engineers and cartographers used these kinds of lines to portray the underwater and then the dry-land configuration. These lines on land, *Höhenlinien* (German) or *courbes de niveau* (French), came to be known in English as *contour lines*. During the first half of the nineteenth century they were widely used, but not until relatively late in the century did contouring become a common method of depicting the terrain on large-scale, topographic maps.

After the development of lithography in the early part of the nineteenth century, it became possible easily to produce continuous tonal variations or shading to simulate the appearance of the irregular land surface. It was not until late in that century, however, that "hill shading" as a type of area symbol was widely used for the represen-

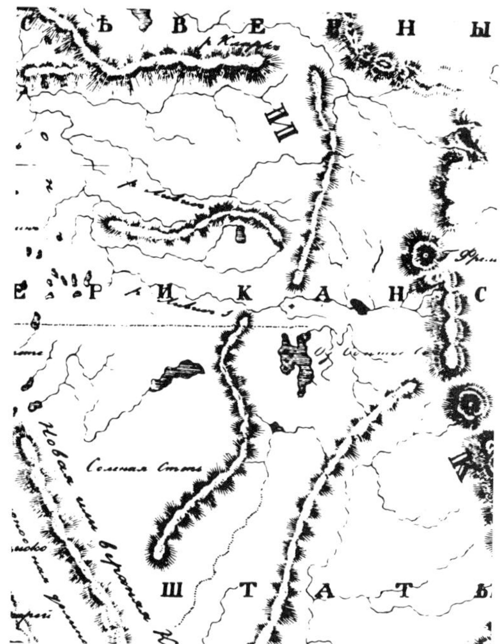

FIGURE 15.5 Genus *Hachure*, species *caterpillar*. (From an old Russian atlas of western North America.)

tation of the terrain, the shading applied being a function of the slope.

By the beginning of the twentieth century, the basic methods of presenting terrain on large-scale topographic maps (contouring, hachuring, and hill shading) had been attempted, and the essential incompatibility between metrical quality on the one hand and visual realism on the other was readily apparent. For the last several decades the problem has been how to combine the techniques to achieve both ends. The newer topographic maps which combine hill shading with contours are the most effective yet produced (Fig. 15.6).

The representation of the land surface at large scales is concerned essentially with the three major elements of configuration: the slopes, the heights, and the shape of the surfaces formed by the combinations of elevations and gradients. The various methods outlined above, and their combinations and derivatives, seem to have provided the answer, more or less, for the problem at large scales, but the representation of the land surface at smaller scales has been quite another matter. Here the generalization required is so great that only the higher orders of form, elevation, and slope may be presented. As knowledge of the land surface of the earth grew, so also did the need for a variety of methods of presenting effectively that surface at smaller scales.

No one method could satisfy all the small-scale requirements for land-surface delineation. Consequently, with the growth of reproduction techniques and drawing media, the variety of the ways of depicting the land surface steadily increased.

One early development, which paralleled the use of contours on large-scale maps, was their extreme generalization on small-scale maps, resulting in the familiar range-graded "relief map," together with "layer tinting" or "layer coloring" according to altitude. The colors, of course, serve as area symbols between isarithms. Layer tinting came into wide use during the latter part of the nineteenth century, and it has since been more commonly employed than any other technique, mainly because of its relative simplicity. It leaves much to be desired. The character of the surface is presented only by implications of elevation; green, usually used for low elevation zones, is often wrongly equated with plains (and vegetation); the generalized contours show little except regional elevations, which are ordinarily not very significant; and the problems of color gradation and the visibility of the other geographical data are sometimes difficult.

Hachuring was employed at small scales soon after its introduction, but as was observed, it can easily degenerate to wormlike forms. Hill shading can be very effective at any scale, but in monochrome, it cannot be easily combined with other map information, and it often becomes little more than an uneven background tone, which serves largely to reduce the visibility of the other data.

Hachuring is no longer used on most modern maps as the sole method of portraying the land-surface form, although many maps still in existence depict terrain this way. In theory hachuring was metrical, the various graphic systems being made proportional to the slopes of the land sur-

FIGURE 15.6 A section of a modern hill-shaded topographic map. From Bashbish Falls, Massachusetts, 7.5-minute quadrangle, U.S. Geological Survey, 1:24,000. The original color is much more effective than this monochrome reproduction.

face, but the actual slope values were difficult to ascertain. A nonmetrical form of hachuring, consisting of short, straight lines lying in the direction of steep slopes, is often employed today on large-scale topographic maps to portray small features such as gravel pits, mine dumps, cuts and fills along roads and railroads, levees, and similar small but significant phenomena, which in many cases would not be effectively portrayed by contouring.

Today the two conventional methods for portraying the land-surface form are contouring and hill shading. They are often combined, providing both a quantitative delineation and a realistic portrayal. For more specialized purposes there are many other ways of depicting the characteristics of the land-surface form. We will first review these briefly before treating in more detail the two primary systems.

UNCONVENTIONAL METHODS

After the seventeenth century, concern for the portrayal of landforms increased, partly because of the growth of interest in nature's wonders as one manifestation of the rise of romanticism and because of the realization of the profound effects of elevation upon temperatures, vegetation, and other aspects of physical geography. Various experiments were tried, and out of this variety the methods of contouring and hill shading (its forerunner being hachuring) have emerged as the dominant methods, one metrical and the other impressionistic; both objectives are desirable for effective portrayal. It is readily apparent, however, that strict contouring and hill shading cannot satisfy all the needs for presenting information about landforms on maps, particularly at small scales. Consequently, as early as the first half of the nineteenth century cartographers began to experiment with other ways of portraying data about landforms. Often these are adaptations of other methods, so they are not easy to classify, but in the following discussion we will recognize three general groups: perspective pictorial portrayals, morphometric maps, and terrain unit maps.

Perspective Pictorial Maps

Even on ancient maps the major terrain features were represented pictorially in crude fashion. As artistic capabilities increased, and as knowledge of the character of the earth's surface expanded, pictorial representation became increasingly effective. Within the past 50 years, this method of presenting the land features has progressed rapidly. The capability of the computer to manipulate forms has added greatly to our versatility.

Block Diagram. Many varieties of perspective delineation of the land are in common use today. Most of them stem from the attempts made during the nineteenth century to portray the concepts that were being rapidly developed by the growing science of geology. One such delineation, the block diagram, has today become a standard form of graphic expression.

The geologists of the eighteenth century illustrated their studies and reports with cross sections in order to show the structural relationships of the rock formations. The top edge of a cross section is, when properly plotted, a profile across the land surface, and it was only natural that some pictorial sketching was occasionally added to the profile to add realism to the section. An obvious development was to cut out a "block" of the earth's crust in order to view it as though the observer were looking at it obliquely from above in such a way that two sides as well as the top were visible. Even the simplest block diagram is easily accepted as a reduced replica of reality and is remarkably graphic (Fig. 15.7). A great variety of information may be portrayed with block diagrams, ranging from structural data to the successive stages in the geologic development of an area.

It is not necessary, of course, to show the rock structure on the side of a block or even to provide a block or base on which the oblique view of the land surface can rest, although without a base the undulating surface seems to float in space.

FIGURE 15.7 A simple block diagram prepared for student field trip use. The natural appearance of the surface forms on a perspective block makes the concepts easily understandable.

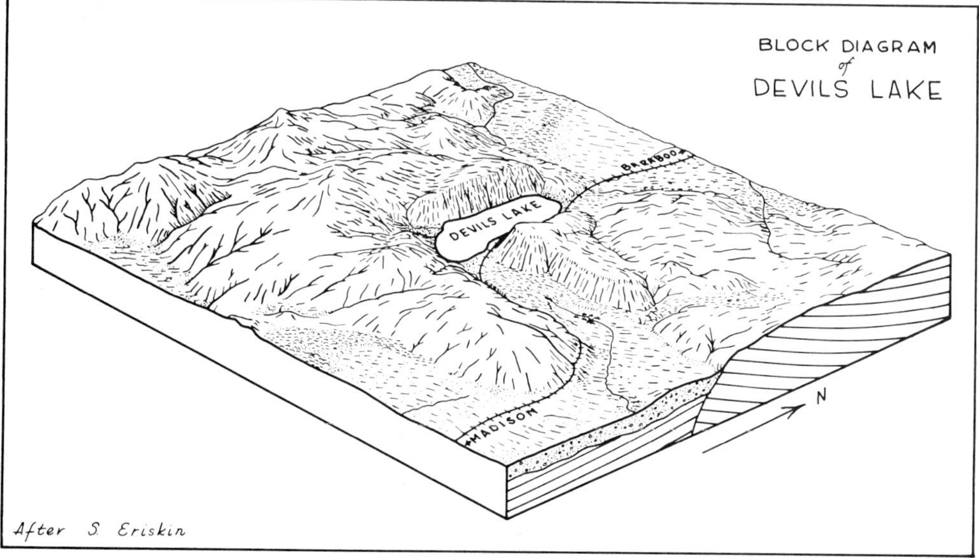

Perspective views are easily produced by computer-driven plotters (see Chapter 13), and consequently almost any kind of surface distribution may now readily be displayed on a perspective landform base. The viewing distance, elevation, and oblique angle of view are easily adjusted.

Block diagrams and similar perspective views are ordinarily restricted to small segments of the earth and do not take into account the curvature of the earth's surface. The extension of the perspective view to larger areas leads to another perspective portrayal, the oblique regional view.

Oblique Regional View. For an oblique regional view one is assumed to be viewing a section of a globe, and it is usually constructed on either an orthographic projection or an "oblique" photograph of an actual globe, to provide the structural base on which the landforms are drawn. Such views are particularly useful as illustrations of national viewpoints and strategic concepts. The modeling is an artistic portrayal, and in addition to the necessary talent, it requires a sympathetic understanding of the role of exaggeration in realistic landform portrayal. Remarkably graphic effects can be created by this method of pictorial representation (Fig. 15.8). Although computer-driven plotters can produce the base materials, the rendering of the surface forms is still more efficiently done by manual means.

Because of scale reduction, pictorial portrayals of the terrain must, of course, be greatly simplified from reality, but more important, the apparent elevations must be markedly exaggerated. Departures of the earth's surface up or down from the spheroid are actually very small, relative to horizontal distances, and they can hardly be shown at all at most medium and small scales. For example, if the highest mountain on the earth, Everest (8,848 m) in the Himalayas, were represented accurately on a 3-D model of Asia at a scale of 1:10,000,000 (a map about 1.2 m square), it would be less than 1 mm high! On the other hand, people are tiny in comparison to landforms, which magnifies our perception of the size of mountains and hills, so that in order to match our subjective impressions, almost all pictorial terrain representation must greatly exaggerate the relative heights of terrain features.

Physiographic Diagram–Landform Map. The block diagram and oblique regional view are not conventional maps in the sense that a map is viewed orthogonally and its scale relationships

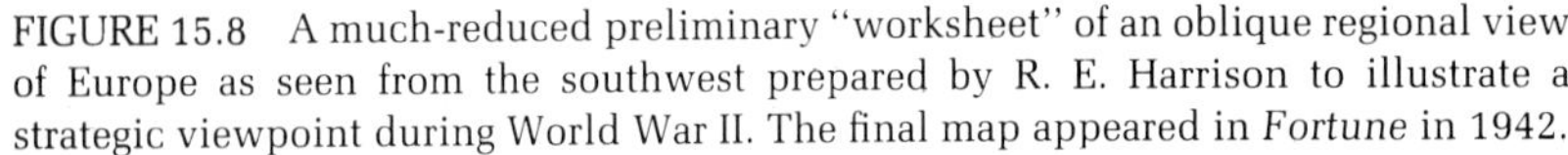

FIGURE 15.8 A much-reduced preliminary "worksheet" of an oblique regional view of Europe as seen from the southwest prepared by R. E. Harrison to illustrate a strategic viewpoint during World War II. The final map appeared in *Fortune* in 1942.

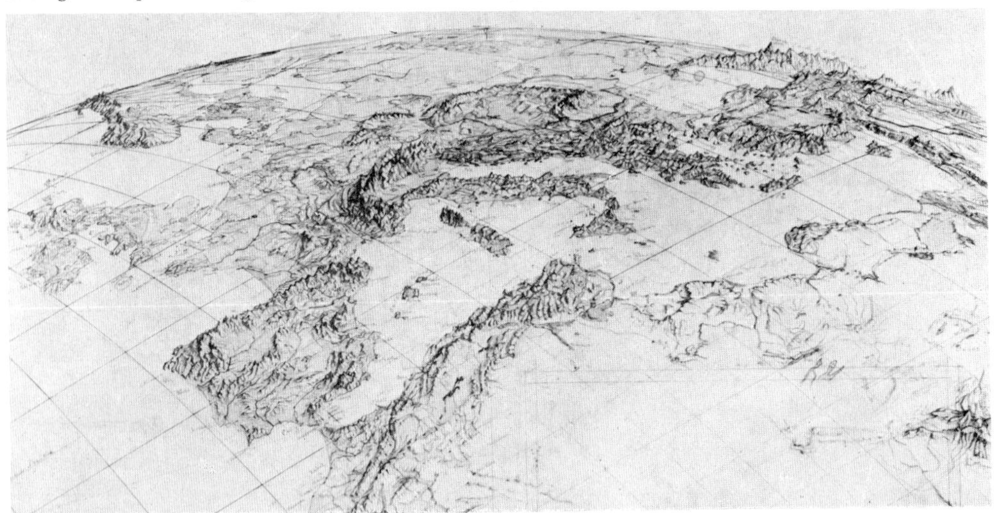

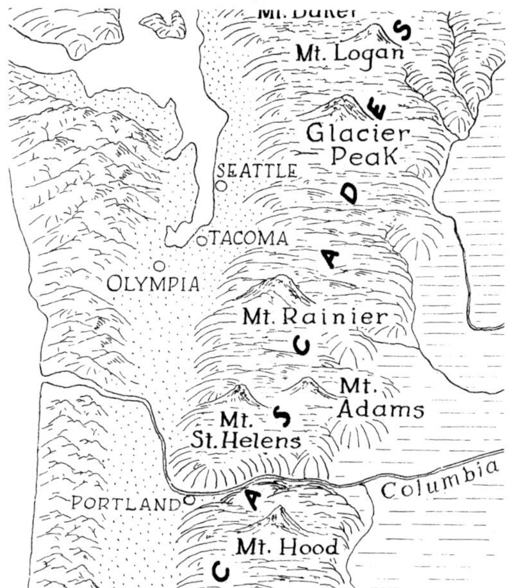

FIGURE 15.9 A portion of a small, generalized physiographic diagram, which clearly shows the planimetric displacement caused by positioning oblique views on a planimetrically correct map. (From A. K. Lobeck, *Things Maps Don't Tell Us*. New York: The Macmillan Company, 1956).

are systematically arranged on a plane projection. From the block diagram and the oblique view, however, have come other map types, which to some extent combine the perspective view of undulations of the land and the planimetric (two-dimensional) precision of the map. These are the physiographic diagram and the landform map.

These maps are made on a conventional, planimetrically correct map base, but the symbols themselves are derived from their oblique appearance. All physiographic diagrams and landform maps have one major defect in common: the side view of a landform having a vertical dimension requires horizontal space, and on a conventional map horizontal space is reserved for planimetric position. For example, if a single mountain is drawn on a map as seen from the side or in perspective, the peak or base and its profile will be in the wrong place planimetrically. This is well illustrated in Figure 15.9. This fundamental defect in physiographic diagrams and landform maps has, of course, been recognized by the cartographers who draw them, but it has, properly, been justified on the ground that the realistic appearance more than outweighs the disadvantages of planimetric displacement. This is especially true on small-scale drawings in which the planimetric displacement is not bothersome to the reader and only occasionally causes the cartographer concern, for example, when some feature of significance is "behind" a higher area. At much larger scales, however, the conflict of the perspective view and the consequent error of planimetric position make it desirable to adopt other techniques, such as the use of inclined traces as described in Chapter 13.

Schematic maps in which the pictorial treatment of the landform is treated more systematically are usually called physiographic diagrams or landform maps. The former attempts to relate the forms to their origin. In these, by varying darkness and textures, the major structural and rock-type differences which have expression in the surface forms are suggested. They do not have a particularly realistic appearance, and their common name, "physiographic diagram" is appropriate (see Fig. 15.10).

Landform or land-type maps are those in which more emphasis is placed on the character of the surface forms, with less attention to their genesis. This type of map often uses schematic symbols such as those developed by Erwin Raisz,* to represent various classes of the varieties of landforms and land types (see Fig. 15.11). There is, of course, no sharp distinction between the physiographic diagram and the landform map. All possible combinations of attention may be paid to the underlying structures, rock types, and geomorphic processes.

Morphometric Maps

Morphometric analysis of the land-surface form is concerned with such measures as average elevations, slope categories, relative relief, degrees of dissection, and so on. Some of these statistics

*Erwin Raisz, "The Physiographic Method of Representing Scenery on Maps," *The Geographical Review* 21 (1931): 297–304.

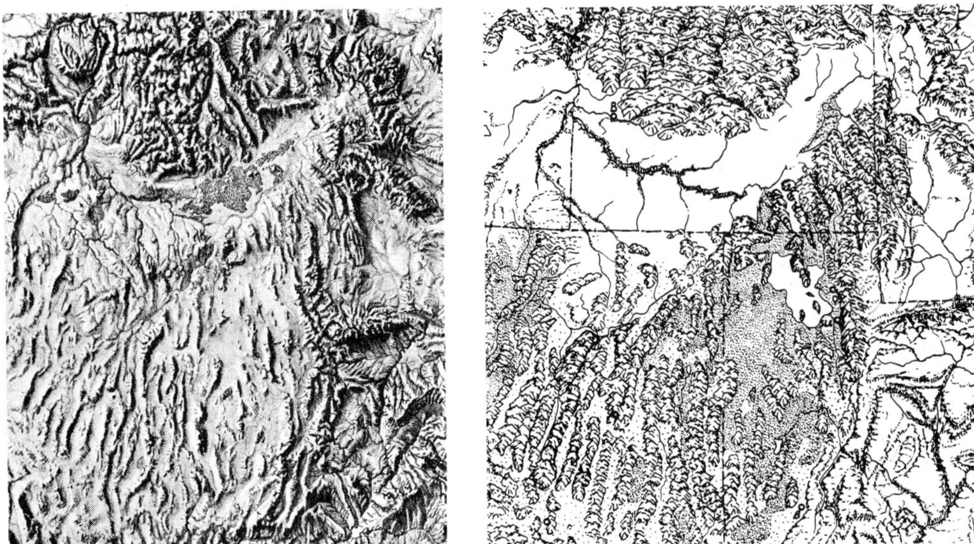

FIGURE 15.10 Physiographic diagrams are indeed diagrammatic. On the left is a relatively realistic portrayal of the region around Great Salt Lake (just to the right of center) and the Snake River valley, drawn by R. E. Harrison for the *National Atlas of the United States* (Courtesy of U.S. Geological Survey). On the right, the same area from A. K. Lobeck, "Physiographic Diagram of the United States," which employs a schematic treatment to emphasize the geomorphic characteristics (Courtesy Geographical Press, Hammond Company).

FIGURE 15.11 A portion of a small-scale landform map. Compare this with Figure 15.10. Note the inclusion of descriptive terms (from E. Raisz, "Landforms of Arabia," 1:3,600,000).

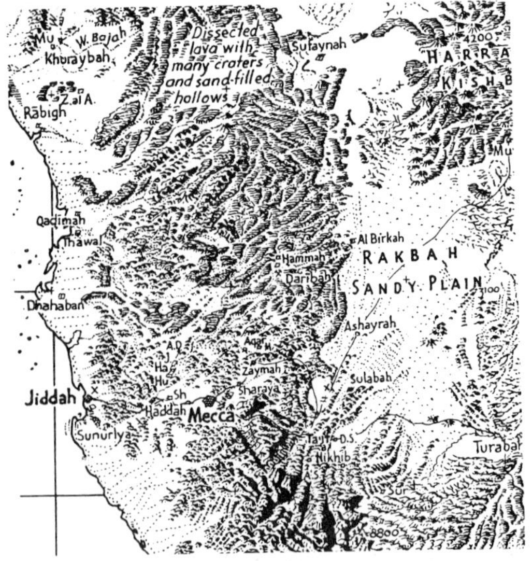

show strong correlations with variations in human activity and might be expected to be used as "background data" in place of the simple elevation shown on most general maps. This has not occurred, and such treatments of the terrain are generally reserved for the specialist.

The portrayal of such data is straightforward; area symbols are used to reinforce either isarithms or dasymetric lines. The major problem inherent in these methods is the determination of what to present, not how to present it. Consequently, to use these methods, the cartographer must be essentially a specialist about the distribution being mapped or else must simply present the work of others.

In both Europe and the United States, the concept of relative relief, as opposed to elevation above a datum, has been tried. Relative, or local, relief is the difference between the highest and lowest elevations in a limited area, for example, a 5-minute quadrangle. These values are then portrayed with isarithms or by a simple choropleth map. The relative relief method is valuable

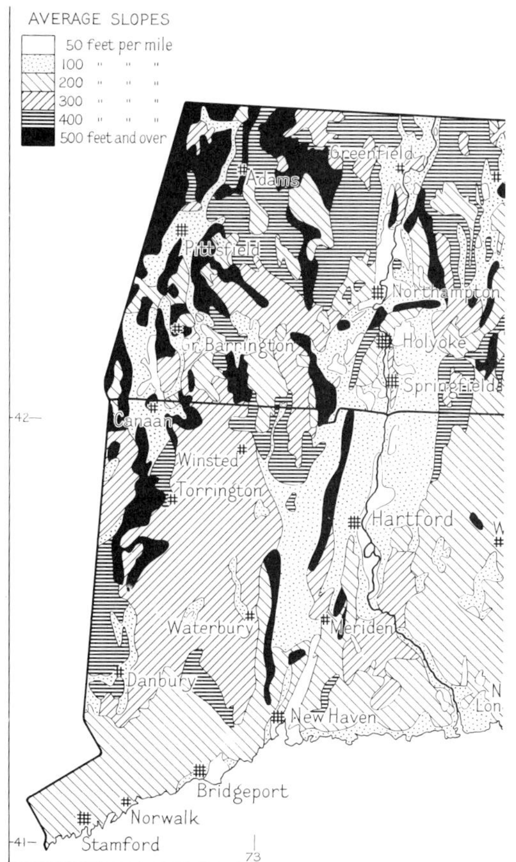

FIGURE 15.12 A portion of a slope zone map of part of southern New England. The areas of similar slope were outlined on topographic maps by noting areas of consistent contour spacing. (Courtesy of *The Geographical Review*).

when applied to areas of considerable size because basic features and divisions of the distribution are emphasized, but it seems to be unsuited for differentiating important details too small to extend beyond the confines of the unit area chosen for statistical purposes. It is best adapted to relatively small-scale representation.

Slope zone maps show categories of slopes either in terms of average gradients, as degrees of inclination, as percentage slopes, or by areas designated by such terms as flat, gently sloping, steep, and so on. They may be developed at relatively small scales (Fig. 15.12), but they are likely to be more useful at larger scales. In such cases the relation between various kinds of soil erosion, such as mud slides or gullying, runoff rates, and other critical factors can be evaluated more easily with the aid of maps showing the different categories of slopes rather than from the contour maps themselves. Figure 15.13 is an example of an experimental slope zone map prepared by the U.S. Geological Survey. These maps are prepared only for special needs.

The slope values may be readily obtained from digitized contour data, and they have also been prepared by photomechanical techniques.

Terrain Unit Maps

The terrain unit method employs descriptive terms that range from the simple "mountains," "hills," or "plains" designations to complex, structural, topographic descriptions such as "maturely dissected hill land, developed on gently tilted sediments." This system is essentially a modification of the dasymetric system (see Chapter 14), and the lines bounding the area symbols have no meaning other than being zones of change from one kind of area to another. This method of presenting landforms has been found useful in textbooks and in regional descriptions for a variety of purposes, ranging from military terrain analysis to regional planning. Its basic limitations are the regional knowledge of the maker and the geographical competence of the map readers.*

CONTOURING

Portrayal of the land surface by contours is the most metrical system yet devised. As described in Chapter 13, contours are isarithms; they are the traces that result from passing parallel, equally spaced, horizontal surfaces through the three-dimensional land surface and projecting these traces orthogonally to the map surface.** The vertical distance between the parallel surfaces is called

*First-class examples of terrain unit maps are "Classes of Land-Surface Form" by Edwin H. Hammond in *The National Atlas of the United States of America* (Washington, D.C.: U.S. Geological Survey, 1970), pp. 62–64.

**The horizontal surfaces that produce the contours are in fact curved and parallel to the ellipsoidal mapping surface. After that surface has been transformed to a plane, the horizontal surfaces may be thought of as parallel planes.

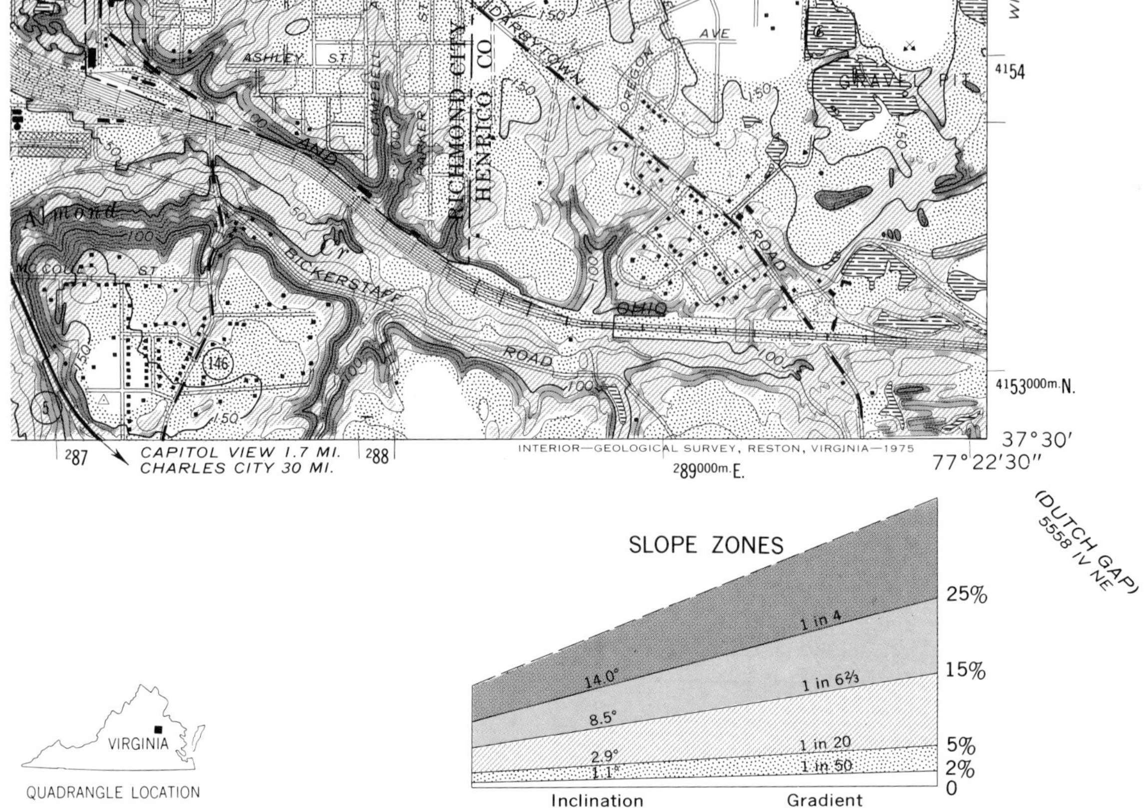

RICHMOND, VA.

EXPERIMENTAL PRINTING–1975

FIGURE 15.13 A portion of the experimental Richmond, Virginia, slope map, 1:24,000, prepared by the U.S. Geological Survey.

the *contour interval.* Natural contours may occur occasionally as strandlines left on a shore as water levels have receded; generally, however, contours are not visible on the land in the normal experience of the average person. Figure 15.14 shows the usual way we see form—by the interplay of light and dark—and, for comparison, a contour representation of the identical form.

Most series of topographic maps conventionally employ several widths and styles of contour lines to facilitate interpretation. The United States Geological Survey uses six kinds of contour lines as follows:

1. *Index Contours.* Every fourth or fifth line (depending upon the contour interval) is made noticeably thicker than the other contours. These are usually labeled with the elevations they represent.
2. *Intermediate Contours.* The three or four contours between adjacent index contours, spaced at the basic contour interval, are about half the width of the index contours.
3. *Supplementary Contours.* Dashed or dotted lines may be spaced at one-half, one-fourth, or one-fifth the basic contour interval.
4. *Depression Contours.* In the portrayal of the land surface by contouring, the area inside a closed contour line is assumed to be higher than the area outside, but when in fact the area inside is lower, as in a basin or sink, this fact is denoted by right-angle ticks on the downside—the inner side—of the closed line.
5. *Carrying Contours.* These are single lines representing several contour lines as would occur in portraying vertical or near-vertical features, such as cliffs, cuts, and fills.
6. *Approximate Contours.* Sometimes called form lines, these are dashed lines, with no specific elevational value, which are used to show

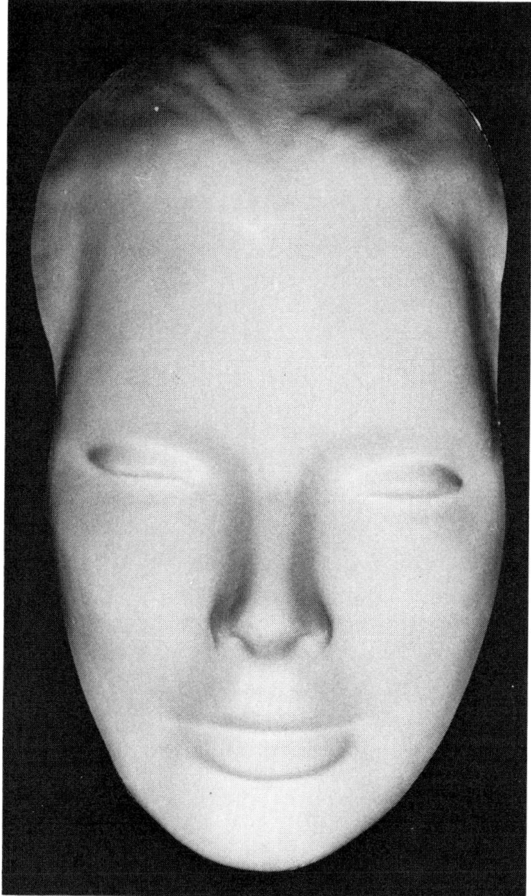

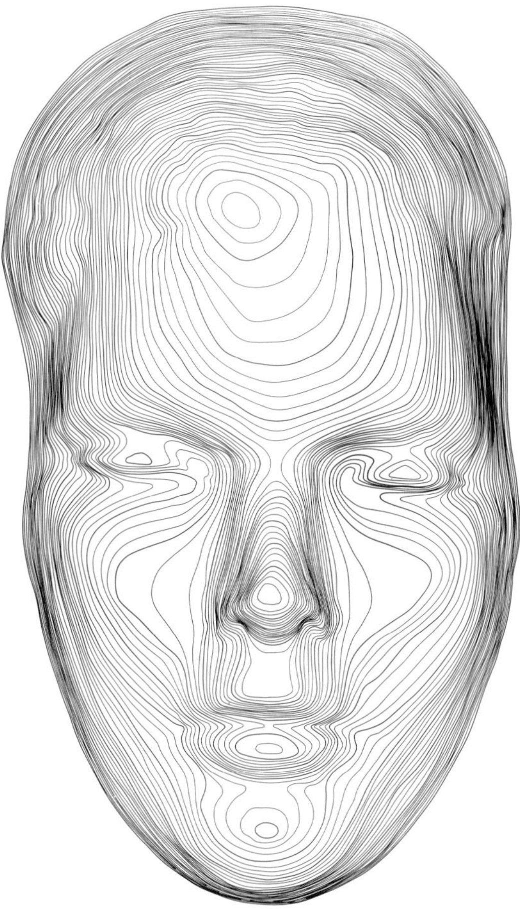

FIGURE 15.14 A vertically lighted plaster model and a precise contour map with a 1 mm interval derived from it. The contours were obtained by photogrammetric methods. (Courtesy G. Fremlin.)

general shapes where accurate contours are not feasible.

Although contours do not present as clear a visual picture of the surface as hill shading, the immense amount of information that may be obtained by careful and experienced interpretation makes contouring by far the most useful way of portraying the land on topographic maps. We must remember, however, that not all contour maps are of the same order of accuracy. Before the acceptance of the air photograph as a device from which to derive contours, the lines were drawn in the field with the aid of a scattering of "spot heights" or elevations. Consequently, they were often not precisely located.

It is apparent that much of the usefulness of contours depends on their vertical spacing, and the choice of a contour interval is not an easy task. The portrayal of the slopes (gradients) is one of the major objectives of contour maps; consequently, contour intervals are almost always equal-step progressions. When an uneven contour interval is employed, slopes are difficult to visualize and calculate, and misleading impressions are likely.

As the contour interval is increased, the amount of surface detail lost between the contours becomes correspondingly greater. If, because of a lack of data or scale, the contour interval must be excessive, other methods of presentation, such as hill shading, are likely to be more useful.

Contour Accuracy

The quality of the delineation of the land surface by contours involves two kinds of accuracy, absolute accuracy and relative accuracy. Absolute accuracy is specified by the United States National Map Accuracy Standards as:

> *Vertical accuracy, as applied to contour maps on all publication scales, shall be such that not more than 10 percent of the elevations tested shall be in error more than one-half the contour interval. In checking elevations taken from the map, the apparent vertical error may be decreased by assuming a horizontal displacement within the permissible horizontal error for a map of that scale.**

The standards may be converted to a mathematical equivalent by employing the concept of the standard deviation, in this application usually called the root-mean-square error (RMSE).**

Relative accuracy refers to the relationship of the contours with one another, for it is the association among the lines which portrays the shapes of the terrain features. The absolute accuracy standard specifies that contour lines must be positioned within a band representing one-half the contour interval above or below the true elevation. This is adequate for most engineering purposes but not for showing the true slopes and configurations of the terrain.† For example, given a constant contour interval, a uniform slope should be delineated by evenly spaced contours on the map. If, however, adjacent contours were spaced successively nearly a half interval above and below their correct elevations, the resulting contour pattern would imply terraces instead, although the map would meet the absolute map accuracy standard (Fig. 15.15). Relative accuracy is as important for portraying shape as is absolute accuracy for portraying elevation. Fortunately, modern photogrammetric methods are sufficiently precise so that under normal circumstances both objectives can be obtained. Although absolute accuracy is not difficult to assess, relative accuracy is, and consequently present accuracy standards do not include any requirement for it. Relative accuracy can be checked only by comparing the delineation with a stereo model.

The horizontal planes that produce the contours are measured from a defined datum such as mean sea level, and where they intersect the three-dimensional land surface is strictly a matter of geometry. Whether a particular terrain feature is portrayed adequately, that is, with good relative accuracy, is a matter of chance. Sometimes it is desirable to shift contours, within the limits of absolute accuracy, in order to portray more effectively a feature that is prominent in the landscape but not adequately revealed by the accident of geometry (Fig. 15.16). The smaller the contour interval, the less likely are such conditions to occur.

Contour Intervals

The choice of the contour interval is extremely important and is based on a number of factors. Foremost, of course, is the accuracy and completeness of the data, which for large-scale mapping today with photogrammetric methods is not a great problem. For compiled maps at smaller scales it continues to be a significant element. Another major factor is the purpose of the map, which can range from the very large-scale topographic map to be used for engineering and planning, which requires a very small interval, to a small-scale regional map on which only the major lineaments of the land-surface form are needed and for which a large contour interval would be adequate. Map scale itself, regardless of purpose, is an important factor, since an interval that is too small for the scale will result in unwanted

*Revised and adopted by the U.S. Bureau of the Budget, June 17, 1947, and still in effect. The permissible horizontal error is 1/30 in. at scales larger than 1:20,000 and 1/50 in. at smaller scales.

**See M. M. Thompson, *Maps for America,* 2d ed. (Washington, D.C.: U.S. Geological Survey, 1981), pp. 102–7; also M. M. Thompson, "A Current View of the National Map Accuracy Standards," *Surveying and Mapping* 16 (1960):449–57.

†*Topographic Instructions of the United States Geological Survey* (Reston, Va.: U.S. Geological Survey, 1979), Chapter 4B8, "Relief Features."

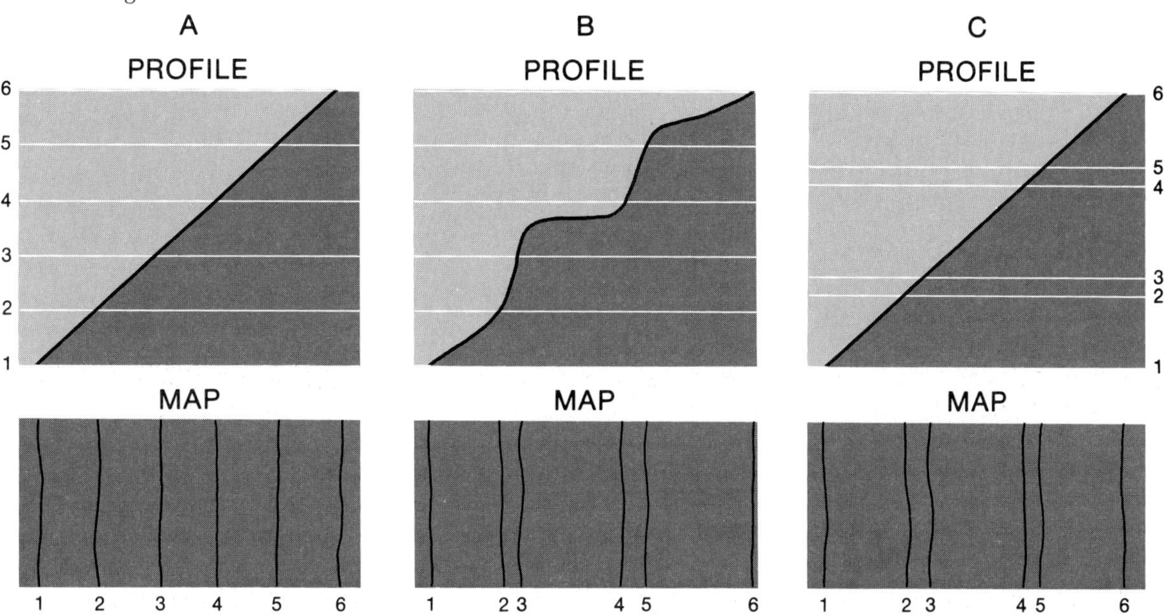

FIGURE 15.15 How displacement of contours can affect portrayal. The profiles in (A), smooth slope, and in (B), benches or terraces, are characteristically portrayed by contours on their respective maps when the contours are accurately located. If, as in (C), contours 2, 3, 4, and 5 were displaced as shown—less than half the contour interval, which meets map accuracy standards—then the resulting map would suggest benches, as in (B), instead of the true smooth slope.

crowding. It is worth reiterating that, except in unusual circumstances, the contour interval should be uniform; that is, all the assumed horizontal planes should be the same distance apart for easy recognition of slopes and shapes.

From a theoretical point of view, if our objective is solely to portray the surface configuration on a map unencumbered by other data, then the best contour interval would be the smallest possible, since that would provide the most elevational data and the most accurate portrayal of the forms. Although that theoretical ideal is rarely possible, since one of the primary functions of the maps is to show the geographical association of features, the concept of the minimum possible contour interval is a useful way to approach the topic of the appropriate interval.* The minimum possible contour interval is a function of scale $(\frac{1}{S})$, the maximum inclination of slope (a), and the greatest possible number of lines which will be distinguishable, that is, "legible," per unit of horizontal distance on the map (k). A width of 1 mm on a map at a scale of $\frac{1}{S}$ corresponds to a distance $\frac{S}{1000}$ meters, and with a slope angle of a in degrees, this distance represents a vertical difference in meters of $\frac{S}{1000} \times \tan a$. The minimum contour interval in meters, therefore, is

$$\frac{S \tan a}{1000\,k}$$

A value of 2 for k calls for lines about 0.1 mm thick separated by about 0.4 mm space, which

*Eduard Imhof, *Cartographic Relief Representation,* ed. H. J. Steward (Berlin and New York: Walter de Gruyter & Co., 1982) pp. 113–22.

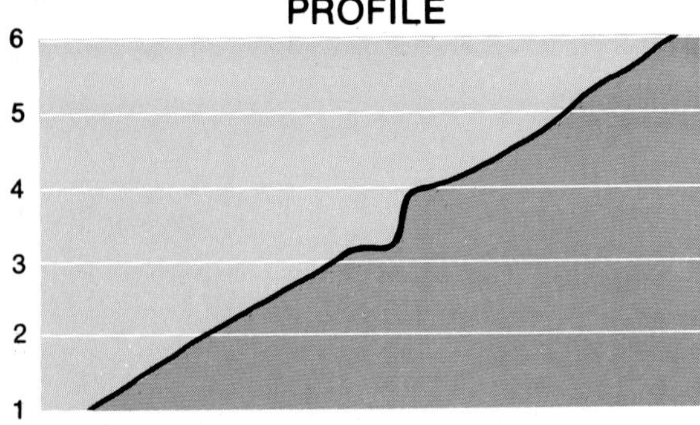

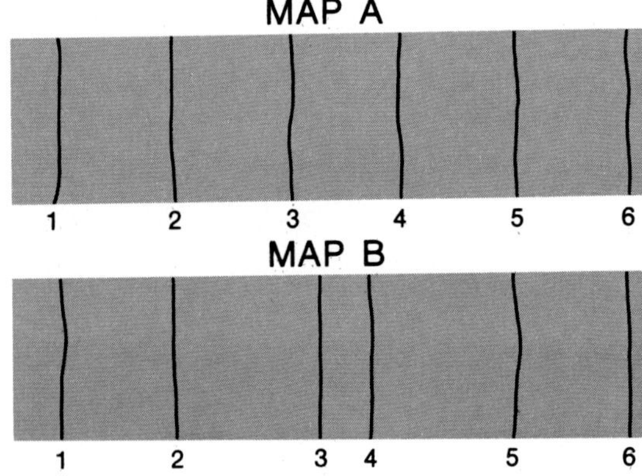

FIGURE 15.16 The prominent bench or cliff in the profile does not show in (A) because it occurs between two contour planes. It is desirable to shift contours 3 and 4 closer together on the map (each less than one-half the contour interval), as in (B), to improve the relative accuracy of the portrayal. (Based upon *Topographic Instructions of the United States Geological survey.*)

seems reasonable for persons with average vision. Table 15.1 lists the contour intervals at various values of *a*, with $k = 2$ at a scale of 1:25,000. The values for other map scales are inverse multiples, that is, for a scale of 1:50,000 the values would be twice those in the table.

The use of such odd contour intervals as listed in Table 15.1 would be very inconvenient both for counting on maps and for employing index contours, such as every fifth one on maps with an interval of 20 m or ft. Consequently, contour intervals are normally set at easily added and divisible spacings, such as 5, 10, 20, 30, 50, 100, and so on, m or 5, 10, 20, 25, 40, 50, 100, 200,

TABLE 15.1 Minimum Contour Interval (rounded to nearest whole number) at a Scale of 1:25,000 for Slope Angles 5° to 45° with $k = 2$

$a°$	Meters	Feet
5°	1	4
10°	2	7
15°	3	11
20°	5	15
25°	6	19
30°	7	24
35°	9	29
40°	11	34
45°	13	41

and so on, ft.** At very large scales, that is, larger than 1:10,000, depending upon the requirements for the map, the interval may be very small, such as 0.5 or 1.0 m (or 1 or 2 ft.).

Differences in terrain make it impossible to have a consistent contour interval for an entire map series regardless of scale. For example, a contour interval of 40 ft. would be quite appropriate for the standard U.S. 1:24,000, 7.5-minute topographic series in a high relative relief area in western United States, but that interval would allow only eight contour planes for the entire state of Florida (maximum elevation 345 ft.). On the other hand, it is very inconvenient to have different contour intervals on adjacent maps of a series, so it is desirable to maintain regional consistency in contour intervals. Such consistency is aided by the use of *supplementary contours* (dotted or screened lines on U.S. Geological Survey maps). They are usually established at a simple fraction of the basic interval. They are particularly useful on flood plains and areas of similarly

* In the United States it is the intent of the National Mapping Program to convert to the use of the International System of Units (SI). Metrication is complex and includes expressing contours, contour intervals, elevations, soundings, and so on, in meters as well as employing "metric" map scales, such as 1:25,000. The conversion will take time and will be accomplished in steps. A brief description is contained in Morris M. Thompson, *Maps for America,* 2d ed. (Washington, D.C.: U.S. Geological Survey, 1981), pp. 231–32.

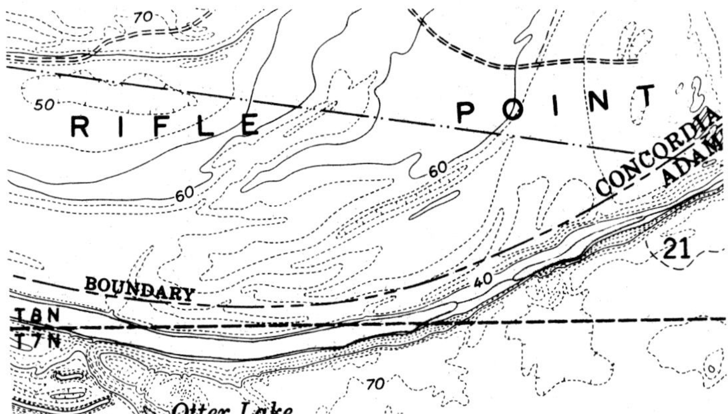

FIGURE 15.17 Supplementary contours at a 5-ft. interval used to show detail on a section of the Lower Mississippi flood plain. From Natchez, Mississippi-Louisiana, 7.5-minute quadrangle, U.S. Geological Survey, 1:24,000, for which the basic interval is 20 ft.

low relief when the basic interval for the rest of the map is too large to delineate minor forms, such as natural levees and banks, which are very significant locally (Fig. 15.17). A similar use is for flat-topped ridges, mesas, and the flatter areas of mountainous regions; they are also used in areas where the basic contours are widely spaced and where there are extreme differences in steepness (Fig. 15.18).

On smaller-scale maps the functions of contouring shift from having the dual role of portraying both elevation and slope angles to showing elevation and suggesting only general shapes. For example, the contour intervals on the International Map of the World (IMW) series are not constant but instead are chosen primarily to show elevations. The general specifications call for "principal contours" at 100, 200, 500, 1000, 1500, 2000, 3000, 4000, 5000, and 6000 m. Additional (auxiliary) contours at regular intervals of 10, 20, 50, or 100 m may be added where desirable.

Generalization of Contours

Contours on maps are lines that have thickness, which in turn limits their radius of curvature. Consequently, it is literally impossible for a contour to reflect every variation in elevation. Thus, all contour maps are generalized, that is, simplified, to some extent, regardless of their scale.

The manner in which the generalization is car-

FIGURE 15.18 Supplementary contours at a 10-ft. interval used to show detail in an area where there are great differences in steepness. From North Ogden, Utah, 7.5-minute quadrangle, U.S. Geological Survey, for which the basic interval is 40 ft.

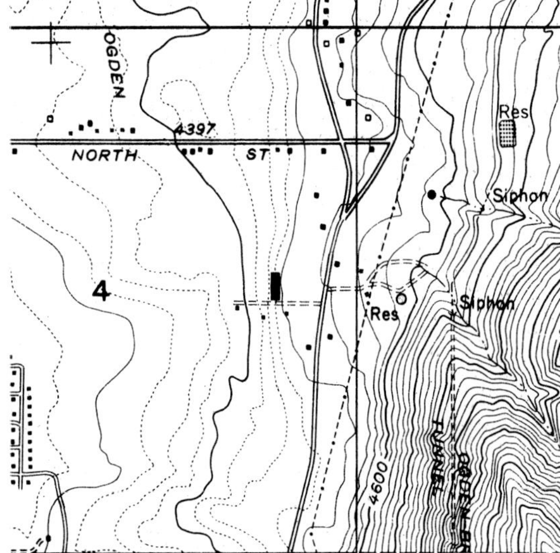

ried out depends upon the terrain and the scale of the map. The primary objectives are to locate the contours as accurately as possible, to show as much detail as is consistent with the scale of the map, and to portray effectively the character of the terrain. Unlike a coast, which is a single line outlining the shape of the land-water boundary, a single contour line shows only elevation. The surfaces of the land are shown by patterns or arrangements of contours, so that it is the *set* of lines which is generalized, not the individual lines. Consequently, the major masses, ridges, valleys, and depressions should retain their intrinsic character, and when contours are generalized, they should neither merely be smoothed into more rounded forms nor simply made more jagged.

At most scales some of the information that could be included must be left out in order to display the remainder more clearly, but in all cases any sizable, sharply defined breaks in slope, such as crestlines and scarps, should be retained.

In some areas the terrain is characterized by such closely spaced erosional or depositional forms, such as in badland topography or sand dunes, that the simplification for even large-scale maps must take on some of the character of sampling. When there are larger features, they can be shown, but if there is little difference in size, the selection of those to be shown is made arbitrarily. At small scales, where reduction has caused similar intricacies, as in the representation of a complex mountain area, such features as well-known routes, major valleys, and regional divides are some of the factors taken into consideration in the choice of the forms to retain.

The production of generalized contour maps from a digital data base is an involved undertaking because it requires a series of roundings and transformations. Perhaps the best way to illustrate the kinds of operations involved is to quote the explanation of the procedures involved in the production of a "Digital Terrain Map of the United States," by Richard H. Godson at a scale of about 1:7,500,000.

> *The Digital Terrain Map of the United States was generated from a U.S. Geological Survey (USGS) digital data base created from data obtained from the National Geodetic Survey (NGS). NGS received these data from the Defense Department's Electromagnetic Compatibility Analysis Center (ECAC), whose primary data source was the Defense Mapping Agency Topographic Center (DMTC). Data for a few degree squares, especially around the coastline, were obtained by ECAC from other sources.*
>
> *The original data from DMATC were generated by digitizing contour lines, spot elevations, and stream and ridge line data from the 1:250,000 scale series of maps, which have contour intervals ranging from 15 m to 61 m. These data were then converted to a rectangular grid of values, producing elevations spaced every .254 mm on each 1:250,000 scale map (approximately 63 m on the ground).*
>
> *ECAC used the DMATC data and data from other sources to obtain point elevations, rounded to the nearest ten meters, for every 30 seconds of latitude and longitude (approximately every .83 km on the ground). The USGS transformed this 30-second point elevation data to average elevations for cells 30 seconds on a side by a simple average of the elevations of the four corners of each cell. The 30 second cell data were then converted to three minute cell elevations by averaging the 30 second cell data contained in each three minute block.*
>
> *The three minute geodetic coordinates were then changed to map coordinates by using an Albers projection with a central meridian of 96 degrees. Because the machine contouring method requires equally spaced data, elevations were determined at intervals of 6 km by using a spline under tension technique. . . . The interpolation method created some values slightly less than zero near the coastline and is the reason for some white dots seen on the map. The final interpolated data set was contoured using Applicon Incorporated proprietary software and plotted on an Applicon color plotter.**

Layer Tinting

Perhaps the most widely used method of presenting land-surface information on wall maps,

*Miscellaneous Investigations Series, Map I-1318, U.S. Geological Survey, Reston, Va., 1981. Contours on the map are actually shown only by a change from one colored and patterned layer tint to another, and these edges are not very precise. A contour interval of 100 m (330 ft.) is employed, resulting in thirty-two layer tints, which run the gamut of the spectrum from a violet at the low end to a red at the high.

in atlases, and on other "physical" maps is the method variously called layer tinting, hypsometric coloring, or altitude tinting. This is the application of different area symbols (hue, pattern, or value) to the areas between contours. On small-scale maps the simplification of the chosen contours must, of necessity, be large. Consequently, the contour arrangements on such maps are not particularly meaningful, and the system degenerates into a presentation of large categories of surface elevation. It should be emphasized that the larger the scale, assuming a reasonable degree of simplification, the more useful is the layer system. When the scale has been increased to the point at which the character of the individual contours and their relationships become meaningful, the representation graduates to being a most useful map.

Colored layer tints at small scales on terrain maps, when combined with pictorial terrain or shading, nearly satisfy most of the landform requirements of the general map reader. They present the major structures as well as some of the details of form. Some of the modern atlases are excellent examples of the effectiveness of this combination.*

HILL SHADING

Representation of the surface configuration with small-interval contours and spot heights provides a highly metrical portrayal of elevations and slopes, but it does not look very realistic. To be sure, one trained in the "interpolation" of contour maps can visualize the major and minor forms symbolized by the arrangements of the line symbols, but the representation of objects by means of projecting to a flat map the traces of a series of equally spaced horizontal planes is not the way objects ordinarily appear to us. Instead, we usually recognize shapes primarily by the interplay of light and dark, which creates patterns recognizable because they correspond with those in our previous experience. The most realistic portrayal of the land surface is thus a pictorial method that duplicates as far as possible the normal way we see shapes. There are, of course, other elements besides light and shade involved in the artistic creation of surface form and arrangement, such as color, clarity, texture, perspective, and so on, but by far the most important is the interrelation of dark and light. The portrayal of the land-surface form by this "natural" method is called hill shading, plastic shading, or loosely just shading.*

There is nothing new about the artistic modeling of shapes by variation in light and shade. Chiaroscuro is a standard art term derived from the Italian words *chiaro* (lightness) and *oscuro* (darkness) to describe any graphic method that derives its effect primarily from the extensive gradation of light and dark in the attempt to attain realism. It was not applied very early in cartography because of the lack of interest in portraying the land-surface form and the difficulty of reproducing smooth tones. The shaded hachuring (see Fig. 15.4) introduced in the nineteenth century is a form of hill shading obtained by closely spaced lines. A great variety of techniques, from manual chalk lithography to machine-produced vertical profiles laid side by side on the paper were tried out during the nineteenth century.** Not until the end of the nineteenth century, however, after the various photographic, engraving, lithographic, and halftone processes had been brought together, could the reproduction of smooth tonal variations from light to dark be easily obtained. Those developments, helped along by the perfection of the airbrush, led to a surge in interest in hill shading.†

*Two of many examples are *The International Atlas* (Chicago: Rand McNally and Company), and *de Grote Bosatlas* (Groningen, The Netherlands: Walters-Nordhoff).

*The adjective "plastic" as it is used here derives from the German *plastik* (from the Greek *plastikos*), which refers to the modeled, three-dimensional effect; it has nothing to do with the modern synthetic materials for which the same term is used.

**E. M. Harris, "Miscellaneous Map Printing Processes in the Nineteenth Century," in *Five Centuries of Map Printing*, ed. D. Woodward (Chicago: University of Chicago Press, 1975).

† The airbrush is a small, pencil-shaped, precise ink atomizer which uses compressed air to apply intricate detail and smooth tones.

FIGURE 15.19 A portion of a small-scale, vertically viewed map incorporating hill shading by pencil on Ross board. Drawn by R. E. Harrison. (From J. Polter, *The Flying North*, 1947, courtesy The Macmillan Company.)

For the most part manual methods rely on the techniques of the artist adapted to the constraints of cartography, namely adherence to standards of horizontal position and, so far as possible, relative elevation. Until the adoption of plastics as drawing surfaces, registration was a major problem. Continuous tone can be rendered in various ways: by airbrush, or pencil and eraser on various drawing surfaces, such as illustration board, gray painted metal or plastic, or pebbled paper, such as Ross board (Fig. 15.19); by wash drawing; or by gouache painting.* Effective hill shading by manual methods requires considerable skill, not only in handling the tools and materials, but also in interpreting the characteristics of the terrain from the source data, usually contour maps and air photographs. It is tedious and time-consuming, but in the hands of a competent artist, manual methods can produce the most effective portrayals. Individual arrangements of forms can be handled to best advantage, and since all but the very largest-scale maps must be significantly generalized, important and unimportant elements can be dealt with effectively. A well-done manually shaded map is, in effect, a portrait of the landforms of an area that captures its "character" as distinguished from any kind of mechanical rendering. On the other hand, since each artist will produce a different "interpretation," adjacent sheets of a series hill-shaded by different persons are likely to look different.

Since manual hill shading is so dependent upon skill, the question naturally arises, why not simply use photographs of the subject itself, the earth, since with modern orthophotographic techniques, perspective distortions can be dealt with? There are several reasons, and their consideration brings into focus the basic objective of hill shading.

Obviously clouds are an important factor, but there are other problems involved in using photographs of the earth. If the earth's surface were barren, like that of the moon, some large-scale vertical photographs could be used for hill shading. Unfortunately, in many areas and in various seasons the surface is masked by varieties of vegetation, field crops, snow, and other things, all of which cause considerable differences in the reflection of incident light and the electromagnetic radiations (albedo), which affect photographic emulsions, spectral receptors, and our eyes. Consequently, in many cases the land-surface forms are simply not visible. Hill shading assumes a denuded land surface.

The surface of the earth is not generalized, and except for some areas at very large scale, simplification is required to produce an effective "portrait." As scale is reduced, the myriad slope variations tend to coalesce and become merely a sort of pattern-like symbol, which camouflages the major or more important lineaments. Furthermore, at smaller scales the forms tend to appear flatter and cease to seem as visually significant as we humans think they should.

*Wash drawing is watercolor painting in translucent grays, while gouache painting uses opaque pigments in water and gum.

Natural lighting is often ineffective, because it is either from too high an angle or from the wrong direction. For example, in the northern hemisphere, much of the time the land surface is illuminated from a southerly direction which is a poor direction for portraying landforms on maps with north at the top.

Many of the problems associated with trying to use the earth itself as the direct source of hill shading can be lessened by constructing a suitably generalized 3-D model and photographing it. This has been done, sometimes effectively, as an alternative to drawing and painting by an artist. Instead of using a physical 3-D model, one can employ a numerical digital terrain model and program a computer to direct an output device, which can produce a range of tones. Both these options will be discussed in following sections.

Whatever the method of preparing a hill-shaded map, the production of effective light-and-shade relationships depends upon various assumptions concerning the illumination of the irregular surface.

Location of the Light Source. In its simplest definition, hill shading is a map of the brightness differences that result from incident light from a constant azimuth and elevation being reflected to an observer from the various orientations and gradients of slopes with an ideal reflecting surface. The observer is assumed to be viewing the map orthogonally (from directly overhead at all points), so that different brightness maps result from changing the azimuth and elevation of the light source. If the surface is illuminated from directly above, then the tones that result will be a function only of gradient, resulting in the "steeper-the-darker" relationship that Lehmann first proposed for his hachure system. The result is less than satisfactory, since the illumination direction is abnormal and is not efficient in the production of clues to form (Fig. 15.20). Every other elevation and azimuth for the light source will produce a different map of brightness variations.

Therefore, one of the problems facing the cartographer who plans to use shading is the direction from which the light is to come. A curious phenomenon is that with light from some directions, depressions and rises will appear reversed. When the light comes from a direction in front of the viewer, elevations generally appear "up" and depressions "down." In addition to providing the correct impression of relief, the direction of lighting is important in effectively portraying the surface being presented. Many areas have a "grain" or alignment of features that would not show up effectively if illuminated from a direction parallel to it. For example, a smooth ridge with a northwest-southeast trend would result in about the same illumination on both sides if the light were to come from the northwest.

The elevation of the light source also has a significant effect on the map of brightness differences. Lowering the elevation of the light will emphasize low relief forms and increase the apparent relief by producing greater contrasts, but at the same time it will decrease the visibility of detail in the darkened areas. On lighted 3-D models lowering the light source will also lengthen any shadows.

With manual methods of hill shading, variations in the elevation of the light source can be regulated by the artist as needed. This is impossible in the photography of a single model, but it can be incorporated to some extent in the computer generation of brightness maps.

Model Photography

Three-dimensional models of the land surface have been prepared at all scales, ranging from dioramas of small localities to the representation of terrain on both large and small globes. Some 3-D models have been prepared expressly for photography, such as Figure 15.21, as a base on which to portray other data, or Figure 15.22, which was prepared specifically to produce wall maps. Most such model preparation for photography has been at small scales.

In its simplest form, hill shading on a map attempts to evoke the impression we think we ought to receive when viewing a photograph of a 3-D model. Our initial reaction is that a 3-D portrayal

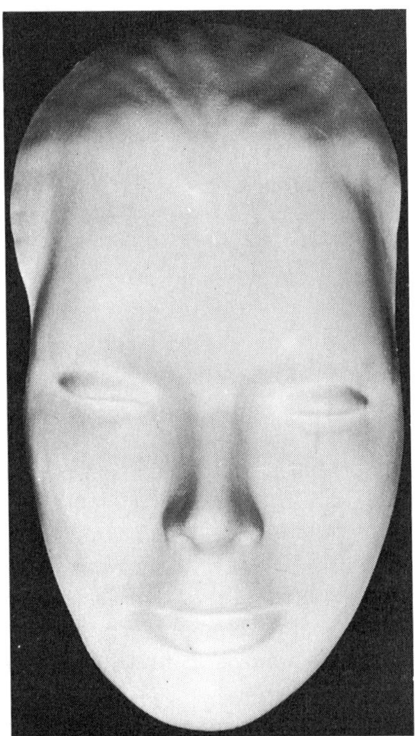

FIGURE 15.20 Two illuminations of the same plaster model. The one on the left is lighted from the direction the model is viewed (Lehmann's system). The right hand model is lighted from the upper left. (Courtesy of G. Fremlin.)

of an area ought to be the most realistic of all maps, and in some respects it is, especially at large scales. On the other hand, to photograph it to provide the hill shading for a map involves a great many complications.

The preparation of a 3-D model is not easy. Details of their construction are beyond the scope of this book. Suffice it to say that it involves such operations as preparing an adequate contour map, carving terrain with a digitally controlled milling machine cutting out successive contour layers of cardboard, stacking them to form a step-like negative mold, preparing a cast from the mold, modeling it with some pliable substance to smooth the stepped surface realistically, preparing a negative mold of the modeled surface, and then

FIGURE 15.21 A portion, much-reduced, of a small-scale 3-D model of Greece and adjacent areas prepared expressly for photography.

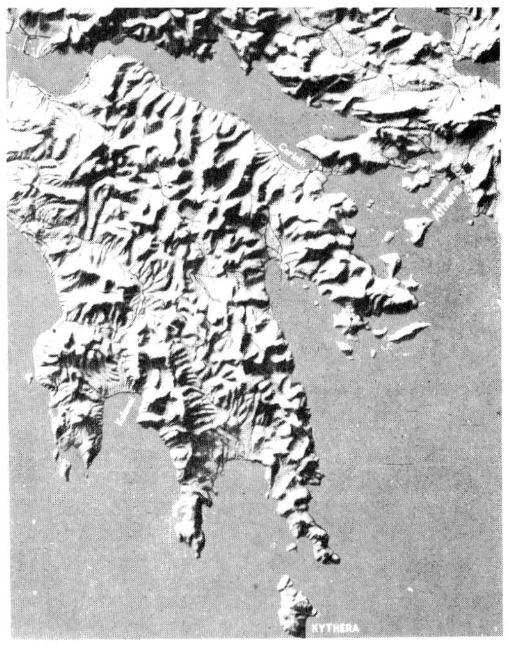

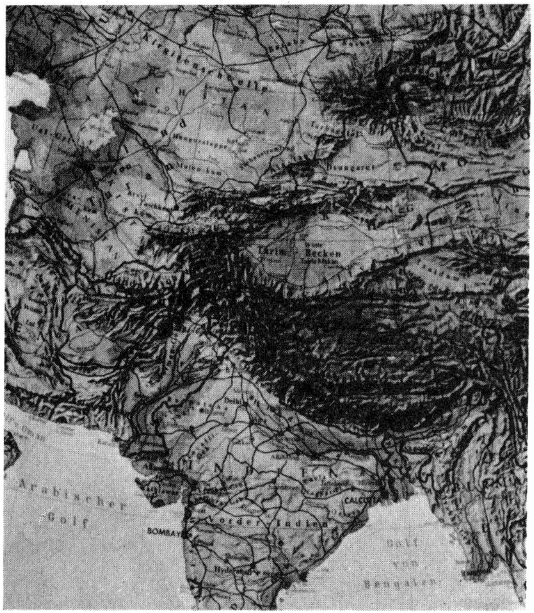

FIGURE 15.22 A much-reduced monochrome print of a portion of a modern color wall map emphasizing surface. The detailed terrain is derived from photographing a carefully made, three-dimensional model. The map is reproduced by complex color printing analogous to process color. (Map by Wenschow, courtesy Denoyer-Geppert Company.)

making a cast from it.* At that point, except for the surface configuration, the model has no visible character, that is, it is a uniform color and it all looks alike except for whatever shadows appear; there are no coastlines or rivers, lakes are like flat plains, and so on.

The lighting of such a model to obtain an adequate photograph is very troublesome, and often next to impossible, since contrasts, highlights, and shadows are difficult to manage. Proper highlights in one area may produce shadows that mask out the detail in the shadow area. This is not an unmanageable problem in photographing familiar objects, such as the face in Figure 15.20, since our familiarity with such forms allows us to "fill in" the detail in the shadowed area, but this does not extend to the complexities of surface configuration.*

Other problems are involved. Such a large-scale model may give the appearance of being a replica of a section of the earth's surface, but it is usually not. As pointed out earlier, our conceptions of the significance of elevational differences and steepnesses of slopes do not accord with their real horizontal and vertical relationships. Generally, if we think there is a marked difference in elevation or there are notable slopes in an area (else why make a model?), it is usually necessary to incorporate an exaggeration in the vertical dimension compared to the horizontal in order for the display to appear realistic. If, in the modeled area, there is a wide range in the local relief, and if the elevations in the areas of lower local relief are exaggerated to look correct, then unless the vertical exaggeration ratio is changed from place to place, the areas of higher local relief will appear unrealistic and perhaps even needle-like.

Manual Hill Shading

Manual hill shading is an art, but it is obviously not entirely subjective, since its aim is to create in the eyes of the viewer the "appropriate" impression of how the land-surface form would appear if seen from above. The image, therefore, must reflect reality in that every form should be in its correct place and the various slopes and

*See Charles Curtis Ryerson, "Improved Methods of Reproducing Large Relief Models," *The American Cartographer* 10(1983):151–57. If the model is to portray other data, the various characteristics to be shown, such as transportation routes, hydrography, settlements, various boundaries, field lines, vegetation, and so on must be compiled and somehow made to adhere to the 3-D surface. In "custom-made" models this is normally done by painting on the data, but flat maps previously printed on plastics may, with reasonable success, be made to conform by heat and pressure to a previously prepared mold. A great deal of ingenuity has been devoted to the preparation of 3-D maps.

*One solution to this problem has been literally to paint the highlights and shadows on a neutral gray surface by airbrushing white and black paints from opposite directions and then to do the photography with the model lighted from above. This was done to produce Figure 15.21.

elevations should be in proper relation with one another. On the other hand, as mentioned earlier, human conceptions of land-surface forms are not at all in accord with their actual 3-D geometry, and this combined with the necessity to generalize at most scales, means that hill shading is not the same as an "objective" overhead view of the denuded land. Furthermore, for proper portrayal of the complexities and often awkward arrangements of forms, it is necessary to adjust the intensities and directions of the assumed incident illumination, which creates the lights and darks of the portrayal. The line between a realistic portrayal and a faulty one is very fine. Like the eyes, ears, nose, and mouth of a recognizable portrait, no matter how well drawn the individual features are, they must be fitted together as they are in nature. It takes the eye and the hand of an artist first to visualize and then to create a meaningful chiaroscuro rendering of forms from the essentially geometric patterns of a contour map.

The usefulness of hill shading to portray surface configuration has long been appreciated, but its nonmetrical character rendered it much less appealing to the military and the engineer who had to work with grades and volumes for transport and construction. Nevertheless, hill shading was so effective visually that it was only natural that there would be attempts to make the representation measurable as well as realistic appearing. The first of these attempts was systematic hachuring, first introduced by Lehmann. Although the range of shading in the various systems was methodically based on slope values and was widely used in the nineteenth century, in its various forms it proved to be more visual than metrical, and within the century it was superseded by contours for large-scale mapping.

Illuminated contours with flat area tones and shadowed contours have both been employed to make depiction by metrical contours appear more realistic. For illuminated contours the lines are systematically thickened according to their angular relationship with the direction of light, usually from the upper left. The lines facing the assumed light source are made lighter than the gray background and those facing away are made darker. An effective display of the system for a volcanic area of Japan was prepared by K. Tanaka, and it has become known as the Tanaka Method (Fig. 15.23).* For shadowed contours the lines are systematically thickened only on the side away from the light source. This gives the illusion of looking at an obliquely lighted, stepped surface.** Various other possibilities of combining a degree of shading with the measurable characteristics of slope or elevation have been suggested. None has progressed beyond the experimental stage, nor is it likely that any will, since with modern color printing, hill shading now can be effectively combined with contouring.

FIGURE 15.23 Map with illuminated contours of a volcanic landscape in Japan by K. Tanaka. Scale ca. 1:175,000 from 1:100,000 original. Contour interval 20 m. (Courtesy of *Geographical Review*.)

*K. Tanaka, "The Relief Contour Method of Representing Topography on Maps," *The Geographical Review* 40 (1950):444–56.

**Pinhas Yoeli, "Shadowed Contours with Computer and Plotter," *The American Cartographer* 10 (1983):101–10.

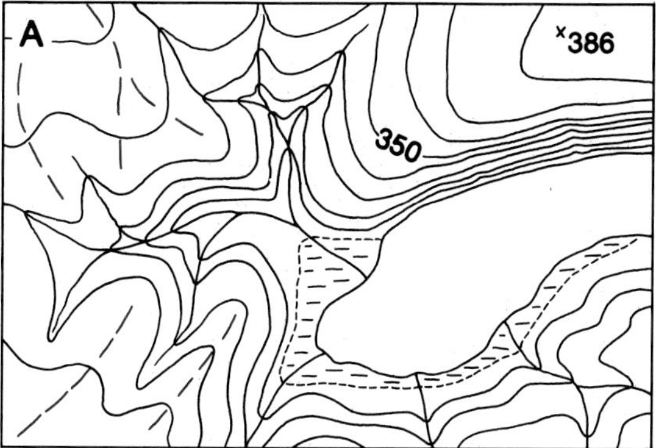

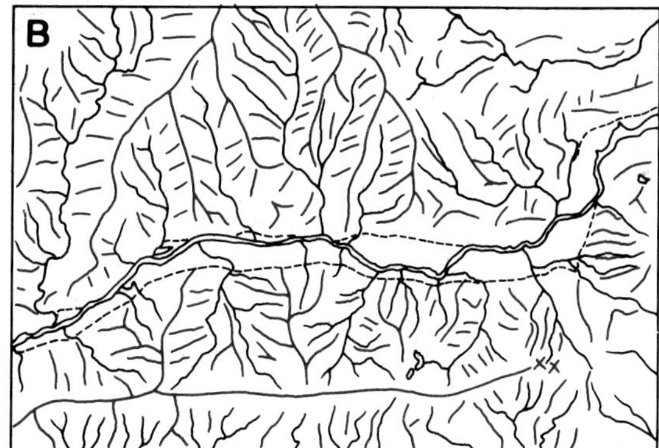

FIGURE 15.24 Compilation guides for hill shading: (A) large-scale; (B) small-scale.

Compilation Guides. In order to obtain correct form relationships and accuracy, hill shading is done over (or on) a carefully prepared guide drawing. For large-scale maps the guide is a contour map with the closest possible contour interval to which have been added streams, various crest or ridge lines, and other significant elements, such as flat or level areas (Fig. 15.24). In areas where significant forms, such as terraces or levees, are not well shown by the standard contours, supplementary contours or form lines are added. Aerial photographs are a useful supplementary aid, since they often show details and characteristics that contours do not reveal.

For hill shading small-scale maps, contours on a compilation guide are of little use, since they are so greatly generalized that they do not reveal form. Instead, a network of structural lines consisting of valley and crest lines and the like is prepared as a guide (Fig. 15.24). On guides for both scales the streams and valley bottoms are usually shown with solid lines and the divides with dashed lines. Distinguishing them by color is helpful. In much humid-land terrain, valley bottoms are more sharply defined than are the divides between them.

Depending upon the drawing surface being used, these guides may be placed on the sheets in nonphotographic blue. If the hill shading is to be done on a translucent medium, the guide may be in black and colors and placed beneath the drawing surface.

Shading Systems. In theory the light source that produces the variations in reflectance can be located: (1) at the zenith, that is, where the light source is at the vertical viewing position, or (2) at any other position, which will provide oblique illumination. Ordinarily, as we have seen, oblique illumination is from the upper left, that is, northwest if the map is oriented with north at the top.

Vertical illumination produces a pattern of reflectance that leads to what we can call Type A or slope shading, in which horizontal surfaces are white and in which the effect is "the steeper the darker" (Fig. 15.3). The first systematized hachuring adopted this method, using lines (hachures) to create the shading and making the widths of the lines strictly proportional to the slope angle. Lehmann chose a slope of 45° as the limit, so that for any slopes over 45° the hachures would coalesce and create solid black. Other systems and limits were devised; in some of them a solid black did not occur.

Light from any direction other than the zenith produces a completely different pattern of reflectance, which leads to what we can call Type B or oblique shading. In oblique illumination the radiance of a surface depends upon the relation of a surface normal to the direction from which the incident light comes. A surface normal is a perpendicular to a tangent to the slope. When the surface normal parallels the illumination, that is "points" exactly in that direction, then radiance will be at a maximum. The more the surface normal differs from that direction, the less the radiance and therefore the more the shading. If we assume a normally oriented map with the illumination from the northwest, then a surface facing northwest with a slope perpendicular to the illumination will have the maximum radiance, and accordingly the least shading ("white"). Any other northwest-facing surface having a greater or lesser slope would be darker in proportion to the difference in slope. Similarly the more the azimuth of the surface normal departs from northwest the darker the surface, with the maximum occurring where the steepest slope faces southeast. A horizontal surface would have a radiance approximated by a medium gray shading.

Slope shading (Type A) and oblique shading (Type B) are both less than completely satisfactory. Type A is unnatural in that we do not usually see forms lighted from the direction in which we are looking, and that direction of illumination does not provide the clues to depth with which we are familiar. Type B is entirely normal and provides a realistic impression, but if strictly applied, the lights are too light and the darks too dark, resulting in excessive contrast among similar slopes which happen to face in different directions. Furthermore, flat areas are relatively dark, which obscures other data in valleys. A more appropriate hill shading seems to be a combination of slope and oblique shading which we can label Type C. In Type C hill shading similar slopes are made to appear more related to one another, thus taking on some of the characteristics of Type A shading, while the contrasts of the oblique lighting of Type B shading are mitigated. In Type C shading the darkness of higher, partly illuminated surfaces is lightened somewhat over that which would be produced by the usual oblique light source. The combination of the two types of shading results in an effect similar to that produced in the nineteenth century by the obliquely lighted hachuring in which the basic elements of Type A shading were employed ("the steeper the darker") but which simply employed a set of thinner lines on the light-facing slopes (Fig. 15.4). The contrasts among the three types of shading are illustrated in Figure 15.25.*

Type C shading, although combining contradictory elements of both pure slope shading and pure oblique shading, is based on the considerable cartographic heritage of the hachuring of the past. The combination enables us to portray effectively details and relationships that would not be so apparent if we were to proceed entirely systematically. One should always keep in mind that a hill-shaded map is not at all a reproduction of the real world; it is a generalized and exaggerated portrayal of what we think our impressions might be of a denuded land surface. Only in large-scale portrayals of relatively coarse-grained mountain areas, where the magnitudes of the forms overwhelm the characteristics of the coverings, or of terrain in barren arid or frozen areas, do the representations approach reality.

The visual logic of acceptable hill shading is extremely complex. For example, unidirectional lighting from the northwest at a zenith angle of 45° would cause any southeast-facing slope of more than 45° to be in complete shadow. Furthermore, any low area immediately to the southeast of a high region would be in a cast shadow. Both are undesirable, partly because the excessive contrast confuses the recognition of unfamiliar forms, and because the darkness would lower the visibility of other mapped data. For these and other reasons, a diffuse light is assumed, which allows a "softer" surface without excessive darks. Similarly, highlights, that is, spots or areas of intense radiance, are inadvisable in hill shading—which is why a model to be photographed is given a matte surface.

---

*Much more extensive discussion of this and all other aspects of manual hill shading by the authority on the subject will be found in Eduard Imhof, *Cartographic Relief Presentation*, ed. H. J. Steward (Berlin and New York: Walter de Gruyter, 1982).

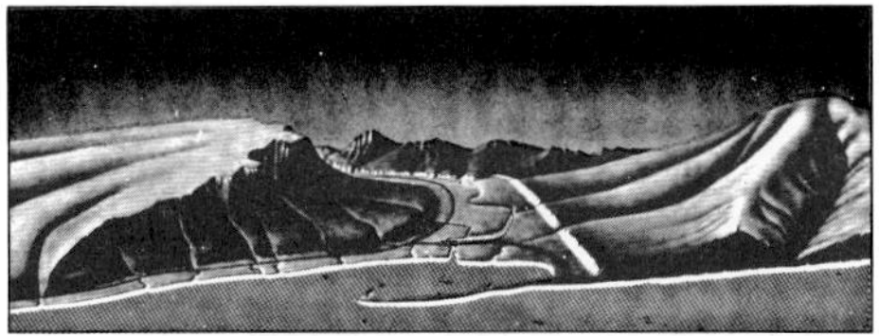

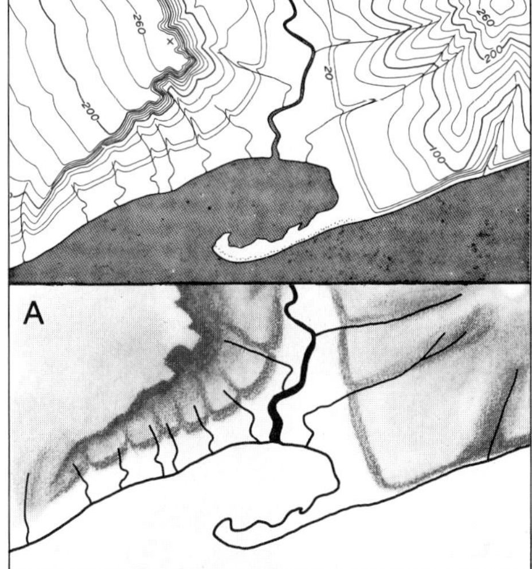

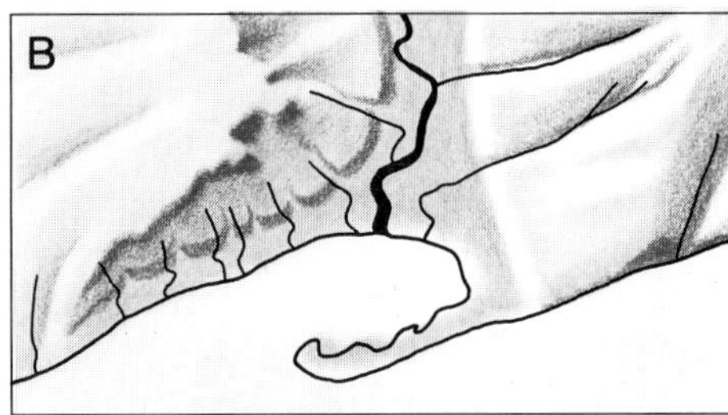

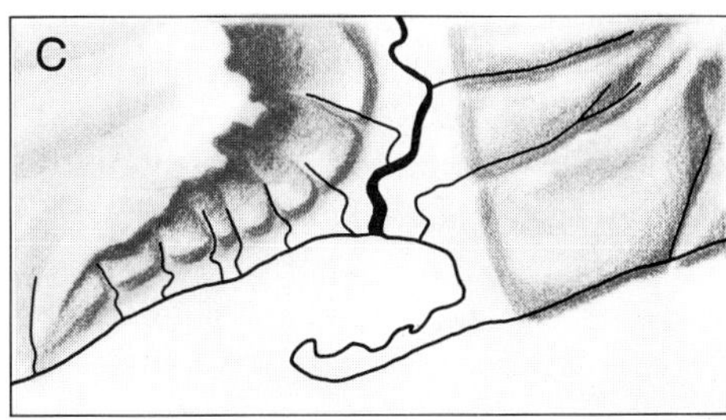

FIGURE 15.25 Sketches to illustrate the three basic methods of manual hill shading: Type A—slope shading; Type B—oblique shading; and Type C—combined slope and oblique shading. The contour map used as a base and the perspective view are the standard diagrams used by the U.S. Geological Survey to illustrate the fundamentals of contouring.

The impression of relief is greatly dependent upon the position of the light source (azimuth and elevation), but rigorous adherence to one position is not useful. There are two circumstances where it is desirable to shift the position of the light source. The first of these has already been mentioned, namely the incorporation in Type C shading of some of the characteristics of the darkness-slope relationship resulting from strict slope shading (Type A). With a given light-source position, the surface on which the light falls perpendicularly will be the brightest, but steeper surfaces, less sloping surfaces, and those progressively turned away from the light source (in azimuth) will all appear darker. This can make it difficult to interpret forms; to correct this problem and to enhance minor but important relief forms, it is advisable to vary the elevation of the light source.

A second problem that can result from rigid adherence to a light source position occurs when smooth, linear uplands, ridges, or valleys trend parallel to the azimuth of the light source. This relationship results in equal reflectance from both sides of the form. The result is that it appears essentially expressionless and upsets the viewer. This can be corrected by employing local variations in the azimuth of the light source, and the fact that the reasonable portrayal was obtained by varying the azimuth of the light will be unnoticed. The variations within adjacent areas of the map must be small, however, although differences of as much as 25° or 30° from one section of a map as compared to another can occur.

Another variation that may be incorporated in hill shading of high relief areas is the phenomenon known as aerial perspective. In normal viewing circumstances atmospheric haze lowers contrasts, which tends to obscure distant features, while those closer to us appear sharper. The 3-D effect can be somewhat enhanced by increasing slightly the sharpness of edges and the contrasts among the lights and darks of the higher elements, but only where there are great differences in local relief.

Hill shading is the attempt to create a believable portrayal of the land-surface form, and there is a tendency to think of it as being reality itself rather than being a symbolization of it. In actuality hill shading employs areas symbols—tones—to produce a 3-D effect, and it is as much symbolization as all other maps representations. Often it is desirable to try to incorporate distinctions among such things as surface material, glacier ice, bare rock, sand, and so on, but this is difficult to do in monochrome, since variation in tone has been appropriated for representing form, and thus only textural variations are available. Such representations must be made with great care so as not to ruin the 3-D effect. When color is available, the task is easier, since for example, one can give glacial forms a blue tint or sand areas a buff. The employment of full color for "realistic" portrayal of the land-surface form incorporating vegetational, seasonal, and other variations in surface coloring, is an art form well beyond the scope of this book.

Much hill shading is destined for monochrome reproduction, but *light* tints of green, brown, yellow, and so on, incorporated by film manipulation, can be used to incorporate elevation data or to enhance the naturalistic effect. The drawing must either be prepared on some special surface, such as pebbled drawing paper for line reproduction, or it must be halftoned, since it employs continuous tone (see Chapter 17). Whatever the process, the contrasts incorporated in the drawing generally will be lessened in the reproduction so that tones must be somewhat exaggerated in the original.

Materials and Tools. The best drawing surfaces for hill shading are smooth, white papers that erase well, but they have the disadvantage of not being translucent. Consequently, the compilation guide must somehow be transferred to them. Translucent papers or plastics with a matte surface are more convenient, since the compilation guides can be placed beneath them, but their drawing surfaces are not as good.

Hill shading is best done with high quality pencils of varying hardness, which give a very black, nonshiny mark and which do not smudge when erased. Smoothing can be accomplished with a stump (a short, thick roll of paper). The use of watercolors with brushes or the airbrush requires great familiarity with them. The airbrush is useful for laying in flat tones, but when used for modeling, it tends to produce soft, rounded forms which often need sharpening. Removal of shading to lighten tones and to provide detail can be done with an eraser or by scraping. It is important to keep the drawing surface clean and "untouched by human hands." The drawn areas should be protected from accidental marking especially when using an eraser. Protection of the final drawing is best done by covering it, rather than by using a fixative spray; if the drawing is to be photographed, fixatives add a sheen to the surface.

For smaller-scale hill-shaded maps the intermediate gray tones should preferably be relatively uniform over the entire map. This is difficult to do when modeling on a white sheet with pencil. One way to help attain this effect is to start with a uniform gray tone to which the darker shading can be added in the usual way. The lighter areas can be erased or scraped, if the materials allow it, or can be added with white paints or chalk pencil. This process, known as the dark-plate or gray-tone method, allows somewhat faster production of simplified drawings, but its main advantage is the ease of obtaining uniformity in the middle gray tones.

The combinations of tools, materials, and processes that may be used to produce hill shading are as varied as artists' media, and their successful use generally requires uncommon skill. On the other hand, some terrain can be portrayed in a reasonable manner, especially as a background for other data, in somewhat more stylized form by methods that require less artistic skill. One of the more recent suggestions involves outlining on a compilation guide areas of relative uniformity in up to eight categories of darkness according to their directions and slopes. These "facets" are then made to appear as distinct flat tones by using tint screens with open-window negatives prepared from peelcoats (see Chapter 18).*

Automated Hill Shading

The highest quality hill shading is manual, but such quality can only be produced by a skillful artist who has an uncommon familiarity with geomorphology. It is probably fair to state also that some of the worst hill shading has been manually produced by persons who either do not have the artist's eye for form or have no understanding of the character of land-surface forms. Since hill shading has been considered very desirable for centuries, it is not surprising that there have been numerous attempts to make the process more mechanical and thus minimize the need for the talents of the artist. These attempts have ranged from enhancing contours to photographing terrain models, but except for the latter, most forms of mechanical assistance have not been very effective.

A century ago, scientists worked out the equations for determining the intensity of light reaching an observer from a point on a three-dimensional surface under illumination by parallel rays. Their work was largely of theoretical interest, because of the immense amount of calculation needed to make use of the equations, as well as the practical problem of producing any graphic result. Now, however, the computer makes possible quick and easy computation, and as a consequence hill shading now can be produced by automated methods.

Automated hill shading requires elevation data in the form of a digital elevation (terrain) model, consisting of a square array of spot heights. Digital elevation models have been generated by interpolation from contour maps, but where good quality aerial photographs are available, the digital elevation data can be obtained essentially automatically.

*A. J. Karssen, "Mask Hill Shading: A New Method of Relief Representation," *ITC Journal*, 1982-2, pp. 160–69.

In its simplest terms automated hill shading begins by estimating (calculating) the gradient at each data point of the array. This involves determining the slope in the west-to-east direction and in the south-to-north direction. This calculation may be done in various ways, such as, for example, by taking into account more than just the adjacent elevations. Then, given an azimuth and a zenith angle for a light source, a reflectance (gray level) for each point is calculated. These numbers can then be fed into a graphic output device, which can produce continuous tone or halftone film transparencies.*

The general procedure is quite straightforward, and any system of hill shading can be automated. Furthermore, numerous variations can be incorporated in the programs, including those described in the previous section on manual hill shading: Gradients can be smoothed by using more complex gradient estimators; apparent resolution can be increased by calculating and plotting gray tones for the four intermediate positions around each value in the digital elevation model; shading contrasts can be increased or decreased by changing the gradients by a constant factor or by changing the elevation of the light source depending upon slope values and changing its direction depending upon the general orientation of ridge and stream lines. Furthermore, various models of reflectance may be incorporated in systems depending upon assumptions as to the character of the reflecting surface, such as smooth, glossy, rough, or somewhat light-absorbing, and so on. Some examples of shaded images prepared from digital elevation models are shown in Figure 15.26.

The technology for automatic hill shading is now readily available, but there are some aspects which render it a less-than-perfect answer to the realistic portrayal of the land-surface form. Ob-

*A very thorough study of automated hill shading, with all relevant equations and a full bibliography, is Berthold K.P. Horn, "Hill Shading and the Reflectance Map," *Proceedings of the IEEE* 69, no. 1 (1981):14–47; also in *Geo-Processing* 2, no. 1, (1982):65–146.

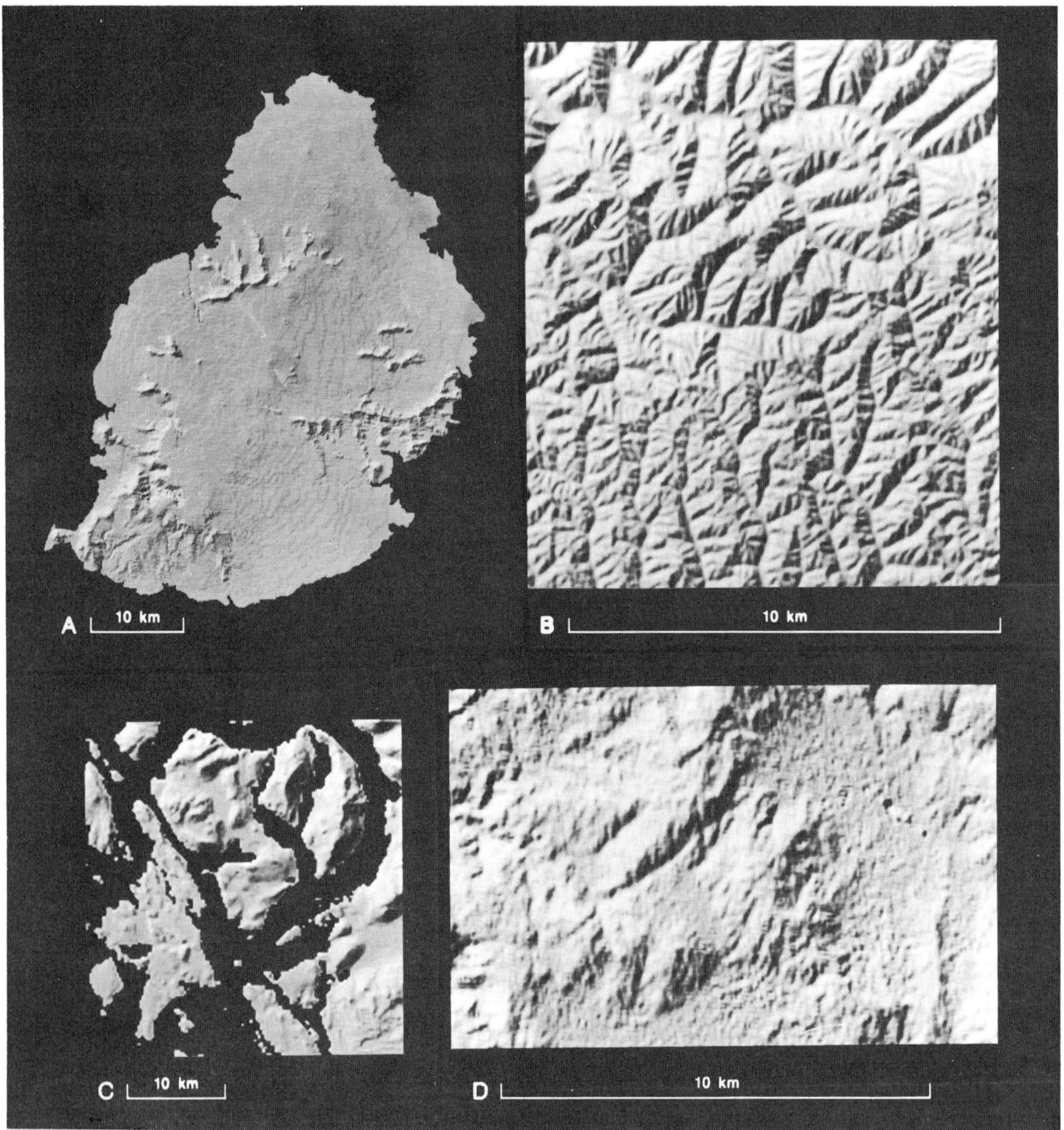

FIGURE 15.26 Shaded images of four digital terrain models: (A) The island of Mauritius, Indian Ocean; (B) Jewell Ridge area, Virginia; (C) Gulf Islands, British Columbia; (D) Tehachapi Mountains, California. (Courtesy of and by permission from Berthold K. P. Horn, "Hill Shading and the Reflectance Map," *Proceedings of the IEEE* 69 (1981). Copyright 1981, IEEE.)

viously it depends upon the availability of accurate and detailed digital elevation data. The square array of spot heights must be on or fitted to the appropriate map projection. The fact that it is a mechanical process means that interaction to incorporate variations in the system from place to place to allow modifications due to unfortunate arrangements of reality or to enhance or suppress elements as may be desirable, tend to be expensive and time consuming. Furthermore, the operator must be as sensitive to the need to make such modifications as would a competent manual hill shader. A formidable and so far intractable problem is how to simplify a large-scale digital elevation model in a nonlinear fashion in order to obtain an appropriate generalization for smaller scales, that is, one which will retain essential features and discard the nonessential. Nevertheless, automated hill shading has several important characteristics. It is fast and relatively inexpensive, given the equipment. Considerably greater detail is possible compared to shading based only on contours. A particular advantage is that the several maps in a series will appear to have been made by the same system and the adjoining sheets will match.

SELECTED REFERENCES

Brassel, K., "A Model for Automated Hill Shading," *The American Cartographer* 1 (1974): 15–27.

Castner, H. W., and R. Wheate, "Re-assessing the Role Played by Shaded Relief in Topographic Scale Maps," *The Cartographic Journal* 16 (1979):77–85.

Dickinson, G. C., *Maps and Air Photographs,* 2d ed. New York: John Wiley & Sons, 1979.

Gilman, C. R., C. W. Richter, and F. S. Brownworth, "Slope Maps—A New USGS Product," *Proceedings of the American Society of Photogrammetry,* 1972 Fall Convention, pp. 384–97.

Horn, B.K.P., "Hill Shading and the Reflectance Map," *Proceedings of the IEEE,* 69 (1981): 14–47; also in *Geo-Processing* 2 (1982): 65–146.

Imhof, E., *Cartographic Relief Presentation,* ed. H. J. Steward. Berlin and New York: Walter de Gruyter, 1982.

Irwin, D., "The Historical Development of Terrain Representation in American Cartography," *International Yearbook of Cartography* 16 (1976):70–83.

Keates, J. S., "Techniques of Relief Representation," *Surveying and Mapping* 21 (1961): 459–63.

Lobeck, A. K., *Block Diagrams,* 2d ed. Amherst, Mass.: Emerson-Trussel Book Company, 1958.

Means, R., *Shaded Relief Technical Manual,* Part I, ACIC Technical Manual RM-895. St. Louis, Mo.: Aeronautical Chart and Information Center, 1958; reprinted 1962.

Monkhouse, F. J., and H. R. Wilkinson, *Maps and Diagrams.* London: University Paperbacks, Methuen and Co., Ltd., 1964.

Raisz, E., *General Cartography,* 2d ed. New York: McGraw-Hill Book Co., 1948.

Robinson, A. H., and N.J.W. Thrower, "A New Method for Terrain Representation," *The Geographical Review* 47 (1957):507–20.

Ryerson, C. C., "Improved Methods of Reproducing Large Relief Models," *The American Cartographer* 10 (1983): 151–57.

Tanaka, K., "The Relief Contour Method of Representing Topography on Maps," *The Geographical Review* 40 (1950):444–56; also in *Surveying and Mapping* 11 (1951).

Thompson, M. M., *Maps for America* 2d ed. Washington, D.C.: U.S. Geological Survey, 1981.

U.S. Geological Survey, *Topographic Instructions of the United States Geological Survey.* Washington, D.C., U.S. Geological Survey. 1979, "Relief Features," Chapter 4B8.

Yoeli, P., "Shadowed Contours with Computer and Plotter," *The American Cartographer* 10 (1983): 101–10.

PART FOUR

The Practice of Cartography: Production and Reproduction

16 Compilation and Credits

COMPILATION

In cartography the term "compilation" refers to the assembling and fitting together of the diversity of geographical data that will be included in a map. The "fitting together" means locating the various data in their proper relative horizontal positions (planimetry) according to the map projection system and the map scale being used.

The objective of the compilation process is the preparation of a composite that contains all the base reference data, the lettering, the geographical distribution(s) being mapped, and everything else that will appear. This becomes the guide for the construction of the map, either manually by scribing or ink drawing or automatically by printing or plotting. The mechanical processing or electronic manipulation of the compilation is not of concern in this chapter (see Chapter 18), but it is important that the cartographer proceed at this stage in a manner that will make the ultimate map construction as easy as possible.

The Compilation Process

The compiling of data may require the use of other maps, text and tabular sources, and digital records for the desired information. The maps (or the digital records that represent the maps) may be on different projections; they may differ markedly in level of accuracy; the dates of publication may vary; their scales will probably be different; and they are likely to have different forms (photo versus line imagery). Text and tabular data may exhibit an equal variety of characteristics.

The cartographer must pick and choose, discard this, and modify that, all the while placing the selected data on the new map, locating each item precisely.

The first rule of compilation is to work from larger to smaller scales because even the largest scale maps, whether made up of photographic or line imagery, show data that have been generalized. These data (representing linear features, such as coasts, rivers, roads, or boundaries) may be "accurate" for the scale at which they are presented, but the generalization made necessary by scale and purpose would usually not be "accurate" for a larger scale. If we compile from smaller to larger scale, we may be building inappropriate error into the compilation.

The techniques of compilation (that is, the means employed to position the data on the new map) range from those that are largely electronic to those that consist of the transfer of data by eye. Between these extremes lies a variety of useful mechanical and photomechanical methods.

The Worksheet

The composite that results from the compilation process is called the *worksheet*. The worksheet may be constructed by hand or machine on a sheet of dimensionally stable (see below) drafting material, or it may consist of nothing more than an electronic image on a display screen. For a simple map the worksheet contains everything, but complex maps may need more than one worksheet, each carefully registered (accurately fitted and positioned) with its companions (see below). When completed, a worksheet is comparable to a corrected, "rough draft" manuscript. All that is then necessary is for the final flaps (the artwork) to be prepared by drafting, scribing, or some other production method.

It is a common experience for a cartographer to find that many of the elements included in one map might easily be used for another. General features such as boundaries, hydrography, and even lettering may not vary much, if at all, from one to another in a series of maps. Consequently, the cartographer can save considerable future effort by anticipating possible subsequent use of the compilation efforts; the cartographer should prepare the worksheet and plans with these possibilities in mind. Modern reproduction methods make it relatively easy to combine different separation drawings or images, even when printing in one color (see Chapter 17).

SOURCE DATA AND MAP TYPE

As suggested in Chapter 2, one of the more important distinctions in mapmaking is the one between the processes employed to produce large-scale topographic and special reference maps and the methods used for small-scale general and thematic cartography. One of the major differences lies in the methods and techniques used to acquire and compile the data to be portrayed.

Large-Scale Mapping

Compilation for large-scale maps, generally considered to include scales larger than about 1:75,000, is ordinarily done by photogrammetric methods, field survey, or a combination of the two. The planimetric accuracy of such maps is controlled as carefully as possible, and within the limits of definition, map scale, and human error, they are correct. The basic principles involved in photogrammetry and the use of remotely sensed images were treated in Chapter 10. The disciplines involved in field survey are plane and geodetic surveying, and except for an occasional reference where the principles overlap or impinge on the cartographic process, they will not be treated in this textbook.

It is important to appreciate that there are different kinds of surveys according to the definitions and assumptions made by the survey organization. Plane (or cadastral) survey is commonly done for a limited area; because the curvature of the earth's surface is relatively insignificant over a small area, it may not even need to be taken into account. The lines of such a plane survey are determined from ground observations and are normally mapped as observed instead of first being referred to a spheroid. Topographic maps, on the other hand, are based on a framework developed by geodetic survey and the ground observations referred to a spheroid; consequently, the two kinds of surveys usually do not match.

Most large-scale plane survey and cadastral maps do not show much physical environmental data. If compilation requires the union of physical and cultural data, the cartographer may be hard put to resolve the differences. For example, if we wish to make a map showing up-to-date information concerning (1) the streams, lakes, and swamps, and (2) the landownership of a region, we may find the first category on topographic maps (of different dates!) but not the second; ownership data will probably be available from the county surveyor's maps, but these may not show the drainage details. The two sources will be essentially "accurate" according to the definitions used for their mapping, but they will not match one another. In general, the practical significance of these kinds of problems varies according to the scale of the map being compiled; the smaller the scale, the less the difficulties, since positional discrepancies diminish and the desirability of generalization increases with reduction in scale.

Medium-Scale Mapping

Compilation for medium-scale maps, generally considered to include scales between 1:75,000 and 1:1,000,000, is accomplished in several ways. Selected data may be traced from the larger-scale maps and the result reduced manually, mechanically, photomechanically, or electronically. The degree to which interpretation and generalization take place in this process will vary with the technique. "Eyeball" generalization has long been practiced. Strictly mechanical and photomechanical processes involve relatively little interpretation and generalization. Electronic methods, on the other hand, routinely involve complex mathematical generalization procedures.

Medium-scale maps are being compiled more and more frequently from remotely sensed imagery taken from high-altitude aircraft and space vehicles. Compilation may involve creating a graphic image from digital sensor data, or it may involve deriving a line image from a photographic record. Recent interest in land use and land cover mapping from satellite imagery has followed this latter approach.

Small-Scale Mapping

The compilation methods employed in preparation of small-scale general and thematic maps are quite different from those used in large-scale mapping. Often, small-scale mapping (scales smaller than 1:100,000,000) involves maps of relatively large areas; this means that data may need to be acquired from diverse sources. Additionally, thematic mapping very often requires that the primary subject matter of the map be presented against a background of locational information, which is called the base data. This base material is ordinarily compiled first, and the accuracy with which it is done largely determines the accuracy of the final map. This is because of the practical requirement that the thematic cartographer must compile much of the subject-matter data by using the base material as a skeleton on which to hang it. These data, usually consisting of coasts, rivers, lakes, and political boundaries, are generally available from large-scale general reference maps.

A large amount of base data useful for small-scale mapping have been converted to digital records and stored in cartographic data banks. More data are being added all the time. Statistical data are also being stored with increasing frequency in digital data files. If the cartographer has access to these records and appropriate computer hardware and software are available, compilation work for small scale-maps can often be done in electronic form. Of course, cartographers who lack access to suitable base data files are not forever prevented from electronic map compilation procedures, but they may have to create their own digital records.

THE SIGNIFICANCE OF BASE DATA TO THEMATIC MAPS

The importance of including on the finished map an adequate amount of base data cannot be overemphasized. Nothing is so disconcerting to a map reader as to see a large amount of detail presented on a map and then be confronted with the realization that there is no "frame" of basic geographic information to which the distributions can be related. The most important objective of thematic mapping is to communicate geographical relationships. Since few people are able to conjure up an acceptable "mental map" and project it on the thematic map, it is incumbent on the cartographer to provide it.

The amount and detail with which the base data are shown will, of course, vary from map to map. The usual thematic map must have on it the coastlines, major rivers and lakes, and at least basic civil divisions. The graticule, in most cases, should also be indicated in some fashion. The objectives of the map will dictate the degree of detail required, but it is a rare map that can be made without these kinds of information to aid the reader in understanding the relationships presented.

Coastlines

The compiling of coasts for very small-scale maps is not much of a problem, since they usually require so much simplification that detail is of little consequence. This is not the case when compiling at medium scales where considerable accuracy of detail is necessary.

Perhaps the major problem facing the cartographer is the matter of source material. It is well to bear in mind that some coasts will be shown quite differently on different maps; yet both may be correct. Hydrographic charts are made with a datum, or plane of reference, of mean low water, whereas topographic maps are usually made with a datum of mean sea level. The two are not the same elevation, and it is to be expected that there will therefore be a difference in the resulting outline of the land. In parts of the world that experience high tidal ranges or where special planes of reference are used, the differences will be greater. Another difficult aspect of dealing with such source materials is that the coloring of the various charts and maps of the same area may be inconsistent. Marshland, definitely not navigable, is likely to be colored as land on a chart, and a compiler would assume it to be land by its appearance. On the other hand, low-lying swamp on a topographic map is likely to be colored blue, as water, and only a small area may be shown

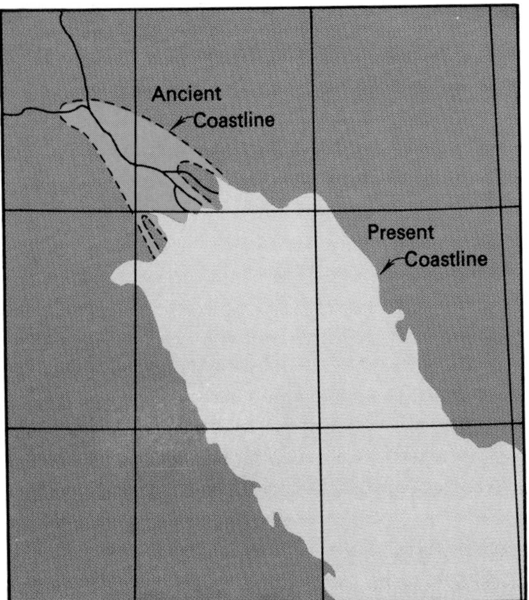

FIGURE 16.1 Major changes in coastlines occur over long periods of time, and these changes may be significant even on small-scale maps. A portion of the Persian Gulf is shown.

as land. On many low-lying coasts the cartographer may be faced with a decision as to what is land; the charts and maps do not tell him.

Through the years some coasts change outline so much that it makes a difference even on medium-scale maps. Figure 16.1 shows the north coast of the Persian Gulf in the past and at present. If we were making maps of a historical period, we would endeavor to recreate the conditions at the period of the map. This problem is particularly evident on coastal areas of rapid silting, which in many parts of the world seem to be important areas of occupancy.

In a number of areas of the world, notably in the polar regions, the coastlines, like many other elements of the map base, are not well known, and they vary surprisingly from one source to another (Fig. 16.2). On some maps a particular region may appear as an island; on others, as a series of islands; and on still others, as a peninsula. On simple line maps a broken or dashed line suffices for unknown positions of coastlines, but it becomes a larger problem when the water is to be shaded or colored. No matter what type of line is used to delineate the coast, the tonal or color change outlines it clearly.

Administrative Boundaries

Compiling political boundaries is sometimes a remarkably complex problem, since the boundaries must be chosen for the purpose and date of the map. The problem becomes more difficult as the area covered by the map increases. Almost all humanly created boundaries change from time to time, and it is surprising how difficult it is to search out the minor changes. The problem is compounded in some kinds of thematic maps. For example, a population map of the distribution of people in central Europe prior to World War II, but also showing present boundaries, raises

FIGURE 16.2 There are marked differences in the planimetry of features shown on the best medium-scale maps of the Antarctic coastline and modern remote sensing imagery. Here the Smyley Island portion of the USGS Bryan Coast map (1:500,000) is compared with landsat images E-1536-12563 and -12570 (Courtesy James W. Schoonmaker, Jr.)

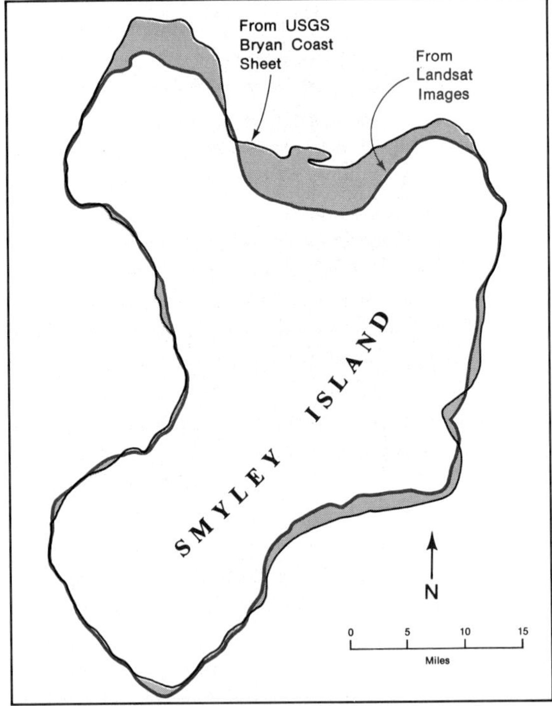

these problems: (1) international boundaries today, and (2) census division boundaries as of the dates of the enumerations in the various countries. The major difficulties are twofold. The first is finding maps that show enumeration districts and that also show latitude, longitude, and other base data so that the boundaries may be transferred to the worksheet in proper planimetric position. The second is placing later international boundaries in correct relation to the earlier enumeration boundaries.

It is not uncommon for official civil division boundary maps to lack any other base data, even projection lines. Such a condition should impress on the cartographer who uses such maps the need to provide base data for the users of the maps.

At any one time there are numerous disputes concerning jurisdiction over territories, ranging from boundaries between states to considerable areas claimed by two countries, such as the Amazon valley boundary between Ecuador and Peru or the Kashmir region in dispute between India and Pakistan. There are also usually several wars being waged over the territories, which call for a decision to be made whether to show the *de facto* (in fact) situation or the *de jure* (by law) jurisdiction.

Hydrography

Compiling rivers, lakes, and other hydrographic features as part of base data is very important. These elements of the physical landscape are in some instances the only relatively permanent interior geographical features on many maps; they provide helpful "anchor points" both for the compilation of other data and for the map reader's understanding of the "place correlations" being communicated.

The selection of the rivers and lakes depends, of course, on their significance to the objective of the map. On some maps the inclusion of well-known state or lower-order administrative boundaries makes it unnecessary to include any but the larger rivers. Maps of less well-known areas require more hydrography, since the drainage lines are sometimes a better known phenomenon than the internal boundaries. Care must be exercised to choose the "main stream" of rivers and major tributaries. Often this depends not on the width, depth, or volume of the stream, but on some economic, historical, or other significant element.

Just as coastlines have characteristic shapes, so do rivers, and these shapes help considerably on the larger-scale maps to identify the feature. The braided streams of dry lands, intermittent streams, or meandering streams on flood plains are examples. On small-scale thematic maps it frequently is not possible to include enough detail to differentiate among stream types, but the larger sweeps, angles, and curves of the stream's course should be faithfully delineated. Likewise, the manner in which a stream enters the sea is important. Some enter at a particular angle, some enter bays, and some break into a characteristic set of distributaries. Swamps, marshes, and mud flats are also commonly important locational elements on the map base.

THEMATIC DATA

We make thematic maps to show the distribution or structure of some data. There is no limit to the kinds or classes of data that can be called "thematic." They can range from a shaded relief or an otherwise general map to a map depicting the incidence of mortality from cancer. Consequently, the possible sources of thematic data are practically unlimited, but some general types are likely to be productive and helpful. National and state governments have vast numbers of bureaus which compile and publish data on an enormous array of subjects. Many libraries maintain separate collections of government publications, and the catalogs (and the librarians) are a rich source of information. Professional societies, of which there are hundreds, will usually respond to requests for information by suggesting sources or persons who might help. There are hundreds of national, state, and regional atlases, which contain diverse kinds of data and which usually list sources. There are also scores of specialized "thematic" atlases, which deal with almost any subject imaginable, from medicine to electrical distribution and from roads to agriculture. Information is not in short supply, and a careful search is usually successful.

ANALOG COMPILATION WORKSHEET

When a map is to be produced by traditional manual or photomechanical methods, a worksheet of the compilation is prepared before any construction begins on the final artwork. It should be viewed as an essential planning document. If it is well thought out and prepared, it will guide the cartographer through subsequent production and reproduction stages in a smooth, straightforward manner. But if it is poorly conceived and executed, there is no end to the frustration that may result in trying to carry the artwork through subsequent production and reproduction steps.

Ideally, the worksheet should be a clear, accurate representation of everything to be shown on the finished map. It also should be laid out, designed, and formatted in such a way that subsequent production and reproduction procedures are best facilitated. To be effective and efficient in these endeavors, a number of compilation factors must be considered, including the transparency and dimensional stability of base materials, methods of forming the guide image, possible separations of the artwork, registry of the worksheet to separate flaps and the finished artwork, image scale and geometry, and potential uses of the worksheet. Each of these factors warrants special consideration.

Base Materials

The first step in preparing a compilation worksheet is the selection of an appropriate base material. Two considerations are involved in making this decision. The first is the degree of *dimensional stability* needed to attain the desired fidelity in artwork construction. As a rule, drafting papers are less stable than drafting films (polyesters). Consequently papers contract and expand more with changes in temperature and humidity, altering the dimensions of the artwork in the process. If all artwork for a small map is being constructed on a single sheet of material, and it will not have to be precisely registered (see below) with any other image, then paper may suffice, and it is cheaper. But if a very large, accurate map is needed, or if precise registry (that is, the matching of several compilation or final artwork sheets) is a requirement, a dimensionally stable material such as plastic should be used. In fact, the added cost of drafting film over paper is such a small part of overall mapping costs that it is good practice to use plastic base materials for all compilation chores. In the long run the benefits achieved by so doing will be substantial.

The second consideration in choosing a base material concerns its level of transparency. Perhaps nothing helps the compiling procedure so much as a translucent material with which to work. When translucent material is used, tracing is easier, and if desired, the worksheet also can be used to make contact negatives or to contact images to other sensitized materials.

A tracing medium of some sort enables the compiler to accomplish a number of things in addition to the convenience of being able to trace some data. The compiler may lay out lettering for titles and the like, and move the layout around under the compilation worksheet. For drawing a series of parallel lines, lettering at an angle, or placing dots regularly, the cartographer need only place some cross-section (graph) paper for a guide under the material.

Forming the Guide Image

It is usually advantageous to prepare the image on the worksheet at the scale planned for final drafting or scribing. Since it then can be directly traced or scribed, a potential scale-changing step can be eliminated. Furthermore, working at production scale gives the cartographer a better impression of how much feature generalization is needed.

For positive artwork, the worksheet normally is done on one sheet of material. The various marks on the worksheet may be done in pencil, ball-point pen, or any satisfactory medium. It will be of great assistance in the subsequent drafting stage if each feature to be shown differently on the final map is compiled with a different color. Lettering, if its positioning is no problem, may be roughly done. But if the positioning is important, the lettering should be laid out with approximate size and spacing. If the first try does not work, it may be erased and done over. Borders and obvious linework need only be suggested by ticks.

Separations

There are a number of situations in which it is appropriate for the worksheet to take the form of more than one sheet of base material. This is generally the case when the artwork is to be reproduced in multiple colors. It also may be required when the artwork is to undergo complex photomechanical manipualtions to obtain special effects (see Chapter 18). Finally separation artwork may be necessary when images are to be contacted to sensitized materials, say for subsequent scribing.

The artwork can be separated in two ways. One approach is to place different categories of information, such as roads, rivers, political boundaries, and so forth, on separate sheets, called *feature separations.* Alternatively, various features to be reproduced eventually in the same color may be placed together on a separate sheet.

Of course, whether two or ten worksheets are prepared, it is important to use dimensionally stable material to minimize potential registry problems. It may still be helpful to use different colored pencils to distinguish between categories of symbols that are drawn on each separation. The exception might be when a separation is to be contacted to sensitized materials, in which case dark linework (black is recommended) provides for the best image transfer. In this latter instance the visual form of lines representing different categories of features should be varied, for example by changing width, dotting, dashing, dot-dashing, and so forth.

Registry

Multiple worksheets or artwork flaps require some means of registration to ensure that the final map composite is accurate. Two methods of registration are commonly employed: pins and graphic marks. Each technique serves a special purpose in subsequent map production and reproduction.

The most precise method of registering the various flaps is to fasten them together with registry pins (Fig. 16.3). These are available in two forms (Fig. 16.4). The most common type consists of a flat piece of material (metal or plastic) to which one or more machined ¼-inch round studs have been attached.* Studs of various heights are marketed. The shorter pins will accept several sheets of material in overlay and can be used in vacuum frames and camera copy boards without danger of breaking the glass. The longer pins cannot be safely used in camera or contact work, but they may be handy during compilation or artwork construction when it is necessary to overlay many flaps.

In order to use these registry pins, precisely matched holes are required in the worksheet and each proposed flap before any artwork construction begins. Special precision punches are manufactured in a variety of sizes especially for this purpose (Fig. 16.5). Large, precision punches are quite expensive, however, and their cost may be hard to justify. Fortunately, for maps of relatively small size we can easily make our own. With an ordinary two- or three-hole adjustable office punch, on which the punches have been *welded into a fixed position,* we can produce perfectly matching holes on a series of sheets. For greater

*Available from the Chester F. Carlson Company, 2230 Edgewood Avenue, Minneapolis, MN 55426; Brown Manufacturing Company, 2000 Dempster Street, Evanston, IL 60204; and Arkay Corporation, 228 South First Street, Milwaukee, WI 53204.

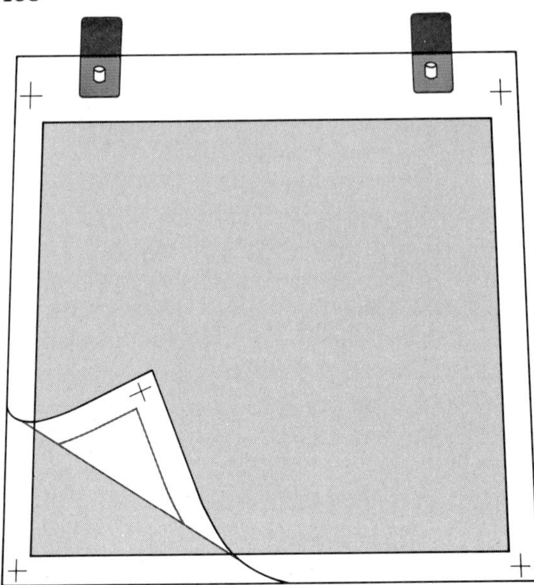

FIGURE 16.3 Each separate sheet of material used in the preparation of artwork for a map can be referred to as a flap. The 1/4-inch pins hold the flaps in register and the small crosses are used for positioning the printing plates on the press for successive colors.

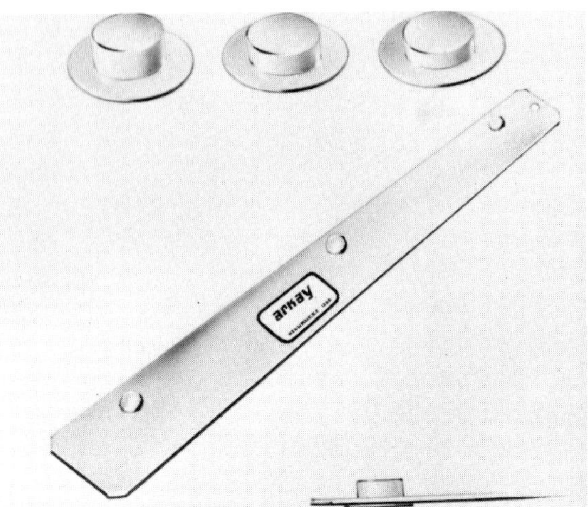

FIGURE 16.4 Registry pins are marketed in several forms. Single flared-base pins and a multiple pin bar are illustrated. (Courtesy ARKAY/HULEN.)

distances between holes, we can fasten separate punches to lengths of metal or wood.

Another way to produce properly spaced holes is to use prepunched tabs. These are available commercially in gummed and ungummed form.* If gummed tabs are used, special care should be taken because they tend to move (or slip) under changing temperature and humidity conditions. Normally two of these tabs are stuck or taped along the top of the worksheet, and pins are then stuck through their holes. Each flap is then successively placed onto the worksheet, the holes in another set of tabs are carefully positioned over the pins, and the tabs are rigidly fixed to the material.

When all flaps are properly punched or tabbed before artwork construction begins, each sheet will fit precisely on the pins throughout the entire production and reproduction process (see Film Registration in Chapter 18). Interchanging flaps can then be accomplished quickly and accurately. Two pins, however, may not provide the rigidity needed for large sheets of material. An extra pin may be placed at the bottom of the sheet, or a pin may be used at the midpoint on each of the four sides.

Using three slots instead of ¼-inch circular holes results in more precise registry and is a more satisfactory system for large maps. This system consists of a flat table surface with three movable punch assemblies that can be locked into position with the axes perpendicular. The rectangular pins used in this system are the same width as the slots but slightly shorter. This lengthwise clearance permits resetting the punch to conform with previously punched sheets by only approximate visual location of the sliding punches. The arrangement of the slots and the use of the rectangular pins keep the center of the map in register and allocate proportionately on both sides of the center axis any misregister caused by contraction or expansion.

Pin registration may not be sufficient when multiple-flap artwork is to be reproduced by plate-based methods. Although pins normally will provide the printer with an adequate means of registry, to achieve the highest quality printing, it

*Various kinds of tabs are available from Berkey Technical Corporation, 25-15 50th St., Woodside, NY 11377.

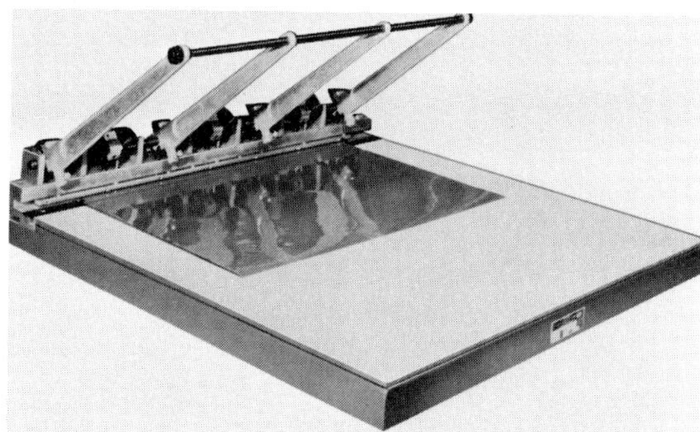

FIGURE 16.5 Pin registry of graphic arts materials requires that precisely punched holes be made in each flap. The use of specially designed punch machines for this purpose can save time and simplify operations, as well as improve registry. (Courtesy ARKAY/HULEN.)

may be necessary to make small adjustments in the paper feed to the press. To serve this purpose, a second method of registering the negatives and the printing plates is needed; it can be used in conjunction with register pins.

Graphic *registry marks* are commonly used for this purpose. Marks such as small crosses (crosshairs) are drafted or stuck in each of the four margins outside the map border (see Fig. 16.3). During map preparation, these registry marks can be used visually to place each flap in the same position on the worksheet. The marks are retained when the negative is prepared and are transferred along with the map image to the printing plate. When the printer is satisfied that the plate is in the proper position on the press, the registry marks are removed with an abrasive. The crosses should be made with very fine lines and should be placed far enough from the map proper to enable the printer to remove them easily without damaging the map. On the other hand, if they are placed too far from the map, much more film is necessary to record their images.

Image Geometry

The process of compilation often requires a change in the geometry of the source materials. In this case it may be necessary to transfer much of the data by eye to the worksheet. The transformation geometry established by the graticules shown on the source map and worksheet constitute the guidelines from which the positions of all features must be estimated. The eye is remarkably discriminating and, with practice, can position data with all the precision that is ordinarily necessary. Very often the use of mechanical means for reduction is prevented by a change in transformation geometry. Do not be too concerned about the accuracy of such a process. Before the advent of computer-assisted cartography, probably 90 percent of all small-scale maps were compiled in this manner, and a large number still are because the desired data are not available in a form that can be handled mechanically.

When the projections differ between the sources being used and the map being compiled, the cartographer must become adept at imagining the shearing and twisting of the graticule from one projection to another and modify the positioning of features accordingly. The difficulties occasioned by map projection differences between the sources and the compilation can be largely eliminated by making the graticules comparable, that is to say, having the same interval on each will greatly facilitate the work (Fig. 16.6).

Compilation is most easily undertaken by first outlining on the new projection the areas covered by the source maps. This outlining is similar to the index map of a map series. The sheet outlines may be drawn, and the special spacing of the graticule on each source (5°, 2°, and so on) may be lightly indicated.

Scale

Much of the compiler's work is in one way or another associated with map scale. The scale of the compilation worksheet must be determined on the basis of the area to be covered and the size of the final map. Once the worksheet scale has been determined, it is often found to differ from the scale of available source maps. Any scale changing that may be necessary at this point can

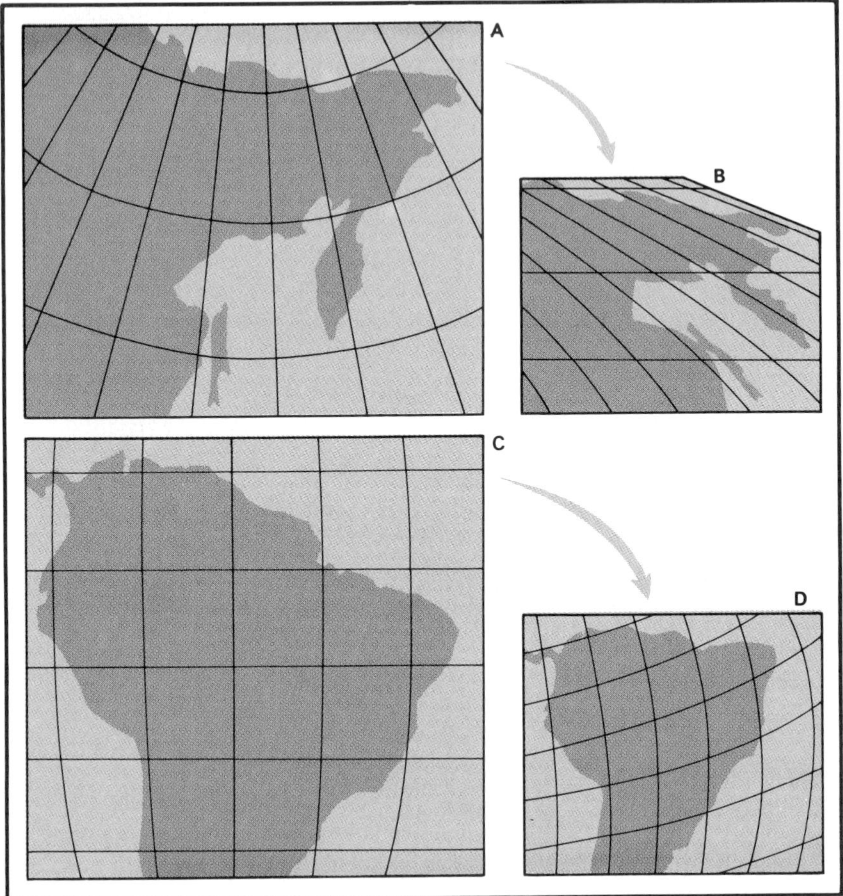

FIGURE 16.6 Shearing and shape changes in the compilation procedure. Maps (B) and (D) are derived from (A) and (C).

be either tedious or quick, depending on the procedure that is used.

Determining the Map Scale. All maps are, of course, constructed to a scale. In actual practice many thematic maps are prepared in a size to fit a prescribed format, the format being the size and shape of the sheet on which the map will appear. The format may be a whole page in a book or atlas, a part of a page, a separate map requiring a fold, a wall map, or a map of almost any conceivable shape and size. Whatever the format may be, the map must fit within it.

The shapes of earth areas may vary considerably when mapped, depending on the projections on which they are plotted. Hence, an important concern of the cartographer regarding map scale may be the projection on which the map will be made. When the projection choice has been narrowed to those that are suitable, the variations in the shapes of the mapped area on the different projections can be matched against the format to see which will provide the best fit and maximum scale.

The easiest way to do this is to establish the vertical and horizontal relationship of the format shape on a proportion basis and then compare the proportion against representations on the various projections. In this instance the actual dimensions as to length and breadth are not the critical matter. It is the proportion between them in relation to the format that is important. When the projection has been selected that best fits the purpose and format of the proposed map, the scale of the finished map may be determined. Normally it is good practice to use either a round number RF or a simple stated scale (map distance to earth distance).

Compilation Scale in Relation to Finished Scale. If the map is to be scribed, normal practice is to compile at the finished scale. If carefully done, the compilation itself will be the basis for the image that can be made to appear in the scribe coat. Similarly, if the map is to be reproduced from an original drawing on plastic or paper by a process that does not allow changing scale, compilation will have to be done at the finished scale. On the other hand, if reproduction is to be by a process that allows reduction, it is desirable to compile and draft at a larger scale and reduce the artwork to the finished scale.

Sometimes the complexity of the map or other characteristics make it desirable to compile at a larger scale even for a map to be scribed. In this case reduction would be used to place the image on the scribe coat.

In general an ink drawing can be reduced with advantage: at the larger scale drafting is easier; reduction "sharpens up" the lines, and so on. The amount of reduction will depend on the process of reproduction and on the complexity of the map. It will also depend on whether definite specifications have been determined for the reproduction, as may be the case with maps of a series. In general, ink-drafted maps are made for from one-quarter to one-half reduction in the linear dimensions. It is unwise to make a greater reduction because the design problem then becomes difficult.

FIGURE 16.7 Optical projection devices are handy for changing the scale of relatively simple map artwork and for transferring details from one map to another when their scales differ. (Courtesy of Artograph, Inc.)

Changing Map Scale. The scale of a base map or a source map may be changed by optical projection, photography, or similar squares.*

Optical Projection. Various projection devices are available, and most cartographic establishments have one. They may operate by projection from overhead onto an opaque drawing surface or from underneath onto a translucent tracing surface (Fig. 16.7). A source map is placed in the projector, and by adjusting the enlargement or reduction, the scale of the projection of the image to the drawing surface may be changed to that of the map being compiled. When properly adjusted, the desired information may then be traced. This task is rather simple if only a few image details need to be transferred, but it can be a major chore if a complex, detailed map im-

*The pantograph, a mechanical device for changing scale used in past centuries, is rarely used today because it is quite cumbersome and slow.

age is involved. Furthermore, if the transformation geometry of the source map and the map being compiled are very different, a projector will not be much help.

Occasionally, a cartographer is called on to produce an oversized chart or map that involves great enlargement. In most cases, extreme accuracy is not required. If the outlines cannot be sketched satisfactorily because of their intricacies, it is possible to accomplish an adequate solution by projection onto a wall. A slide or film transparency may be projected to paper affixed to the wall and the image traced thereon.

Photography. The photographic process can produce good quality copies at desired sizes on dimensionally stable materials, making it an ideal process for compilation stages of map construction. Existing maps can be enlarged or reduced through the use of the conventional film process, and either negatives or positives can be used for tracing detail to a new map.

It is necessary to specify the reduction or enlargement by a percentage ratio, such as 25 percent reduction or 150 percent enlargement. Care should be exercised in specifying the percentage change. A 50 percent enlargement (called "50 percent up") in the linear dimension of the original is obtained by setting the camera to 150 percent. Conversely, 25 percent reduction in the linear dimension is obtained by setting the camera to 75 percent, and a 75 percent reduction in the linear dimension of the original is achieved by setting the camera to 25 percent. In other words, an enlargement *of* 50 percent is an enlargement *to* 150 percent, and a reduction *of* 75 percent is a reduction *to* 25 percent (see Fig. 7.17).

Photographic enlargement or reduction has several advantages over projection techniques that require hand tracing of the image. For one thing, the photographic process is not affected by the level of image detail. Moreover, the photographic process is more precise than hand copying. The image may be either projected or viewed in the camera, making it possible to scale the dimensions of the image more exactly. All the cartographer needs to do is to specify a line on the piece to be photographed and then request that it be reduced to a specific length. The ratio can be worked out exactly, and the photographer needs only a scale to check the setting. Any clearly defined line or border will serve as the guide. If none is available, the cartographer may place one on the drawing with light blue pencil so that it will not be apparent after subsequent photographic processing.

It is possible by photography or with the help of a projector to transform a scale difference of as much as four or five times, but much larger changes are difficult. Any larger reduction or enlargement may have to be done by repeating the process (that is, by making one or more intermediate copies). If the image gets too large, it may have to be done in sections and later pieced together. Another problem is found in the relation between linear and areal scales: the size (area) of maps of the same region will vary as the square of the ratio of their linear scales (see Fig. 7.17). For example, a region shown in 1 cm^2 at a scale of 1:50,000 will occupy only 1 mm^2 ($\frac{1}{100}$ cm^2) at a scale of 1:500,000. Therefore, although positions of individual points or simple lines could be easily determined, anything that involves complications in areal spread would be reduced to an almost indecipherable complexity.

The photographic and xerographic processes can sometimes be combined to real advantage. An example might be when the source map occurs as a figure in a book. The problem then is that most books are too thick to fit under the glass in the camera copy board. But the book illustration can be readily copied at scale xerographically and the resulting electrostatic image placed in the camera for necessary enlargement or reduction. Care must be exercised, however, since some xerographic copiers change the scale of the image slightly while others stretch the image somewhat in one dimension. Although some copiers can also reduce or enlarge, on most machines only a limited number of preset reduction or enlargement factors is available.

Similar Figures. When appropriate photographic equipment is not available, we may be forced to change scale by a method called "similar squares." This involves drawing a grid of squares on the original and drawing the "same" squares, only larger or smaller, on the compilation. The lines and positions may then be transferred by eye from one grid to the other (see Fig. 16.8). With care it is quite an accurate process, although it is the most tedious of the various forms of compilation.

Use of the Worksheet

When the worksheet and the compilation have been completed, the image may be transferred to a scribe sheet for processing or ink drafting may be done on translucent material directly over the worksheet. If the map is simple and the artwork is to be prepared by the cartographer, things such as the character of the lines will be planned. But if the artwork is to be constructed by someone else, the cartographer must prepare a sample sheet of specifications as a guide in map production. Even for the cartographer-produced map, it is wise to prepare such a sample sheet. The task is simple if each category is in a different color or otherwise clearly distinguished. Separation drawings for small maps may be made easily from a single worksheet, and they will register.

DIGITAL COMPILATION WORKSHEET

The introduction of electronic technology into the mapping process has provided an alternative method of preparing a compilation worksheet. The end result, the worksheet, serves the same purpose regardless of how it is prepared, of course, but there are major differences between the two methods. Many of the factors that were critical in manual and photomechanical compilation are not relevant in digital compilation. Conversely, digital compilation introduces factors not found in analog compilation. Still other factors are relevant in both methods.

Digital methods generally provide the compiler with greater flexibility than analog methods. The compiler has much more freedom to try alternative designs, to merge data from diverse sources, to change scale and geometry of the worksheet at will, and to skip costly steps that manual and photomechanical techniques normally require. But these advantages are gained only at a price. Obviously, digital records must be available, along with the software to manipulate them and the computer hardware to carry out data entry, processing, and display functions. The cost of these items has been and remains high, but is progressively declining. Digital com-

FIGURE 16.8 Changing scale by similar squares. (B) has been compiled from (A).

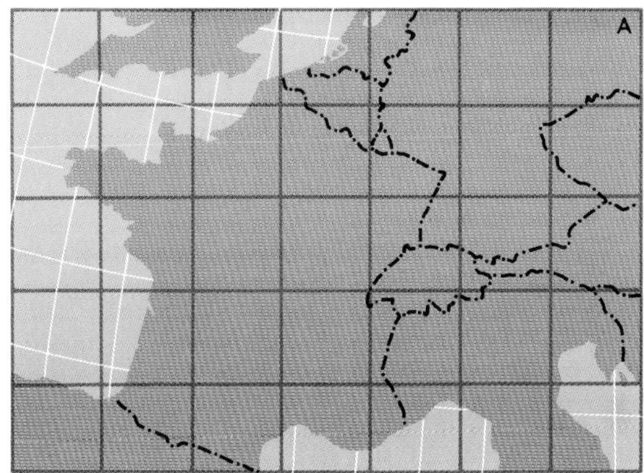

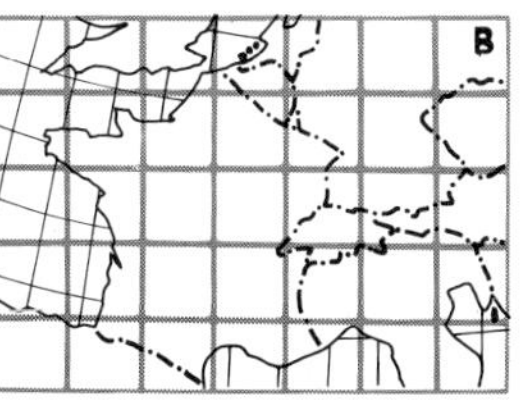

pilation still lies out of reach of many cartographic compilers.

Thus, in spite of its promise, digital compilation is still a long way from replacing analog methods completely. The cartographer merely has been provided with a second option and must carefully weigh the costs and benefits of each before making a commitment to one method or the other. Both methods have their place in the cartographic environment.

In trying to determine whether digital compilation is a viable option, the compiler must answer a number of specific questions. For instance, is access to existing data bases sufficient, or will digitizing be required? If digitizing will be necessary, what is the best means of carrying it out and what provisions must be made for post-digitizing file manipulation in order to clean up, edit, and transform the data to a usable form? Is software available for formatting a projection base and for converting digital records to a common (preferably geographical) reference system, or will it have to be written? Are equipment and software available for constructing the actual worksheet and for taking advantage of the potential benefits of having such a worksheet in subsequent mapping operations? If not, is the cost of purchasing or creating these items manageable or prohibitive? Will there be timely and adequate access to back-up equipment in case of project equipment failures? In sum, are the reserves at hand to make digital compilation a practical, working means of accomplishing the compilation goals?

Data Bases

A large amount of geographical information already exists in the form of digital records called *data bases.* A primary objective of these data bases is to systematize access to the data elements. Several types are available. *Locational base files* (sometimes called cartographic data bases) are most common. They include coordinate records, usually of the vector type, for physical features (coastlines, hydrography), cultural features (roads, railroads, population centers), and boundaries (political, economic, administrative). For the most part these locational data bases form a scale-related hierarchy. The earliest files were created from very small-scale maps, and successive generations of files have generally been done from larger and larger scale source materials. Thus, the first world outline map was done in the early 1960s by the Naval Weapons Laboratory from a 1:12,000,000 scale map base, while the latest was completed in the early 1980s using 1:1,000,000 to 1:4,000,000 scale maps. For the United States itself, the USGS provides complete digital line graph (DLG) coverage (taken from National Atlas sheets) at 1:2,000,000 in its U.S. GeoData files (Color Fig. 17). The same agency is currently working toward the goal of complete DLG coverage at 1:24,000, using the topographic quadrangles as base information.

Thematic base files make up the second category of digital records that are available for mapping purposes. A great deal of statistical census data, for example, has been *geocoded* (that is, given locational reference) and built into digital records. Massive amounts of remote sensing data, such as that from the LANDSAT program (see Chapter 10), are also available in the form of digital data tapes. Finally, digital elevation models for the land surface and the sea bottom are available by one-degree quadrilaterals (geographical coordinate system) for the entire world and at considerably larger scales for the North American and European continents. Extremely high resolution digital elevation models are also produced as a by-product of analytical stereoplotting and orthophotomapping.* Most of these thematic data bases can be purchased or access to them gained by some other means.

The term *geographical base files* refers to a third type of data base. The name normally is given to

**Digital terrain model* (DTM) is the generic term for numerical terrain data of all sorts. These may include lists or grids of spot heights, elevation contours, ridge and valley lines, and so on. The term *digital elevation model* (DEM) refers specifically to a regular grid of spot heights.

merged files, which have locational and thematic (or statistical) data bases combined. For the most part geographical base files are not widely available because they generally are created to serve local and regional needs and therefore are not of widespread interest. Their availability locally, however, may constitute an important resource for the map compiler and should be investigated.

One can expect all these data files to be augmented regionally and locally with larger-scale bases for states, regional planning districts, and metropolitan areas. It is unlikely that these more localized files will be compatible with the more comprehensive files, but standardization of data file structure will be a part of every cartographer's future. The first step in this direction is being attempted by the National Cartographic Information Center at the USGS in Reston, Virginia.

Digitizing

Although a large number of computer-compatible data bases are already available, the cartographic compiler for some time yet will encounter situations where existing digital records are inadequate to serve the needs at hand. In such cases the compiler can decide either to create (or have created) the necessary files or to forget about them altogether and rely instead on a nondigital compilation procedure. Both options have their place.

Before beginning a digitizing project, the costs as well as the benefits must be weighed against those of alternative compilation methods. As a rule, unless a data file is to be used a number of times, the costs of producing it may be difficult to justify. Files representing phenomena that change rapidly might meet this multiple-use criterion, but would have to be updated so frequently (maybe even completely redigitized) that digital compilation again may be of questionable value. Hand compilation might make better sense in these situations.

If, after the alternatives have been carefully considered, the compiler still feels that digitizing source materials is the desired approach, then it is necessary to determine the best way to go about the task. There are several possible approaches. Analog-to-digital conversion of image detail can be done by hand, or it can be carried out with the assistance of a stand-alone digitizing machine. In either case, the digital records can be structured in either a vector or a raster format.

Vector Format Digitizing. Throughout their history maps have consisted of graphic analog images made up of point, line, and area symbols (plus lettering), which have been produced by various manual drafting or scribing techniques, and by sticking up preprinted symbols. Regardless of the approach taken, however, the elements of the component artwork have a definite graphic form. Vector digitizing preserves the form of point, line, and area features and, therefore, is functionally similar to what cartographers have come to regard as the standard way of going about the mapping process.

What constitutes point, line, and area features in digital terms takes some getting used to, however (Fig. 16.9). A point is specified by a coordinate pair, a line by a series of coordinate pairs, and an area by a closed series of coordinate pairs (that is, the first and last coordinate pair in the string being the same). The thing to remember here is that all the information contained in the graphic analog image is also preserved in the digital record; it is only in a different state.

Vector digitizing was initially done by hand, but the task was found to be so laborious that special digitizing equipment was soon developed. Today the great bulk of vector digitizing is done with the aid of automated devices. Although all of these machines involve human operators, some are more labor intensive than others.

Manual Digitizing. Strictly manual vector digitizing is generally carried out with the source material overlaid on graph paper. The coordinate locations of point, line, and area features are then visually determined, the results entered on a coding sheet, and the coded data keypunched or

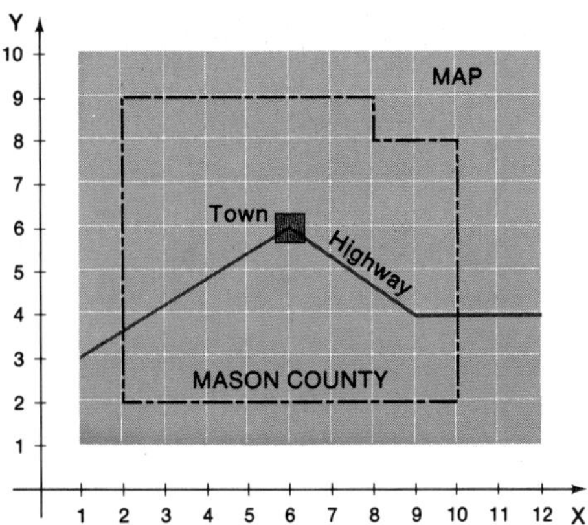

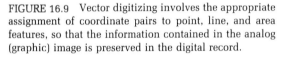

Point (Town) = (6,6)

Line (Highway) = (1,3), (6,6), (9,4), (12,4)

Area (County Boundary) = (2,9), (8,9), (8,8), (10,8), (10,2), (2,2), (2,9)

FIGURE 16.9 Vector digitizing involves the appropriate assignment of coordinate pairs to point, line, and area features, so that the information contained in the analog (graphic) image is preserved in the digital record.

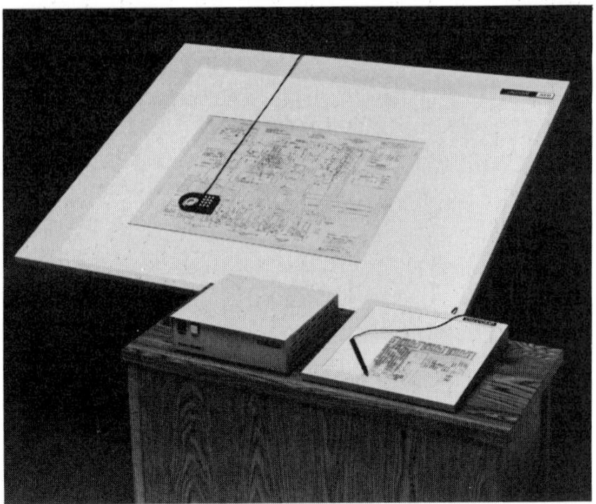

FIGURE 16.10 Electro-mechanical digitizers serve the function of electronic graph paper. As the operator traces a cursor over the map image, its location is sensed electronically and automatically translated into x and y coordinates. (Courtesy California Computer Products.)

typed into a computer terminal. As you might expect, the process is slow, tedious, and expensive. As the number of coordinates to be recorded increases, the labor involved quickly becomes overwhelming. Furthermore, due to the nature of the manual digitizing process, numerous errors tend to creep into the digital records that are produced. For all of these reasons, manual digitizing should if possible be avoided. The cost of electronic digitizing equipment usually can be justified for projects of even relatively small size.

Electro-Mechanical Digitizers. The most widely used digitizing machines are manually operated electro-mechanical devices. These machines consist of a number of components, including a coordinate table (equivalent to electronic graph paper), a movable cursor usually containing a cross-hair, a system control unit, and a data output/storage device (Fig. 16.10). As the operator moves the cursor by hand over material that has been mounted to the table surface, its position is sensed by the equipment's electronics and the trace of the cross-hair over the image is translated into arbitrary *x-y* coordinates (Fig. 16.11). These table coordinates are automatically recorded, normally on punch cards, disc, or magnetic tape.

Most electro-mechanical digitizers can be operated in two modes: point and line or stream. In *point mode,* coordinates are only recorded when the operator gives a special signal, such as a verbal* or a push button command. This mode is used in recording point features, and can also be effective in digitizing line features if only a few points (called nodes) are involved (i.e., shapes are made up of straight line facets or are geometrically rather simple).

*A recent innovation with some electro-mechanical digitizers is to place plot annotation or tagging operations under control of a voice entry device, thereby freeing the operator's hands for other activity. Indeed, a two-way exchange is possible, with the machine prompting the operator for the information needed, and the operator speaking the correct labels or values to the machine. This machine-operator interaction reduces errors as well as greatly speeds up the vector digitizing process.

FIGURE 16.11 The sixteen-button Keypad Control cursor provides all control and record functions for ALTEK's AC 90 SM SUPER-MICRO advanced microprocessor digitizing system. (Courtesy ALTEK Corporation.)

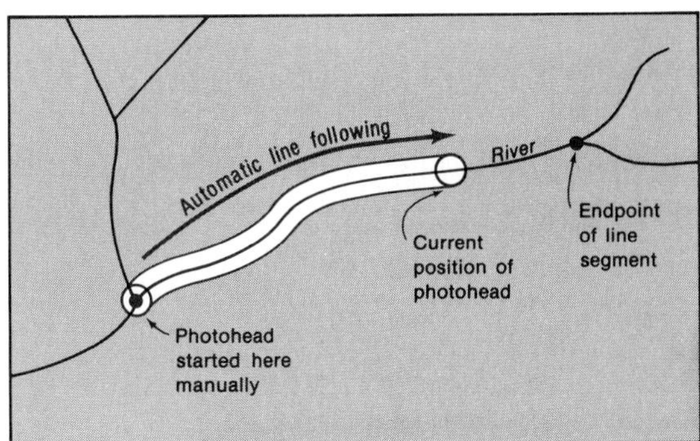

FIGURE 16.13 Line-following digitizers record coordinates as they automatically trace out line features. However, operator intervention is required to position the optical head on each line and to resolve ambiguous situations that may arise in the course of line following.

In *line mode* the machine is preset to record coordinates automatically at given time or distance intervals. These increments, in conjunction with the scale of the source map, determine the spatial resolution of the resulting digital record. The tighter the curves and the more the detail in the original linework, the greater the digitizer resolution required to capture or preserve the form of features in the digital file (Fig. 16.12). More detailed linework also requires more operator time, data storage capacity, and computer processing time, all of which translate into greater compilation costs. This is the reason why existing data bases tend either to be of rather low resolution or of rather limited coverage.

FIGURE 16.12 Resolution in vector digitizing is defined by changes in the size of the minimum increments between adjacent data points in the x and y directions. Greater resolution is achieved by specifying finer increments.

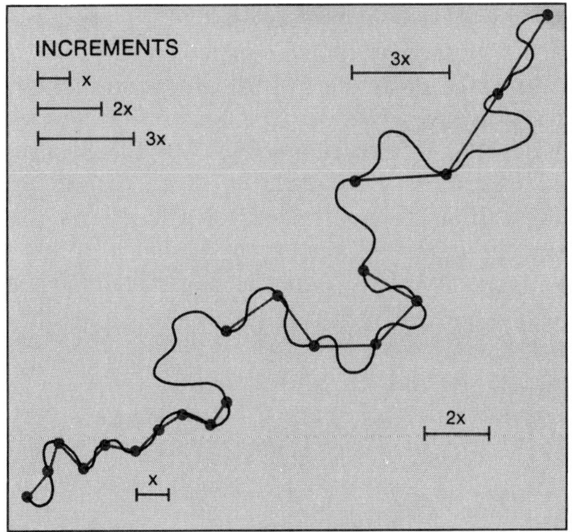

Semiautomatic Line Followers. Another type of vector format digitizing machine has an automatic line-following capability. With one such machine, the operator begins by "locking" a servo-controlled photo sensor onto one end of a line. The sensor then follows the line automatically to its end, or to where it intersects another line or goes off the edge of the sheet (Fig. 16.13). As the photohead moves along the line, its changing position is recorded as a string of *xy* table coordinates. The process is then repeated for the next line, and so forth.

The FASTRAK line-following digitizer system that is marketed by Laser-Scan Limited* works in a somewhat different fashion. In this case a fine laser beam probe is steered at high speed (500 scans/second) in either a vertical or horizontal scan pattern along a selected feature (Fig. 16.14). Information generated by this local raster

* Laser-Scan Limited, Cambridge Science Park, Milton Road, Cambridge CB4 4BH, England.

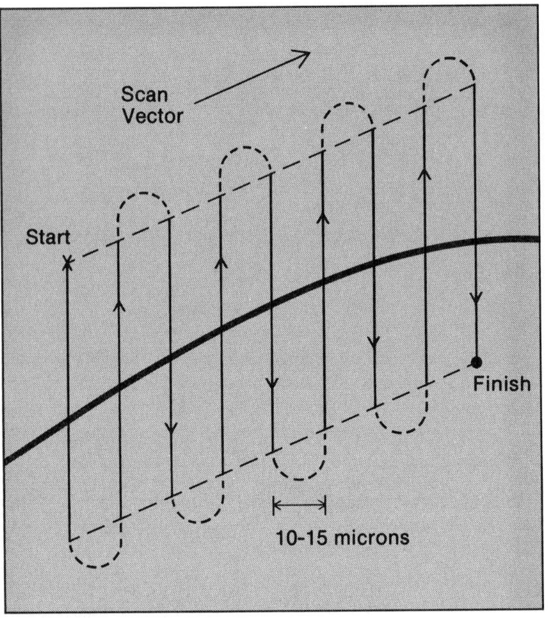

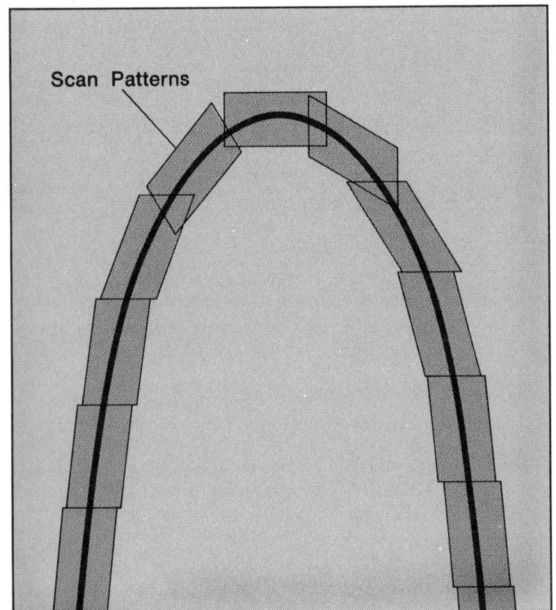

FIGURE 16.14 The FASTRAK line-following digitizer is a hybrid device in the sense that it operates locally in raster mode to produce a scan vector but globally in a vector mode to produce a scan pattern.

scan is processed by an integral computer facility to derive vector (coordinate) data and to provide guidance for the laser beam's next scan pattern.

Neither type of line-following digitizer is fully automatic, of course. As with manual line-following digitizing, the more detailed the map, the more tedious and time-consuming the human task. Operator intervention is necessary or desirable to correct mistakes; to clear up problems that arise in ambiguous situations requiring interpretation, such as with crossed line features (like a road and river) or closely spaced lines (such as contours on a steep slope); and to make annotations.

Raster Format Digitizing. We have already discussed the collection of field data in raster format in our consideration of remote sensing (see Chapter 10). The digitizing of source maps in the laboratory follows many of the same principles. But in the laboratory there is much greater control over the variables involved. An extremely intense and narrowly focused light source can be used. This means that the scan line pattern can be very fine (actually microscopic), and there can be great spectral sensitivity. Scan digitizers currently in use vary widely in speed, resolution, and the image size they can accommodate.

The format of scanner data may be very strange to the cartographer who has practiced traditional vector methods. It must be remembered, however, that the raster record at least in theory can contain the same information that is found in the analog (graphic) source material image (Fig. 16.15). The form of point, line, and area features can all be preserved or, at least, recovered through processing of the raster record. What may seem strange is the sequential structure of coding, whereby adjoining pixels between scan lines may be found greatly separated in the digital record.

In practice, the format and resolution of the raster record may be such that considerable image generalization has taken place. But this need not be so. The range in resolution presently attainable varies from the microscopic to the grossly abstracted (see Color Fig. 18). The finer the resolution, the more scanning time (measured in square inches or centimeters of image processed

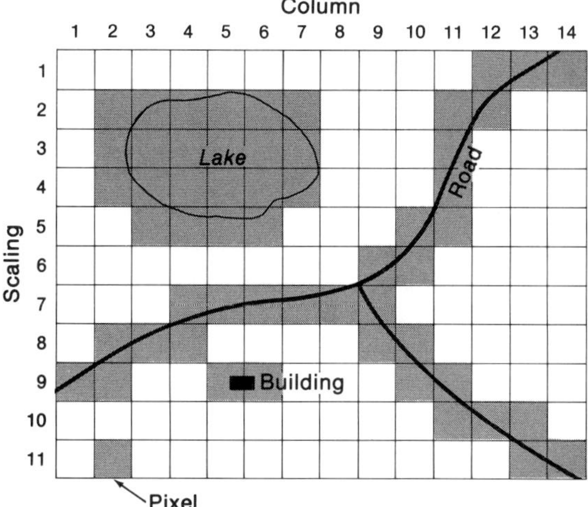

FIGURE 16.15 Raster scanning technology can be used to decompose an analog image into a matrix of discrete picture elements (called pixels). Increasing the number of rasters increases resolution, and if desired, vector elements of the image can be recovered through subsequent software manipulation of the raster record.

per minute) and the more data handling, of course. Thus, the resolution sought should take into consideration subsequent compilation needs.

Clearly, raster digitizing eliminates much of the human labor that is normally associated with vector data input methods. But raster data files may also present the compiler with certain drawbacks. For one thing, vast amounts of data are generated nonselectively. A scan of a 48 cm by 60 cm (19 inch by 24 inch) map at 25 microns yields over 400 million pixels!* A full matrix of raster data is collected even if the only feature on the map image is a sparse river system. Secondly, annotations on the original image may cause problems, and they are difficult to add to the raster files after the fact as well. Finally, the data structure may not be compatible with subsequent mapping needs. Raster coordinates are not in cartesian (*xy*) form; thus, conversion may be needed. Vectorization of the raster records may also be necessary to make the data compatible with available data processing or mapping software. Fortunately, data structure incompatibilities are becoming less of a problem because appropriate conversion routines are being integrated into new cartographic software.

Both continuous tone and line (binary) images can be scan-digitized. In the case of continuous tone photos, a range of numbers (say from 0 to 255) is used to represent the tonal density values. With line maps, quantitizing generally takes a binary form, with 0's signifying empty cells and 1's representing occupied cells. Some automated scanners work from transmission source material, while others work from reflection copy. Some require miniature images (such as 35 mm film), while others work with large format images (up to several feet). Through use of special techniques, even multicolored originals can be scan-digitized into separate numerical records for each color.

Scanning Densitometers. Rotating-drum microdensitometers can be used to raster-digitize systematically small-format source materials. The

FIGURE 16.16 The C-4500 Colormation is a rotating drum color scanning microdensitometer and color film recorder combination. (Courtesy Optronics International, Inc.)

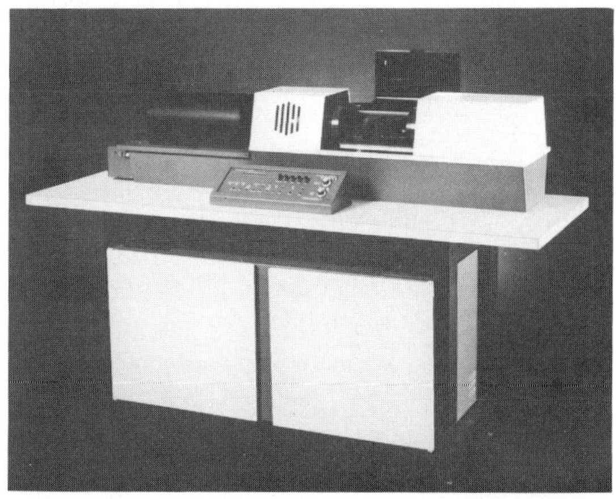

*The space in memory needed to store and reproduce the original line map at the scanned resolution can be greatly reduced by *run-length encoding*. This technique involves recording only line or feature edge crossings along each raster stroke. This eliminates the potentially large amount of data that might be used to record blank spaces, although achievable data compaction decreases as the map detail increases. Strategies such as this trade a reduction in the computer memory requirements for increased calculation when reconstructing the image, of course.

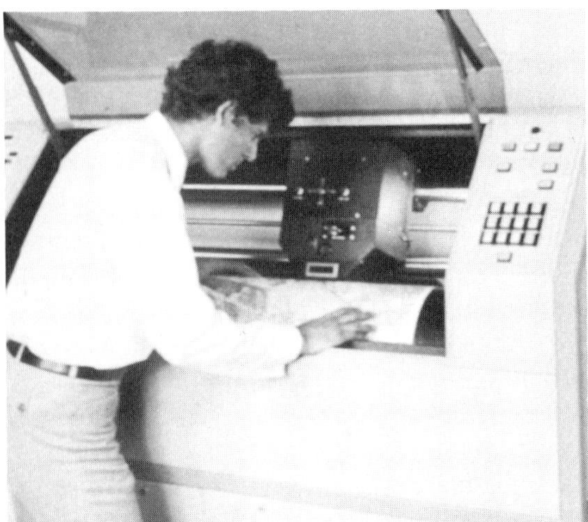

FIGURE 16.17 The Scitex Response-250 color scanner converts analog source maps into raster data at extremely high speed and resolution. The scanner accepts source material up to 36 by 36 inches and can automatically recognize and separate up to twelve precalibrated colors per scan. (Photo courtesy Scitex America Corporation.)

actual scanning is accomplished rapidly and precisely by density sampling pixel after pixel in a circumferential path as an image that is mounted over a window in the drum rotates past the scanner optics (Fig. 16.16). At the end of each drum revolution, the photodetector is advanced (incremented) one raster width along the drum axis, and the scanning process is repeated. The data are processed and then recorded on magnetic tape, or put directly on tape.

Both transmission and reflection source maps up to approximately 35 by 43 cm (14 by 17 inches) may be processed on these electromechanical devices. Up to 64 hue levels or, in black-and-white operation, 256 gray levels, can be differentiated. Operator-selectable spatial resolutions up to a 12.5 micron (2000 lines per inch) maximum can be achieved. Color originals are separated by rescanning the image through automatically operated color filters. Practical operating speeds up to 30,000 pixels per second can be attained. And, in addition to these factors, the equipment can be conveniently operated under normal room lighting conditions.

Scanning densitometer technology has also been extended to large-format raster digitizers. For example, the scanner employed in the Response-250 System* (Fig. 16.17) that is currently being used by both the USGS and the U.S. Defense Mapping Agency (DMA) can accommodate both transmission and reflection source maps up to 90 by 90 cm (36 by 36 inches). The operator calibrates the equipment by pointing the scanner head at each of the source map's colors. The scanner can then recognize up to twelve colors per scan automatically, making it ideal for digitizing a multicolored input map such as a USGS topographic sheet. The source maps are read at approximately 40 lines per mm (1000 lines per inch) and at a rate of 130 lines per minute.** By computer standards the scan rate is rather slow, but by manual digitizing rates the scan rate is extremely fast. Digitized data are taken directly to tape or disk.

Electronic Video System. An electronic video system can also be used for raster digitizing. Components of such a system normally include a video (that is, television) camera, a raster cathode ray tube monitor, and a processing unit to make the analog to digital conversion. Images can be digitized from either reflection or transmission (requires backlighting) imagery. The results may be displayed directly on a raster monitor, or transferred to disk or magnetic tape for future processing. Up to 256 shades of gray can be differentiated in a black-and-white image, so the system is far more sensitive than the human eye. Electronic video systems are far less expensive than scanning densitometers, but they cannot achieve the extremely high resolution of the more costly equipment. The compiler must weigh these factors against mapping requirements.

---

*Manufactured by Scitex America Corporation, 75-D Wiggins Avenue, Bedford, MA 01703.

**The X4040 Large Format Systems marketed by Optronics International, Inc. (7 Stuart Road, Chelmsford, MA 01824) has an even greater capability. It operates at a speed of 1300 lines per minute with materials up to 1 by 1 meter (40 by 40 inches) in size at resolutions as high as 25 microns (1000 lpi).

Data Structures. The differences between vector and raster formatted data are indeed striking. But other aspects of the data base are of equal if not greater importance in computer mapping and spatial analysis. This is particularly true of the arrangement or structure of data within a data base. In addition to being concerned with the implications associated with the handling of vector or raster data, we must also consider such issues as data base flexibility, ease of access to individual data elements, data storage/processing efficiency, and the comparability of records holding different statistical and geographical information. In response to changing attitudes toward these critical factors, there has been a dramatic evolution in the development of data structures over the past several decades.

Early data bases were developed primarily with data input convenience as the overriding concern. It is simplest to encode features on an *entity-by-entity* basis, so this procedure was followed in these first-generation data structures. A pair of coordinates defines a point (node), a line (arc or chain) is defined by a sequential list of coordinate pairs representing the nodes, and lines closing on themselves define the boundaries of regions or polygons (Fig. 16.18A). Unfortunately, when dealing with polygon boundaries, this procedure is not very efficient in terms of data handling or mapping. The reason is that many points end up being digitized twice, and some points may be recorded three or more times.

The redundancy inherent in an entity-by-entity coding scheme can be reduced considerably by replacing the sequential list format with a *locational dictionary*. In this procedure each node on the source map is given an identifying label and then digitized only once. Any number of line or polygon boundary lists can then be compiled from the master coordinate dictionary (Fig. 16.18A). Although some advantages are achieved over simple entity-by-entity coding by using locational dictionaries, applications of the file structure are still limited because no regard is given to spatial adjacencies, overlap, connectivity, and similar aspects of the environmental context. The consequence of ignoring geographical context is that editing, error detection, chaining of line segments into polygons, and analytical processing of the data prior to mapping are all tedious and time-consuming.

In order to make applications of data bases more flexible and efficient than is possible with strict entity-by-entity encoding, it was found necessary to incorporate explicit *topological structure* into the data arrangement. This led to a second generation of data structures in which attention is given to neighboring geographical entities when coding line segments (Fig. 16.18B). For instance, the topological description of each line segment would include not only the beginning and ending bounding nodes but also the left and right designations of the cobounding regions. Each polygon record might also include the identification of any interior islands or, if the polygon is itself an interior island, the identification of the single exterior polygon. This file structure not only accommodates the natural pattern of mapped features efficiently, but also provides the information needed to facilitate automated editing and manipulation of the data base.

The desire for better access to elements in the data base eventually led to still another generation of data structures. Two types of network or graph-based data arrangements are representative of this third class. When environmental features form natural nested hierarchies, a *tree structure* can be employed. Administrative units, for example, are often arranged in a tree in which a path can be traced from a root at the national level through successively lower orders at the regional, state, county, and minor civil division levels (Fig. 16.18C). More complex circuit-type networks might be required when it is desirable to establish direct linkages between a large number of data elements both within and between levels in a hierarchy. Data bases which exhibit cross-referencing of this latter type are commonly referred to as *relational data structures* (Fig. 16.18D). They represent the most sophisticated and powerful class of data structures that have been developed to date.

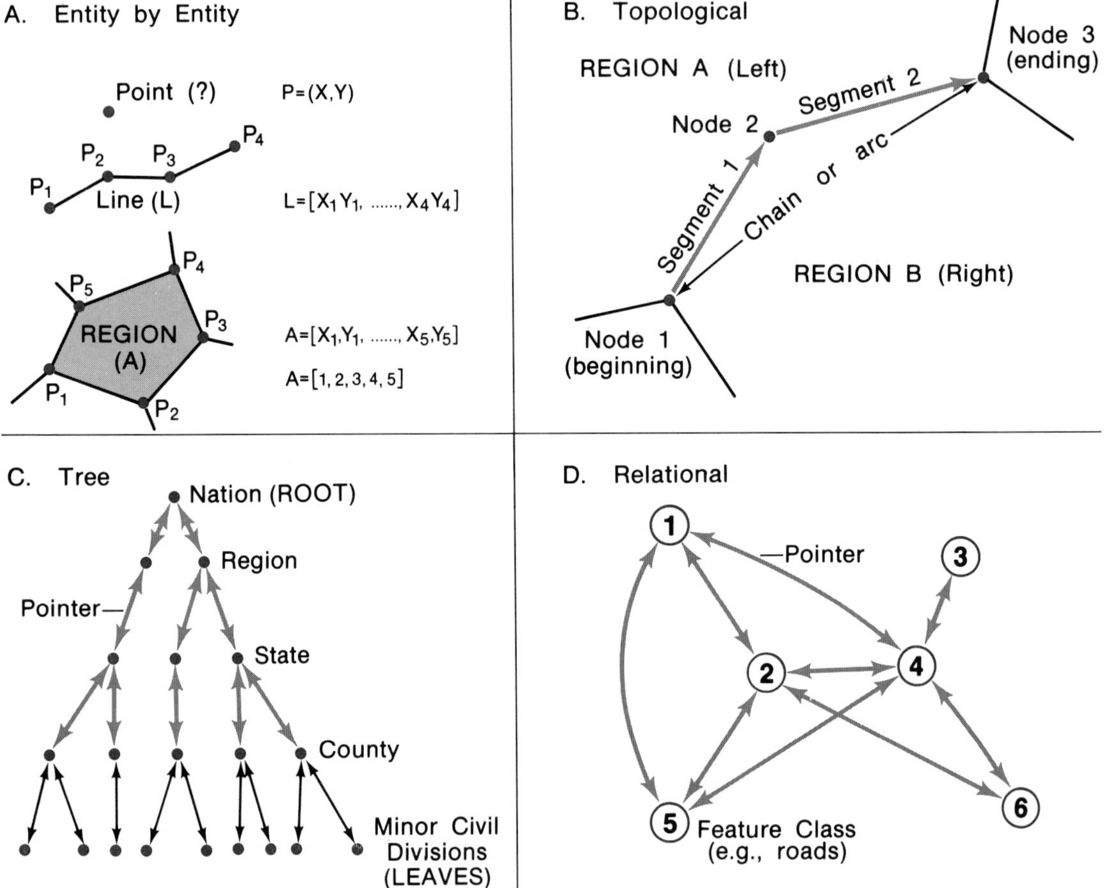

FIGURE 16.18 The arrangement of information within a digital data base has a critical effect on use of the data in computer mapping and spatial analysis. As a rule the effort involved in structuring a data base is directly related to the flexibility of its use.

The key to access efficiency within network-type data bases lies in the use of *pointers* which specify the linkages or paths between data elements. If administrative units are ordered hierarchically, for example, pointers directed upward toward the root (note reverse of normal tree) make it possible to aggregate data from lower levels to higher levels. Conversely, by providing downward pointers, data for a particular area can easily be retrieved by specifying the appropriate path. Pointers serve a similar function in a relational data structure, although the linkages are generally much more complex than they are with tree structures. The more features that have to be linked directly together, the more detailed the system of pointers becomes, of course.

Data structures in current use span all three developmental generations. But there has been a steady trend toward building more geographical context into data bases. In most cases, input stage convenience is no longer the primary objective. Indeed, the development of second and third generation data structures can become so detailed that elaborate tables are required to guide the encoding and data entry processes. This form of data input is actually very tedious and inconvenient. But if efficiency of mapping and spatial analysis are the goals, there seems to be no substitute for devoting this extra effort to the initial stage of data base construction.

Data File Manipulation. Raw digital records as they come from digitizing operations may require considerable processing and manipulation to bring them up to a standard that is cartographically acceptable. To begin with, the digitizing process is usually far from perfect, requiring a variety of clean-up chores before the data can be put to use in compilation. Additionally, general editing of the digital records is usually necessary to correct errors, rectify omissions, and make deletions. Finally, data structure conversions may be required if the format used in digitizing is not compatible with that needed in subsequent mapping operations.

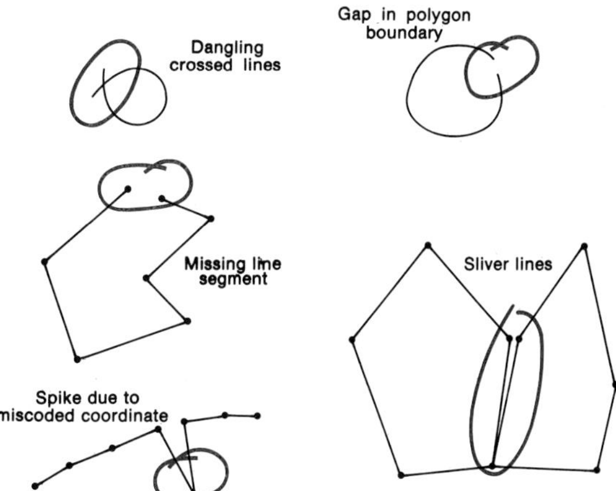

FIGURE 16.19 Cleanup and editing of digital vector files is usually necessary in order to obtain a suitable plot from a newly created digital record. Errors, omissions, and deletions must be taken care of.

Cleanup and Editing. Even with careful planning and execution, things may go wrong in the digitizing operation. Both machine and human errors occur. Additionally, problems with the source material may have been overlooked. In any case, file cleanup and editing are routine chores for the map compiler who is involved with newly created digital records.

The errors that need correction can take several forms. When working with files generated by manually operated vector digitizers, lines may have crossed when they were not supposed to, region (polygon) boundaries may not have closed, features may have been misidentified (such as a river coded as a road), and so forth. Some features may have been digitized twice (usually with slightly different positioning), or they may not have been done at all. All of these errors, omissions, and deletions should be taken care of during this cleanup and editing stage (Fig. 16.19).

Files generated on semiautomated line-following digitizers may also need some work to bring them up to an acceptable standard. Much of this work normally involves correcting errors that result from the line follower not tracking along a line properly, especially in ambiguous situations where lines come close together or cross. A great deal of manual coding may also be necessary to identify the digital records properly so that they are usable.

Scanner-generated files exhibit a whole different set of cleanup and editing problems. Much of the work that must be performed is associated with finding the center of line and point features, and region boundaries, so that the digital files can be used in compiling maps accurately at different scales. Unless elaborate color coding by feature was used initially in conjunction with color sensitive scanners, a great deal of identification and coding work may also be required to put the digital record in a usable form.

The processes of cleanup and editing normally involve first plotting out or displaying the digital records. This portrayal will generally reveal problems, especially if it is presented at larger-than-digitizing scale. At this point the compiler has several options. The necessary changes can be made in the digital records by essentially manual techniques, a process that can be so laborious and time-consuming that it is suitable for only the smallest projects. Alternatively, cleanup and editing can be done interactively. This second approach is much preferred, of course, even though it requires the aid of rather expensive CRT

FIGURE 16.20 The Intergraph high performance work station with local processing power is well suited for mapping applications. The screen on the left can be used for reference while the one on the right displays the results of interactive editing or analysis. In this case, land use polygons are being analyzed. (Courtesy Intergraph Corporation.)

devices (Fig. 16.20) and software that are especially suited for the purpose. But without an interactive capability, it would be extremely difficult to tackle major digital compilation projects in an efficient and effective manner.

Updating Files. Regardless of how complete a digital record might be when it is created, it will become progressively more outdated as time passes. Some environmental features change more quickly than others, of course. Thus we can expect that files of base data (coastlines, rivers, political boundaries) generally become out-of-date more slowly than files of thematic information (land use, socioeconomic factors). Some digital records will require so few changes that it is worthwhile to update them in piecemeal fashion. Others may require such extensive changes that it makes more sense to do them over entirely.

File updating may be far from easy, and it represents a major ongoing problem in computer-assisted mapping. As in cleanup and editing, the task of file updating is greatly simplified if the compiler has the instrumentation and software necessary to work interactively with the digital records. In this case the old map image can be displayed on a video display terminal; features to be added, deleted, or modified can be identified; and the necessary changes can be made in the digital files with the aid of various data input devices (keyboard, light pen, stylus and tablet, digitizer).

If an interactive capability does not exist, these same updating chores can become a major burden for the compiler. In fact, if updating is too difficult, the digital record may actually be abandoned. This if preferable, of course, to going ahead with compilation in spite of known deficiencies in the digital records. Perhaps, in the future major sources of primary cartographic information, such as the USGS, will assume responsibility for keeping their digital records updated, so that compilers involved in mapping at the secondary or lower levels will be relieved of the bulk of the file updating burden.

Data Format Conversions. What is most convenient in digitizing may not always be most convenient in other mapping operations. Thus, there are situations in which it may be desirable to convert data that have been recorded in a vector format to a raster format. Conversely, we may want to convert raster records to a vector data format before going ahead with other mapping tasks. Software programs are available (or can be written) to perform both types of transformations. At this time most of these programs are used in stand-alone fashion. In other words, data format conversions are performed as a separate operation. With time, however, we can expect the capacity for making data conversions to become an integral part of standard mapping programs. Indeed, as we gain the ability to convert digital records routinely from one data format to another, the compiler will become less and less restricted by the coding scheme that was employed in the initial digitizing stage. The digitizing scheme will become far less obvious to the user (i.e., user-transparent) than it is today.

Projections and Coordinate Systems

An initial consideration when compiling a base map is to choose the projection or coordinate system on which feature compilation will take place. The impact of computer assistance has been particularly large in this area of the mapping process. The result has been greatly increased flexibility in the use of different projections and coordinate systems with unconventional orientations and origins. Consequently the type of information that a cartographer must now know about projections has changed.

Most commonly used projections are now available in software packages which, when used in conjunction with cartographic data bases, can produce plots of graticule, coastlines, boundaries, and so on at almost any desired scale (Fig. 16.21). It is no longer necessary, except in very unusual circumstances, for a cartographer to perform the calculations, plot the graticule, and compile general base data.

FIGURE 16.21 Even microcomputers are being used for the map compilation task. This oblique Lambert Azimuthal Equal Area projection was programmed for an Atari-800 microcomputer and prepared on a dot matrix printer by John P. Snyder, a cartographic researcher with the U.S.G.S. Shorelines were taken from a 9,000-point data file in the Atari "Mapware" Software Program©, written by Koons and Prag.

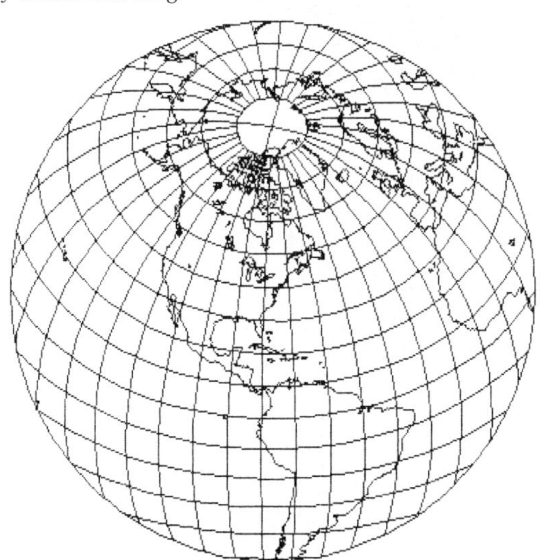

Cartographers now need concentrate only on the design aspects and the fundamental purposes for which the map is being prepared. For example, it is literally possible to make the map fit the format that is available. For special effects cartographers can easily produce elongated, sheared, or unconventionally rotated projections and overlay them with coordinate systems. They must be thoroughly schooled in the use of projections and their properties and be able to select the projection possessing the property that is needed and that best fits the required format (see Chapter 5).

Establishing a Common Base

As we have seen, vector digitizers and raster scanners both code data with respect to a machine coordinate system. Yet, knowing how many inches a location is in the *x* and *y* direction from an arbitrarily established origin on a vector digitizer table, or how many rows and columns a pixel is away from the beginning point (row 1, column 1) in a scan-digitized matrix, is insufficient information for many compilation purposes (Fig. 16.22). There is no real problem if all source information has been quantitized within the same format and with respect to the same origin. But

FIGURE 16.22 Machine reference systems inherent in digitizer table and raster scanner records need to be converted to a common geographical reference system if subsequent mapping flexibility is to be facilitated.

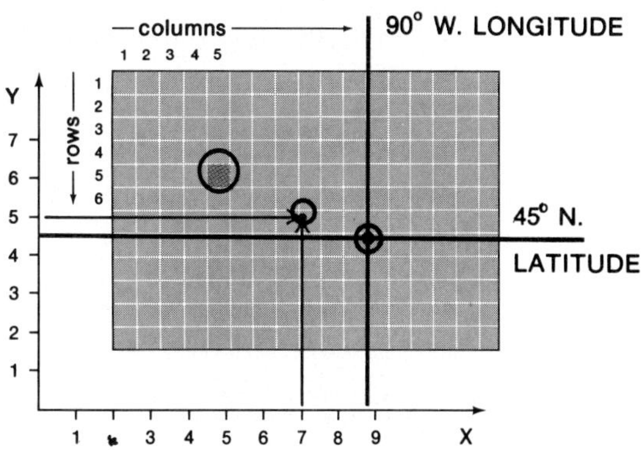

machine coordinates are not very useful when compiling from digital records that have been coded with respect to different formats and origins.

A prerequisite of digital compilation is that the digital records representing all source materials be coded with a common reference system. This usually means converting local machine coordinates to a common geographical reference system such as latitude-longitude. Existing cartographic data bases in most cases will already have had their data converted to standard geographical coordinates. In addition to latitude-longitude, state plane coordinates, Universal Transverse Mercator coordinates, and the United States Public Land Survey parcel descriptions have all been used for this purpose.

Unfortunately, not all existing data bases have their coordinates given with respect to a geographical reference system, or the system used may be different from the one the cartographer desires. Additionally, of course, any direct involvement on the part of the compiler in the conversion of analog source material to digitial form will generate files of information that lack geographical coordinates. In each instance, further manipulation of the digital records will be required prior to using them in preparation of the compilation worksheet.

The task of converting digital records to a common reference base would be most onerous if carried out solely by hand for anything more than a few coordinates. Fortunately, the software appropriate for handling this coordinate transformation chore is available (or could be written). Available routines take advantage of the fact that a map projection establishes a systematic mathematical relationship between ground and map locations. In fact, the relationship works both ways. Thus, just as we can move freely from ground reference to map reference, we can reverse the process and move from map reference to ground reference. The mathematical routines that make this possible require as input pertinent information concerning key locations on the source map sheet (for example, several corners) and the position of the map sheet within its general projection framework (for example, with respect to origin, central meridians or parallels, and so on). This information is provided along with the coordinate file of geographical features, and the computer program does the rest. The result is a new file that is coded with respect to the desired geographical reference base.

Preparation of the Worksheet

When working with computer assistance, the preparation of the compilation worksheet is under software control. Someone has already taken responsibility for the program algorithm and for providing options and defaults* for its use. The cartographer is of course still responsible for making most of the same decisions that are required of the manual compiler, but in this case these decisions are implemented through software instructions to the computer and computer-driven peripheral devices. If all needed digital records are compatible with respect to their data structure format and their locational reference base, preparation of the worksheet can usually proceed with great speed and efficiency.

The role of the map compiler is to define the individual cartographic elements that are associated with the cartographic processes of selection, classification, simplification, and symbolization. Among a host of other things, including choice of projection, area of coverage, scale, and features to be mapped, it involves choosing a mapping technique (dot, choropleth, isoline, and so forth) and specification of mapping parameters (class intervals, number of classes, and so on). In other words, the compiler prepares the worksheet with the aid of the computer but bears much the same responsibility in this endeavor as normally with manual compilation.

Once all the information necessary for constructing the worksheet has been communicated to the computer, the compiler can call for a plot,

*Most mapping programs have built-in specifications for such things as number of classes, class interval scheme, shading patterns, and so on. Unless the user instructs the program to do otherwise, these standard or default specifications will be utilized in map production.

print, or display of the selected data. This portrayal of the compilation material may be done in the form of a composite image, or as a set of registered separations. The latter may involve merging several data files. Before the worksheet is finalized, considerable interaction between the compiler and the data files and program controls may occur.

Uses of the Worksheet

Digital compilation worksheets can be put to a variety of uses. Cartographers always have used the worksheet to test layout and design concepts, but the time and expense involved kept this function to a minimum. The great speed and relatively low cost of electronic compilation have encouraged cartographers to expand greatly their use of the worksheet as a means of previewing design alternatives. Indeed, a number of rough worksheets might be made, especially if the compiler is working interactively with a video display screen image, before settling on the one that is to be carried through into production.

Worksheets compiled electronically may also be used as a basis for subsequent production by nonautomated methods. In this case the worksheet is treated no differently from those compiled by hand procedures. In fact, it is becoming quite common to compile part of the map artwork by hand and part by electronic methods. An example might be a city street map that was compiled by hand being used as a base upon which to overlay digitally compiled thematic information, such as the sites of traffic accidents, schools, fire stations, and other data.

A third use of worksheets that have been compiled from digital records is to provide a preproduction test of data files, mapping program controls, and graphics equipment. In this case the worksheet serves as an intermediate step prior to going ahead with electronic map reproduction. Commonly, for example, a map that is to be produced finally in liquid ink on drafting film is first plotted in preliminary form on paper in ballpoint pen.

Finally, the digitally produced worksheet may constitute the one and only map that is constructed. In other words, the mapping process may actually end with the compilation step. This tends to occur when a researcher or planner is using cartographic output as an aid in tackling environmental problems. A quick look at the rough map image may be all that is needed. The essence of the cartographic portrayal, not a finished product, is what is important in such a case.

ACCURACY AND RELIABILITY

In most situations "accuracy" is a relative term. In cartography it can refer to the closeness with which data on the map are located with respect to defined geographical positions, to the detail with which elevations are indicated, or to the degree to which a mapped distribution portrays the "real" distribution. The concept can only be dealt with in terms of "accuracy for what purpose," but that does not in any way change the fact that the accuracy of source data must always be a matter of primary concern for the map compiler. Errors can easily be propagated and exaggerated through the compilation process. Unfortunately, the quality of source data is not always readily evident. For this reason the cartographer needs to develop sensitivity to potential sources of error.

The age of the source material is one thing to watch closely. As a rule, methods for acquiring data have improved over the years. With time, a more complete picture also builds up. Thus, we can expect that the scope and depth of environmental knowledge continually improves. Newer sources are generally preferred over older ones. In any case the date that source data were collected or last revised should be noted and, preferably, passed along to the map user.

The scale of source material must also be carefully weighed. Progressive generalization with smaller scales is an inevitable aspect of the mapping process. For this reason compilation should always be from larger-scale sources rather than smaller. The temptation to enlarge a smaller-scale source map is bad enough. But it would be even worse to blow up a smaller-scale map of one feature (such as soils categories) to be overlaid on a compilation worksheet containing other features (such as landownership parcels) that were compiled from larger scales. The different degree of cross-variable generalization on the resulting map could quite easily invalidate it with respect to its intended use.

A third concern regarding the quality of source data concerns the level of responsibility that can be associated with the data collection or mapping agency. The locations of landownership boundaries found in County Plat Books produced by many commercial vendors in the United States are highly generalized and therefore notoriously "inaccurate" in terms of any coordinate reference system (Fig. 16.23) in spite of the potential hardship that this may cause those who do not know that these cadastral maps are designed more for topological than planimetric accuracy. The standard topographic quadrangles that are published by the United States Geological Survey are produced to strict map accuracy standards and are relatively reliable within the limits of the map scale, date, and so forth. Clearly, the compiler is obliged to pay careful attention to this responsibility factor.

Many relatively large-scale compiled maps have been put together from a variety of sources in a kind of patchwork fashion. Instead of individual phenomena being compiled from several sources that cover the whole map area, one part of the map may be derived from modern topographic maps, another part from older compiled maps, and so on. If there is any significant difference among the quality of the sources, then a coverage diagram should be prepared. An example of such a diagram is shown in Figure 16.24. Many kinds of information can be shown on coverage diagrams, such as the quality of source maps, dates of censuses, or names of compilers or investigators.

FIGURE 16.23 Map accuracy standards are not widely defined or adopted by the various private and government mapping establishments. As a consequence there may be marked variations in the planimetry of features from one map to the next, as is illustrated with this stream segment traced from a county plat book (1:53,350) and a U.S.G.S. topographic quadrangle (1:24,000).

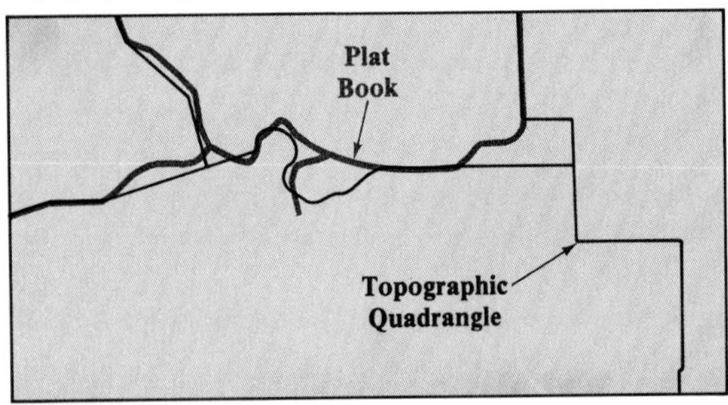

SOURCES AND CREDITS

The sources to which a cartographer may turn to obtain geographical information are legion, and it is not our purpose to suggest and evaluate them. It is useful, however, to divide the great variety into two classes with respect to the necessity for citation.

A considerable share of the geographical information compiled as base data—things such as coastlines, rivers, and boundaries—may be classed as general knowledge in the sense that they are well-known facts. They have been mapped many times, and it is quite unnecessary to cite the source of such information. On the other hand, if the delineation differs significantly from the usual way the phenomenon has been mapped because some very new source has been used, such as a new survey, remotely sensed imagery (such as LANDSAT, see Chapter 10), or as a result of recent boundary changes, it is helpful to the map user for that to be indicated.

In all instances of thematic maps and large-scale specialized reference maps, it can be assumed that the data being mapped will not be common knowledge. Consequently, the map user should be given the source and, if it is important, the date of the special information. Indications such as, "U.S. Bureau of the Census, 1980," "Land use data from provisional 1982 map by Special Project Committee . . . , etc.," or "Zoning boundaries as approved by the County Board in 1976," will provide the user with necessary information.

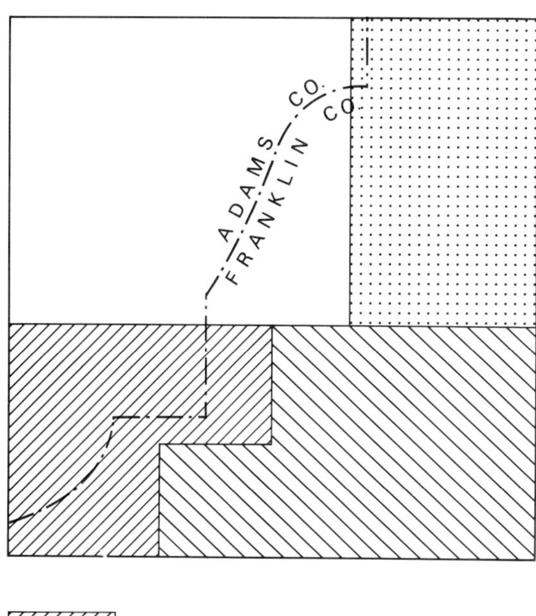

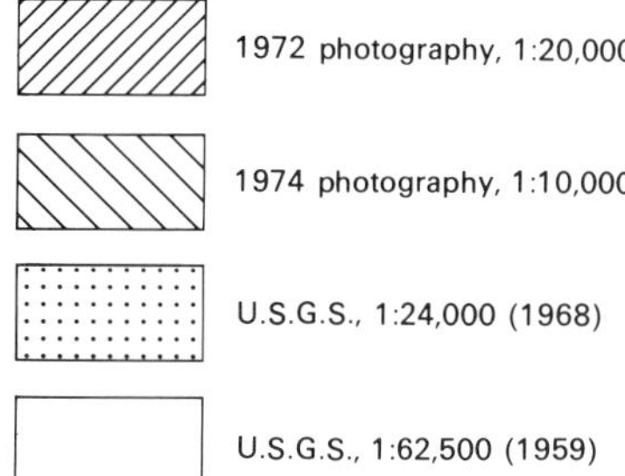

FIGURE 16.24 Coverage diagram showing the kind and date of sources used in compilation.

Whenever specialized information has been obtained from a nonpublic source, such as a research foundation paper, a scholarly publication, or an industrial report, it is common courtesy to acknowledge the source. Judgment is always involved. The information may be widely known and may appear in substantially the same form in several sources. In that case citation may be unnecessary, but whenever there is any doubt, the source should be acknowledged.

Copyright

A large proportion of published material is protected by copyright, the name given to the legal right of a creator to prevent others copying his or her original work. The application of statutory copyright to maps is not very well understood because the law is aimed primarily at the protection of literary works, but it encompasses a large number of classes of materials ranging from musical compositions to photographs, and it includes maps. We cannot, of course, offer legal advice, but a few general observations are provided so that you will appreciate both the complexity of the situation and the necessity for complying with the provisions of the law.

There are two aspects to the copyright law. Both have to do with the protection afforded the copyright holder, but one is concerned with the degree to which a map, or a base data file, is considered "original" and thus can properly be subject to copyright. Over the years it appears that the judicial view of originality has been very restrictive with regard to maps. They have often been seen only as collections of facts and thus not being very original. The other aspect has to do with the responsibility of the mapmaker to refrain from copying the work of another. The outcome of a court case is never predictable, but whatever might happen, it is bound to be upsetting and expensive, one way or another. Consequently, as a rule, no copyrighted map should be *copied* without obtaining permission from the copyright holder. The term "copied" is used here in its literal sense, meaning "reproduced exactly." One must not, therefore, either through photography or tracing, copy exactly the way some other mapmaker has represented geographical data. Although the exact position of the coastline of the United States, for example, cannot be copyrighted, presumably the way someone has generalized it can. Furthermore, the exact manner in which some cartographer has processed and symbolized statistical data for an area is protected.

Most publications, including maps, that are produced by agencies of the U.S. government are in the public domain; that is, they are not copyrighted and can be used freely. One ought always to give credit where it is due, and in the case of specialized publications containing judgments and opinions of named authors, it is not only courteous but wise to request permission from the issuing office, since the material may have been copyrighted by the authors separately or some of it may have come from copyrighted sources.

Although governmentally produced survey maps in the United States of the topographic variety, census materials, and the like may be used freely as sources of data, this is most definitely *not* the case with such materials produced by other governments. In most instances reproduction of such materials without permission is strictly prohibited. Since the United States is a party to the international Universal Copyright Convention, and since this and other agreements make it obligatory for citizens to comply with the copyright laws of the agreeing nations, one must be very careful.

Generally, whenever one wants to copy or reproduce material copyrighted by others for any purpose other than private study, term papers, theses, and the like, written permission to do so must be obtained from the holder or holders of the copyright.* The request should specify (1) the work in which the wanted material is to be used, (2) the publisher, (3) the exact reference to the material (author, book or periodical, pages, figures, tables, and so on), and (4) how the material is to be used (as an illustration, a quotation, or whatever). It is important to direct the request to the copyright holder; in many instances this is the publisher, not the author. Sometimes a fee is demanded, but in all cases the credit line or acknowledgment of the source should be given exactly as specified by copyright holder. Obtaining permissions can be a tedious and lengthy operation, involving much letter writing, so it is wise to start early.

*There are many complications involved, such as the so-called fair-use doctrine, which applies to certain scholarly uses of copyrighted materials, but it is difficult for individuals to obtain clear legal opinions on copyright matters. The best advice is always to request written permission; it is usually given.

SELECTED REFERENCES

Antill, P. A., "The National High-Altitude Photographic Data Base, *Technical Papers,* ACSM Fall Convention, Hollywood, Florida (1982), pp. 28–37.

Bond, B. A., "Cartographic Source Material and its Evaluation," *The Cartographic Journal* 10 (1973):54–58.

Boyle, A. R., "Scan Digitizing and Drafting," *Technical Papers,* ACSM Annual Meeting, St. Louis (1980), pp. 136–42.

Cerney, James W., "Awareness of Maps as Objects of Copyright," *The American Cartographer* 5, no. 1 (1978):45–56.

Davies, J., "Copyright and the Electronic Map," *The Cartographic Journal* 19 (1982): 135–36.

Jenks, G. F., "Lines, Computers, and Human Frailties," *Annals of the Association of American Geographers* 71 (1981):1–10.

Light, D. L., "Mass Storage Estimates for the Digital Mapping Era," *Technical Papers,* ACSM Annual Meeting, Washington, D.C. (1983), pp. 152–59.

McEwen, R. B., and H. W. Calkins, "Digital Cartography in the USGS National Mapping Division: A Comparison of Current and Future Mapping Processes," *Cartographica* 9 (1982):11–26.

Nagy, G., "Optical Scanning Digitizers, *Computer* 16, no. 5 (1983):13–24.

Peucker, T. K., "Digital Terrain Models: An Overview," *Proceedings,* Auto Carto IV, vol. 1, Reston, Va. (1979), pp. 97–107.

Peucker, T. K., and N. Chrisman, "Cartographic Data Structures," *The American Cartographer* 2, no. 1 (1975):55–69.

Pierce, R. J., "Evolution of the Defense Mapping Agency's Digital Data Specifications," *Technical Papers,* ACSM Fall Convention, Hollywood, Florida (1982), pp. 320–29.

Rogers, A., and J. A. Dawson, "Which Digitizer?" *Area* 11 (1979):69–73.

Snyder, J. P., "Efficient Transfer of Data Between Maps of Different Projections," *Technical Papers,* ACSM Annual Meeting, Washington, D.C. (1983), pp. 332–40.

17

Map Reproduction

Most maps are made to be duplicated because copies are needed to get the map to its potential users. To satisfy this need, cartographers have a long tradition of emphasizing faster, less costly, and higher quality means of map reproduction. To a large degree this has meant involving proportionately more capital (for equipment and supplies) and less labor in the duplicating process. It has also involved the development of more versatile (flexible) procedures, and expanding the degree to which the cartographer is integrated with and in control of the reproduction process.

As might be expected under these circumstances, the fundamental processes used to duplicate maps have changed dramatically over the years. Traditional *manual processes* were in large part replaced by *mechanical processes* with the development of printing in the fifteenth century. These mechanical processes were then supplemented by *photomechanical processes* in the late 1800s. In the mid-twentieth century, existing methods were further augmented by *electronic processes,* which now show signs of growing dominance.

Historically, as each new technology was introduced, the dominance of the older technologies diminished. But the shift from one technology to the next was not abrupt or complete. A long transitional phase between dominant technologies was more the rule. During these mixing and melding periods some of the old ways were salvaged intact while others were adapted or modified to be compatible with the new processes. Thus the imprint of each of the earlier technologies is more or less evident in each of the modern techniques of map reproduction. The present state of integration between manual, mechanical, photomechanical, and electronic processes merely represents the latest evolutionary stage in the continuing development of map duplication processes.

These fundamental processes, taken individually and in various combinations, form the basis of a wide variety of map duplication methods. Although the actual procedures are given different names or commercial labels, they all involve the transfer of the map image from one form (state) or material to another. Sometimes this is done physically by using inks, toners, or dyes. At other times it is done through a photochemical reaction (as when silver salts are transformed into metalic silvers). It can even be done through electrochemical reactions, as when phosphor compounds are made to glow on a cathode ray tube screen through the stimulation of an electron beam.

The various reproduction processes seem to be supporting a number of trends. One is to make obtaining fewer copies of a map more economical. Another is to eliminate intermediate steps in the preparation for the reproduction stage and thereby save time and materials. A third trend is from wet to dry processing of materials and the related movement away from materials that require darkroom processing. Finally, in large part because of the high cost of silver, there is a growing interest in materials that do not contain forms of this metal in their emulsions.

When duplicate copies of maps were made by hand, before the advent of mechanical, photomechanical, and electronic reproduction methods, the cartographer was free to prepare the map by any one of several procedures. Even today, if a map is not to be reproduced, the cartographer is under few constraints as to design or the kinds of materials that may be used. On the other hand, since most maps are made to be duplicated, several factors, such as cost and requirements for artwork, place significant restrictions on the cartographer. The means of reproduction each have special requirements. Unfortunately, maps are sometimes prepared by people who are unfamiliar with these restrictions and who give no thought to how the map is to be duplicated. In many cases it is discovered, too late, that there is no process that can adequately or economically reproduce the map.

It must be emphasized that the proper sequence in preparing artwork is first to determine the process by which it will be duplicated,* and

*Very often there is no choice. One then needs to know how to proceed in light of what the duplicating process will do.

then to plan how the artwork can best be prepared to fit that method. In the selection of a process, we consider the kinds of copies we need, how many copies are required, the quality desired, and the size. To make intelligent choices and to be able to plan for the proper artwork, a knowledge of common duplicating processes is a necessity.

In this chapter we describe several processes to enable you to choose the best one for a particular situation. An understanding of the possibilities and limitations of the processes will be needed as a background for Chapter 18, which deals with the preparation of artwork.

CLASSIFICATION OF METHODS

A source of considerable confusion for the beginner is the terminology employed in the printing and duplicating businesses. Over the years, cartographers have adopted some of this terminology to describe materials that they furnish for duplication. Because of the kinds of materials currently being prepared by cartographers, many of the terms that have been used in the past no longer seem appropriate. For example, the word "copy" can mean either text material (words), artwork (drawings), or both. It also can be interpreted to mean one sheet of material, and it also seems to suggest that the material is in positive form. Referring to a duplicate of the copy as a "copy" or to duplicates as "copies" only adds to the confusion.

We will refer here to all map materials furnished for duplication as *artwork, art, original,* or *original drawing.* Negative materials will be called *open-window negatives, peelcoats, negative scribe coat,* or *scribe coat,* and the term *negative** will be used to describe the film or other materials that can be used in a duplicating process that requires transmitted light to make copies. A *positive* is a print made from a negative, and a *film positive* is a print made on transparent or translucent film. A *mask* is a positive or negative that is used as a blocking device when compositing or screening negatives.

Each year new products are added to the growing list of materials and systems that can be used to duplicate maps. Some of the methods are simple and require little equipment; others may require a substantial investment in processing apparatus. Selection of a particular process may depend on considerations such as quality of the end product (probably the most important criterion), cost, physical nature of the copies, and availability of a given system. The cartographer should know enough about some of the more commonly used methods to be able to make intelligent choices and to prepare artwork in the most efficient manner.

It is difficult to classify reproduction methods in a satisfactory manner because many techniques require more than one process, and the intermediate processes in one may be an end in themselves in another, or they may be intermediate in several different techniques. For example, photography is a step in the printing process, but it can also be considered a separate reproduction process. Perhaps the most practical manner of classification is to group the reproduction methods on the basis of whether or not they involve a decreasing unit cost with increasing numbers of copies. Thus, we can group methods according to whether they are more appropriate for reproducing a limited number of copies or many copies.

In the past it so happened that segregation on this basis also separated the common techniques according to whether or not they require printing plates and printing ink, that is, whether they were printing or nonprinting processes. But the development of new procedures, especially those associated with electronic technology, has already blurred this distinction to the point where it is no longer meaningful. Currently, to decide which method of reproduction will produce copies most economically, we must not only consider both the size of the map and the number of copies

*A negative is something in which the dark and light tones of normal artwork are in reverse.

required, but also take into account the form of the artwork that is to be duplicated.

In the following descriptions only the widely used and generally available techniques are considered. It should, however, be pointed out that there are a number of other methods that produce excellent results in specific "requirement situations." These processes range from stencil reproduction and silk screen to gravure and collotype. These and others of the same category (not widely used in cartography) are not considered here, but the interested student can find abundant information about them in the graphic arts literature.

GRAPHIC ARTS PHOTOGRAPHY

The use of the photographic process in cartography was previously discussed in relation to environmental remote sensing (Chapter 10). The photographic process is also an integral part of map reproduction, in which case it is called *graphic arts photography*. Field and laboratory applications of the photographic process have many similarities but also some major differences. With respect to cartographic applications, it is convenient to organize further discussion of the topic under the headings of recording medium, exposure control, and material processing.

Recording Medium

Normally we consider photography to include those films and papers that are sensitized to the visual part of the electromagnetic spectrum. The recording medium consists of several layers (Fig. 17.1). In the case of film, the base material can be acetate or some other plastic material such as polyester or polystyrene. Since acetate is not dimensionally stable, it is usually unsuitable for cartographic work. Commonly the film base is sandwiched between two coatings. On the front is one (for black-and-white rendition) or more (for color rendition) layers of light-sensitive emulsion. Traditionally, this has been composed of minute crystals or grains of silver halide (a salt) suspended in a solidified gelatin matrix.* The back of the film base may be coated with an antihalation material that absorbs any light rays that penetrate the base during exposure, thereby preventing reflection back to the emulsion.** Also, this coating tends to minimize curling and compensates for distortion or dimensional changes caused by changes in the emulsion when it absorbs moisture.

FIGURE 17.1 This exaggerated cross section of black-and-white film shows the position of the sensitized emulsion and the antihalation layer relative to the film base.

Color Sensitivity. The color sensitivity of films is built in during manufacture and is indicated in film data specifications by a diagram called a *wedge spectrogram* (Fig. 17.2). The height and extent of the colored area in each spectrogram indicates the sensitivity of a particular film to the various colors. Although only the bands of blue, green, and red wavelengths are labeled in Figure 17.2, the film sensitivity to other colors can be determined. For instance yellow, an additive mixture of red and green, falls where those two colors merge. Since films of different sensitivity serve different purposes, a variety is used in map reproduction.

Blue-Sensitive Emulsions. These emulsions record high negative densities for blue areas of an original, whereas greens, yellows, and reds will appear as low negative densities. On a contact positive, the tones are reversed, and the blue areas

*Numerous emulsions in current use or under development are sensitive to radiant energy but do not contain silver halides. Although, strictly speaking, these are nonphotographic materials, they may perform the same function and have similar handling characteristics. In fact, clear distinctions are becoming more difficult each year.

**Not all films used by cartographers have antihalation backing, however.

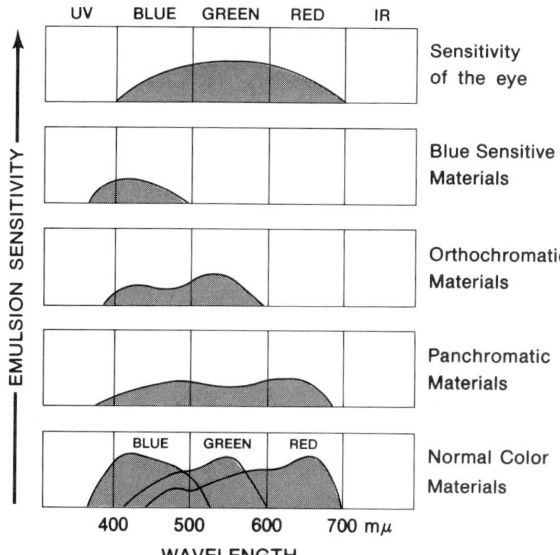

FIGURE 17.2 Wedge spectrograms in which the sensitivity of each film to the various colors is indicated by the height and extent of the shaded area. (Yellow falls where red and green merge.)

will be transparent while the greens, yellows, and reds will be opaque. Since the film is blind to most of the spectrum, a bright yellow or red safelight can be used in the darkroom without exposing (or fogging) the film.

Orthochromatic Emulsions. These emulsions are sensitive only to blue, green, and yellow, and are blind to red. This permits use of a red safelight and also causes red to register the same as black on the film. Both positive- and negative-acting emulsions are available. On standard *reversal film,* both red and black produce transparent areas (low densities) on a negative, while blue, green, and yellow cause the film to be opaque.

In contrast, *duplicating film* is positive acting. It can be used to make duplicate copies of positive artwork or film. That is, a negative can be made directly from a negative, or a positive can be made from a positive. If a sheet of this film is developed with no exposure to light, it will be completely opaque. Exposure to white light, however, removes density, so that light passing through transparent areas of a negative will cause transparent areas on the duplicating film. To produce duplicate negatives or positives that have the same orientation (emulsion-image-reading relationship) as the original, the base (nonemulsion side) of the original should be placed in contact with the emulsion of the film. With a somewhat longer exposure, the same result can be achieved by placing the emulsion of the original in contact with the base of the duplicating film.

Orthochromatic emulsions are by far the most important sensitized materials used in map reproduction. Their high contrast makes them ideally suited for processing line artwork (inked or scribed) and for photomechanical manipulation of separations, masks (positive and negative), and screens in preparation for plate-based printing (such as offset lithography). Several types of orthochromatic emulsions are available. *Regular lithographic* (lith) *film* has long been the industry standard, but *stabilization lith film* (also referred to as *direct access film*) is becoming extremely popular. In practical terms, the primary difference between these two materials lies in the way they are processed after exposure.

Regular lith film is highly sensitive to development conditions, including the temperature and strength of the developer, and the time in the chemical bath. This gives the cartographer control of product quality during both exposure and development. This can be a disadvantage as well as an advantage, of course. Stabilization film, on the other hand, is not very sensitive to development conditions and, therefore, must be controlled through exposure alone. Although this may reduce the chance of making mistakes, it also restricts the special effects that can be achieved. Thus, the choice of one material over the other may have a direct effect on the cost of production and the quality of the final product. To complicate matters even further, materials from competing manufacturers exhibit variations in speed, thickness, dimensional stability, quality, exposure/development latitude, cost, and so forth.

Panchromatic Emulsions. These emulsions are sensitive to all visible colors and render colored artwork in tones of gray. A dark area on the artwork, which reflects little light, will appear more translucent than a lighter area that reflects more light. Making a print from the negative reverses the relationship and produces a replica of the original. Because of the wide range of sensitivity, this film must be handled and processed in complete darkness. When compared with the widespread usefulness of orthochromatic emulsions, the cartographic applications of panchromatic emulsions seem rather limited. Black-and-white aerial photography is the major application. Laboratory use of panchromatic emulsions is appropriate when duplicating continuous tone imagery or artwork (such as shaded relief portrayals).

Color Film. This type of film has three separate layers of continuous-tone (panchromatic) emulsion, each formulated during manufacture to be sensitive to a particular portion of the visible spectrum (Color Fig. 19). The two kinds of color film, color reversal and color negative, differ in that the former produces a color transparency that has the same tones and values as the original scene, while the latter yields a negative in which the tones and values are reversed. The color negative can be used to produce either a positive transparency or a color print on paper.

Reversal (color) processing makes use of the residual-silver phenomenon* to form a positive image directly from the original film exposed in the camera. In this process the film is developed in black and white developer, producing a negative silver image in each of the three emulsion layers. Next the film is exposed to white light or is chemically treated to fog the silver halide that was not used during the first development. The film is then placed in the color developer, where the most recently exposed silver halide oxidizes the developer to form positive images of yellow, magenta, and cyan. Combinations of varying amounts of these dyes produce the colors of the original scene. The final step is a bleaching process in which the silver is converted to salts that are soluble in hypo. Since the dye is insoluble, it remains to record the colored image.

Negative color films incorporate the dyes in the emulsion at the time of manufacture. After development, the silver image, which is formed along with the dyed image, is removed by bleaching. The remaining dyes are complimentary to the colors and negative in relation to the tonal gradations of the original scene. To produce a color print, white light is directed through the negative onto a three-layer emulsion on paper. The dyes permit selected wave-lengths to pass and expose the appropriate layers on the paper, producing an image in the proper colors.

Contrast. Sometimes it is useful to categorize films on the basis of their *characteristic curves,* which are defined by the relation between the exposure time and resulting film density. A characteristic curve for a particular film is obtained by plotting the exposure time on a horizontal logarithmic scale and the corresponding densities produced on a vertical arithmetic scale (Fig. 17.3). The central section of the graph is of special interest. In this straight-line portion of the curve there is a constant relationship between the resulting densities and the exposure lengths. The tangent of the angle between this part of the curve and the horizontal is a measure of the steepness of the curve and is referred to by the Greek letter gamma (γ).

The characteristic curve for *low-contrast* film has a gently sloped straight line and is said to have a low gamma. Small changes in exposure cause small variations in density. Both panchromatic and color film fall into this category. For this reason, they can faithfully copy variations in tones of gray or color to reproduce continuous tone photographs or shaded drawings. Low-contrast films are not designed for copying sharp line originals, however, because they produce soft (transitional) breaks between image and non-image areas.

The characteristic curve for *high-contrast* film has a steep straight line and is said to have high gamma. Small changes in exposure cause great

*Development of a negative leaves residual silver halide that was not used to form the negative image and that has the gradation of a positive.

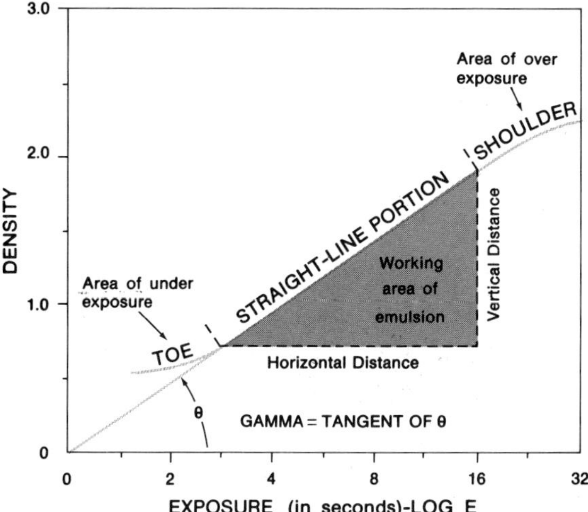

FIGURE 17.3 The straight-line portion of the characteristic curve shows a constant relationship between the resulting densities and the length of exposure. The tangent of the angle between the straight-line portion and the horizontal is a measure of the steepness of the curve and is referred to by the Greek letter gamma.

variations in density. Indeed, there are virtually no intermediate tones. The result is a clean break between the image and nonimage areas. Thus, high contrast films are not designed for copying continuous tone originals.

High-contrast film is very popular in cartographic work, not only because it can to some degree sharpen up messy originals, but also because many map reproduction techniques depend upon exposure to negatives that have areas that are either transparent or completely opaque. High-contrast films are the only ones suitable for platemaking in conjunction with plate-based printing.

Continuous Tone and Halftone. The continuous tone negative produced with panchromatic film or color film cannot be used directly with many methods of map reproduction for two reasons. First, many sensitized materials used in proofing and map reproduction work like high-contrast films. This means that they will not record tonal differences faithfully. Secondly, the plates used in conventional printing (lithography, letterpress) must be entirely divided into two kinds of surfaces, one that takes ink and one that does not (see below). If the printing plates are to be made photographically, the negative must be composed of either opaque areas or transparent areas with nothing intermediate.

The way cartographers get around this problem of reproducing continuous tone artwork is to change the color shading or gray tones into varying sized dots on the negative. The goal is to produce a negative on which the size of the clear dots relative to the opaque areas between them is directly related to variations in the darkness and lightness of tones on the original copy (Fig. 17.4 and Color Fig. 20). To understand how this is done, it is convenient to think of the original image as consisting of three main areas: *highlights* (0 to 30 percent density), midtones (30 to 70 percent density), and shadows (70 to 100 percent density). The immediate aim is to produce a negative in which the lightest highlights are represented by small clear openings in an other-

FIGURE 17.4 How a halftone looks under magnification. The illustration shows clearly that the number of dots per unit area remains uniform; only their sizes vary. Compare with any photograph in this book under a strong magnifying glass.

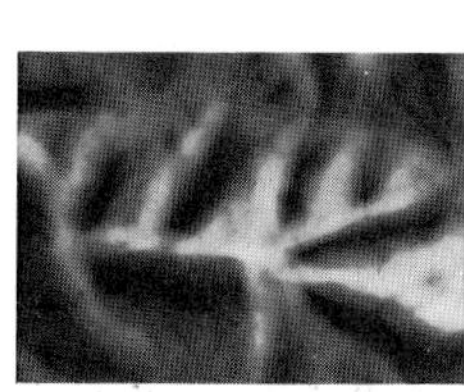

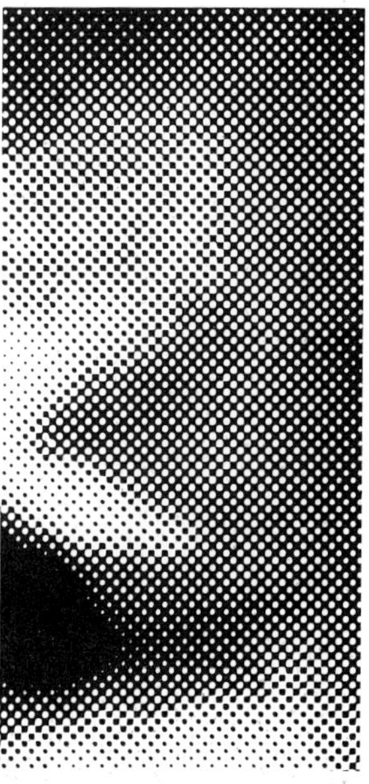

wise solid area, while the darkest shadows are represented by tiny black dots in an otherwise clear area (Fig. 17.5). Dots that just begin to touch or connect, forming a checkerboard pattern, should occur at the middle midtone values. The eventual aim is to have a printed result on which approximately 5 percent of the lightest highlight areas is covered by tiny black dots and about the same proportion of the darkest shadow areas is covered by clear dots.

Traditionally, this conversion of a continuous tone image to a reproducible dot structure has been done by using a special mechanical screen to produce what is called a *halftone negative*. Orthochromatic (high-contrast) film is required for this purpose. Both glass and contact halftone screens are used to produce the desired dot structure.

Initially, *glass halftone screens* were used. To produce a screen of this type closely spaced parallel lines are engraved on the surface of two panes of glass and filled with an opaque pigment. The two pieces of glass, with the engraved surfaces in contact, are glued together with a transparent cement. Lines on one pane are positioned perpendicular to the lines on the other, forming a network of opaque lines and tiny transparent squares. When the screen is placed between the film and the lens of the camera, light reflected from different tones on the artwork passes through the openings in the screen and produces dots on the emulsion of the film that vary in size with the amount of light transmitted. Because all the dots are small and closely spaced, they blend when viewed from normal reading distance and create the illusion of continuously changing tone.

FIGURE 17.5 A good halftone negative will exhibit a range of density from very small clear dots in the highlight areas to tiny black dots in the shadow areas. All other tones will be represented by dot sizes falling between these extremes.

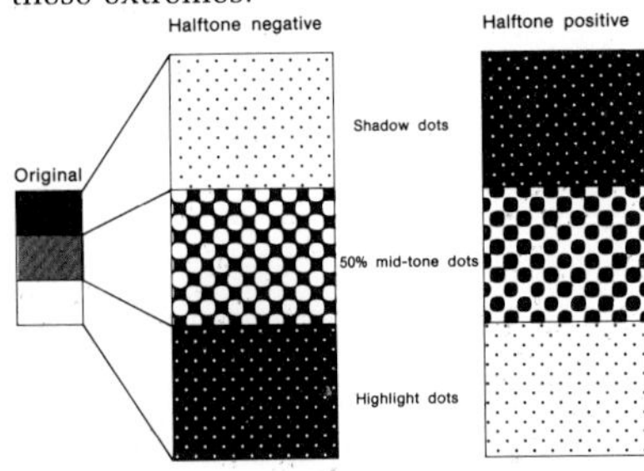

In the last few decades the *contact halftone screen* has largely replaced the glass halftone screen. Halftone screens of this second type are made on flexible film and are used in contact with the exposed emulsion of the photographic material. Thus, they are much less expensive to make and easier to use than glass halftone screens. Contact halftone screens produce the varying sized dots* through modulation of light by the optical action of a vignetted dot pattern of the screen acting on the film emulsion (Fig. 17.6). The vignetted dots are produced by a dye that is thicker near the centers of the dots, and the size of the halftone dot depends on the amount of light that is able to pass through the dye. Magenta screens are generally preferred for reproducing black-and-white originals, while gray screens are used for color reproduction.

The latest and most dramatic development in halftone production is *electronic screening*. Conventional mechanical screens are replaced by a laser scanner that is controlled by a computer algorithm that arranges clusters of microscopic points to form the individual screen dots. As the electronic record of the map image is exposed to film or sensitized plates, each digitally specified intensity is translated into an appropriate dot size and shape, according to the screen ruling specified.

The number of opaque lines per inch on halftone screens varies from less than 100 to more than 300. All other things being equal, the closer the lines, the smaller and closer together the dots will be. The closer together they are, the more

*The "dots" on these halftone screens may be round, square, or elliptical. Each of these dot structures produces slightly different effects in the final product.

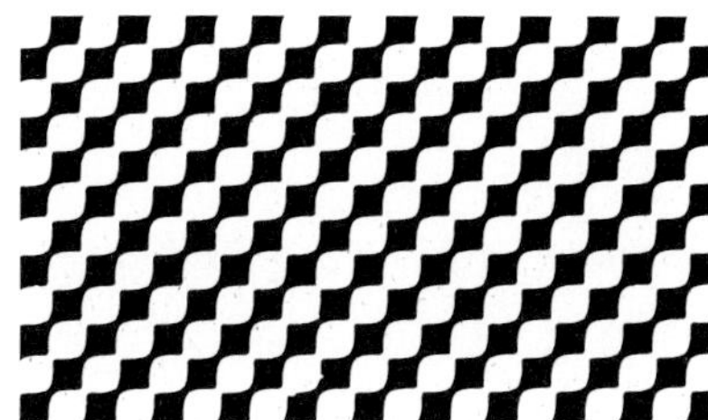

FIGURE 17.6 The vignetted dots on a contact halftone screen are produced by a dye that is thicker near the centers of the dots. The size of the halftone dot depends on the amount of light that is able to pass through the dye. (Enlarged and reproduced in halftone.)

difficult it is for the eye to see them individually, and the smoother and more natural the result will appear. Finer screens demand higher quality reproduction media, procedures, and equipment, of course.

Different visual effects can be created by printing halftones with their lines of dots oriented at different angles.* For this reason, care should be taken when orienting screens for halftone work. With monochrome reproduction of continuous tone copy, for example, the least visual disturbance seems to occur when the halftone negative is made with the screen oriented at 45° to the horizontal.

Even greater care must be exercised in positioning the screens when making halftones for process color reproduction. The reason is that the lines of dots for the different printed colors normally must have an angular separation of 30° in order to prevent a usually undesirable moiré effect (see Color Fig. 21 and Fig. 18.21). In three-color process work, the usual practice is to place one screen at 45° from the horizontal, one at 75°, and one at 105°. If a fourth screen must be used, it should be assigned to the lightest color and rotated to an angle of 90°. Thus, in four-color process, it is customary to place the black screen at 45°, the magenta screen at 75°, the yellow screen at 90°, and the cyan screen at 105° (see Color Fig. 21).

Electronic screening simplifies the problem of producing halftones with different screen angles. Desired screen angles need merely to be specified prior to activating the laser plotter. The process is a little more involved when mechanical screens are being used, however. Two approaches can be taken. Either a single screen can be rotated to the desired angle for each exposure, or a set of pre-angled screens can be used. Since a set of pre-angled glass halftone screens would be prohibitively expensive, these screens are usually manufactured in a circular form so that they can be rotated the desired angle while mounted in the camera. Since contact (vignetted) halftone screens are considerably less expensive than glass (cross-line) halftone screens, pre-angled sets of screens of rectangular shape are customarily used. However, a single contact screen may be used, usually with the aid of a special screen angle guide (template).

Exposure Control

In photography for many cartographic purposes the goal is to produce an image (either negative or positive) at the desired reproduction scale. Since the originals and their resulting copies on photographic media may each be large in size, the equipment required for controlling exposure to the emulsion must be correspondingly large. Devices must be available for handling both reflection and transmission copy, and for making precise enlargements or reductions if that is necessary. Devices must also be available for handling digital records. These needs are served by process (graphic arts) cameras, vacuum frames, platemakers, and electronic imaging systems. In order to accommodate emulsions ranging widely in sensitivity, a variety of light sources are used in conjunction with this specialized equipment.

Projection Copying. The most versatile piece of photographic equipment for the cartographer

* Screen angle is defined by the angle that the rows of dots or lines hit the edge of the screen.

is the large-format *process camera.** It is designed to copy large pieces of artwork onto equally large sheets of film and can make precision enlargements or reductions if needed. The parts of the process camera are fairly standard, although in some machines they are arranged vertically and in others horizontally. The components include a copy frame (copyboard) that holds the artwork in place and flat under pressure or vacuum, a lens (or lens system),** a vacuum-activated film or plate holder, a light source for frontlighting reflection copy or backlighting transmission copy, and a set of controls for changing scale and adjusting exposure (Fig. 17.7). Both the copyboard and the lens are movable, making it possible to vary the distance between the lens and the film and the lens and the copyholder to produce the desired enlargements or reductions while maintaining sharp focus.

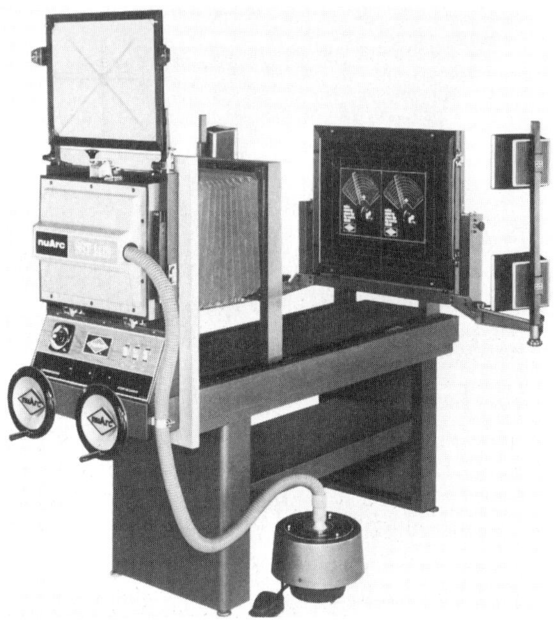

FIGURE 17.7 The basic components of a camera are a film holder, which keeps material flat, a lens, and a copyholder, which can be moved perpendicular to the plane of the film. The positions of the lens and copyholder control the amount of enlargement or reduction. (Courtesy nuArc company, Inc.)

Contact Copying. Although a camera is necessary when a change in scale is desired, one-to-one reproduction is usually best accomplished by making contact copies. They can be made on either film or paper from negatives or positives or from translucent artwork. During exposure, it is necessary that the unexposed film be in tight contact with the piece being reproduced.

Vacuum Frame. Contact photography is usually accomplished in a *vacuum frame* that is equipped with a rubber pad, a glass lid, and a pump to remove air from inside the frame. As air is exhausted, atmospheric pressure holds the pieces of film snugly together and against the glass. The length of the exposure is controlled by an accurate timing device. Light sources vary depending on exposure requirements for different emulsions. For low-intensity exposures a point light source is usually suspended over a horizontal contact frame (Fig. 17.8). Filters and a rheostat can be used to regulate wavelength and brightness characteristics.

Platemaker. When a high-intensity light source is needed, a *platemaker* is commonly used. With this device the light source and a contact frame are enclosed in a box. Commonly the light source is positioned in the bottom of the box and the photographic material is held in a downward-facing vacuum frame at the top (Fig. 17.9). Because of the short distance between the light source

*Devices known as *enlargers* are sometimes used in cartographic laboratories to make enlargements from small-format negative artwork (such as 35 mm slides). Although use of these machines has been limited by a lack of appropriate map artwork in the past, increasing involvement with remote sensing imagery and photographs of maps displayed on electronic screens may generate more interest in enlargers in the future.

**The lens normally used in a process camera is color corrected and specially designed for copying flat originals. Generally speaking, the longer the focal length, the less geometric distortion in the result.

FIGURE 17.8 A vacuum frame is commonly used for contact photography when a relatively low intensity exposure to a point light source is sufficient. (Courtesy nuArc company, Inc.)

FIGURE 17.9 A platemaker consists of a box that contains a vacuum frame and a high-intensity source of illumination. It is widely used with graphic arts materials and emulsions that can be safely handled under normal daylight conditions. (Courtesy nuArc company, Inc.)

and the photographic material, an array of lights rather than a single bulb is customarily used to give even illumination of the full copy area. The lights themselves are commonly high intensity sources rich in ultraviolet rays. Among other things, the enclosing box helps prevent potentially harmful exposure of the operator.

In *flip-top platemakers* the contact frame can be rotated around a central axis so that it can either face upward or downward. The arrangement facilitates loading and unloading the vacuum frame. It also makes it possible to gain added exposure flexibility by suspending a point light source over the platemaker. When this is done, the flip-top platemaker can be used as a standard vacuum frame when in one position (facing up), and as a platemaker when in the other position (facing down).

Electronic Imaging System. Laser or electronic cameras have been around since the mid-1970s and now seem to be coming of age. The more complete laser cameras are more appropriately characterized as electronic imaging systems, since they consist of such components as a job-planning station, data entry station, laser scanner, output laser film recorder, and a film processor (see also Photo Plotting in Chapter 18 and Electronic Video System in Chapter 16). These devices convert images to digits and then store them on discs, magnetic tape, or whatever. At that point the machines can be used to perform all manner of electronic manipulations of the image, including the capability of up to 150-line screening.

Although a small film format in laser cameras in the past severely limited cartographic applications, technical advances in equipment design have solved this problem except for the largest maps. In contrast to standard process cameras where exposure is given in seconds, input and output exposures with an electronic imaging system are stated in terms of inches or square feet per minute. Output can be in the form of negatives or positives, and multicolor images can be processed electronically as black-and-white separations.

Film Registration

When working with separation artwork (see Chapter 16), registration needs to be maintained throughout graphic arts photography. The principles and procedures are similar to those used in handling the registration of original artwork. Filmwork, particularly when it is carried out under darkroom conditions, does introduce some special problems, however. Since camera and contact registration require different procedures, they are treated separately.

Camera Work. A high-quality graphic arts camera in good condition is necessary if separation artwork is still to be in accurate registry at the film stage. Poorly maintained or inexpensive equipment may lack the precision necessary to make repeated matching exposures. Thus, unless one has access to a precision camera that can return to the same exact setting time after time, it is best to do all the photography of multiple-flap artwork in a single session. The actual procedure for maintaining registry through the camera varies with the type of registration system that is used on the original artwork.

When graphic registration marks are the only means of registration used with multiple-flap artwork, there are two choices. The flaps could be photographed one after another, with no concern for registry, and the processed film sheets later registered visually. This registration chore grows more difficult and inaccurate as the size of the material increases and becomes tedious as the number of flaps increases. Furthermore, the map reproduction process may require pin registration, which means that registration still must be transferred to prepunched strips or tabs of stable material that are attached to the film sheets. Since this *strip registration,* or *stripping,* process is also tedious, it is better avoided, if possible.

The preferred alternative to graphic mark and strip registration of film is to maintain pin registry through the camera. Since this requires pin-registered separation artwork, materials that are registered solely by graphic marks should first be converted to this form. The next step is to punch the sheets of film that will be used so that the holes match those of the original artwork.* Registry pins should then be attached (taped) to the copy board so that they will accept the artwork separations, as well as to the filmboard, so that they will accept the prepunched film sheets. It may take some juggling of the pins to get the image area to fall properly on the film, especially if you're not using a piece of film of generous size. Finally, each flap is photographed with both film and artwork positioned on its respective pins. After processing and remounting on pins, the film sheets should be in proper registry.

Contact Work. Film registration in contact photography is far less complicated than it is in camera work. Punch registration is again the preferred system. The first step is to see to it that the material containing the unexposed emulsion and the materials being used to control the exposure are identically punched and properly registered.** Appropriately spaced pins should then be taped to the bed or mat of the vacuum frame or platemaker. Finally, prepunched materials should be superimposed on the pins in the desired order and the exposure made. The result will be a sheet of exposed material that is fully registered with the other materials.

Light Sources

The terms *light source* and *light sensitive emulsion* are used rather loosely in cartography. Radiation generated by so-called light sources often extends beyond the visible spectrum into the longer infrared and shorter ultraviolet wave-

*As an alternative, punch registry may be achieved by stripping prepunched tabs or strips of stable material to the film sheets. But care must be taken to ensure that the stripped materials do not fall off during the wet film processing that is to follow. Tape containing a special adhesive is available for this purpose.

**Sheets of material that are used to control exposure in contact photography are called *masks*. They are discussed in detail in Chapter 18.

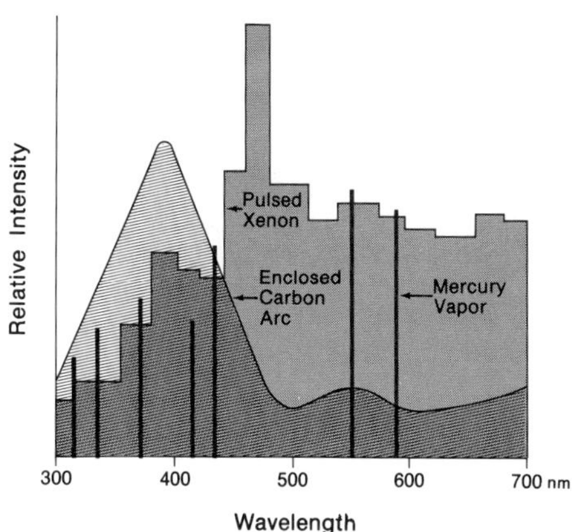

FIGURE 17.10 The different sources of illumination used in graphic arts photography exhibit very different wavelength characteristics. This makes it possible to take advantage of specialized emulsions possessing a wide range of sensitivities.

lengths. Similarly, the emulsions of many reprographic materials used routinely by cartographers are affected by infrared and ultraviolet energy as well as by visible light wavelengths. This nonliteral use of terms may be confusing, but otherwise it is of little practical consequence. What is important is for the cartographer to be able to match the particular wavelength composition of different light sources to the photographic characteristics of the original maps to be reproduced and the sensitivity of the material used in their reproduction. Two types of artificial light sources are used in cartography: flood lamps and laser beams.

Flood Sources. Two general types of flood light sources are used in cartographic applications of graphic arts photography. Their wavelength characteristics are illustrated in Figure 17.10. *Incandescent lamps* are one common type. They generally exhibit a continuous spectrum of emission and range from relatively low intensity tungsten filament lamps (including the "photoflood" type) to the high intensity tungsten-halogen (or quartz-iodine) lamp.

The second, more specialized, category of light sources includes *electric* or *vapor discharge lamps*. Carbon arc lamps (enclosed or open), mercury-vapor lamps (including the fluorescent lamp modification), and pulsed xenon arc lamps all fall in this second category. Although some of these artificial sources of illumination (carbon arc, fluorescent) exhibit a continuous spectrum of emission, others (mercury vapor and pulsed xenon arc) generate energy only at certain wavelengths or within limited wavebands of the spectrum. To some degree cartographers can tailor the energy discharge characteristics of the various light sources through use of appropriate filters.

Laser Sources. Ordinary light sources emit light in many different colors (wavelengths). In contrast, lasers produce an intense beam of light of a very pure single frequency or color. Many types of lasers have been built since the first was constructed in 1960 (Fig. 17.11). Cartographers have taken advantage of the fact that light produced by lasers is in general far more monochromatic, directional, and powerful than that generated with other light sources. Current applications include laser scanners for transforming graphic artwork

FIGURE 17.11 Lasers exhibit a wide range of power and wavelength characteristics, which permits them to be matched to the variety of sensitized graphic arts materials currently used by cartographers.

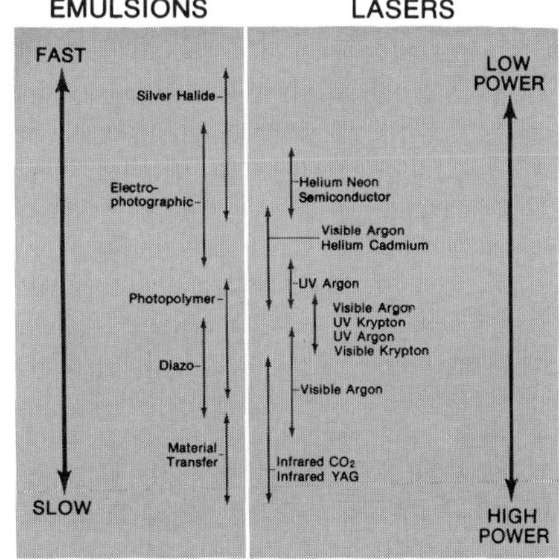

to digital form and laser recorders for transforming digital records to graphic form. Since many new laser-based products are being introduced to the marketplace each year, there is little doubt that lasers will be used extensively throughout the high-quality graphic arts industry (including cartography) during the coming decade.

Processing Photographic Materials

When photographic materials are exposed to light, a photochemical reaction takes place within the emulsion, creating an invisible *latent image*. Careful chemical processing is then required to bring out (develop) and give some degree of permanency (stability) to this image. With emulsions containing silver halide crystals, the visible image is formed by reducing the exposed silver salts to grains of metalic silver that appear black. Unexposed silver salts are removed, leaving the film transparent in nonimage areas.

Both wet and dry processes are used in handling photographic materials. In *dry processing* the necessary chemicals are built into the photographic material or a material sandwich that is subsequently separated. The process is convenient but expensive. Since the cartographer has little control over the processing, high-quality results are sometimes difficult if not impossible to obtain.

In contrast, *wet processing* normally entails a four-step procedure of developing, fixing, washing, and drying. Traditionally these steps were carried out manually with the aid of a special developing sink and chemical trays. Automated processors now perform the same chores with greater speed and consistency (Fig. 17.12). Regardless of the approach to wet processing, the function of each step is the same.

Developing consists of immersing the exposed sheet of photographic material in a temperature-controlled chemical bath. With standard *lithographic films* development time is critical and depends on the strength and temperature of the lith developer. Material that is taken out of the developer too soon will be underdeveloped (details will be fuzzy and the background will be gray rather than black). Material that is left in the developer too long will be overdeveloped (fine details will be lost). Careful development can produce an extremely high-quality image, however. Furthermore, to some degree development time can be altered to compensate for exposure errors.

FIGURE 17.12 Automated wet film processors perform the sequence of developing, fixing, washing, and drying operations with great speed (normally in 60 seconds) and consistency. (Photo courtesy Eastman Kodak Company).

With *stabilization* or *rapid access* films and chemistry, time in the developer and the strength and temperature of the chemical bath are not so critical. Material placed in the chemical bath develops up to a point and then stops. Furthermore, the chemical bath can be reused a number of times before its strength is depleted and it must be replaced. To gain this convenience, the cartographer must give up something, however: the ability to create special effects through careful control of the development time and procedure.

Once development of the photographic material is complete, the material is taken out of the developing bath and dipped into an acid solution called a *stop bath*. The function of the stop bath is to neutralize the development chemicals (which are basic), and thereby halt further change in the image.

The next step is to clear the unexposed silver salts from the nonimage area of duplicating film or the image area of reversal film. When this is completed, high-contrast film will be opaque in places and transparent in others, and high-contrast paper will be either black or white. Low-contrast materials will exhibit a full range of gray tones. The process is called *fixing,* and the solution is called *fixer.*

The fourth step is to wash the fixed material in fresh tap water to remove all traces of processing chemicals that may in time lead to a deterioration of the photographic image or material.

The final step in wet processing of photographic materials is drying. Depending upon the characteristics (hardness or softness) of the emulsion at this stage and the nature of the base material, drying can take one of several forms. The material may be squeezed, hung to dry, blot-dried, blow-dried,* or dried through some combination of these procedures. This may be done by hand, or it may be carried out automatically in a mechanical processor or a film dryer.

*Blow-drying, especially if heat is involved, may damage some materials. If in doubt, run a test on a piece of scrap material.

FIGURE 17.13 Before being stripped into position for platemaking, pinholes or unwanted words or symbols are removed by applying a water-soluble opaque or retouching fluid to the film.

Retouching and Alterations

Touching up or manipulating (making additions, deletions, or corrections to) film negatives or positives to improve the quality of the image is one of the more important steps in the cartographic technique, since some results can be obtained better at this stage than in the drafting stage. No matter what processing or modification is to take place, however, the negative or positive must be brought to perfection, so that the image is left sharp and clear (Fig. 17.13). To do this, the film is first placed on a light table where all pinholes and blemishes in the dark emulsion area are removed by covering them with a water-soluble paint called *opaque* or with a special retouching fluid.* Dark spots and blemishes in areas on the film that are supposed to be clear can be removed by scraping with a sharp blade. If the film is to be composited with a screen, opaquing and scraping should be done first and screening second because the film is more difficult to touch up if the emulsion already contains a fine pattern of lines or dots in addition to other graphic marks.

Deletions are easily made on positives by scraping and on film negatives by covering transparent areas with opaque or retouching fluid, or by covering the areas with red lithographer's tape. This is normally done on the base side of the film to prevent accidental damage to the emulsion. Additions of names or symbols are more difficult.

*Refillable, fiber-tipped retouching pens are available in several widths and fluid colors (opaque black and visually transparent ruby). The availability of fast-drying, high-density retouching fluid in pens is a great convenience. A representative brand name is *Kimoto Pake* (marketed by Kimoto USA, Inc. (East), 1116 Tower Lane, Bensenville, IL 60106).

The emulsion may be removed from the areas requiring repair or change by cutting or scraping it away from the film base. A new piece of emulsion or emulsion-like material containing the artwork to be added may then be "stripped" into place on the negative. Alternatively, artwork can sometimes be added by "engraving" it into the negative emulsion. The emulsion is quite brittle, however, and tends to chip off, often resulting in ragged lines. Finally, type and symbols that have been laid out on a transparent base material can sometimes be stuck up as a unit in clear areas on the film.

METHODS FOR A FEW COPIES

Limited-copy methods of reproduction are normally used when a relatively small number of copies is needed, since the unit cost does not decrease rapidly with increased numbers of copies. The cost of producing a second or third copy is as much or nearly as much as producing the first. Limited-copy methods also tend to be most economical when the size of the map to be duplicated is large.

Limited-copy methods include some materials that the cartographer can process with rather inexpensive equipment. Other methods require certain kinds of graphic arts equipment such as process cameras, platemakers, and darkroom facilities. Ordinarily these processes are not used to produce great numbers of copies, and often only one color is possible. With some materials, however, color can be added to a copy for display, for teaching, or for research. Maps for which there is only an occasional need for a copy or maps that must be revised often can be reproduced most economically by these limited-copy methods.

In recent years, a number of new sensitized materials have been developed that can be used under ordinary lighting conditions. Some of these systems are invaluable for the cartographic process. Intermediate duplicating steps in the production of a map and the proofing process can be accomplished with a minimum of equipment. Use of the new materials and a trend toward having more duplicating apparatus in most cartographic establishments permits the mapmaker to be much more involved in the process of preparing the materials to produce a printing plate. It is not uncommon for the cartographer to furnish the printer with materials that have been proofed and are ready for platemaking (see Lithography, below).

The widely used limited-copy techniques are: diazo (for example, Ozalid), xerography (for example, Xerox), photography, various others that fall in the category of proofing methods, and a group of computer display devices (printers, plotters, display screens). This listing is unsystematic in several respects, since there are many similarities and differences among these techniques that would enable one to group them quite differently. On the other hand, these categories are usually used in "the trade," so it is convenient not to deviate.

The organization of the remainder of this section is based on a twofold logic. On one hand, it is possible to argue that one may approach the task of duplication with a map in hand asking what options are available for reproduction and what are the relative advantages and disadvantages of each method. The idea is to get to the point where it is possible to say, "If I had this, then I could do the following."

On the other hand, we have already seen that it is important for the cartographer to produce maps so that they can be most efficiently and effectively duplicated. Understanding what can be done with different forms of artwork is an important consideration for the cartographer prior to launching into map production. For this reason map reproduction is introduced in this chapter before discussing map production in Chapter 18.

When a map is in hand, or when deciding how best to produce a map, it is useful to organize your thinking about map reproduction around the topics of black-and-white reflection (opaque) originals, color reflection artwork, negative and positive film separations, and digital records. These general topics form the headings for the following discussion. Under each heading the more commonly used techniques will be treated in some detail.

Opaque Black-and-White Original

Often it is desirable to reproduce a map that exists only in the form of reflection or opaque black-and-white artwork. Maps of this sort could have been produced manually, either by ink drafting or shading, or by sticking up preprinted point and line symbols, type, tints, and patterns. On the other hand the map may be a black-and-white photograph or electronic scanner image generated through environmental remote sensing, or it could be a photograph of a video terminal. More and more frequently black-and-white maps are the end product of electronic mapping. In this case it may be a temporary image on a display screen, or it could be available in the form of hard-copy output from a character (line) printer, dot matrix printer, electrostatic printer/plotter, laser plotter, ink jet printer, or a line plotter. Finally it is quite possible that the opaque map image represents a composite print that was created from separation negatives or positives.

Various methods permit the direct reproduction of these forms of opaque black-and-white artwork, including xerography, diffusion transfer, and instant photography. Each method permits reduction of the original. Opaque black-and-white maps are being duplicated with ever-greater frequency for projection rather than for direct viewing. In part this is done to overcome size limitations, but it also helps to get around the cost of providing each map user with a personal copy of a map. For projection purposes film positives are made in the form of slides or transparencies. Upon projection, of course, actual viewing size is generally much enlarged from that of the original artwork.

Xerography. Reproduction of black-and-white originals can be achieved in a dry photomechanical process called electrophotography or xerography (for example, Xerox). Rather than using chemical reactions or solutions as is common among other photographic processes, xerography is an electrostatic process that involves photoconductivity and surface electrification.

Xerography begins by giving a special metal plate a positive charge, which upon exposure to the original through a lens, disappears in those nonimage areas subjected to light. The remaining positively charged surface constitutes a latent image, which is "developed" with a negatively charged black powder toner that adheres to the image area (recall that opposite electrical charges attract). Paper, or a similar material, which has been given a positive charge, is placed against the plate. It receives the image by transfer of toner and it is fixed onto the material with heat.

Xerography can provide excellent copies at a reasonable cost. Because the xerographic process does not involve chemicals, time of immersion in solutions and subsequent drying are saved, and copies can be provided in just a few seconds. The image can be placed on a variety of papers, plastics, or other materials (cloth, metal, wood), since the transfer is physical. Since the original is projected to the plate through a lens, scale changes are possible.

Diffusion Transfer. Black-and-white originals can be duplicated directly by using a photomechanical process called *diffusion transfer.** A diffusion transfer print is made by first exposing a sheet of light-sensitive carrier material (which serves as a negative) to the original in a process camera. Both line and continuous tone materials are available. The carrier sheet is then sandwiched with a sheet of positive print material (paper or film), and processed by running it through a machine containing a reusable activating solution (Fig. 17.14). A few seconds after the lam-

*Diffusion transfer materials are marketed under several trade names, including COPYPROOF (AGFA-GEVAERT, Inc., Graphic Systems Division, 150 Hopper Ave., Waldwick, NJ 07463) and PMT (Eastman Kodak Co., Rochester, NY 14650).

FIGURE 17.14 Diffusion transfer materials are processed through an activator bath and then sandwiched together between a set of rollers in a simple desk-top machine. (Courtesy AGFA-GEVAERT, Inc.)

inated materials emerge from the processor rollers, the transfer of the image from the carrier to the print is complete. The two sheets of material can then be separated, leaving a positive (film or paper) print. Halftones can be created by using a special contact screen during exposure.

For a variety of reasons cartographers are showing a growing interest in the diffusion transfer process. The results may not be quite as good as those obtained with photographic prints, but the quality of diffusion transfer prints is still high enough for it to be used as a basis for subsequent line and halftone printing operations. Diffusion transfer is relatively inexpensive, especially for reproducing continuous tone copy (for example, aerial photos), since there is no need for a film negative (the carrier with its latent image is discarded) and the developer can be reused many times before it is exhausted. The process is also fast, since in addition to eliminating a separate photomechanical step, there is no need for fixation or drying steps. The possibility of making scale changes is another attractive feature, particularly for purposes of compilation or pasteup reformatting. Still another useful trait of the diffusion transfer process is that the line quality of original artwork that lacks sharpness and density can actually be improved to some extent in direct contrast to the normal image degradation that occurs with most photographic processes.

Instant Photography. Instant photography is another means of map reproduction, both for duplicating hard-copy originals and for copying the image displayed on video screens.* The popularity of the technique can be traced to its convenience. A finished, dry, positive print or 35 mm slide is produced within a few minutes of exposure, and no bulky or messy tanks, chemical solutions, or water baths are required.

Instant photography has some limitations, however. Exposure is more critical than it is with conventional photography, since there is no way to control development to compensate for exposure problems. The film or paper for instant photography is also rather expensive. Furthermore, image size for the most part is limited to 35 mm slides or page size (8 by 10 inch) or smaller prints by the availability of film and special cameras. Yet, for many applications, the convenience and relatively high quality of instant photography more than compensate for these limitations.

*Standard video screens can be photographed with relatively inexpensive computer graphics cameras, but the quality of reproduction is usually rather poor. This problem is largely overcome by using special computer graphic film recorders that incorporate a separate flat screen monitor for displaying the map image. Even with these more expensive film recorders the quality of the image is still limited by the screen resolution, however.

Opaque Color Original

Sometimes it is desirable to reproduce a map that exists only in the form of reflection or opaque colored artwork. Maps of this sort could have been produced manually, either by hand coloring with ink or paint, or by sticking up preprinted flat colors. They may represent color photography or scanner images generated through environmental remote sensing. Color proofs of separation artwork made by photomechanical or xerographic methods also fall into this category of opaque colored artwork. And, finally, the map may be the end product of electronic map production, consisting either of an ephemeral display screen image or of hard-copy output from an incremental plotter, laser or ink jet printer, or a computer graphics camera. In any case, the problem is producing a suitable reproduction of the opaque color original. There are several solutions, each of which is relatively expensive when compared with reproducing opaque black-and-white originals.

Methods that permit the direct reproduction of opaque colored artwork for direct viewing include color xerography, instant color photography, and color diffusion transfer. Due to the high cost of equipment and materials, reproduction size using these methods is usually rather limited. Page size or smaller is most common. Relatively small maps can be duplicated at scale; otherwise they must either be reduced or copied in sections, which later can be stripped together.

With increasing frequency, opaque colored maps are being reproduced for projection rather than for direct viewing. This is a way to get around part of the high cost of duplicating large colored maps for direct viewing. For projection purposes film positives are made in the form of slides or transparencies. When these maps are projected, of course, actual viewing size is generally much enlarged from that of the original.

Xerography. Whereas only a single plate and toner are required for monochrome xerography (see above), full-color xerographic reproduction of opaque artwork involves the use of separation plates. The standard procedure is to use a different plate for each of the subtractive primaries (cyan, magenta, yellow). Color separation is accomplished through a lens and filter (red, green, blue) arrangement. Line artwork and continuous tone imagery can both be copied.

Color xerography is a little slower and more expensive than monochrome xerography, but its price is very competitive when compared to other procedures for duplicating color originals. Reproduction color fidelity has been a serious problem in the past, but seems to be steadily improving. The lack of a black plate (and toner) to give density and sharp definition to the image remains a problem, however. Yet, on balance, the advantages of xerographic reproduction for many cartographic purposes now seem to outweigh the disadvantages. Thus, we can expect that use of this method of color reproduction in cartography will increase substantially in the near future.

Diffusion Transfer. High-quality color duplication can also be accomplished using recently developed diffusion transfer materials. The process is identical to black-and-white diffusion transfer reproduction, except that a special color print material is employed. A camera is used, which makes scale changes possible. No screens or other exposure devices are needed. Although these color materials are still too new to be properly evaluated, their high quality and relative inexpensiveness give them great promise for cartographic applications. They could become particularly important in duplicating remote sensing imagery where control of color balance by the cartographer is advantageous.

Diffusion transfer color print materials from different manufacturers, such as COPYCOLOR* and EKTAFLEX PCT,** have similar handling

*A product of AGFA-GEVAERT, Inc., Graphic Systems Division, 150 Hopper Ave., Waldwick, NJ 07463.

**A product of Eastman Kodak Company, Rochester, NY 14650.

characteristics. The prints are as simple to make as black-and-white prints and can be produced from various forms of reflection copy. The first step in making a print is to expose a sheet of light-sensitive color diffusion transfer film to the original in a process camera. The exposed transfer film is then run through a chemical activator bath in a special table-top processor, which also laminates the exposed film with a special sheet of paper or film. During lamination, the color image transfers from the transfer film to the film or paper. After a few minutes, the film/paper or film/film sandwich can be peeled apart, yielding a color print that air-dries in several minutes. There is no need for exact temperature control, chemical mixing, print washing, or other treatment.

Instant Photography. Instant photographic reproduction of colored artwork does not differ significantly from the same method for duplicating black-and-white artwork (see above), except that the film or paper used has several emulsion layers instead of one. Specifically, instant color film has three layers of colored dyes, three light-sensitive layers, and filters that assure that each emulsion layer reacts to just one of the primary colors of light. An appropriate color dye will appear wherever a layer of light-sensitive emulsion is exposed and developed.

As with instant black-and-white film or paper, instant color film or paper has a few drawbacks. For instance, it is fussier than color negative film about correct exposure. Since its development process is fixed and beyond the control of the cartographer, errors made in exposure cannot be corrected in the darkroom as they might with conventional film. Instant color film is also expensive. Yet, in spite of these problems, instant color film is growing in popularity as a means of reproducing opaque colored maps.

Transmission Original

Some cartographic artwork is produced directly in the form of transmission copy, for example, when artwork is constructed on a translucent or transparent drafting medium or is created by negative scribing and electronic film plotting or recording. Reflection artwork can also be readily converted to film negatives or positives through photographic processing (see above). Regardless of how the artwork came to be in transmission form, further processing of the image involves light transmission through the material rather than light reflection off the surface as is the case when processing opaque copy. The transmission artwork may be monochrome (black-and-white) or multiple color, and the artwork may either consist solely of linework or may be continuous tone (at least in part).

Limited-copy reproduction methods suitable for handling transmission artwork take on many different forms, depending on the nature of the image. For the most part all of these methods produce extremely high-quality results. For convenience, we can classify the various methods into those that duplicate the original with a single sensitized coating (diazo, photographic, or electrostatic), and those that may use multiple sensitized coatings to build up an image progressively in different colors. Many of these techniques fall into the category of *map proofing systems* (see Proofing in Chapter 18).

Single Sensitized Layer. Diazo, xerographic, and photographic processes are useful for reproducing maps in limited numbers from film negatives or positives. Of the three methods, photographic techniques are overwhelmingly dominant. A wide variety of printmaking papers that are coated with a standard photographic emulsion can be used in map reproduction, either in contact or through projection. Some are restricted to the production of black-and-white duplicates, while others provide full-color capability. Prints can be made on any of a variety of special films that provide good drafting surfaces and are dimensionally stable. Of the many materials available, three that exhibit special properties are singled out for further discussion.

Diazo. The diazo process (for example, Ozalid) relies on the reaction of diazo compounds with other substances to produce colored dyes. Although purple (blue) dye is standard, the process colors as well as a wide variety of vivid hues are also available. Papers, films, and cloth have all been sensitized with diazo emulsions. Reproductions are made by exposing these materials to an intense (mainly ultraviolet) light while in contact with a positive opaque image (map) on transparent or translucent material. Light passing through the nonimage areas of the map decomposes the sensitized emulsion, while the opaque materials on the artwork protect the image areas by blocking the light. When the exposed material is developed in ammonia fumes, the unexposed diazo forms a colored dye.

It is not possible when making direct contact diazo prints to enlarge or reduce. Consequently, the artwork must be designed for "same-size" reproduction. Since the process depends on the translucence of the drawing surface, painting out imperfections with white is not possible because the paint is as opaque as the ink and would appear as a dark spot in the print. Creases on tracing paper or plastic and heavy erasures that affect the translucence are also frequently visible on the print. Prints are commonly obtained by feeding the original drawing into a machine against a roll of the diazo-sensitized paper; the exposure and dry developing take place rapidly within the machine. The drawing is returned and may be immediately reinserted for another exposure and copy. The image is not absolutely permanent; it fades especially rapidly when exposed to sunlight and is not acceptable in circumstances where permanence is required.

The ultraviolet light source of diazo equipment furnishes a light of constant intensity. Exposure time is controlled by adjusting the speed with which the sensitized material passes through the machine. Original drawings on material of greater opacity require longer exposures in order to decompose the nonimage area so that it will appear colorless. Unless the drawing material is quite translucent and the image area is opaque, the resulting print is likely to suffer from lack of contrast.

Shading films, type stickup, and similar products on drawings produce best results in the direct contact process when they are placed on the back of the drawing to prevent "shadowing." To use stickup type in this fashion, the adhesive would have to be on the upper surface. An additional hazard with the standard forms of these materials is present in the *Ozalid* diazo process because the drawing must be fed around a roller, and such materials tend to curl off the paper if it is heated and rolled. New types of preprinted sheets and stickup lettering with heat-resisting adhesive are available to obviate this difficulty.

Another way of circumventing this problem is to convert the drawing into a direct reading film positive. If the positive is prepared so the emulsion is on the bottom when it reads correctly, it can be fed onto the drum of an Ozalid machine with good results. It is not absolutely necessary to have the emulsion arranged as suggested, but the result will be more satisfactory.

Ordinary diazo paper has a poor drafting surface, so this process is not much used for obtaining copies of base maps on which other information is to be added. However, a great variety of sensitized diazo materials such as tracing cloth, cloth-backed paper, and various types of drafting films that do have adequate drafting surfaces are available.

Color diazo films* are ideal when used singly for overhead projection or as map overlays. Additionally, by compositing several sheets of different colored material a multiple overlay sandwich can be formed (see Overlay Systems, below). This can serve as an inexpensive color proof of multiflap (separation) artwork. Dimensionally

*Representative brand names include Technifax Diazochrome (James River Graphics, Inc., South Hadley, MA 01075) and Chromatic Diazo Film (GAF Corporation, 140 West 51st St., New York, NY 10020).

stable polyester film should be used, of course, when accurate registration is important.

Kodagraph. Kodagraph Wash-Off Contact Film, produced by Kodak,* provides the cartographer who lacks darkroom facilities with a capability for contacting film positives or negatives. Only an ultraviolet source and a pressure or vacuum frame plus a flat tray for the developer are necessary. After a very short exposure, the film is developed for 1 minute and then rinsed under warm water until the unwanted emulsion has been washed away. An added advantage is that the material has a good drafting surface enabling additions and corrections.

The film also has the ability to accept bichromated emulsions, so colors can easily be added to a film positive of a base map (see Color Emulsion Solutions below).

Dylux. Dylux 608 Registration Master Film** is available on paper or dimensionally stable plastic. It produces a blue line image immediately upon exposure to ultraviolet light through a negative or positive. The image requires no chemical developing and self-fixes under normal room lighting conditions. A pale yellow background disappears within an hour, but the process can be speeded by exposing the film to pulsed xenon or carbon arc lamps that are filtered with an ultraviolet blocking material.

The light source for the initial exposure must be exclusively ultraviolet to produce the maximum contrast. Best results are obtained with black light fluorescent lamps that have suitable phosphors, yet screen out nearly all visible light. The amount of density of the image is directly related to the length of exposure. Separation negatives can be given different exposures to produce images in different tones. Screens and masks can be used as in normal film compositing.

Since ultraviolet light forms an image and visible light deactivates the emulsion, positive images can be made from positive film. With the film in contact with the positive, an exposure is made with a bright light, which removes the sensitivity in the nonimage areas. After removing the positive, the sheet is exposed to ultraviolet to produce the image.

Multiple Sensitized Layers. Film separations can also be reproduced in positive, directly viewable form with a variety of materials that use separate emulsions for each color to be copied. With one such class of materials the image is constructed by superimposing one layer of emulsion after another directly upon an opaque base material. A second class of materials permits the overlaying of transparent sheets containing the separate emulsions.

Superimposition Systems. The various superimposition systems for map reproduction from film separations are very different in their operation. Contact xerography, for example, is an electrostatic process involving no photographic emulsion in the conventional sense. In contrast, a variety of photosensitive emulsions are available in liquid form, which can be wiped on base materials prior to exposure. Finally, images can be built up by laminating thin emulsion layers onto a base sheet one after another. Unfortunately, each of these superimposition systems suffers from several drawbacks. One is that when using several colors, problems with any one of the emulsion layers can easily ruin the quality of previous work and the overall result. Secondly, success in making one copy of a map through superimposition methods in no way guarantees that subsequent copies will be of comparable quality.

Standard projection xerography (see above) can be used to reproduce color transparencies as well as reflection copy. But a new system of *contact*

* Eastman Kodak Company, Rochester, NY 14650.

**A product of E.I. Dupont de Nemours and Co., Wilmington, DE 19898. The material is available in sheets up to 30 by 40 inches, and in roll stock 40 inches by 100 feet. Shelf life appears to be indefinite.

xerography, called KIMOFAX,* shows even greater promise as a means for short-run, multiple-color map reproduction. With this transfer-type system a transparent positive separation is required for each color desired (eight color toners are available). A single plate is reused by cleaning and reexposing between each successive color application. Contact originals up to 61 by 85 cm can be accommodated with present equipment. Since a wide variety of reproduction materials can be used, the process is relatively inexpensive. Yet, the resulting composite artwork is of extremely high quality (see Color Fig. 22).

Colored emulsion solutions that can be applied to special plastic materials (including scribecoat) are available. A variety of colors are marketed in each of several product lines.** Additionally, by mixing proper amounts of the primary colors, any desired hue can be created, although in practice this is a rather tedious, trial-and-error process.

The smoothest application of color and therefore the best results are obtained with a mechanical whirler. Unfortunately, few small mapping establishments have access to such a device. With sufficient practice, however, the sensitizer can be applied by hand evenly over a fairly wide area. The three-step procedure involves first taping the plastic material to be coated to a smooth, flat surface, next pouring the correct amount of the color solution onto the center of the sheet and, finally, using a small block applicator covered with a soft pad to spread the color evenly in alternating series of horizontal and vertical strokes (Fig. 17.15). The trick is to spread just the right amount of emulsion over a clean, dry surface with soft, even strokes until the emulsion is dry.

Exposure is made in contact with negative artwork with light in the ultraviolet end of the spectrum. Developing the image consists of rinsing the exposed material under a water tap and then drying thoroughly. Once the material is completely dry, another color can be added over the first and the process repeated. As many colors as desired can be applied in this manner, and if well done, the result can be comparable to a map that has been printed in several colors by separate press runs.

FIGURE 17.15 Wipe-on emulsions must be spread evenly over the surface if good results are to be achieved. To accomplish this by hand, a block applicator covered with a soft pad is used in alternating series of horizontal and vertical strokes. (Courtesy Direct Reproduction Corporation).

Great care must be taken to ensure high-quality results, however. Not only must the coating be uniform, but such environmental factors as dust, temperature, and humidity must also be rigidly controlled. For these reasons, wipe-on emulsions tend to be more suitable for generating a rough, quick preview copy of a map, or for transferring a guide image from one surface to another (as in preparation for scribing), than for repro-

*This electrostatic color reproduction system is marketed by Kimoto USA, Inc. (East), 1116 Tower Lane, Bensenville, IL 60106.

**Brite-Line, Camden Products Company, Box 2233 Gardner Station, St. Louis, MO; Kwik-Proof and Watercote, Direct Reproduction Corporation, 835 Union St., Brooklyn, NY 11215.

FIGURE 17.16 Lamination systems for color map reproduction normally require the use of a special mechanical laminator if consistent, high-quality results are to be achieved. (Courtesy 3M Company).

ducing maps from film separations in any numbers.

Color map reproduction of potentially high quality in a wide range of colors can be produced on a single sheet of material through use of several *lamination processes.* By employing successive laminations of different colors, full-color maps can be created. The two systems that are discussed here involve quite different procedures.

With the *3M Transfer Key System** a clear liner material that has been factory coated with a color emulsion is pressure laminated to a special opaque base material (Transfer-Base). Although the lamination can be done with a hand roller, the process is plagued with problems. For consistent, high-quality results a special mechanical laminator should be employed (Fig. 17.16). In either case, once the lamination is completed, the clear liner is peeled away, leaving the color emulsion adhered to the Transfer-Base. Exposure is then made in a contact frame with a light source rich in ultraviolet. With negative-acting Transfer Key, development is carried out by hand with 3M Brand Color Key/Transfer Key developer. The positive-acting material combines low polluting aqueous development with rapid machine processing and yields a high resolution product. Once the material has thoroughly dried, additional colors can be added by repeating the process with other Transfer Key films of the chosen hues.

The *Cromalin Color Display System** is a dry proofing process. The Cromalin material consists of a tacky photopolymer layer (light sensitive) sandwiched between a protective polypropylene sheet and a mylar cover sheet. During machine lamination, the lower protection sheet is discarded under heat and pressure, and the colorless sensitized layer is made to adhere to an opaque base sheet. After exposure to a light source high in ultraviolet, the top protection cover is stripped away, leaving the clear emulsion on the base unprotected. Finely powdered color toner gently applied at this time clings to the sticky emulsion in the image area. After all loose toner has been cleaned from the surface with a soft cloth, the base can again be run through the laminator to prepare it for another color. The process is repeated until all colors are in place. A protective covering can then be added by exposing a final lamination without any masking material in place.

Both negative- and positive-acting Cromalin Color Display films are available. Unfortunately,

*3M Transfer Key is a product of the Printing Products Division, Minnesota Mining and Manufacturing Company, 3M Center, St. Paul, MN 55101. The material is available in sheet stock up to 25 inches by 38 inches and 24 inches by 100 feet rolls in a limited color selection. Shelf life seems to be several years.

*Cromalin is a product of E.I. Dupont de Nemours and Company, Wilmington, DE 19898. The material is available in widths up to 50 inches in 300-foot roll stock. Shelf life appears to be very short (3 to 6 months).

laminators suitable for one type of emulsion are not suitable for the other. Negative-acting material is generally more convenient because it does not require prior compositing of flaps to produce film positives. Both negative- and positive-acting film suffer from rather short shelf life and high cost, although when properly used, color representations of extremely high quality can be attained.

Overlay Systems. A variety of transparent overlay materials containing colored emulsion coatings have useful cartographic applications. The clear sheets or foils of film are factory-coated with presensitized colored emulsions in the standard hues, including the regular process colors (yellow, cyan, and magenta). The materials can be used under daylight conditions, and the only equipment needed is a pressure or vacuum frame* and a source of light that is rich in high-intensity ultraviolet.

Overlay methods share several features that are of special significance to cartographers. One advantage is that each color flap can be inspected independently from the others. A second desirable feature is that the factory-applied emulsion coatings assure precise color fidelity and even color density, as well as permanent pigment colors. To achieve results that most closely resemble the appearance of final printed maps, the sandwich of foils should be superimposed on a white base, but even then the order with which the foils are overlaid will influence the overall color effects. If the overlay sandwich is viewed over a light table or projected onto a screen, still different color effects will occur. Indeed, to some degree these variable color renditions under different viewing conditions may represent a disadvantage of overlay systems. In addition to the colored diazo materials that have already been discussed (see above), several other product lines warrant special mention.

*Enco NAPS (Negative Acting Proofing System)** uses a material that is exposed emulsion to emulsion in contact with a negative, with the colored emulsion facing the source of illumination. A special developer is then spread gently over the exposed emulsion using a piece of cotton and a light sweeping motion. After the background emulsion is removed (dissolved), rinsing under a water tap and blow or blot drying completes the operation. The product is a positive print in the chosen color.

A series of positives in different colors, of separate flaps of a map, can be prepared and then overlaid to form a full color proof of the map. To aid this process the Enco NAPS film can easily be punched for pin registration. Tint screens and halftones are faithfully reproduced, and colors combine in the same manner as on a printing press.

*3M Color-Key Proofing Film*** is a negative-acting material that is similar to NAPS in handling and results. One difference, however, is that the 3M product is exposed with the color emulsion facing away from the source of illumination. Another difference is that Color-Key, unlike Enco NAPS, cannot easily be punched, so to obtain pin registration, prepunched strips or tabs of some other material normally are taped to the 3M product. A special feature of Color-Key is that the orange emulsion is actinically opaque and therefore can be used as you would employ film negatives in subsequent processing of materials.

Another special feature of 3M Color-Key is that sheets of material with a positive-acting emulsion are also available. After this material is exposed in contact with a film positive, the image is de-

*Color diazo materials (see above) are an exception, since they also require special machine processing.

*NAPS is a product of Azoplate, Division of American Hoechst Corporation, 558 Central Ave., Murray Hill, NJ 07974. The material is available in sheet sizes to 25 inches by 38 inches, and has a shelf life of several years if properly stored.

**3M Color-Key Proofing Film is a product of Minnesota Mining and Manufacturing Company, 3M Center, St. Paul, MN 55101. The material is available in sheet sizes to 25 inches by 38 inches and has a shelf life of several years if properly stored.

veloped normally. A bleaching solution is then applied to complete the process.

*General Color-Guide** is a negative-acting material that is similar to NAPS in that the film carrying the colored emulsion can easily be punched and is similar to 3M Color Key in that exposure is made with the emulsion facing away from the light source. It differs from both in the way the image is developed, however. With General Color Guide the development procedure involves first immersing the exposed material in a special salt bath to fix the latent image. Next, instead of rubbing away the unexposed emulsion by hand as with the other overlay materials, the Color Guide emulsion is removed by rinsing with free running warm water. Flushing briefly with cold water then sets the remaining emulsion, although it remains extremely soft and easily scratched until thoroughly dried. The material is thicker than Color-Key and has a slightly gray cast; therefore, color is not as good. There is no shelf-life limitation with General Color Guide, and if sheets are accidently or improperly exposed, they may be returned undeveloped to a light-tight box and reused after 72 hours.

Digital Original

With rapidly growing frequency the information required to construct and reproduce a map image is stored in computer-readable form. The actual production of maps for direct viewing from these digital records can take several forms. One group of graphic output devices generates *hard-copy* maps on printers or plotters. Another class of machines produces *soft-copy* (ephemeral) images on graphic display terminals (screens). Each class of output device plays a special but different role in the realm of map reproduction. In doing so, each also involves special considerations for the cartographer.

Hard Copy. We have come to expect that map reproduction techniques will produce a tangible, physical product. Understandably, then, a variety of computer output devices fall in this category. These include raster devices such as character printers (like ordinary computer line printers) and a variety of dot printers (matrix, electrostatic, laser, ink jet). Monochrome (usually black-and-white) artwork can be generated on all of these machines, although electrostatic, laser and ink jet technology also provide a multiple-color capability. Another class of hard-copy output devices is made up of vector plotters, which are capable of constructing linework with pen and ink in a variety of colors.

As we have seen in previous discussion, all these forms of computer output graphics can be reproduced for direct viewing using various non-computer methods. It is also possible, of course, to use the production device itself as a reproduction machine. Thus, by repeating the printing or plotting step as many times as desired, multiple copies of a map can be generated from the digital record in control of mapping software.

Since each additional copy of a map produced on a computer graphics device will cost almost as much as the previous one, this approach is generally appropriate only when a few copies of a map are needed. If very many duplicates are required, it will often make better economic sense to use the computer output artwork as a master for xerographic, diffusion transfer, or some other form of reproduction, particularly for artwork generated on vector plotters because plotting time can be significant and increases in direct proportion to the amount of detail in the image.

Soft Copy. In spite of the long tradition of reproducing maps in tangible form, the popularity of electronic map displays is growing dramatically and now adds a new dimension to the way we think about map reproduction. Map users are finding that in some situations they are able to get the information that they desire from a temporary map display. By so doing, they can avoid many of the problems and some of the expense

*General Color-Guide is a product of General Photo Products Company, 10 Patterson Ave., Newton, NJ 07860. The material is available in sheet sizes to 28 inches by 50 inches and in roll stock to 100 feet by 42 inches.

normally associated with obtaining, storing, and retrieving individual map sheets. What this means is that a map need only to be produced once and then sent out on demand to dozens, even thousands, of viewers who have linked themselves together in a telecommunications network. In a sense this is the electronic version of reproducing maps in the form of slides or transparencies, except that the map image is sent out to the individuals rather than bringing them together for a group showing.

Map reproduction by means of electronic display could provide the viewer with several extra benefits. It might be possible for the viewer to center the map on any chosen geographical location and display artwork within a surrounding "window." This would help to overcome frustrating problems that occur when the location of interest falls at the edge of a conventional map sheet. The viewer could also zoom in on a portion of a map in order to get a blown-up picture of fine image detail. Possibly best of all, it would be easier for the custodian agency to keep the master artwork current, since reproduction costs could be minimized. This should mean that when the viewer called up the map image on a display screen, it would be more up-to-date than we have come to expect conventional sheet maps to be.

Growing interest in the ability to reproduce maps upon demand on a display screen is doing more than merely changing the way we go about duplicating cartographic artwork. It is also forcing us to rethink what we mean by the term "map reproduction," because if, in the future, potential map users are provided with easy-to-use mapping software and ready access to digital records (cartographic and statistical data bases), they could produce their own unique cartographic portrayals directly. Rather than spending their time producing maps themselves, the professional cartographers' role might then at least in part shift to keeping the digital records current, to working on improving map design software, and to raising cartographic literacy to the point where potential users could take full advantage of their newfound mapping capability. Since under these conditions there may not be a need to produce the artwork for even a master map, except possibly for preview or data editing purposes, there might also be far less need for map reproduction in the traditional sense.

METHODS FOR MANY COPIES

Often large numbers of copies of a map are needed. In this case reproduction considerations usually differ somewhat from those associated with duplication methods involving only a few copies. In particular, since initial preparation costs can be spread over many more copies of the map, they can be quite high compared to those acceptable when using limited-copy methods, especially if low-cost reproduction materials and highly automated processing techniques could be employed. The result would be low-cost duplication at a unit cost that decreases with the number of copies run.

Given these factors, it might be expected that methods appropriate for reproducing many copies of a map may not be appropriate for very small runs. This is indeed the case. Modern high-speed printing presses produce the first thousand copies of a map almost as cheaply as the first copy. In fact, as a rule, many-copy methods become more and more economical as the number of copies required grows larger.

Generally the most economical way to reproduce a map in large numbers is from a special *plate* using one of the mechanical printing processes. Three distinct approaches have been taken in creating these plates (Fig. 17.17). The earliest mechanical printing was done using a process known as *relief printing* or *letterpress,* in which the paper receives the ink directly from surfaces standing in relief. The initial method of forming raised features on the plate by mechanically cutting away the nonprinting areas of a smooth block of wood was in time replaced by photographic image transfer and chemical etching of plastic or metal plates.

A second development in mechanical repro-

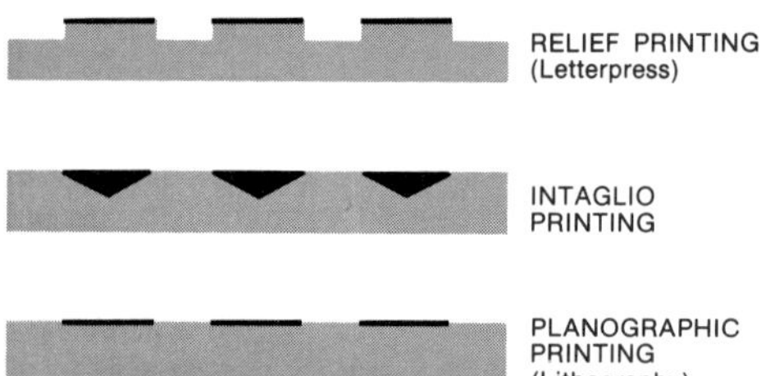

FIGURE 17.17 The basic processes of letterpress, intaglio, and planographic printing operate in different ways to produce a surface from which ink may be transferred to paper.

duction, *intaglio printing,* also involved ink and an uneven surface. Grooves were cut in a flat metal plate, usually copper, and then filled with ink. The surface of the plate was then cleaned off and the plate, with its ink-filled grooves, was squeezed against a dampened sheet of paper. The paper "took hold" of the ink and, when removed from the metal plate, the pattern of grooves appeared as ink lines. In a sense, this process of printing from an intaglio surface is just the opposite of letterpress printing, since the inking area is "down" instead of "up."

A third important development in mechanical printing occurred in 1798 with the invention of *lithography*. This new printing process was based on the incompatibility of grease and water. In contrast with printing from relief or an intaglio surface, the surface of the lithographic printing plate was a plane, with no significant difference in elevation between the inked and noninked areas. Although the early lithographic plates were made by drawing with a greasy ink or crayon directly on the smooth surface of a particular kind of limestone (which gave the process its name), this bulky medium has long since been replaced by thin metal, plastic, or paper plates.

All three of these plate-based printing techniques are based on an all-or-nothing inking principle. In other words, ink is either deposited in full amount (density) or not deposited at all. The image area is printed in solid black (or some other color), and the nonimage area receives no ink. This ink/no ink dichotomy causes no problem in reproducing simple line artwork. But the reproduction of continuous tone artwork, such as an aerial photograph or a shaded relief map, is another story. Truly effective printing of this type of artwork requires special processing of the image prior to printing. Specifically some means for decomposing the tonal variations in the image into fine, discrete marks that, when printed, give the illusion of continuous tone. During the 1880s the mechanical halftone screen was devised for this purpose (see above). Although these screens are still widely used, the same effect can now be achieved electronically under the control of a digital record of the artwork. Either a film recorder can be used to produce a halftone negative, or a laser platemaker can be used to place the halftone image directly on the printing plate.

Lithography

In recent years modern offset photo-lithography (hereafter referred to simply as lithography) has largely displaced other plate-based methods in the printing industry. With respect to cartography, lithography probably now accounts for over 99 percent of the maps that are printed from plates. Although other plate-based methods are of historical interest, they will not be discussed in the remainder of this chapter.

The overwhelming dominance of lithography can be attributed to the economical, high-quality results that can be achieved. Colors and fine lines can be reproduced with great fidelity. In addition, lithography is particularly well suited for printing large maps. For convenience, single and multiple color lithography are considered separately in the following discussion. If the cartographer can determine in advance the eventual printing strategy and the grade of paper to be used, the map can be designed effectively and efficiently for that particular combination.

Single Color. The lithographic reproduction of maps in a single color (usually black) takes several steps. The printing process includes the basic operations of processing the original artwork (if necessary), exposing and developing (processing) the plate, and presswork.

Processing the original may take several forms. Basic operations may involve changing scale, reversing artwork, and conversions from analog to digital form, or vice versa. The map image can be transferred to a lithographic plate directly in serial fashion from a digital record through the use of laser and electron beam plotters, or the transfer of the image can be indirect through simultaneous exposure to a flood-light source in conjunction with a negative or positive mask.

In cases where a digital record is available and exposure to the plate is to be made directly, processing the original consists of deciding upon an appropriate set of machine instructions. The digital record can also be used to create masks. When a microfilm recorder is used for this purpose, the result is called computer output microfilm (COM). Due to the restricted film format (size) of microfilm, subsequent photographic processing will normally be required to attain a suitable printing scale. This problem can be avoided if a large format laser film plotter is used; in this case additional preparation for printing is minimal.

Similarly, if negative artwork is provided in the form of scribed linework and symbols and open-window peelcoats (masks), processing of the artwork by the printer is virtually eliminated. Since negative artwork is generally made at the desired printing scale, enlargement or reduction is rarely necessary.

If, on the other hand, artwork is provided to the printer in positive form, it generally has to be transformed into a negative prior to printing. The most modern approach is to use an electronic scanner to convert the original (analog copy) to a digital record, and then process the digital data with a computer. The more conventional procedure is to photograph the original with a standard graphic arts camera and then process the film negative by hand. In this case relatively slow film is used in order to give the photographer greater control over the quality of the negative. The resulting negative must be composed of either opaque areas or transparent areas, and nothing intermediate. Grays are not permitted on the negative, as they are in conventional (for example, 35 mm) photography.

This means, of course, that if the printer is provided with continuous-tone copy, such as relief shading or a photographic image, it must be converted to a halftone negative prior to printing.* This processing step involves the use of a graphic arts camera in conjunction with a glass or contact halftone screen (see above). With high-quality lithographic plates and presswork, halftone screens of up to 300 lines per inch can be reproduced on coated stock, although the use of 130- to 170-line screens is most common.

Multiple Color. The lithographic reproduction of maps in multiple colors does not differ fundamentally from black-and-white lithography except that a variety of colored inks are used. Each separate ink requires a separate printing plate, of course, and thus also requires a complete duplication of the steps in the whole lithographic process. Thus, generally, the cost of reproducing multiple-color maps by lithography is proportionately greater than that of single-color (usually black) reproduction. There are two basically different procedures for multiple-color lithography. One is called process color printing and the other flat color printing.

Process Color. Process color printing is the method used to reproduce natural colored photographic imagery and artwork that has been prepared using color media in all hues on a single drawing. This method is based on the fact that almost all color combinations can be obtained by varying mixtures of the subtractive primaries (magenta, yellow, cyan) and black (color Fig. 3). Inks in these colors are called *process inks.*

---

*Experiments are being conducted with a variety of lithographic plates that employ the natural grain of the plate surface to produce a random dot pattern. A resolution equal to 600 lines per inch has been achieved with these screenless lithographic plates. This compares with a maximum of 300 lpi for the best-quality halftone prints. The added level of detail this increase in resolution makes possible is dramatic. But further development work is needed before screenless lithography is an operational procedure in map reproduction. Short press runs and lack of consistency are two problems being encountered.

If artwork is prepared by hand, two flaps are commonly made. A black flap usually contains features such as the border, lettering, graticule, and special annotations (scale, direction indicator, and the like). On a second flap all color work is done by painting, airbrush, preprinted color tints, and similar methods. If the original map is an aerial photograph, the colors are rendered directly by the dyes in the separate emulsion layers. A second (black) flap is not so common with photos, but one may be made to enhance certain features or provide special annotations.

The varying colors on the original traditionally were color-separated by using a camera equipped with green, red, and blue filters. Four exposures are made on separate pieces of film. By using the proper filters and by rotating a halftone screen to the proper angle for each exposure, the resulting separation negatives record the three component additive primary colors (blue, green, red) of the original as black-and-white densities.

The modern method of process color separation involves the use of an electronic scanner. With these devices, light reflected from or transmitted through the image is picked up by photocells covered with red, green, and blue separation filters. These photocells generate three electrical signals proportional to the intensity of the light they have received. These signals in turn are converted to digits and entered into a computer-compatible storage medium, from which the information can be retrieved and processed in ways that mimic the full range of manual photomechanical operations (including screening). The output signal from the computer is then used to control a glow lamp or laser beam focused on a sheet of photographic film (or a printing plate). Some scanners produce one separation at a time, while others can produce the four separations simultaneously.

Understanding how process color reproduction works involves a quick review of color theory (see Chapter 8). Begin with the fact that a filter transmits light of its color and absorbs the light rays of most other colors. Thus, when white light strikes the *blue filter,* the blue wavelengths are allowed to pass through to expose the film, and the other wavelengths are blocked. The resulting negative is least dense in the green and red (yellow light) areas. The positive printing plate made from this negative will print most ink in the green and red areas and, therefore, it is inked with yellow. Similarly, the *green filter* transmits green light and blocks blue and red (magenta light). Since most ink will be deposited in the blue and red areas, magenta ink is used. Finally, the *red filter* passes red light and absorbs green and blue (cyan light). The resulting plate deposits cyan ink in the areas of green and blue.

The three (positive) printing plates are halftones of varying amounts of the appropriate subtractive primaries. When these plates are printed together in *three-color process,* the halftone dots and transparent inks perform their subtractive assignments, merge, and recreate the full color range of the original photo or drawing. Yellow ink mixed with magenta results in reds, yellow combined with cyan forms greens, magenta combined with cyan produces blues, and so forth.

Unfortunately, three-color process does not produce a true black. To gain the extra definition provided by a black ink a fourth printing plate is needed. The procedure is then called *four-color process printing*. The black plate includes the line drawing (if any) and the halftone that is used as a toner to increase the shadow densities and improve the overall contrast. The halftone negative for the black plate can be made by using three partial exposures with each of the three color filters.

Flat Color. Flat color printing is the method most often used for reproducing line maps. Artwork for flat color reproduction is prepared in black and white and requires at least one separate flap for each colored printing ink. Of course, many combinations of line and halftone effects are possible (see Chapter 18).

Flat color reproduction takes two different forms.

The modern approach is to separate the artwork so that a different color can be created for each map category or feature using the four process inks in combinations and tone values. The tone values are achieved by use of appropriate tint screens (see Chapter 18).

Because the whole gamut of colors can be created with the four process inks, the procedure is sometimes called fake process color printing.* Most printers have color charts available that illustrate a wide selection of these color combinations (Color Fig. 9). This flat color printing technique is especially recommended if more than a half dozen or so areal tints are needed, as might be the case in reproducing a soils or land use map. The reason is that a total of four press runs suffices regardless of the number of distinct colors needed on the map. But since all colors other than the subtractive primaries are created by superimposition of some combination of the four process inks, extremely precise registration is required for high-quality color reproduction of fine linework or type. This degree of registration is usually not attainable or practical unless bi-angle screens and a modern four-color press is used. For this reason the use of process inks to copy a map containing fine lines or type may not be the best choice.

The traditional approach to flat color reproduction and one that still is widely used today is to color-separate the artwork so that each category or feature can be printed with a special preblended ink of a *conventional color.* A wide selection of printing colors, each of which is uniquely identified by descriptive numbers, is available for this purpose. In 1963 these colors were standardized in a coordinated matching system for printing inks, called PANTONE®** Matching System. PANTONE colors are now a printing industry standard.*

The PANTONE Matching System comprises over 500 standard PANTONE colors, all of which were produced by blending eight basic colors, plus black and transparent white. For the convenience of the user, swatches of the PANTONE colors are printed with transparent inks on coated and uncoated paper and displayed in the handy PANTONE Color Formula Guide. The PANTONE color identification name or number and the complete blending formula for the color are provided with each color swatch. Since the inks and paper stock used in producing the PANTONE Color Formula Guide are representative of those found throughout the commercial printing industry, the PANTONE colors can be readily reproduced in map printing.

Since a conventional color is printed directly (by a preblended ink of that color), the technique is especially well suited for color reproduction of fine lines and type (Table 17.1). Use of conventional colors is also convenient when a limited number of discrete areal tints are to be printed. But if the desired number of conventional colors grows much beyond four, it might be better to use process inks. Indeed, there is a practical limit on how many conventional colors can be printed economically. This is determined in large part by the fact that each additional ink requires another printing plate and another run of the map through the press. Extra time and materials are involved, and both cost money.

*The term "fake" process color printing is used because it simulates the full-color effects of process color.

**Pantone, Inc.'s check-standard trademark for color reproduction and color reproduction materials.

*The Defense Mapping Agency (DMA), which is the largest map producer in the United States, uses an in-house system of *Standard Printing Colors* (SPC), which serves the same function for the government that the PANTONE colors do for the commercial printing industry. In the SPC system each conventional color is uniquely identified by a five-digit number that is composed of the significant digits of dominant wavelength, purity, and luminous reflectance measures in the CIE system (see Chapter 8). Eight basic inks, plus black and transparent white, provide the colors from which all of the other inks are blended (Color Fig. 23).

TABLE 17.1 Ink Colors for Features on U.S. Government Maps

USGS $7\frac{1}{2}$-minute Quadrangles (Topographic)		U.S. DMA Aeronautical Charts	
Conventional Ink	Feature	Process Ink	Feature
Brown	Contour lines	Cyan	Heavy type
			Military Grid
Green	Forest tint		Aeronautical annotations
			Rivers
Blue	Water bodies		Forest symbol
	Rivers	Magenta	Aeronautical information
	Grid ticks		Roads
Red	Roads		High elevation tints
	Urban areas		Heavy type
	USPLS grid		City tint
			Boundary tint
Magenta	Photo revisions	Yellow	Roads
Black	Type		Elevation tints
	Boundaries		Boundary tint
	Railroads		
	Buildings	Black	Relief shading
	Feature outlines		Railroads
	Point symbols		Roads
	Roads		Boundaries
	Grid ticks		Grid
			Fine type
			Contour lines
			Point symbols

There is no need to rely exclusively on either conventional inks or process inks, of course. To some degree the limitations of each approach can be overcome by combining the two methods. Thus, it is not uncommon to find the four process inks plus one or two conventional inks used on the same map. In this situation the cartographer can choose from the full range of colors made possible by the few process inks while still having the advantage of using a preblended color or two for fine line features or type. The use of modern five-color presses encourages this sort of hybrid color use in mapping.

Platemaking. Lithographic plates must, of course, be divided into image and nonimage areas. The image areas hold the greasy ink, while the nonimage areas attract water and repel the ink. In modern offset lithography (see Presswork, below) the image on the plate must be *right reading*.

Lithographic plates may or may not be presensitized. Plates that have no sensitized coatings are known as *direct-image plates* or *masters*. The base material of the plate is often paper, but it can be plasticized paper, acetate, aluminum, or aluminum foil. Images can be placed on the plate by any of the following methods.

1. By use of a typewriter with a special ribbon or drawn by hand with pencil, ink, lithographic crayon, or ball-point pen. These materials must be slightly greasy to produce images that attract printing ink.
2. By printing using lithography.

3. By sensitizing plates with wipe-on solutions such as diazo. When coated, they are handled like presensitized plates and exposed through a negative or with a laser beam.

Until quite recently most lithographic plates were "manufactured" in the printing shop. A sheet of metal, usually zinc, was first treated so it would accept and hold a thin film of water. Next it was coated with bichromated albumin and dried. After exposure through a negative, a developing ink was applied that produced the image. The unhardened, nonimage area was then washed away, and the plate was ready for the press.

Today the most popular surface plate has a diazo emulsion applied by the manufacturer. These presensitized plates are surface coated with a light-sensitive emulsion during manufacture. Most sensitizers are of the nonphotographic variety, that is, the emulsions are slow and can be handled under normal room lighting conditions or under bright yellow lights.* To prepare this plate for presswork, it is only necessary to expose it to an ultraviolet light source, through a negative or with a scanning laser beam, and then process it for a few minutes to develop the image.

On a surface plate, the image tends to wear away with long press runs. Dots in a halftone or in a screened area become ragged and smaller. To overcome these problems several forms of printing were developed in which multilevel plates are used. These methods, involving deep-etch or bimetal plates, have the advantage of more faithful reproduction and much larger press runs (as many as several million impressions).

Regardless of the type of presensitized lithographic plate that is used, overexposing the emulsion does not affect its printing qualities. Therefore, a single plate can be successively exposed to several negatives. The process, called *double burning* or *double printing,* permits the printer to composite two or more separate flaps on one plate. For example, line and halftone negatives can easily be combined. Similarly, line negatives can be screened by interposing the mechanical screen between the negative and the sensitized plate. This accomplishes the same end result on the printing plate as drafting or sticking up tints or patterns on the original artwork, or making composite negatives from open-window negatives and negative screens. However, better results are obtained with extra-fine patterns of lines or dots by applying them at this last stage, after photography, than by making them "stand up" through extra steps of the photographic process, particularly if precision graphic arts photography is beyond reach of the cartographer.

A difficulty with double burning of plates is that registry marks are not visible on the plate until it has been exposed separately to each negative and the composite image has been developed. Thus, to make certain that each negative is registered properly, a mechanical system, usually incorporating the use of registry pins, is used (see Chapter 16).

After proper exposures have been made, the plate is developed, placed on the press, and is ready to print. At this stage, most corrections on the plate are impractical if not impossible and are generally limited to deletions. If errors are found on the relatively inexpensive plate, it is usually discarded, the map artwork revised, a new negative made, and a new printing plate prepared.

Presswork. Most lithographic presses incorporate an *offset arrangement* whereby the impres-

*Some plates are coated with silver halides and can be used in a camera to produce plates directly from positive artwork. Modern "instant printing," for example, uses inexpensive plastic plates that can be exposed in cameras specially equipped with a high-intensity light source. With the very high-contrast emulsion used on these plates there is an upper limit of eighty-five lines per inch in reproducing patterns. The general results are also slightly fuzzier than those achieved with metal plates. However, with properly designed artwork and careful platemaking and printing, the quality of copies can be quite good. The real attraction of the system are its ability to reduce artwork at the platemaking stage, its speed (intermediate steps are eliminated), and its extremely low cost. For these reasons black-and-white maps of rather simple design, such as those included in reports and similar documents, are commonly duplicated through instant printing.

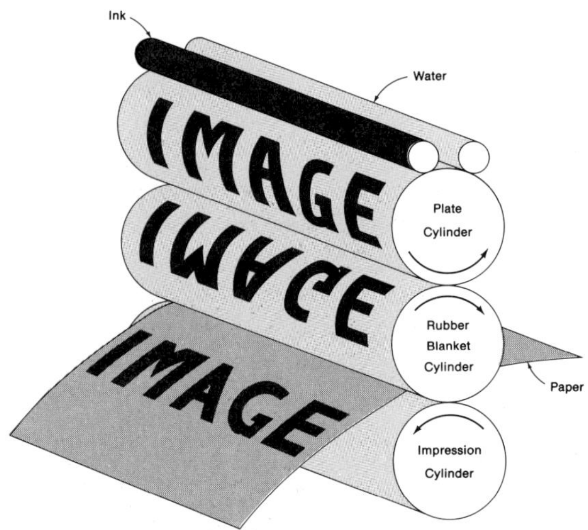

FIGURE 17.18 In offset lithography the image is passed from the plate to a rubber blanket and from the blanket to the paper. Thus, the plate must be right-reading.

sion from the plate is transferred to another cylinder covered with a rubber mat, called a *blanket,* and the image is then transferred from the blanket to the paper (Fig. 17.18). Actually, the offset arrangement is so standard when using a lithographic plate that the terms "offset" and "lithography" have become almost synonymous. The soft rubber blanket that deposits the ink image on the paper in the offset process does not crush the paper fibers, and fine lines and dots can be faithfully reproduced, even on soft, absorbent paper. Of course, better results are usually achieved with finished or coated paper stock.

No advantage would be gained by attempting to describe various types of printing presses in detail. But there are some fundamental differences that are worth mentioning. One way to classify lithographic presses is according to how many printing units they contain and, therefore, how many plates can be mounted simultaneously. Older presses contained one unit which restricted them to printing a single ink in a press run. A press of this type is quite appropriate for reproducing maps in a single color, for example, black. But the machine is handicapped if it is necessary to print another ink on the same map. To do this the press has to be cleaned, re-inked, a new plate mounted, and the process repeated.

Modern presses, in contrast, may contain up to a half dozen printing units arranged in tandem and therefore can print as many colors in sequence in a single press run. As might be expected, these multiple-unit presses are especially suited to reproducing multiple-color maps. Two- and four-color machines seem to be most common.

SELECTED REFERENCES

Bach, D. N., "Non-Press Proofing Methods," *Proceedings of the American Congress on Surveying and Mapping,* Fall Convention, Lake Buena Vista (1973), pp. 32–41.

Cuff, D. J., and M. T. Mattson, *Thematic Maps: Their Design and Production.* New York: Methuen, 1982.

Defense Mapping Agency, *Standard Printing Color Catalog for Mapping, Charting, Geodetic Data and Related Products,* March 1972.

Defense Mapping Agency, *Standard Printing Color Catalog (Process) for Mapping, Charting, Geodetic Data and Related Products,* June 1975.

Defense Mapping Agency, *Standard Printing Color Identification System,* ACIC Technical Report No. 72-2. February 1972.

E. I. Du Pont de Nemours and Company, Inc., *The Contact Screen Story.* Wilmington, Del.: E. I. Du Pont de Nemours and Company, 1972.

Eastman Kodak Company, *Halftone Methods for the Graphic Arts.* Rochester, N.Y.: Eastman Kodak Co., 1982.

Eastman Kodak Company, *Basic Photography for the Graphic Arts.* Rochester, N.Y.: Eastman Kodak Co., 1982.

Graves, F. W., and D. L. Des Rivieres, "Cartographic Applications of the Diffusion Transfer Process," *The American Cartographer* 6 (1979): 107–15.

International Paper Company, *Pocket Pal: A Graphic Arts Production Handbook,* 12th ed. New York: International Paper Co., 1980.

Keates, J. S., *Cartographic Design and Production*. London: Longman Group Ltd., 1973.

Kidwell, R., et al., "Experiments in Lithography from Remote Sensor Imagery," *Technical Papers,* ACSM Annual Meeting, Washington, D.C. (1983), pp. 384–93.

Kuhn, L., and R. A. Meyers, "Ink-Jet Printing," *Scientific American* 240 (April 1979):162–78.

Maxwell, W. C., *Printmaking: A Beginning Handbook*. Englewood Cliffs, N.J.: Prentice-Hall, Inc., 1977.

Mertle, J. S., and Gordon L. Monsen, *Photomechanics and Printing*. Chicago: Mertle Publishing Company, 1957.

Moore, Lionel C., *Cartographic Scribing Materials, Instruments and Techniques,* 2d ed., Technical Publication No. 3. Washington, D.C.: American Congress on Surveying and Mapping, Cartography Division, 1968.

Ovington, J. J., "An Outline of Map Reproduction," *Cartography* 4 (1962):150–55.

Poush, B., and M. Magee, "Major Development Paves Way for Multicolor Field Maps," *ACSM Bulletin,* May 1980, pp. 17–18.

Rich, J., "Want Special Color Effects? Fake It!," *Graphic Arts Monthly,* June 1982, pp. 113–16.

Rosenthal, R. L., "Digital Screening and Halftone Techniques for Raster Processing," *Technical Papers,* ACSM Annual Meeting, St. Louis (1980) pp. 126–35.

Schlemmer, Richard M., *Handbook of Advertising Art Production,* Englewood Cliffs, N.J.: Prentice-Hall, Inc., 1966.

Stefanovic, P., "Digital Screening Techniques," *ITC Journal,* Special Cartography Issue (1982-2): 139–44.

Urbach, J. C., T. S. Fisli, and G. K. Starkweather, "Laser Scanning for Electronic Printing," *Proceedings of the Institute of Electrical and Electronic Engineering* 70 (1982):597–618.

18

Map Production

MAP CONSTRUCTION

Whether maps are printed from plates or are duplicated by some other method, the quality of the copies is closely related to the nature of the original artwork. With the possible exception of image reduction by photographic or electronic means, duplicating processes cannot improve much the appearance of lines, type, or other map symbols. Flaws or irregularities in the artwork will usually be apparent on copies. The preparation of proper artwork involves choosing the most efficient and economical procedures and using the materials and instruments best suited for a particular job.

The duplicating industries have made great strides during the past few years in perfecting new products and techniques. Some of these have solved problems with which cartographers have lived for years. New sensitized materials are available for intermediate steps in the cartographic process that permit more efficient map construction and facilitate the preparation of trial runs or proofing. Rapid advances have also been made in computer-assisted map construction. Automated graphics devices can be used to preview design alternatives quickly and cheaply in an interactive environment prior to actual map production. In many respects, however, the only significant difference is that computer-driven plotting and printing devices are performing tasks previously done by hand. Thus, for the most part, the principles and processes discussed in this chapter are applicable regardless of whether the map construction is by people or machines.

The cartographer is no longer forced to rely on the printer to process artwork. Dimensionally stable materials are available that can be presented to the printer ready for platemaking. If desired, all the artwork necessary to produce multicolor maps can be produced by the cartographer without assistance from the printer. Printers may be avoided altogether by constructing maps on computer-driven graphics devices.

In order to make use of new techniques and plan work properly, the cartographer must become acquainted with the new materials and processes. No one procedure is the best for all maps, and the production plan should be tailored for each situation. In a book of this size we cannot cover all the information that is available in the form of trade manuals, books, and materials and equipment catalogs. This presentation is built on the background that was developed in the chapters on design, color, compilation, typography, and reproduction. With the exercise of some imagination, you should be able to devise workable plans for the construction of well-designed maps that can be reproduced efficiently.

In the discussion of positive artwork, a variety of materials and equipment are mentioned along with information about possible uses and limitations. The list is not exhaustive but, on the other hand, those mentioned have been used and tested. The same holds true for the subsequent section, which deals with negative artwork. When reading both these parts, keep in mind that there is a good deal of overlap, since negative and positive materials are often used in combination for one map. Finally, the section on complex artwork and proofing should provide a basis for devising particular plans for solving complex construction and reproduction problems.

Map Artwork

Maps can be prepared in positive or negative form or in combinations of both. Preparing positive artwork involves constructing the point, line, and area symbols, along with the lettering, on a relatively white or translucent surface. Execution is done in ink or with a variety of preprinted materials. During the reproduction process, negatives of the artwork may be prepared for use in making a special printing plate. The copies produced by a variety of reproduction methods are replicas of the original map.

Negative materials are prepared in a different manner. The symbols are produced either by cutting or scribing them into a coating that is actinically opaque,* or by exposing photographic material through symbol templates or with a directed light beam (such as a laser). These prepared negatives can be used in the same manner as a conventional film negative in duplicating processes.

Before beginning the map, the cartographer should decide which form of artwork would be most suitable and select the one that will produce adequate copies at an economical cost. The cartographer must consider things such as the type of reproduction to be used, size of the copies, amount of detail, kinds of symbols, registration procedures, amount of reduction, access to specialized equipment, and the desired quality of the finished product.

Whether artwork is prepared in negative or positive form or a combination of the two, it often consists of more than one sheet of material, called a *flap* (see Fig. 16.3). If a map is to be printed in one color, only one sheet of material may be necessary. More than one color or certain special effects such as areas of smooth gray tones usually require separate flaps. These are then combined at some stage in the production or reproduction process to produce the map on a single sheet.

POSITIVE ARTWORK

Methods for constructing positive artwork vary considerably. Traditionally, manual drafting was used. This technique has recently been adapted for use with automated plotted devices. Computer technology has introduced several additional methods for producing positive artwork as well. Printers, for example, are now widely used in map production. They are of two basic types: character printers that work on principles made familiar through the standard typewriter, and electronic machines (dot matrix, electrostatic, ink jet, and xerographic) that build up an image by printing large numbers of tiny dots. In addition to these hard-copy printers, maps are now routinely being produced on video display screens, some of which can be electronically duplicated on photographic materials. Each of these methods warrants special consideration.

Before discussing these different methods of producing positive artwork, however, it should be pointed out that such artwork is constructed with one of several goals in mind. For instance, the artwork may be prepared on translucent material at the proposed reproduction size, with the idea that a negative (or separation negatives) can subsequently be made by contact methods in a vacuum frame. Normally these negatives would then be photomechanically manipulated to produce a composite print, or they would be used to make a printing plate. Alternatively, the artwork may be prepared at the desired final size on an opaque (usually white) medium. This high-contrast artwork will normally be suitable for direct viewing. By constructing positive artwork at scale using either approach, the cartographer gains the advantage of being able to monitor image complexity and relationships among different symbols readily. But there is also the potential disadvantage that any irregularities in constructing or registering the artwork may also be apparent in the finished product.

If, on the other hand, negatives are to be produced from positive artwork, or if it eventually is to be viewed at a different size, a process camera is usually required. Actually the reduction that use of a camera permits can be put to real advantage. By preparing artwork somewhat larger than the proposed reproduction size, symbols are easier to construct and more detail can be conveniently placed on the map. Furthermore, reduction tends to sharpen the image, make slight irregularities less apparent, and improve registry.

When preparing artwork that is to be reduced, there is a tendency, however, to add so much detail or such fine detail that it is either lost or difficult to read at the reduced size. It is not easy

*The negative need not be visually opaque, but opaque only to the wavelengths of light that affect light-sensitive emulsions.

to anticipate how certain artwork will appear with reduction of much over 50 percent.* For this reason artwork is normally prepared at no more than twice reproduction size (200 percent) and construction at 125 to 150 percent is most common.

Positive artwork displayed on a video screen clearly does not fall neatly into either of these categories because the resulting image does not have a physical quality. These images are intended for direct viewing, not as something that would be carried around by a map user. If a hard copy of a displayed map is required, it can be produced on one of the automated plotting or printing devices that were mentioned in the previous chapter.

Drafting

Many of the advantages of cartographic drafting can be traced to the high degree of flexibility it provides in artwork preparation. The ease and naturalness of working in positive is a major factor. Additionally, the availability of different materials, pens, and inks make it possible to achieve a wide range of useful effects. The options of using prepared materials and mechanical guide devices provide still further drafting alternatives. Drafting technology has proven as adaptable to computer-age plotting devices as to traditional manual execution. In order to put the power of cartographic drafting to best use, the potential contribution made by each of these factors toward mapping flexibility needs to be understood. The individual topics will, therefore, be more closely inspected in the following sections.

*Artwork intended for great reduction has to be obviously exaggerated at construction (up scale or size) if it is to look good after being reduced (down scale). Experience helps when it is necessary to make the map several or more times larger. But, even better, a sample area should be constructed according to the proposed design. A reduction of this sample will indicate whether or not the proposed design is acceptable. This can be done quickly and inexpensively using photomechanical transfer (PMT) materials.

Drawing and Printing Surfaces. The sheet materials to be used in the construction and reproduction of the map may sometimes be selected on the basis of a particular attribute. More often, however, they are chosen on the basis of some overall qualities. Economy is sometimes a consideration, but the cost of materials is usually an insignificant part of the total cost of producing a map. The following brief listing includes the more important qualities for cartographic use.

1. *Dimensional stability.* This refers to the degree of shrinkage or expansion in a material with changes in temperature and humidity. Probably the most stable materials used in cartography today are flexible polyester-base films. Even these materials will undergo a relatively small change in dimension as temperature and humidity vary. Temperature and humidity increases or decreases are additive in their effect on dimensional change. An increase in one and a decrease in the other, however, will have a canceling effect.
2. *Ink adherence.* A good drafting surface not only "holds on to" the ink well, but it accepts enough ink to provide an opaque image. Some special inks have been developed for special surfaces; but most of them tend to clog drawing pens. The material that will accept standard drafting and printing inks is usually most desirable.
3. *Translucence.* This refers to the ease with which it is possible to see through the material, and the ability of the material to transmit light for the purpose of exposing sensitized materials. Translucence is of special concern in cartographic drafting. Not only is a considerable amount of tracing usually done, but also much drafting for reproduction is done on separate flaps. Obviously it is desirable to be able to see what is where on other flaps when they are all in register. Translucent material makes it possible to produce contact negatives and enables the cartographer to make proofs or produce other

materials for the cartographic process without the use of a camera.

4. *Erasing quality*. The ability to remove ink without damaging the drafting surface is necessary both for making corrections and for revising artwork.
5. *Strength*. Some drawings must withstand repeated rolling and unrolling or even folding, and some receive wear from certain duplicating processes. For such drawings a durable material is required.
6. *Reaction to wetting*. Many maps call for painting with various kinds of paints and inks. A material that curls excessively when wet is inappropriate for such a purpose.

There are many possible types of drawing and printing surfaces, but only a few are useful for cartographic purposes. These fall into two categories: plastics (particularly polyesters) and papers. The special and very desirable qualities of *plastics* have made them the prime drafting medium of the modern age. Their translucence permits tracing even without the use of a light table. The surface of these materials has been given a special coating or has been roughened in some way to cause ink to adhere. Indeed, these plastics take ink so freely that a pen literally glides over the surface and the lines produced with any given pen are slightly wider than those drawn on prepared tracing paper (see below). As a rule plastics are relatively expensive, but this higher cost is far outweighed by their ease of use and improved line quality.

Although space does not permit a comparison of the various kinds of polyester drafting materials, a representative product called *Cronaflex* will be described.* Cronaflex is available in rolls or in separate sheets and in varying thicknesses and sizes. Either side of the material provides an excellent drafting surface. The surface can be cleaned of oil or grease with a soft cloth dampened with trichlorethylene.* Such cleaning can be accomplished before drafting begins or at any time during production without affecting inked portions. Ordinary drawing ink can be completely removed at any time by use of a Q-Tip or cloth dampened with water without altering the drafting surface, which greatly facilitates correcting and revising artwork. The high dimensional stability of Cronaflex is another important quality. The material can be used in conjunction with very stable photographic films or scribe sheets and maintain registry. This is an especially valuable trait when preparing large maps where precise registry of separations is crucial.

For maps that do not require great dimensional stability, a *prepared tracing paper* provides an excellent and economical drafting surface for ink.** Its translucence also permits tracing even without the use of a light table, and good diazo copies can be made from it. It reacts to wetting, however, and it will curl or ripple if large areas are covered with ink or paint. Corrections can be made by gently scraping most of the ink away with a knife or razor blade and then rubbing gently with a medium-hard eraser. If you are extremely careful and do not excessively loosen the fibers of the paper, further drafting may sometimes be done in the corrected area. But as a rule corrections and revisions of artwork are much more difficult on paper than on plastic.

---

*Cronaflex UC-4 Drafting Film is manufactured by E.I. Du Pont de Nemours and Company, Photo Products Department, Wilmington, DE. Cartographic Drafting Film developed by Keuffel and Esser Company has essentially the same qualities as Cronaflex.

* Trichloroethylene is a highly toxic substance and should not be used unless extreme care is taken to avoid direct contact with the skin and breathing of fumes. Standard rubbing alcohol works almost as well as a general cleaning agent and is potentially far less hazardous. Alcohol will remove ink from the surface, however, so it is not suitable for general post-drafting cleanup.

**Albanene Tracing Paper Number 107155 can be purchased from Keuffel and Esser Company. A standard tracing paper is very thin tissue and is generally not suitable for cartographic work.

Tracing paper is not particularly stable and tends to shrink or expand with changes in temperature and humidity. For small drawings, a foot or so square, it can be used successfully, even if several flaps need to fit precisely. If, for some special reason, it is necessary to make separation drawings for large maps on tracing paper, we must take special precautions. Tracing paper does not change dimensions equally in both directions. The sheets, which should all be cut from the roll at one time, should have the grain of the paper running in the same direction. During the drafting, all the flaps should be kept at the same location. If some of the sheets are exposed to different temperature and humidity conditions, size changes will be unequal and registration will become a problem.

Ink. When you first think about drafting ink, you might focus on the different colors that might be available. Although this may be natural, ink color has little relevance in most cartographic drafting projects because most artwork is executed in black, regardless of the eventual reproduction color desired. Black ink has many attributes that make it the first choice for drafting.

When choosing an ink, the cartographer should consider a number of factors that may be less obvious than its color. For example, the ability of an ink to adhere to different drafting surfaces is critical. Ink that will produce a dense line of constant width on paper may skip or produce gray lines on plastic. The reverse may also be true. Ink that flows well in some brands or types of pens may tend to clog in others. Inks should dry relatively quickly on the drawing medium to minimize the chance of smearing and speed up the drafting operation, but they should not dry so quickly that they stop up the pen nibs. Cured ink should not end up so brittle that it easily chips or flakes when the drafting medium is bent or handled. As a rule, problems resulting from these factors can be minimized or avoided by carefully following the recommendations of ink and pen manufacturers.

Ink has a definite shelf life. It contains special agents that deteriorate or dissipate with age. To keep ink as fresh as possible, a good practice is to keep its container tightly capped and stored at a cool temperature when not in use. Since the replacement cost of ink is an insignificant part of total drafting costs, and since bad ink can lead to major drafting problems, if there is any doubt about its condition, the ink should be discarded.

Drawing Pens. The cartographer has a wide selection of pen types and styles from which to choose. Each has certain capabilities and limitations, and some are far easier to use than others. Which pen is best depends on the skill of the cartographer and the purpose for which the pen is intended. For an average project the cartographer may use two or three different kinds. Common to all pens, however, is the need to keep them clean and well maintained for proper operation. Ragged lines, gray lines instead of black, and lines of inconsistent width often result from dirty, clogged, or damaged pens. Equipment must be cleaned even during use, since bits of lint or other debris are constantly being picked up from the drafting surface. Pens must also be carefully handled in both use and storage if damage to delicate parts is to be avoided.

Technical Pens. The invention of the technical pen in the mid-1950s revolutionized ink drafting techniques.* Technical pens, as since refined, have been universally adopted for mechanical as well as manual drafting (Fig. 18.1). For the most part they have made other types of drafting pens obsolete. They are equally suitable for drafting straight and irregular lines, and can be used freehand or with a mechanical guide (see below).

*Technical pens (sometimes called reservoir pens) are available from a number of manufacturers. The better known are: Faber-Castell, KOH-I-NOOR Rapidograph, K & E Leroy, and MARS. Most graphic arts supply stores stock several of these brand names.

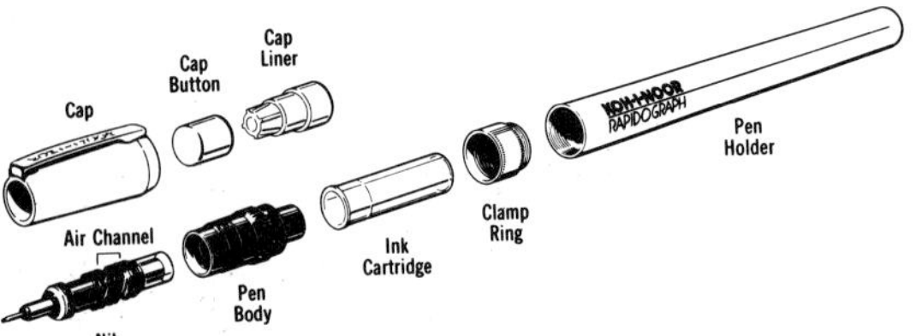

FIGURE 18.1 Technical reservoir pens have become an industry standard, both for manual drafting and automated machine plotting. (Courtesy Koh-I-Noor Rapidograph, Inc.)

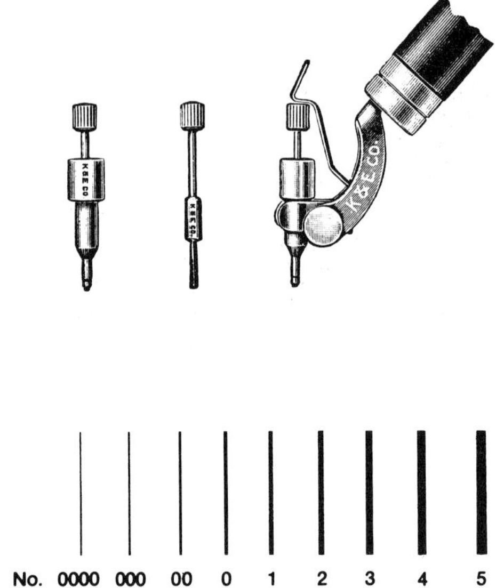

FIGURE 18.2 Leroy pens and penholder, and widths of lines made by various sizes of pens. (Courtesy Keuffel and Esser Company).

Their overwhelming popularity can be attributed to their design for quality output and ease of use. Technical features include a large capacity (usually cartridge type) ink reservoir and a wide selection of point sizes (from 0.1 mm to 6 mm) made of wear-resistant materials such as stainless steel, jewel, and tungsten carbide.

Functionally, technical pens can be handled as conveniently as a pencil. Their tip and ink-flow mechanism are engineered to minimize clogging and ink blobs. Through use of a special cap, the ink held throughout the nib and pen body is kept ready for instant startup after weeks, even months, of storage. Although initially rather expensive, technical pens quickly pay for themselves in reduced drafting time and frustration.

Other Pens. Although technical pens are easier to use and produce more precise results, several older, less sophisticated types of drafting pen are still used widely enough in manual drafting to warrant special mention. The *Leroy pen* stands out among this group. It was designed for use with Leroy lettering templates (see Chapter 9). But once a person learns to handle the pen consistently, it can be used to produce excellent freehand lines. The pen is made up of a cylinder containing a pin for a point, and ink is fed between these two components from a small, open reservoir (Fig. 18.2). A set of points of different sizes is required to produce a range of line widths.

Another pen that after some practice can be used to produce excellent freehand lines is the *Pelican-Graphos pen*. A set of nibs, which have been machined to produce lines of specific widths, can quickly be changed on a fountain-type pen. The "A" nib set is most practical for cartographic purposes, although the wider "T" nibs can come in handy for heavy linework and the "R" nibs are convenient for producing round, uniform dots.

A drafting instrument of somewhat different design is the *ruling pen*. Cartographers still may find occasions when they must resort to this dated instrument, such as when using an old drafting set to make special symbols like a circle. The ruling pen consists of two blades whose spacing is adjustable with a small screw on the side, so that a continuous range of line widths may be made with the same pen. Because they are hard to use, they have become unpopular. First, it takes skill and patience to make the fine adjustment of the blades required to produce lines of desired widths. Secondly, it takes equal skill to get the ink to flow at all, let alone to flow evenly to produce a line of constant width.

Quill pens made of metal are needed to produce special kinds of lines and can also be useful in cartographic drawing. A large variety are obtainable, and it is helpful to have a good selection on hand. Some are hard and stiff and make uniform lines; others are very flexible and are used for lines, such as rivers, that require a changing width on the drawing. A favorite is the one called a "crow quill," a relatively stiff pen, which requires a special holder. Quill pens of any type may be dipped in the ink bottle, but a better practice is to use the dropper to apply a drop of ink to the underside of the pen. This procedure helps to produce a finer line and allows frequent cleaning without excessive waste of ink.

Pen Guides. Although a large proportion of cartographic drafting involves irregular linework, there are also many instances where regular lines must be drawn. Borders, reference grids, geometric symbols, and legend boxes all fall in this latter category. Fortunately, a number of rather simple but highly useful mechanical aids are available to assist the cartographer in producing this regular linework. Straight edges, curves, templates, compasses, and plotting heads all can be of immense help if properly used.

Straight Edges. Mechanical guides for drafting straight lines take several forms. The edge of the drawing board or table can establish a fixed, straight edge from which a T-square can be used to produce parallel straight lines. By employing triangles of various angles (and in various combinations) in conjunction with a T-square, parallel straight lines can be constructed at other angles as well. Mechanical drafting machines perform the function of both triangles and T-squares with much greater convenience.

The drafting material itself may contain grids of fine blue lines, which are not reproduced when photographed with orthochromatic film. Alternatively, drafting material may be registered over a worksheet that has been compiled on gridded (graph) paper. Finally, the worksheet and drafting material may be taped down on a drafting table or board that has been covered with a gridded material.

Smooth Curves. Curved lines generally are more difficult to construct than straight lines. Mechanical aids are available commercially in a wide selection of sizes, shapes, and styles. But even if the cartographer has a large collection of such devices handy, there still will be many curves that cannot be readily fitted. Flexible curve guides that can be tailor-bent to suit the situation are handy, but they by no means solve the problem completely. What this means is that a curved line normally must be done in stages. Each segment of the line will require use of a different place on the guide device, or different guides altogether. To construct such a line without the individually constructed segments being apparent is the ultimate test of drafting skill. It is usually better to start from a drawn segment than to try to meet it. High-quality curve construction takes patience, a good eye, and a great deal of practice.

Templates. A variety of symbol templates are commercially available that can be used in conjunction with technical and Leroy pens. These are particularly useful when constructing small geometric symbols such as squares, rectangles, circles, ellipses, and so forth. Even more than with the previously discussed guide devices, care must be taken to prevent ink from flowing under the template because when drawing a closed figure involving pen movements in a variety of directions, it is difficult to hold the pen at a constant angle against the guide.

Compasses. Compasses are used for drawing arcs or circles. Usually they are designed so a pencil or pen can be interchanged. Older compasses (for example those found in commercial drafting sets) usually have blades comparable to the standard ruling pen, while newer designs may have special attachments that permit the use of modern technical pens or Pelican-Graphos pens (Fig. 18.3). For circles or arcs with long radii, a special *beam compass,* which can be extended by the insertion of extra linkage, is necessary. The *drop compass,* on the other hand, is used for making small circles. The pen is loose on the

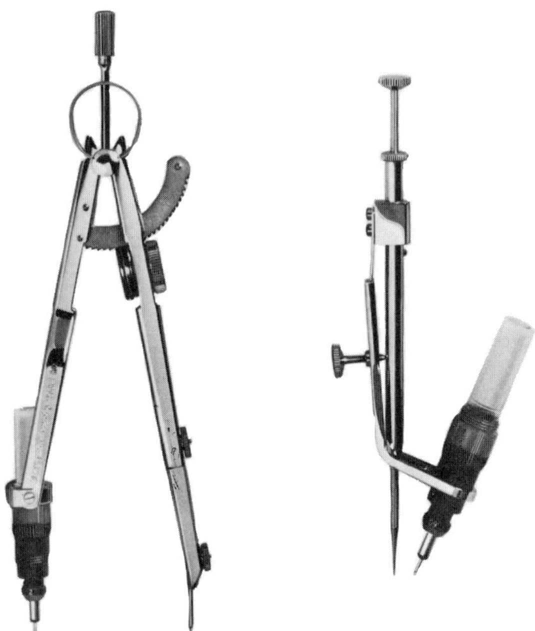

FIGURE 18.3 Drafting compasses are available in several designs to serve special needs. In recent years some manufacturers have replaced the conventional ruling pen tip with a more convenient-to-use technical pen. (Courtesy, Koh-I-Noor Rapidograph, Inc.)

pointed shaft; when the center has been located, the pen is dropped to the drafting surface and twirled.

Plotter Heads. Possibly the ultimate pen guide is the program-controlled drafting head found on automated plotting devices (Fig. 18.4). In this case the digital record of the artwork acts as an invisible "electronic template" in directing the pen around the drafting surface. Since mechanical guides that might smear ink and soil the drafting surface are eliminated, and since the pen is always oriented at the same angle with the drafting surface, the quality of automated drafting can be very high. Through simple program instructions special effects such as double, dotted, dashed, or dot-dashed lines are readily achieved. These devices are discussed in more detail later.

FIGURE 18.4 The drafting head on modern line-plotting devices normally holds from four to eight pens representing a selection of line weights and colors. (Courtesy, The Gerber Scientific Instrument Company).

Preprinted Materials. Much of the artwork that in the past was constructed by laborious drafting procedures is now commonly done by applying preprinted stickup materials. These easy-to-use materials save much of the time that was formerly necessary to construct a map and drastically reduce the demands for artistic and drafting skill on the part of the cartographer. Preprinted materials are marketed in a variety of forms and styles sufficient to satisfy most drafting needs. Point symbols, lines, tints (flat shadings), patterns, and flat colors are all available from most graphic arts supply houses. Catalogs may be obtained from the local outlet or from the manufacturer.* If you

*Para-Tone, Inc., 512 W. Burlington Avenue, La Grange, IL 60525, manufactures Zip-A-Tone; Craftint Manufacturing Company, 1615 Collamer Avenue, Cleveland, OH 44110, manufactures Craf-Tone; Chart-Pak, Inc., Leeds, MA 01053, manufactures Contak as well as an adhesive tape useful as a substitute for drafted lines; Artype, Inc., 127 S. Northwest Highway, Barrington, IL 60010, manufactures dot and line screens with a regular progression of tonal values; Graphic Products Corporation, 3810 Industrial Avenue, Rolling Meadows, IL 60008, manufactures Format. One word of caution: The quality of prepared materials varies from one brand to another with respect to blackness, sharpness, consistency of dot sizes, and adhesion characteristics.

cannot find a special symbol, many of these same firms will tailor-make one according to your specifications (you supply an example or master). Alternatively, you can make up your own stickup symbols in house by putting them on *stripping film* or by using a special material called *Image-N-Transfer.** The big advantage of making your own symbols is that you can change the texture and size of available prepared media according to your own preference to fit your own design.

Point and Line Symbols. Point and line symbols are widely used in mapping and therefore are widely available in preprinted form. The transfer of the preprinted image to the flap is accomplished in one of several ways. One type of material consists of a carrier sheet of thin, transparent film containing *dry transfer* or *rub-on* symbols. With the desired symbol properly positioned over the map, it is rubbed with a burnishing tool to transfer the symbol to the drafting surface. These materials are convenient to use, but the symbol material is fairly thin and rather brittle, and the image can easily be distorted or chipped during the transfer process. Dry transfer materials also seem to have a relatively short shelf life, and some drafting surfaces seem to resist the transfer.

A second type of material consists of point symbols that have been preprinted on thin, transparent film with an adhesive backing. The adhesive may either be wax (especially older materials) or a heat-resisting coating that is protected by a translucent backing sheet. In this case the symbols are cut out of the material, stuck up on the map at the appropriate position, and burnished down. Although the cut-out materials may be somewhat slower to use for certain purposes than the rub-on type, the quality is generally higher and the results more uniform.

Straight-line symbols that can substitute for inked linework are available in the same two forms as point symbols. Rub-on lines of any length are not recommended, however, because it is extremely difficult to perform the dry transfer without distorting or chipping the line. Cutout symbols are available in a wide variety of widths and styles. For broken lines (dashed, dot-dash, and the like) in particular, these stickup materials can be used efficiently to make a high-quality product. Line symbols are also marketed in the form of both rigid and flexible tapes in many widths and styles. Both forms are suitable for constructing straight lines, although as a rule the rigid types (printed on a clear carrier film up to $\frac{1}{4}$ inch wide) tend to be easier to use. Flexible tapes can also be used to produce curved linework, providing changes of direction are rather gentle. Problems with tapes can usually be traced to the fact that they seem to have a memory for their original length. Thus, stretched tapes will try to shrink back to their former length, causing straight lines to shorten and curved lines to creep across the drafting surface. Temperature and humidity changes make stretching and shrinking problems even worse.

Patterns and Tints. Indispensable to many maps are the patterns and tints used to differentiate one area from another. Since the repetitive drafting of the fine details that make up these areal symbols is tedious and time consuming and perfection is very difficult to attain, preprinted materials are extensively used. Patterns, consisting of visually discernible point and line symbols, are available in different arrangements, orientations, and textures (Fig. 18.5). Tints, on the other hand, are available in different densities (percentages of blackness) and textures (numbers of lines per inch). Most commercial brands are marketed in densities ranging from 10 to 70 percent black and in textures ranging from 27.5 to 85 lines per inch (Fig. 18.6).

When choosing tint screens, the cartographer should keep in mind some of the hazards that characterize map production and reproduction. A 20 percent dot screen, for example, is supposed to produce a dot size that will in the aggregate constitute 20 percent of the area on the final map. This ideal is not often attained for several reasons. The tint screen usually has to be processed photographically at least once, maybe

* Image-N-Transfer (I.N.T.) is a 3M product. It is a rub-on sheet rather than a stickup medium. Stripping film is a 3M product of the stickup variety.

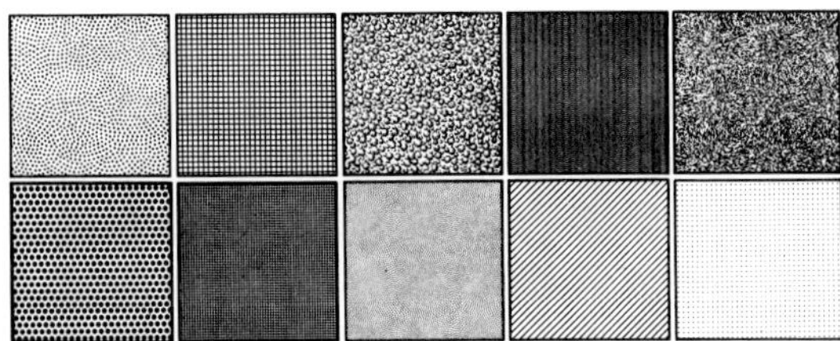

FIGURE 18.5 Examples of preprinted symbols, in this case Zip-A-Tone. Many of the maps and diagrams in this book have been prepared with the aid of these kinds of symbols.

twice, in reproducing the artwork. The process involves light sources, exposure times, emulsion sensitivity, and developing, all of which can affect the dot size. The resulting image can easily depart from the initial value by several percent in either direction. In the case of plate-based printing, other problems exist because of variations in printing inks, press functioning, and paper characteristics, which can cause the dots to be larger or smaller than those on the printing plate. In general, the printed dots tend to be larger than the screen dot diameters for black dots (up to 50 percent) and the white dots (over 50 percent) tend to be smaller.

Both patterns and tints are stuck up on the map in the same fashion. The material may be stripped from its protective backing sheet, placed over the area to be covered on the map, and cut along the boundary of the area with a sharp needle or blade (X-acto knife or razor). The excess material is then stripped away and the remaining pattern or tint burnished to the drawing surface. Extreme care must be exercised when cutting the adhesive film over inked lines on the drawing, however. The cutting tool may damage the linework and ink will occasionally be pulled off the drawing when stripping away excess material. For these reasons, the material is sometimes positioned over the map and cut out as before, but with its backing sheet in place. The cut out section must then be positioned on the map and burnished down. This second approach takes great care if the material is to be matched precisely to the area outline and, as a rule, is not recommended for regions that have complex shapes.

The use of pattern symbols creates special problems for the cartographer, because the texture, form, and orientation of the pattern must to some degree be coordinated with the shape characteristics of the region being covered with the areal symbol. The size, shape, and orientation of the region may interact with certain pattern attributes to produce special effects that usually are undesirable. As a rule, patterns should be selected that have a fine enough texture to prevent visual interaction between the border of the region and pattern details, and to ensure that all

FIGURE 18.6 Tint screens are available in different densities and textures. Finer textures give the illusion of gray when applied to artwork, making it possible to create a range of values on the map.

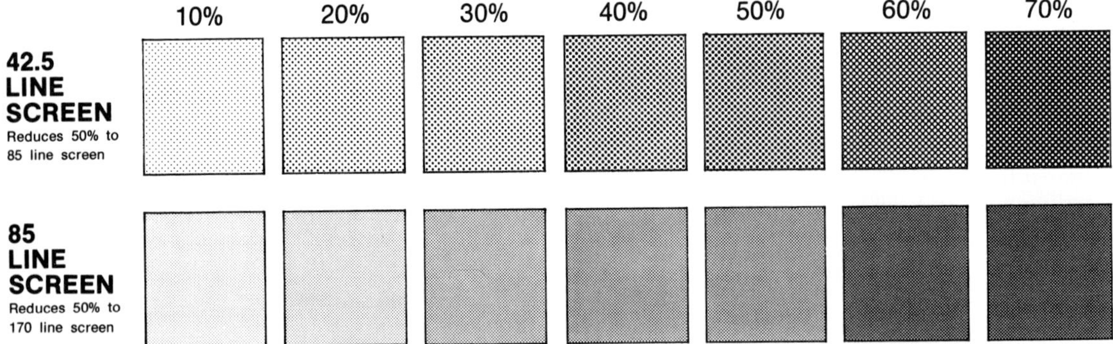

parts of the region (particularly narrow extensions) receive sufficient pattern elements to be readily identifiable.

Knowledge of the relation of tints to reduction is critical to the preparation of graded series of tones or values. The cartographer must work within the limits set by reproduction processes at one extreme and by the human perceptual mechanism at the other. As a rule, reproduction scale tints of much more than 200 lpi are difficult to print consistently. Due to the many variables involved when using preprinted materials, it is best to leave a safe margin for error. Thus a maximum in the neighborhood of 150 lpi is probably a more practical working limit.

At reproduction scale a tint with less than 75 lpi will generally be seen by most people as a pattern rather than a flat tone, although 100 lpi represents a more certain lower limit. To be safe, therefore, the cartographer should generally aim for a reproduction scale tint ranging from 100 to 150 lpi. If these general limits are accepted, it would mean that on a map designed for 50 percent reduction, tints of more than 75 lpi and less than 50 lpi should not be used; on a map designed for 25 percent reduction, tints of 75 lpi and 60 lpi are the upper and lower limits and so forth.*

Flat Colors. Sheets of preprinted flat color materials also can serve several useful cartographic functions. Maps can be produced for direct viewing with these materials just as they would with preprinted patterns or tints. This may be done to explore design alternatives or to produce a map in color that will not be duplicated. Preprinted colors may also be used to prepare a map for subsequent photographic reproduction, or for photographic color separation prior to plate-based printing.

A quite different use of preprinted color sheets involves a special application of the red material. When large areas are to appear ultimately as a solid color, say black, the traditional approach is laborously filling in the area with black ink. This task may be more efficiently performed by using an actinically opaque but visually translucent solid red film that is cut to fit the area.* This is especially efficient when the region to be covered has a smooth boundary or a simple geometric shape. It can also be effective when working with a medium such as paper that may become distorted when large areas are wetted. This technique is commonly used when constructing artwork for producing open-window negatives (see below).

Drafting the Map. Drafting traditionally has been a manual activity that placed large demands on the cartographer's time, energy, patience, and skill. Fortunately, mechanical plotting devices driven by electronic computers can now mimic most manual drafting procedures. Not only does this reduce the need for manual drafting skills, but in theory it can free the cartographer from much of the burden of manual drafting. At the least, plotters change the nature of the cartographer's chores and require a somewhat different mix of skills. In practice, however, the cartographer will still encounter enough situations that can best be handled by manual methods to warrant their being learned. Appropriate plotting equipment, digital records, and mapping software simply may not be sufficiently accessible to justify their use. Alternatively, it may be more cost-efficient to draft rather than to plot a simple map, or to touch up by hand (rather than to replot) a mechanically plotted map that contains some easily corrected errors or defects. In sum, the goal should be to gain sufficient skill to be able to draft a map manually or plot it mechanically, and to know enough about each method

*Exact upper and lower working limits will depend on a number of factors, including the quality of the following: artwork construction, photographic processing, printing medium, and the printing process.

*Specially prepared red color materials (transparent blocking films) are marketed under several trade names. *Parapaque* (manufactured by Para-Tone, Inc., 512 Burlington Ave., La Grange, Il 60525) is a representative material.

to be able to merge or mix the two techniques as effectively as possible during execution of a mapping project.

When positive artwork is to be reproduced by plate-based methods, it will be converted by the printer into negative form, either by use of a camera or by direct contact in a vacuum frame. In either case, good, sharp contrast between the symbols of the map and the drafting surface is important to produce quality negatives. The line artwork should be opaque black or red. Gray lines cause difficulty in determining exposure times, and the result is usually broken or weak line work on the printed copies.

Making a separate flap or flaps for lettering is standard procedure when a map involves the use of color. On single-color maps, however, the lettering is usually incorporated into a flap that contains other lines and symbols. In any case, there are usually situations when the lettering must fall over other map features. If colored areas or lines are not too dark, black lettering can usually be superimposed; however, dark or black lines should be interrupted to accommodate the names. A good practice is to leave spaces for the lettering when the linework is drafted.

Manual Drafting. Learning drafting skill is a long, involved process. Many months of practice are usually required before techniques are mastered to the point where professional results can be obtained. The quality and detail of even a skilled draftperson's work is always limited by the nature of the medium, ink, and the instruments available. Details concerning how drafting skill can be acquired fall beyond the scope of this book. Those interested in an in-depth treatment of this subject should consult a manual on mechanical drafting. What will be presented here are a few hints that should help someone who wants to learn to draft quickly, accurately, and consistently. All three traits are very desirable in cartographic drafting.

1. Begin by laying out materials, tools, and other items that will be needed to complete a project. Then make sure that everything required is available and in working order.
2. Clean yourself, the work area, and the drafting surface prior to actual drawing. Clean drafting instruments and surfaces frequently during the drawing process.
3. Know the line weights each pen size produces on different drafting media, as well as the relative visual weights these lines exhibit at various reductions. A simple grid can be constructed for this purpose (see Fig. 7.3).
4. Keep your hands off the drafting surface. Ever-present skin oils will cause ink pens to skip or produce low-density lines. Two ways to solve these problems are either to wear thin, soft gloves or to rest your hand on a small sheet of drafting material that can be slid around the map area as required.
5. When starting to use a new pen, always test its functioning on a scrap piece of material before working directly on the map.
6. Draw only in the most comfortable direction. This may be toward the body for some, and away from the body for others. In either case, proper body position can be maintained by moving around the drafting table, or by rotating a drafting board upon which the compilation sheet has been taped. Artwork that has been taped or pinned to a free-floating compilation sheet can also be readily oriented and reoriented for ease of drafting, and permits use over a light table.
7. Handle pens in a consistent fashion. The pen tip should always form the same angle (close to 90°) with the drafting surface, and a constant pressure (or touch) should be maintained.
8. Before using guide devices, first run your finger along the edge to feel for any cuts or notches that might lead to an irregular line. When using these devices, take care to keep the pen tip away from the bottom edge in order to avoid having ink flow (creep) under the guide. It is most convenient to use the newer acrylic guides that are manufactured with beveled edges. However, if these are

not available, placing several layers of drafting tape on the underside of standard (unbeveled) guides will raise them slightly off the drafting surface and thereby minimize the problem of ink blobs.

9. Work as systematically as possible. If pens are difficult to clean between uses, try to draw all lines of the same weight before moving on to the next pen. If ink is not drying rapidly, proceed from one side of the map to the other (usually from top to bottom) drawing parallel or near parallel lines as you go.
10. Make corrections or changes that involve scraping the material as the last drafting step so that damage to the drafting surface will not be a serious matter. With care, marvelous things can be accomplished with a sharp blade and white opaque. If the artwork is prepared specifically for direct contact negatives, however, white opaque should not be used for deletions because it registers on the film the same as opaque black.
11. Clean and clearly label all separations, being sure that sufficient registry marks are provided.
12. Carefully add preprinted materials, taking special precautions to avoid damaging existing inked linework. When preprinted materials are burnished onto the map, the adhesive tends to be squeezed out along the edges and can become distributed over the drafting surface, causing difficulty for further drafting. If ink must be added after this has happened, it is necessary to clean the area to be inked. Wax can be effectively removed by swabbing with trichloroethylene.*
13. Check the artwork carefully and systematically and then place it between layers of protective material to minimize the potential for damage during future handling and storage.

*Care must be exercised to prevent the trichloroethylene from coming in contact with preprinted materials, since damage may occur. Also keep in mind that the fluid is extremely toxic to humans.

Mechanical Plotting. Anything that can be drafted by hand can be plotted mechancially in the same vector fashion under the control of mapping software. Indeed, many intricately detailed maps, particularly those portraying statistical data, that manual draftspeople would avoid if at all possible are now being routinely plotted by machine.

Cartographers who want to work with a mechanical plotter either must find appropriate mapping software (programs), have someone create such software, or write the program instructions themselves. Once this is done, and a machine-readable digital record of cartographic and statistical data is available, the map is produced with the help of user-supplied plotting instructions, which are generally outlined in a program user's manual. Decisions to be made might include the width (11 or 30 inches being most common) and type (plastic or paper) of plotting medium, scale of the plot, choice of line weights (pen sizes), type of pen (ball point, fiber tip, fluid ink), color of ink, desired artwork separations (for example, latitude and longitude grid on one sheet, hydrography on another, and so forth), and location of registry marks for multiple separations. Beyond that, the biggest concern is to ensure that the pen is working properly, a concern that is all too familiar in manual drafting as well.

As a rule ballpoint pens are used when testing or working out problems with a new plot even though the resulting linework is of relatively poor quality and there is little (if any) selection of line widths. There are two reasons. Ballpoint pens are relatively easy to maintain and keep working, so operator intervention in the plotting process is minimized. Secondly, ballpoint plotting is relatively inexpensive (due both to the low cost of the pens and less need for operator time). These factors can become significant if the plot has to be done a number of times to work out the bugs. Then, if higher-quality artwork is desired, fluid pens can be used on the final plot.

Another consideration when plotting a map is that many of the plotters being used are of the "incremental" type, meaning that as the pen moves to draw a diagonal line of the drafting medium, the pen must be stepped or incremented in the horizontal and vertical direction from one end to the other. The steplike or zigzag increments along diagonal lines produced on low-resolution plotters, an effect known as *aliasing,* can be quite noticeable (Fig. 18.7). Plotting at up scale for photo reduction to final size usually is of some help. But if high-quality linework is needed, a better approach is to use a higher-resolution plotter or one of the newer models that can plot diagonal lines without incrementing.

Printers

In contrast to vector plotters, printers operate on a raster principle. As a result, the detail or complexity of the artwork being produced has no effect on the speed of production. Each cell or printing element is printed regardless of whether it is empty or full. Some printers are essentially mechanical in design with the image created by physically striking an inked ribbon against the paper. Other printers are electronic in design, producing artwork through a variety of nonimpact methods. These may involve electrostatic, thermographic, electromagnetic, and ink-jet technologies.

FIGURE 18.7 An enlargement of a map section produced on an incremental plotter. The "steps" are readily apparent, especially along the bounding graticule lines.

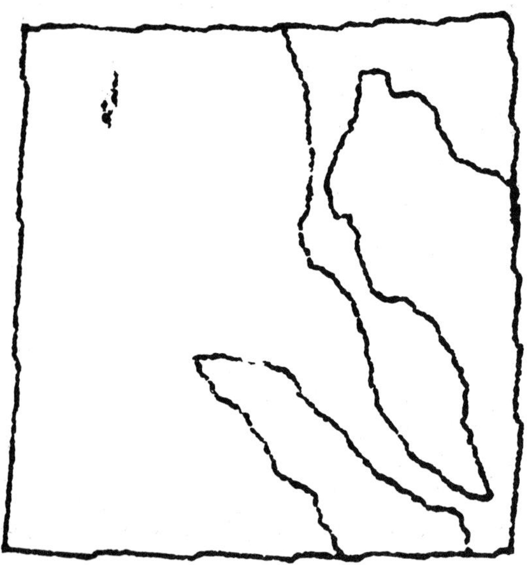

Another way to classify printers is by the way they build up an image. Some devices print whole characters, just as the standard typewriter does. These character printers may work serially like a typewriter, or print a whole line of characters simultaneously. In contrast to character printers, a number of computer graphics output devices build up images in raster fashion by printing rows or matrices of tiny dots. These machines are called dot printers.

Character Printers. Soon after digital computers were invented, cartographers began producing rather crude maps using the standard typewriter characters (numbers, letters, special symbols) found on ordinary computer *line printers.** Cartographic output from these line printers is relatively familiar in large part because of the widespread use of the SYMAP software program from the Harvard Laboratory for Computer Graphics, which initially relied heavily on the line printer (Fig. 18.8).

Point, line, and areal symbols can all be created on these raster devices, although the results tend to be rather coarse. The reason is that line printers have a maximum resolution of ten characters (printing positions) per inch horizontally and only six or eight characters per inch vertically. Line-printer maps also lack density and contrast because even with an overprinting capability the maximum density that can be achieved is about 55 percent area inked.* Thus, if more than a few density levels are used, they become extremely difficult to differentiate visually.

Maps can be made at large scales on several sheets, taped together, and photographically reduced, with some lessening of visual coarseness. But since this does nothing to improve density or

*Ordinary computer line printer maps can be greatly improved by the use of special chains that employ symbols different from the common alphanumeric symbols. These devices have not become very popular, however.

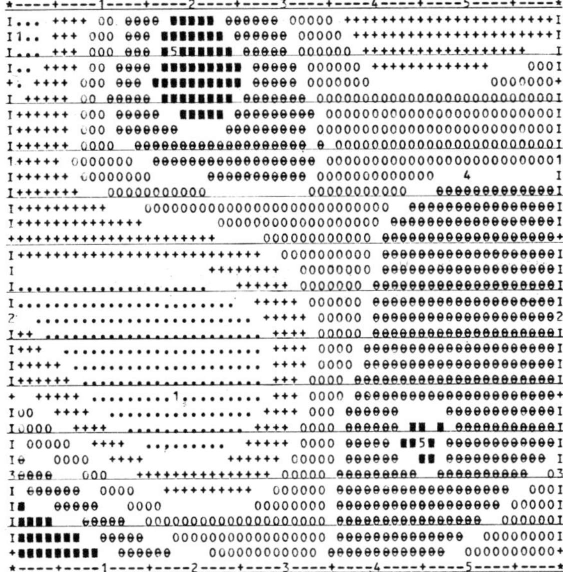

FIGURE 18.8 Printout from the software program SYMAP. Such output is readily obtainable because of the widespread availability of this mapping package, but it is crude. Reduction of the printer output does enhance the visual appeal somewhat; this illustration has been reduced from an original that measured 30 cm^2.

contrast, the overall effect is little improved. As a result line printer maps currently are not very important in map production. In large government agencies, they simply are not used anymore. Instead, they are used primarily for map preview, to test design concepts, to provide a rough but economical picture of the nature of a distribution for use in planning and research, or to serve as a compilation worksheet for other, more refined production methods.

Dot Printers. Dot printers vary substantially in their resolution and in the way that they produce their dots. The lower-resolution printers produce artwork that lacks density and has rather fuzzy or ragged edges, but the highest-resolution printers produce artwork of extremely high quality.

Dot printers all work in a raster format. Some, such as the matrix printer, operate in serial fashion and use the impact principle. In contrast, electrostatic printer/plotter, ink-jet printer, and laser printer are all line printers that function on nonimpact principles. As a rule the higher-cost machines operate with the best resolution and greatest speed.

Matrix Printers. Matrix printers work on much the same principle as a standard electric typewriter, except that a single printing head containing a matrix of pins (say 7 by 9) takes the place of the full alphanumeric character set (Fig. 18.9). The printing head moves serially from left to right across the platen to print the first line, the paper advances, the process is repeated for the second row, and so forth. At each printer position any combination of the dots (maybe none) within the matrix may be activated. Pins striking the ribbon against the paper print the dots.

By activating different patterns (arrangements, textures, orientations) of dots, a wide variety of graphic effects can be produced (Fig. 18.9). The machines are noisy and slow but inexpensive enough to have become a standard graphic output device with personal computer buffs (see Fig. 16.21). If low-quality artwork is not considered a serious drawback, matrix printers can be a useful map production device.

Electrostatic Printer/Plotter. Electrostatic printer/plotter devices are line-printing machines that work on a nonimpact principle. Small dots are printed a row at a time as a specially coated conductive

FIGURE 18.9 Dot matrix printers are slow and limited in the paper size that they can handle, but they are inexpensive and popular graphic output devices for microcomputers. (Courtesy, Richard Smith).

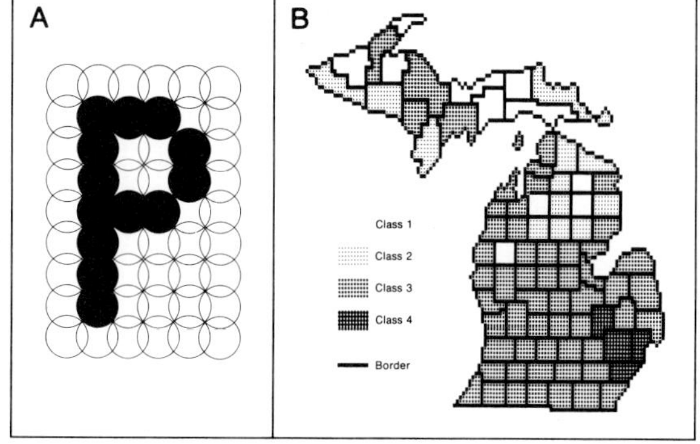

paper or film (up to 72 inches wide) moves at a constant rate beneath one or more linear arrays of conducting nibs (Fig. 18.10). Resolution up to 400 dots per inch and rates up to 3 inches per second can be achieved. Each nib (actually an electrode) is independently controlled by the digital record to give the paper or film a small round dot of negative electrical charge. This latent electrical image is "developed" by passing a positively charged toner (normally black) over the paper or film. The toner adheres to the medium where the charges were deposited to form the positive image dots.

The primary advantage of electrostatic printer/plotters over matrix printers is their much greater speed. They also can accommodate wider paper, are quieter to operate, and have fewer mechanical parts to malfunction. Full-color electrostatic plotting using magenta, cyan, yellow, and black toners is also possible. Similar to matrix printers, limited tints and patterns can be achieved by defining small texture components with different numbers of dots (a procedure known as *dithering*).

FIGURE 18.10 Electrostatic plotters are non-impact devices that utilize one or more linear arrays of printing nibs to build up an image a row of dots at a time. By offsetting the printing heads the dots can be made to overlap, producing high quality linework. (Courtesy, Benson).

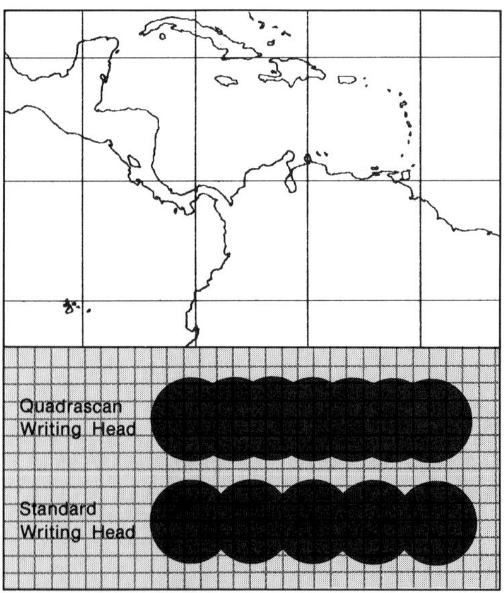

A possible disadvantage of electrostatic printing/plotting is that the special image medium is relatively expensive, although still only a minor part of total production costs. In addition, particularly on the lower resolution machines, the image may lack density, contrast, and sharpness. Yet for many purposes, the great speed and low cost of producing positive artwork on an electrostatic printer/plotter far outweigh the possible disadvantages associated with image quality.

Ink-Jet Printers. Ink-jet printers literally spray-paint the artwork onto the image medium (uncoated papers or clear films). A stream of ink is broken up into a stream of tiny droplets (as small as .0025 inch), and shot from a nozzle at a rate of over 100,000 per second. Under the control of the digital record the droplets acquire an electrical charge. Then, passing between charged plates, they are deflected according to the charges they carry to the appropriate places on the paper. By using multiple nozzles, ink-jet printers can simultaneously print in several colors of ink (see Color Fig. 24). If process inks are used, the result looks very much like a process color halftone printing from lithographic plates.

The droplets of ink used in ink-jet printers play the same role as electrons in a cathode ray tube. Unfortunately, ink droplets are more troublesome to work with than electrons. The mechanical system of ink reservoirs, pumps, tubes, and nozzles is subject to clogging and other failures. Furthermore, the discrete ink droplets can be placed imprecisely, potentially compromising the quality of the print. But if colored artwork (including transparencies) needs to be done quickly and relatively inexpensively, ink-jet printing might be a good choice. As with electrostatic methods, however, the image is not of as high quality as might be produced by photomechanical means. But for many mapping purposes it is perfectly adequate.

Laser Printers. Laser printers work much like a standard xerographic copier, which has led some to refer to this printing process as electronic xerography. The main difference between the two is that the image is drawn from a digital record by a laser rather than being reflected to the photosensitive drum from a sheet of brightly illuminated artwork. The laser beam is routed through a modulator that regulates its intensity (that is, turns it on and off) according to the digital information that is provided. A series of mirrors then leads the modulated beam to a scanning prism that sweeps it across the photosensitive drum line by line in raster fashion. The image on the drum, now a pattern of positive charges, is next dusted with negatively charged toner particles. A positively charged sheet of paper rolled over the drum attracts the toner particles, which then are fused on to the surface under heat and pressure (Fig. 18.11).

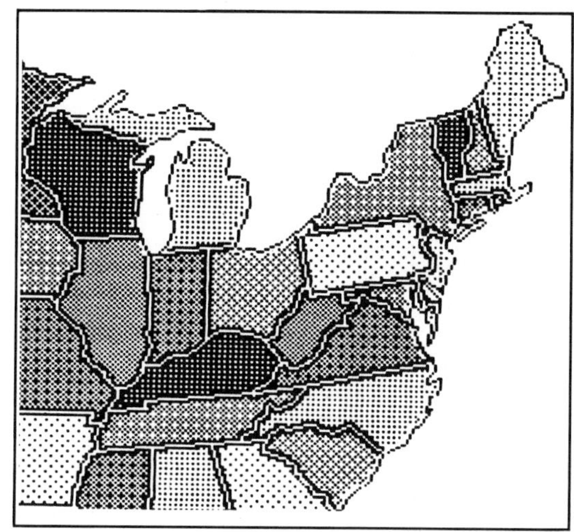

FIGURE 18.11 Laser printers with resolutions of 300 rasters per inch can produce maps of fairly high graphic quality at very high speed. (Courtesy, DataPoint).

Laser printers can work only about one-third as fast as ink-jet printers, and they cost about twice as much. What they lack in speed and economy, however, they more than make up for in quality. Machines with a resolution of 300 rasters per inch (90,000 dots per square inch) are currently in use. Indeed, the quality of graphic output from laser printers approaches that of the best offset presses.

Display Screen

With ever-increasing frequency maps are being produced for direct viewing on display screens rather than in hard-copy form. The display device is usually a CRT, which is comparable to a standard television screen, and will probably be so for some time, although a variety of new display devices are attracting attention.* The CRT displays we use today had their origins in the early 1960s and can be classified into two types, depending upon the structure of the displayed image.

Vector displays (also called random-scan displays) create a map image line by line in much the same way that it might be done with drafting or scribing. Various parts of the map (line segments) can be depicted in any order. To do so an electron beam is moved across the phosphor-coated screen from endpoint to endpoint of successive line segments. Points, lines, and characters can all be plotted at screen resolutions up to about 4000 by 4000 points, which is extremely high resolution for a display screen (Fig. 18.12).

Early vector displays were of the *refresh* type, meaning that the entire image must be rewritten every $\frac{1}{30}$ to $\frac{1}{60}$ second in order to avoid visual flicker. The problem is that as the image grows complex, this may not be enough time for a complete redrawing of the image, and the image tends to flicker. This problem was circumvented with the development of the *direct-view storage tube* (DVST). With this type of display screen the image stays on the screen (until erased) once it is written. The electron beam plots rather slowly on these devices, however. DVST's still tend to be

*Display screens based on technologies other than cathode ray tubes might be more common in the future. Possible technologies include gas discharge, plasma panels, light emitting diodes (LED), and liquid crystal displays (LCD).

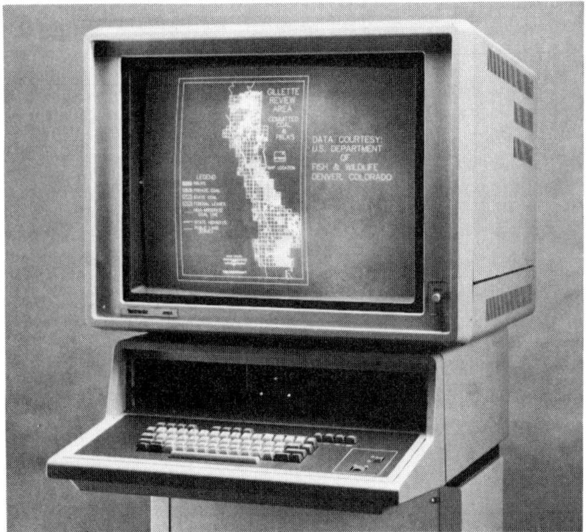

FIGURE 18.12 A random-scan image on a vector display terminal is in many respects comparable to a conventional line map that has been drafted or scribed one line segment or feature at a time. (Courtesy, Tektronix, Inc.).

popular for applications that require the plotting of tens of thousands of high-precision lines but that do not need dynamic picture manipulation.

Raster-scan displays are patterned after familiar television technology. The map image is divided into horizontal lines. Then all parts of the first line are depicted in left-to-right order, the process is repeated for the second line, and so forth. During the left-to-right sweep the electron beam intensity is modulated to create different shades of gray or colors. The raster structure of the image makes possible the display of solid areas, typically in color (see Color Fig. 25). The refresh process is independent of image complexity, so that flicker is not a problem. Since data must be stored for each picture element (pixel) regardless of whether it is filled or empty, the data-processing requirements grow rapidly with greater screen resolution. The highest resolution raster-scan displays currently in use (approximately 1000 columns and rows) fall far short of the resolution of the best vector displays. Even then, smooth image motion (dynamics) is not possible. The reason is that a million or so pixels would have to be altered in less than $\frac{1}{30}$ second, and this rate of operation is not yet practical. A refresh vector screen is more suitable for image animation because only the endpoints of the line segments have to be transformed between pictures.

One of the greatest benefits of using a display screen is that it enables the cartographer to interact with the map that is being produced. Vast improvements in the technology make this interaction easier. Initially light pens were pointed at the screen, but they were clumsy and fragile and have been superseded by other devices. One is a thin stylus that can be moved over a data tablet or menu. Another is a transparent, touch-sensitive panel that is mounted on the screen. Possibly the ultimate development is audio communication, which permits hands-free interaction with the map. Using these devices, the cartographer can in one way or another identify picture components to be operated upon, specify operations to be performed, or make additions to the image. Actually, cartographers working in an interactive environment commonly are in control of several input devices, gaining tremendous flexibility and power in performing editing, data manipulation, design, and other mapping tasks.

NEGATIVE ARTWORK

Use of negative artwork generally improves cartographic quality and may reduce reproduction costs as well. The capability for the cartographer to produce very precise lines in negative form came with the *scribing process,* which has come into wide use in the past few decades. Recently, many kinds of materials have appeared that can be used in conjunction with the scribe coat and permit the cartographer to prepare artwork in negative form more easily. Computer-driven vector plotters can also be outfitted with scribing tools and materials, greatly reducing the need for manual scribing—assuming, of course, that the information to be mapped is available in the form of a digital record. Electronic technology has also made it possible to plot directly on film with a small spot of light or laser or electron beams. Some of these film plotters work on a vector principle, while others employ raster technology.

Negative artwork is normally made at reproduction size. But advances in microfilm technology have also made it possible for cartographers to plot map images economically at much reduced scale for later enlargement to reproduction size. The printer can then go directly from this negative artwork to the platemaking stage. By skipping the photographic step that is required to reverse positive artwork, the cartographer potentially has control of the materials much further into some (especially plate-based) reproduction processes than would otherwise be possible.

Scribing

In light of the drawbacks inherent in the drafting method that we pointed out in the previous section, not surprisingly, cartographers have adopted a more efficient technique for producing line drawings. The alternative is called *scribing*. In contrast to drawing with pen and ink where the desired marks are applied to a drafting surface, in scribing the desired marks are obtained by removing material. The cartographer starts with a sheet of hard plastic film to which a soft, translucent coating has been applied. Then, working over a light table, the coating is removed by cutting and scraping with special engraving instruments to produce the lines and symbols. Scribing can also be done on a flatbed plotter directed by a computer. When it is finished, the sheet has the same general appearance as a negative made photographically.

Because of its many advantages, scribing has been widely adopted as a replacement for conventional drafting by commercial and government mapping establishments. Scribed artwork requires fewer processing steps than drafted artwork during certain reproduction processes. Furthermore, correction and revision are simpler than with ink drafting. Scribing is also easier than drafting for some people to master, possibly reducing training time. Since there is no ink to spill or to keep flowing evenly through fine pen points, the quality of scribed linework can be measurably improved as well. All these factors contribute to substantial time savings in preparing artwork, which means that scribing can under some conditions be far less costly than drafting.

Materials. The scribing medium consists of a clear base material that is coated with an actinically opaque, scribable material. A variety of base materials and coatings have been tried over the years. Although some products proved more useful than others for specialized applications, the industry has evolved to the point where a few products are now standard.

Glass, thermoplastic, vinyl, and polyester film have all been tried as scribe base materials. Polyester film is clear, extremely tough, flexible, and exhibits excellent dimensional stability under a wide range of thermal and humidity conditions, and is the most common base in use today. The film is available in thicknesses varying from .0025 to .01 inch, with the .0075 inch thickness in most general use.

The requirements for a good scribe coating are even more exacting than those for a suitable scribe base. The zone of optimal quality with respect to the hardness, thickness, and translucence of the coating, and the adhesion of the coating to the base material seems to be very limited. Years of experimental work were required to arrive at the optimal combinations of these factors that today's high-quality scribe coats exhibit.

Different types of scribe coat are available to suit different applications. If it will be necessary to overlay the scribe sheet on the guide image, then it would be desirable to use material with a transparent dye coating because they can be used without a light table. Overlay scribing is made difficult, however, by light diffusion through the coating. As a rule it is advantageous to have the guide image placed on the surface of the scribe coat, so that it can be viewed directly rather than indirectly through the material. A variety of relatively opaque scribe coats have been specifically designed for this purpose. The pigment-type coatings are generally preferred to transparent dye coatings for most scribing operations because guide images and scribed lines are easier to see.

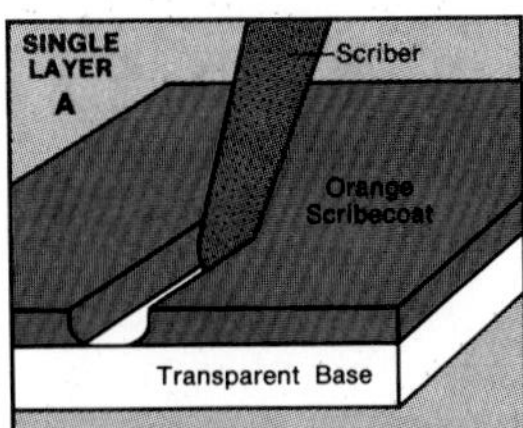

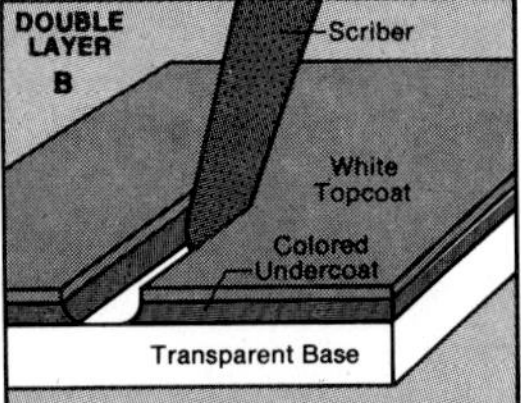

FIGURE 18.13 Single-layer scribe coat is most popular and is generally used over a light table. If a light table is not available, special double-layer scribe coat may be more convenient, however.

Single-layer scribe coats of the pigment type are available in such colors as rust (most common), red, and yellow (Fig. 18.13A). Scribing with these coatings is usually done over a light table to make the scribed lines stand out and to review the quality of the scribed work. If a special white pigment coating is used, the scribed sheet can be viewed and used as positive artwork by sandwiching black material beneath the scribe sheet.

Special double-layer scribe coats are sometimes used when it is necessary to scribe without a light table (Fig. 18.13B). White on rust, green, or black are the most common colors. Lines scribed through the white top layer are outlined by the darker second layer, making it easy to follow the progress of the scriber.

Guide Image. As with positive artwork, a worksheet should be compiled from which the final artwork is prepared. Although the translucent quality of the scribe coat permits tracing the worksheet on a light table, this method is seldom used because tracing an image through the scribe coat is hindered by light diffusion. Instead, scribing is more easily and accurately done if an image of the worksheet is first placed directly on the scribe coat. Even then, however, most scribing is still done on a light table.

The worksheet is usually produced in positive form on translucent material at the proposed scribing scale. This permits transfer of the image to the scribe coat base without the use of a camera or darkroom facilities. If, for some reason, it is necessary to prepare the worksheet at a larger scale, it can be reduced to the desired size by photography and then contacted to the scribe base.

A guide image can be placed on a pigment-type scribe coat in two ways. The simplest procedure is to obtain one of the commercially available scribe sheets with a surface that is precoated with a sensitized emulsion. Scribe sheets sensitized with a diazo coating require that a positive of the line image worksheet be exposed to an ultraviolet source in contact with the scribe coat and the guide image developed in ammonia fumes. The result will be a positive guide image.

Scribe sheets sensitized with a photographic emulsion for use with line image worksheets are available in both negative- and positive-acting form. The worksheet is exposed in a platemaker in contact with the scribe coating to transfer the guide image. Regardless of whether you start with positive or negative artwork, by choosing the proper photographic emulsion, it is possible to place either a negative or positive guide image on the scribe coat. The worksheet is exposed in a platemaker in contact with the scribe coating to transfer the guide image. Since some scribers find a negative guide image easiest to work with over a light table, you may want to experiment to find the procedure that best suits your needs.

If it is desirable to scribe from a continuous tone guide image, such as an aerial photograph, a special emulsion is needed. K & E produces one such product with the brand name *Stabilene Contone Film*. This material is made up of a photomechanical emulsion on the surface of a double-layer (white/rust) scribe coat. When it is necessary to produce a line negative in exact registry with a photographic image map, this product is superior.

Although presensitized scribe sheets are convenient, they have several drawbacks. They are expensive, and their shelf life is limited, which means that it is difficult for the low-volume user to obtain consistent results. In addition, their emulsion can only be developed once, limiting the scriber to a monochrome guide image. We can overcome these drawbacks to some degree by purchasing regular (unsensitized) scribe coat and applying the sensitizer ourselves.

Do-it-yourself sensitizing may be done with either a diazo or bicromate emulsions. Although bicromate emulsions are negative-acting, thus requiring a negative worksheet if a positive guide image is desired, they do tend to be more versatile than a diazo sensitizer because bicromate emulsions are available in several colors, which permits a color-separated guide image to be built up through successive applications and developments of the colored sensitizers. To accomplish this end, the worksheet artwork would, of course, have to be separated on the appropriate flaps.

Separation Flaps. Before beginning the worksheet, we should consider whether separation scribe coat negatives will be necessary. If linework is to appear on two or more scribe coats, it may be advisable to separate the information on separate flaps of the worksheet. For example, if drainage on the final map is to be printed in blue and the roads in red, each will be scribed on a separate sheet. If the information is separated on two flaps of a worksheet, a separate image can be placed on each scribe sheet, or both images can be contacted to the scribe sheets in different colors. Actually, contacting all the flaps in different colors to each scribe coat can reveal discrepancies in the worksheet flaps. A road, for example, may have been plotted in error so that it crosses a small lake. The multiple image would show this, and the proper position of the road would need to be established.

For very detailed and complicated maps, a modification of the multiple printing of the worksheet flaps might be used. Assume that the hydrographic features for the map were compiled from very accurate large-scale topographic sheets, roads were selected from a somewhat smaller-scale, more generalized map, and lines such as soil boundaries had been originally plotted on aerial photographs. Since the hydrographic features are likely to be in the most accurate planimetric position, that image is transferred to the scribe coat and the lines scribed. The soil boundaries are related to the hydrographic features and, in some cases, might end abruptly at lakes or streams. The second scribe coat will include the image of the worksheet of the soil boundaries and the scribed image of the hydrographic features. The soil boundaries are then scribed with the hydrographic detail as a guide. On the last scribe coat, the already scribed images of the hydrography and soil boundaries are combined with the worksheet flap of the roads. The delineation of the roads can then be adjusted slightly to fit the previously scribed features.

Image Orientation. Scribing procedures initially may seem strange because the scriber must learn to "think in reverse." The reason is that a negative image is created directly, rather than indirectly through a camera step as is the case in positive ink drafting. Since the camera reverses positive map copy, the scriber must perform this same function. Working in negative requires some special thought. The image orientation of the scribe sheet should be coordinated with the reproduction process that will ultimately be used to duplicate the map (Fig. 18.14). If, for example, the lithographic printing process is to be used, the image placed on the scribe coat should be wrong reading because in the lithographic process the plate is prepared by placing the emulsion of the film or the coating of the scribe sheet face down in contact with the plate. If the emulsion side is "up," light tends to creep because of the thickness of the film, and the image on the plate might not be as sharp as it could be. A lithographic negative should, therefore, have a wrong-reading image when viewed from the emulsion side. Then when the negative is placed emulsion side against the plate, the resulting image on the plate will be right-reading.*

*In photoengraving a letterpress plate, a very thin film is used so it can be "flopped" to produce a wrong-reading plate. For letterpress printing, then, the image should be scribed right-reading on the scribe sheet, so that the scribe coat emulsion can be in contact with the plate during exposure.

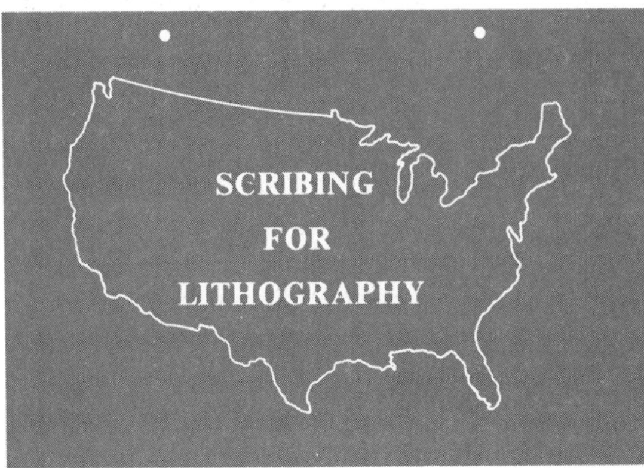

FIGURE 18.14 For best results guide image orientation on the scribe coat should be wrong-reading for reproduction by offset lithography.

Scribing Instruments. Although scribing tools are available in many design variations, there are only a few basic kinds of instruments. Regardless of design, each of these instruments has two components: a cutting point and a point holder called a graver or scriber.* The type of point and the way it is held (or guided) are crucial in scribing. To become a skilled scriber, you must learn to match properly the choice of cutting point and point holder to the different scribing tasks.

Cutting Points and Blades. Scribing points are designed for two types of cutting action. A *round* (needle) or *conical* point is most commonly used (Fig. 18.15). No special holder is required to guide these points because sideways pressure in any direction will remove the scribe coat. Round points are designed for use in freehand point holders with which it is difficult to hold the point vertical to the surface of the scribing medium. Conical points, on the other hand, require some form of rigid holder to keep the tip perpendicular to the scribing surface.

*Although the terms graver and scriber are often used interchangeably, in order to avoid confusion in this book they are given different definitions. Graver will be used in reference to the point holder and the term scriber will refer to the person doing the scribing.

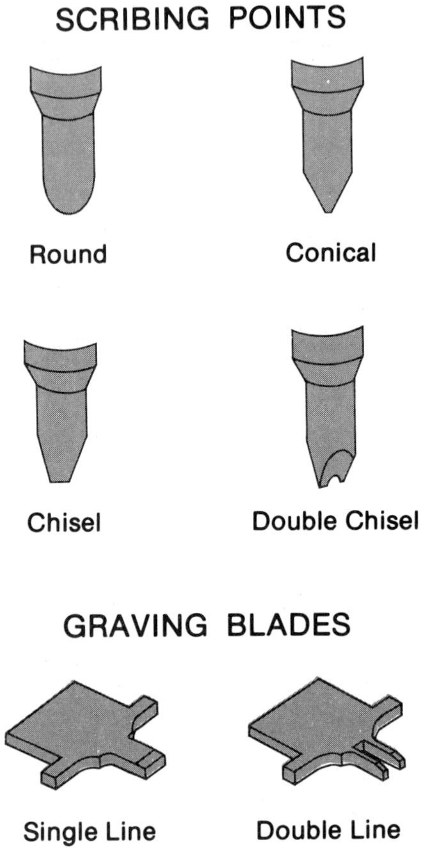

SCRIBING POINTS

Round Conical

Chisel Double Chisel

GRAVING BLADES

Single Line Double Line

FIGURE 18.15 Scribing points are designed for two types of cutting action. To be used effectively, they require point holders of different designs.

Round and conical points have several drawbacks. For one thing, the pressure required to move the point through the scribe coat increases quite dramatically as the size (diameter) of the point increases. Furthermore, a round or conical cutter is not suitable for scribing special point symbols such as small dots or squares.

The limitations of a round or conical scribe point are overcome to a large degree with either a *chisel-shaped cutting point* or a chisel-tipped *graving blade* (Fig. 18.15). These relatively blunt-tipped cutters remove the scribe coating more cleanly than a round or conical point, although to do so the direction of movement must be at a right angle to the edge of the cutter. To ensure that the cutter is always in proper orientation to the movement of the point, a rather elaborate *swivel-head* point holder is required (see below). Multiple-tipped chisel points and graver blades are available for cutting double-line and triple-line symbols, as are special cutters for making point symbols.

The choice of material used to make scribing points is critical because it determines how they must be handled and whether or not they will eventually need sharpening. A sharp point is necessary to cut a clear, crisp line and to keep scribing effort at a minimum.

Steel points are the least expensive and can be handled rather roughly without breaking. Because the base material is abrasive, however, steel points do tend to wear with use. The result will be a change in line width. Steel points can be sharpened periodically with the use of a special jig, but this tends to be an exacting, tedious task. In the long run it is often more efficient simply to replace these rather inexpensive points when they become worn.

Jewel scribing points (generally artificial sapphire) are more expensive and more brittle than steel points; thus, they must be handled with great care. Jewel points are easier to use than steel points on some scribe coatings. Their greatest advantage, however, is that they strongly resist wear.

Both steel and jewel points come in a variety of sizes and styles. Point sizes are measured in tenths of a millimeter or thousandths of an inch. A set of points with a broad range of sizes and styles is generally required for mapping purposes.

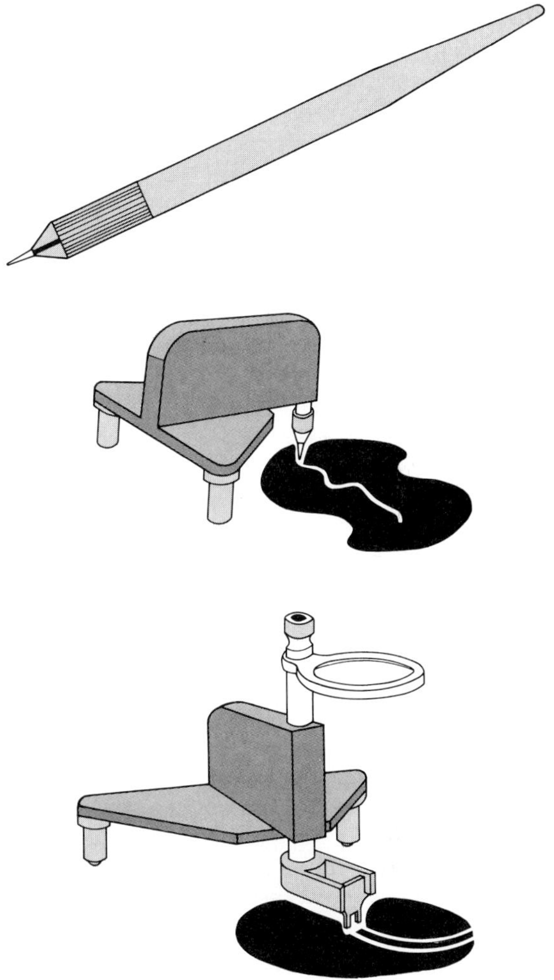

FIGURE 18.16 A variety of tools are used to hold scribing points in proper orientation. Each design serves a special purpose.

Point Holders. Simple hand-guided gravers are sometimes used for scribing fine lines and for doing final "touch-up" work. These *fineline gravers* may be nothing more than a vice or chuck to hold the scribing point (commonly a steel phonograph needle) and a pen-type handle (Fig. 18.16). For this reason this type of instrument is sometimes referred to as a *penholder graver.* More elaborate freehand models incorporate such items as a magnifying glass or an offset point vice, which holds the point in a near-vertical position and keeps the handle out of the scriber's line of sight when the tool is held in normal writing position.

Fineline gravers are relatively difficult to use. Considerable practice may be required to develop the proper "touch." Too much pressure makes the fine point tend to gouge the scribe base. Too little pressure means that the desired line weight is not obtained. Variable pressure leads to inconsistent line weight. If the scribe point is not held in a near-vertical position, all three problems may occur. Because of the difficulties of freehand scribing, fineline gravers are best used sparingly, at least until sufficient skill has been attained in their use.

The problems associated with freehand scribing are largely overcome by using a tripod-type point holder. The *rigid graver* is the simplest instrument of this type. The standard version consists of a horizontal plate with a vertical center rib for a handgrip supported by two legs and the cutting tip which functions as a third leg (Fig. 16.18B). The scribing point is held in an adjustable chuck which permits the use of different cutter sizes and can be raised or lowered to level the instrument (to keep the cutting point or blade vertical).

On some rigid gravers a third supporting leg is used and the chuck is spring-loaded. This arrangement makes it possible to preset the desired scribing pressure by adjusting the spring tension. On still other models a turret head holding a selection of point sizes is incorporated into the instrument.

Both conical and chisel-edged scribing points can be mounted in a rigid graver. If a conical point is used, the graver can be moved in any direction. Unfortunately, scribing difficulty increases in direct relation to the size of a conical scribing point. In practice, conical points larger than .006 inch are rarely used. We can get around this problem by mounting a chisel-edged point in the rigid graver. However, when we do so, the instrument is good only for straight-line scribing along a straight edge because the cutting edge must be kept at a right angle to the direction the point is traveling.

Some of the problems associated with rigid gravers are overcome by using a tripod graver that incorporates a swivel chuck (Fig. 16.18C). This *swivel graver* arrangement permits both conical points (needles) and chisel-edged points (both single and multiple tipped points or blades) to be used in the instrument. Since the chuck "swivels" into proper position as the instrument is moved, the direction of movement is not a concern once the point is correctly aligned. The scriber does have to take care, of course, that the point is properly aligned before cutting pressure is applied to the graver. For the expert scriber, the greater cost of a swivel head and the extra effort required to master a swivel cutter are more than compensated by the higher quality of the resulting scribe work. Swivel gravers are particularly suited to scribing double irregular lines, such as roads.

Several semiautomatic gravers are also available to perform specialized scribing tasks. One example is the *dot graver,* which operates by rotating the cutter tip. Another form is the building graver. This instrument permits movement of the cutter tip a preset distance in one direction, producing squares or rectangles.

Scribing Technique. Scribing is sufficiently different from ink drafting to require its own special techniques to take full advantage of the method. It is necessary to know how to handle the special instruments, how to correct or alter scribed artwork, and how to clean up the final negative image. Since scribing is essentially a technique for constructing lines in negative form, it is also necessary to learn how to use special techniques for dealing with type, patterns, and point symbols.

Holding and Handling Instruments. If the hard, outer layer of the base material is penetrated, the scribing point will gouge the material, and controlling the path of the scribing tool becomes difficult, resulting in a ragged, irregular line. Thus the *scribing touch* is rather critical. Once you acquire a "feel" for the material and instrument, however, gouging presents no real problem. With the proper touch and correctly sharpened points, the scribe coating is removed smoothly and cleanly.

Each scribing instrument and each point style and size requires its own special handling technique. Most people find it easiest to start with a rigid graver in the .004 to .006 inch point size range. As you gain skill, you can then work to smaller and larger sizes. Each point size will take a different amount of pressure, which may be adjusted by bearing down on the instrument with different force, or by moving the fingers closer or farther from the point. It is difficult to start cutting lines over .02 inch with a round point unless the graver is tilted slightly in the direction of the intended cut when the line is started. Once the rigid graver has been mastered, go on to the harder-to-use swivel gravers, and repeat the process. Finally, go to the pen-type gravers, which are still harder to learn to use.

The direction a scribing instrument is moved relative to the body position is not as critical in producing high-quality linework as it is in ink drafting. Some people prefer to pull the instrument toward them, while others find that pushing the instrument away works best. Most scribers like to have the scribe sheet loose (not fixed) on the light table, so that it can be freely rotated and moved about during scribing. An exception might be when scribing long straight lines, where a fixed scribe sheet tends to be more convenient.

Corrections and Alterations. Sometimes it is desirable or necessary to alter or correct a scribed image. Changes are made by filling in the scribed line in question with an opaque material that can itself be scribed. Specially prepared liquid and crayon-type opaques are available for this purpose. Unfortunately, scribing lines through hand-applied opaque is generally less satisfactory than working with the original scribe coat.

Cleanup. Bits of loose scribe coat are a natural by-product of the scribing process. Since these particles of material will block light and cause broken or ragged lines in subsequent photomechanical processing stages, they must be removed. The bulk of the debris may be removed by systematically brushing the finished scribe sheet with a soft-bristle brush. If particles still remain in the scribed lines, they can be removed by touching them with a tacky gum eraser. Another technique that works just as well is to press the sticky side of a piece of drafting tape against the scribed line. When the tape is lifted, the unwanted bits of material adhere to it and are removed from the base sheet. It often helps to check the quality of scribed lines and to look for loose bits of scribe coat under a shop microscope.

Point Symbols, Patterns, and Type. Lettering, areal tints or patterns, and complex point symbols create special problems for the scriber. Although special scribing devices have been developed and a variety of scribing templates and guides are available, their practical use tends to be limited. If only a few names, point symbols, or small patches of areal symbols are needed, they can be drafted or stuck-up (preprinted) in positive form, converted to negative form photographically, and then stripped into areas cleared of scribe coat.

Usually when a map is scribed, however, lettering, point symbols, and areal patterns are best handled with a separate positive flap. The normal procedure is to place a sheet of transparent or translucent material in registry over the right-reading scribe sheet. If scribing has been done wrong-reading for subsequent lithographic printing, it will first have to be flipped over. Once the artwork has been stuck up on the positive flap, the scribe sheet can be placed in registry over it to opaque as necessary to interrupt lines. Finally, the positive flap is photographically reversed to create a negative, which can be combined with the scribed negative artwork on the film or printing plate.

Mechanical Scribing. Mechanical scribing is carried out on a flatbed plotter in much the same way as drafting (see above). The difference is that the drawing surface and pen are replaced by a scribe sheet and scribing points. What has been said previously about the quality of output from incremental plotters holds true for mechanical scribing as well. But if suitable data are contained in the digital record and the plotter is capable of high resolution, the results can be more accurate than those achieved with manual scribing. Mechanical scribing is also faster than manual work.

The person who is operating mechanical scribing equipment must initially mount the proper size scribe sheet and select the appropriate scribing point size and style. Later, during the actual plotting the operator may also have to intervene to mount a new scribe sheet (when separation artwork is being plotted) or to change scribing points (when artwork is being constructed with different line weights and forms). Otherwise, the plotting operation is under the control of the mapping software and the digital record and therefore is automatic.

Photo Plotting

Map images can be plotted directly on photographic film or paper with devices called photo plotters (the terms photorecorders and photowriters are also used). Some of these machines work at map reproduction scale, while others operate in microfilm format. Both types of photo recorders have versions capable of handling either vector or raster data. Regardless of their design, however, photo plotters work at extremely high speed and produce precise, finished negative artwork.

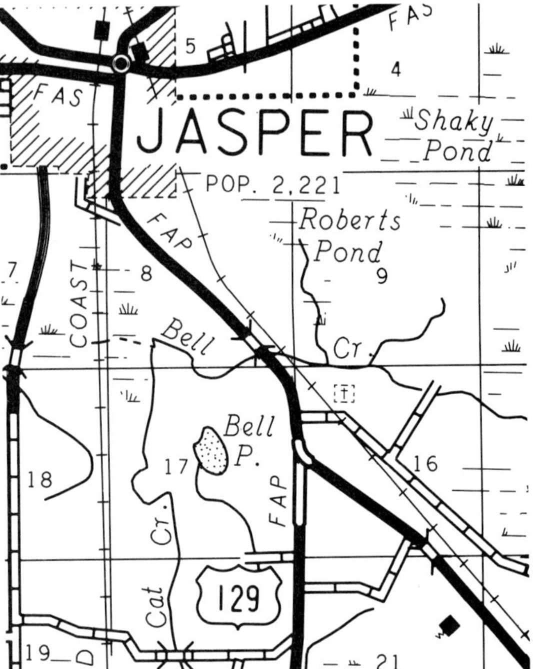

FIGURE 18.17 Large-format photo plotting can produce extremely high-quality film positives or negatives exhibiting a wide range of symbol types. (Plot courtesy, the Gerber Scientific Instrument Company).

Large Format. Large-format photo plotters are of two types. One type operates on a vector principle, while the other type operates in a raster mode. Vector photo plotters are actually modified versions of standard flatbed incremental plotters. The main difference is that an optical exposure head (photohead) is substituted for the pen or scribe point holder. The photohead is designed to produce a beam of light, which can expose light-sensitive material (paper or film) that has been mounted on the table surface. Most such machines are operated in a photographic darkroom.

The light beam is controlled in both size and intensity and for continuous or flash-exposing under program control of an aperture turret that might contain up to several dozen templates. Some devices even have a variable aperture photo exposure head, which provides a fast and accurate means of producing a variety of rectangular shapes, repetitive patterns, and line widths.

The quality of artwork constructed by these photo plotters can be outstanding (Fig. 18.17). Aperture sizes can be as small as .002 inch, resolutions go as high as .0001 inch (2.5 microns), and plotting speeds up to 60 inches per second have been attained. Add to these features the extremely high reliability that can be achieved with a machine that has so few moving parts and it becomes apparent why vector photo plotters are gaining importance in map production.

The second major class of large-format photo plotters includes those that work in raster form (Color Fig. 26). These machines convert digital density data from magnetic tape or disk files to continuous tone analog imagery on film or paper. Machines are available that can perform high-resolution reconstruction of composite color imagery directly on color film as well as produce black-and-white images.

Large-format photo plotters generally operate

with a laser light source. Up to 256 discrete modulation levels of the laser beam are standard. Raster (spot) size and turret-mounted apertures are usually selectable. The best of current machines can plot film sheets up to 30 by 30 inches at resolutions of 1000 lines per inch at a rate of a million pixels per second. Furthermore, the machines can be conveniently operated under normal light conditions.

Small Format. By taking advantage of microfilm technology, cartographers can produce negative artwork that is much smaller than the desired final map size. What makes this possible are the extremely fine resolution of the film that is used and the precision operation of the computer output on microfilm (COM) unit. When good equipment and materials are used, reproduction-size products that have been enlarged many times will show no evidence that they were produced on microfilm.

Several types of microfilm plotters are currently in use. One type operates on a vector plotting principle. In this case an electron beam traces out the linework image on a CRT screen from the coordinate descriptions of the lines that are held in the digital record. A physical image is created by exposing the screen traces to microfilm, which subsequently can be enlarged to reproduction size. Artwork such as contour maps and base map projection grids have been successfully plotted in this fashion (Fig. 18.18).

Computer output on microfilm can also be used for the high-speed production of open-window negatives (Fig. 18.19). These are then directly usable for making plates for color map printing (Color Fig. 10). In this case a raster plotter is employed. A computer program first has to convert the coordinate descriptions of the data areas to a raster format. The program next has to determine which film separations should have each region opened to produce a window. Finally, an electron or light beam sweeps across the microfilm line by line, turning on (to expose) or off as directed by the raster scan file. Except for its small size, the resulting film separation looks identical to one that might be made by more conventional manual and photomechanical methods.

COMPLEX ARTWORK

The cartographer with a good knowledge of duplicating and printing methods can take advantage of their capabilities to produce special effects for improving design and legibility. To use the techniques of duplicating trades most efficiently, the artwork must be carefully planned and the use of particular materials is often necessary. The following section describes some of these materials and provides some examples of their use. It is hoped that students will be able to adapt and expand these techniques and procedures for solving their particular problems.

Negative Masks

The cartographer has several choices when desiring to produce tints, patterns, or colors within given regions on a map. We have already discussed how this can be done through positive artwork. An alternative that generally produces higher quality results is to place the areal symbols in a given area by photomechanical means. To do so a negative is needed that is actinically opaque except in that area in which the areal symbol is desired. These are called *negative masks* or *open-window negatives.* The idea is to make the border of the open area fall exactly on the bounding line of the region to be filled. The required negatives can be constructed manually, mechanically, photomechanically, or electronically.

Ink and Stickup. Preparing an open-window negative completely by hand can be a tedious job, especially if the area boundary is complicated. The most convenient procedure is to ink a wide border carefully around the region using a fairly small pen size so that high accuracy can be achieved, and then fill the center of the region with a restickable masking film such as Para-

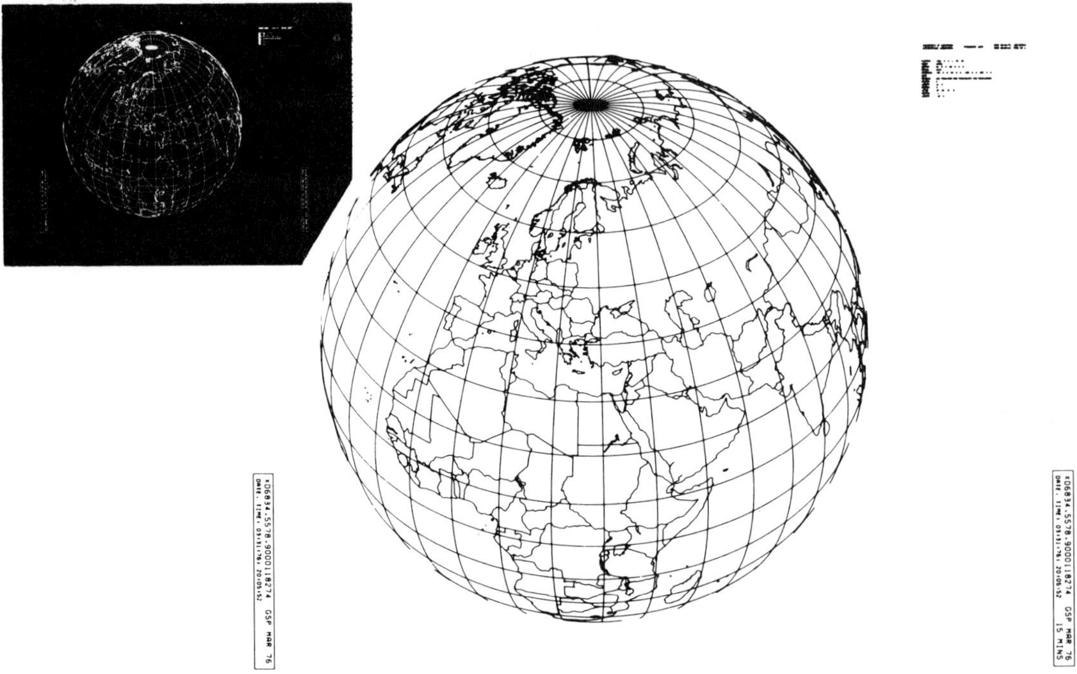

FIGURE 18.18 An enlargement of a line map produced from the microfilm strip shown at the top left.

paque or Kimoto Mask.* Alternatively, the central area could be inked with a heavy pen size, although caution must be exercised, since large areas of ink may tend to crack and peel after drying. In any case, artwork done at scale on a translucent base can be used directly in a contact frame. Artwork done at other scales requires camera work to produce the proper size open window.

Cut and Peel. Open-window negatives can also be prepared by using special materials consisting of a thin masking film laminated to a clear polyester base.** The window is created by using a sharp blade to cut around the border of a region and then peeling the masking layer away from the base. The cut-and-peel method is fast and accurate if regions have simple geometric shapes or rather smooth curved boundaries (or thick-lined borders). But it becomes progressively more tedious and less satisfactory as a region becomes more detailed or complex in shape. Furthermore, it is generally easier and more accurate to ink than to cut material along a guide line.

Etch and Peel. Open-window negatives can also be prepared photochemically. Several kinds of sensitized materials are available for this purpose.* They are alike in appearance, each having a thin ruby film adhered to a transparent polyester base sheet. No mechanical cutting is involved. Instead, the lines bounding areas are ex-

*Parapaque is a product of Para-Tone, Inc., 512 Burlington Ave., LaGrange, IL 60525. Kimoto Mask is a product of Kimoto USA, Inc. (East), 1116 Tower Lane, Bensenville, IL 60106.

**Rubilith and Amberlith are manufactured by Ulano Graphic Arts Supplies, Inc., 610 Dean St., Brooklyn, NY 11238. *Kimoto Strip Coat* is produced by Kimoto USA, Inc. (East), 1116 Tower Lane, Bensenville, IL 60106.

*Striprite is produced by Direct Reproduction Corporation, 811–13 Union Street, Brooklyn, NY 11215. Peelcoat Film is available from the Keuffel and Esser Company. *Kimoto Peel Coat* is produced by Kimoto USA, (East), 1116 Tower Lane, Bensenville, IL 60106.

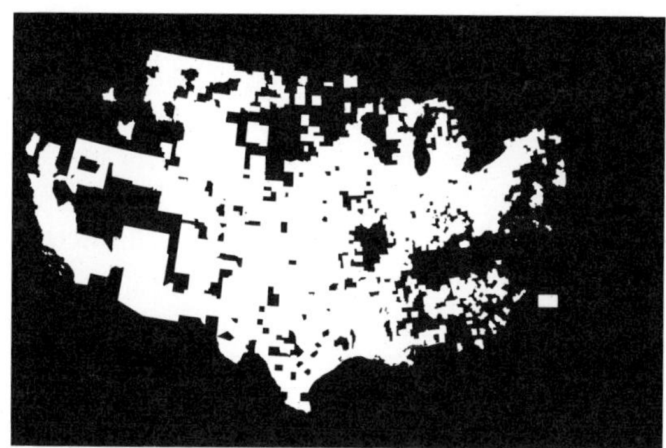

FIGURE 18.19 One of a number of open-window negatives that were produced as computer output on microfilm in conjunction with the reproduction of a complex colored map. (Courtesy, U.S. Census Bureau).

posed to the material in a vacuum frame with a source rich in ultraviolet light, after which the sheet is developed and then chemically etched.

After etching is complete, but before peeling begins, unwanted lines on the processed sheet must be blocked-out. This is done with a special orange, water soluble opaquing material.* The opaque can be applied by hand with a squeegee or by mechanical whirler. Once the opaque has dryed, the peeling step can be carried out as it is with other peelable masking films. For easy peeling, lift a corner of the material to be removed with a knife and strip it off with a piece of adhesive tape. As a final sep, the opaque *trap line* that remains along the edges of an open window after peeling may be removed using a moist cotton swab (a Q-Tip is ideal).

Peel coat materials may be processed under normal room lighting conditions and can, therefore, be used in cartographic drafting rooms that have limited equipment. All materials are processed using negatives; however, *Striprite* is also available in a form that permits the use of positives. All materials produce good-quality negatives that can be registered almost perfectly. A pin registry system is used throughout the process, of course, to insure registry of each open-window negative with the lines on the scribe coat.

Film Plotting. A fourth way to produce open-window negatives is to create them directly from a digital record using a film plotter (see above). In this case a spot of light or laser beam is scanned back and forth across the film in raster fashion, while it is being modulated by the electronic record of the map data. The film is exposed line by line until the image is completed. The plotting may be done at reproduction scale, or it may be done on microfilm and later enlarged for use as an open window in print or platemaking. Once the film has been plotted, the negative is handled in the same way as those produced by other means.

Negative Screens

Patterns and tonal values can be added to maps by using a contact screen in conjunction with an open-window negative. This can be done on paper or film in a vacuum frame or at the platemaking stage. It is necessary only to insert the appropriate screen between the carefully registered negative and the sensitized material. Since any reduction of the artwork has already been accomplished, there is not the danger of the details of the tints and patterns disappearing as there is when preprinted areal symbols are placed on the original artwork (Fig. 18.20). This means that contact tint and pattern screens generally produce higher quality results than are achieved with use of the equivalent positive areal symbols.

Pattern Screens. Patterns contact screens are the same as their preprinted pattern counterparts, except that they are in negative rather than positive form. Otherwise the comments that were made in previous discussion of positive patterns is equally appropriate with respect to contact pattern screens. Furthermore, since handling procedures for pattern screens are much the same as those outlined for tint screens in the following sections, no further discussion of pattern screens is provided here.

*A Keuffer and Esser Company product called *Mask Kote* is widely used for this purpose.

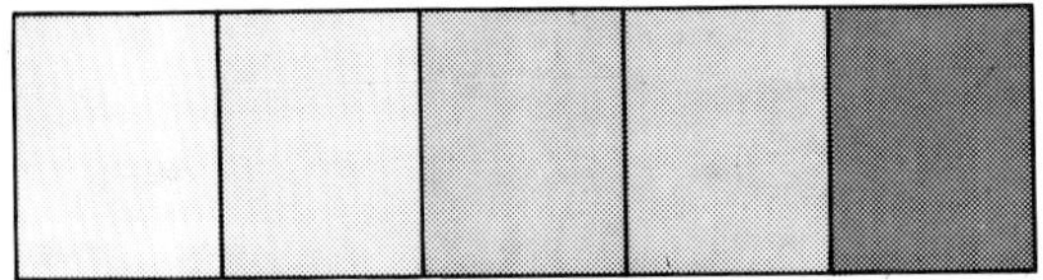

FIGURE 18.20 Preprinted dot patterns can be applied to the artwork to give an illusion of gray; however, the effect can be destroyed if too much reduction is necessary. When tint screens are introduced at the film compositing or platemaking stage, the amount of reduction of the original artwork need not be a consideration. The tints shown above were labeled as 10, 20, 30, 40, and 50 percent.

Tint Screens. These screens are composed of fine dots or closely spaced lines that produce the impression of gray on the printed copies. As with positive tints (see above), contact tint screens are specified by two measures: texture in lines per inch and density as percent blackness. Since the texture of contact tint screens is defined in the same fashion as it is with positive tints, the topic needs no further discussion. It should be noted, however, that it is possible to use finer contact tint screens than positive tints because normal image degradation associated with an extra photographic step has been eliminated.

The density rating of contact tint screens, on the other hand, warrants a few comments. The rating of the tone produced by tint screens is designated in percentages of ink coverage; therefore, a 10 percent screen will produce a very light tone and an 80 percent screen will give a very dark tone. Tint screens are considered to be in negative form; therefore, an 80 percent screen actually has very small opaque dots, with most of the film transparent, whereas a 10 percent screen is mostly opaque with small transparent openings.*

Tint screens graded by 1 percent intervals from 0 to 100 are not available. The usual set of percentage calibrated screens includes 10, 20, 30, 40, 50, 60, 70, 80, and 90 percent, and some series also include 5, 7½, 15 and 25 percent. Less common are screens calibrated throughout at a 5 percent interval.* Since this limited selection of screen values may not match those called for on the basis of good design principles, the cartographer often finds it necessary to work within available options. Combining screens to form composite tones may be the only way to approximate a desired value.

Multiple Screens. In monochrome printing, up to three tint screens can be printed on one area. In flat color printing, as in full-color halftone printing, a fourth screen is often added (see Chapter 17). Whenever more than one screen is used extreme care must be exercised in positioning the screens to prevent usually undesirable moiré effect (Fig. 18.21 and Color Fig. 21). Superimposing screens within a single color is a satisfactory method only if the screens are rotated so that the orientations of the lines or lines of dots are separated by angles of 30 degrees. If more than one color is to be used to print a map, each color is normally assigned a screen angle 30 degrees from the others (see Chapter 17). Screens of different colors can then be superimposed.

Normally a mechanical screen angle gauge or a graphic device is used to format standard screens so that they can be readily positioned at the proper angle (Fig. 18.22). The alternative to using guide devices is to use preangled screens, which can be purchased from companies that manufacture standard tint screens, or they can be made ahead of time by the cartographer from standard tint screens.**

*Custom-made tint screens can be produced by the cartographer by using a contact halftone screen in a vacuum frame. Exposing orthochromatic film through the screen produces uniform dots, the size of which are determined by the length of exposure. Since contact screens are available in lines, dots, and other patterns, numerous kinds of tint screens can be produced.

*Brochures detailing product lines available can be obtained from the manufacturers, including Borrowdale (250 West 83rd St, Chicago, IL 60620); LogEtronics, Inc. (7001 Loisdale Rd, Springfield, VA 22150); or ByChrome Co. (P.O. Box 1077, Columbus, Ohio 43216).

**J. M. Olson, "A Simple Technique for Pre-Angling Screens," *The American Cartographer* 9 (1982): 81–83.

FIGURE 18.21 Tint screens should be aligned with an angular separation of 30 degrees. Other angles can produce various moiré patterns such as the one shown.

The ability to superimpose screens means that it may not be necessary to prepare a separate flap for each tone. If, for example, we wanted values of 20, 40, and 60 percent, only two flaps would be required. On flap number one, the areas of 20 and 60 percent would be opaqued and a 20 percent screen specified. A second flap would include the areas of 40 and 60 percent and a 40 percent screen specified. Since the two flaps overlap, the 60 percent area would receive 20 plus 40 percent ink (Fig. 18.23). The cost would be reduced by one less negative and one less plate exposure.

A possible drawback of this method of creating hybrid tints by superimposing screens is that the resulting blackness (percent area inked) will actually be somewhat less than the sum of the screen percentages involved. Thus in the previous example, the combined 20 and 40 percent screens do not actually produce 60 percent area inked. The reason for this discrepancy is that some of the dots from one screen will overlap with some of the dots from the other screen. Kimerling has analyzed the cross-screening relationship and his values are given in Table 18.1. Fortunately, the deviation is normally small when compared with the cartographer's design thresholds and therefore may not be a serious problem.

Most often when gray tints are abutted, a black line is needed between them because making them join exactly is not usually possible. If, however, the tones are a result of the superposition of

FIGURE 18.22 A graphic template can be used to trim standard tint screens so that they can be quickly and accurately positioned to achieve proper angular separation between successive screen exposures.

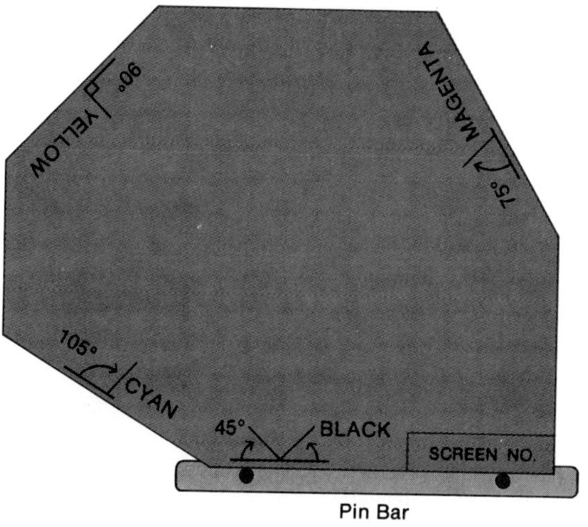

FIGURE 18.23 The ability to superimpose screens means that a separate flap may not have to be made for each tone desired on the map. In this example only two flaps are required to produce three values.

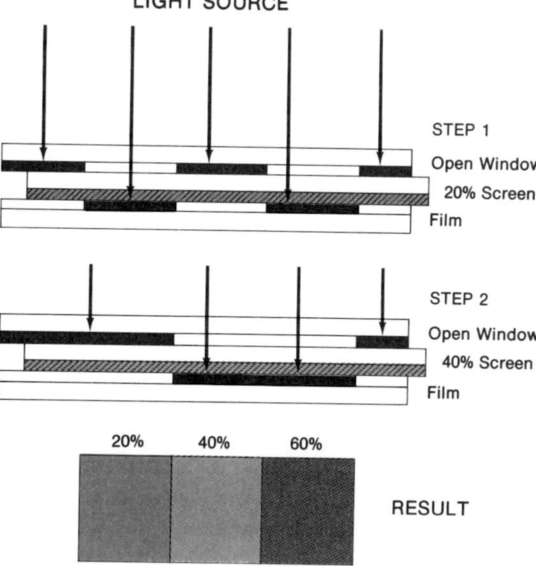

TABLE 18.1 Percent Area Inked for Various Combinations of Dot Tint Screens

Screen 1 (%)										
	10	19								
	20	28	36							
	30	37	44	51						
	40	46	52	58	64					
	50	55	60	65	70	75				
	60	64	68	72	76	80	84			
	70	73	76	79	82	85	88	91		
	80	82	84	86	88	90	92	94	96	
	90	91	92	93	94	95	96	97	98	99
		10	20	30	40	50	60	70	80	90
		Screen 2 (%)								

Source: A. Jon Kimerling, "Visual Value as a Function of Percent Area Inked for the Cross-Screening Technique," *The American Cartographer* 6, no. 2, (1979):141–48.

screens, the black line may not be necessary (Fig. 18.24). When a screened area is adjacent to a solid area, the artwork should always carry the tint into the solid. If we attempt to make them just join, white spaces are almost sure to occur.

Screened Symbols. A very useful technique is to screen the areas of preprinted patterns that have been applied to the artwork, which often tends to look rather harsh when printed in a solid color (Fig. 18.25). Also, superimposed lettering or symbols that might otherwise be lost can show clearly through the screened patterns. Again, a separate flap is prepared for screening. An open-window negative of this flap is then contacted with both the pattern screen (above) and the tint screen (below) in a vacuum frame (Fig. 18.26). The result will be patterned artwork that lacks strong contrast and looks rather gray in appearance.

When linework, type, or other symbols are screened, the tint screen dots can reduce the sharpness of feature edges. This situation can be improved somewhat by using finer textured screens

FIGURE 18.24 It is extremely difficult to make two tint screens match precisely (left). A tint superimposed on another (right), however, causes no difficulty.

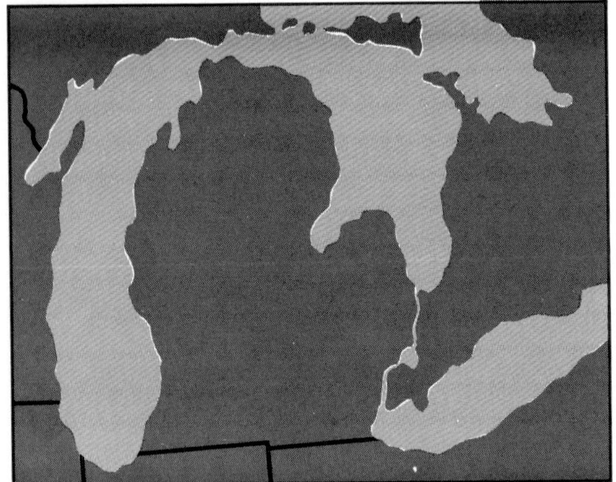

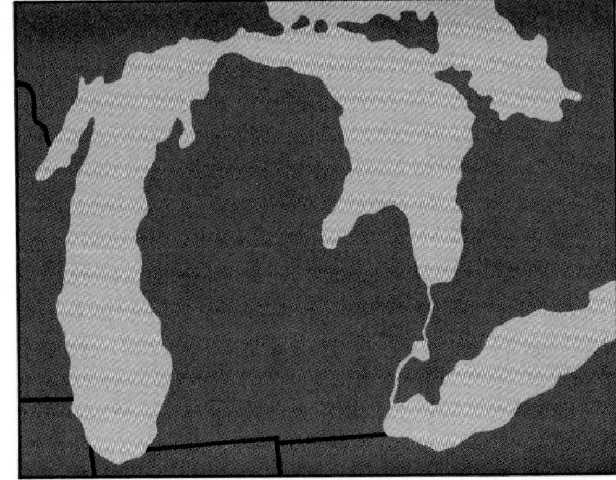

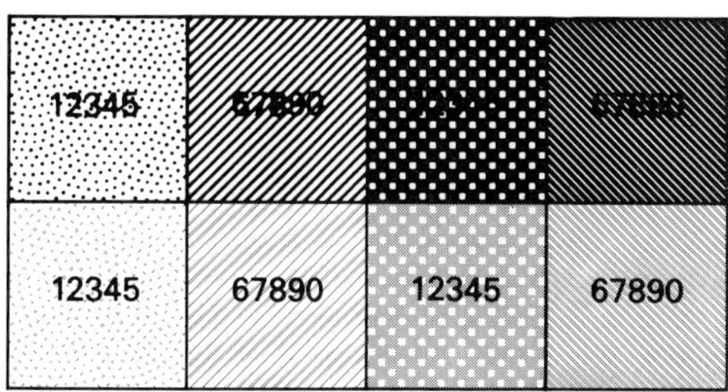

FIGURE 18.25 The patterns in the lower row have been screened to 30 percent of black. The patterns appear less harsh, and lettering or symbols applied remain legible.

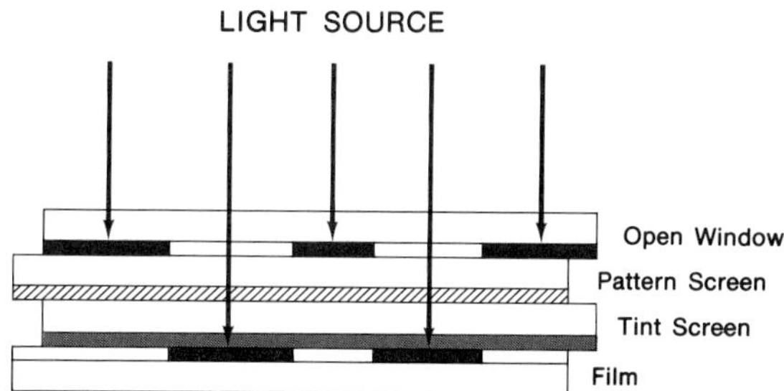

FIGURE 18.26 Patterns can be screened in a vacuum frame by using this arrangement of open-window negative, pattern screen, tint screen, and film. Note the emulsion orientations.

or bi-angled screens,* or by choosing coarser patterns. It is also possible for a moiré to develop when screening patterns are made up of fine dots or certain other marks. As a precaution, test the orientation of patterns and tint screens by superimposing them on a light table.

Lines, lettering, and symbols can also be screened, but very narrow lines and thin serifs on letters may be lost. Lettering is probably most successfully screened with line screens, which do not produce the ragged edges caused by dot screens. Bold lines and sans-serif lettering, when screened, can add much to the utility and overall appearance of a map.

Positive Masks

The opposite of an open-window negative (or negative mask) is a *positive mask*. Open windows are used to create special effects by permitting light to expose particular parts of a sensitized material. In contrast, positive masks block light from reaching certain parts of a sensitized material. The positive mask can be constructed in positive on a translucent surface for direct contact work. Alternatively, it can be obtained by exposing positive artwork to duplicating film or by exposing an open-window negative to standard reversal film.

Positive masks can be used in making vignettes, open or reverse artwork, and chokes and spreads. These applications are demonstrated in the following sections. The point to remember is that positive masks can be used to achieve map design attributes that in some cases would be extremely difficult to realize by any other means.

Vignettes. With the use of a contact halftone screen, it is possible to produce dots of a diminishing size or a vignette (see Fig. 18.27). This can be a useful technique when, for example, we wish to indicate an indefinite zone instead of the sharp boundary produced by a line or the unaltered edge of a pattern. Quite commonly, a vignette is used along a shoreline or coastline instead of employing a constant tint over the entire water body.

To produce this effect on water bodies, we need an open-window negative of the feature and a reverse or positive mask. In the contact printer we use a sandwich of film, screen negative, spacer, and positive, as shown in Figure 18.28. Light cannot reach the film where the open-window negative, in contact with the screen, protects the land area. Spacers, consisting of clear sheets of film between the positive and the open-window negative, permit the light to spread into the edge of the area of the water body. The amount of light reaching the film decreases with increasing distance from the edge, producing smaller and

*Bi-angled screens are created by superimposing two standard dot screens and thus exhibit an irregular or tandom or dot rosette pattern. Whereas a single (standard) screen will pick up parts of lines oriented in some directions better than in others, bi-angled screens have a better chance of preserving sharp edges on features oriented in all directions.

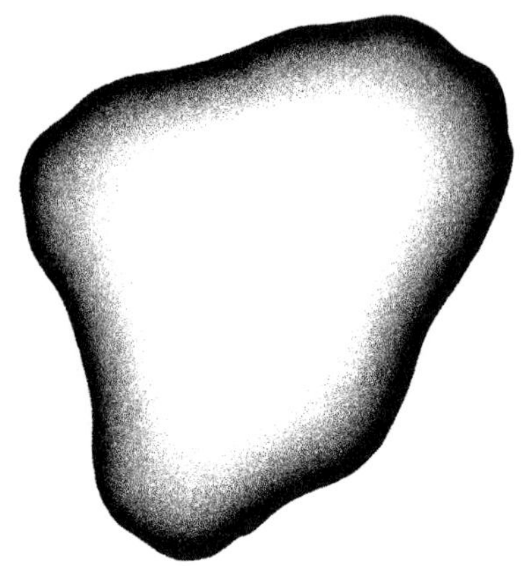

FIGURE 18.27 When it is more realistic to portray an indefinite zone than to show a sharp boundary, a vignette may create the desired effect. In this example the vignette is made up of diminishing size dots.

smaller dots. The size of the dots is related to the length of exposure. The width of the vignette band is determined mainly by the amount of space between the negative and the positive and, to some extent, by the length of exposure.

A positive prepared this way can also be used as a mask to vignette the edges of area tints and patterns. The positive blocks light in decreasing amounts away from the area boundary while an open-window negative is used to restrict the light to the image area. The area symbol must be in the form of a film negative.

FIGURE 18.28 A vignette can be produced using the above arrangement of film, contact screen, open-window negative, spacers, film positive, and diffuser with a contact frame.

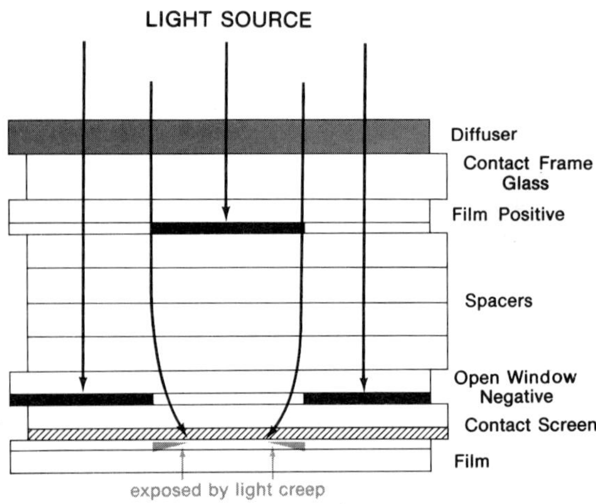

Relief Shading. Rendering the surface of the land in continuous shading requires special production as well as reproduction techniques. The conceptual basis of hill shading along with several manual and computer-assisted methods were covered in Chapter 15. Since automated procedures are under the control of mapping software and the concepts underlying several such programs were already outlined, they will not be considered further.

Instead, this section will focus on manual relief shading procedures. The idea is to create a suitable visual impression of undulating landforms by using a range of tones in combination with clear highlights. In doing so, there are several important things to remember. One is that good registry between the hydrography and the shading is critical. Secondly, if the shaded map is to be reproduced with plate-based methods, a halftone of the image will be required. If that is the case, the shading has to be done in such a way that the effects of the halftoning process on the quality of the image are taken into consideration.

One way to produce relief shading is to shade with graphite (pencil) or charcoal on a sheet of translucent material that has been registered over a contour map of the landform. This ability to separate the guide image from the shaded relief flap has several advantages. First, it is convenient because no special manipulations or extra steps are required prior to shading. In addition, there is no worry that the guide image will carry over through subsequent photomechanical processing of the image.

When an opaque medium such as illustration board is used, the guide image must be placed directly on the construction surface. This is commonly done in fine blue lines which are not supposed to affect the orthochromatic film used in making halftones. In practice, extreme care must be taken or the guide image will indeed show.

Opaque materials can be handled in several ways. One method is to use a mechanical airbrush. This device literally spray-paints tones (in the form of tiny dots) on a white background. Alternatively, a gray background is sometimes used. With this "dark-plate" method, shadow areas are further darkened with graphite, charcoal, or airbrush, and highlight areas are lightened with white pencil or chalk.

When the artwork for a map consists partly of line and partly of continuous tone, they should be prepared on separate flaps. In the halftone process even solid black areas are converted to dots and will, therefore, appear as dark grays. Lettering and other line symbols have a fuzzy instead of a sharp black appearance (Fig. 18.29). When separate flaps are prepared, the line work is photographed separately to produce a regular line negative and a halftone screen is used only to produce the negative of the continuous tone flap. The two negatives are then combined by exposing the one printing plate to each negative separately. The linework will be solid and its edges will be less fuzzy.

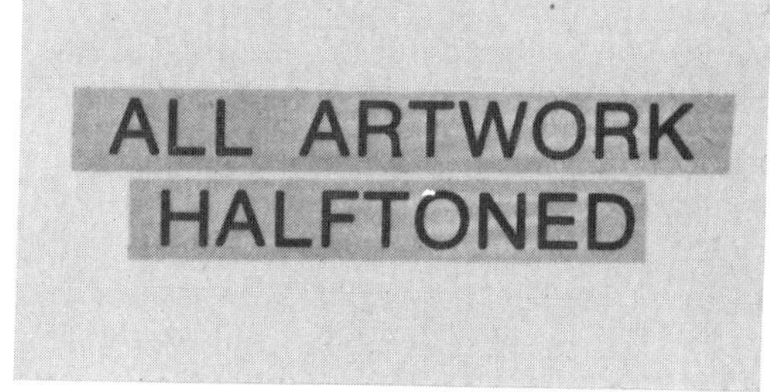

FIGURE 18.29 Lettering used in conjunction with a halftone should be placed on a separate flap and burned separately into the plate. When halftoned, the lettering (or any fine linework) tends to be fuzzy.

Open or Reverse Artwork. It is useful for the map designer to have the option of producing white lines or type on a dark background. Unfortunately, white lines and type are difficult to produce directly on the artwork. Opaque white ink will not flow properly in most pens, and it tends to flake off when the artwork is handled. White stickup materials work somewhat better, but still leave much to be desired. The most satisfactory method is to reverse artwork at the print (film) or platemaking stage. Reversing lines and type produces sharp images from artwork that is conveniently prepared with the usual black opaque materials.

The artwork that is to be open is produced on a separate flap and prepared in the same way as any other flap. One of the other flaps will, of course, be an open-window negative to produce either a tone or solid color for the background. Instructions to the printer indicate that a given flap is to be reversed from another. When the plate is exposed, a film positive of the flap to be reversed is placed, in registry, between the open-window negative and the plate (Fig. 18.30). The opaque positive artwork blocks the light and produces a nonprinting area on the plate. The white paper then shows through the tone or solid color, and white lines or type is the result.

The positive used to reverse can be referred to as a mask and indeed is useful in many cases to block out unwanted portions of images. Suppose a shaded relief map is to be prepared for an area in which numerous lakes are to have a smooth gray tone. Shading the terrain can be done on the first flap without regard for the lake outlines and actually should overlap into the lake area. A second flap, on which all the lakes are opaqued, will provide an open-window negative to be screened to produce an even tint (Fig. 18.31). When making the plate, a film positive of the

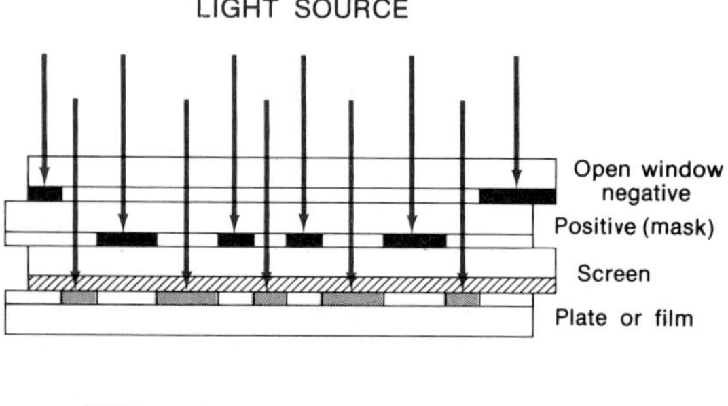

FIGURE 18.30 White (or reverse) lettering and symbols on a dark background can be produced by using a positive mask during the film compositing or platemaking stage.

second (lake) flap is placed over the halftone negative made from the terrain drawing, which will leave the plate unexposed in the lake areas.

FIGURE 18.31 The shaded relief was done without regard for the lake. A film positive (mask) of the flap that is to lay the tint in the lake has been used to reverse the halftone dots out of the shaded relief (step 1). An open window negative (mask) of the lake flap then is screened and lays an even tint in the lake (step 2).

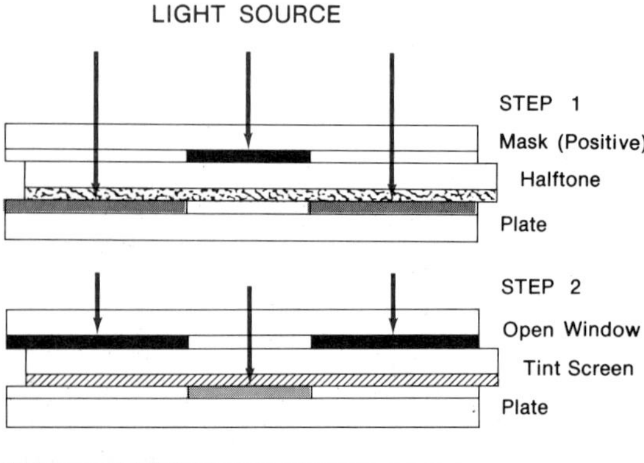

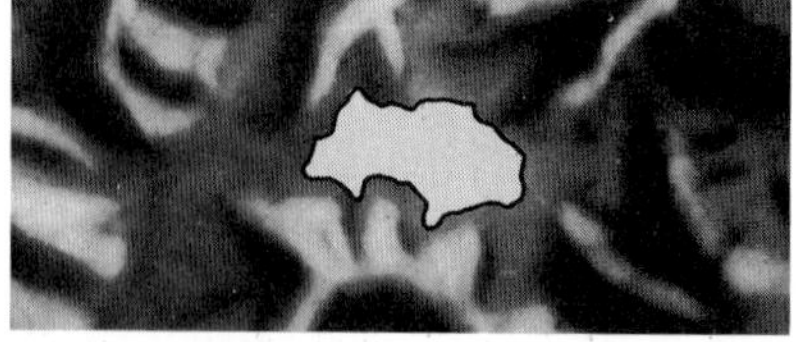

The open-window negative of the lake flap with an attached screen is then used to lay the even tint in the masked areas.

The background surrounding open artwork has a critical effect on the success of the technique. The same is true of solid artwork, of course. If the background is too light, white areas will not be distinguishable. If the background is too dark, the solid color will not be discernible. In both cases the problem is that the background will not provide enough contrast. Any dark background tint down to about 30 percent is adequate for reversing, while any tint up to about 60 percent is satisfactory for overprinting artwork in solid black (Fig. 18.32). These percentages change somewhat when the background tint or the solid artwork is printed in colors other than black, however.

When the background is variable in tone, special problems can arise. Reversed type or linework on continuous tone aerial photographs or shaded relief maps, for example, may exhibit visual contrast in some areas but not enough in others. The result is that type or lines may fade in and out from one part of the image to another, especially if they are rather thin in the first place. The same is true when printing special symbols in solid colors over existing background artwork. In the case of type, this problem is sometimes overcome by using a positive mask to open up rectangles within which solid lettering is placed. Unfortunately, this not only destroys background detail excessively, but is not very attractive aesthetically. A better method is to outline the reverse or solid artwork using a method described in the next section.

Chokes and Spreads. Special design effects can sometimes be created by making lines or other symbols wider or narrower instead of trying to maintain accurate widths. For example, we may wish to create outlines for more contrast with surroundings (see previous section). This can be done by using chokes and spreads produced from the original artwork, and then sandwiching them

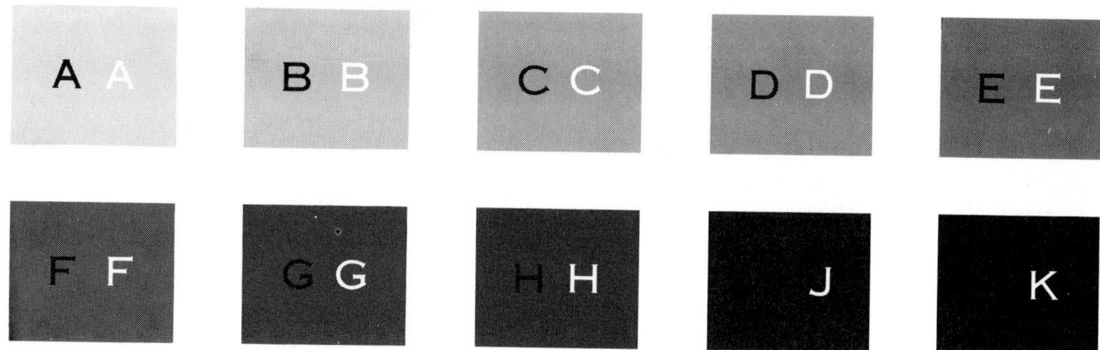

FIGURE 18.32 Lettering has been superimposed and reversed from tint screens ranging in equal value increments from 10 percent to solid black. We can determine from this illustration approximately the point where reversing or overprinting can be unsatisfactory.

together in manner similar to that used in making vignettes (see above).

The chokes and spreads are produced in a vacuum frame by separating the emulsions of a positive or negative and the film with clear spacers (acetate or film) during illumination (Fig. 18.33). The exposure is made through a sheet of frosted material, such as drafting film, that is placed on top of the contact frame glass.* This diffuses the light and allows it to undercut the image after it passes through the spacers. If reverses of the resulting chokes or spreads are needed, they can be produced by contact methods.

An outline of type or other symbols can then be made from solid artwork by contacting with a choke positive and a spread negative in register. Special effects, such as shadowing and highlighting, can be produced by shifting the positive and negative slightly out of register. A solid outline around a reverse on a tinted background can be made by first contacting the film with a positive choke in combination with a negative of the tint background, and next contacting a spread negative in register with the positive choke (Steps 1A and 1B in Fig. 18.34). In contrast, an open outline around solid artwork can be made by first contacting the film with the spread positive and a negative of the background tint, and next contacting it with a choke negative (Steps 2A and 2B in Fig. 18.34).

*An alternative method is to place the light source off to the side at an angle of 20 to 45 degrees with the vertical and then spin the vacuum frame to undercut the image.

Image Editing and Updating

Often it is necessary to make a few changes to an otherwise acceptable map. A new road may

FIGURE 18.33 Chokes and spreads are produced in a vacuum frame by separating the emulsions of the positive or negative and the film with clear spacers that encourage light creep during illumination.

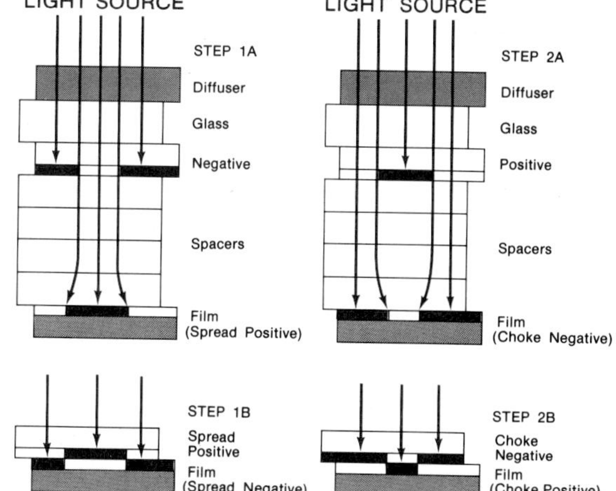

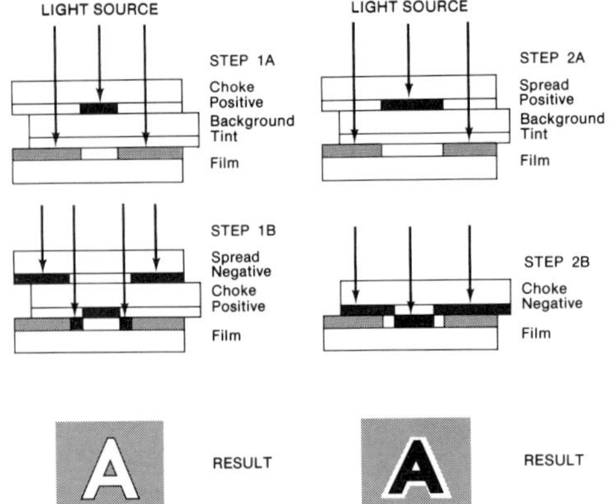

FIGURE 18.34 Special effects such as outlines, shadowing, and highlighting can be created by using different combinations of chokes and spreads as masks.

have been built, a railroad abandoned, buildings constructed, a reservoir created, and so forth. A material called sensitized or photo *Cronaflex** is especially versatile for these map updating or correction chores. Not only has this material been provided with a sensitized photographic emulsion, but it has an excellent drafting surface as well.

To use sensitized Cronaflex, negatives of base maps can be photographically contacted to the material in a vacuum frame. When developed and dried, existing image elements can be removed by scraping, and ink or preprinted materials can be added. The artwork can then be rephotographed with high-quality results.

Sensitized Cronaflex is especially useful when a series of thematic maps is to be made on the same base. The base needs to be constructed only once, and a negative of the base contacted to produce the required number of high-quality prints. Each of these base images can then in turn be completed by adding special information in the form of inked or preprinted symbols.

*E.I. DuPont de Nemours and Co., Photo Products Department, Wilmington, DE 19898.

Artwork for Plate-Based Color Printing

Colored maps are usually produced from artwork that is prepared in black and white. The color results from separate press runs using different colored inks. At least one flap—and often more than one—is necessary for each color. The cartographer furnishes a series of registered flaps to the printer with specific instructions for processing and printing.

A two-color map usually involves black and one other color, although it can be two colors other than black. Three-color maps ordinarily would have black and any other two colors. A four-color map, however, most often uses the subtractive primary colors, magenta, yellow, and cyan, along with black. Theoretically, any other hue can be produced by combinations of varying amounts of these primary colors. In plate-based color printing the combining is accomplished by successive press runs, and the amount of each color is controlled by proper screens. Charts showing the hues produced by superimposing different percentages of magenta, yellow, cyan, and black are prepared by most printing firms (Color Fig. 9). When choosing colors, the cartographer should use a chart prepared by the same printer who will reproduce the map. If another chart must be used, a copy of it should be furnished to the printer to indicate the precise color desired.

To demonstrate the procedure that we might follow in planning and preparing the artwork for a colored map, consider an actual case of a map that requires separate colors for ten categories of information and light blue for numerous small lakes (Table 18.2). From this listing we can prepare a chart that will reveal the number of flaps necessary to produce these colors and indicate which areas are to be opaqued on each flap (Fig. 18.35). Other flaps that are necessary will include a solid black flap with the base information, lettering, and area boundaries, a solid blue flap for the hydrographic features, and a 20 per-

TABLE 18.2

Area	Colors	Result
1	Yellow solid, magenta 60%	Light orange
2	Yellow 60%, magenta 60%, cyan 10%	Reddish orange
3	Yellow 60%, magenta 10%	Medium yellow
4	Yellow solid, magenta 20%	Dark yellow
5	Yellow solid, magenta solid, cyan 10%	Dark orange
6	Yellow solid, magenta 10%, cyan 20%	Light green
7	Yellow 40%, magenta 40%, cyan 20%	Beige
8	Yellow 40%	Light yellow
9	Yellow 60%, magenta solid, cyan 10%	Red
10	Yellow solid, magenta 10%, cyan 80%	Dark green
Lakes	·Cyan 20%	

cent blue to lay a tint on the lakes. A separate flap for the lakes is also prepared to provide a film positive that can be used as a mask when exposing the plates for all other colors. This will prevent the other colors from falling within the lakes.

To produce the various flaps in positive form, it is customary to secure copies of the base map on a stable material with the image in light blue. Using the blue lines of the areas as a guide, black opaque is applied to the appropriate areas on each flap. When the open-window negatives are prepared, the light blue images will disappear. A specified tint screen is then attached to the negatives and the plates are prepared according to a special instruction sheet that is sent to the printer along with the artwork (Fig. 18.36).*

To produce a map on which linework is to be in different colors, another method can be used in which all linework is placed on one flap. When the art is photographed, multiple negatives are made, one for each proposed color. Then each negative is opaqued to leave only that part of the image that is to be printed in a particular color. Since most film has relatively good dimensional stability, excellent registry can be maintained.

*Screen percentages on printed maps depend on a variety of factors, including platemaking, the type of ink, the nature of the press, and the characteristics of the printing medium (such as paper). What this means is that the printer should be told what screen percentage is wanted on the final map, not what percent screen to use. The printer then has the flexibility to manipulate and control the various factors in the best way possible to achieve the desired result.

FIGURE 18.35 When complicated color combinations are necessary, a chart of this sort can be a helpful guide for the cartographer. Numbers in the boxes represent the area designations from Table 18.2.

	YELLOW	RED	BLUE
Solid	1 4 5 6 10	5 9	
80%			10
60%	2 3 9	1 2	
40%	7 8	7	
20%		4	6 7 Lakes
10%		3 6 10	2 5 9

UNIVERSITY OF WISCONSIN CARTOGRAPHY LAB
PRINTING INSTRUCTIONS

MAP NUMBER ______ PRINTING SIZE ______
NUMBER OF FLAPS ______ COLORS ______

Flap (a) - Color ______, Solid or ____%, Reverse from ____, Halftone ______
Flap (b) - Color ______, Solid or ____%, Reverse from ____, Halftone ______
Flap (c) - Color ______, Solid or ____%, Reverse from ____, Halftone ______
Flap (d) - Color ______, Solid or ____%, Reverse from ____, Halftone ______
Flap (e) - Color ______, Solid or ____%, Reverse from ____, Halftone ______
Flap (f) - Color ______, Solid or ____%, Reverse from ____, Halftone ______
Flap (g) - Color ______, Solid or ____%, Reverse from ____, Halftone ______
Flap (h) - Color ______, Solid or ____%, Reverse from ____, Halftone ______
Flap (i) - Color ______, Solid or ____%, Reverse from ____, Halftone ______
Flap (j) - Color ______, Solid or ____%, Reverse from ____, Halftone ______
Flap (k) - Color ______, Solid or ____%, Reverse from ____, Halftone ______
Flap (l) - Color ______, Solid or ____%, Reverse from ____, Halftone ______

Additional Comments:

FIGURE 18.36 To minimize the potential for map reproduction problems, a complete set of instructions should accompany clearly labeled artwork when it is sent to the printer. A standard instruction sheet is handy for this purpose.

Multiple Use of Negatives and Positives

With the capabilities of the printer to produce tints and reverse linework, the cartographer has many opportunities to enhance the map by various combinations of white, grays, and black. The decisions concerning the use of such combinations are made at the design stage, so that before drafting is begun, the artwork has been planned in a fashion that will produce the best results with a minimum of effort. A good knowledge of the possibilities and limitations of the printing processes is necessary for us to visualize how the various pieces of film are to be used. We must keep in mind, for example, that a print or a printing plate cannot be exposed to two negatives at the same time; that is, negatives require successive exposures.* On the other hand, more than one film positive, or film positives and a negative, can be *sandwiched* together and used for a single exposure. On one print or printing plate, then, there can be an image that resulted from a series of exposures, any of which can incorporate more than one piece of film.

*The use of contact tint screens to screen contact pattern screens is an exception to this rule, since both contact screens being used are normally in negative form (see above).

The following example will demonstrate that cartographers are not as limited in one-color printing as one might believe. Suppose a design program specifies that the hydrographic features and section lines of a map are to appear white, while the swamp areas, county boundaries, and township lines are to appear solid black on a background flat tone of 40 percent black (Fig. 18.37). Hydrographic features must be reversed from the background tint and from the swamp areas, but the black boundaries are to print over the white.

Four flaps are used to produce the map.

Flap A All lowland areas.

Flap B Entire area of map opaque to provide open window negative for overall tint screen.

Flap C Hydrographic features and section lines

Flap D County boundaries and township lines.

All the flaps can be prepared in positive form with black ink and *Zip-A-Tone*.

The following are the instructions given to the printer.

1. Lay a 60 percent tint screen on a negative of flap B, and, using a positive of flap A as a mask, contact to a new film.
2. Double print on the plate:
   a. The new film with a positive of flap C as a mask.
   b. The negative of flap D.

When a 60 percent screen is used with the negative of flap B and contacted to the new film with a positive of flap A, the screen is reversed to 40 percent. It will not cover the low wet areas, however, since the mask holds the light back and transparent areas result. If the plate were to be exposed to the new film, a 40 percent tint would

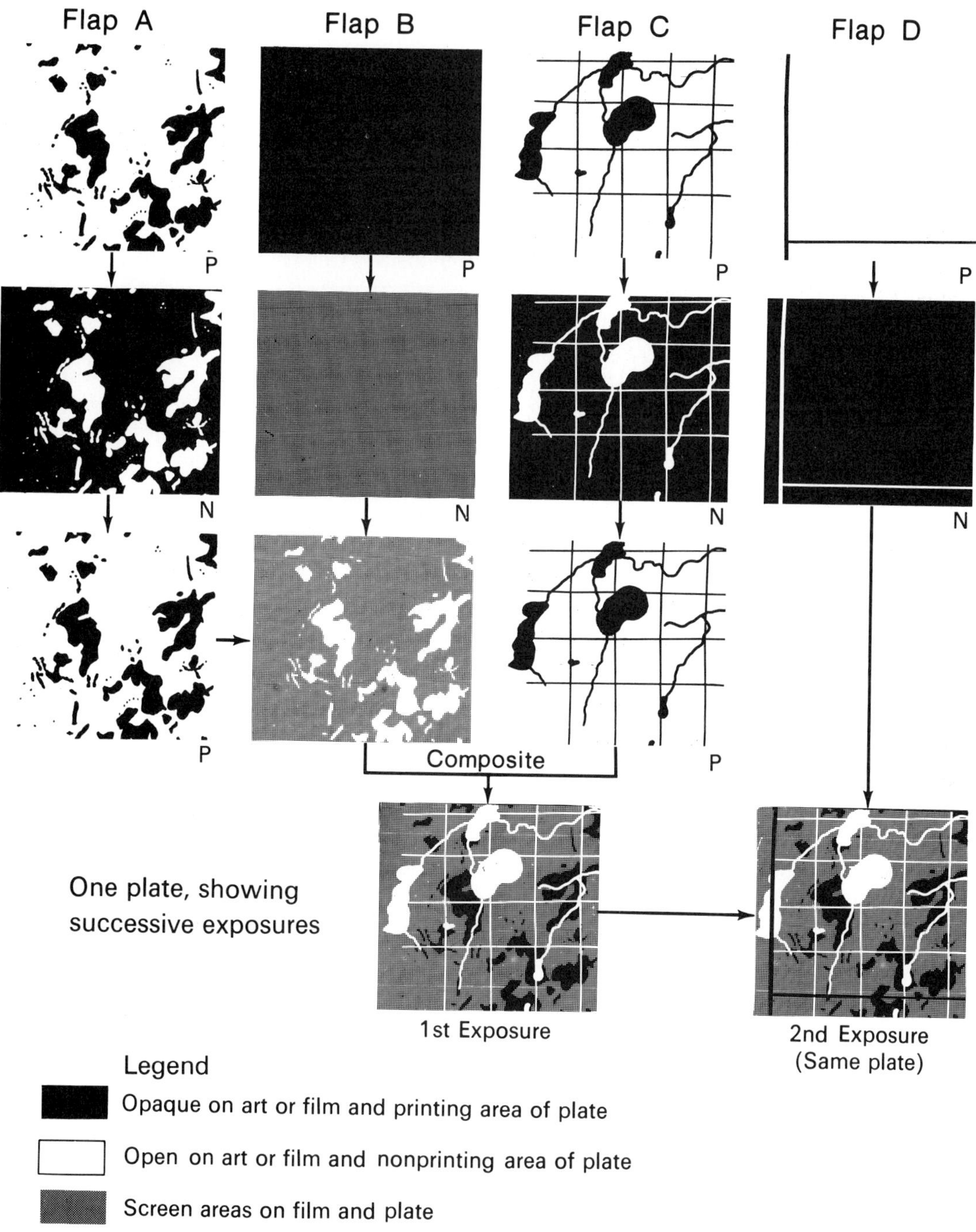

FIGURE 18.37 Possibilities for use of black and white are unlimited. Through imaginative design, many of the advantages of multiple-color map production can be achieved.

result over the whole map except in the wet areas, which would be solid black. But, by combining a positive of flap C with the new film, the hydrographic features and the section lines block the light and produce a nonprinting area in the background tint and in the low wet areas.

PROOFING

When artwork is completed, it must be carefully and thoroughly checked before being sent for duplication. When facilities and materials are available, it is wise to proof a map if it consists of more than one flap. Even one-color maps can consist of as many as ten flaps or more, greatly increasing the chances for error and making editing difficult. A properly prepared proof will closely resemble the copies that will be printed from the artwork and can therefore evaluate the success of the design as well as reveal errors.

Artwork prepared on translucent material can be used to prepare proofs in the drafting room, even though darkroom facilities are not available, as described in Chapter 17. The various proofing processes, such as *General Color Guide Film, 3M* proofing materials, and the use of bichromate sensitizers, require the use of negatives for exposure. These direct contact negatives can be made using *Kodagraph Contact Film* or with *Orange 3M Color Key,* both of which are actinically opaque; either of these negatives can be used to make the proofs with any of the proofing systems. During each step of the proofing operation, the pin registration system must be employed. Because of the nature of the 3*M* material, it cannot be easily punched. It is necessary, therefore, to use the gummed prepunched tabs or to punch strips of discarded film and tape them to the proofing sheets.

If map artwork is available in the form of digital records, the full range of computer-driven printers, plotters, and display screens can also be used in proofing a map. Color screens are particularly well suited for this task. With appropriate hardware, software, and data files, the digital records can be altered interactively, permitting the corrected image to be viewed almost instantaneously. This not only reduces proofing time and effort, but also avoids the high materials cost normally associated with photochemical proofing methods. There is a trade-off in programming, computer charges, and capital costs, of course.

If a map is to be reproduced by plate-based methods and facilities are not available for proofing, we can request that the printer furnish proofs after making the negatives but before making the printing plates. Errors found at this stage can often be corrected on the negative. If the artwork must be revised, we have lost only the cost of the negatives and not the cost of the plates, paper, and press time.* For complicated colored maps we may wish to request *press proofs,* in which case the actual plates that will be used to print the final copies are used to print a small number of copies of the map. Errors found at this point save the cost of paper and press time. A final and important precaution is for the cartographer to be present while the map is actually being printed to inspect such things as registry and color quality.

SELECTED REFERENCES

Adams, R. V., "Electron Printing System," *Graphic Arts Monthly,* March 1982, pp. 68–76. Also see graphic arts and printing industry trade journals, such as *American Printer* and *Graphic Arts Monthly*.

Bell, P. A., and P. A. Woodsford, "Use of the HRD-1 Laser Display for Automated Cartography," *The Cartographic Journal* 14 (1971):128–34.

Cuff, D. J., and M. T. Mattson, *Thematic Maps: Their Design and Production*. New York: Methuen, 1982.

*The cost of the negative is not very much different from the cost of the plates these days. What is expensive is the cost of labor to remake the plate.

Edwards, K., and R. M. Batson, "Preparation and Presentation of Digital Maps in Raster Format," *The American Cartographer* 7 (1980):39–49.

Foley, J. D., and A. VanDam, *Fundamentals of Interactive Computer Graphics*. Reading, Mass.: Addison-Wesley Publishing Co., 1982.

Groop, R. E., and P. Smith, "A Dot Matrix Method of Portraying Continuous Statistical Surfaces," *The American Cartographer* 9 (1982):123–30.

Groop, R. E., and R. E. Smith, "Matrix Line Printer Maps," *The American Cartographer* 9 (1982):19–24.

Irwin, D., "Black and White Maps from Color," *The Canadian Cartographer* 8 (1971):137–42.

Johnston, P., "Graphics by Computer," *Graphic Arts Monthly*. March 1982, pp. 54–67.

Keates, J. S., "Developments in Non-Automated Techniques," *Transactions of the Institute of British Geographers,* New Series 2 (1977):37–48.

Kers, A. J., "Flow Diagrams in Map Production," *ITC Journal,* Special Cartography Issue, 1982-2 (June), pp. 37–48.

Kimerling, A. J., "Visual Value as a Function of Percent Area Inked for the Cross-Screening Technique," *The American Cartographer* 6 (1979):141–48.

Logsdon, T., "High-Speed Printers," *Technology Illustrated* 2 (December/January 1983):101–4.

Muehrcke, P. C., "An Integrated Approach to Map Design and Production," *The American Cartographer* 9 (1982):109–22.

Shearer, J. W., "Cartographic Production Diagrams: A Proposal for a Standard Notation System," *The Cartographic Journal* 19 (1982):5–15.

Stolle, H. J., "Notes on the Single Peelcoat Method for Color Separation of Textbook Maps," *The Canadian Cartographer* 9 (1972):141–45.

Smith, R. M., "Improved Area Symbols for Computer Line-Printed Maps," *The American Cartographer* 7 (1980):51–57.

Turner, E., "New Imaging Materials Assist in Map Drafting," *The American Cartographer* 10 (1983):73–76.

Wray, J. R., "New Precision in Mapping Possible with Laser Plotter," *Computer Graphic News,* January/February 1983, pp. 8–9, 20.

Yoeli, P., *Cartographic Drawing with Computers,* Computer Applications, 8. Nottingham, England: Department of Geography, University of Nottingham, 1982.

Appendix A

Useful Dimensions, Constants, Formulas, and Conversions

Equatorial circumference of earth	40 075.1 km	24 901.5 mi. (statute)
Area of earth (approximate)	510 064 500 sq km	196 937 000 sq. mi.
Radius of the sphere of equal area (approx.)	6 371 km	3 959 mi.
Circumference to the diameter of a circle (π)	3.141 593	
Area of a circle	πr^2	
Area of a sphere	$4\pi r^2$	

CONVERSIONS

Note. SI refers to the International System of Units (metric), USCS to United States Customary System. The values in the table are based on the generally employed conversion ratio that 1.0 inch of the International Foot equals 25.4 millimeters exactly. Note, however, that in the United States, for surveying and large-scale mapping and charting purposes, the slightly different ratio (established in 1866) that 1.0 inch of the "U.S. Survey Foot" equals 25.40005 millimeters (1.0 meter = 39.37 inches exactly) has been retained. The difference is about 2 parts per million.

SI A	USCS B	To Convert from A to B, Multiply A by	To Convert from B to A, Multiply B by
Meters	Feet	3.280 840	0.304 8*
Meters	Yards	1.093 613	0.914 4*
Centimeters	Inches	0.393 700 8	2.54*
Centimeters	Feet	0.032 808 40	30.48*
Kilometers	Miles (U.S. statute)	0.621 371 1	1.609 344*
Kilometers	Miles (international nautical)	0.539 956 8	1.852*
Square meters	Square feet	10.763 91	0.092 903 04*
Square meters	Square yards	1.195 990	0.836 127 36*
Square centimeters	Square inches	0.155 000 3	6.451 6*
Square centimeters	Square feet	0.001 076 39	929.030 4*
Square kilometers	Square miles (U.S. statute)	0.386 102 1	2.589 988
Hectares	Acres	2.471 054	0.404 685 6
Hectares	Square miles	0.003 861 02	258.998 8

*Figures identified with an asterisk are exact. The others are generally given to seven significant figures.
From Units of Weights and Measures, National Bureau of Standards Miscellaneous Publication 286, Supt. of Documents, U.S. Government Printing Office, Washington, D.C., 1967.

Trigonometric Functions of an Acute Angle in a Right Triangle

$$\text{sine} = \frac{\text{opposite side}}{\text{hypotenuse}} \qquad \text{cotangent} = \frac{\text{adjacent side}}{\text{opposite side}}$$

$$\text{cosine} = \frac{\text{adjacent side}}{\text{hypotenuse}} \qquad \text{secant} = \frac{\text{hypotenuse}}{\text{adjacent side}}$$

$$\text{tangent} = \frac{\text{opposite side}}{\text{adjacent side}} \qquad \text{cosecant} = \frac{\text{hypotenuse}}{\text{opposite side}}$$

Appendix B
Geographical Tables

TABLE B.1 Lengths of Degrees of the Parallel

Lat.	Meters	Statute Miles	Lat.	Meters	Statute Miles	Lat.	Meters	Statute Miles
° ′			° ′			° ′		
0 00	111 321	69.172	30 00	96 488	59.956	60 00	55 802	34.674
1 00	111 304	69.162	31 00	95 506	59.345	61 00	54 110	33.623
2 00	111 253	69.130	32 00	94 495	58.716	62 00	52 400	32.560
3 00	111 169	69.078	33 00	93 455	58.071	63 00	50 675	31.488
4 00	111 051	69.005	34 00	92 387	57.407	64 00	48 934	30.406
5 00	110 900	68.911	35 00	91 290	56.725	65 00	47 177	29.315
6 00	110 715	68.795	36 00	90 166	56.027	66 00	45 407	28.215
7 00	110 497	68.660	37 00	89 014	55.311	67 00	43 622	27.106
8 00	110 245	68.504	38 00	87 835	54.579	68 00	41 823	25.988
9 00	109 959	68.326	39 00	86 629	53.829	69 00	40 012	24.862
10 00	109 641	68.129	40 00	85 396	53.063	70 00	38 188	23.729
11 00	109 289	67.910	41 00	84 137	52.281	71 00	36 353	22.589
12 00	108 904	67.670	42 00	82 853	51.483	72 00	34 506	21.441
13 00	108 486	67.410	43 00	81 543	50.669	73 00	32 648	20.287
14 00	108 036	67.131	44 00	80 208	49.840	74 00	30 781	19.127
15 00	107 553	66.830	45 00	78 849	48.995	75 00	28 903	17.960
16 00	107 036	66.510	46 00	77 466	48.136	76 00	27 017	16.788
17 00	106 487	66.169	47 00	76 058	47.261	77 00	25 123	15.611
18 00	105 906	65.808	48 00	74 628	46.372	78 00	23 220	14.428
19 00	105 294	65.427	49 00	73 174	45.469	79 00	21 311	13.242
20 00	104 649	65.026	50 00	71 698	44.552	80 00	19 394	12.051
21 00	103 972	64.606	51 00	70 200	43.621	81 00	17 472	10.857
22 00	103 264	64.166	52 00	68 680	42.676	82 00	15 545	9.659
23 00	102 524	63.706	53 00	67 140	41.719	83 00	13 612	8.458
24 00	101 754	63.228	54 00	65 578	40.749	84 00	11 675	7.255
25 00	100 952	62.729	55 00	63 996	39.766	85 00	9 735	6.049
26 00	100 119	62.212	56 00	62 395	38.771	86 00	7 792	4.842
27 00	99 257	61.676	57 00	60 774	37.764	87 00	5 846	3.632
28 00	98 364	61.122	58 00	59 135	36.745	88 00	3 898	2.422
29 00	97 441	60.548	59 00	57 478	35.716	89 00	1 949	1.211

Tables B.1 and B.2 are from U.S. Coast and Geodetic Survey; Table B.3 is from *Smithsonian Geographical Tables*. The values are based on the Clarke (1866) spheroid. When full tables are available for GRS 80 (see Chapter 4), they will differ only slightly.

TABLE B.2 Lengths of Degrees of the Meridian

Lat.	Meters	Statute Miles	Lat.	Meters	Statute Miles	Lat.	Meters	Statute Miles
°			°			°		
0–1	110 567.3	68.703	30–31	110 857.0	68.883	60–61	111 423.1	69.235
1–2	110 568.0	68.704	31–32	110 874.4	68.894	61–62	111 439.9	69.246
2–3	110 569.4	68.705	32–33	110 892.1	68.905	62–63	111 456.4	69.256
3–4	110 571.4	68.706	33–34	110 910.1	68.916	63–64	111 472.4	69.266
4–5	110 574.1	68.707	34–35	110 928.3	68.928	64–65	111 488.1	69.275
5–6	110 577.6	68.710	35–36	110 946.9	68.939	65–66	111 503.3	69.285
6–7	110 581.6	68.712	36–37	110 965.6	68.951	66–67	111 518.0	69.294
7–8	110 586.4	68.715	37–38	110 984.5	68.962	67–68	111 532.3	69.303
8–9	110 591.8	68.718	38–39	111 003.7	68.974	68–69	111 546.2	69.311
9–10	110 597.8	68.722	39–40	111 023.0	68.986	69–70	111 559.5	69.320
10–11	110 604.5	68.726	40–41	111 042.4	68.998	70–71	111 572.2	69.328
11–12	110 611.9	68.731	41–42	111 061.9	69.011	71–72	111 584.5	69.335
12–13	110 619.8	68.736	42–43	111 081.6	69.023	72–73	111 596.2	69.343
13–14	110 628.4	68.741	43–44	111 101.3	69.035	73–74	111 607.3	69.349
14–15	110 637.6	68.747	44–45	111 121.0	69.047	74–75	111 617.9	69.356
15–16	110 647.5	68.753	45–46	111 140.8	69.060	75–76	111 627.8	69.362
16–17	110 657.8	68.759	46–47	111 160.5	69.072	76–77	111 637.1	69.368
17–18	110 668.8	68.766	47–48	111 180.2	69.084	77–78	111 645.9	69.373
18–19	110 680.4	68.773	48–49	111 199.9	69.096	78–79	111 653.9	69.378
19–20	110 692.4	68.781	49–50	111 219.5	69.108	79–80	111 661.4	69.383
20–21	110 705.1	68.789	50–51	111 239.0	69.121	80–81	111 668.2	69.387
21–22	110 718.2	68.797	51–52	111 258.3	69.133	81–82	111 674.4	69.391
22–23	110 731.8	68.805	52–53	111 277.6	69.145	82–83	111 679.9	69.395
23–24	110 746.0	68.814	53–54	111 296.6	69.156	83–84	111 684.7	69.398
24–25	110 760.6	68.823	54–55	111 315.4	69.168	84–85	111 688.9	69.400
25–26	110 775.6	68.833	55–56	111 334.0	69.180	85–86	111 692.3	69.402
26–27	110 791.1	68.842	56–57	111 352.4	69.191	86–87	111 695.1	69.404
27–28	110 807.0	68.852	57–58	111 370.5	69.202	87–88	111 697.2	69.405
28–29	110 823.3	68.862	58–59	111 388.4	69.213	88–89	111 698.6	69.406
29–30	110 840.0	68.873	59–60	111 405.9	69.224	89–90	111 699.3	69.407

TABLE B.3 Areas of Quadrilaterals of Earth's Surface of 1° Extent in Latitude and Longitude

Lower Latitude of Quadrilateral °	Area Sq km	Sq Mi (U.S. statute)	Lower Latitude of Quadrilateral °	Area Sq km	Sq mi (U.S. statute)
0	12 308.04	4 752.16	45	8 686.85	3 354.01
1	12 304.39	4 750.75	46	8 533.26	3 294.71
2	12 297.08	4 747.93	47	8 377.03	3 234.39
3	12 286.15	4 743.71	48	8 218.14	3 173.04
4	12 271.57	4 738.08	49	8 056.65	3 110.69
5	12 253.34	4 731.04	50	7 892.65	3 047.37
6	12 231.50	4 722.61	51	7 726.14	2 983.08
7	12 205.99	4 712.76	52	7 557.20	2 917.85
8	12 176.88	4 701.52	53	7 385.82	2 851.68
9	12 144.17	4 688.89	54	7 212.13	2 784.62
10	12 107.83	4 674.86	55	7 036.14	2 716.67
11	12 067.87	4 659.43	56	6 857.90	2 647.85
12	12 024.36	4 642.63	57	6 677.48	2 578.19
13	11 977.24	4 624.44	58	6 494.91	2 507.70
14	11 926.56	4 604.87	59	6 310.30	2 436.42
15	11 872.30	4 583.92	60	6 123.61	2 364.34
16	11 814.52	4 561.61	61	5 934.98	2 291.51
17	11 753.18	4 537.93	62	5 744.44	2 217.94
18	11 688.36	4 512.90	63	5 552.05	2 143.66
19	11 620.01	4 486.51	64	5 357.86	2 068.68
20	11 548.19	4 458.78	65	5 161.95	1 993.04
21	11 472.90	4 429.71	66	4 964.36	1 916.75
22	11 394.13	4 399.30	67	4 765.16	1 839.84
23	11 311.95	4 367.57	68	4 564.41	1 762.33
24	11 226.35	4 334.52	69	4 362.16	1 684.24
25	11 137.39	4 300.17	70	4 158.54	1 605.62
26	11 045.03	4 264.51	71	3 953.51	1 526.46
27	10 949.33	4 227.56	72	3 747.22	1 446.81
28	10 850.31	4 189.33	73	3 539.71	1 366.69
29	10 748.01	4 149.83	74	3 331.04	1 286.12
30	10 642.42	4 109.06	75	3 121.27	1 205.13
31	10 533.61	4 067.05	76	2 910.50	1 123.75
32	10 421.57	4 023.79	77	2 698.74	1 041.99
33	10 306.34	3 979.30	78	2 486.13	959.90
34	10 187.95	3 933.59	79	2 272.69	877.49

TABLE B.3 Areas of Quadrilaterals of Earth's Surface of 1° Extent in Latitude and Longitude

Lower Latitude of Quadrilateral °	Area Sq km	Area Sq mi (U.S. statute)	Lower Latitude of Quadrilateral °	Area Sq km	Area Sq mi (U.S. statute)
35	10 066.43	3 886.67	80	2 058.50	794.79
36	9 941.82	3 838.56	81	1 843.63	711.83
37	9 814.14	3 789.26	82	1 628.17	628.64
38	9 683.45	3 738.80	83	1 412.17	545.24
39	9 549.75	3 687.18	84	1 195.69	461.66
40	9 413.10	3 634.42	85	978.83	377.93
41	9 273.56	3 580.54	86	761.66	294.08
42	9 131.11	3 525.54	87	544.21	210.12
43	8 985.81	3 469.44	88	326.60	126.10
44	8 837.71	3 412.26	89	108.88	42.04

Appendix C

Glossary of Technical Terms

The terms included in this glossary are primarily from other disciplines related to cartography. Many of the terms are important for all cartographers to learn, and a large number are used in the text. Numerous widely used cartographic terms are not included, since they are defined in the text and the reader can locate the definition by looking for the terms in the index.

Absorption (re: Energy) The process by which radiant energy is absorbed and converted into other forms of energy.

Accuracy The degree to which a measurement is known to approximate a given value: correctness: usually refers to computation.

Active system A remote sensing system that transmits its own electromagnetic emanations and then records the energy reflected or refracted back to the sensor.

Address A numeric or symbolic designation of a memory location or peripheral unit where information is stored.

Algorithm A set of rules of procedure for solving a specific problem.

Aliasing The "jaggy" or "staircase" appearance of lines and edges on raster displays.

Alphanumeric (1) A character set of letters, integers, punctuation marks, and special symbols. (2) Term used to denote a combination of letters and numbers.

Analog The representation of a numerical quantity by a physical variable, e.g., graphic marks or electric voltages; contrasted with digital.

Anchoring stimuli Legend values strategically selected from the mapped data to aid the map reader in correctly estimating the values of the symbols.

Angstrom (Å) Unit of measurement. 10^{-10}m.

Array The arrangement of a series of terms in a geometric pattern, like a matrix.

Artwork Material, drawings, etc., prepared for reproduction; copy.

Assembler A computer program that translates computer instructions written in a source language directly into machine language instructions.

Atmospheric windows Those wavelength ranges in which radiation can pass through the atmosphere with relatively little attenuation. Commonly used atmospheric windows occur at approximately 0.3–2.5, 3.0–4.0, 4.2–5.0, and 7.0–15.0 μm.

Attenuation In physics, any process in which the flux density (or power, amplitude, intensity, illuminance) of a "parallel beam" of energy decreases with increasing distance from the energy source.

Attitude The angular orientation of a remote sensing system with respect to a geographical reference system.

Autocorrelation A statistic calculated as the correlation of an ordered series of observations with the same series of observations offset by a stated interval (lag).

Backlight Passing light through transmission copy from behind the image in photochemical processing.

Backscatter The scattering of radiant energy into the space bounded by a plane perpendicular to the direction of the incoming radiation and lying on the same side of the plane as the incoming ray; the opposite of forward scatter.

Band (1) A selection of wavelengths. (2) Frequency band. (3) Absorption band. (4) A group of tracks on a magnetic drum. (5) A range of radar frequencies, such as X-band, Q-band.

Band-pass filter A wave filter that allows only a specified band of wavelengths to pass while blocking all others.

Base map A map containing geographical reference information on which attribute data may be plotted for purposes of comparison or geographical correlation.

Batch processing A method whereby items are coded, collected in groups, and processed sequentially.

Binary A numbering system based on two. Only the digits 1 and 0 are used.

Bit Abbreviation for binary digit.

Blackbody The theoretical material that transforms heat energy into radiant energy with the maximum rate permitted by thermodynamic laws.

Block A group of records or words treated as a logical unit of information.

Buffer A storage register or section of memory where information from one part of the computer system may be held temporarily until another part is ready to receive it.

Bug An error in a program or routine: an equipment malfunction.

Byte A group of binary digits handled as a unit.

CAI Computer-assisted / computer-aided instruction.

Card image A binary representation in storage of the hole patterns of a Hollerith card.

Cathode ray tube A vacuum tube in which slender beams of high-speed electrons are projected onto a fluorescent screen to produce luminous spots.

CCT Computer compatible tape containing digital LANDSAT data.

Central processing unit The part of the computer that contains the circuits that control and perform the execution of computer instructions.

Chain (1) The printing part of a line printer: a set of printing heads, like those of a typewriter, that strikes an inked ribbon to leave an impression. (2) A synonym for string, as in "a string of coordinates."

Character A basic symbol used to convey information, e.g., letters, numbers.

Chroma (Munsell) The range from neutral gray to full saturation of hue, with value held constant, with neutral gray specified as 0. The third designation in the Munsell specification.

Chromatic coordinates (CIE) The designated fractional amounts of the three primary colors that compose a given color in the CIE system. The coordinates are designated x, y, and z.

Chromaticity diagram (CIE) A graph for plotting the chromatic coordinates x and y from which measures of dominant wavelength and purity can be obtained. Also called a Maxwell diagram.

CIE Commission International de l'Éclairage.

CIE color system A method for specifying the physical characteristics of a color numerically by measures of dominant wavelength, luminosity, and purity devised by the Commission International de l'Éclairage.

Clinographic curve A graph in two dimensions formed by arraying the z values of the data in numerical order on the y axis against the cumulative areas to which they refer on the x axis. The x axis is scaled with a square root scale of percentage of cumulative area.

Clustering The analysis of a set of measurement vectors to detect their inherent tendency to form clusters in a multidimensional measurement space.

Coefficient A number or quantity serving as a

measure of some attribute, usually a constant quantity in any given mathematical operation.

Color separation A positive or negative representing one of the colors to be used in process color printing.

COM Computer output on microfilm.

Command The portion of an instruction word that specifies the operation to be performed.

Commensurable Having a common measure, specifically, divisible by a common measure or unit.

Compilation Preparation of a new or revised map or chart, or a portion thereof, from field surveys, remote sensing, census data, existing maps, and other sources.

Compiler (1) A computer software program that converts a source language program into object language. (2) One who compiles data for a map.

Computer compatible tapes Tapes containing digital LANDSAT data. These tapes are standard 19 cm ($7\frac{1}{2}$ in.) wide magnetic tapes in nine-track or seven-track format. Four tapes are required for the four-band multispectral digital data corresponding to one LANDSAT scene.

Computer network Several CPUs linked together by a telecommunications system.

Configuration The specific arrangement of computer hardware and peripherals making up a system.

Conjugate principal point The position on a photograph of the principal point of the adjacent photograph along a flight line.

Contact print A photographic reproduction made from a negative or positive in contact with sensitized paper, film, or printing plate.

Continuous tone Smooth and continuous transition of tones, such as on an ordinary photograph, in contrast to halftone and line images.

Contrast stretching Expanding the original range of recorded digital values to the full contrast range available on the recording film or display device. This is done to increase the amount of contrast on the image.

Coordinate An ordered set of data values that specifies a location: may be absolute or relative.

Copy Used in the printing industry to refer to any material (e.g., pictures, artwork) to be reproduced.

Core The most accessible information storage unit(s) of a digital computer.

Correlation The interdependence between two sets of numbers; a relation between two quantities such that when one changes, the other does. Simultaneous increasing or decreasing is positive correlation; one increasing and the other decreasing is negative correlation.

CPU Central processing unit.

CRT Cathode ray tube.

CRT display Graphic output using the screen of a cathode ray tube as the viewing element.

Cumulative frequency graph An arithmetic graph of the ranked values of a geographical distribution on the Y axis against the accumulated areas of the regions to which the values refer. The resulting curve is also called a hypsometric curve.

Cursor A movable part of an instrument (e.g., a digitizer or CRT) that indicates a position visible to the operator and whose x-y coordinates are known by the machine.

Data bank or data base A store of information (usually in digital form) organized such that retrieval can be done on a selective basis.

Data encoding The process of converting data to machine-readable form.

Data structure The arrangement and interrelation of records in a file.

Datum Any level surface, line, or point used as a reference for measurement of another quantity.

Densitometer An instrument for measuring the amount of light transmitted or reflected.

Density slicing The process of converting the continuous gray tone of an image into a series of distinct gray tones, each corresponding to a specific digital range.

Dependent variable In a function, the variable whose value (in the range) is determined and specified by the associated value of the independent variable.

Dichroic grating A grating in a multispectral scanning sensor system that separates the incoming reflected wavelengths from the incoming emitted wavelengths.

Digital The representation of a quantity by a number code; contrasted with analog.

Digital incremental plotter A device used for drawing line segments. The length and direction of the line segments are specified by a series of commands in machine-usable form.

Digitizer An instrument that converts graphically represented information into digital form.

Digitizing The data entry process whereby graphic images are converted to digital coordinates.

Dimensional stability The ability of paper or film to resist dimensional distortion with changes in temperature and humidity.

Disc/disk A magnetic storage device for digital information.

Discriminant function One of a set of mathematical functions used to divide measurement space into decision regions.

Displacement Any shift in the position of an image on a photograph due to tilt or scale change in the photograph or to the relief of the objects photographed.

Dithering The printing or display of different arrays of constant-area dots on graphic output devices in an attempt to produce tones that approximate the variable-area dots of halftone reproduction.

Dominant wavelength (CIE) The monochromatic light which, when mixed in suitable proportion with the illuminant light, will match the color in question.

Drum plotter A digital incremental plotter in which the output material is mounted on a rotating drum or cylinder.

Edge enhancement The use of analytical techniques to emphasize selected adjacent pixel differences in imagery.

Electromagnetic radiation (EMR) Energy propagated through space in the form of an advancing interaction between electric and magnetic fields. Also commonly referred to as simply radiation or electromagnetic energy.

Electromagnetic spectrum The ordered array of known electromagnetic radiations including the shortest cosmic rays, as well as gamma rays, X rays, ultraviolet radiation, visible radiation, infrared radiation, microwave, and all longer wavelengths.

Emissivity The ratio of the radiation given off by a surface to the radiation given off by a blackbody at the same temperature.

EMR Electromagnetic radiation.

Emulsion (photography) A suspension of a light-sensitive substance, such as a silver halide, in a colloidal medium (usually gelatin), which is used for coating photographic films, plates, and papers.

Enumeration unit Any area used as the region for the collection of data about that region.

Factor analysis A statistical procedure for reducing a set of data variables to a lesser number of variables, each of which is a function of one or more of the original variables.

False color film Film that does not register the blue radiation being reflected by an object, but that does register the green, red, and infrared radiation being reflected. This produces an image that appears abnormal to the human eye.

Feature code A numeric label attached to a line or point.

Feature separation Process of preparing a separate drawing, engraving, or negative for selected types of data in the preparation of a map or chart.

Fiducial marks Index marks (usually four), rigidly connected with the camera lens through the camera body, which form images on the negative. The marks are adjusted so that the intersection of lines drawn between opposite fiducial marks defines the principal point of a photograph.

File An organized collection of records.

Filter (1) Any material, which, by absorption or reflection, selectively modifies the radiation transmitted. (2) To remove a component(s) of electromagnetic radiation, usually with the aid of a filter.

Flap Separate sheet on which a map or selected features of a map are prepared. (Other terms that are used to refer to these separate sheets are overlay and separate.) One map may require several flaps.

Flat In offset lithgraphy, the assembled, stripped composite of film materials ready for platemaking.

Flatbed plotter A digital incremental plotter in which the output material is mounted on a fixed or movable plane.

Flight line The path that an aircraft, or other airborne sensor, follows when sensing or collecting data. *See* Ground track.

Focal length of a camera The distance measured along the optical axis from the optical center (rear nodal point) of the lens to the plane of critical focus.

Frequency curve A graph in two dimensions relating the number of items observed, on the one axis, to the category or value of the observation, on a perpendicular axis.

Frontlight Bouncing light off the surface of an image in photochemical processing.

Function An association of a certain object (or objects) from one set (the range) with each object from another set (the domain).

Gamma rays A term used to denote very short wavelengths in the range of 10^{-6} micrometers on the electromagnetic spectrum.

Geocode The process of attaching a geographic reference to a phenomena, such as assigning a latitude/longitude designation to a water well site.

Geographic mean The areally weighted mean in which the sum of the x values multiplied by the areas they represent is divided by the total area. Also, the height, measured on the y axis, of a rectangle equal in area to the area below the curve and above the x axis on a cumulative frequency curve.

Geographic median The x value above which and below which half the total area occurs. Also, the y value, read off the cumulative frequency curve, at a point obtained by erecting a line perpendicular to the x axis at the midpoint of the cumulated areas plotted along the x axis.

Gradient The rate of change, with respect to distance, of a variable quantity.

Ground control Accurate positional data for identified horizontal and/or vertical points on the ground.

Ground range The horizontal distance from the ground track (nadir) to a given object.

Ground resolution cell The area on the ground that is covered by the instantaneous field of view of a detector.

Ground track The orthogonal projection of the actual flight path of a vehicle carrying the remote sensing system onto the surface of the earth or other body.

Halftone A continuous-tone image in which the gradations in tone have been converted to tiny, discrete dots or lines of varying size for purposes of plate-based printing.

Hard copy Any physical map, chart, or graphic presentation that has some degree of permanence.

Hardware The physical components of a computer and its peripheral equipment.

High pass filter A filter which blocks the low-frequency (long-wave) component of transmitted radiation.

Histogram The graphic display of a set of data, which shows the frequency of occurrence (along the vertical axis) of individual measurements or values (along the horizontal axis).

Hue The attribute of color associated with wavelength, e.g., blue, red, green.

Hue (Munsell) One of the 100 pigments which constitute the full range of the system. The first designation in the Munsell specification.

Hypsometry Vertical control in map making; referring to elevation relative to an established datum.

IFOV Instantaneous field of view.

Independent variable In a function, the controlling variable. The domain of the function contains all values of the independent variable.

Infrared Energy in the 0.7–100.0 μm wavelength region of the electromagnetic spectrum. For remote sensing, the infrared wavelengths are sometimes divided into near infrared (0.7–1.3 μm), middle infrared (1.3–3.0 μm), and far infrared (7.0–15 μm). Far infrared is sometimes referred to as thermal infrared. (Authorities differ on the exact bounds of these arbitrary divisions.)

Input Information or data prepared for use by a computer.

Instantaneous field of view A term specifically denoting the narrow field of view whose electromagnetic radiation is being recorded by the detector at any one instant. While as much as 120° may be under scan, the electromagnetic radiation from only a small area is recorded at any one instant.

Instruction A code that defines an operation to be performed as well as the data or unit of equipment to be used.

Integrated circuit A small piece of semiconductor material containing a number of circuits.

Intensity (color) The richness or brilliance of a color. A general term which encompasses the attributes precisely defined by such terms as chroma, saturation, or purity in systems of color specifications.

Interactive mode A method of operation that allows on-line communication between a person and a machine; commonly used to enter data or to direct the course of a program.

Interface The common boundary between individual components of a system.

Interval (1) A set containing all numbers between two given numbers and one, both, or neither end point. (2) A scale of measurement upon which numbers of a set are differentiated by relative amount.

Iteration Repetition. A repeated application of a process or manipulation.

Iterative solution A solution arrived at through a series of iterations.

Label A group of characters used as a symbol to identify an item of data, an area of memory, a record, or a file.

Lag An interval with respect to space or time; a set distance or period.

Layover Displacement of the top of an elevated feature with respect to its base on a radar image.

Library routine A prewritten standard algorithm for use in computer programs.

Light pen A cursor the size of a ballpoint pen; used for pointing to a location on a CRT screen.

Linear array A type of scanning sensor consisting of numerous small detectors arranged in a straight line. A line of data is obtained by sampling the response of the arrayed detectors.

Line artwork or copy Artwork or other material composed of discrete marks as distinguished from continuous tone.

Line map Map composed of discrete marks (i.e., point, line, and area symbols) as distinguished from continuous tone imagery.

Line printer A peripheral computer device like a typewriter that prints one line at a time. A set of alphanumeric characters is available at each position along the line.

Low pass filter A filter that blocks the high-frequency (short-wave) component of transmitted radiation.

Luminosity (CIE) The lightness or darkness of a color as specified by tristimulus value Y.

Machine language Instructions written in machine code that can be obeyed by a computer without further translation.

Magnetic tape Ferrous-coated tape. Selective polarization of the surface permits the sequential storage of digital data.

Mask A sheet of material used to block exposure on particular parts of sensitized material. Positive masks are used to reverse lines and symbols; negative masks (open-window negatives) are used when exposing relatively large areas.

Matrix A rectangular arrangement into columns and rows of the elements of a set.

Maxwell diagram *See* Chromaticity diagram.

Mean A single number or quantity representing a set of numbers and having a value intermediate between the values of other numbers, e.g., an average.

Memory An organization of storage units retained primarily for the retrieval of information.

Method of least squares A method based upon the principle that the most appropriate value for a quantity that can be deduced from a set of measurements or observations is that value for which the sum of the squared deviations between observations and their theoretical expected values is a minimum.

Micrometer (abbr. μm) A unit of length equal to one millionth (10^{-6}) of a meter or one thousandth (10^{-3}) of a millimeter.

Micron (abbr. μ) Equivalent to and replaced by micrometer; 10^{-6} m.

Microprocessor A CPU consisting of a single integrated circuit.

Microwave A very short electromagnetic wave; any wave between 1 meter and 1 millimeter in wavelength or 300 GHz to 0.3 GHz in frequency. Passive systems operating at these wavelengths are sometimes called microwave systems. Active systems are called radar. The exact limits of the microwave region are not defined.

Millimicron *See* Nanometer.

Mimetic Exhibiting mimicry: imitating or resembling closely.

Minicomputer A relatively low-cost CPU with limited core capacity.

Modem A device that transforms electrical signals into audio tones for transmission over telephone lines.

Moiré A geometric pattern of light and dark tones which occurs when one tint screen is superimposed on another at any but a particular angle.

Monochromatic Pertaining to a single wavelength or, more commonly, to a very narrow band of wavelengths; a one-color graphic.

Moving average A series of arithmetic averages obtained by averaging subsets of successive equal-intervaled (with respect to distance or time) terms.

MSS Multispectral scanner.

Multispectral (line) scanner A remote sensing device that is capable of recording data in the ultraviolet, visible, and infrared portions of the spectrum.

Multivariate analysis A set of data-analysis procedures that use multidimensional interrelations and correlations within the data to enable more effective discrimination among the data.

Munsell color system A method for designating a color by finding its position in a color solid (i.e., matching it with a pigment color) as specified by the perceptual dimensions of hue, value, and chroma. Named after its originator, A. H. Munsell.

Nadir (1) That point on the celestial sphere vertically below the observer, or 180° from the zenith. (2) In aerial photography, that point on the ground vertically beneath the perspective center of the camera lens.

Nanometer A unit of length equal to 1×10^{-9} m; one billionth of a meter. Also called a millimicron.

Near infrared The preferred term for the shorter wavelengths in the infrared region extending from about 0.7 micrometers (visible red), to around two or three micrometers (varying with the author). The longer wavelength end grades into the middle infrared. The region emphasizes the radiation reflected from plant materials, which peaks around 0.85 micrometers. It is only available for use during the daylight hours.

Neat line Line separating the body of a map from the map margin.

Negative An image in which the tones of the original are reversed so that light areas appear dark and vice versa. The reverse of positive.

Node A point common to two or more line segments.

Noise Unwanted variations introduced into data during measurement or display.

Nominal A scale of measurement upon which members of a set are differentiated by kind only.

Normal distribution A theoretical frequency distribution characterized by a bell-shaped curve, symmetrical about the mean with points of inflection at ± one standard deviation from the mean.

Object language A machine language resulting from the output of a compiler; the target language to which a source language program is converted by the assembler or compiler.

Off-line Processing of information that is not directly under the control of the CPU.

On-line Processing of information that is directly under the control of the CPU.

Opaque Impervious to the rays of light. Opaque also refers to any of a variety of substances, whether white, black, or red, that prevent transmission of light.

Opaque, actinically Opaque to the wavelengths of light by which sensitized materials are affected.

Open-window negative *See* Mask.

Ordinal A scale of measurement upon which members of a set are differentiated by order.

Orthogonal Right angled; pertaining to or depending upon the use of right angles or perpendiculars.

Orthophotograph Image derived from a conventional perspective photograph by simple differential rectification so that image displacements caused by camera tilt and terrain relief are removed.

Oscillate To move to and fro, as a pendulum; to fluctuate regularly.

Output Computer results, e.g., answers to mathematical and statistical problems or graphic plottings.

Overedge Any portion of a map lying outside the nominal map border.

Overlay Printing or drawing on a transparent, translucent medium intended to be placed in register with other base material on a map.

Overprint New material printed on a map or chart to show data of importance or special use, in addition to those data originally printed.

Panchromatic film Film that is sensitive to the entire visible part of the spectrum.

Parallax The apparent change in the position of one object, or point, with respect to another, when viewed from different angles, e.g., the base and top of a hill.

Parallel Equidistant apart; extending in the same direction equidistant at all points and never converging or diverging.

Parameter An arbitrary constant or a variable in a mathematical expression which distinguishes a specific case.

Passive system A sensing system that detects or measures radiation emitted by the target.

Peripheral The input/out equipment used to transmit data and receive data from the CPU.

Perpendicular Vertical, orthogonal; meeting a given line or surface at right angles.

Photobase The distance between the principal points of two adjacent prints in a series of vertical aerial photographs.

Photogrammetry The art or science of obtaining reliable measurements by means of photography.

Photographic interpretation The act of examining photographic images for the purpose of identifying objects and judging their significance. Photo interpretation and image interpretation are other widely used synonyms.

Photolithography Photomechanical process for converting an object that can be photographed into any of the various kinds of images on lithographic plates.

Photomap (photographic map) Map made by adding marginal information, descriptive data, and a reference system to a photograph or assembly of photographs.

Photon detector A type of detector used in remote sensing systems. Direct interaction between photons of incoming radiation and energy levels of electrical charge carriers in the detector produce a measurable and recordable response.

Pins, registry Machined metal or plastic studs that are inserted into matched holes in flaps of a map to keep the flaps registered.

Pixel A sensed unit of a digital image, usually a very small area (e.g., 25, 50, or 100 microns square) on which is recorded the radiation from the area of instantaneous field of view (IFOV).

Plane of best fit A geometric planar surface fit by the method of least squares to a data set defined most often in three dimensions.

Planimetric map Map that represents only the horizontal positions for features represented; distinguished from a topographic map by the omission of relief in measurable form.

Planimetry The horizontal control in mapmaking obtained by the measurement of only horizontal relationships.

Plan position indicator A radar system characterized by a circular display screen on which a radial sweep indicates the position of returning signals. Rarely used in mapping applications.

Polarization The direction of vibration of the electrical field vector of electromagnetic radiation. In SLAR systems, polarization is either horizontal or vertical.

Position-related variable Any data set the members of which carry a measure of their location (position) as an integral part of their numeric values.

Positive An image in which the light and dark tones are the same as those of the original. The reverse of negative.

Potential A type of function for which the intensity of a field or distribution may be derived.

PPI Plan position indicator.

Precision A statistical measure of repeatability, usually expressed as a variance or standard deviation (root mean square, RMS) of repeated measurements.

Primary color(s) One of three hues which, when mixed in suitable proportions, can match any color. Additive (or light) primaries are blue, green, and red. Subtractive (or pigment) primaries are cyan, magenta, and yellow.

Principal point The point of intersection of straight lines connecting opposite fiducial marks on a photograph.

Process color(s) The colors of the inks used in printing to duplicate the appearance of polychromatic originals: black, cyan, magenta, and yellow.

Program A series of logical statements or instructions written in source language or object language to specify what operations the machine is to perform.

Programming language A language used to prepare computer programs, e.g., FORTRAN, ALGOL, PASCAL, PL1, BASIC.

Progression A sequence of terms in which there is a constant relation between each term and the one succeeding it.

Proof Any of various kinds of preliminary copies of a map that are used to check the accuracy, registration, and legibility of a map or to give an indication of its appearance in final printed form.

Purity (CIE) The degree of saturation of a given color (designated by its dominant wavelength) ranging from 100 percent with full color to 0 percent.

Pushbroom scanning A process of scanning using linear arrays. The sampling of the detectors in the linear arrays is synchronized with the speed of the airborne platform holding the sensing system. Successive lines of ground coverage are therefore obtained as the sensor moves overhead.

Radar Acronym for radio detection and ranging. A method, system, or technique, including equipment components, for using beamed, reflected, and timed electromagnetic radiation to detect, locate, and (or) track objects, to measure altitude or to acquire a terrain image.

Radar shadow A dark area of no return on a radar image that extends in the direction away from the sensor and the object on the terrain that intercepted the radar beam.

Radar, synthetic aperture (SAR) A radar in which a synthetically long apparent or effective aperture is constructed by integrating multiple returns from the same ground cell.

Radial displacement The shift in position of the image of an object on a photograph measured along a line drawn from the center of the photograph through the image of the object. The shift is caused by a difference between the elevation of the object and the elevation at the center of the photograph.

Radiation The emission and propagation of energy through space in the form of waves, e.g., the emission and propagation of electromagnetic waves, or of sound. The process of emitting radiant energy.

Radiometer An instrument for quantitatively measuring the intensity of electromagnetic radiation in some band of wavelengths. Usually used with a modifier, such as an infrared radiometer or a microwave radiometer.

Radio waves A term used to denote long wavelengths in the range of 10^8 micrometers on the electromagnetic spectrum.

RAM Random access memory.

Random (1) Processing without definite aim or purpose. (2) A process of selection in which each item of a set has an equal probability of being chosen.

Random access memory Memory in a computer to which there is equal access to any location irrespective of the previous registers consulted.

Raster The screen area of luminescent material in a CRT where an image is produced.

Rastering *See* Aliasing.

Ratio (1) A scale of measurement upon which members of a set are differentiated by absolute amount. (2) The quotient of two quantities.

Ratioing A method in the processing of remotely sensed images where the analyst calculates the ratio from the spectral reflectances recorded in two different bands of the electromagnetic spectrum.

RBV Return beam vidicon.

Read only memory A semiconductor memory containing fixed data that can be read by the computer but cannot be changed in any way.

Record (1) A collection of related items of data, treated as a unit. (2) One segment of a file; for example, the outline of an area may form a record, and a complete set of such records may form a file.

Reflectance The ratio of the radiant energy reflected by a body to that incident upon it.

Reflection artwork or copy An opaque image that is viewed and photographed by light reflected from its surface. Requires frontlighting.

Reflection (EMR theory) Electromagnetic radiation neither absorbed nor transmitted is reflected.

Register (1) A storage location in a hardware unit. (2) To fit properly, e.g., when several pieces (flaps) of artwork are used to prepare a map, they must be in register.

Register marks Special marks such as crosses that are applied to the original artwork prior to photography and are used for positioning negatives or color flaps.

Regression (linear) In general, any equation which represents a trend in a set of data is a regression equation. Most often, the equation of the relationship between two data sets is obtained by applying the method of least squares to a given set of observations from both data sets. One data set is designated as the independent variable, the remaining, the dependent variable.

Regression (multiple) An extension of linear regression where more than one independent variable is used to establish the relationship between the independent variables and the dependent variable.

Residual A term used in statistics to denote the difference between a result obtained by observation and that obtained by a formula.

Resolution A measure of the ability of a device to differentiate a value; high resolution means fine differentiation. It may be expressed as lines per millimeter or in many other units such as temperature or other physical properties being measured.

Return beam vidicon A modified vidicon television camera tube in which the output signal is derived from the depleted electron beam reflected from the tube target. RBVs provide the highest resolution television imagery and are used in the LANDSAT series.

Reverse Changing the relationship of the transparent areas and opaque areas on photographic film. For example, if film is exposed to light passing through the transparent areas of a negative, those areas will be opaque on the new film and the remainder of the new film will be transparent.

Reverse lettering Light lettering on a dark background.

Right reading The image, such as type, is "readable" in the normal left to right situation. Whether the image-carrying side (emulsion) is up or down should be specified.

ROM Read only memory.

Scan line The narrow strip on the ground that is recorded by the instantaneous field of view of a detector in a scanner system.

Scanner Any device that systematically decomposes a sensed image or scene into pixels and then records some attribute of each pixel.

Scatter diagram A diagram showing the occurrence of paired point values of two variables. The values of one variable are indicated along the x axis and those of the other along the y axis.

Scattering The process by which small particles suspended in a medium of a different index of refraction diffuse a portion of the incident radiation in all directions.

Screen angle The angle at which a screen is placed with relation to others to avoid usually undesirable moiré patterns.

Screen, contact A variable-opacity screen that can be used in a camera or a vacuum frame to convert continuous tone to varying sized dots. (Patterns other than dots are also available.)

Screen, halftone A close network of perpendicular lines etched into glass and filled with an opaque pigment. When used at the photographic stage, it converts continuous tone to varying sized dots.

Screen, tint A film that is interposed between a negative and a sensitized surface to produce a pattern. Those with closely spaced small dots or narrow lines produce the illusion of solid gray.

Scribe coat An actinically opaque plastic coating (on a transparent base) into which lines and other symbols can be cut. The finished scribe sheet can be used as a photographic negative in duplicating processes.

Sensor Any device that detects energy, converts it into a signal and transmits it in a form suitable for obtaining information about the environment.

Sensor platform The vehicle or structure that holds a sensing system allowing it to operate. Platforms can range from hot air balloons to space stations.

Separate *See* Flap.

Sidelooking radar An all-weather, day or night remote sensor which is particularly effective in imaging large areas of terrain. It is an active sensor, as it generates its own energy which is transmitted and received to produce a photo-like picture of the ground. Also referred to as sidelooking airborne radar; abbr. SLAR.

Skew Asymmetry in a distribution.

Slant range For radar images this term represents the distance measured along a line between the antenna and the target.

Smoothing The removal of high frequency components from data by the application of moving averages or other filtering processes.

Software The already established program and routines used in computer operations.

Source language The language used by a programmer to communicate computational requirements to the compiler for that source language.

Spatial filter An image or digital transformation usually a one-to-one operator used to lessen noise or enhance certain characteristics of the image or map. For any particular (x,y) coordinate on the transformed image, the spatial filter assigns a gray shade or a numerical value on the basis of the gray shades or numerical values of particular recorded values near the coordinates (x,y).

Spatial variable Any variable whose numeric values vary over an area.

Spectral reflectance The reflectance of electromagnetic energy at specified wavelength intervals.

Spectral signature Quantitative measurement of the sensed properties of an object at several wavelength intervals.

Spectrophotometer An instrument which measures the proportions of the various wavelengths of light in a color.

Standard deviation The square root of the arithmetic mean of the squares of the deviations from the mean. It serves as a measure of the dispersion from the mean in a frequency distribution.

Standard error of the mean A measure of the representativeness of a calculated mean value from a sample of data. It is obtained by dividing the standard deviation of the sample mean by the square root of the number of samples values.

Stenciling A procedure in image analysis which allows the analyst in effect to "zero" out all data not of interest for the analysis of a given set of features. Analogous to masking in photo lithography.

Stereocompilation Production of a map or chart manuscript from aerial photographs and geodetic control data by means of photogrammetric instruments.

String A consecutive sequence of characters or physical elements.

Stylus A cursor or device in a recording machine used to designate data to be digitized or stored.

Subroutine A set of computer instructions to carry out a predefined computation; usually exists as part of a larger computer program.

Surface fitting A statistical procedure, usually a form of regression analysis, which approximates a surface in three or more dimensions from a set of data values.

Telecommunications The process of transmitting data between geographically separate pieces of electronic equipment.

Thematic mapper An advanced multispectral scanning system aboard LANDSAT 4. It has seven channels of operation, six returning 30 m resolution and the thermal infrared channel returning 120 m resolution images.

Thermal band A general term for infrared wavelengths which are transmitted through the atmospheric window at 8–14 μm. Occasionally also used for the windows around 3–6 μm.

Thermal infrared The preferred term for the wavelength range of the infrared region, extending roughly from 3 μm to about 15 or 20 μm. In practice the limits represent the envelope of energy emitted by the earth behaving as a gray body with a surface temperature around 290°K.

Throughput The total computer processing of data from input to output.

Tilt The angle between the optical axis of the camera and a plumb line.

Time sharing Allocating in turn small divisions of a total time period to two or more functions so that machine time can be used to peak efficiency.

Transmission copy A map image through which light must pass in order for the image to be seen and photographed. Requires backlighting.

Transmittance The ratio of the radiant energy transmitted through a body to that incident upon it.

Tristimulus values (CIE) The summations of the amounts of the red (X), green (Y), and blue (Z) primary colors needed to match each wavelength (or band of wavelengths) of a given color.

Ultraviolet radiation Electromagnetic radiations of shorter wavelengths than visible radiation; roughly, radiation in the wavelength interval between 10 and 4000 Å.

Update The modification of data records as necessary.

Value (color) The sensation of relative darkness (low) or lightness (high) of a color or tone compared to another or as measured on a gray scale.

Value (Munsell) The range of chromatic grays from 0 (black) to 10 (white) based upon a perceptual scale. The second designation in the Munsell specification.

Variable A quantity which can take on any of the numbers of a set.

Variance The square of the standard deviation. Variance of a random variable is the expected value of the square of the deviation between that variable and its expected value. It is a measure of the dispersion of the observed values about their mean.

Vector A quantity that has magnitude, direction and may be assigned a meaning.

Visible wavelengths The radiation range to which the human eye is sensitive, approximately 0.4–0.7 μm.

Wavelength (symbol γ) Wavelength equals velocity divided by frequency. In general, the average distance between maxima (or minima) of a roughly periodic pattern. Specifically, the least distance between particles moving in the same phase of oscillation in a wave disturbance.

Window A band of the electromagnetic spectrum which offers maximum transmission and minimal attenuation through a particular medium when used with a specific sensor.

Word An ordered set of characters handled as a unit by a computer.

Word length The number of bits in a computer word.

Wrong reading The image, such as type, is "unreadable" in that it is a mirror image and thus reads from right to left. Whether the image-carrying side (emulsion) is up or down should be specified.

X rays A term used to denote short wavelengths in the range of 10^{-4} micrometers on the electromagnetic spectrum.

Index